D0142274

Modern Digital and Analog Communication Systems

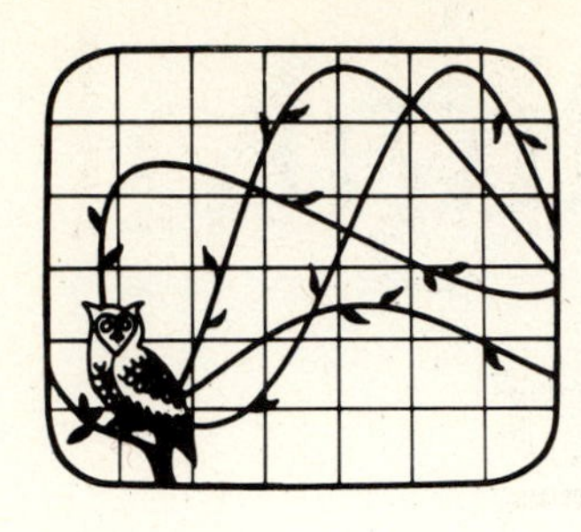

HRW
Series in
Electrical and
Computer Engineering

M. E. Van Valkenburg, Series Editor

L. S. Bobrow ELEMENTARY LINEAR CIRCUIT ANALYSIS
C. H. Durney, L. D. Harris, C. L. Alley ELECTRIC CIRCUITS: THEORY AND ENGINEERING APPLICATIONS
G. H. Hostetter, C. J. Savant, Jr., R. T. Stefani DESIGN OF FEEDBACK CONTROL SYSTEMS
S. Karni and W. J. Byatt MATHEMATICAL METHODS IN CONTINUOUS AND DISCRETE SYSTEMS
B. C. Kuo DIGITAL CONTROL SYSTEMS
B. P. Lathi MODERN DIGITAL AND ANALOG COMMUNICATION SYSTEMS
A. Papoulis CIRCUITS AND SYSTEMS: A MODERN APPROACH
A. S. Sedra and K. C. Smith MICROELECTRONIC CIRCUITS
M. E. Van Valkenburg ANALOG FILTER DESIGN

Modern Digital and Analog Communication Systems

B. P. LATHI

CALIFORNIA STATE UNIVERSITY, SACRAMENTO

Holt, Rinehart and Winston

New York Chicago San Francisco Philadelphia
Montreal Toronto London Sydney Tokyo
Mexico Rio de Janeiro Madrid

Copyright © 1983 CBS College Publishing
All rights reserved.
Address correspondence to:
383 Madison Avenue New York, NY 10017

Library of Congress Cataloging in Publication Data

Lathi, B. P. (Bhagwandas Pannalal)
Modern digital and analog communication systems.

(HRW series in electrical and computer engineering)
Bibliography: p.
Includes index.
1. Telecommunications systems 2. Digital communications 3. Statistical communication theory. I. Title II. Series.
TK5101.L333 1983 621.38'0413 82-23224
ISBN 0-03-058969-X

Printed in the United States of America

Published simultaneously in Canada
2 3 4 5 039 9 8 7 6 5 4 3 2 1

CBS COLLEGE PUBLISHING
Holt, Rinehart and Winston
The Dryden Press
Saunders College Publishing

Contents

Preface

The study of communication systems can be divided into two distinct areas:

1. How communication systems work.
2. How they perform in the presence of noise.

The study of these two areas, in turn, requires specific tools. To study the first area, the students must be familiar with signal analysis (Fourier techniques), and to study the second area, a basic understanding of probability theory is essential.

For a meaningful comparison of various communication systems, it is necessary to have some understanding of the second area. For this reason most instructors feel that the study of communication systems is not complete unless both of the areas are covered reasonably well. As one of my colleagues put it, "I cannot imagine teaching communication systems without teaching their behavior in the presence of noise." Most of us will agree with this sentiment. However, it poses one serious problem: the material to be covered is enormous. The two areas along with their tools are overwhelming; it is difficult to cover this material in depth in one course.

The current trend in teaching communication systems is to study the tools in early chapters and then proceed with the study of the two areas of communication. Because too much time is spent in the beginning in studying the tools (without much motivation), there is little time left to study the two proper areas of communication. Consequently, teaching a course in communication systems has been a real dilemma. The second area (statistical aspects) of communication theory is a degree harder than the first area, and can be properly understood only if the first area is well assimilated. One of the reasons for this problem is our attempt to cover both areas at the same time. The students are forced to grapple with the statistical aspects while also trying to become familiar with how communication systems work. This practice is most unsound pedagogically because it violates the basic fact that one must learn to walk before one can run. The ideal solution would be to offer two courses in sequence, the first course dealing with how communication systems function and the second course dealing with statistical aspects and noise. But in the present curriculum, with so many competing courses, it will be difficult to squeeze in two basic courses in the communications area.

There is, however, a way out of this difficulty. A careful examination shows that it is really not necessary to go into probabilistic aspects (at least in the first course) in order to study comparative behavior in the presence of noise. In analog communication systems, the noise can be treated as an interference, and using a Rayleigh model to represent noise as a sum of sinusoids, it is possible to find the noise output resulting from these interfering sinusoids. This model uses the frequency domain description (the power spectral density) of noise and permits the derivation of noise power outputs and signal-to-noise ratios, and the discussion of the relative merits of various systems without requiring any statistical description of noise. The additional advantage of this approach is that, here, the power of a signal is a time average rather than an ensemble average. Time averages are much more direct and easier for the students to understand in a first course. The concept of ensemble averages is rather confusing to a beginner. Its appreciation requires a level of maturity that is unreasonable to expect from an average undergraduate in the very first course in communication systems. Also, in digital communication systems, if we consider threshold detection, the error probability depends only on the difference of strengths of pulses to be distinguished. The relative performance of various schemes (such as on-off, polar, bipolar, etc.) can be determined with ease. For example, for the same noise immunity (that is, the same error probability), we can show, without any recourse to statistical concepts, that on-off or bipolar requires two times the power needed for the polar scheme. Even the determination of error probability, which amounts to determining the probability that the noise amplitude will exceed some value, requires only a modest discussion of the probability density function. Thus to study the comparative behavior of communication systems (digital and analog) in the presence of noise, it is possible, by and large, to avoid statistical concepts and ensemble averages. I have found this to be the most appropriate way of dealing with the dilemma mentioned earlier.

The first four chapters in this book follow precisely this philosophy. These chapters treat in depth how digital and analog communication systems work and how they behave in the presence of noise in the manner discussed above. Thus, they form a sound, well-rounded, comprehensive survey course in communication systems that is within the reach of an average undergraduate and that can be taught in three to four semester hours. Once the students have mastered the first four chapters, they are ready for an in-depth treatment of statistical concepts in communication theory. Chapters 5 through 9 provide such a treatment and are appropriate for advanced undergraduates or graduate students.

Chapter 1 introduces the students to a panoramic view of communication systems. All the important concepts of communication theory are explained qualitatively in a heuristic way. This gets the students deeply interested to the point where they are anxious to study the subject. Because of this momentum, they willingly submit to the discipline of studying the tool of signal analysis in Chapter 2. Signal distortion caused by various types of channel imperfections is also discussed in this chapter. Chapter 3 deals with digital communication systems, including the digital transmission of analog signals (PCM and DM). Chapter 4 discusses linear and exponential (or angle) modulation of a carrier by analog as well as digital signals. Chapter 5 is a reasonably thorough treatment of the theory of probability and random processes. This is the second tool required for the study of communication systems. Every attempt is made to motivate the students and sustain their interest through this chapter by providing applications to communications problems wherever possible. Chapter 6 discusses the behavior of communication systems in the presence of noise—this time using ensemble averages. Optimum signal detection is discussed in Chapter 7 and information theory is introduced in Chapter 8. Finally, error-control coding is discussed in Chapter 9.

Analog pulse modulation systems such as PAM, PPM, and PWM are deemphasized in comparison to digital schemes (PCM and DM) because the applications of the former in communications are hard to find. The digital schemes are used widely now and will be used even more widely in the future. Tone-modulated FM also receives its share of deemphasis for a sound reason. Since angle modulation is nonlinear, the conclusions derived from tone modulation cannot be blindly applied to modulation by other baseband signals. In fact, these conclusions are misleading in many instances. For example, in the literature PM gets short shrift as being inferior to FM, a conclusion based on tone-modulation analysis.* It is shown in Chapters 4 and 6 that PM is, in fact, superior to FM for all practical cases (including voice). For this reason, tone-modulated FM is deemphasized and more space is devoted to PM and the comparison of FM with PM.

In my earlier books (*Signals, Systems and Communication,* Wiley, 1965 and *Communication Systems,* Wiley, 1968), I had devoted a great deal of space and effort to signal-vector analogy. In terms of the time available, signal-vector analogy, despite its charm, is a luxury which we can now ill afford due to the addition of several new areas to communication. Moreover, it unnecessarily prolongs the study of the tools (signal analysis) and is therefore distracting in a survey course. For these reasons, in the present book, the signal-vector analogy is omitted and the study of signal space is postponed to Chapter 7.

Chapters 2, 4, 5, and 6 each have two parts. Each of these chapters could easily have been split into two separate chapters. I have avoided this temptation for a good reason. When a subject is fragmented into too many chapters, beginning students are confused and bewildered by the proliferation of seemingly endless topics and they fail to see the interrelationship between them. On the other hand, it is easy to see the whole when it is divided into fewer but well-defined parts. The situation is similar to a visit to a big mansion. When a first-time visitor sees many rooms, he is confused and bewildered and fails to see the wholeness of the mansion. But if a guide carefully divides the mansion into fewer but more well-defined areas such as living area, sleeping area, recreation area, etc., it is much easier to grasp the purpose of each room and its relationship to the whole structure.

One of the aims in writing this book has been to make learning a pleasant or at least a less intimidating experience for the student by presenting the subject in a clear, understandable, and logically organized manner. Every effort has been made to give an insight—rather than just an understanding—as well as heuristic explanations of theoretical results wherever possible. Many examples are provided for further clarification of abstract results. Even a partial success in achieving my stated goal would make all my toils worthwhile.

It is a pleasure to acknowledge the assistance received from several individuals during the preparation of this book. Many students have helped me to prepare illustrations and to proofread. I would particularly like to mention Dave Carpenter, Dave Lewis, A. Sohrofiroozani, Ron Taylor, and Tao Zen. I wish to thank Professors

*Another reason given for the alleged inferiority of PM is that the phase deviation has to be restricted to a value less than π. It has been shown in Chapter 4 that this is simply not true of bandlimited analog signals.

Ronald C. Houts and S. C. Kwatra and two anonymous reviewers for many helpful comments, and Professor Yuzo Yano for his help in preparing the illustrations. The help of Professors L. H. Gabriel and C. G. Nelson in preparing the manuscript is appreciated. Thanks are also due to the secretaries who typed the successive drafts of the manuscript. I would like to mention Carma Kuhl, Beverly Barnes, Maureen Reed, and especially Vilma Pavanatti whose dedication to the effort was beyond the call of duty.

Finally, I owe a debt of gratitude to my family: my wife Rajani, my children Anjali and Pandit for their patience and understanding. A mere "thank you" really cannot make up for the hardships they suffered on account of this book.

B. P. Lathi

A Note to Instructors

This book can be used for a variety of undergraduate and graduate level courses in communications systems and theory. With a judicious selection of topics, the book can be used for a two-semester or a three-quarter sequence in the area of communication. Sufficient material is included to allow flexibility in adopting the book for a variety of courses. Some possible options are suggested below:

I. Undergraduate courses which do not require probability theory
  1. A survey course in communication systems: Chapters 1, 2, 3, and 4 (3 to 4 semester-hours or 4 to 6 quarter-hours). Sections dealing with noise (Secs. 2.10, 3.6, 4.7, and 4.16) can be omitted, if so desired, without loss of continuity.
  2. Modulation theory and noise calculations: Chapters 1, 2, and 4 (3 semester-hours or quarter-hours).
  3. Digital communications: Chapters 1, 2, and 3, and possibly 9 (2 to 3 semester-hours or 3 to 4 quarter-hours).

II. Advanced undergraduate or graduate-level courses
  1. Statistical theory of communication: Chapters 5 and 6 or 5, 6, and 7.
  2. Analog communication in noise: Chapters 1, 2, 4 (review), 5, and Part I of 6.
  3. Digital communications: Chapters 1, 2, 3 (review), 5, Part II of 6, and 7.
  4. Information theory and coding: Chapters 8 and 9.

1

Introduction

This book examines communication by electrical signals. In the past, messages have been carried by runners, carrier pigeons, drum beats, and torches. These schemes were adequate for the distances and "data rates" of the age. In most parts of the world, these modes of communication have been superseded by electrical communication systems* that can transmit signals over much longer distances (even to distant planets and galaxies) and at the speed of light.

Electrical communication is reliable and economical; communication technology will alleviate the energy crisis by trading information processing for a more rational use of energy resources. Some examples: Important discussions now mostly communicated face-to-face in meetings or conferences, often requiring travel, will increasingly use "teleconferring." Similarly, teleshopping and telebanking will provide services by electronic communication, and newspapers will be replaced by electronic news services.

---

*With the exception of the postal service!

1.1 COMMUNICATION SYSTEM

Figure 1.1 shows three examples of communication systems. A typical communication system can be modeled as shown in Fig. 1.2. The components of a communication system are as follows:

The *source* originates the message, such as a human voice, a television picture, a teletype message, or data. If the data is nonelectrical (human voice, teletype message, television picture), it must be converted by an *input transducer* into an electrical waveform referred to as the *baseband signal* or the *message signal.*

The *transmitter* modifies the baseband signal for efficient transmission.*

Figure 1.1 Some examples of communication systems.

* The transmitter consists of one or more of the following subsystems: a preemphasizer, a sampler, a quantizer, a coder, and a modulator. Similarly, the receiver may consist of a demodulator, a decoder, a filter, and a deemphasizer.

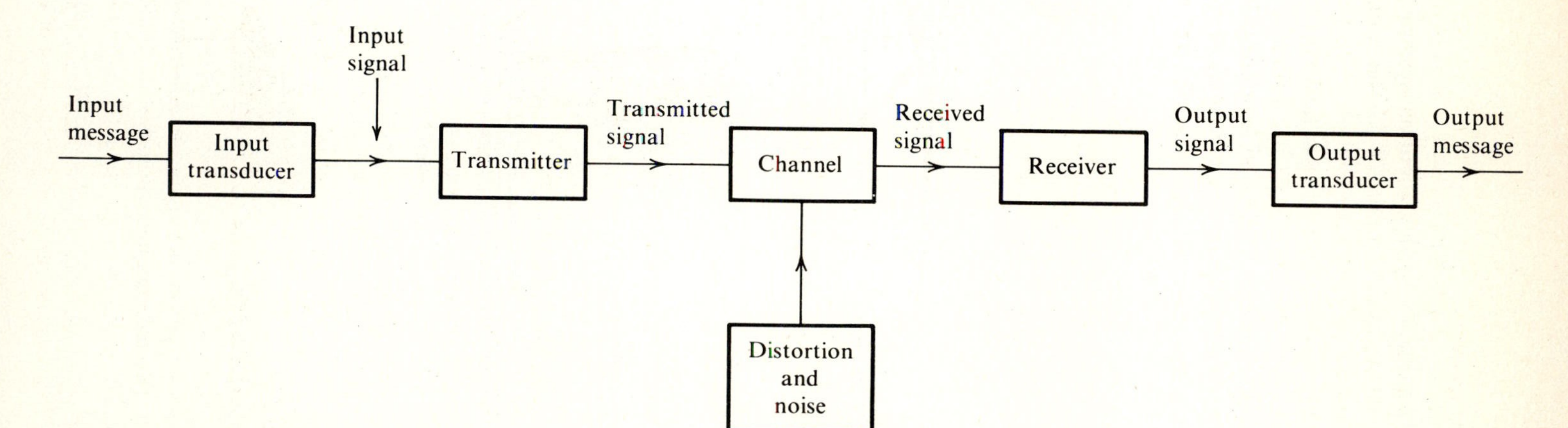

Figure 1.2 A communication system.

The *channel* is a medium—such as wire, coaxial cable, a waveguide, an optical fiber, or a radio link—through which the transmitter output is sent.

The *receiver* reprocesses the signal from the channel by undoing the signal modifications made at the transmitter and the channel. The receiver output is fed to the *output transducer,* which converts the electrical signal to its original form—the message.

The *destination* is the unit to which the message is communicated.

A channel acts partly as a filter, to attenuate the signal and distort its waveform. The length of the channel increases attenuation, varying from a few percent for short distances to orders of magnitude for interplanetary communication. The waveform is distorted because of different amounts of attenuation and phase shift suffered by different frequency components of the signal. For example, a square pulse is rounded or "spread out" by the process. This type of distortion, called *linear distortion,* can be partly corrected at the receiver by an equalizer with gain and phase characteristics complementary to those of the channel.

The channel may also cause *nonlinear distortion* through attenuation that varies with the signal amplitude. Such distortion can also be partly corrected by a complementary equalizer at the receiver.

The signal is not only distorted by the channel, but it is also contaminated along the path by undesirable signals lumped under the broad term *noise* which are random and unpredictable signals from causes external and internal. External noise includes interference from signals transmitted on nearby channels, man-made noise generated by faulty contact switches for electrical equipment, by automobile ignition radiation, fluorescent lights, and natural noise from lightning, electrical storms, solar and intergalactic radiation. With proper care, external noise can be minimized or even eliminated. Internal noise results from thermal motion of electrons in conductors, random emission, and diffusion or recombination of charged carriers in electronic devices. Proper care can reduce the effect of internal noise but can never eliminate it. Noise is one of the basic factors that sets a limit on the rate of communication.

The *signal-to-noise ratio* (*SNR*) is defined as the ratio of the signal power to the noise power. The channel distorts the signal, and noise accumulates along the path. Worse yet, the signal strength decreases while the noise level increases with distance from the transmitter. Thus, the SNR is continuously decreasing along the channel. Amplification of the received signal to make up for the attenuation is of no avail because the noise will be amplified in the same proportion, and the SNR remains, at best, unchanged.*

1.2 ANALOG AND DIGITAL MESSAGES

Messages are digital or analog; digital messages are constructed with a finite number of symbols. For example, printed language consists of 26 letters, 10 numbers, a "space," and several punctuation marks. Thus, a text is a digital message constructed from about 50 symbols. Human speech is also a digital message, because it is made

* Actually, amplification further deteriorates the SNR because of the amplifier noise.

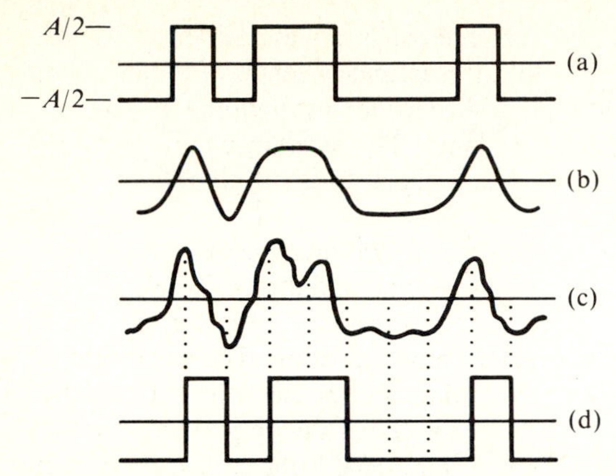

Figure 1.3 (a) Transmitted signal. (b) Received distorted signal (without noise). (c) Received distorted signal (with noise). (d) Regenerated signal (delayed).

up from a finite vocabulary in a language.* Similarly, a Morse-coded telegraph message is a digital message constructed from a set of only *two* symbols: mark and space. It is, therefore, a *binary* message, implying only two symbols. A digital message constructed with M symbols is called an *M-ary* message.

On the other hand, analog messages are characterized by data whose value varies over a continuous range. For example, the temperature or the atmospheric pressure of a certain location can vary over a continuous range and can assume infinite possible values. Similarly, a speech waveform has amplitudes that vary over a continuous range. Over a given time interval, infinite possible different speech waveforms exist, in contrast to only a finite number of possible digital messages.

Digital messages are transmitted by using a finite set of electrical waveforms. For example, in the Morse code, a mark can be transmitted by an electrical pulse of amplitude $A/2$, and a space can be transmitted by a pulse of amplitude $-A/2$. In an M-ary case, M distinct electrical pulses (or waveforms) are used; each of the M pulses represents one of the M possible symbols. The task of the receiver is to extract a message from a distorted and noisy signal at the channel output. Message extraction is often easier from digital signals than from analog signals. Consider a binary case: two symbols are encoded as rectangular pulses of amplitudes $A/2$ and $-A/2$. The only decision at the receiver is the selection between two possible pulses received, not the details of the pulse shape; the decision is readily made with reasonable certainty even if the pulses are distorted and noisy (Fig. 1.3). Hence a digital communication system can transmit messages with greater accuracy than an analog system in the presence of distortion and noise.

The possibility of using *regenerative repeaters* is a further advantage for digital

* Here we imply the written text of the speech rather than its details such as pronunciation of words and varying inflections, pitch, emphasis, etc. The speech signal from a microphone contains all these details. This signal is an analog signal and its information content is more than a thousand times the information in the written text of the same speech.

communication. A repeater station detects pulses and transmits new clean pulses, thus combating accumulation of distortion and noise and enabling information transmission over longer distances with greater accuracy.

In contrast to digital messages, the waveform in analog messages is important, and even a slight distortion or interference in the waveform will cause an error in the received signal. A further difficulty: a regenerative repeater is not viable for analog signals because the noise and distortion, no matter how small, cannot be cleaned up from a signal.

As a result, the distortion and the noise interference are cumulative over the entire transmission path. To compound the difficulty, the signal is attenuated continuously over the transmission path; thus with increasing distance, the signal becomes weaker, whereas the distortion and noise become stronger. Ultimately, the signal, overwhelmed by the distortion and the noise, is mutilated. Amplification is of little help, because it enhances the signal and the noise in the same proportion. Consequently, the distance over which an analog message can be transmitted is limited by the transmitter power. Yet analog communication is being used widely and successfully despite these problems. A tendency exists, however, to replace analog systems with digital systems as the latter become more economical because of a dramatic cost reduction achieved in the fabrication of digital circuitry.

Analog-to-Digital Conversion

A meeting ground exists for analog and digital signals: conversion of analog signals to digital (A/D conversion). The frequency spectrum of a signal indicates relative magnitudes of various frequency components. The *sampling theorem* (to be proved in

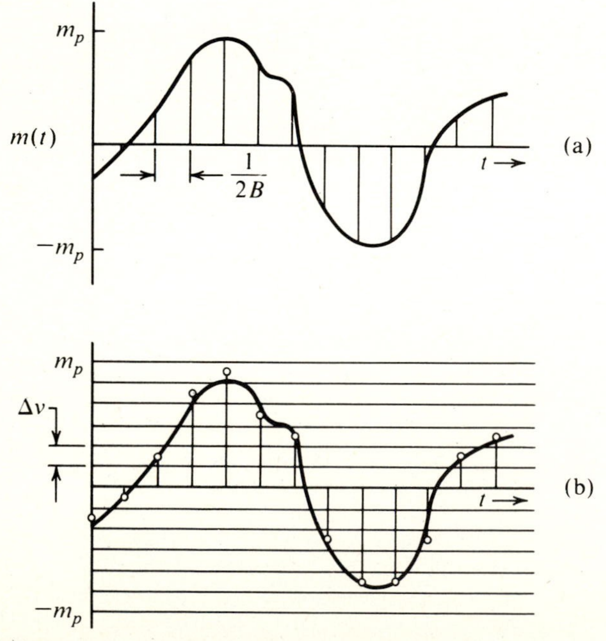

Figure 1.4 Signal sampling and quantizing.

Chapter 2) states that if the highest frequency in the signal spectrum is B (in Hz), the signal can be reconstructed from its samples, taken at a rate not less than $2B$ samples/second. This means that in order to transmit the information in a continuous signal, we need only transmit its samples (Fig. 1.4*a*). Unfortunately, the sample values are still not digital because they lie in a continuous range and can take on any one of the infinite values in the range. We are back where we started! This difficulty is neatly resolved by what is known as *quantization,* where each sample is approximated, or "rounded off," to the nearest quantized level, as shown in Fig. 1.4*b*. Amplitudes of the signal $m(t)$ lie in the range $(-m_p, m_p)$, which is partitioned into L intervals, each of magnitude $\Delta v = 2m_p/L$. Each sample amplitude is approximated to the midpoint of the interval in which the sample value falls. Each sample is now approximated to one of the L numbers. The information is thus digitized.

The quantized signal is an approximation of the original signal. We can improve the accuracy of the quantized signal to any desired degree by increasing the number of levels (L). For intelligibility of voice signals, for example, $L = 8$ or 16 is sufficient. For commercial use, $L = 32$ is a minimum, and for telephone communication, $L = 128$ or 256 is commonly used.

During each sampling interval T_o, we transmit one quantized sample, which takes on one of the L values. This requires L distinct waveforms, each of duration T_o. These may be constructed, for example, by using a basic rectangular pulse of amplitude $A/2$ (Fig. 1.5) and its multiples (for example, $\pm(A/2)$, $\pm(3A/2)$, $\pm(5A/2)$, . . . , $\pm[(L-1)A/2]$) to form L distinct waveforms to be assigned to the L values to be transmitted. Amplitudes of any two of these waveforms are separated by at least A to guard against noise interference and channel distortion. Another possibility is to use fewer than L waveforms and form their combinations (codes) to yield L distinct patterns. As an example, for the case $L = 16$ we may use 16 pulses, $\pm(A/2)$, $\pm(3A/2)$, . . . , $\pm(15A/2)$, each of duration T_o. The second alternative is to use only two basic pulses, $A/2$ and $-A/2$, each of duration $T_o/4$. A sequence of four such pulses gives $2 \times 2 \times 2 \times 2 = 16$ distinct patterns, as shown in Fig. 1.6. We can assign one pattern to each of the 16 quantized values to be transmitted. Each quantized sample is now coded into a sequence of four binary pulses. This is the so-called binary case, where signaling is carried out by means of only two basic pulses (or symbols).

An intermediate case exists where we use four basic pulses (quaternary pulses) of amplitudes $\pm(A/2)$ and $\pm(3A/2)$, each of duration $T_o/2$. A sequence of two quaternary pulses can form $4 \times 4 = 16$ distinct levels or values.

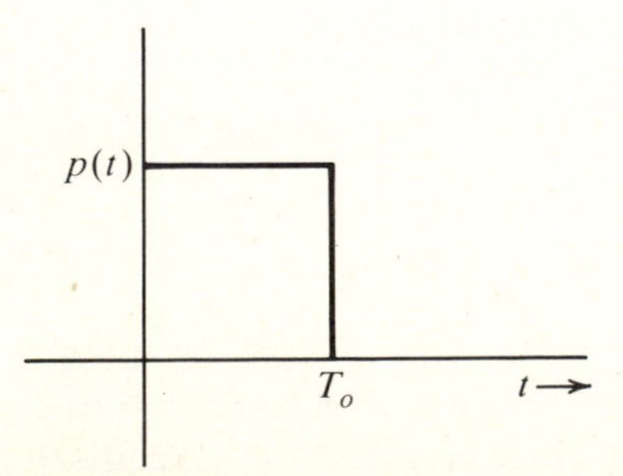

Figure 1.5

Digit	Binary equivalent	Pulse code waveform
0	**0000**	
1	**0001**	
2	**0010**	
3	**0011**	
4	**0100**	
5	**0101**	
6	**0110**	
7	**0111**	
8	**1000**	
9	**1001**	
10	**1010**	
11	**1011**	
12	**1100**	
13	**1101**	
14	**1110**	
15	**1111**	

Figure 1.6 An example of a binary pulse code.

The binary case is of great practical importance because of its simplicity and ease of detection. Virtually all digital communication today is binary. The above scheme of transmitting data by digitizing and then using pulse codes to transmit the digitized data is known as *pulse-code modulation* (*PCM*).

Noise Immunity of Digital Signals

A typical distorted binary signal with noise acquired over the channel is shown in Fig. 1.3. If A is sufficiently large compared to typical noise amplitudes, the receiver can still correctly distinguish between the two pulses. The pulse amplitude is typically 5 to 10 times the rms noise amplitude. For such a high SNR, the probability of error at the receiver is less than 10^{-6}; that is, on the average, the receiver will make less than one error per million pulses. The effect of random channel noise and distortion is thus practically eliminated. Hence, when analog signals are transmitted by digital means, the only error, or uncertainty, in the received signal is that caused by quantization. By increasing L, we can reduce the uncertainty, or error, caused by quantization to any desired amount. At the same time, because of the possibility of regenerative repeaters, we can transmit signals over a much longer distance than would have been possible for the analog signal. As will be seen later in this chapter, the price for all these benefits of digital communication is paid in terms of increased bandwidth of transmission.

From all this discussion, we arrive at a rather interesting (and by no means obvious) conclusion—that every possible communication can be carried on with a minimum of two symbols. Thus, merely by using a proper sequence of a wink of the eye, one can convey any message, be it a conversation, a book, a movie, or an opera star's singing. Every possible detail (such as various shades of colors of the objects and tones of the voice, etc.) that is reproducible on a movie screen or on the best quality color television can be conveyed with no less accuracy, merely by a wink of an eye.*

Although PCM was invented by P. M. Rainey in 1926 and rediscovered by A. H. Reeves in 1939, it was not until the early sixties that Bell Laboratories installed the first communication link using PCM. The cost and size of vacuum tube circuits was the chief impediment to the use of PCM in the early days. It was the transistor that made PCM practicable.

1.3 THE SIGNAL-TO-NOISE RATIO, THE CHANNEL BANDWIDTH, AND THE RATE OF COMMUNICATION

The fundamental parameters that control the rate and the quality of information transmission are the channel bandwidth B and the signal power S. The appropriate quantitative relationships will be derived later. Here we shall demonstrate these relationships qualitatively.

Bandwidth of a channel is the range of frequencies that it can transmit with reasonable fidelity. For example, if a channel can transmit with reasonable fidelity a signal whose frequency components occupy a range from 0 (dc) up to a maximum of 5000 Hz (5 kHz), the channel bandwidth B is 5 kHz.

To understand the role of B, consider the possibility of increasing the speed of information transmission by time compression of the signal. If a signal is compressed in time by a factor of two, it can be transmitted in half the time, and the speed of transmission is doubled. Compression by a factor of two, however, causes the signal to "wiggle" twice as fast, implying that the frequencies of its components are doubled. To transmit this compressed signal without distortion, the channel bandwidth must also be doubled. Thus, the rate of information transmission is directly proportional to B. More generally, if a channel of bandwidth B can transmit N pulses per second, then

*Of course, to convey the information in a movie or a television program in real time, the winking would have to be at an inhumanly high speed.

to transmit KN pulses per second we need a channel of bandwidth KB. To reiterate, the number of pulses/second that can be transmitted over a channel is directly proportional to its bandwidth B.

Signal power S plays a dual role in information transmission. First, S is related to the quality of transmission. Increasing S, the signal power, reduces the effect of the channel noise, and the information is received more accurately, or with less uncertainty. A larger signal-to-noise ratio (SNR) also allows transmission over a longer distance. In any event, a certain minimum SNR is necessary to communication.

The second role of the signal power is not as obvious, although it is very important. We shall demonstrate that the channel bandwidth B and S are exchangeable. That is, to maintain a given rate and accuracy of information transmission, we can trade S for B and vice versa. Thus, one may reduce B if one is willing to increase S, or one may reduce S if one is willing to increase B. The rigorous proof of this will be provided in Chapter 8. Here we shall give only a "plausability argument."

Consider the PCM scheme discussed earlier, with 16 quantization levels ($L = 16$). Here we may use 16 distinct pulses of amplitudes $\pm (A/2)$, $\pm (3A/2)$, . . . , $\pm (15A/2)$ to represent the 16 levels (16-ary case). Each sample is transmitted by one of the 16 pulses during the sampling interval T_o. The amplitudes of these pulses range from $-15A/2$ to $15A/2$. Alternately, we may use the binary scheme, where a group of four binary pulses is used to transmit each sample during the sampling interval T_o. In the latter case, the transmitted power is reduced considerably because the peak amplitude of transmitted pulses is only $A/2$, as compared to the peak amplitude $15A/2$ in the 16-ary case. In the binary case, however, we need to transmit four pulses in each interval T_o instead of just one pulse required in the 16-ary case. Thus, the required channel bandwidth in the binary case is four times as great as that for the 16-ary case. In both cases, the minimum amplitude separation between transmitted pulses is A, and we therefore have about the same error probability at the receiver. This means the quality of the received signal is about the same in both cases. In the binary case, the transmitted signal power is reduced at the cost of increased bandwidth. We have demonstrated here the exchangeability of the S with B. It will be noticed that relatively little increase in B enables significant reduction in S. In conclusion, the two primary communication resources are the bandwidth and the transmitted power. In a given communication channel, one resource may be more valuable than the other, and the communication scheme should be designed accordingly. A typical telephone channel, for example, has a limited bandwidth (3 kHz) but a lot of power is available. On the other hand, in space vehicles, infinite bandwidth is available but the power is limited. Hence, the communication schemes required in the two cases are radically different.

Since the SNR is proportional to the power S, we can say that SNR and bandwidth are exchangeable. It will be shown in Chapter 8 that the relationship between the bandwidth expansion factor and the SNR is exponential. Thus, if a given rate of information transmission requires a channel bandwidth B_1 and a signal-to-noise ratio SNR_1, then it is possible to transmit the same information over a channel bandwidth B_2 and a signal-to-noise ratio SNR_2 where

$$\text{SNR}_2 \simeq \text{SNR}_1^{B_1/B_2} \tag{1.1}$$

Thus, if we double the channel bandwidth, the required SNR is only a square root of the former SNR, and tripling the channel bandwidth reduces the corresponding SNR to only a cube root of the former SNR. Thus, a relatively small increase in channel bandwidth buys a large advantage in terms of reduced transmission power. But a large increase in transmitted power buys a meager advantage in bandwidth reduction. Hence, in practice, the exchange between B and SNR is usually in the sense of increasing B to reduce transmitted power and is rarely the other way around.

Equation (1.1) gives the upper bound on the exchange between the SNR and B. Not all systems are capable of achieving this bound. For example, frequency modulation (FM) is one scheme that is commonly used in radio broadcasting for improving the signal quality at the receiver by increasing the transmission bandwidth. We shall see that an FM system does not make efficient use of bandwidth in reducing the required SNR, and its performance falls far short of that in Eq. (1.1). PCM, on the other hand, comes close (within 10 dB) to realizing the performance in Eq. (1.1). Generally speaking, transmission of signals in digital form comes much closer to the realization of the limit in Eq. (1.1) than does transmission of signals in analog form.

The limitation imposed on communication by the channel bandwidth and the SNR is dramatically highlighted by Shannon's equation.*

$$C = B \log_2 (1 + \text{SNR}) \qquad \text{bits/second} \tag{1.2}$$

Here C is the rate of information transmission per second. This rate C (known as the channel capacity) is the maximum number of binary symbols (bits) that can be transmitted per second with a probability of error arbitrarily close to zero. In other words, a channel can transmit $B \log_2 (1 + \text{SNR})$ binary digits, or symbols, per second as accurately as one desires. Moreover, it is impossible to transmit at a rate higher than this without incurring errors. Shannon's equation clearly brings out the limitation on the rate of communication imposed by B and SNR. If there were no noise on the channel ($N = 0$), $C = \infty$, and communication would cease to be a problem. We could then transmit any amount of information in the world over a channel. This can be readily verified: if noise were zero, there would be no uncertainty in the received pulse amplitude, and the receiver would be able to detect any pulse amplitude without ambiguity. The minimum pulse-amplitude separation A can be arbitrarily small, and for any given pulse, we have an infinite number of levels available. We can assign one level to every possible message. For example, the contents of this book will be assigned one level; if it is desired to transmit this book, all that is needed is to transmit one pulse of that level. Because an infinite number of levels are available, it is possible to assign one level to any conceivable message. Cataloging of such a code may not be practical, but that is beside the point. The point is that if the noise is zero, communication ceases to be a problem, at least theoretically. Implementation of such a scheme would be difficult because of the requirement of generation and detection of pulses of precise amplitudes. Such practical difficulties would then set a limit on the rate of communication.

---

*This is true for a certain kind of noise—the white gaussian noise.

In conclusion, we have demonstrated qualitatively the basic role played by B and SNR in limiting the performance of a communication system. These two parameters then represent the ultimate limitation on a rate of communication. We have also demonstrated the possibility of trade or exchange between these two basic parameters.

Equation (1.1) can be derived from Eq. (1.2). It should be remembered that Shannon's result represents the upper limit on the rate of communication over a channel and can be achieved only with a system of monstrous and impracticable complexity and a time delay in reception approaching infinity. Practical systems operate at rates below the Shannon rate. In Chapter 8, we shall derive Shannon's result and compare the efficiencies of various communication systems.

1.4 MODULATION

Baseband signals produced by various information sources are not always suitable for direct transmission over a given channel. These signals are usually further modified to facilitate transmission. This conversion process is known as *modulation*. In this process, the baseband signal is used to modify some parameter of a high-frequency carrier signal.

A carrier is a sinusoid of high frequency, and one of its parameters—such as amplitude, frequency, or phase—is varied in proportion to the baseband signal $m(t)$. Accordingly, we have amplitude modulation (AM), frequency modulation (FM), or phase modulation (PM). Figure 1.7 shows a baseband signal $m(t)$ and the corresponding AM and FM waveforms. In AM, the carrier amplitude varies in proportion to $m(t)$, and in FM, the carrier frequency varies in proportion to $m(t)$.

At the receiver, the modulated signal must pass through a reverse process called *demodulation* in order to retrieve the baseband signal.

As mentioned earlier, modulation is used to facilitate transmission. Some of the important reasons for modulation are given below.

Ease of Radiation

For efficient radiation of electromagnetic energy, the radiating antenna should be of the order of one-tenth or more of the wavelength of the signal radiated. For many baseband signals, the wavelengths are too large for reasonable antenna dimensions. For example, the power in a speech signal is concentrated at frequencies in the range of 100 Hz to 3000 Hz. The corresponding wavelength is 100 km to 3000 km. This long wavelength would necessitate an impracticably large antenna. Instead, we modulate a high-frequency carrier, thus translating the signal spectrum to the region of carrier frequencies that corresponds to a much smaller wavelength. For example, a 1 MHz carrier has a wavelength of only 300 meters and requires an antenna whose size is of the order of 30 meters. In this aspect, modulation is like letting the baseband signal hitchhike on a high-frequency sinusoid (carrier). The carrier and the baseband signal may be compared to a stone and a piece of paper. If we wish to throw a piece of paper, it cannot go too far by itself. But by wrapping it around a stone (a carrier), it can be thrown over a longer distance.

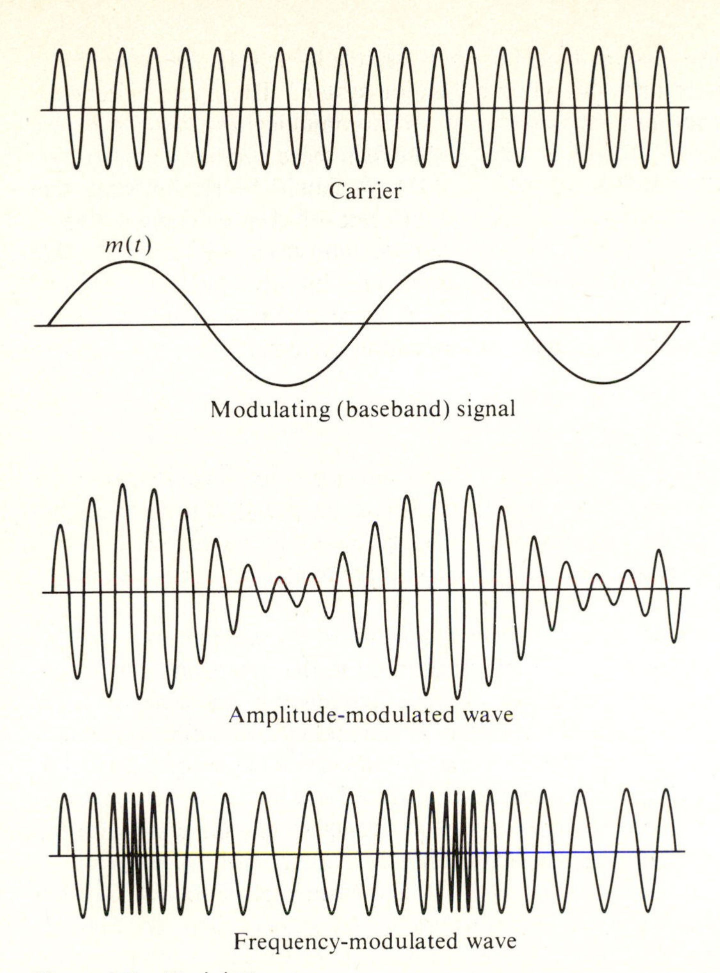

Figure 1.7 Modulation.

Simultaneous Transmission of Several Signals

Consider the case of several radio stations broadcasting audio baseband signals directly, without any modification. They would interfere with each other because the spectra of all the signals occupy more or less the same bandwidth. Thus, it would be possible to broadcast from only one radio or TV station at a time. This is wasteful because the channel bandwidth may be much larger than that of the signal. One way to solve this problem is to use modulation. We can use various audio signals to modulate different carrier frequencies, thus translating each signal to a different frequency range. If the various carriers are chosen sufficiently far apart in frequency, the spectra of the modulated signals will not overlap and thus will not interfere with each other. At the receiver, one can use a tuneable bandpass filter to select the desired station or signal. This method of transmitting several signals simultaneously is known as *frequency-division multiplexing* (*FDM*). Here the bandwidth of the channel is shared by various signals without any overlapping.

Another method of multiplexing several signals is known as *time-division multiplexing* (*TDM*). This method is suitable when a signal is in the form of a pulse train (as in PCM). The pulses are made narrower, and the spaces that are left between pulses are used for pulses from other signals. Thus, in effect, the transmission time is shared by a number of signals by interleaving the pulse trains of various signals in a specified order. At the receiver, the pulse trains corresponding to various signals are separated.

Effecting the Exchange of SNR with *B*

We have shown earlier that it is possible to exchange SNR with the bandwidth of transmission. Frequency or phase modulation can effect such an exchange. The amount of modulation (to be defined later) used controls the exchange of SNR and the transmission bandwidth.

1.5 RANDOMNESS, REDUNDANCY, AND CODING

Randomness plays an important role in communication. As noted earlier, one of the limiting factors in the rate of communication is noise which is a random signal. Randomness is also closely associated with information. Indeed, randomness is the essence of communication. Randomness means unpredictability, or uncertainty, of the outcome. If a source had no unpredictability, or uncertainty, it would be known beforehand and would convey no information. Probability is the measure of certainty, and information is associated with probability. If a person winks, it conveys some information in a given context. But if a person were to wink continuously with the regularity of a clock, it would convey no meaning. The unpredictability of the winking is what gives the information to the signal. What is more interesting, however, is that from the engineering point of view, also, information is associated with uncertainty. The information of a message, from the engineering point of view, is defined as a quantity proportional to the minimum time needed to transmit it. Consider the Morse code, for example. In this code, various combinations of marks and spaces (code words) are assigned to each letter. In order to minimize the transmission time, shorter code words are assigned to more frequently occurring (more probable) letters (such as e, t, and a) and longer code words are assigned to rarely occurring (less probable) letters (such as x, q, and z). Thus, the time required to transmit a message is closely related to the probability of its occurrence. It will be shown in Chapter 8 that for digital signals, the overall transmission time is minimized if a message (or symbol) of probability P is assigned a code word with a length proportional to $\log (1/P)$. Hence, from an engineering point of view, the information of a message with probability P is proportional to $\log (1/P)$.

Redundancy also plays an important role in communication. It is essential for reliable communication. Because of redundancy, we are able to decode a message accurately despite errors in the received message. Redundancy thus helps combat noise. All languages are redundant. For example, English is about 50 percent redundant; that is, on the average, we may throw out half of the letters or words without destroying the message. This also means that in any English message, the speaker or the writer has free choice over half the letters or words, on the average. The remaining half is determined by the statistical structure of the language. If all the redundancy of English were removed, it would take about half the time to transmit a telegram or telephone conversation. If an error occurs at the receiver, however, it would be rather difficult to make sense out of the received message. The redundancy in a message, therefore, plays a useful role in combating the noise in the channel. This same principle of redundancy applies in coding messages. A deliberate redundancy is used to combat the noise. For example, in order to transmit samples with $L = 16$ quantizing levels, we may use a group of four binary pulses, as shown in Fig. 1.6. In this coding scheme, no redundancy exists. If an error occurs in the reception of even one of the pulses, the receiver will produce a wrong value. Here we may use redundancy to eliminate the effect of possible errors caused by channel noise or imperfections. Thus, if we add to each code word one more pulse of such polarity as to make the sum of all positive pulses even, we have a code that can detect a single error in any place. Thus to the code words **0001** we add a fifth pulse, of positive polarity, to make a new code word, **00011.** Now the sum of positive pulses is two (even). If a single error occurs in any position, this parity will be violated. The receiver knows that an error has been made and can request retransmission of the message. This is a very simple coding scheme. It can only detect that an error is made but cannot locate it. Moreover, it cannot detect an even number of errors. By introducing more redundancy, it is possible not only to detect but also to correct errors. For example, for $L = 16$, it can be shown that properly adding three pulses will not only detect but also correct a single error occurring at any location. This subject of error-correcting codes will be discussed in Chapter 9.

2

Analysis and Transmission of Signals

PART I Signal Analysis

This chapter deals with the basic tools required in the analysis and transmission of signals. In Part I, a different way of signal specification, the so-called *frequency-domain* description, will be studied.

When a signal is described as a function of time, such as $\cos(\omega_o t + \theta)$ or e^{-at}, it is a *time-domain* description of the signal. We shall show in this chapter that any signal that can be generated in a laboratory can be expressed as a sum* of sinusoids of various frequencies. Each signal, therefore, has a frequency spectrum represented by amplitudes (and phases) of various frequency components and is completely specified by its spectrum. A signal can therefore be described as a function of time (time-domain description) or by its frequency spectrum (frequency-domain description).

In Part II, we shall discuss the effects of channel characteristics on a signal during its transmission. The frequency-domain description of a signal facilitates the understanding of these effects.

---

* The sum may be a discrete sum or an integral (sum over a continuum).

2.1 PERIODIC SIGNAL REPRESENTATION BY FOURIER SERIES

Figure 2.1 shows an example of a periodic signal. A periodic signal $g(t)$ is defined by the property

$$g(t) = g(t + T_o) \qquad T_o \neq 0 \tag{2.1}$$

The smallest value of T that satisfies Eq. (2.1) is called the *period*. The repetition frequency f_o is $1/T_o$. A periodic signal remains unchanged by a positive or a negative shift of any integral multiple of the period T_o. This means a true periodic signal must begin at $t = -\infty$ and go on forever until $t = \infty$.

Let us consider a signal $g(t)$ formed by adding sinusoids of frequencies 0, f_o, $2f_o, \ldots, kf_o$.

$$\begin{aligned} g(t) &= a_0 + a_1 \cos 2\pi f_o t + a_2 \cos 2(2\pi f_o)t + \cdots + a_k \cos 2\pi k f_o t \\ &\qquad + b_1 \sin 2\pi f_o t + b_2 \sin 2(2\pi f_o)t + \cdots + b_k \sin 2\pi k f_o t \\ &= a_0 + \sum_{n=1}^{k} a_n \cos 2\pi n f_o t + b_n \sin 2\pi n f_o t \end{aligned} \tag{2.2a}$$

It can readily be seen that this signal is periodic with period $T_o = 1/f_o$. This follows from the fact that

$$\begin{aligned} g(t + T_o) &= a_0 + \sum_{n=1}^{k} a_n \cos 2\pi n f_o(t + T_o) + b_n \sin 2\pi n f_o(t + T_o) \\ &= a_0 + \sum_{n=1}^{k} a_n \cos (2\pi n f_o t + 2\pi n) + b_n \sin (2\pi n f_o t + 2\pi n) \\ &= a_0 + \sum_{n=1}^{k} a_n \cos 2\pi n f_o t + b_n \sin 2\pi n f_o t \\ &= g(t) \end{aligned}$$

Therefore, any combination of sinusoids of frequencies 0, f_o, $2f_o, \ldots, kf_o$ is a periodic signal with period T_o. This is true regardless of the values of amplitudes (a_n's and b_n's) of these sinusoids in Eq. (2.2a). It is clear that by changing the values of a_n's and b_n's in Eq. (2.2a) we can construct a variety of periodic signals.

The converse of this result is also true. Any periodic signal $g(t)$ with period T_o can

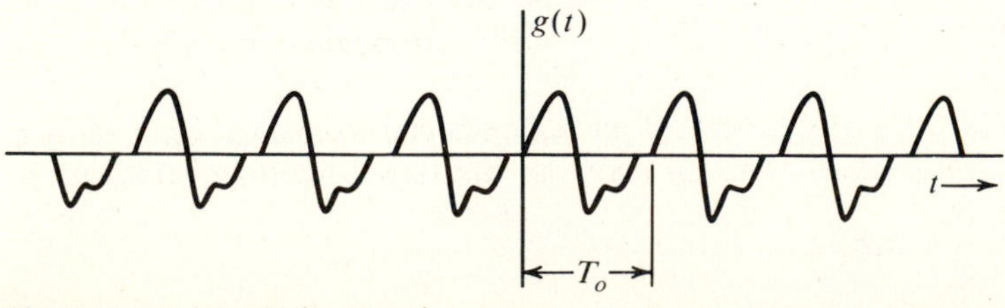

Figure 2.1 A periodic signal.

be expressed as a sum of sinusoids of frequencies f_o and all its integral multiples* ($f_o = 1/T_o$) as in Eq. (2.2a). Note that frequency 0 is dc. The frequency f_o is known as the *fundamental frequency* and the frequency nf_o is the nth harmonic frequency. For convenience, we let $\omega_o = 2\pi f_o$ in Eq. (2.2a) to yield

$$g(t) = a_0 + \sum_{n=1}^{\infty} a_n \cos n\omega_o t + b_n \sin n\omega_o t \tag{2.2b}$$

To determine coefficients a_n and b_n, we first observe that

$$\int_{t_o}^{t_o+T_o} \cos k\omega_o t \cos n\omega_o t \, dt = \begin{cases} 0 & k \neq n \\ T_o/2 & k = n \end{cases}$$

$$\int_{t_o}^{t_o+T_o} \sin k\omega_o t \sin n\omega_o t \, dt = \begin{cases} 0 & k \neq n \\ T_o/2 & k = n \end{cases} \tag{2.3}$$

$$\int_{t_o}^{t_o+T_o} \sin k\omega_o t \cos n\omega_o t \, dt = 0$$

Integrating both sides of Eq. (2.2b) over one period t_o to $t_o + T_o$, we obtain a_o:

$$a_0 = \frac{1}{T_o} \int_{t_o}^{t_o+T_o} g(t) \, dt \tag{2.4a}$$

Next we multiply both sides of Eq. (2.2b) by $\cos k\omega_o t$ and integrate over one period $(t_o, t_o + T_o)$. Every term on the right-hand side, except the one for $n = k$, will vanish because of Eq. (2.3) giving a_k or a_n as:

$$a_n = \frac{2}{T_o} \int_{t_o}^{t_o+T_o} g(t) \cos n\omega_o t \, dt \qquad \omega_o = 2\pi f_o = \frac{2\pi}{T_o} \tag{2.4b}$$

In a similar way, by multiplying both sides of Eq. (2.2b) by $\sin n\omega_o t$ and integrating, we get

$$b_n = \frac{2}{T_o} \int_{t_o}^{t_o+T_o} g(t) \sin n\omega_o t \, dt \tag{2.4c}$$

where t_o is arbitrary.

The trigonometric Fourier series in Eq. (2.2) can be written in a more compact and meaningful way as follows:

$$g(t) = C_0 + \sum_{n=1}^{\infty} C_n \cos (n\omega_o t + \theta_n) \qquad \omega_o = 2\pi f_o \tag{2.5}$$

---

* To be more precise, the right-hand side of Eq. (2.2b) converges to $g(t)$ in the mean; that is, the mean square error (averaged over one period) approaches zero, where the error signal $\epsilon(t)$ is defined as[1—3]

$$\epsilon(t) = g(t) - [a_0 + \sum_{n=1}^{\infty} a_n \cos n\omega_o t + b_n \sin n\omega_o t]$$

From the trigonometric identity* it follows that

$$C_0 = a_0$$

$$C_n = \sqrt{a_n^2 + b_n^2}$$

and

$$\theta_n = -\tan^{-1}\frac{b_n}{a_n} \tag{2.6}$$

From the compact Fourier series it follows that $g(t)$ consists of sinusoidal signals of frequencies $0, f_o, 2f_o, \ldots, nf_o, \ldots$. The nth harmonic, $C_n \cos(n\omega_o t + \theta_n)$, has amplitude C_n and phase θ_n. We can plot the magnitude spectrum (C_n vs. ω) and the phase spectrum (θ_n vs. ω) for a given periodic signal.

EXAMPLE 2.1

Find the trigonometric Fourier series for the periodic signal $g(t)$ in Fig. 2.2. In this case:

$$T_o = \tfrac{1}{2},\; f_o = 2, \text{ and } \omega_o = 2\pi f_o = 4\pi$$

Hence the Fourier series consists of frequencies 0, 2, 4, 6, . . . or angular frequencies $0, 4\pi, 8\pi, 12\pi, \ldots$.

$$g(t) = a_0 + \sum_{n=1}^{\infty} a_n \cos 4\pi nt + b_n \sin 4\pi nt \tag{2.7}$$

where†

$$a_0 = 2\int_0^{1/2} e^{-t}\,dt = 0.79 \tag{2.8a}$$

$$a_n = 4\int_0^{1/2} e^{-t}\cos 4\pi nt\,dt = 0.79\left(\frac{2}{1 + 16\pi^2 n^2}\right) \tag{2.8b}$$

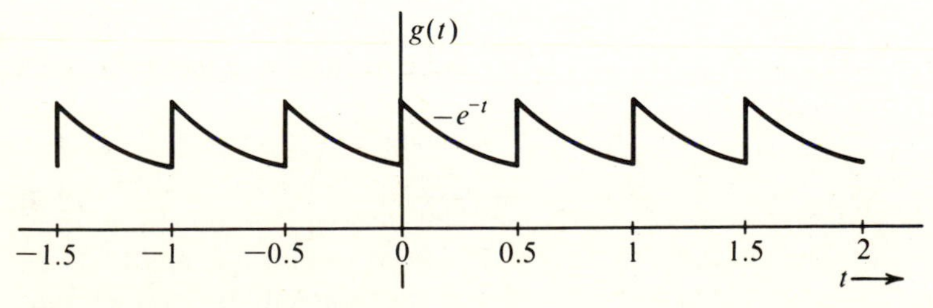

Figure 2.2 A periodic signal.

---

*$a\cos x + b\sin x \equiv \sqrt{a^2 + b^2}\cos(x - \tan^{-1} b/a)$

†Here we are using $t_o = 0$. Any other value of t_o would yield the same result.

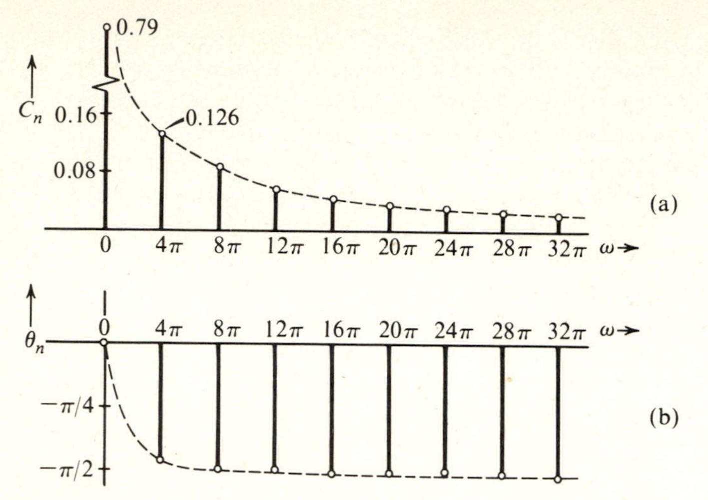

Figure 2.3 Magnitude and phase spectra of the periodic signal in Fig. 2.2.

Similarly,

$$b_n = 4 \int_0^{1/2} e^{-t} \sin 4\pi nt \, dt = 0.79 \left(\frac{8\pi n}{1 + 16\pi^2 n^2} \right) \tag{2.8c}$$

The compact Fourier series is given by

$$g(t) = C_0 + \sum_{n=1}^{\infty} C_n \cos (4\pi nt + \theta_n) \tag{2.9}$$

where, from Eqs. (2.6) and (2.8),

$$C_0 = 0.79$$

$$C_n = 0.79 \left(\frac{2}{\sqrt{1 + 16\pi^2 n^2}} \right) \tag{2.10}$$

$$\theta_n = -\tan^{-1} (4\pi n)$$

The magnitude and the phase spectra of $g(t)$ are plotted as a function of ω in Fig. 2.3. ■

In our discussion, we shall use the angular frequency ω instead of f in order to avoid carrying the factor 2π, which appears in numerous relationships [e.g., Eq. (2.2a)]. Many texts and much of the advanced literature in journals, however, use f instead of ω. The use of f does simplify certain relationships by eliminating the factor 2π and making time-domain and frequency-domain relationships more symmetrical. Unfortunately, it accomplishes this at the cost of adding the factor 2π to many other relationships and making them rather unwieldy. Moreover, students are much more at home with the variable ω because it is widely used in all areas of systems analysis. The author feels that the use of ω has much to offer over that of f, at least in a first course in communication. Actually, ω should be considered as an abbreviated notation for $2\pi f$. In this text we shall move from ω to f and from f to ω whenever convenient, thus making the best of both worlds. Although both ω and f will be referred to as frequency, it should be understood that ω refers to angular frequency.

A signal $g(t)$ may be specified as a function of time, as in Fig. 2.2 (time-domain specification), or it may be specified by its spectra, as in Fig. 2.3 (frequency-domain specification). The frequency spectra of a periodic signal exist only at frequencies 0, f_o, $2f_o$, . . . and are therefore *discrete*. We shall soon see that the spectra of non-periodic signals exist over all frequencies in a given range and are continuous spectra.

The signal $g(t)$ has a jump discontinuity at $t = 0$ with $g(0^-) = e^{-0.5} = 0.606$ and $g(0^+) = 1$. The Fourier series, however, converges to a point midway between these values, viz., $(1 + 0.606)/2 = 0.803$. This may be verified by substituting $t = 0$ on the right-hand side of Eq. (2.9).

■ EXAMPLE 2.2

Find the Fourier series for the rectangular pulse train $k(t)$ shown in Fig. 2.4*a*. Here the period is T_o, $f_o = 1/T_o$, and $\omega_o = 2\pi f_o = 2\pi/T_o$

$$k(t) = a_0 + \sum_{n=1}^{\infty} a_n \cos n\omega_o t + b_n \sin n\omega_o t$$

$$a_0 = \frac{1}{T_o}\int_{-T_o/2}^{T_o/2} k(t)\,dt = \frac{1}{T_o}\int_{-\tau/2}^{\tau/2} A\,dt = \frac{A\tau}{T_o} \tag{2.11a}$$

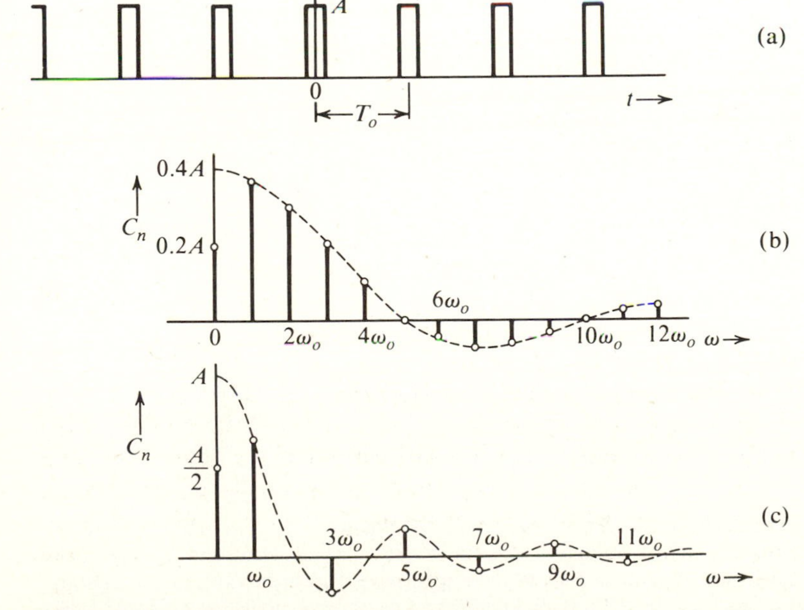

Figure 2.4 A rectangular pulse train and its spectrum.

$$a_n = \frac{2}{T_o}\int_{-\tau/2}^{\tau/2} A \cos n\omega_o t \, dt = \frac{2A}{\pi n} \sin \frac{n\pi\tau}{T_o} \tag{2.11b}$$

$$b_n = \frac{2}{T_o}\int_{-\tau/2}^{\tau/2} A \sin n\omega_o t \, dt = 0$$

Hence

$$k(t) = C_0 + \sum_{n=1}^{\infty} C_n \cos (n\omega_o + \theta_n) \tag{2.12a}$$

where

$$C_0 = \frac{A\tau}{T_o} \tag{2.12b}$$

$$C_n = \frac{2A}{\pi n} \sin \left(\frac{n\pi\tau}{T_o}\right) \tag{2.12c}$$

$$\theta_n = 0 \tag{2.12d}$$

The magnitude spectrum* C_n is shown in Fig. 2.4*b* for the case $\tau = T_o/5$.

A square pulse train is a special case of the rectangular pulse train, when $\tau = T_o/2$. For this case

$$C_0 = \frac{A}{2}$$

$$C_n = \begin{cases} \dfrac{2A}{\pi} \dfrac{(-1)^{(n-1)/2}}{n} & n, \text{ odd} \\ 0 & n, \text{ even} \end{cases} \tag{2.13a}$$

The Fourier series in Eq. (2.12a) becomes

$$k(t) = \frac{A}{2} + \frac{2A}{\pi} \sum_{n=1,3,5,\ldots}^{\infty} \frac{(-1)^{(n-1)/2}}{n} \cos n\omega_o t \tag{2.13b}$$

$$= \frac{A}{2}\left[1 + \frac{4}{\pi}\left(\cos \omega_o t - \frac{1}{3}\cos 3\omega_o t + \frac{1}{5}\cos 5\omega_o t - \frac{1}{7}\cos 7\omega_o t + \cdots\right)\right] \tag{2.13c}$$

Figure 2.4*c* shows the magnitude spectrum for this case.

Another special case of interest is when $\tau \to 0$ and $A \to \infty$ such that $A\tau = 1$. In this case, each pulse in Fig. 2.4*a* becomes an impulse of unit strength, and $k(t)$ is

* Strictly speaking, a magnitude spectrum is always positive. Whenever $\theta_n = 0$, as in Eq. (2.12), the values of C_n may be positive as well as negative. The negative sign may be accounted for by a phase shift of π or $-\pi$ (in fact, any odd multiple of $\pm\pi$), because a negative C_n can be expressed as $|C_n|e^{\pm jk\pi}$ (k, odd). The magnitude spectrum for $k(t)$ (Fig. 2.4*a*) will be a rectified version of the spectrum in Fig. 2.4*b* and the phase spectrum will be $\theta_n = \pi$ or $-\pi$ for those frequencies where C_n is negative. A single magnitude spectrum in Fig. 2.4*b*, however, is sufficient and more convenient.

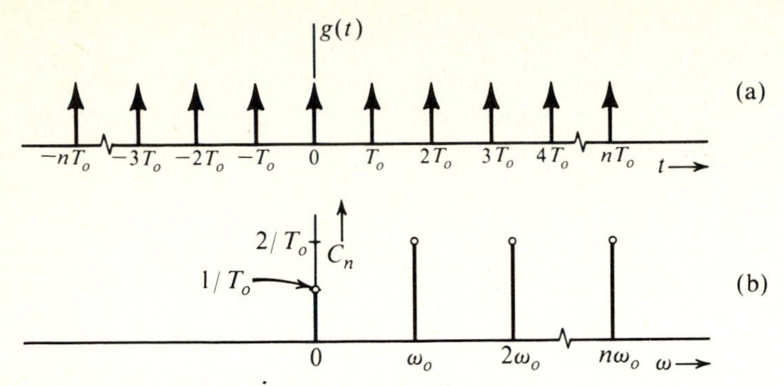

Figure 2.5 An impulse train and its spectrum.

simply a uniform train of unit impulses, as shown in Fig. 2.5*a*. The results of Example 2.2 are valid here with the appropriate limits. Thus,

$$\sum_{n=-\infty}^{\infty} \delta(t - nT_o) = C_0 + \sum_{n=1}^{\infty} C_n \cos n\omega_o t$$

Using $A\tau = 1$ and $\tau \to 0$, we have from Eq. (2.12),

$$C_0 = \frac{A\tau}{T_o} = \frac{1}{T_o} \tag{2.14a}$$

$$C_n = \lim_{\tau \to 0} \frac{2A}{\pi n} \sin\left(\frac{n\pi\tau}{T_o}\right)$$

$$= \lim_{\tau \to 0} \frac{2}{T_o} \frac{\sin\left(\frac{n\pi\tau}{T_o}\right)}{\left(\frac{n\pi\tau}{T_o}\right)}$$

$$= \frac{2}{T_o} \tag{2.14b}$$

The last result follows from the use of L'Hospital's rule to obtain

$$\lim_{x \to 0} \frac{\sin x}{x} = 1$$

Hence

$$\sum_{n=-\infty}^{\infty} \delta(t - nT_o) = \frac{1}{T_o}[1 + 2(\cos \omega_o t + \cos 2\omega_o t + \cdots + \cos n\omega_o t + \cdots] \tag{2.14c}$$

where $\omega_o = 2\pi/T_o$.

The frequency spectrum is shown in Fig. 2.5*b*. Note that amplitudes of all frequency components except dc are the same, viz., $2/T_o$. The dc-component amplitude is $1/T_o$. ■

Periodic Signal Representation by Exponential Fourier Series

Because a sinusoidal signal of angular frequency $n\omega_o$ can be expressed in terms of exponential signals $e^{jn\omega_o t}$ and $e^{-jn\omega_o t}$, a periodic signal $g(t)$ with a period T_o can also be represented by a series consisting of exponential components as

$$g(t) = G_0 + G_1 e^{j\omega_o t} + G_2 e^{j2\omega_o t} + \cdots + G_n e^{jn\omega_o t} + \cdots$$
$$+ G_{-1} e^{-j\omega_o t} + G_{-2} e^{-j2\omega_o t} + \cdots + G_{-n} e^{-jn\omega_o t} + \cdots \tag{2.15a}$$

$$= \sum_{n=-\infty}^{\infty} G_n e^{jn\omega_o t} \qquad \omega_o = 2\pi f_o = \frac{2\pi}{T_o} \tag{2.15b}$$

The coefficients G_k in this series can be obtained by multiplying both sides of Eq. (2.15b) by $e^{-jk\omega_o t}$ and integrating over one cycle $(t_o, t_o + T_o)$ for any value of t_o. This yields

$$\int_{t_o}^{t_o+T_o} g(t) e^{-jk\omega_o t}\, dt = \sum_{n=-\infty}^{\infty} G_n \int_{t_o}^{t_o+T_o} e^{j(n-k)\omega_o t}\, dt \tag{2.16}$$

In evaluating integrals on the right-hand side of Eq. (2.16), we use the result

$$\int_{t_o}^{t_o+T_o} e^{j(n-k)\omega_o t}\, dt = \begin{cases} 0 & n \neq k \\ T_o & n = k \end{cases} \tag{2.17}$$

Substitution of Eq. (2.17) in Eq. (2.16) yields

$$G_k = \frac{1}{T_o} \int_{t_o}^{t_o+T_o} g(t) e^{-jk\omega_o t}\, dt \tag{2.18}$$

The representation in Eq. (2.15) is called the *exponential Fourier series* representation of a periodic function.

In conclusion, a periodic waveform $g(t)$ with a period T_o can be expressed by an exponential Fourier series*:

$$g(t) = \sum_{n=-\infty}^{\infty} G_n e^{jn\omega_o t} \qquad \omega_o = 2\pi f_o = \frac{2\pi}{T_o} \tag{2.19a}$$

The coefficients G_n are complex in general and are given by

$$G_n = \frac{1}{T_o} \int_{t_o}^{t_o+T_o} g(t) e^{-jn\omega_o t}\, dt \tag{2.19b}$$

The constant t_o is arbitrary and, if properly chosen, can simplify integration in Eq. (2.19b). The trigonometric and the exponential Fourier series are not two different series but represent two different ways of writing the same series. The coefficients of one series can be obtained from those of the other. From Eqs. (2.4) and (2.19) it

---

* To be more precise, the right-hand side of Eq. (2.15) converges to $g(t)$ in the mean; that is, the mean square of the error (averaged over one period) approaches 0, where the error ϵ is defined as[1]

$$\varepsilon = g(t) - \sum_{n=-\infty}^{\infty} G_n e^{jn\omega_o t}$$

follows that

$$a_0 = G_0$$

$$\left.\begin{aligned} a_n &= G_n + G_{-n} \\ b_n &= j(G_n - G_{-n}) \end{aligned}\right\} n \neq 0$$

and

$$\left.\begin{aligned} G_n &= \frac{1}{2}(a_n - jb_n) \\ G_{-n} &= \frac{1}{2}(a_n + jb_n) \end{aligned}\right\} n \geq 1 \tag{2.20}$$

For real $g(t)$, a_n and b_n are real numbers [Eq. (2.4)], and coefficients G_n and G_{-n} are conjugates:

$$G_{-n} = G_n^* \tag{2.21a}$$

Thus, if

$$G_n = |G_n|e^{j\theta_n}$$

$$G_{-n} = |G_n|e^{-j\theta_n} \tag{2.21b}$$

$|G_n|$ is the magnitude and θ_n is the angle or phase of G_n. Hence, for a real $g(t)$, $|G_{-n}| = |G_n|$, and the magnitude spectrum $|G_n|$ vs. ω is an even function of ω. Similarly, the phase spectrum θ_n vs. ω is an odd function of ω because $\theta_{-n} = -\theta_n$.

■ EXAMPLE 2.3

Find the exponential Fourier series and plot the corresponding frequency spectra for the periodic signal $g(t)$ in Fig. 2.2. Here $T = 1/2, f_o = 2, \omega_o = 4\pi$, and from Eq. (2.19a and b), we have

$$g(t) = \sum_{n=-\infty}^{\infty} G_n e^{j4\pi nt}$$

where

$$\begin{aligned} G_n &= 2\int_0^{1/2} e^{-t}\, e^{-j4\pi nt}\, dt \\ &= 2\int_0^{1/2} e^{-(1+j4\pi n)t}\, dt \\ &= \frac{0.79}{1 + j4\pi n} \end{aligned}$$

Therefore,

$$\left.\begin{aligned} |G_n| &= \frac{0.79}{\sqrt{1 + 16\pi^2 n^2}} \\ \theta_n &= -\tan^{-1}(4\pi n) \end{aligned}\right. \tag{2.22}$$

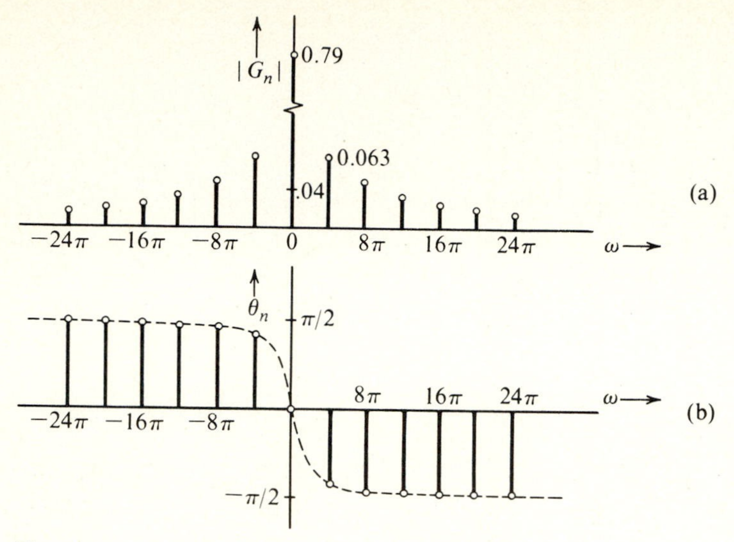

Figure 2.6 Magnitude and phase spectra of the periodic signal in Fig. 2.2.

Compare these results with those in Eq. (2.10) for the trigonometric Fourier series.

The magnitude spectrum ($|G_n|$ vs. ω) and the phase spectrum (θ_n vs. ω) are shown in Fig. 2.6. As expected, the magnitude spectrum is an even function and the phase spectrum is an odd function of ω. Compare these spectra with those in Fig. 2.3*a* and *b*. ■

■ EXAMPLE 2.4

Find the exponential Fourier series and plot the corresponding frequency spectra for the rectangular pulse train $k(t)$ in Fig. 2.4*a*.

We have

$$k(t) = \sum_{n=-\infty}^{\infty} K_n e^{jn\omega_o t} \qquad \omega_o = 2\pi f_o = \frac{2\pi}{T_o} \tag{2.23a}$$

where

$$\begin{aligned} K_n &= \frac{1}{T_o}\int_{-\tau/2}^{\tau/2} A e^{-jn\omega_o t}\, dt \\ &= \frac{A}{\pi n} \sin \frac{n\omega_o \tau}{2} \\ &= \frac{A}{\pi n} \sin\left(\frac{n\pi\tau}{T_o}\right) \end{aligned} \tag{2.23b}$$

Compare this with Eq. (2.12c). The magnitude spectrum for the case $\tau = T_o/5$ is shown in Fig. 2.7*a*. For the case of a square pulse train, $\tau = T_o/2$, $K_n = (A/\pi n) \sin (n\pi/2)$, and the spectrum K_n vs. ω is shown in Fig. 2.7*b*. Compare this

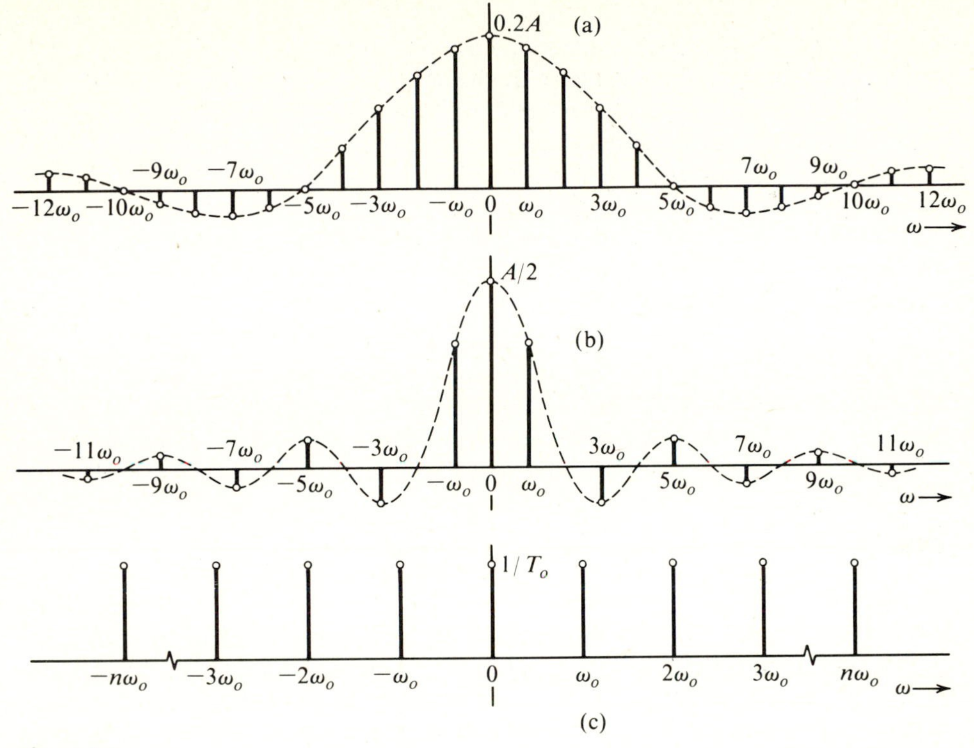

Figure 2.7

spectrum with that in Fig. 2.4*c*. For the case where $\tau \rightarrow 0$ and $A\tau = 1$ (unit impulse train in Fig. 2.5*a*), we have from Eq. (2.23b)

$$K_n = \lim_{\tau \rightarrow 0} \frac{1}{T_o} \frac{\sin\left(\frac{n\pi\tau}{T}\right)}{\left(\frac{n\pi\tau}{T}\right)} = \frac{1}{T_o} \tag{2.24a}$$

and

$$\sum_{n=-\infty}^{\infty} \delta(t - nT_o) = \frac{1}{T_o} \sum_{n=-\infty}^{\infty} e^{jn\omega_o t} \qquad \omega_o = \frac{2\pi}{T_o} \tag{2.24b}$$

The spectrum has components of frequencies $n\omega_o$, n varying from $-\infty$ to ∞, including 0, all with an equal strength of $1/T_o$, as shown in Fig. 2.7*c*. Compare this spectrum with that in Fig. 2.5*b*. ■

Exponential Fourier Spectrum

In the case of the exponential Fourier series, we have the components of frequencies $0, \pm\omega_o, \pm 2\omega_o, \ldots, \pm n\omega_o, \ldots$, with the spectrum ranging over positive as well as negative frequencies. The idea of a negative frequency, at a first glance, is rather disturbing, because by definition frequency (number of repetitions per second) is a positive quantity. How shall we interpret a negative frequency? The confusion arises in the first place because here the term frequency is used in a loose sense rather than in its conventional sense (that is, number of repetitions per second). We must remember that in the exponential Fourier series, the basic functions are exponentials, $e^{jn\omega_o t}$, not sinusoids. Its spectrum represents the amplitudes of these components, and we loosely call $n\omega_o$ the frequency of the signal $e^{jn\omega_o t}$. In this case, the idea of negative frequency is inevitable. For example, a signal $\cos n\omega_o t$ can be represented as

$$\cos n\omega_o t = \frac{1}{2}[e^{jn\omega_o t} + e^{-jn\omega_o t}]$$

Thus, a signal of frequency $n\omega_o$ (in the conventional sense) is expressed as a sum of two signals of frequencies $n\omega_o$ and $-n\omega_o$ (in the exponential sense). We must remember that in the exponential Fourier series, the components of frequencies $n\omega_o$ and $-n\omega_o$ combine to yield the component of frequency $n\omega_o$ (in the conventional sense). This point can be further clarified by rearranging the exponential Fourier series in Eq. (2.19a) as

$$g(t) = G_0 + \sum_{n=1}^{\infty} G_n e^{jn\omega_o t} + G_{-n} e^{-jn\omega_o t} \tag{2.25a}$$

The use of Eq. (2.21b) in Eq. (2.25a) yields

$$g(t) = G_0 + \sum_{n=1}^{\infty} 2|G_n| \cos (n\omega_o t + \theta_n) \tag{2.25b}$$

Comparison of this with the compact trigonometric Fourier series in Eq. (2.5) brings out the kinship of the two series. It follows that

$$\begin{aligned} C_0 &= G_0 \\ C_n &= 2|G_n| = |G_n| + |G_{-n}| \qquad n \geq 1 \end{aligned} \tag{2.26}$$

The magnitude C_n of the component of frequency $n\omega_o$ (in the conventional sense) is the sum of magnitudes of components $n\omega_o$ and $-n\omega_o$ (in the exponential sense). It looks as if the trigonometric spectrum C_n vs. ω is formed by the exponential spectrum $|G_n|$ vs. ω by folding the latter about the vertical axis and adding the superimposed components at frequencies $n\omega_o$ and $-n\omega_o$. This is easily verified by comparing Fig. 2.3*a* with Fig. 2.6*a*, Fig. 2.4*b* with Fig. 2.7*a*, and Fig. 2.5*b* with Fig. 2.7*c*.

In conclusion, when we plot a spectrum of exponential components, the spectrum exists over negative as well as positive frequencies because we have components of the form $e^{-jn\omega_o t}$ and $e^{jn\omega_o t}$, and we loosely refer to the exponential index as the frequency of that component. But in the sinusoidal (or conventional) sense, frequencies can only be positive, and the spectrum of the trigonometric Fourier series exists only for positive ω. Because it is more convenient to use exponential representation rather than trigonometric, we shall be using exponential spectra in the rest of the book. It is therefore important that the reader thoroughly understand what it means.

In this section, we have discussed a way of representing a periodic signal in terms of sinusoidal or exponential signals of the form $e^{jn\omega_o t}$. A periodic signal by definition begins at $t = -\infty$, and the sinusoids or exponentials in the Fourier series also begin at $t = -\infty$. For convenience, signals that start at $t = -\infty$ will be called *eternal* signals.

We would like to extend the results derived so far to $g(t)$ that is nonperiodic. Such an extension is indeed possible by what is known as the *Fourier transform* representation of $g(t)$.

2.2 EXPONENTIAL REPRESENTATION OF NONPERIODIC SIGNALS: THE FOURIER TRANSFORM

The representation of nonperiodic signals by eternal exponential signals can be accomplished by a simple limiting process, and we shall show that nonperiodic signals, in general, can be expressed as a continuous sum (integral) of eternal exponential signals.

We desire to represent the nonperiodic signal $g(t)$ shown in Fig. 2.8*a* by eternal exponential signals. Let us construct a new periodic signal $g_p(t)$ consisting of the signal $g(t)$ repeating itself every T_o seconds, as shown in Fig. 2.8*b*. The period T_o is made long enough so that there is no overlap between the repeating pulses. This new signal $g_p(t)$ is a periodic signal and consequently can be represented by an exponential Fourier series. In the limit, if we let T_o become infinite, the pulses in the periodic signal repeat after an infinite interval, and

$$\lim_{T\to\infty} g_p(t) = g(t)$$

Thus, the Fourier series representing $g_p(t)$ will also represent $g(t)$, in the limit $T_o \to \infty$.

The exponential Fourier series for $g_p(t)$ is

$$g_p(t) = \sum_{n=-\infty}^{\infty} G_n e^{jn\omega_o t} \qquad \omega_o = \frac{2\pi}{T_o} \tag{2.27a}$$

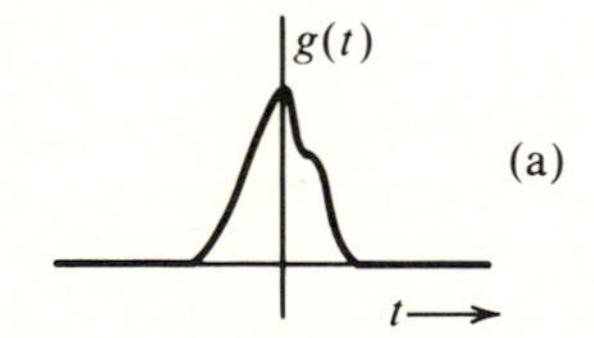

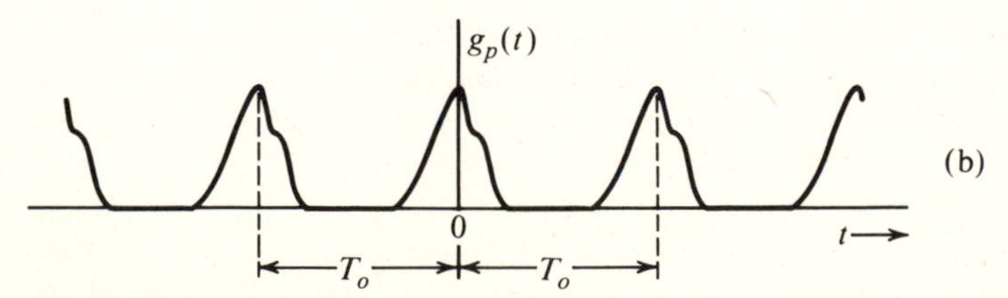

Figure 2.8

and

$$G_n = \frac{1}{T_o} \int_{-T/2}^{T/2} g_p(t)\, e^{-jn\omega_o t}\, dt \tag{2.27b}$$

As T_o becomes larger, ω_o (the fundamental frequency) becomes smaller and the spectrum becomes denser. As seen from Eq. (2.27b), the amplitudes of the individual components become smaller, too. The shape of the frequency spectrum, however, is unaltered. In the limit as $T_o \to \infty$, the magnitude of each component becomes infinitesimally small, but now there exist an infinite number of frequency components ($\omega_o \to 0$). The spectrum exists for every value of ω and is no longer a discrete but a continuous function of ω. To illustrate this point, let us make a slight change in notation. Because in the limit as $T_o \to \infty$ $\omega_o \to 0$, the quantity ω_o is infinitesimal and may be denoted by $\Delta\omega$.

$$T_o = \frac{2\pi}{\omega_o} = \frac{2\pi}{\Delta\omega}$$

From Eq. (2.27b),

$$T_o G_n = \int_{-T/2}^{T/2} g_p(t) e^{-jn\Delta\omega t}\, dt \tag{2.28a}$$

The quantity $T_o G_n$ is a function of $n\Delta\omega$. Let

$$T_o G_n = G(n\Delta\omega) \tag{2.28b}$$

From Eq. (2.27a and b),

$$g_p(t) = \sum_{n=-\infty}^{\infty} \frac{G(n\Delta\omega)}{T_o} e^{jn\Delta\omega t}$$

$$= \sum_{n=-\infty}^{\infty} \left[\frac{G(n\Delta\omega)\Delta\omega}{2\pi}\right] e^{jn\Delta\omega t} \tag{2.29a}$$

Here $g_p(t)$ is expressed as a sum of eternal exponentials of frequencies 0, $\pm\Delta\omega$, $\pm 2\Delta\omega$, $\pm 3\Delta\omega$, ... The strength of the component of frequency $n\Delta\omega$ is $G(n\Delta\omega)\Delta\omega/2\pi$, which is infinitesimal and approaches zero in the limit as $T_o \to \infty$. We have a strange situation of having a component of every possible frequency ($\Delta\omega \to 0$), but the strength of each component is approaching zero. We shall soon see that this is not such a strange case after all, and in fact we come across many such situations in other areas. Note that although the amplitudes of all frequency components approach zero, their relative strengths are given by $G(n\Delta\omega)$.

Because in the limit as $T_o \to \infty$, $g_p(t) \to g(t)$,

$$g(t) = \lim_{T_o \to \infty} g_p(t) = \lim_{\Delta\omega \to 0} \frac{1}{2\pi} \sum_{n=-\infty}^{\infty} G(n\Delta\omega) e^{jn\Delta\omega t} \Delta\omega \tag{2.29b}$$

The right-hand side,* by definition, is the integral

$$g(t) = \frac{1}{2\pi}\int_{-\infty}^{\infty} G(\omega)e^{j\omega t}\, d\omega$$

Also, from Eq. (2.28a and b),

$$\begin{aligned} G(n\Delta\omega) &= \lim_{T\to\infty}\int_{-T_o/2}^{T_o/2} g_p(t)e^{-jn\Delta\omega t}\, dt \\ &= \int_{-\infty}^{\infty} g(t)e^{-jn\Delta\omega t}\, dt \end{aligned}$$

Hence

$$G(\omega) = \int_{-\infty}^{\infty} g(t)e^{-j\omega t}\, dt$$

To recapitulate, we have shown that†

$$g(t) = \frac{1}{2\pi}\int_{-\infty}^{\infty} G(\omega)e^{j\omega t}\, d\omega \tag{2.30a}$$

where

$$G(\omega) = \int_{-\infty}^{\infty} g(t)e^{-j\omega t}\, dt \tag{2.30b}$$

We have succeeded in representing a nonperiodic signal $g(t)$ in terms of eternal exponential signals. Equation (2.30a) represents $g(t)$ as a continuous sum of eternal exponential functions with frequencies lying in the interval $(-\infty < \omega < \infty)$. The amplitude of the component at any frequency ω is proportional to $G(\omega)$. Therefore, $G(\omega)$ represents the frequency spectrum of $g(t)$. The frequency spectrum is continuous and exists at all values of ω.

Equation (2.30a and b) are usually referred to as the *Fourier transform pair,* with $G(\omega)$ being the *direct Fourier transform* of $g(t)$, and $g(t)$ as the *inverse Fourier transform* of $G(\omega)$. Symbolically, these transforms are also written as

$$G(\omega) = \mathscr{F}[g(t)] \quad \text{and} \quad g(t) = \mathscr{F}^{-1}[G(\omega)]$$

---

* Equation (2.29b) can also be expressed as

$$g(t) = \lim_{T_o\to\infty} g_p(t) = \lim_{\Delta\omega\to 0}\frac{\Delta\omega}{2\pi}\left[G(0) + \sum_{n=1}^{\infty} 2|G(n\Delta\omega)|\cos[(n\Delta\omega)t + \theta_n]\right]$$

where $\theta_n = \angle G(n\Delta\omega)$. Hence a nonperiodic signal can be expressed as a sum of sinusoids of frequencies $n\Delta\omega$ $(n = 0, 1, 2, \ldots)$ in the limit as $\Delta\omega \to 0$.

† To be more precise, the right-hand side of Eq. (2.30a) converges to $g(t)$ in the mean; that is, the mean square error approaches zero, where the error ϵ is[1]

$$\varepsilon = g(t) - \frac{1}{2\pi}\int_{-\infty}^{\infty} G(\omega)e^{j\omega t}\, d\omega$$

These relationships can also be expressed symbolically by a Fourier transform pair as

$$g(t) \leftrightarrow G(\omega)$$

This means $G(\omega)$ is the Fourier transform of $g(t)$, and $g(t)$ is the inverse Fourier transform of $G(\omega)$.

The Fourier transform is a linear operator. This means if

$$g_1(t) \leftrightarrow G_1(\omega) \qquad \text{and} \qquad g_2(t) \leftrightarrow G_2(\omega)$$

then

$$a_1 g_1(t) + a_2 g_2(t) \leftrightarrow a_1 G_1(\omega) + a_2 G_2(\omega) \tag{2.31}$$

The proof is trivial. The result can be extended to the sum of any number of signals. If $g(t)$ is a real function of t, then from Eq. (2.30b) it follows that

$$\begin{aligned} G(-\omega) &= \int_{-\infty}^{\infty} g(t)e^{j\omega t}\, dt \\ &= G^*(\omega) \end{aligned} \tag{2.32}$$

Hence, if

$$G(\omega) = |G(\omega)|\, e^{j\theta_g(\omega)} \tag{2.33}$$

then

$$G(-\omega) = |G(\omega)|\, e^{-j\theta_g(\omega)}$$

This means for real $g(t)$

$$|G(-\omega)| = |G(\omega)|$$

and

$$\theta_g(-\omega) = -\theta_g(\omega) \tag{2.34}$$

Hence, for real $g(t)$, the magnitude spectrum $|G(\omega)|$ is an even function of ω, and the phase spectrum $\theta_g(\omega)$ is an odd function of ω.

Comments About the Fourier Transform

We have expressed a nonperiodic signal $g(t)$ as a continuous sum of exponential signals with frequencies in the interval $-\infty$ to ∞. The amplitude of a component of any frequency ω is infinitesimal but is proportional to $G(\omega)$.

The concept of a continuous spectrum is sometimes bewildering because we generally picture the spectrum as existing at discrete frequencies and with finite amplitudes. The continuous spectrum concept can be appreciated by considering an analogous concrete phenomenon. Consider a beam loaded with weights of G_1, G_2, G_3, . . . , G_n units at uniformly spaced points x_1, x_2, x_3, . . . , x_n, as shown in Fig. 2.9*a*. The beam is loaded at n discrete points, and the total weight W_T on the beam is

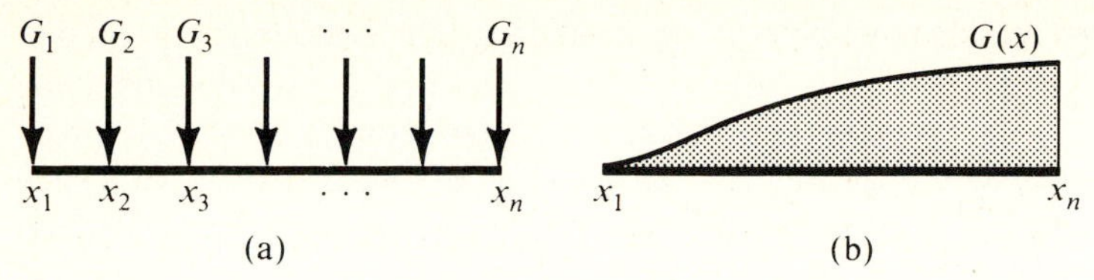

Figure 2.9 (a) A beam loaded at discrete points. (b) A continuously loaded beam.

given by

$$W_T = \sum_{i=1}^{n} G_i$$

Now, consider the case of a continuously loaded beam, as shown in Fig. 2.9*b*. The loading density $G(x)$, in kilograms per meter, is a function of x. The total weight on the beam is now given by a continuous sum of the weight—that is, the integral of $G(x)$ over the entire length:

$$W_T = \int_{x_1}^{x_n} G(x)\, dx$$

In the former case (discrete loading), the weight existed only at discrete points. At other points there was no load. On the other hand, in the continuously distributed case, the loading exists at every point, but at any one point the load is zero. The load in a small distance dx, however, is given by $G(x)\, dx$. Therefore, $G(x)$ represents the relative loading at a point x. An exactly analogous situation exists in the case of a signal and its frequency spectrum. A periodic signal can be represented by a sum of discrete exponentials with finite amplitudes:

$$g(t) = \sum_{n=-\infty}^{\infty} G_n e^{j\omega_n t} \qquad (\omega_n = n\omega_o)$$

For a nonperiodic signal, the distribution of exponentials becomes continuous; that is, the spectrum exists at every value of ω. At any one frequency ω, the amplitude of that frequency component is zero. The total contribution in an infinitesimal interval $d\omega$ is given by $G(\omega)e^{j\omega t}\, d\omega/2\pi$, and the function $g(t)$ can be expressed in terms of the continuous sum of such infinitesimal components.*

$$g(t) = \frac{1}{2\pi}\int_{-\infty}^{\infty} G(\omega)e^{j\omega t}\, d\omega$$

Another comment about the Fourier transform is that it is a tool for representing a signal as a sum (continuous sum) of eternal exponential components. This leads to a

---

* The factor 2π in the above equation can be removed if the integration is performed with respect to the variable f instead of ω. We have $\omega = 2\pi f$, $d\omega = 2\pi\, df$ and

$$g(t) = \int_{-\infty}^{\infty} G(2\pi f)e^{j2\pi ft}\, df$$

rather fascinating picture when we try to imagine how these eternal exponential components sum to $g(t)$, particularly when $g(t)$ is a signal that is zero outside some finite interval (a, b). All the exponential (or sinusoidal) components in the spectrum are eternal; that is, they begin at $t = -\infty$ and continue forever. How can such components add to a pulse $g(t)$, which exists only over a finite interval (a, b) and is zero outside this interval? The answer is that these components, which are infinite in number, have such amplitudes and phases that they sum to zero everywhere except over the interval (a, b), where they sum to $g(t)$. Such a perfect and delicate balance of magnitudes and phases would truly tax human ingenuity. Yet the Fourier transform does it routinely without much thinking on our part. Indeed, we get so involved in the mathematics that we fail to notice this marvel.

Existence of the Fourier Transform

From Eq. (2.30b), we have

$$|G(\omega)| \leq \int_{-\infty}^{\infty} |g(t)|\, dt$$

Hence, if the right-hand side is finite, the existence of the Fourier transform is guaranteed. Thus, if a function $g(t)$ is absolutely integrable, that is

$$\int_{-\infty}^{\infty} |g(t)|\, dt < \infty \tag{2.35a}$$

the Fourier transform of $g(t)$ exists. It can also be shown that if

$$\int_{-\infty}^{\infty} |g(t)|^2\, dt < \infty \tag{2.35b}$$

the existence of the Fourier transform is guaranteed.[2] Thus, either of the two conditions above guarantee the existence of the Fourier transform. If $g(t)$ satisfies either of these two conditions, it is said to be *Fourier transformable*. Note that either one of these conditions is sufficient for the existence of the Fourier transform. They are not necessary conditions for the function to be Fourier transformable. Several functions exist that do not satisfy these conditions yet have Fourier transforms in the limit. Functions such as a step function or a sinusoidal function violate the conditions in Eq. (2.35a and b) yet have Fourier transforms in the limit when one admits the use of generalized functions (or distributions) such as impulse functions.

The topic of the existence of the Fourier transform is rather involved and is beyond our scope. We may remember that any function $g(t)$ that can be generated in a laboratory is Fourier transformable.

The integral in Eq. (2.35b) is very important and warrants further discussion. If a voltage $g(t)$ is applied across a 1 ohm resistor (or if a current $g(t)$ is passed through a 1 ohm resistor), the dissipated energy E_g is given by*

---

* For a complex signal $g(t)$, the energy E_g is defined as

$$E_g = \int_{-\infty}^{\infty} |g(t)|^2\, dt$$

$$E_g = \int_{-\infty}^{\infty} g^2(t)\, dt \tag{2.36}$$

For this reason, the integral on the right-hand side of Eq. (2.36) is defined as the *energy* of a real signal $g(t)$. In terms of this definition, we can say that finite energy is a sufficient condition for a signal to be Fourier transformable.

EXAMPLE 2.5

Find the Fourier transform of the single-sided exponential pulse $e^{-at}u(t)$ shown in Fig. 2.10*a*.

We are given

$$g(t) = e^{-at}u(t)$$

$$G(\omega) = \int_{-\infty}^{\infty} e^{-at}u(t)e^{-j\omega t}\, dt$$

$$= \int_{0}^{\infty} e^{-(a+j\omega)t}\, dt$$

$$= \frac{1}{a + j\omega} \qquad a > 0$$

$$= \frac{1}{\sqrt{a^2 + \omega^2}} e^{-j\tan^{-1}(\omega/a)}$$

Here

$$|G(\omega)| = \frac{1}{\sqrt{a^2 + \omega^2}} \qquad \text{and} \qquad \theta_g(\omega) = -\tan^{-1}\frac{\omega}{a}$$

The magnitude spectrum $|G(\omega)|$ and the phase spectrum $\theta_g(\omega)$ are shown in Fig. 2.10*b*.

Note that the integral for $G(\omega)$ converges only for $a > 0$. For $a < 0$, the Fourier transform does not exist. This also follows from the fact that for $a < 0$, $g(t)$ is not

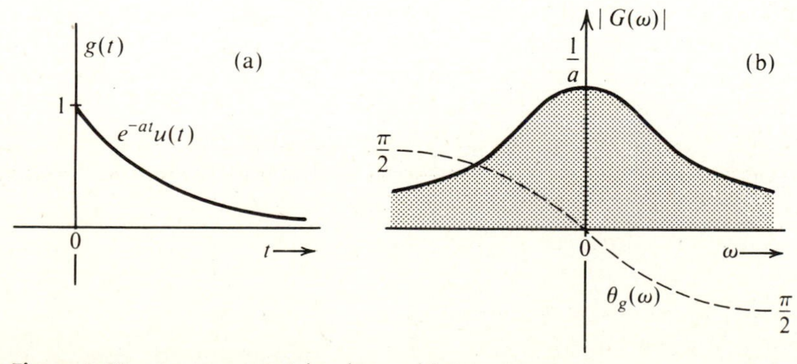

Figure 2.10 An exponential pulse and its spectrum.

absolutely integrable. Also note that $|G(\omega)|$ is an even function and $\theta_g(\omega)$ is an odd function of ω. ■

■ EXAMPLE 2.6

Find the Fourier transform of a gate function $\Pi(t)$ defined by (Fig. 2.11*a*)

$$\Pi(t) = \begin{cases} 1 & |t| < \frac{1}{2} \\ 0 & |t| > \frac{1}{2} \end{cases} \tag{2.37}$$

Because a function $g(t/a)$ is the same as $g(t)$ expanded by a factor a in time scale for $a > 1$, $\Pi(t/\tau)$ is a gate function of width τ centered at the origin (Fig. 2.11*b*).

$$\begin{aligned} \mathcal{F}\left[\Pi\left(\frac{t}{\tau}\right)\right] &= \int_{-\tau/2}^{\tau/2} e^{-j\omega t}\, dt \\ &= \frac{1}{j\omega}\left[e^{j\omega\tau/2} - e^{-j\omega\tau/2}\right] \\ &= \tau \frac{\sin(\omega\tau/2)}{\omega\tau/2} \end{aligned} \tag{2.38}$$

The frequency spectrum of $\Pi(t/\tau)$ is shown in Fig. 2.11*c*. The function with the form $\sin x/x$ in Eq. (2.38) plays an important role in communication theory. For com-

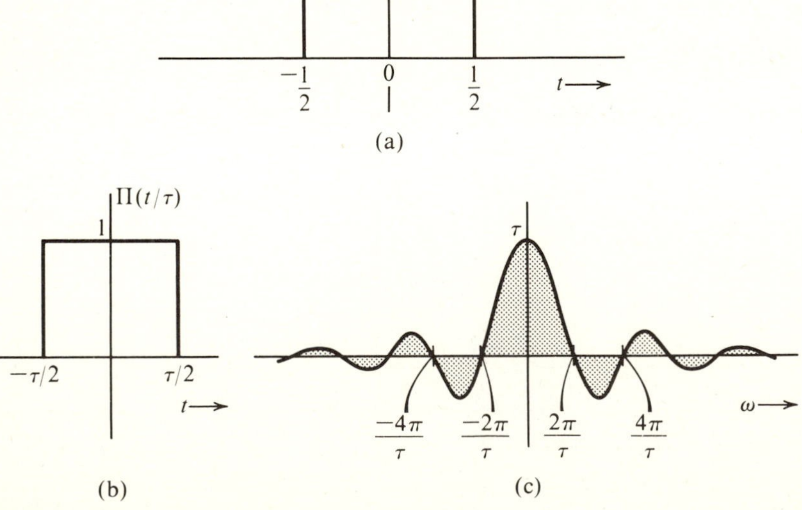

Figure 2.11 A gate function and its spectrum.

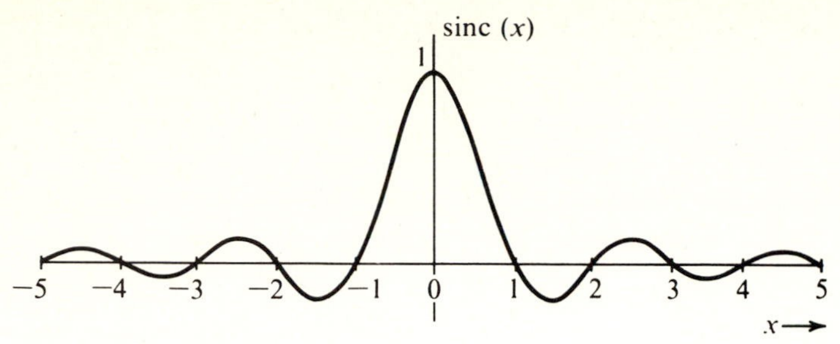

Figure 2.12 The function sinc (x).

pactness, a special notation is used for this function. We define

$$\text{sinc}\ (x) = \frac{\sin \pi x}{\pi x} \tag{2.39}$$

and, from Eq. (2.38),

$$\Pi\left(\frac{t}{\tau}\right) \leftrightarrow \tau\ \text{sinc}\left(\frac{\omega\tau}{2\pi}\right) \tag{2.40}$$

The function sinc (x) is shown in Fig. 2.12. Using L'Hospital's rule, we find that sinc $(0) = 1$. Also, sinc $(x) = 0$ for all integral values of x. Thus, sinc $(\omega\tau/2\pi) = 0$ at all integral values of $\omega\tau/2\pi$, or at $\omega = \pm(2\pi/\tau), \pm(4\pi/\tau), \pm(6\pi/\tau) \ldots$. ■

■ EXAMPLE 2.7

Find the Fourier transform of the signum function sgn (t), defined by

$$\text{sgn}\ (t) = \begin{cases} 1 & t > 0 \\ -1 & t < 0 \end{cases} \tag{2.41}$$

The transform of this function (Fig. 2.13) can be obtained by considering the function as a limit.

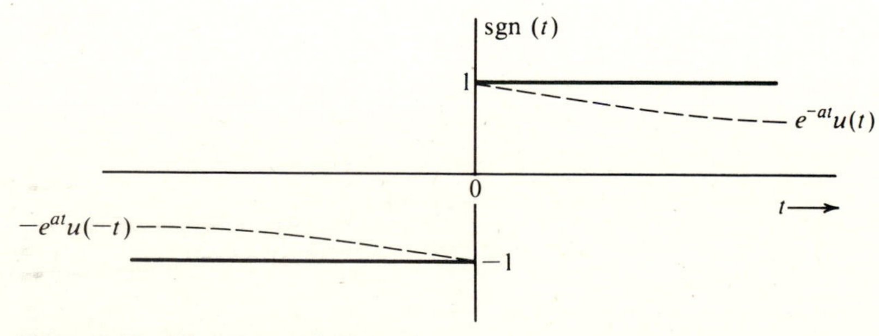

Figure 2.13 The signum function.

$$\operatorname{sgn}(t) = \lim_{a\to 0} [e^{-at}u(t) - e^{at}u(-t)]$$

$$\mathcal{F}[\operatorname{sgn}(t)] = \lim_{a\to 0} \int_0^{\infty} e^{-at}e^{-j\omega t}\,dt - \int_{-\infty}^{0} e^{at}e^{-j\omega t}\,dt$$

$$= \lim_{a\to 0} \left[\frac{1}{a+j\omega} - \frac{1}{a-j\omega}\right] = \frac{2}{j\omega} \tag{2.42}$$

■

■ EXAMPLE 2.8

Find the Fourier transform of the unit impulse function $\delta(t)$. Dirac defined the unit impulse function as:

$$\int_{-\infty}^{\infty} \delta(t)\,dt = 1 \qquad \text{and} \qquad \delta(t) = 0 \text{ for } t \neq 0 \tag{2.43}$$

The unit impulse can be considered as a narrow pulse of unit area in the limit as its width $\varepsilon \to 0$ and its height $1/\varepsilon \to \infty$ (Fig. 2.14*a*).

The Fourier transform of the unit impulse can be obtained by using the so-called *sampling* (or *sifting*) property of the impulse function

$$\int_{-\infty}^{\infty} \varphi(t)\,\delta(t)\,dt = \varphi(0) \tag{2.44a}$$

where $\varphi(t)$ is a continuous function at the origin. The sampling property is proved by multiplying $\varphi(t)$ by the narrow pulse shown in Fig. 2.14*a* that represents $\delta(t)$ in the limit $\varepsilon \to 0$. The product $\varphi(t)\,\delta(t)$ is $\varphi(0)/\varepsilon$ over the interval $[-(\varepsilon/2), (\varepsilon/2)]$, and

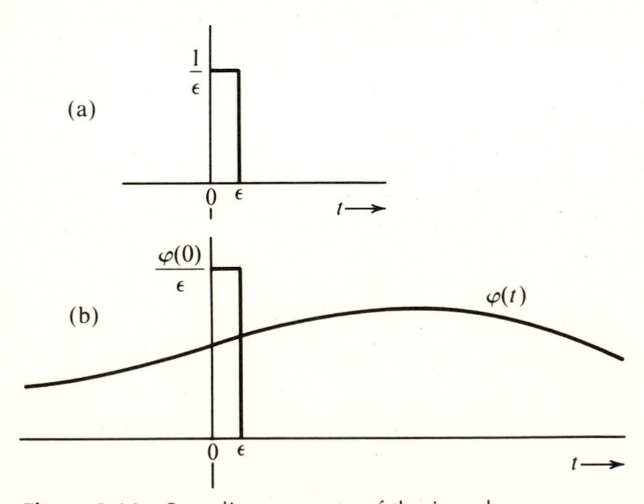

Figure 2.14 Sampling property of the impulse.

$$\int_{-\infty}^{\infty} \varphi(t)\, \delta(t)\, dt = \lim_{\varepsilon \to 0} \frac{\varphi(0)}{\varepsilon} \int_{-\varepsilon/2}^{\varepsilon/2} dt = \varphi(0)$$

Using a similar argument, the above result is generalized as

$$\int_{-\infty}^{\infty} \varphi(t)\, \delta(t - t_o)\, dt = \varphi(t_o) \tag{2.44b}$$

and

$$\int_{-\infty}^{\infty} \varphi(t - t_1)\, \delta(t - t_2)\, dt = \varphi(t_2 - t_1) \tag{2.44c}$$

The impulse function definition in Eq. (2.43) leads to a nonunique function.[3] Secondly, it is not very rigorous because $\delta(t)$ is zero everywhere except at $t = 0$, where it is undefined. In a more rigorous approach, the impulse function is defined not as an ordinary function but as a *generalized* function (also known as a *distribution*). A generalized function $\delta(t)$ is defined by Eq. (2.44). In other words, an impulse function $\delta(t)$ is such a function that the area under the product $\varphi(t)\, \delta(t)$ is $\varphi(0)$. A more detailed discussion of this subject is beyond our scope. A reasonable and short discussion can be found in Papoulis.[3]

Using the sampling property, Eq. (2.44a), we have

$$\mathscr{F}[\delta(t)] = \int_{-\infty}^{\infty} \delta(t) e^{-j\omega t}\, dt = 1 \tag{2.45}$$

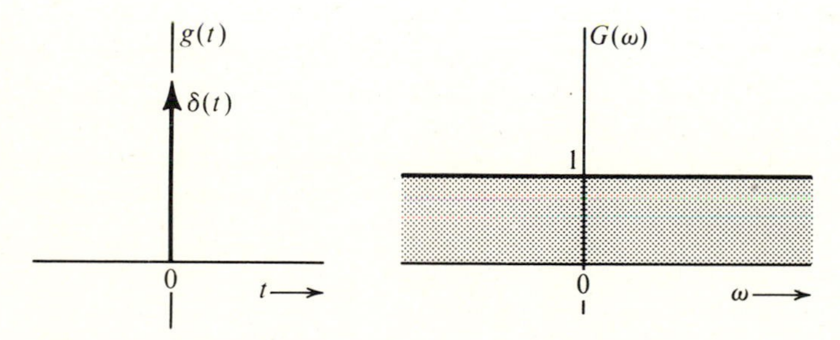

Figure 2.15 An impulse function and its transform.

Figure 2.15 shows $\delta(t)$ and its spectrum (the Fourier transform). Because $\delta(t)$ is the inverse transform of 1, we have

$$\delta(t) = \frac{1}{2\pi} \int_{-\infty}^{\infty} e^{j\omega t}\, d\omega$$

By letting $\omega = -x$, we have

$$2\pi\, \delta(t) = \int_{-\infty}^{\infty} e^{-jtx}\, dx \tag{2.46}$$

■

■ EXAMPLE 2.9

Find the Fourier transform of (1) $g(t) = 1$, and (2) $g(t) = u(t)$.

$$\mathscr{F}[1] = \int_{-\infty}^{\infty} e^{-j\omega t}\, dt = \int_{-\infty}^{\infty} e^{-j\omega x}\, dx$$

Use of Eq. (2.46) immediately yields (Fig. 2.16)

$$1 \leftrightarrow 2\pi\, \delta(\omega) \tag{2.47}$$

The spectrum of a constant is a single impulse located at $\omega = 0$. This is logical because a constant is a dc signal and has only one frequency component ($\omega = 0$). Because

$$1 + \operatorname{sgn}(t) = 2u(t)$$

$$\mathscr{F}[u(t)] = \frac{1}{2}[\mathscr{F}[1] + \mathscr{F}[\operatorname{sgn}(t)]]$$

$$= \pi\, \delta(\omega) + \frac{1}{j\omega} \tag{2.48}$$

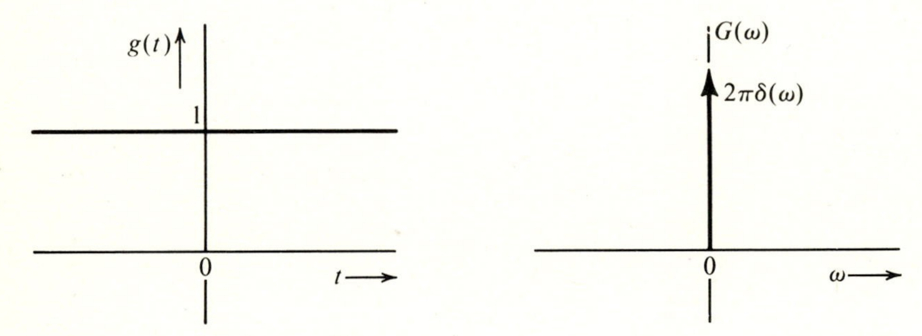

Figure 2.16 A constant (dc) and its spectrum.

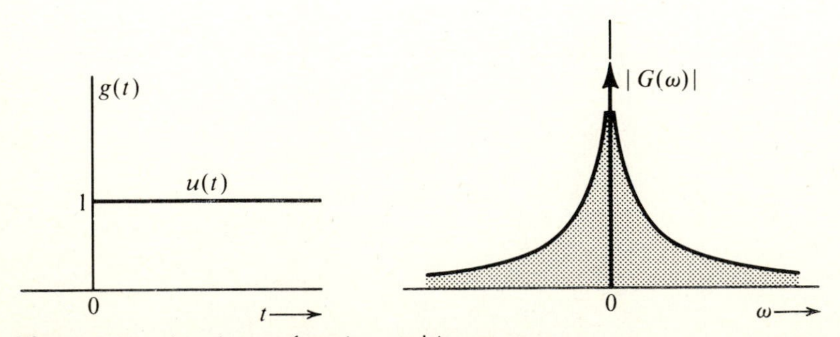

Figure 2.17 A unit step function and its spectrum.

This is shown in Fig. 2.17. Note that $u(t)$ is not a true dc signal. Hence it has a continuous spectrum consisting of all possible frequency components in addition to dc. ■

Table 2.1 lists several useful Fourier transform pairs.

Table 2.1 Fourier transform pairs

$g(t)$	$G(\omega)$
1. $e^{-at}u(t)$	$\dfrac{1}{a + j\omega}$
2. $te^{-at}u(t)$	$\dfrac{1}{(a + j\omega)^2}$
3. $\|t\|$	$\dfrac{-2}{\omega^2}$
4. $\delta(t)$	1
5. 1	$2\pi\,\delta(\omega)$
6. $u(t)$	$\pi\,\delta(\omega) + \dfrac{1}{j\omega}$
7. $\cos\omega_o t$	$\pi[\delta(\omega - \omega_o) + \delta(\omega + \omega_o)]$
8. $\sin\omega_o t$	$j\pi[\delta(\omega + \omega_o) - \delta(\omega - \omega_o)]$
9. $\cos\omega_o t\, u(t)$	$\dfrac{\pi}{2}[\delta(\omega - \omega_o) + \delta(\omega + \omega_o)] + \dfrac{j\omega}{\omega_o^2 - \omega^2}$
10. $\sin\omega_o t\, u(t)$	$\dfrac{\pi}{2j}[\delta(\omega - \omega_o) - \delta(\omega + \omega_o)] + \dfrac{\omega_o}{\omega_o^2 - \omega^2}$
11. $e^{-at}\sin\omega_o t\, u(t)$	$\dfrac{\omega_o}{(a + j\omega)^2 + \omega_o^2}$
12. $2B\,\mathrm{sinc}\,(2Bt)$	$\Pi\left(\dfrac{\omega}{4\pi B}\right)$
13. $\Pi\left(\dfrac{t}{\tau}\right)$	$\tau\,\mathrm{sinc}\left(\dfrac{\omega\tau}{2\pi}\right)$
14. $\begin{cases} 1 - \dfrac{\|t\|}{\tau} \cdots \|t\| < \tau \\ 0 \cdots \|t\| > \tau \end{cases}$	$\tau\,\mathrm{sinc}^2\left(\dfrac{\omega\tau}{2\pi}\right)$
15. $e^{-a\|t\|}$	$\dfrac{2a}{a^2 + \omega^2}$
16. $\sum_{n=-\infty}^{\infty}\delta(t - kT)$	$\omega_o\sum_{n=-\infty}^{\infty}\delta(\omega - n\omega_o) \qquad \omega_o = \dfrac{2\pi}{T}$
17. $e^{-t^2/2\sigma^2}$	$\sigma\sqrt{2\pi}\, e^{-\sigma^2\omega^2/2}$

2.3 SOME PROPERTIES OF THE FOURIER TRANSFORM

The Fourier transform is a tool for expressing a signal in terms of its exponential components of various frequencies and is just another way of specifying the signal. We therefore have two descriptions of the same function: the time-domain and the frequency-domain descriptions.

It is instructive to study the effect in one domain caused by certain operations on the function in the other domain. We may ask, for example: If a signal is differentiated in the time domain, how is the spectrum of the derivative signal related to the spectrum of the signal itself? How does the time shift affect the signal spectrum? And so on.

It is important to point out at this stage that a certain amount of symmetry exists in the equations defining the two domains. This can be easily seen from the equations defining the Fourier transform.

$$G(\omega) = \int_{-\infty}^{\infty} g(t)e^{-j\omega t}\, dt$$

and **(2.49)**

$$g(t) = \frac{1}{2\pi}\int_{-\infty}^{\infty} G(\omega)e^{j\omega t}\, d\omega$$

We should therefore expect this symmetry, or duality, to be reflected in the properties. For example, we expect that the effect on the frequency domain caused by differentiation in the time domain should be similar to the effect on the time domain caused by differentiation in the frequency domain. We shall now verify that this indeed is the case.

Symmetry or Duality

If

$$g(t) \leftrightarrow G(\omega)$$

then

$$G(t) \leftrightarrow 2\pi g(-\omega) \tag{2.50}$$

☐ **Proof:** From Eq. (2.49) it follows that

$$2\pi g(-t) = \int_{-\infty}^{\infty} G(x)e^{-jxt}\, dx$$

Now change t to ω and the result follows.

■ EXAMPLE 2.10

Show that

$$\frac{1}{a + jt} \leftrightarrow 2\pi\, e^{a\omega}\, u(-\omega) \tag{2.51}$$

This follows immediately from application of Eq. (2.50) to

$$e^{-at}u(t) \leftrightarrow \frac{1}{a + j\omega}$$ ■

■ **EXAMPLE 2.11**

Show that

$$\frac{j}{\pi t} \leftrightarrow \text{sgn}\,(\omega) \tag{2.52}$$

Application of Eq. (2.50) to Eq. (2.42) yields

$$\frac{2}{jt} \leftrightarrow 2\pi\ \text{sgn}\,(-\omega)$$

Because sgn $(-\omega) = -\text{sgn}\,(\omega)$, we get Eq. (2.52). Another example of symmetry or duality is shown in Fig. 2.18. ■

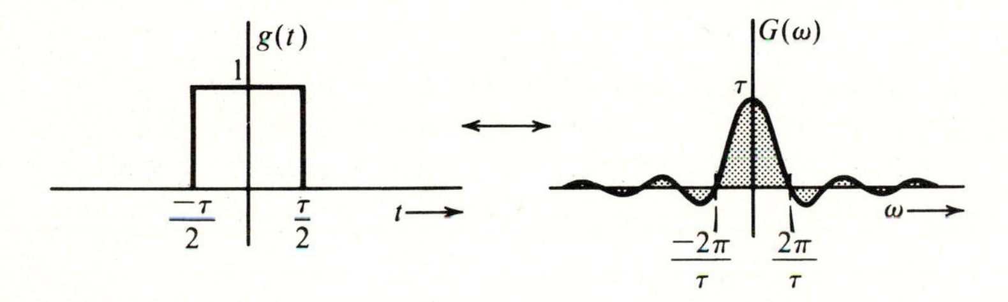

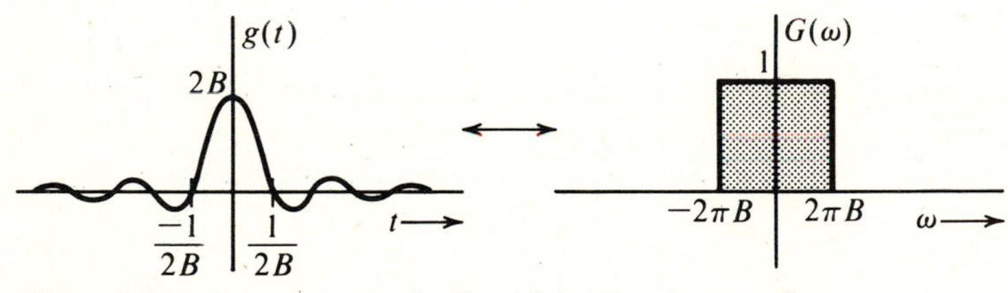

Figure 2.18 Symmetry or duality of the Fourier transform.

Scaling

If

$$g(t) \leftrightarrow G(\omega)$$

then for a real constant a,

$$g(at) \leftrightarrow \frac{1}{|a|}G\left(\frac{\omega}{a}\right) \tag{2.53}$$

☐ **Proof:**

$$\mathcal{F}[g(at)] = \int_{-\infty}^{\infty} g(at)e^{-j\omega t}\, dt$$

Let $x = at$. Then for $a > 0$

$$\mathcal{F}[g(at)] = \frac{1}{a}\int_{-\infty}^{\infty} g(x)e^{(-j\omega/a)x}\, dx = \frac{1}{a}G\left(\frac{\omega}{a}\right)$$

Similarly, it can be shown that if $a < 0$,

$$g(at) \leftrightarrow \frac{1}{-a}G\left(\frac{\omega}{a}\right)$$

Hence follows Eq. (2.53).

Significance of the Scaling Property. The function $g(at)$ represents the function $g(t)$ compressed in the time scale by a factor of a. Similarly, a function $G(\omega/a)$ represents a function $G(\omega)$ expanded in the frequency scale by the same factor a. The scaling property, therefore, states that compression in the time domain is equivalent to expansion in the frequency domain and vice versa. This result is also obvious intuitively, because compression in the time scale by a factor a means that the function is varying more rapidly by the same factor, and, hence, the frequencies of its components will be increased by the factor a. We therefore expect its frequency spectrum to be expanded by the factor a in the frequency scale. Similarly, if a function is

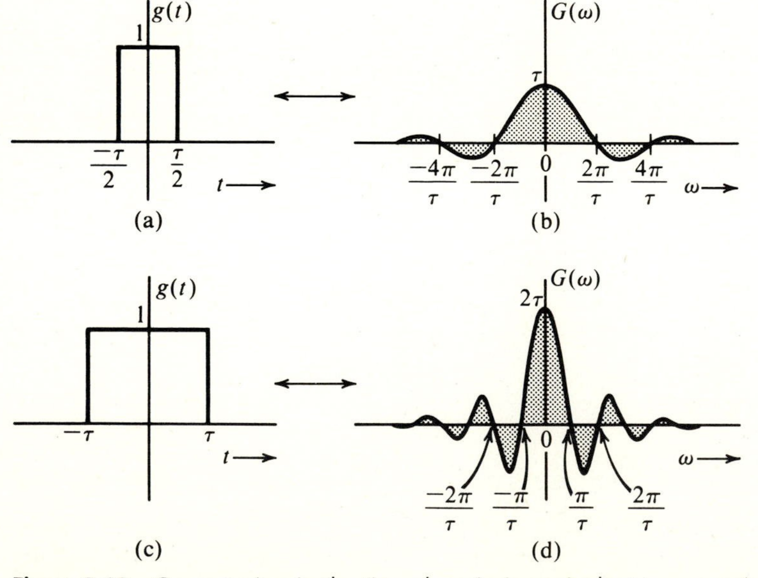

Figure 2.19 Compression in the time domain is equivalent to expansion in the frequency domain and vice versa.

expanded in the time scale, it varies more slowly, and, hence, the frequencies of its components are lowered. Thus the frequency spectrum is compressed. As an example, consider the signal $\cos \omega_o t$. This signal has frequency components at $\pm\omega_o$. The signal $\cos 2\omega_o t$ represents compression of $\cos \omega_o t$ by a factor of two, and its frequency components lie at $\pm 2\omega_o$. Here the frequency spectrum has been expanded by a factor of two. The effect of scaling is demonstrated in Fig. 2.19.

■ EXAMPLE 2.12

Show that

$$g(-t) \leftrightarrow G(-\omega) \tag{2.54}$$

This follows directly by letting $a = -1$ in Eq. (2.53). ■

■ EXAMPLE 2.13

Show that (Fig. 2.20)

$$e^{-a|t|} \leftrightarrow \frac{2a}{a^2 + \omega^2} \tag{2.55}$$

Because

$$e^{-a|t|} = e^{at}u(-t) + e^{-at}u(t)$$

and

$$e^{-at}u(t) \leftrightarrow \frac{1}{a + j\omega}$$

therefore

$$e^{at}u(-t) \leftrightarrow \frac{1}{a - j\omega}$$

and the result follows. ■

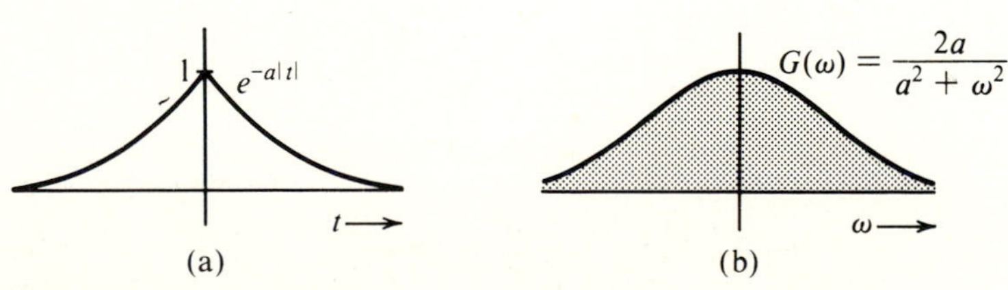

Figure 2.20 A two-sided exponential pulse and its spectrum.

■ EXAMPLE 2.14

The Fourier transform of a pulse $p(t)$ (Fig. 2.21a) is $P(\omega)$. Find the Fourier transform of $g(t)$ (Fig. 2.21b) in terms of $P(\omega)$.

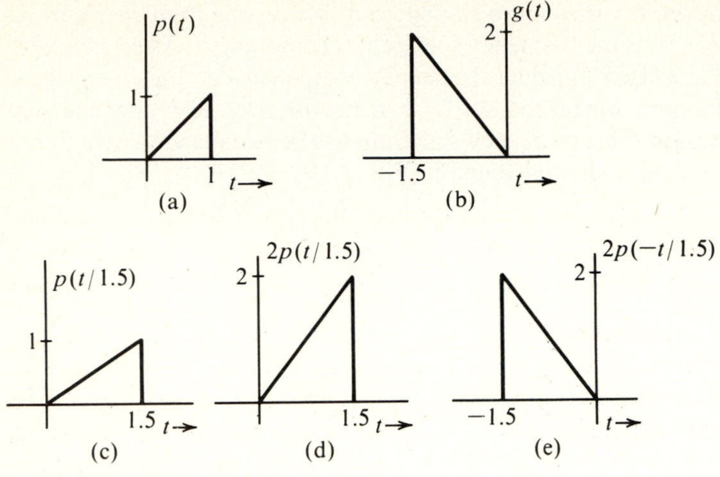

Figure 2.21 Scaling.

We can write $g(t)$ in terms of $p(t)$ as

$$g(t) = 2p\left(-\frac{t}{1.5}\right)$$

Figure 2.21*c, d,* and *e* shows the successive transformation from $p(t)$ to $g(t)$.
Because $p(t) \leftrightarrow P(\omega)$

$$p\left(\frac{t}{1.5}\right) \leftrightarrow 1.5P(1.5\omega)$$

$$2p\left(\frac{t}{1.5}\right) \leftrightarrow 3P(1.5\omega)$$

Hence

$$g(t) = 2p\left(-\frac{t}{1.5}\right) \leftrightarrow 3P(-1.5\omega)$$ ■

Time Shifting

If

$$g(t) \leftrightarrow G(\omega)$$

then

$$g(t - t_o) \leftrightarrow G(\omega)e^{-j\omega t_o} \tag{2.56a}$$

☐ **Proof:**

$$\mathcal{F}[g(t - t_o)] = \int_{-\infty}^{\infty} g(t - t_o)e^{-j\omega t}\, dt$$

Letting $t - t_o = x$, we have

$$\mathcal{F}[g(t - t_o)] = \int_{-\infty}^{\infty} g(x)e^{-j\omega(x+t_o)}\,dx$$
$$= G(\omega)e^{-j\omega t_o}$$

Note that if

$$G(\omega) = |G(\omega)|\,e^{j\theta_g(\omega)}$$
$$g(t - t_o) \leftrightarrow |G(\omega)|\,e^{j[\theta_g(\omega) - \omega t_o]} \tag{2.56b}$$

This result clearly states that a shift of t_o in the time domain leaves the magnitude spectrum unchanged, but the phase spectrum acquires an additional term $-\omega t_o$. The result is also obvious intuitively, because the shifting of a function in the time domain by t_o does not change the magnitudes of its frequency components but shifts each of them by t_o seconds. A shift of t_o for a component of frequency ω is equivalent to a phase shift of $-\omega t_o$. This can be easily verified from the equation

$$\cos \omega(t - t_o) = \cos(\omega t - \omega t_o)$$

■ EXAMPLE 2.15

Find the Fourier transform of $e^{-a|t-t_o|}$. This function (shown in Fig. 2.22*a*) is a time-shifted version of $e^{-a|t|}$ (shown in Fig. 2.20*a*).

Because $e^{-a|t|} \leftrightarrow 2a/(a^2 + \omega^2)$

$$e^{-a|t-t_o|} \leftrightarrow \frac{2a}{a^2 + \omega^2}\,e^{-j\omega t_o} \tag{2.57}$$

The spectrum of $e^{-a|t-t_o|}$ (Fig. 2.22*b*) is the same as that of $e^{-a|t|}$ (Fig. 2.20*b*), except for an added phase shift of $-\omega t_o$. This example clearly demonstrates the effect of the time shift. ■

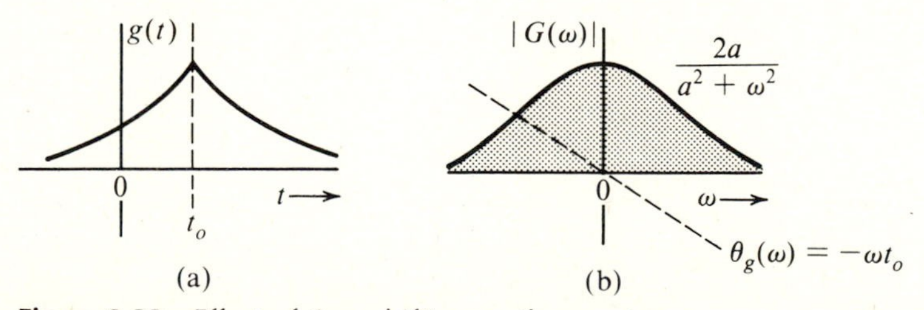

Figure 2.22 Effect of time-shifting on the spectrum.

■ EXAMPLE 2.16

Show that

$$g(t - t_o) + g(t + t_o) \leftrightarrow 2G(\omega)\cos t_o\omega \tag{2.58}$$

This follows directly from Eq. (2.56a). ■

Frequency Shifting

If

$$g(t) \leftrightarrow G(\omega)$$

then

$$g(t)e^{j\omega_o t} \leftrightarrow G(\omega - \omega_o) \tag{2.59}$$

☐ **Proof:**

$$\begin{aligned}\mathscr{F}[g(t)e^{j\omega_o t}] &= \int_{-\infty}^{\infty} g(t)e^{j\omega_o t}e^{-j\omega t}\,dt \\ &= \int_{-\infty}^{\infty} g(t)e^{-j(\omega-\omega_o)t}\,dt \\ &= G(\omega - \omega_o)\end{aligned}$$

The theorem states that a shift of ω_o in the frequency domain is equivalent to multiplication by $e^{j\omega_o t}$ in the time domain. It follows that multiplication by a factor $e^{j\omega_o t}$ translates the whole frequency spectrum $G(\omega)$ by an amount ω_o.

In communication systems, it is often desirable to translate the frequency spectrum. This is usually accomplished by multiplying a signal $g(t)$ by a sinusoidal signal. This process is known as *modulation.** Because a sinusoidal signal of frequency ω_o can be expressed as the sum of exponentials, multiplication of a signal $g(t)$ by a sinusoid (modulation) will translate the whole frequency spectrum. This can be shown by observing the identity

$$g(t)\cos\omega_o t = \frac{1}{2}[g(t)e^{j\omega_o t} + g(t)e^{-j\omega_o t}]$$

Using the frequency-shifting property, it follows that

$$g(t)\cos\omega_o t \leftrightarrow \frac{1}{2}[G(\omega + \omega_o) + G(\omega - \omega_o)] \tag{2.60a}$$

Thus, the process of modulation translates the frequency spectrum up and down in frequency by ω_o. An example of frequency translation caused by modulation is shown in Fig. 2.23. Note the duality between Eqs. (2.58) and (2.60a).

If the signal is multiplied by $\cos(\omega_o t + \varphi)$ rather than by $\cos\omega_o t$, using similar argument we can show that

$$g(t)\cos(\omega_o t + \varphi) \leftrightarrow \frac{1}{2}[G(\omega + \omega_o)e^{-j\varphi} + G(\omega - \omega_o)e^{j\varphi}] \tag{2.60b}$$

Note that the magnitude spectrum of $g(t)\cos(\omega_o t + \varphi)$ is identical to that of $g(t)\cos\omega_o t$, but the phrase spectrum of the former differs from that of the latter by

*This modulation (known as amplitude modulation) is but one of the two basic types of modulation to be discussed in the next chapter.

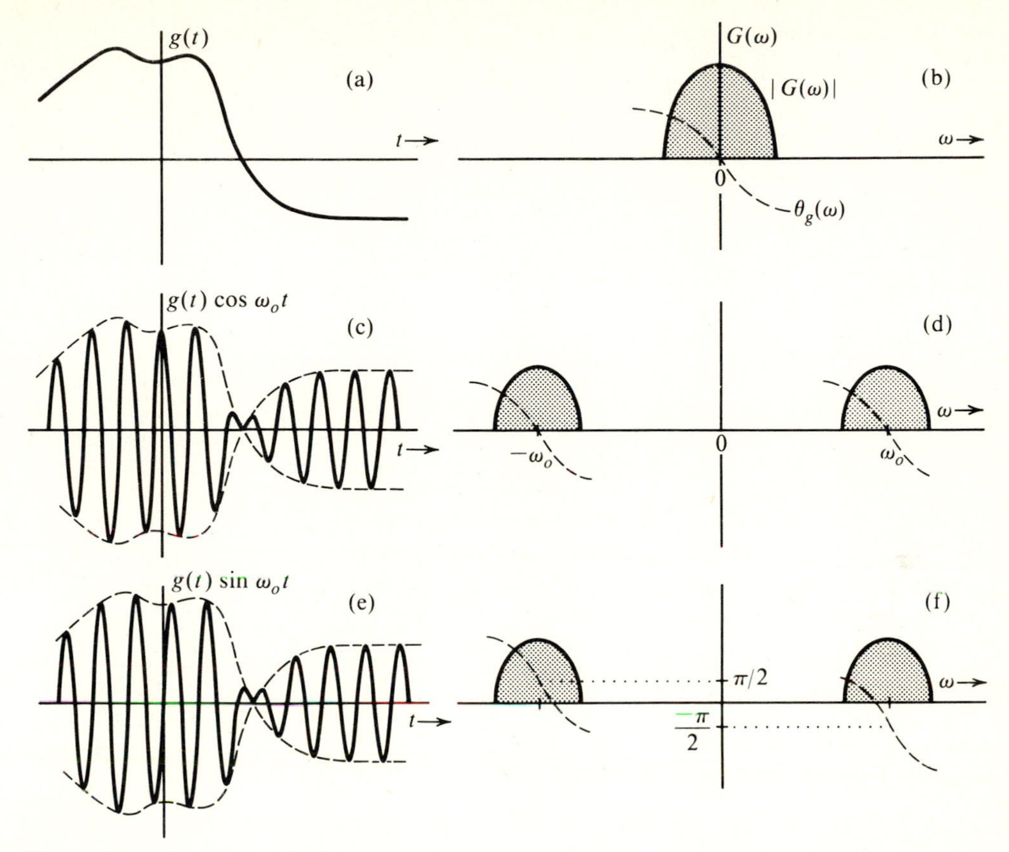

Figure 2.23 Frequency shifting.

a constant φ for $\omega > 0$ and by $-\varphi$ for $\omega < 0$. For a special case when $\varphi = -\pi/2$, Eq. (2.60b) becomes

$$g(t) \sin \omega_o t \leftrightarrow \frac{1}{2}\left[G(\omega + \omega_o)e^{j\pi/2} + G(\omega - \omega_o)e^{-j\pi/2}\right] \quad \textbf{(2.60c)}$$

$$= \frac{j}{2}\left[G(\omega + \omega_o) - G(\omega - \omega_o)\right] \quad \textbf{(2.60d)}$$

The magnitude and the phase spectra of $g(t) \cos \omega_o t$ and $g(t) \sin \omega_o t$ are shown in Fig. 2.23*d* and *f*, respectively.

It should be noted that signals $g(t) \cos \omega_o t$ and $g(t) \sin \omega_o t$ have bandpass-magnitude spectra centered symmetrically at ω_o (as well as at $-\omega_o$). A general bandpass signal $g(t)$ can be expressed as*

$$g_{bp}(t) = g_c(t) \cos \omega_o t + g_s(t) \sin \omega_o t \quad \textbf{(2.61a)}$$

If $g_c(t)$ and $g_s(t)$ are both lowpass signals, each bandlimited to B, then $g_{bp}(t)$ will be a bandpass signal of bandwidth $2B$, and its spectrum will be centered at ω_o (and $-\omega_o$).

* See Sec. 5.10 for a rigorous proof.

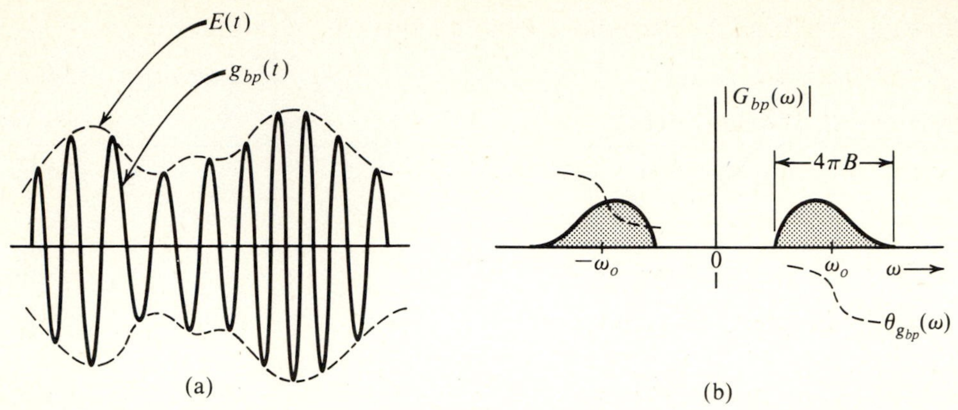

Figure 2.24 A bandpass signal and its spectrum.

Although $g_c(t) \cos \omega_o t$ and $g_s(t) \sin \omega_o t$ both have magnitude spectra that are symmetric about ω_o (and $-\omega_o$), the magnitude spectrum of their sum $g_{bp}(t)$ is *not* symmetric* about ω_o. This is the result of the associated phase spectra. The addition of two spectra is the same as adding two exponentials at each frequency. Because

$$a_1 e^{j\varphi_1} + a_2 e^{j\varphi_2} \neq (a_1 + a_2)e^{j(\varphi_1 + \varphi_2)}$$

magnitudes do not add directly, and the symmetry is destroyed. A general bandpass signal $g_{bp}(t)$ and its spectrum is shown in Fig. 2.24.

Equation (2.61a) can also be written as

$$g_{bp}(t) = E(t) \cos [\omega_o t + \theta(t)] \quad \textbf{(2.61b)}$$

where

$$E(t) = +\sqrt{g_c^2(t) + g_s^2(t)} \quad \textbf{(2.61c)}$$

$$\theta(t) = -\tan^{-1} \frac{g_s(t)}{g_c(t)} \quad \textbf{(2.61d)}$$

Because $g_c(t)$ and $g_s(t)$ are lowpass signals, $E(t)$ and $\theta(t)$ are also lowpass signals. If $2\pi B \ll \omega_o$, $g_{bp}(t)$ is a narrowband signal. Because $E(t)$ is always positive [Eq. (2.61c)], it can be seen from Eq. (2.61b) that $E(t)$ is a slowly varying envelope and $\theta(t)$ is a slowly varying phase of $g_{bp}(t)$ (Fig. 2.24).

■ EXAMPLE 2.17

Find the Fourier transforms of

(1) $e^{j\omega_o t}$
(2) $\cos(\omega_o t + \varphi)$, $\cos \omega_o t$, and $\sin \omega_o t$
(3) $\cos \omega_o t\, u(t)$ and $\sin \omega_o t\, u(t)$

---

* See Sec. 5.10 for a rigorous proof.

From Eq. (2.47) and the frequency-shifting property, Eq. (2.59),

$$e^{j\omega_o t} \leftrightarrow 2\pi\, \delta(\omega - \omega_o) \tag{2.62a}$$

From Eqs. (2.47) and (2.60b),

$$\cos(\omega_o t + \varphi) \leftrightarrow \pi[\delta(\omega + \omega_o)e^{-j\varphi} + \delta(\omega - \omega_o)e^{j\varphi}] \tag{2.62b}$$

Hence

$$\cos \omega_o t \leftrightarrow \pi[\delta(\omega + \omega_o) + \delta(\omega - \omega_o)] \tag{2.62c}$$

$$\sin \omega_o t \leftrightarrow \pi[\delta(\omega + \omega_o)e^{j\pi/2} + \delta(\omega - \omega_o)e^{-j\pi/2}] \tag{2.62d}$$

From Eqs. (2.48) and (2.60a and d),

$$\cos \omega_o t\, u(t) \leftrightarrow \frac{\pi}{2}[\delta(\omega + \omega_o) + \delta(\omega - \omega_o)] + \frac{j\omega}{\omega_o^2 - \omega^2} \tag{2.62e}$$

$$\sin \omega_o t\, u(t) \leftrightarrow \frac{\pi}{2j}[\delta(\omega - \omega_o) - \delta(\omega + \omega_o)] + \frac{\omega_o}{\omega_o^2 - \omega^2} \tag{2.62f}$$

Figure 2.25 shows signals $\cos \omega_o t$ and $\sin \omega_o t$ and their spectra.

The Fourier transform of $e^{j\omega_o t}$ consists of a single impulse at $\omega = \omega_o$. This is an expected result because the Fourier transform is a spectrum of the components of the form $e^{j\omega t}$, and $e^{j\omega_o t}$ consists of only one exponential component of the frequency ω_o. On the other hand, $\cos \omega_o t = \frac{1}{2}[e^{j\omega_o t} + e^{-j\omega_o t}]$, and it consists of two exponential components of frequencies ω_o and $-\omega_o$. Hence its entire spectrum consists of impulses at ω_o and $-\omega_o$. The only difference between $\sin \omega_o t$ and $\cos \omega_o t$ is the phase difference of $\pi/2$. Hence their magnitude spectra are identical, but the phase spectra differ by

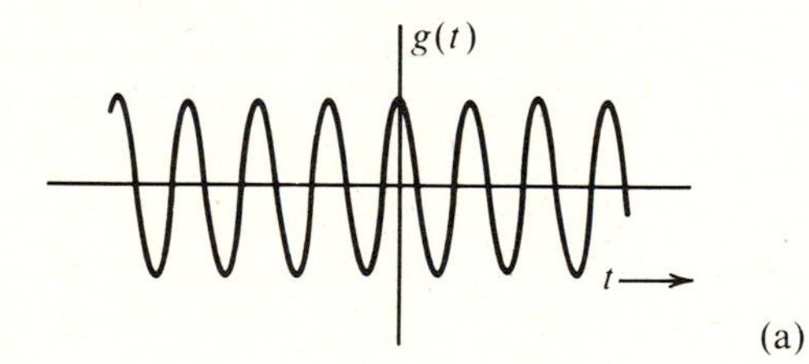

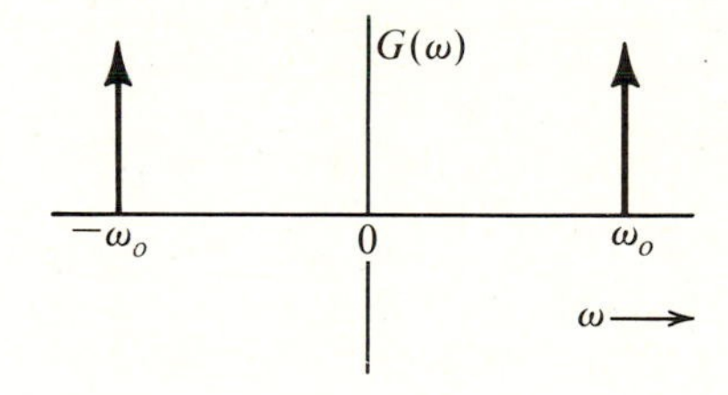

(a)

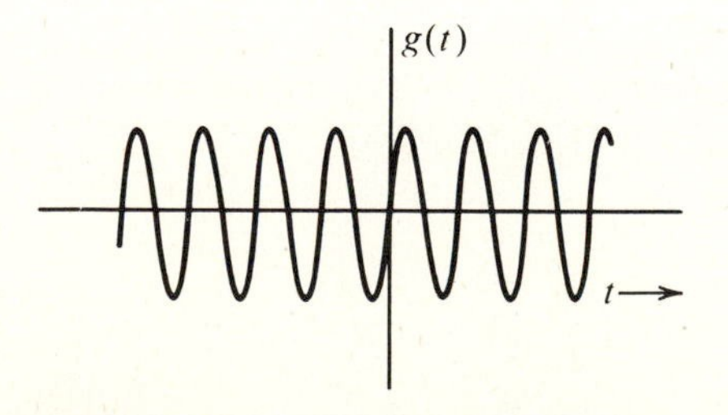

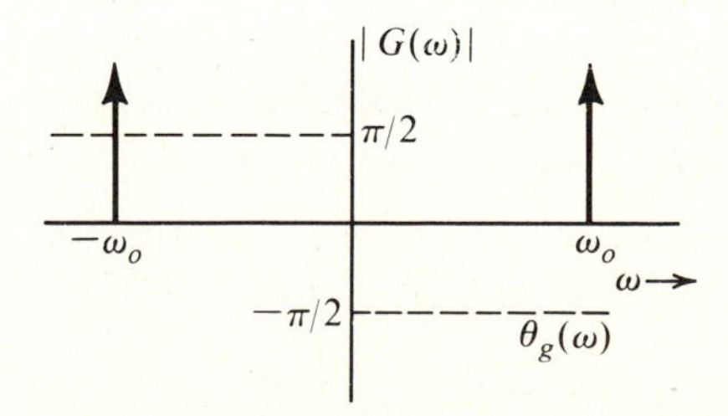

(b)

Figure 2.25 A sinusoid and its spectrum.

$\pi/2$. The signals $\cos \omega_o t\, u(t)$ and $\sin \omega_o t\, u(t)$ are not eternal sinusoids (or eternal exponentials). Therefore, they have continuous spectra in addition to discrete components at $\pm\, \omega_o$. This means such signals consist of all possible frequency components (the continuous spectrum) in addition to discrete components at $\pm\, \omega_o$. At first thought this may sound unreasonable, because these signals seem to be pure sinusoids. We must remember, however, that in the Fourier representation we are talking about representing a signal by eternal sinusoids (or exponentials). If the signals $\cos \omega_o t$ and $\sin \omega_o t$ were to be eternal, it would take only two eternal exponentials to represent each of them. But if they are zero for $t < 0$, we will need an infinite number of components of the proper magnitude and phase that will sum to zero for $t < 0$ and sum to $\cos \omega_o t$ (or $\sin \omega_o t$) for $t > 0$.

Note that the spectrum of $e^{j\omega_o t}$ does not satisfy the condition in Eq. (2.34) because it is not a real function of t. ■

■ EXAMPLE 2.18

Find the Fourier transform of a general periodic signal $g(t)$.

A periodic signal $g(t)$ can be represented by an exponential Fourier series as [Eq. (2.19a)]

$$g(t) = \sum_{-\infty}^{\infty} G_n e^{jn\omega_o t} \qquad \omega_o = \frac{2\pi}{T_o}$$

Therefore

$$g(t) \leftrightarrow \sum_{-\infty}^{\infty} \mathcal{F}[G_n e^{jn\omega_o t}]$$

Hence, from Eq. (2.62a) we have

$$g(t) \leftrightarrow 2\pi \sum_{-\infty}^{\infty} G_n\, \delta(\omega - n\omega_o) \tag{2.63}$$

The Fourier transform of a periodic signal is a sequence of impulses at $\pm\, n\omega_o (n = 0, 1, 2, \ldots)$. ■

■ EXAMPLE 2.19

Find the Fourier transform of the unit impulse train in Fig. 2.26*a*. The exponential Fourier series for this impulse train was found to be [Eq. (2.24b)]

$$g(t) = \sum_{n=-\infty}^{\infty} \delta(t - nT_o) = \frac{1}{T_o} \sum_{n=-\infty}^{\infty} e^{jn\omega_o t}$$

Hence from Eq. (2.63),

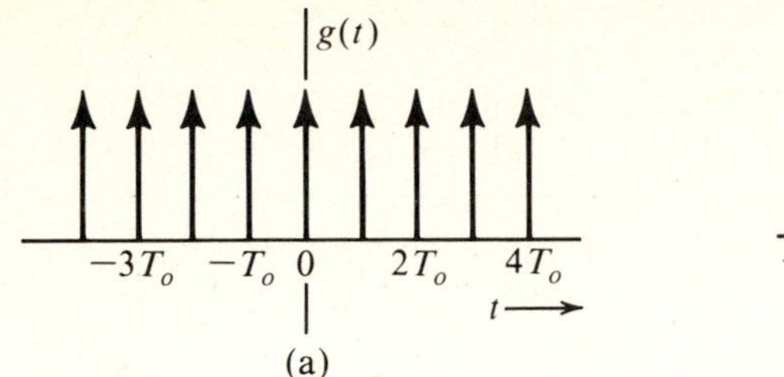

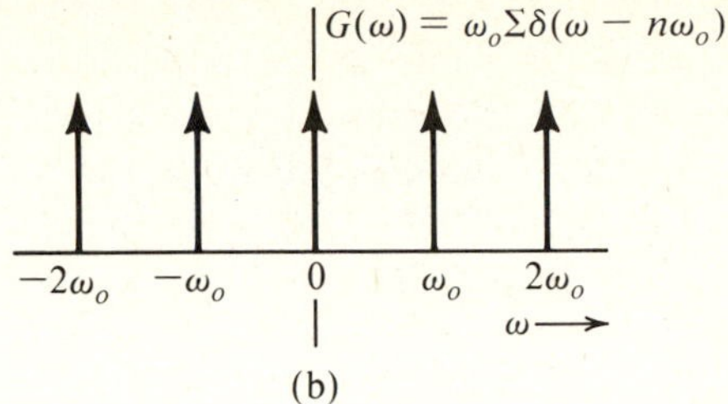

Figure 2.26 An impulse train and its spectrum.

$$\sum_{-\infty}^{\infty} \delta(t - nT_o) \leftrightarrow \frac{2\pi}{T_o} \sum_{n=-\infty}^{\infty} \delta(\omega - n\omega_o)$$

$$= \omega_o \sum_{n=-\infty}^{\infty} \delta(\omega - n\omega_o) \qquad \omega_o = \frac{2\pi}{T_o} \tag{2.64}$$

The spectrum (Fig. 2.26*b*) is also an impulse train. ■

Time Differentiation and Integration

If

$$g(t) \leftrightarrow G(\omega)$$

then*

$$\frac{dg}{dt} \leftrightarrow j\omega G(\omega) \tag{2.65a}$$

and

$$\int_{-\infty}^{t} g(x)\, dx \leftrightarrow \frac{G(\omega)}{j\omega} + \pi G(0)\, \delta(\omega) \tag{2.65b}$$

☐ **Proof:** We have

$$G(\omega) = \int_{-\infty}^{\infty} g(t) e^{-j\omega t}\, dt$$

Integration by parts yields

$$G(\omega) = -\frac{1}{j\omega} g(t) e^{-j\omega t} \Bigg|_{-\infty}^{\infty} + \frac{1}{j\omega} \int_{-\infty}^{\infty} \frac{dg}{dt} e^{-j\omega t}\, dt$$

Because $g(t)$ is Fourier transformable, $\lim_{t\to\pm\infty} g(t) = 0$, so the first term is zero, and we get the desired result.

---

* Equation (2.65a) does not guarantee the existence of the transform of dg/dt. It merely says that if that transform exists, it is given by $j\omega G(\omega)$.

$$j\omega G(\omega) = \int_{-\infty}^{\infty} \frac{dg}{dt} e^{-j\omega t}\, dt$$

Equation (2.65a) is generalized as

$$\frac{d^n g}{dt^n} \leftrightarrow (j\omega)^n G(\omega) \qquad \textbf{(2.65c)}$$

To prove Eq. (2.65b), we note that $u(\cdot)$ is zero for negative values of the argument. Therefore, $u(t - x)$ is 1 for $x < t$ and is zero if $x > t$. Consequently,

$$\int_{-\infty}^{t} g(x)\, dx = \int_{-\infty}^{\infty} g(x)u(t - x)\, dx$$

and

$$\mathcal{F}\left[\int_{-\infty}^{t} g(x)\, dx\right] = \int_{-\infty}^{\infty} e^{-j\omega t}\left[\int_{-\infty}^{\infty} g(x)u(t - x)\, dx\right] dt$$

Changing the order of integration,* we have

$$\begin{aligned}\mathcal{F}\left[\int_{-\infty}^{t} g(x)\, dx\right] &= \int_{-\infty}^{\infty} g(x)\left[\int_{-\infty}^{\infty} u(t - x)e^{-j\omega t}\, dt\right] dx \\ &= \int_{-\infty}^{\infty} g(x)\left[\frac{1}{j\omega} + \pi\,\delta(\omega)\right]e^{-j\omega x}\, dx \\ &= G(\omega)\left[\frac{1}{j\omega} + \pi\,\delta(\omega)\right] \\ &= \frac{G(\omega)}{j\omega} + \pi G(0)\,\delta(\omega)\end{aligned}$$

From Eq. (2.30b), we have

$$G(0) = \int_{-\infty}^{\infty} g(t)\, dt$$

Hence, if the area under $g(t)$ is zero, then $G(0) = 0$, and Eq. (2.65b) becomes

$$\int_{-\infty}^{t} g(x)\, dx \leftrightarrow \frac{G(\omega)}{j\omega} \qquad \textbf{(2.65d)}$$

The time-differentiation and time-integration properties expressed in Eq. (2.65) are also obvious intuitively. The Fourier transform actually expresses a function $g(t)$ in terms of a continuous sum of exponential functions of the form $e^{j\omega t}$. The derivative of

---

* Here we use Fubini's theorem, which states that the order of integration can be changed, that is,

$$\int_{-\infty}^{\infty}\int_{-\infty}^{\infty} g(x, y)\, dx\, dy = \int_{-\infty}^{\infty} dx \int_{-\infty}^{\infty} g(x, y)\, dy = \int_{-\infty}^{\infty} dy \int_{-\infty}^{\infty} g(x, y)\, dx$$

provided any one of these three double integrals with $g(x, y)$ replaced by $|g(x, y)|$ is finite.

$g(t)$ is equal to the continuous sum of the derivatives of the individual exponential components. But the derivative of an exponential function $e^{j\omega t}$ is equal to $j\omega e^{j\omega t}$. Therefore, the process of differentiation of $g(t)$ is equivalent to multiplication by $j\omega$ of each exponential component. A similar argument applies to integration. Thus, the time differentiation enhances whereas the integration suppresses higher frequency components. This, too, is an expected result. The time-differentiation property proves convenient in deriving the Fourier transform of some piecewise continuous signals. This is illustrated by the next example.

■ EXAMPLE 2.20

Find the Fourier transform of the trapezoidal signal $g(t)$ shown in Fig. 2.27*a*.

We differentiate this signal twice to obtain a sequence of impulses. The transform of the impulses is readily found. It is evident from Fig. 2.27*c* that

$$\frac{d^2g}{dt^2} = \frac{A}{(b-a)}[\delta(t+b) - \delta(t+a) - \delta(t-a) + \delta(t-b)] \tag{2.66}$$

Using the time-shifting property, Eq. (2.56a), and the time-differentiation property, the transform of Eq. (2.66) can now be readily written as

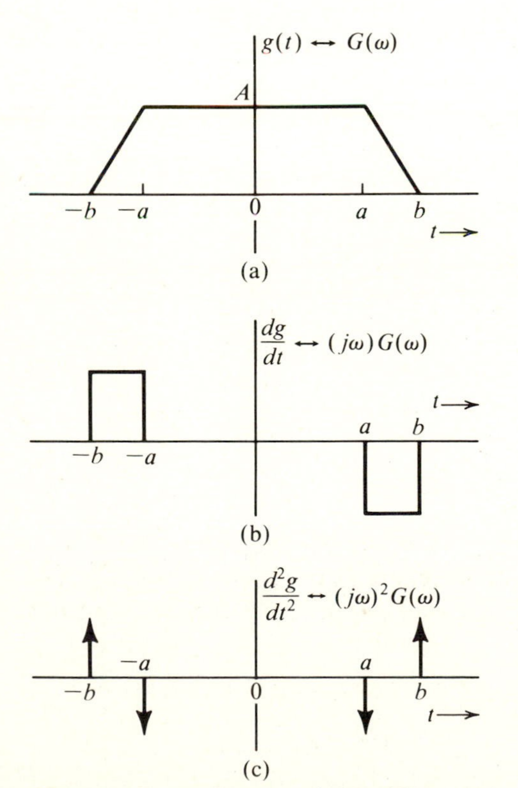

Figure 2.27 Application of time-differentiation property.

$$(j\omega)^2 G(\omega) = \frac{A}{(b-a)}(e^{j\omega b} - e^{j\omega a} - e^{-j\omega a} + e^{-j\omega b})$$

from which we get

$$G(\omega) = \frac{2A}{(b-a)}\left[\frac{\cos a\omega - \cos b\omega}{\omega^2}\right]$$

This problem suggests a numerical method of obtaining the Fourier transform of a function $g(t)$ by approximating it by straight-line segments. ■

Frequency Differentiation

We have a dual of the time-differentiation property

$$-jtg(t) \leftrightarrow \frac{d}{d\omega} G(\omega) \tag{2.67}$$

The proof is similar to that of Eq. (2.65), with the roles of t and ω interchanged.

Convolution

The convolution integral of two signals $g_1(t)$ and $g_2(t)$ is denoted by $g_1(t) * g_2(t)$ and is defined as

$$g_1(t) * g_2(t) = \int_{-\infty}^{\infty} g_1(x)g_2(t-x)\,dx \tag{2.68}$$

By letting $y = t - x$ in Eq. (2.68), we can show that

$$g_1(t) * g_2(t) = g_2(t) * g_1(t) \tag{2.69}$$

The convolution property states that if

$$g_1(t) \leftrightarrow G_1(\omega) \text{ and } g_2(t) \leftrightarrow G_2(\omega)$$

then

$$g_1(t) * g_2(t) \leftrightarrow G_1(\omega)G_2(\omega) \qquad \text{(time convolution)} \tag{2.70a}$$

and

$$g_1(t)g_2(t) \leftrightarrow \frac{1}{2\pi}G_1(\omega) * G_2(\omega) \qquad \text{(frequency convolution)} \tag{2.70b}$$

☐ **Proof:***

$$\begin{aligned}\mathcal{F}[g_1(t) * g_2(t)] &= \int_{-\infty}^{\infty} e^{-j\omega t}\left[\int_{-\infty}^{\infty} g_1(x)g_2(t-x)\,dx\right]dt \\ &= \int_{-\infty}^{\infty} g_1(x)\left(\int_{-\infty}^{\infty} g_2(t-x)e^{-j\omega t}\,dt\right)dx\end{aligned}$$

* See footnote on page 54.

Application of the time-shifting property, Eq. (2.56a), to the inner integral yields

$$\mathcal{F}[g_1(t) * g_2(t)] = \int_{-\infty}^{\infty} G_2(\omega) g_1(x) e^{-j\omega x}\, dx$$

$$= G_1(\omega)\, G_2(\omega)$$

Equation (2.70b) can be proved in a similar way, with the roles of t and ω interchanged.

The convolution property is one of the most powerful tools in Fourier analysis. It permits easy derivation of many important results.

■ EXAMPLE 2.21

Evaluate $g_1(t) * g_2(t)$ for the signals in Fig. 2.28*a* and *b*.

The integrand in Eq. (2.68) is the product $g_1(x)g_2(t - x)$, and the integration is with respect to x (not t). Signal $g_2(-x)$ is obtained by folding $g_2(x)$ about the vertical axis (Fig. 2.28*c*), and $g_2(t - x)$ is obtained by shifting $g_2(-x)$ by t (Fig. 2.28*d*). The area under the product (shown shaded in Fig. 2.28*e*) is the value of the convolution at the instant t. We must repeat this procedure for all values of t (positive as well as negative) to determine the convolution for all values of t.

In the present case, $g_1(x)g_2(t - x) = ae^{-ax}$ over the interval $(0, t)$ and is zero outside this interval. Hence

$$g_1(t) * g_2(t) = \int_{-\infty}^{\infty} g_1(x) g_2(t - x)\, dx = a \int_0^t e^{-ax}\, dx = 1 - e^{-at} \qquad t > 0$$

Note that for $t < 0$, the product $g_1(x)g_2(t - x) = 0$. Hence, the convolution exists only for $t > 0$ and is zero for $t < 0$ (Fig. 2.28*f*).

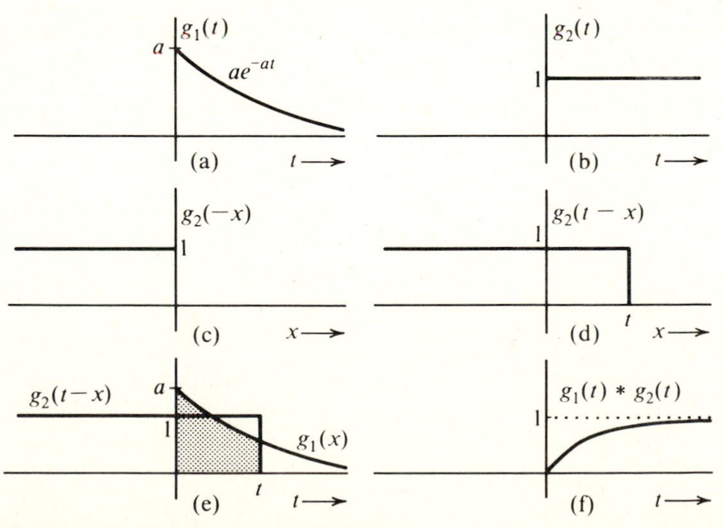

Figure 2.28 Graphical convolution.

We can solve this problem indirectly using the convolution property. We have

$$G_1(\omega) = \frac{a}{a + j\omega} \quad \text{and} \quad G_2(\omega) = \frac{1}{j\omega} + \pi\,\delta(\omega)$$

Also,

$$\begin{aligned} g_1(t) * g_2(t) &= \mathcal{F}^{-1}[G_1(\omega)G_2(\omega)] \\ &= \mathcal{F}^{-1}\left[\frac{a}{j\omega(a + j\omega)} + \frac{\pi a\,\delta(\omega)}{a + j\omega}\right] \\ &= \mathcal{F}^{-1}\left[\frac{1}{j\omega} - \frac{1}{a + j\omega} + \pi\,\delta(\omega)\right] \\ &= (1 - e^{-at})u(t) \end{aligned}$$

Here we used the fact that $a\,\delta(\omega)/a + j\omega = \delta(\omega)$, because this function exists only at $\omega = 0$ where $a/a + j\omega = 1$. The power of the convolution property is evident. ■

■ EXAMPLE 2.22

Determine $g(t) * \delta(t)$

$$g(t) * \delta(t) = \int_{-\infty}^{\infty} g(x)\,\delta(t - x)\,dx$$

From the sampling property, Eq. (2.44a), it follows that

$$g(t) * \delta(t) = g(t) \tag{2.71a}$$

Thus a convolution of a signal $g(t)$ with a unit impulse yields back $g(t)$. This result can also be derived indirectly by noting that $g(t) \leftrightarrow G(\omega)$ and $\delta(t) \leftrightarrow 1$. Hence, from the time-convolution property, Eq. (2.70a), it follows that

$$g(t) * \delta(t) \leftrightarrow G(\omega)$$

and, hence, the result.

Using the generalized sampling property, Eq. (2.44b and c), we can generalize Eq. (2.71a) as

$$g(t) * \delta(t - T) = g(t - T) \tag{2.71b}$$

and

$$g(t - t_1) * \delta(t - t_2) = g(t - t_1 - t_2) \tag{2.71c}$$

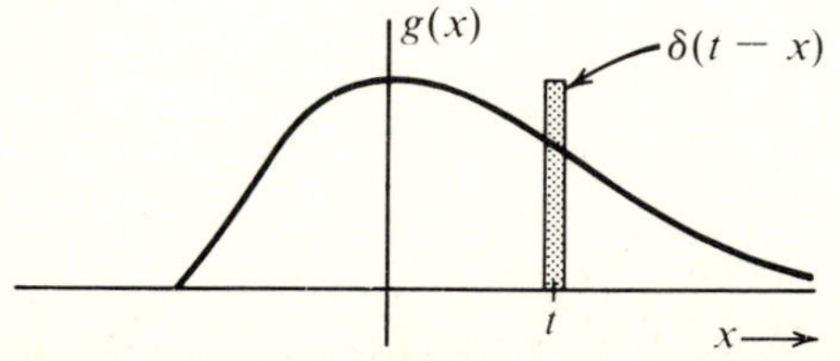

Figure 2.29 Convolution with an impulse.

The result $g(t) * \delta(t) = g(t)$ can also be understood graphically. In convolution, we maintain $g(x)$ fixed (Fig. 2.29) and fold $\delta(x)$ about the vertical axis. We picture $\delta(x)$ as a narrow pulse centered at the origin and having unit area. Also, $\delta(-x) = \delta(x)$, and $\delta(t - x)$ is the same pulse shifted to t. The product $g(x)\ \delta(t - x)$ is also a narrow pulse of area $g(t)$. Hence, as the impulse is shifted by different values of t, it reproduces the function $g(t)$ itself. ■

■ EXAMPLE 2.23

Find $G_1(\omega) * G_2(\omega)$ for the signals shown in Fig. 2.30.

$$G_2(\omega) = k[\delta(\omega - \omega_o) + \delta(\omega + \omega_o)]$$

and*

$$\begin{aligned} G_1(\omega) * G_2(\omega) &= kG_1(\omega) * [\delta(\omega - \omega_o) + \delta(\omega + \omega_o)] \\ &= k[G_1(\omega - \omega_o) + G_1(\omega + \omega_o)] \end{aligned} \tag{2.72}$$

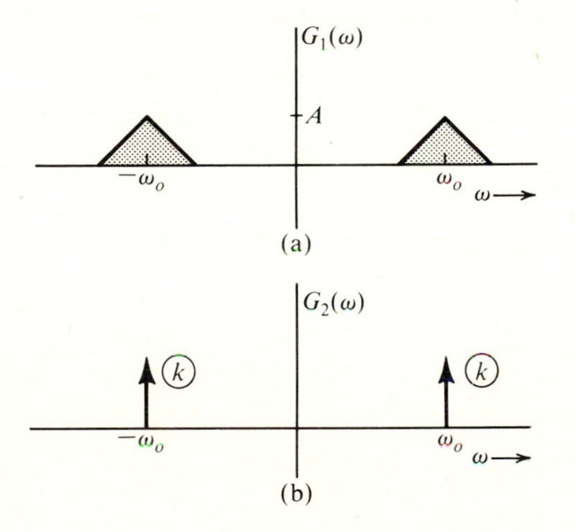

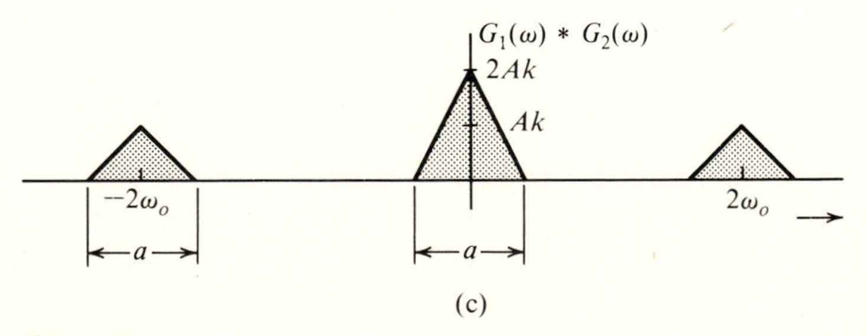

(c)

Figure 2.30

* In deriving Eq. (2.72), we are using the distributive property of convolution:

$$g_1(t) * [g_2(t) + g_3(t)] = g_1(t) * g_2(t) + g_1(t) * g_3(t)$$

The proof is trivial.

Therefore, $G_1(\omega) * G_2(\omega)$ is the sum of $G_1(\omega)$ shifted up and down by ω_o. The result is shown in Fig. 2.30*c*. This result suggests a method of demodulating amplitude-modulated signals. ■

■ EXAMPLE 2.24

A linear time-invariant system (Fig. 2.31) has a unit impulse response $h(t)$. If the input signal is $g(t)$, find the output signal spectrum.

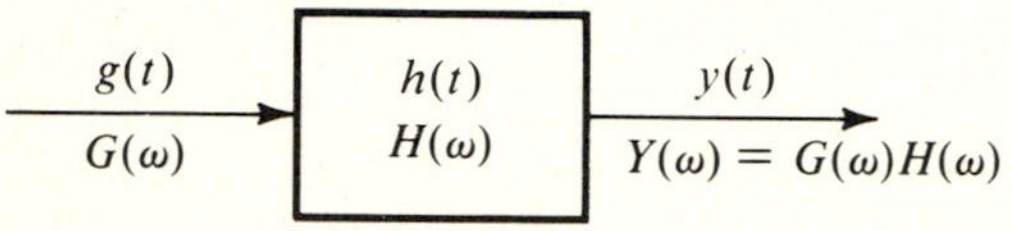

$g(t)$, $G(\omega)$ → [$h(t)$, $H(\omega)$] → $y(t)$, $Y(\omega) = G(\omega)H(\omega)$

Figure 2.31 Signal transmission through a linear system.

It is known that the output $y(t)$ of a linear time-invariant system is the convolution of $g(t)$ and $h(t)$.

$$y(t) = g(t) * h(t) \tag{2.73a}$$

$$Y(\omega) = G(\omega)H(\omega) \tag{2.73b}$$

and

$$y(t) = \mathcal{F}^{-1}[G(\omega)H(\omega)] \tag{2.73c}$$

A linear time-invariant system acts as a filter that changes the spectrum of the input signal from $G(\omega)$ to $G(\omega)H(\omega)$. ■

Time Correlation and Energy

The time-correlation function $\psi_{g_1g_2}(\tau)$ of two real signals $g_1(t)$ and $g_2(t)$ is defined as*

$$\psi_{g_1g_2}(\tau) = \int_{-\infty}^{\infty} g_1(t)g_2(t + \tau)\, dt \tag{2.74a}$$

Letting $t = -x$, we have

$$\psi_{g_1g_2}(\tau) = \int_{-\infty}^{\infty} g_1(-x)g_2(\tau - x)\, dx \tag{2.74b}$$

$$= g_1(-\tau) * g_2(\tau) \tag{2.74c}$$

The function $\psi_{g_1g_2}(\tau)$ is also known as the *time-cross-correlation* function of $g_1(t)$ and $g_2(t)$. In contrast, we define $\psi_g(\tau)$, the *time-autocorrelation* function of a real $g(t)$, as

* A general definition of the correlation that applies to real as well as complex signals is

$$\psi_{g_1g_2}(\tau) = \int_{-\infty}^{\infty} g_1^*(t)g_2(t + \tau)\, dt$$

$$\psi_g(\tau) = \int_{-\infty}^{\infty} g(t)g(t+\tau)\,dt \qquad \textbf{(2.75a)}$$

$$= \int_{-\infty}^{\infty} g(-x)g(\tau - x)\,dx \qquad \textbf{(2.75b)}$$

$$= g(-\tau) * g(\tau) \qquad \textbf{(2.75c)}$$

The application of the time-convolution property to Eqs. (2.74) and (2.75) yields

$$\psi_{g_1 g_2}(\tau) \leftrightarrow G_1(-\omega)G_2(\omega) \qquad \textbf{(2.76a)}$$

$$\psi_g(\tau) \leftrightarrow G(-\omega)G(\omega) = |G(\omega)|^2 \qquad \textbf{(2.76b)}$$

From Eqs. (2.74a) and (2.76a), we have

$$\int_{-\infty}^{\infty} g_1(t)g_2(t+\tau)\,dt = \frac{1}{2\pi}\int_{-\infty}^{\infty} G_1(-\omega)G_2(\omega)e^{j\omega\tau}\,d\omega$$

Letting $\tau = 0$, we have

$$\int_{-\infty}^{\infty} g_1(t)g_2(t)\,dt = \frac{1}{2\pi}\int_{-\infty}^{\infty} G_1(-\omega)G_2(\omega)\,d\omega \qquad \textbf{(2.77a)}$$

Similarly, by considering $\psi_{g_2 g_1}(\tau)$, we can show that*

$$\int_{-\infty}^{\infty} g_1(t)g_2(t)\,dt = \frac{1}{2\pi}\int_{-\infty}^{\infty} G_1(\omega)G_2(-\omega)\,d\omega \qquad \textbf{(2.77b)}$$

If we let $g_1(t) = g_2(t) = g(t)$, we have†

$$\int_{-\infty}^{\infty} g^2(t)\,dt = \frac{1}{2\pi}\int_{-\infty}^{\infty} |G(\omega)|^2\,d\omega \qquad \textbf{(2.77c)}$$

The integral on the left-hand side of Eq. (2.77c) is E_g, the energy of $g(t)$. The right-hand side expresses the signal energy in the frequency domain. This relation, Eq. (2.77c), is known as *Parseval's theorem.*

■ EXAMPLE 2.25

Show that

$$\int_{-\infty}^{\infty} \operatorname{sinc}(2Bt - m)\operatorname{sinc}(2Bt - n)\,dt = \begin{cases} 0 & m \neq n \\ \dfrac{1}{2B} & m = n \end{cases}$$

---

* Equation (2.77a and b) is valid for complex $g_1(t)$ and $g_2(t)$.

† For a complex signal $g(t)$

$$\int_{-\infty}^{\infty} |g(t)|^2\,dt = \frac{1}{2\pi}\int_{-\infty}^{\infty} |G(\omega)|^2\,d\omega$$

This follows from the result in Prob. 2.9 and the fact that Eq. (2.77) is valid for complex signals.

□ **Proof:** Let I represent the integral above. From Eq. (2.77a),

$$I = \frac{1}{2\pi}\int_{-\infty}^{\infty} \frac{1}{4B^2}\left[\Pi\left(\frac{\omega}{4\pi B}\right)\right]^2 e^{j[(m-n)/2B]\omega}\, d\omega$$

Because $\Pi(\omega/4\pi B) = 1$ over $|\omega| \leq 2\pi B$ and is 0 everywhere else,

$$I = \frac{1}{8\pi B^2}\int_{-2\pi B}^{2\pi B} e^{j[(m-n)/2B]\omega}\, d\omega$$

If $m = n$, the integrand is unity, and $I = 1/2B$. If $m \neq n$,

$$I = \frac{1}{8\pi B^2}\left(\frac{-j2B}{m-n}\right) e^{[j(m-n)/2B]\omega}\Bigg|_{-2\pi B}^{2\pi B}$$

$$= \frac{-j}{4\pi B(m-n)}[e^{j(m-n)\pi} - e^{-j(m-n)\pi}] = 0$$ ■

■ EXAMPLE 2.26

Prove the following Schwarz inequality for a pair of real finite energy signals $g_1(t)$ and $g_2(t)$:

$$\left[\int_a^b g_1(t)g_2(t)\, dt\right]^2 \leq \left(\int_a^b g_1^2(t)\, dt\right)\left(\int_a^b g_2^2(t)\, dt\right) \tag{2.78a}$$

with equality only if $g_2(t) = cg_1(t)$, and

$$\left|\int_{-\infty}^{\infty} G_1(\omega)G_2(\omega)\, d\omega\right|^2 \leq \int_{-\infty}^{\infty} |G_1(\omega)|^2 \int_{-\infty}^{\infty} |G_2(\omega)|^2\, d\omega \tag{2.78b}$$

with equality only if $G_2(\omega) = CG_1^*(\omega)$, where c and C are arbitrary constants. We can prove Eq. (2.78a) as follows: For real values of x,

$$\int_a^b [xg_1(t) - g_2(t)]^2\, dt \geq 0$$

or

$$x^2\int_a^b g_1^2(t)\, dt - 2x\int_a^b g_1(t)g_2(t)\, dt + \int_a^b g_2^2(t)\, dt \geq 0$$

Because this quadratic in x is nonnegative for any value of x, its discriminant must be nonpositive, and Eq. (2.78a) follows. If the discriminant equals zero, then for some value of $x = c$, the quadratic equals zero. This is possible only if $cg_1(t) - g_2(t) = 0$, and the result follows.

The proof of Eq. (2.78b) follows from Eq. (2.78a). We have

$$\left[\int_{-\infty}^{\infty} g_1(-t)g_2(t)\, dt\right]^2 \leq \int_{-\infty}^{\infty} g_1^2(-t)\, dt \int_{-\infty}^{\infty} g_2^2(t)\, dt$$

Now, using Eqs. (2.77a) and (2.77c) and noting that $\int g_1^2(-t)\,dt = \int g_1^2(t)\,dt$, we have the desired result. ■

2.4 THE SAMPLING THEOREM

The sampling theorem has a deep significance in communication theory. It states the following:

A signal bandlimited to B Hz (i.e., a signal whose Fourier transform is zero for all $|\omega| > 2\pi B$) is uniquely determined by its values at uniform intervals less than $1/2B$ seconds apart.*

The theorem states that a signal bandlimited to B can be reconstructed from its samples taken uniformly at a rate of not less than $2B$ samples per second.

To prove the sampling theorem, consider a signal $g(t)$ bandlimited to B (Fig. 2.32*a* and *b*). Multiplication of $g(t)$ by a unit impulse train (Fig. 2.32*c*) yields the sampled signal $g_s(t)$ (Fig. 2.32*d*).

$$g_s(t) = g(t) \sum_{n=-\infty}^{\infty} \delta(t - nT_s) \tag{2.79}$$

Using the Fourier series for the impulse train, Eq. (2.24b), we have

$$g_s(t) = \frac{1}{T_s} g(t) \sum_{n=-\infty}^{\infty} e^{jn\omega_s t}$$

$$= \frac{1}{T_s} \sum_{n=-\infty}^{\infty} g(t) e^{jn\omega_s t}$$

Taking Fourier transform of both sides [see Eq. (2.59)], we have

$$G_s(\omega) = \frac{1}{T_s} \sum_{-\infty}^{\infty} G(\omega - n\omega_s) \qquad \omega_s = \frac{2\pi}{T_s} \tag{2.80}$$

The Fourier transform of the sampled signal $g_s(t)$ consists of $G(\omega)$ repeating itself indefinitely at every $\pm n\omega_s$ for $n = 0, 1, 2, 3, \ldots$ (Fig. 2.32*e*). There will be no overlap between successive cycles of $G(\omega)$, provided $\omega_o \geq 2(2\pi B)$, or

$$\frac{2\pi}{T_s} \geq 4\pi B$$

---

*The theorem stated here (and proved subsequently) applies to lowpass signals. A bandpass signal whose spectrum exists over a frequency band

$$f_o - \frac{B}{2} < |f| < f_o + \frac{B}{2}$$

has a bandwith B. Such a signal is uniquely determined by $2B$ samples per second. In general, the sampling scheme is a bit more complex. It uses two interlaced sampling trains, each at a rate of B samples per second (known as second-order sampling). See, for example, Linden.[4]

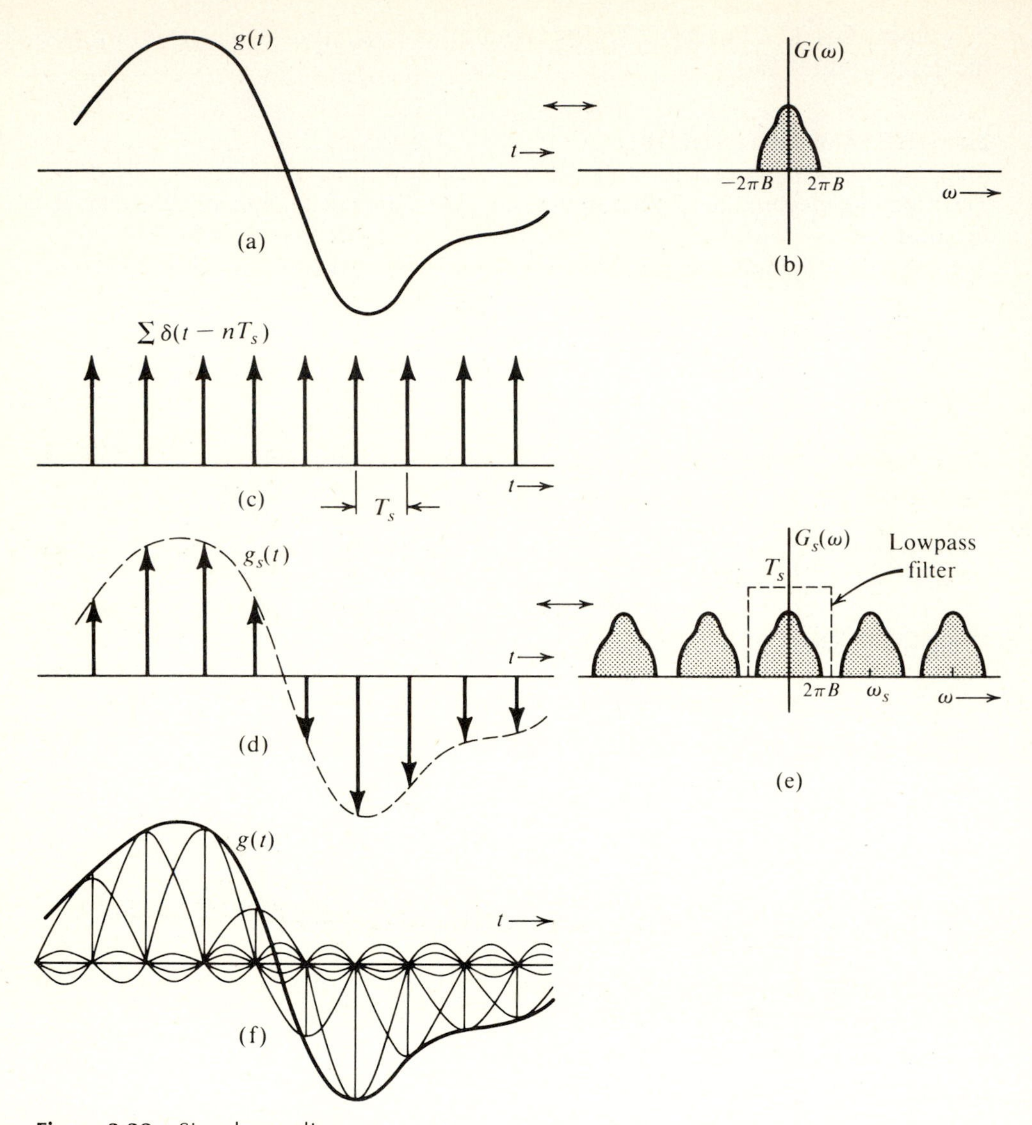

Figure 2.32 Signal sampling.

That is,*

$$T_s \leq \frac{1}{2B} \tag{2.81}$$

Hence, as long as the sampling interval $T_s \leq 1/2B$ or the sampling rate is greater than $2B$ samples per second, $G_s(\omega)$ consists of nonoverlapping repetitions of $G(\omega)$, and,

* In the case where $G(\omega)$ contains an impulse at the highest frequency $\omega = 2\pi B$, Eq. (2.81) modifies to $T_s < 1/2B$. Such is the case when $g(t) = \sin 2\pi Bt$. This is a signal bandlimited to B. But all of its samples taken at a rate of $2B$ samples per second (starting from $t = 0$) are zero, and $g(t)_1$ cannot be recovered from its samples. Hence, to be more general, $T_s < 1/2B$.

consequently, $g(t)$ can be recovered from $g_s(t)$ simply by passing it through a lowpass filter with a transfer function whose magnitude characteristic $|H(\omega)|$ is shown dotted in Fig. 2.32*e*. This proves the sampling theorem. The maximum allowable sampling interval, $T_s = 1/2B$, is known as the *Nyquist interval*, and the corresponding sampling rate ($2B$ samples per second) is known as the *Nyquist sampling rate*.

It is evident from Eq. (2.80) (and from Fig. 2.32*e*) that $g(t)$ can be reconstructed from the sampled signal $g_s(t)$ by passing the latter through a lowpass filter of gain T_s and bandwidth B. The transfer function $H(\omega)$ of such a filter is

$$H(\omega) = T_s \, \Pi\left(\frac{\omega}{4\pi B}\right) \tag{2.82a}$$

The corresponding unit impulse response $h(t)$ is (Table 2.1)

$$h(t) = 2T_s B \text{ sinc } (2Bt) \tag{2.82b}$$

Taking T_s to be the Nyquist interval ($T_s = 1/2B$), we have

$$h(t) = \text{sinc } (2Bt) \tag{2.83}$$

The input $g_s(t)$ to this filter is a sequence of impulses uniformly separated by the interval T_s. The nth impulse, located at $t = nT_s$, has the strength $g(nT_s)$. The filter output from this impulse is $g(nT_s)h(t - nT_s) = g(nT_s) \text{ sinc } 2B(t - nT_s)$. The output $g(t)$ of the filter is the sum of the outputs from all impulses in the sequence at the input. Hence*

$$g(t) = \sum_{n=-\infty}^{\infty} g(nT_s) \text{ sinc } 2B(t - nT_s) \tag{2.84a}$$

$$= \sum_{n=-\infty}^{\infty} g(nT_s) \text{ sinc } (2Bt - n) \tag{2.84b}$$

Hence, $g(t)$ can be reconstructed from its samples $g(nT_s)$ according to Eq. (2.84). The reconstruction procedure is shown graphically in Fig. 2.32*f*.

Aliasing Error

If a signal is undersampled (sampled at a rate below the Nyquist rate), the spectrum $G_s(\omega)$ consists of overlapping repetitions of $G(\omega)$, as shown in Fig. 2.33. Because of the overlapping tails, $G_s(\omega)$ no longer has the complete information about $G(\omega)$, and it is no longer possible to recover $g(t)$ from $g_s(t)$. If the sampled signal $g_s(t)$ is passed through the lowpass filter, we get a spectrum that is not $G(\omega)$ but is a distorted version as a result of two separate causes: 1) loss of the tail of $G(\omega)$ beyond $|\omega| > \omega_o/2$, and 2) this same tail appears inverted, or folded, onto the spectrum at the cutoff frequency. This tail inversion, known as *spectral folding*, or *aliasing*, is shown shaded in Fig. 2.33.

---

* The result can also be obtained directly by observing that the output $g(t)$ is the convolution of $g_s(t)$ and $h(t)$.

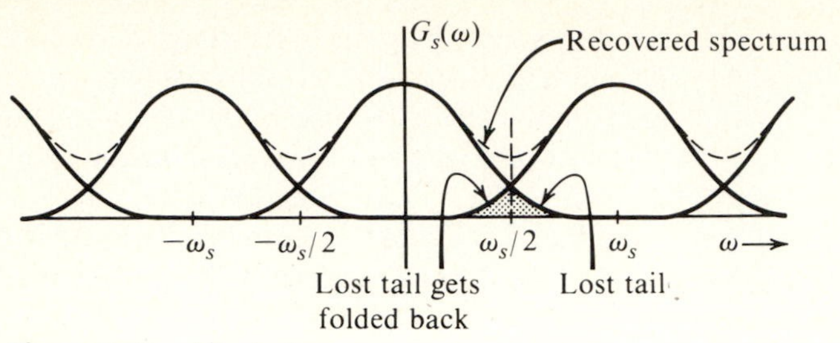

Figure 2.33 Aliasing error.

The aliasing distortion can be eliminated by cutting the tail of $G(\omega)$ beyond $|\omega| = \omega_s/2$ before the signal is sampled. By so doing, the overlap of successive cycles in $G_s(\omega)$ is avoided. The only error in the recovery of $g(t)$ is that caused by the missing tail for $|\omega| > \omega_s/2$. Cutting the tail off reduces the error signal energy by half.

Strictly speaking, a bandlimited signal does not exist in reality. It can be shown that if a signal is time limited (that is, it exists over only a finite time interval), it cannot be bandlimited, and if a signal is bandlimited it cannot be time limited.[3] All physical signals are necessarily time limited because they begin at some finite instant and must terminate at some other finite instant. Hence, all practical signals are nonbandlimited. If a signal is Fourier transformable, however, its energy is finite, and it follows from Eq. (2.77c) that $|G(\omega)|$ must decay at higher frequencies. Most of the signal energy resides in a finite band, and the spectrum at higher frequencies contributes little. The error introduced by cutting off the tail beyond a certain frequency B can be made negligible by making B sufficiently large.

Thus, for all practical purposes, a signal can be considered to be essentially bandlimited at some value B, the choice of which depends upon the accuracy desired.* A practical example of this is a speech signal. Theoretically, a speech signal, being a finite time signal, has an infinite bandwidth. But frequency components beyond 3 kHz contribute a negligible fraction of the total energy. When speech signals are transmitted by PCM, they are first passed through a lowpass filter of bandwidth 3500 Hz, and the resulting signal is sampled at a rate of 8000 samples per second. For a bandwidth of 3500 Hz, the minimum sampling rate (the Nyquist rate) is 7000. Higher sampling rates permit recovery of the signal from its samples using relatively simpler filters. Recovering signals sampled at the Nyquist rate would require sharp cutoff (ideal) filters. (Why?)

When a practical signal $g(t)$ is to be transmitted by its samples (as in PCM), we must first estimate its essential bandwidth B and then cut off its spectrum beyond B. The following example indicates a method of estimating the essential bandwidth of a signal.

---

* In a similar way, if a finite energy signal is strictly bandlimited, its amplitude $\rightarrow 0$ for some $|t| > T$; otherwise, its energy would be infinite. Hence, such a signal is essentially time limited.

■ EXAMPLE 2.27

Estimate the essential bandwidth B of

$$g(t) = \frac{2a}{t^2 + a^2}$$

□ **Solution:** Using the symmetry property, Eq. (2.50), and pair 15 from Table 2-1, we obtain

$$\frac{2a}{t^2 + a^2} \leftrightarrow 2\pi e^{-a|\omega|}$$

The signal $g(t) = 2a/(t^2 + a^2)$ and its Fourier transform $G(\omega) = 2\pi e^{-a|\omega|}$ are shown in Fig. 2.34. The signal $g(t)$ is not bandlimited, and, hence, an aliasing error will occur no matter at what rate this signal is sampled. To avoid the aliasing error, we need to estimate the essential bandwidth B of this signal and then cut off all components beyond B (Fig. 2.34c). The resulting signal $\hat{g}(t)$, is bandlimited to B and can be

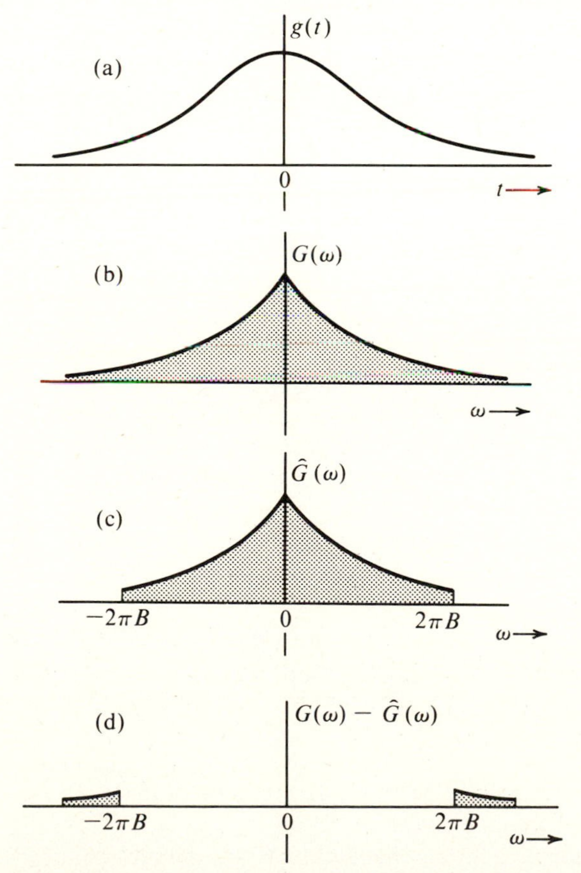

Figure 2.34 Estimation of effective bandwidth of a signal.

sampled at a rate $\geq 2B$ samples per second. From these samples we can reconstruct $\hat{g}(t)$, which is a close approximation of $g(t)$. The error signal $e(t)$ is $g(t) - \hat{g}(t)$, with Fourier transform $G(\omega) - \hat{G}(\omega)$, as shown in Fig. 2.34$d$. To choose B we should stipulate an error criterion. For example, we may choose B such that the energy E_e of the error signal is less than, say, 1 percent of the energy E_g of the signal $g(t)$. From Fig. 2.34d, the energy E_e of the error signal is given by

$$E_e = \frac{1}{2\pi}\int_{-\infty}^{\infty} |G(\omega) - \hat{G}(\omega)|^2 \, d\omega$$

For a real signal $e(t)$, $|E(\omega)|^2$ is an even function of ω, and

$$\begin{aligned} E_e &= \frac{1}{\pi}\int_{0}^{\infty} |G(\omega) - \hat{G}(\omega)|^2 \, d\omega \\ &= \frac{1}{\pi}\int_{2\pi B}^{\infty} |2\pi e^{-a\omega}|^2 \, d\omega \\ &= 4\pi \int_{2\pi B}^{\infty} e^{-2a\omega} \, d\omega \\ &= \frac{2\pi}{a} e^{-4\pi a B} \end{aligned}$$

The energy E_g of $g(t)$ is the same as E_e in the limit $B \to 0$ (see Fig. 2.34d). Hence

$$E_g = \frac{2\pi}{a}$$

We require

$$E_e = 0.01 E_g$$

or

$$\frac{2\pi}{a} e^{-4\pi a B} = 0.01\left(\frac{2\pi}{a}\right)$$

This gives

$$4\pi a B = \ln(100)$$

and

$$B = \frac{0.36}{a}$$

Thus to maintain $E_e \leq 0.01 E_g$, we require $B \geq 0.36/a$. For $B = 0.36/a$, $E_e = 0.01E_g$. For this error, the Nyquist sampling rate is $0.72/a$.

When $g(t)$ is passed through a lowpass filter of bandwidth B, the output signal $\hat{g}(t)$

is bandlimited* to B. From the samples of $\hat{g}(t)$ taken at a rate $\geq 0.72/a$ samples per second, we can reconstruct $\hat{g}(t)$, which is a close replica of $g(t)$. The energy of the error signal $g(t) - \hat{g}(t)$ is only 1 percent of the energy of $g(t)$. ■

Practical Sampling

Sampling a signal $g(t)$ by impulses (known as *instantaneous sampling*) is of theoretical interest only. In practice, sampling is done by narrow pulses of finite width (Fig. 2.35*a*). We can show that $g(t)$ can be reconstructed from the finite-width samples in Fig. 2.35*b* as long as the sampling is done at a rate at least equal to the Nyquist rate (viz., $T_s \leq 1/2B$).

The sampled signal $g_s(t)$ in Fig. 2.35*b* is equal to $g(t)$ multiplied by a rectangular pulse train $k(t)$ (Fig. 2.35*a*), whose Fourier series appears on the right-hand side of Eq. (2.23a).

$$k(t) = \sum_{n=-\infty}^{\infty} K_n e^{jn\omega_s t} \qquad \omega_s = \frac{2\pi}{T_s} \tag{2.85a}$$

where [Eq. (2.23b) with $A = 1$]

$$K_n = \frac{1}{\pi n} \sin\left(\frac{n\pi\tau}{T_s}\right) \tag{2.85b}$$

(a)

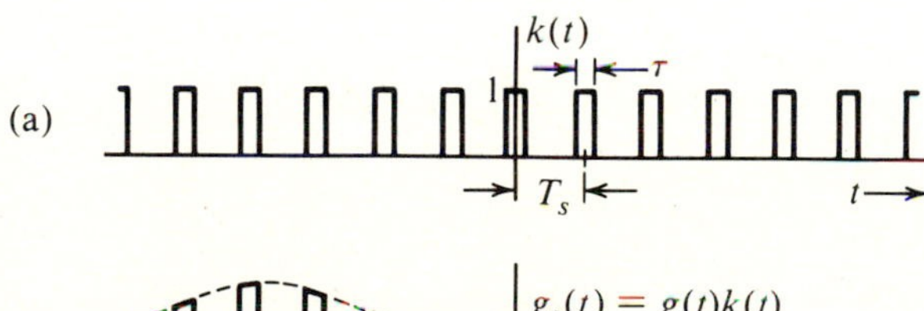

(b)

(c)

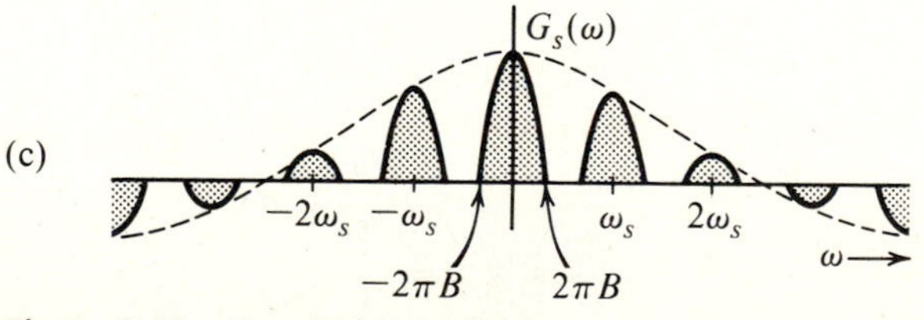

Figure 2.35 Practical sampling.

---

*Actually, a strictly bandlimited signal does not exist, because an ideal lowpass filter is unrealizable (see Sec. 2.5) A practical filter may greatly attenuate but can never completely suppress a band of frequencies. Hence, $g(t)$ will be bandlimited to the extent that the filter used approximates the ideal lowpass filter.

The sampled signal $g_s(t)$ is the product of $g(t)$ and $k(t)$.

$$g_s(t) = g(t)k(t) \tag{2.86a}$$

$$= g(t) \sum_{n=-\infty}^{\infty} K_n e^{jn\omega_s t}$$

$$= \sum_{n=-\infty}^{\infty} K_n g(t) e^{jn\omega_s t} \tag{2.86b}$$

Hence

$$G_s(\omega) = \sum_{n=-\infty}^{\infty} K_n G(\omega - n\omega_s) \tag{2.86c}$$

The spectrum $G_s(\omega)$ consists of the spectrum $G(\omega)$ repeating itself indefinitely every ω_s (Fig. 2.35*c*). The *n*th repetition (at $\omega = n\omega_s$) is multiplied by K_n (shown in Fig.

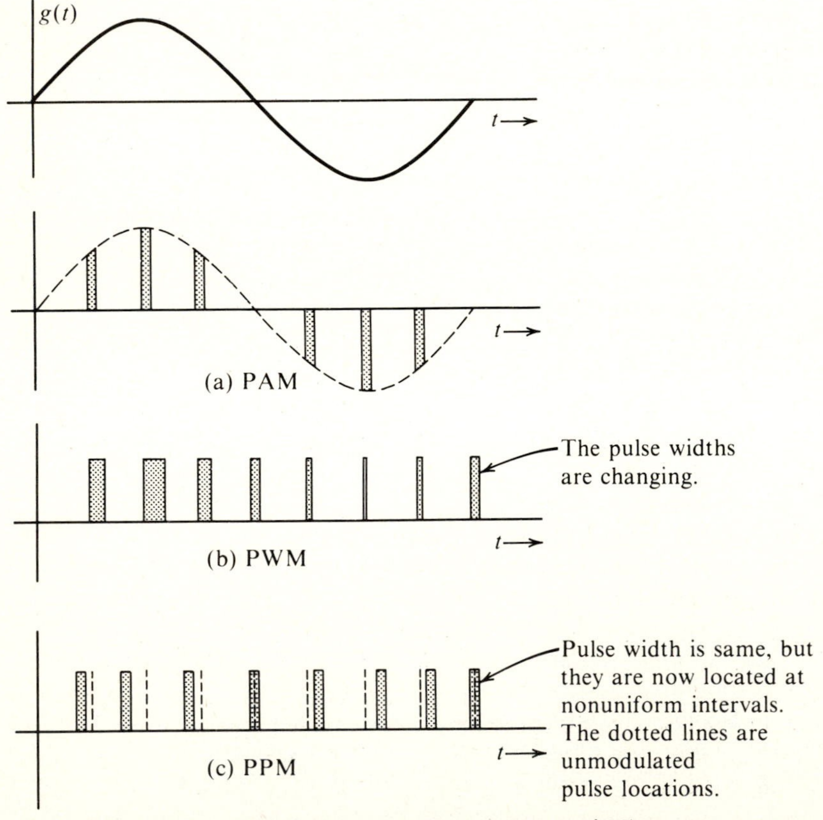

Figure 2.36 Pulse modulated signals: PAM, PWM, and PPM.

2.7*a*). It is evident from Fig. 2.35*c* that $g(t)$ can be reconstructed from $g_s(t)$ by passing the latter through a lowpass filter of bandwidth B.

The derivation here is quite general. We could have used any other periodic signal instead of $k(t)$ to sample $g(t)$. It will only change the values of K_n (exponential Fourier series coefficients of the periodic signal).

The Sampling Theorem and Pulse Communication Systems

The sampling theorem opens a way of communicating analog signals by pulses. The analog signal is sampled, and sample values are used to modify certain parameters of a periodic pulse train. We may vary the amplitudes, widths, or locations of the pulses in proportion to the sample values. Accordingly, we have *pulse-amplitude modulation* (*PAM*), *pulse-width modulation* (*PWM*), or *pulse-position modulation* (*PPM*). Yet another form is pulse-code modulation (PCM), discussed in Chapter 1. Figure 2.36 shows the analog signal $g(t)$ and the corresponding modulated waveforms. Thus, instead of transmitting $g(t)$, we can transmit the corresponding pulse-modulated signal. At the receiver, we must read the information of the pulse-modulated signal and reconstruct the analog signal $g(t)$. The advantage of using pulse modulation is that it permits simultaneous transmission of several baseband signals on a time-sharing basis) time-division multiplexing, or TDM). Because a pulse-modulated signal occupies only a part of the channel time, we can interleave several pulse-modulated signals on the same channel. Figure 2.37*a* shows time-division multiplexing of two PAM signals. We can multiplex several signals, depending upon the channel available.

Figure 2.37*b* and *c* show a block-diagram representation of a transmitter and a receiver, respectively, of a time-division-multiplexed PAM system. At the transmitter, the commutator is switched from channel to channel in a sequence by a timing circuit, which also generates sampling pulses. Thus, the commutator connects different channels in a sequence to the sampling circuit, which samples all of the signals in a sequence by pulses generated by the timing circuit.

The output of the sampling circuit is, thus, a signal that consists of samples of all of the signals interleaved. At the receiver, another timing circuit—which is in synchronism with that at the transmitter—is used to switch the commutator to different channels. The samples of various signals are now properly separated. The desired signal is recovered from each channel by a lowpass filter. The commutators shown in Fig. 2.37 are all electronic switches.

Although Fig. 2.37 refers to time-division multiplexing of PAM signals, the same general concept applies to time-division multiplexing of PCM or any other pulse signals. The Bell System, for example, multiplexes 24 PCM signals on a telephone channel in its T-1 system.

Another way of transmitting several baseband signals simultaneously is frequency-division multiplexing (FDM), mentioned in Chapter 1. In FDM, various signals are multiplexed by sharing the bandwidth of a channel. Each message signal is modified (by modulation, as discussed in Chapter 4) so that it occupies a certain band of the channel not occupied by other signals. The information of various signals is located in nonoverlapping frequency bands of the channel.

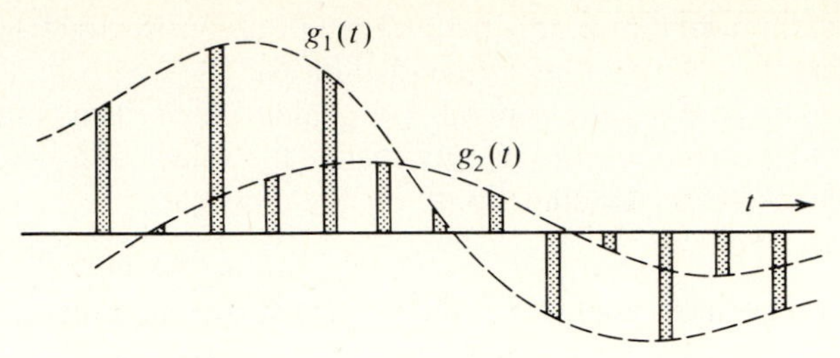

(a) Time-division multiplexing of two signals.

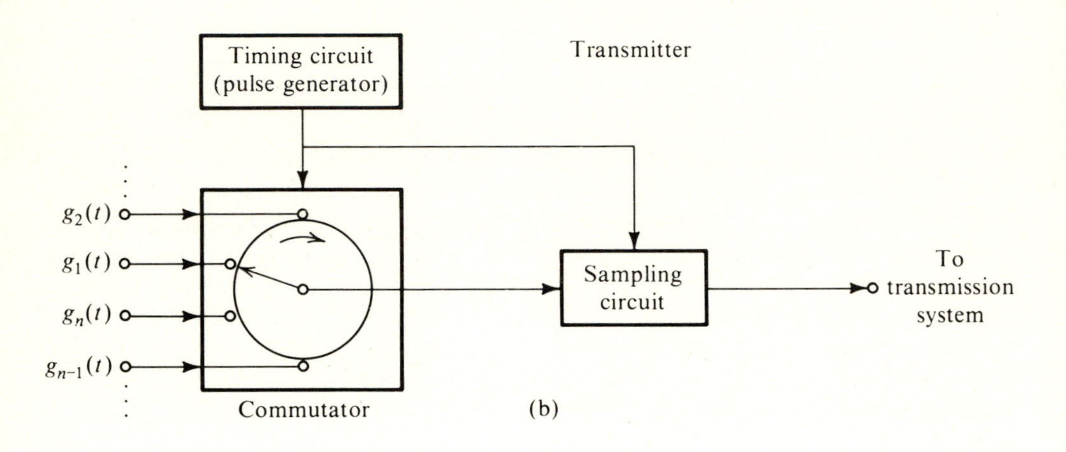

(b)

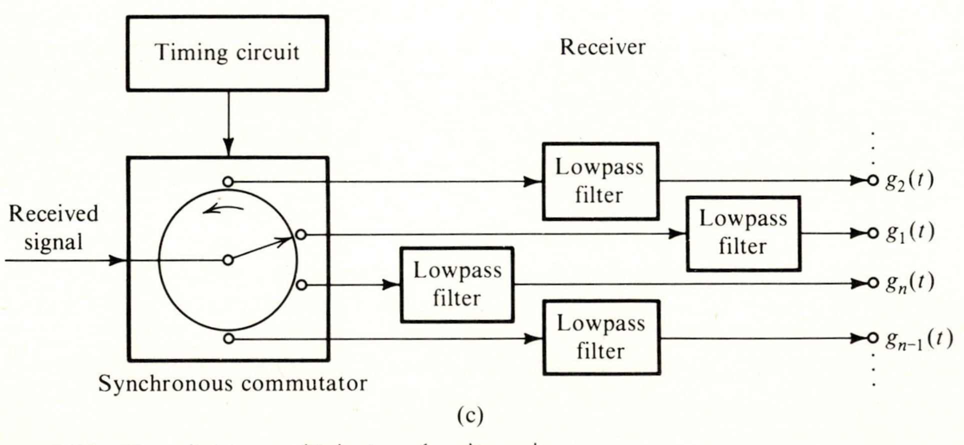

(c)

Figure 2.37 Time-division multiplexing of n channels.

PART II Signal Transmission

2.5 DISTORTIONLESS TRANSMISSION THROUGH A LINEAR SYSTEM

For a given linear system, e.g., a channel, an input $g(t)$ produces an output $r(t)$, thus "processing" the signal $g(t)$ in a way that is characteristic of the system. The input spectrum is $G(\omega) = |G(\omega)|e^{j\theta_g(\omega)}$. If the system transfer function is $H(\omega) = |H(\omega)|e^{j\theta_h(\omega)}$, then the output spectrum $R(\omega)$ is given by

$$R(\omega) = G(\omega)H(\omega) \tag{2.87}$$
$$= |G(\omega)|\,|H(\omega)|e^{j[\theta_g(\omega)+\theta_h(\omega)]}$$

The system thus changes the magnitude spectrum $|G(\omega)|$ to $|G(\omega)|\,|H(\omega)|$ and the phase spectrum $\theta_g(\omega)$ to $\theta_g(\omega) + \theta_h(\omega)$. The magnitude spectrum of $g(t)$ is multiplied by the magnitude spectrum $H(\omega)$, and the phase spectrum is increased by the phase of $H(\omega)$. Thus, the system transfer function $H(\omega)$ modifies the magnitudes and phases of all frequency components of the input signal. Some frequency components may be boosted in amplitude, and others may be attenuated. Relative phases between components are also changed. In general, the output signal may not look like the input. In communication systems, we need to minimize this distortion caused by imperfect channel characteristics. It is therefore important to determine conditions for distortionless transmission.

Transmission is said to be *distortionless* if the input and the output have identical waveshapes within a multiplicative constant. A delayed output that retains the input waveform is also considered distortionless. Thus, in distortionless transmission, the input $g(t)$ and the output $r(t)$ satisfy the condition

$$r(t) = kg(t - t_d) \tag{2.88}$$

Before deriving the conditions for distortionless transmission analytically, it is illuminating to tackle the problem heuristically. For "no distortion," all input frequency components must reach the output undistorted. This means that all frequency components should suffer the same attenuation (or amplification). This implies that

$$|H(\omega)| = k \tag{2.89a}$$

Also, all frequency components must reach the output with the same time delay t_d. For an input $\cos \omega t$, a delay of t_d yields $\cos \omega(t - t_d)$ at the output. But

$$\cos \omega(t - t_d) = \cos(\omega t - \omega t_d)$$

Therefore, a delay of t_d in a component of frequency ω corresponds to a phase lag ωt_d. Hence, for a given time delay t_d, the phase lag is proportional to the frequency ω. For a signal of double the frequency, the phase lag must also be doubled to achieve the same time delay. Hence for distortionless transmission, the phase lag caused by $H(\omega)$ in each frequency component must be proportional to the frequency of that com-

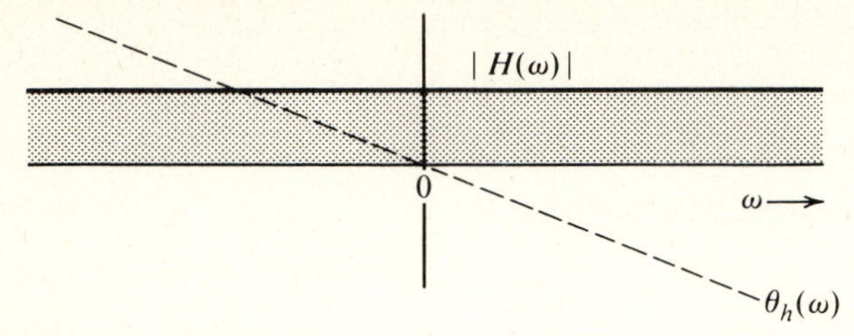

Figure 2.38 Distortionless system characteristics.

ponent. This means*

$$\theta_h(\omega) = -\omega t_d \tag{2.89b}$$

The reader can prove the conditions in Eq. (2.89) analytically from the fact that

$$R(\omega) = H(\omega)G(\omega)$$

and

$$R(\omega) = \mathcal{F}[kg(t - t_d)] = kG(\omega)e^{-j\omega t_d}$$

The ideal magnitude and phase characteristics of a channel for distortionless transmission are shown in Fig. 2.38. Real systems can be, at best, only approximately distortionless.

■ EXAMPLE 2.28

If $g(t)$ and $r(t)$ are the input and the steady-state output, respectively, of a simple RC lowpass filter (Fig. 2.39a), determine the transfer function $H(\omega)$ and sketch $|H(\omega)|$ and $\theta_h(\omega)$. For distortionless transmission through this filter, what is the condition on the bandwidth of $g(t)$? What is the transmission delay? Find $r(t)$ when $g(t) = A \cos 100\, t$.

☐ **Solution:** For this circuit,

$$H(\omega) = \frac{1/j\omega C}{R + (1/j\omega C)} = \frac{1}{1 + j\omega RC} = \frac{a}{a + j\omega}$$

where

$$a = \frac{1}{RC} = 10^6$$

---

* We note at this point that if all frequency components undergo a constant phase shift of $n\pi$ (n integer), the signal will still reach the output undistorted. This is because for n even, the phase shift $n\pi$ in a sinusoid implies the same sinusoid, and the sum of all sinusoids will result in an undistorted signal. When n is odd, the phase shift $n\pi$ in a sinusoid implies a sign change, and the sum of all sinusoids will result in an undistorted signal with a mere sign change. Hence, a more general condition for distortionless transmission is

$$\theta_h(\omega) = n\pi \operatorname{sgn}(\omega) - \omega t_d \qquad (n \text{ integer})$$

The signum function enters here because for real systems, phase is an odd function of ω.

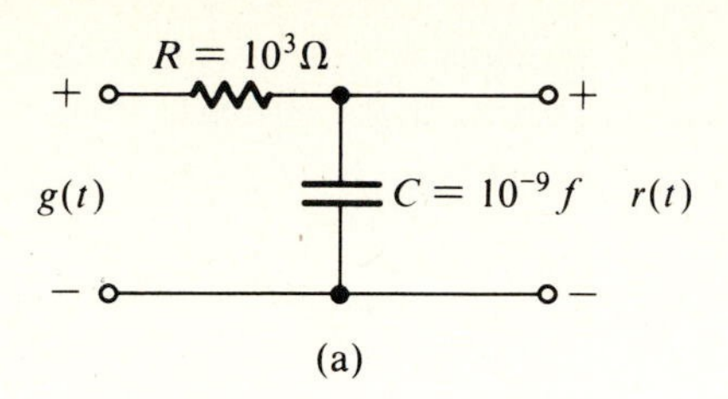

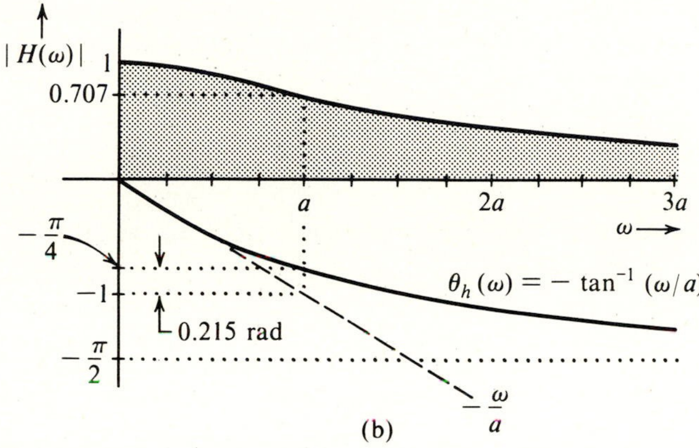

Figure 2.39 A simple lowpass filter and its frequency response characteristics.

Hence

$$\left.\begin{aligned} |H(\omega)| &= \frac{a}{\sqrt{a^2+\omega^2}} \simeq 1 \\ \theta_h(\omega) &= -\tan^{-1}\frac{\omega}{a} \simeq -\frac{\omega}{a} \end{aligned}\right\} \omega \ll a$$

The magnitude and phase are shown in Fig. 2.39*b*. Note that for $\omega \ll a$ ($a = 10^6$), the magnitude and phase characteristics are practically ideal. For example, when $\omega < 200{,}000$ ($\omega < 0.2a$), $|H(\omega)|$ deviates from 1 by less than 2 percent, and $\theta_h(\omega)$ deviates from the ideal linear characteristics by less than 1.5 percent. Hence, for a lowpass signal of bandwidth $B \ll a/2\pi$, the transmission is practically distortionless. The exact value of B will depend upon the amount of distortion that can be tolerated. As we saw, $B = 200{,}000/2\pi \simeq 31{,}847$ Hz gives negligible distortion.

The transmission delay t_d is the negative slope of the phase characteristics [Eq. (2.89b)]. Because for $\omega \ll a$

$$\theta_h(\omega) \simeq -\frac{\omega}{a} \qquad t_d \simeq \frac{1}{a} = RC = 10^{-6}$$

For the input $A \cos 100\, t$, because $100 \ll 10^6$, the transmission is (practically) distortionless with a delay of 10^{-6}. Hence

$$r(t) = A \cos 100(t - 10^{-6}) = A \cos(100\, t - 10^{-4})$$

In practice, variations in $|H(\omega)|$ by a factor of 0.707 are considered tolerable, and the frequency interval over which $|H(\omega)|$ remains within this variation (3 dB) is called the *bandwidth*. For the present case $|H(\omega)| = 0.707$ at $\omega RC = 1$. This gives the bandwidth B as

$$B = \frac{a}{2\pi} = \frac{1}{2\pi RC} = 159.23 \text{ kHz}$$

Over this range of frequencies, the phase deviates from the ideal linear characteristics by at most 0.215 radians. It can be seen from Fig. 2.39*b* that frequencies well below B are transmitted practically without distortion, but frequencies in the vicinity of B will suffer some distortion. ■

■ EXAMPLE 2.29

Show that for a bandpass signal, conditions for distortionless transmission reduce to

$$|H(\omega)| = k$$

$$-\frac{d\theta_h}{d\omega} = t_g \qquad \text{(a constant)} \tag{2.90}$$

over the passband.

A transfer function $|H(\omega)|$ that meets these conditions is shown in Fig. 2.40. Note that the conditions in Eq. (2.90) are more relaxed than the general conditions in Eq. (2.89). It is easy to verify that the conditions in Eq. (2.89) always satisfy Eq. (2.90), but the converse is not true.

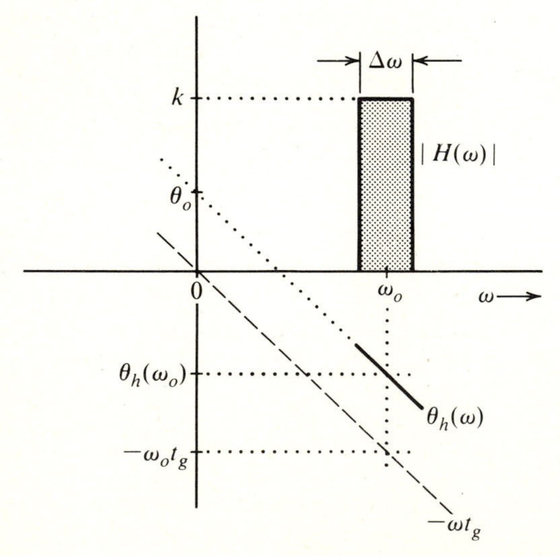

(Distortionless pure delay)

Figure 2.40 Characteristics of a distortionless bandpass system.

To prove these conditions, we consider a general bandpass signal $g_{bp}(t)$ in Eq. (2.61a):

$$g_{bp}(t) = g_c(t) \cos \omega_o t + g_s(t) \sin \omega_o t \qquad \textbf{(2.91a)}$$

where $g_c(t)$ and $g_s(t)$ are lowpass signals of bandwidth $\Delta f/2$ and $g_{bp}(t)$ is a bandpass signal of bandwidth Δf centered at ω_o. The information of $g_{bp}(t)$ is carried by signals $g_c(t)$ and $g_s(t)$. Hence, distortionless transmission in this case means $g_c(t)$ and $g_s(t)$ should remain undistorted. Thus, if the received signal $\hat{g}_{bp}(t)$ is of the form:

$$\hat{g}_{bp}(t) = k[g_c(t - t_g) \cos (\omega_o t - \varphi_o) + g_s(t - t_g) \sin (\omega_o t - \varphi_o)] \qquad \textbf{(2.91b)}$$

with φ_o an arbitrary constant, we have a distortionless transmission.

Let us first consider the transmission of the component $g_c(t) \cos \omega_o t$ through $H(\omega)$ (Fig. 2.40).

The phase $\theta_h(\omega)$ is given by

$$\theta_h(\omega) = -\omega t_g + \theta_o \qquad \omega > 0 \qquad \textbf{(2.92)}$$

The slope of $\theta_h(\omega)$ is $-t_g$. The only difference between this $\theta_h(\omega)$ and that for the distortionless case (Fig. 2.38) is that the former does not pass through the origin.

The distortionless phase function with slope $-t_g$ (shown dashed), is a pure time delay t_g. If the input $g_c(t) \cos \omega_o t$ is applied to this ideal delay filter, the output is $kg_c(t - t_g) \cos \omega_o(t - t_g)$. The phase $\theta_h(\omega)$ of the actual filter differs from the pure delay phase by a constant θ_o. Hence from Eq. (2.60b), it follows that for the same input, the output of the filter with phase $\theta_h(\omega)$ will be $kg_c(t - t_g) \cos [\omega_o(t - t_g) + \theta_o]$. It also follows that when the input is $g_{bp}(t)$ in Eq. (2.91a), the output $\hat{g}_{bp}(t)$ is given by

$$\hat{g}_{bp}(t) = k[g_c(t - t_g) \cos [\omega_o(t - t_g) + \theta_o] + g_s(t - t_g) \cos [\omega_o(t - t_g) + \theta_o]$$

From Eq. (2.92), we have $\theta_h(\omega_o) = -\omega_o t_g + \theta_o$, and

$$\hat{g}_{bp}(t) = k[g_c(t - t_g) \cos [\omega_o t + \theta_h(\omega_o)] + g_s(t - t_g) \sin [\omega_o t + \theta_h(\omega_o)] \qquad \textbf{(2.93a)}$$

Comparison of this equation with Eq. (2.91b) shows that the transmission is distortionless. Equation (2.93a) can be expressed as

$$\hat{g}_{bp}(t) = k[g_c(t - t_g) \cos \omega_o(t - t_p) + g_s(t - t_g) \sin \omega_o(t - t_p)] \qquad \textbf{(2.93b)}$$

where

$$t_p = \frac{\theta_h(\omega_o)}{\omega_o} \qquad \textbf{(2.94)}$$

Rewriting Eq. (2.93b) in terms of envelope [see Eq. (2.61b)],

$$\hat{g}_{bp}(t) = kE(t - t_g) \cos [\omega_o(t - t_p) + \theta_{bp}(t)]$$

t_g and t_p are known as the *envelope (or group) delay* and the *phase delay*, respectively.

As an example, consider once again the lowpass *RC* filter in Fig. 2.39*a*. This filter is distortionless for lowpass signals of bandwidth $B \ll 1/2\pi RC$. Even at frequencies $> 1/2\pi RC$, however, it can transmit bandpass signals with practically no dis-

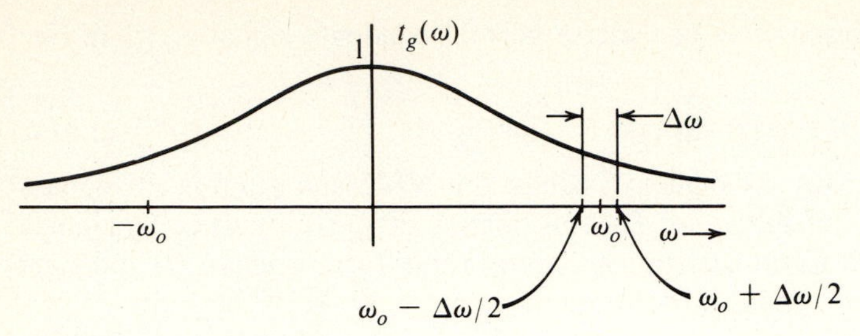

Figure 2.41 Group delay.

tortion, provided the bandpass signal has a narrow bandwidth. To analyze this situation, let us determine the envelope delay t_g at some frequency ω_o.

$$\theta_h(\omega) = -\tan^{-1}\omega RC \qquad \omega > o$$

$$t_g = -\frac{d\theta_h}{d\omega} = \frac{RC}{1 + \omega^2 R^2 C^2}$$

As seen from Fig. 2.41, t_g is not constant but varies continuously with ω. Hence, strictly speaking, this filter cannot transmit a bandpass signal without distortion. But if the bandpass signal has a small bandwidth Δf centered at ω_o, the gain $|H(\omega)|$ and t_g are nearly constant. Therefore, the transmission may be considered practically distortionless.

Let $RC = 10^{-3}$ and $\omega_o = 10^4$. The signal to be transmitted is $g(t) \cos \omega_o t$ where $g(t)$ has a bandwidth $100/2\pi$ Hz.

$$t_g = \frac{10^{-3}}{1 + 10^{-6}\omega^2} = \frac{10^3}{\omega^2 + 10^6}$$

The passband is $\left(\omega_o - \frac{\Delta\omega}{2}, \omega_o + \frac{\Delta\omega}{2}\right)$, or $(10^4 - 50 \text{ to } 10^4 + 50)$.

$$t_g(10^4 - 50) = \frac{10^3}{(10^4 - 50)^2 + 10^6} = 9.99 \times 10^{-6}$$

$$t_g(10^4 + 50) = \frac{10^3}{(10^4 + 50)^2 + 10^6} = 9.80 \times 10^{-6}$$

The variation of t_g over the entire band $\Delta\omega$ is less than 2 percent. Hence

$$t_g \simeq t_g(\omega_o) = \frac{10^3}{10^8 + 10^6} = 9.9 \times 10^{-6}$$

From Eq. (2.94),

$$\omega_o t_p = \theta_o$$

where $\theta_o = -\theta_h(\omega_o) = \tan^{-1} \omega_o RC = \tan^{-1} 10 = 1.47$. Hence

$$t_p = \frac{1.47}{\omega_o} = \frac{1.47}{10^4} = 0.147 \text{ ms}$$

The magnitude $|H(\omega_o)|$ is

$$|H(\omega_o)| = \frac{1}{\sqrt{1 + \omega_o^2 R^2 C^2}} = \frac{1}{\sqrt{101}} \simeq 0.1$$

Thus, for the input $g(t) \cos \omega_o t$, the output is

$$r(t) = 0.1g(t - 9.9 \times 10^{-6}) \cos \omega_o(t - 0.000147)$$

The envelope (or the group) delay is 9.9 μs and the phase delay is 0.147 ms. ■

Ideal Filters

Ideal filters allow distortionless transmission of a certain band of frequencies and suppress the remaining frequencies. The ideal lowpass filter (Fig. 2.42*a*), for example, allows all components below $\omega = 2\pi B$ to pass without distortion and suppresses all components above $\omega = 2\pi B$. Figure 2.43 shows ideal highpass and bandpass filter characteristics.

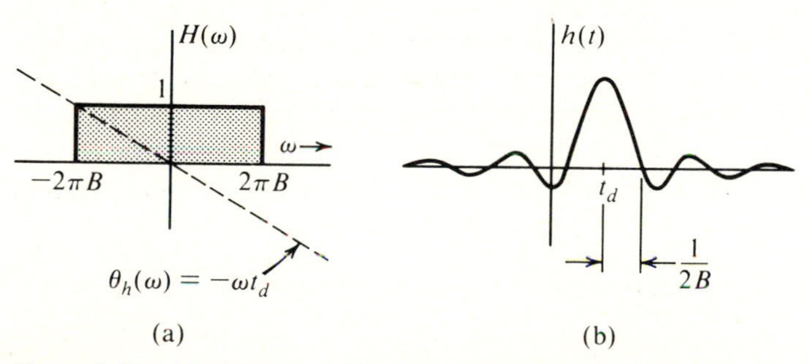

Figure 2.42 Ideal lowpass filter and its impulse response.

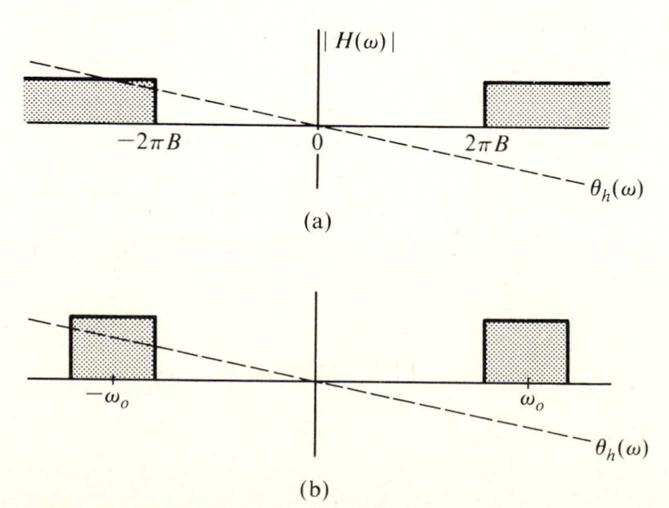

Figure 2.43 Ideal highpass and bandpass filters.

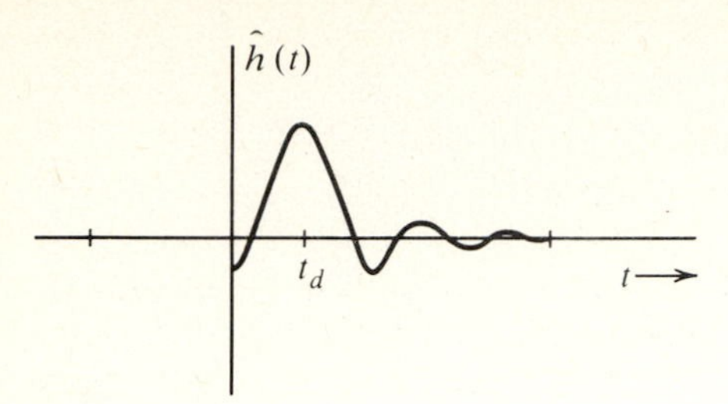

Figure 2.44

The unit impulse response $h(t)$ of the ideal lowpass filter is given by

$$\begin{aligned} h(t) &= \mathcal{F}^{-1}\left[\Pi\left(\frac{\omega}{4\pi B}\right)e^{-j\omega t_d}\right] \\ &= 2B \operatorname{sinc}\,[2B(t - t_d)] \end{aligned} \tag{2.95}$$

This response is shown in Fig. 2.42*b*. Observe that the impulse response exists for $t < 0$. This is a rather strange result, in view of the fact that the impulse input is applied at $t = 0$. The response starts even before the input is applied. The system seems to anticipate the input. Unfortunately, no real system can exhibit such prophetic foresight. Hence, we must conclude that an ideal filter, although desirable, is not physically realizable. One can show, similarly, that other ideal filters (such as ideal highpass or ideal bandpass filters shown in Fig. 2.43) are also not physically realizable.

For a physically realizable system, $h(t)$ must be causal,* that is,

$$h(t) = 0 \qquad \text{for } t < 0 \tag{2.96}$$

The impulse response $h(t)$ in Fig. 2.42*b* is not realizable. But the impulse response $\hat{h}(t)$,

$$\hat{h}(t) = h(t)u(t)$$

is physically realizable because it is causal; that is, it exists only for $t > 0$ (Fig. 2.44). Such a filter, however, will not have ideal filter characteristics and will distort a signal bandlimited to B. But, if we increase t_d sufficiently, $\hat{h}(t)$ will be a close replica of $h(t)$ except for the delay, and the resulting filter $\hat{H}(\omega)$ will be a good approximation of an ideal filter. Hence, an ideal filter can be closely realized at the cost of a delay. This is often true, of unrealizable systems. Theoretically, $t_d = \infty$ is needed to realize the

---

* In the frequency domain, the causality condition, Eq. (2.96), is equivalent to

$$\int_{-\infty}^{\infty} \frac{|\ln |H(\omega)||}{1 + \omega^2}\, d\omega < \infty$$

This is the so-called Paley-Wiener criterion of physical realizability. If $H(\omega)$ does not satisfy this condition, it is unrealizable. Note that if $|H(\omega)| = 0$ over any finite band, $|\ln|H(\omega)|| = \infty$ over that band, and consequently $H(\omega)$ is unrealizable.

ideal characteristics. But a glance at Fig. 2.41*b* shows that t_d = two or three times $1/2B$ will make $\hat{h}(t)$ a reasonably close version of $h(t)$.

At first glance, unrealizable systems seem a bit inscrutable. Actually, there is nothing mysterious about unrealizable systems and their close realization by physical systems with a delay. If one wants to know what will happen in the future, say one year from now, he has two choices: go to a prophet (an unrealizable person) who can give the answer immediately, or go to a wise man and allow him a delay of one year to give the answer! If the wise man is truly wise, he may be able to tell the future very closely with a delay of less than a year by studying the trends. This is exactly the case with unrealizable systems—nothing more and nothing less.

In practice, we can achieve a variety of filter characteristics. The so-called Butterworth filters, for example, are described by transfer functions whose magnitudes are given by

$$|H(\omega)| = \frac{1}{\sqrt{1 + \left(\dfrac{\omega}{2\pi B}\right)^{2n}}} \tag{2.97}$$

These characteristics are shown in Fig. 2.45 for several values of n (the order of the filter). Note that the characteristics approach ideal lowpass behavior as n increases. Of course the filter complexity also increases with n. The characteristic of a real filter is continuous. The half-power bandwidth of a filter is defined as the bandwidth over which the magnitude $|H(\omega)|$ is constant within 3 dB (or a ratio of $1/\sqrt{2}$). For Butterworth filters, the bandwidth is B for all n (Fig. 2.45).

For $n = 4$, the magnitude $|H(\omega)|$, the phase $\theta_h(\omega)$, and the unit impulse response $h(t)$ are shown in Fig. 2.46. Compare these with those for an ideal filter.

It should be remembered that the magnitude $|H(\omega)|$, the phase $\theta_h(\omega)$ of a system are interdependent. That is, we cannot choose $|H(\omega)|$ and $\theta_h(\omega)$ as we please. A

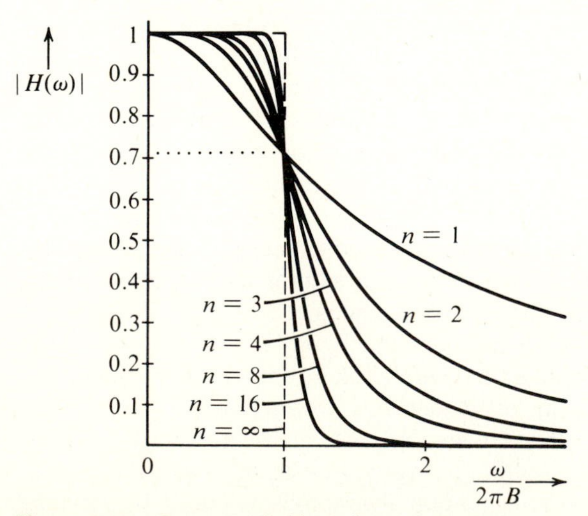

Figure 2.45 Butterworth filter characteristics.

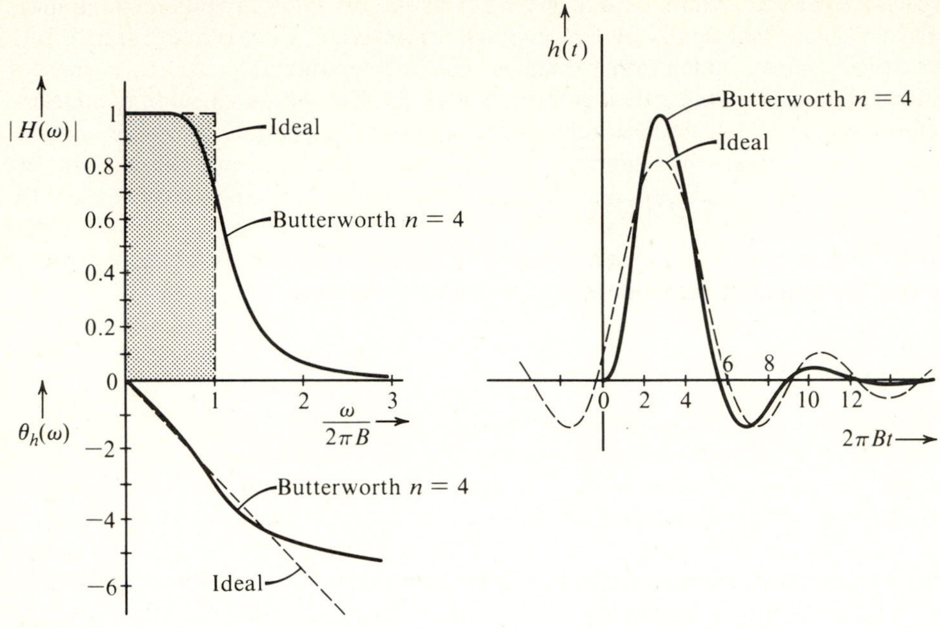

Figure 2.46 Comparison of Butterworth filter ($n = 4$) with an ideal filter.

certain trade-off exists between ideal magnitude and ideal phase characteristics. If we try to perfect $|H(\omega)|$ more, a deterioration in $\theta_h(\omega)$ will occur. For a Butterworth filter, $n = \infty$ gives the ideal magnitude characteristic, but the phase characteristic that goes with it is badly distorted in the vicinity of the cutoff frequency B.

Digital Filters

Analog signals can also be processed by digital means[5,6] (A/D conversion). This involves sampling, quantizing, and coding. The resulting digital signal can be processed by a small, special-purpose digital computer designed to convert the input sequence into a desired output sequence. The output sequence is converted back into the desired analog signal. A special algorithm of the processing digital computer can be used to achieve a given signal operation (e.g., lowpass, bandpass, or highpass filtering).

Digital processing of analog signals has several advantages. A small, special-purpose computer can be time-shared for several uses, and the cost of digital implementation is often considerably lower than that of its analog counterpart. The accuracy of a digital filter is dependent only on the computer word length, the quantizing interval, and the sampling rate (aliasing error). Digital filters employ simple elements, such as adders, multipliers, shifters, and delay elements, rather than *RLC* components and operational amplifiers. As a result, they are generally unaffected by such factors as component accuracy, temperature stability, long-term drift, and so on, that afflict analog filter circuits. Also, many of the circuit restrictions imposed by physical limitations of analog devices can be removed, or at least circumvented, in a digital processor. Moreover, filters of a high order can be realized easily. Finally, digital filters can be modified simply by changing the algorithm of the computer, in contrast to an analog system, which may have to be physically rebuilt.

The subject of digital filtering is somewhat beyond our scope in this course. Several excellent books are available on the subject.[5]

2.6 SIGNAL DISTORTION OVER A CHANNEL

A signal transmitted over a channel is distorted because of various channel imperfections. The nature of signal distortion will now be studied.

Linear Distortion

We shall first consider linear time-invariant channels. Signal distortion can be caused over such a channel by nonideal characteristics of either the magnitude, the phase, or both. We can identify the effects these nonidealities will have on a pulse $g(t)$ transmitted through such a channel. Let the pulse exist over the interval (a, b) and be zero outside this interval. We recall the discussion in Section 2.2 about the marvelous balance of the Fourier spectrum. The components of the Fourier spectrum of the pulse have such a perfect and delicate balance of magnitudes and phases that they add up precisely to the pulse $g(t)$ over the interval (a, b) and to zero outside this interval. The transmission of $g(t)$ through an ideal channel that satisfies the distortionless-transmission conditions also leaves this balance undisturbed, because a distortionless channel multiplies each component by the same factor and delays each component by the same amount of time. Now, if the magnitude characteristic of the channel is not ideal (that is, $|H(\omega)|$ is not equal to a constant), this delicate balance will be disturbed, and the sum of all the components cannot be zero outside the interval (a, b). In short, the pulse will spread out (see the following example). The same thing happens if the channel phase characteristic is not ideal, that is, $\theta_h(\omega) \neq -\omega t_d$. Thus, spreading, or *dispersion,* of the pulse will occur if either the magnitude characteristic or the phase characteristic, or both, are nonideal.

This type of distortion is undesirable in a TDM system, because pulse spreading causes interference with a neighboring pulse and consequently with a neighboring channel (cross talk). For an FDM system, this type of distortion causes distortion (dispersion) in each multiplexed signal, but no interference occurs with a neighboring channel. This is because in FDM, each of the multiplexed signals occupies a band not occupied by any other signal. The magnitude and phase nonidealities of a channel will distort the spectrum of each signal, but because they are all nonoverlapping, no interference occurs among them.

■ EXAMPLE 2.30

A lowpass-filter (Fig. 2.47*a*) transfer function $H(\omega)$ is given by

$$H(\omega) = \begin{cases} (1 + k \cos T\omega)e^{-j\omega t_d} & |\omega| < 2\pi B \\ 0 & |\omega| > 2\pi B \end{cases} \tag{2.98}$$

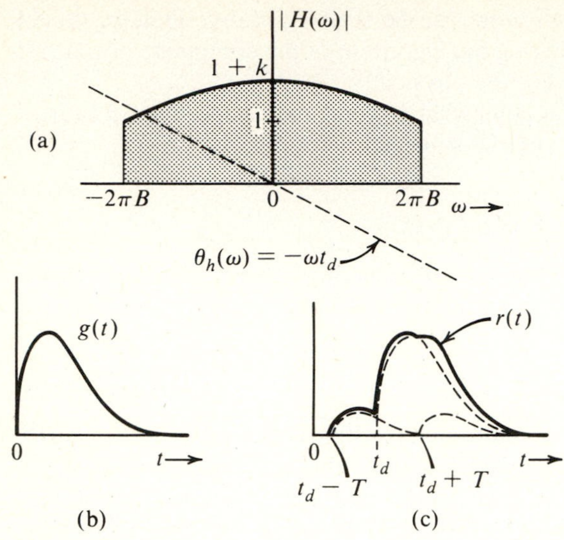

Figure 2.47 Pulse is dispersed when it passes through a system that is not distortionless.

A pulse $g(t)$ bandlimited to B (Fig. 2.47*b*) is applied at the input of this filter. Find the output $r(t)$.

This filter has ideal phase and nonideal magnitude characteristics. Because

$$g(t) \leftrightarrow G(\omega), \qquad r(t) \leftrightarrow R(\omega)$$

$$\begin{aligned} R(\omega) &= G(\omega)H(\omega) \\ &= G(\omega)[1 + k \cos T\omega]e^{-j\omega t_d} \\ &= G(\omega)e^{-j\omega t_d} + k[G(\omega) \cos T\omega]e^{-j\omega t_d} \end{aligned} \tag{2.99}$$

Using the time shifting property, Eq. (2.56a), and Eq. (2.58), we have

$$r(t) = g(t - t_d) + \frac{k}{2}[g(t - t_d - T) + g(t - t_d + T)] \tag{2.100}$$

The output is actually $g(t) + (k/2)[g(t - T) + g(t + T)]$ delayed by t_d. It consists of $g(t)$ and its echoes shifted by $\pm T$. The dispersion of the pulse caused by its echoes is evident from Fig. 2.47*c*.

Ideal magnitude but nonideal phase characteristic of $H(\omega)$ has a similar effect (see Prob. 2.51). ■

Distortion Caused by Channel Nonlinearities

Until now we considered the channel to be linear. This is an approximation valid for small signals. For large amplitudes, nonlinearities cannot be ignored. A general discussion of nonlinear systems is beyond our scope. Here we shall consider a simple case of a memoriless nonlinear channel where the input g and the output r are related by some nonlinear equation.

$$r = f(g)$$

The right-hand side of this equation can be expanded in a McLaurin's series as

$$r = a_0 + a_1 g + a_2 g^2 + a_3 g^3 + \cdots + a_k g^k + \cdots$$

The power-series expansion of this equation allows us to determine the spectrum of the output signal by observing that

$$g^k(t) \leftrightarrow \left(\frac{1}{2\pi}\right)^{k-1} \underbrace{G(\omega) * G(\omega) * \cdots * G(\omega)}_{k\,-\,1 \text{ convolutions}}$$

Hence

$$R(\omega) = 2\pi a_0\, \delta(\omega) + \sum_k \frac{a_k}{(2\pi)^{k-1}} \underbrace{G(\omega) * G(\omega) * \cdots * G(\omega)}_{k\,-\,1 \text{ convolutions}}$$

We can draw some general conclusions from the above equation. The output spectrum consists of an input spectrum plus repeated "autoconvolutions" of the input spectrum. When a spectrum is convolved with itself, the resulting spectrum has twice the bandwidth of the original spectrum. Similarly, $k - 1$ repeated convolutions will increase the bandwidth by a factor of k. This means the output signal will contain new frequency components not contained in the input signal.

This type of distortion causes not only the signal distortion but also interference with neighboring channels because of spreading out (or dispersion) of the spectrum. The spectrum dispersion will cause a serious interference problem in FDM systems. But this effect will not cause interference in TDM systems.

■ EXAMPLE 2.31

A baseband signal $g(t)$ bandlimited to B Hz modulates a carrier of frequency ω_c. The modulated signal $g(t) \cos \omega_c t$ is transmitted over a channel whose input x and the output y are related as

$$y = a_1 x + a_2 x^2 + a_3 x^3$$

Find the received signal, sketch its spectrum, and comment.

The input signal $x(t) = g(t) \cos \omega_c t$. The received signal

$$y(t) = a_1 g(t) \cos \omega_c t + a_2 g^2(t) \cos^2 \omega_c t + a_3 g^3(t) \cos^3 \omega_c t$$

$$= \frac{a_2}{2} g^2(t) + \underbrace{\left[a_1 g(t) + \overbrace{\frac{3}{4} a_3 g^3(t)}^{\text{distortion}} \right] \cos \omega_c t}_{\text{useful component } z(t)} + \frac{a_2}{2} g^2(t) \cos 2\omega_c t$$

$$+ \frac{1}{4} a_3 g^3(t) \cos 3\omega_c t$$

The spectra of the baseband signal $g(t)$, the input signal $x(t)$, and the output signal $y(t)$ are sketched in Fig. 2.48. Observe that in the output signal, there is not only distortion

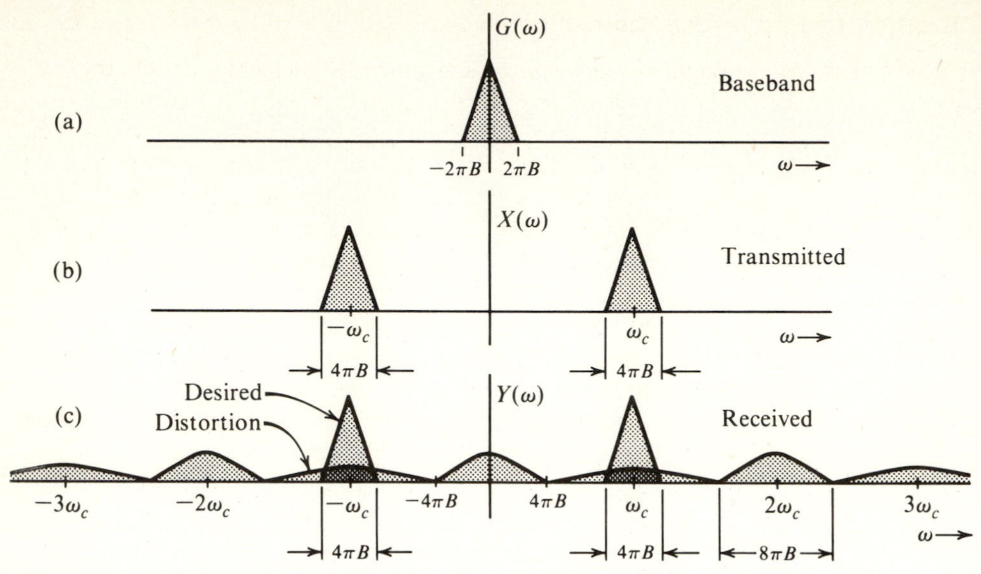

Figure 2.48 Effect of channel nonlinearity on modulated signal.

in the desired signal but also spectrum dispersion, which will interfere with other signals on the channel. ■

Companding. Memoriless nonlinear distortion can be eliminated, or at least reduced, by using a device with characteristics reciprocal of those of the channel (Fig. 2.49). In practice, exact cancellation of nonlinearity is rarely achieved.

In some cases (for example, telephony) a signal is intentionally distorted by passing it through a compressor (Fig. 2.49*a*) that enhances weaker amplitudes in order to better withstand noise. At the receiver, the signal is passed through an expandor (Fig. 2.49*b*) to restore the signal. This process of compression-expansion is known as *companding* and will be discussed later, in Chapters 3 and 6. Compression of an analog signal, in general, increases the signal bandwidth. In PCM telephony, however, the compressed signal sample values—not the compressed analog signal—are transmitted. Because the transmission bandwidth depends only on the number of samples/second, compression does not affect the transmission bandwidth in PCM.

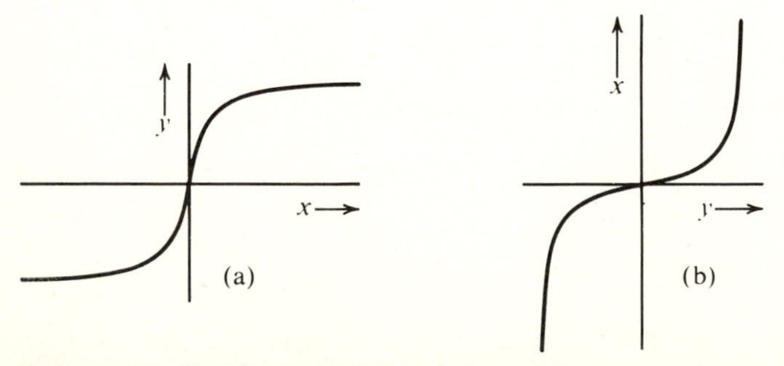

Figure 2.49 Signal compressor and expandor.

Distortion Caused by the Multipath Effect

A multipath transmission takes place when a transmitted signal arrives at the receiver by two or more paths of different delays. For example, if a signal is transmitted over a cable that has impedance irregularities (mismatching) along the path, the signal will arrive at the receiver in the form of a direct wave plus various reflections with various delays. In radio links, the signal can be received by direct path between the transmitting and the receiving antennas and also by reflections from other objects, such as hills, buildings, and so on. In long-distance radio links using the ionosphere, similar effects occur because of one-hop and multihop paths. In each of these cases, the transmission channel can be represented as several channels in parallel, each with a different relative attenuation and a different time delay. Let us consider the case of only two paths: one with a unity gain and a delay t_d, and the other with a gain α and a delay $t_d + \Delta t$, as shown in Fig. 2.50*a*. The transfer functions of the two paths are given by $e^{-j\omega t_d}$ and $\alpha e^{-j\omega(t_d+\Delta t)}$, respectively. The overall transfer function of such a channel is $H(\omega)$, given by

$$
\begin{aligned}
H(\omega) &= e^{-j\omega t_d} + \alpha e^{-j\omega(t_d+\Delta t)} \\
&= e^{-j\omega t_d}(1 + \alpha e^{-j\omega \Delta t}) \qquad \textbf{(2.101a)} \\
&= e^{-j\omega t_d}(1 + \alpha \cos \omega \Delta t - j\alpha \sin \omega \Delta t) \\
&= \underbrace{\sqrt{1 + \alpha^2 + 2\alpha \cos \omega \Delta t}}_{|H(\omega)|}\; e^{-j\underbrace{\left[\omega t_d + \tan^{-1} \frac{\alpha \sin \omega \Delta t}{1 + \alpha \cos \omega \Delta t}\right]}_{\theta_h(\omega)}} \qquad \textbf{(2.101b)}
\end{aligned}
$$

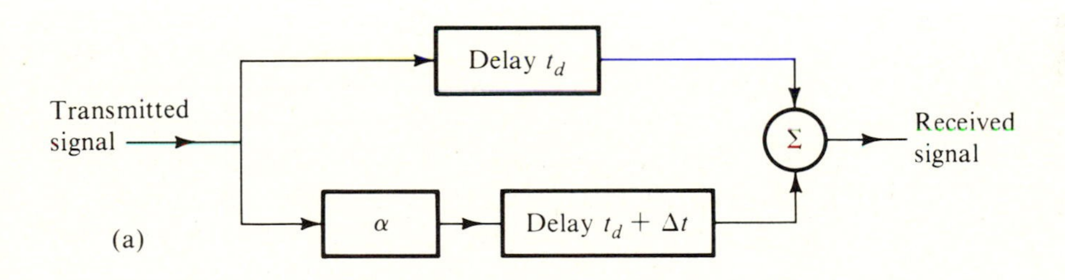

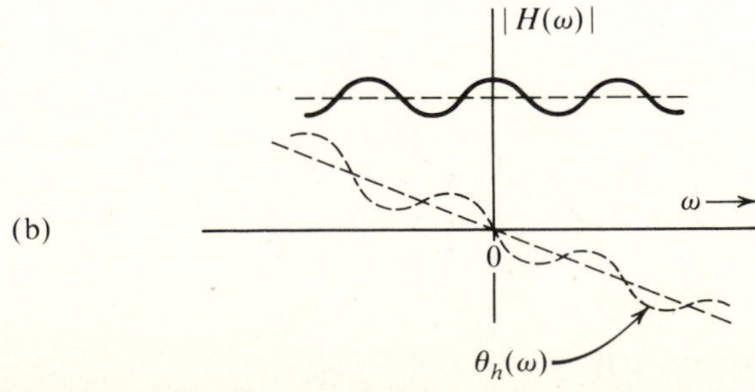

Figure 2.50 Multipath transmission.

Both the magnitude and the phase characteristics of $H(\omega)$ are periodic in ω with a period of $2\pi/\Delta t$ (Fig. 2.50*b*). The multipath transmission, therefore, causes nonidealities in the magnitude and the phase characteristics of the channel and will cause linear distortion (pulse dispersion), as discussed earlier. Such distortion can be partly corrected by using the tapped delay-line equalizer.

Fading Channels

Thus far, the channel characteristics were assumed to be constant with time. In practice, we encounter channels whose transmission characteristics vary with time. These include troposcatter channels and channels using the ionosphere for radio reflection to achieve long-distance communication. The time variations of the channel properties arise because of semiperiodic and random changes in the propagation characteristics of the medium. The reflection properties of the ionosphere, for example, are related to meteorological conditions that change seasonally, daily, and even from hour to hour, much the same way as does the weather. Periods of sudden storms also occur. Hence, the effective channel transfer function varies semiperiodically and randomly, causing random attenuation of the signal. This phenomenon is know as *fading*. One way to reduce the effects of fading is to use *automatic gain control* (AGC).*

Fading may be strongly frequency dependent where different frequency components are affected unequally. Such fading is known as *selective fading* and can cause serious problems in communication.

2.7 THE BANDWIDTH AND THE RATE OF PULSE TRANSMISSION

In digital communication, digital data is transmitted in the form of pulses. The information resides not in the waveform but in the relative strengths of these pulses. Hence, the objective is not the accurate reception of the pulse shape but the accurate reception of the relative pulse strengths. At the receiver, pulses are detected by sampling each pulse at its peak value. From these samples, the decision is made whether a positive or a negative (or a zero) pulse was transmitted. If the transmitted pulses are time limited, their bandwidth is infinity. But because accurate pulse-shape transmission is not the objective, we need a much smaller transmission bandwidth. When pulses are transmitted over a channel of finite bandwidth, their high-frequency components will be suppressed, and the pulses will spread out. For example, a rectangular pulse (Fig. 2.51*a*) transmitted over a finite bandwidth channel will lose its sharp edges, and the pulse will rise as well as decay gradually, causing dispersion, as shown in Fig. 2.51*b*. The dispersion, by causing interference with the neighboring pulses, will cause errors in the reading of relative pulse strengths. The larger the bandwidth, the smaller the dispersion. For a given channel bandwidth, pulses should be spaced far enough apart to reduce the interference caused by dispersion. In short,

* AGC will also suppress slow variations of the original signals.

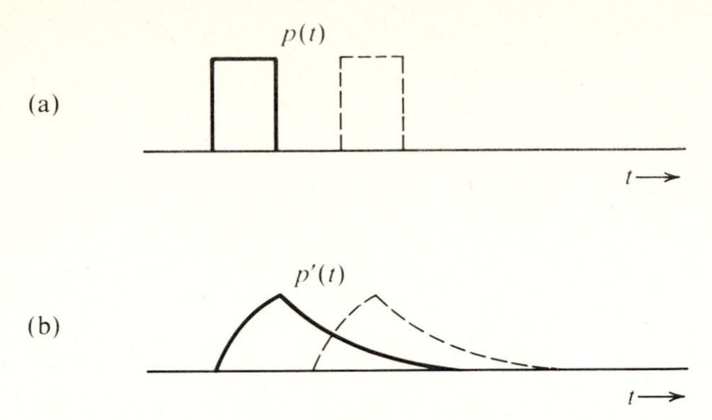

Figure 2.51 Pulse spreading and interference.

the rate of pulse transmission is limited by the channel bandwidth. We shall now show that the rate of pulse transmission is directly proportional to the channel bandwidth B.

The sampling theorem states that a signal bandlimited to B Hz can be specified by $2B$ independent pieces of information per second. Hence, transmission of a signal bandlimited to B is equivalent to transmission of $2B$ independent pieces of information per second. Consequently, an ideal channel of bandwidth B can transmit correctly $2B$ independent pieces of information per second. We can use this corollary to determine the rate of pulse transmission. In case of PAM, PCM, or digital data, the information is transmitted through pulse amplitudes. Transmission of one pulse implies transmission of one piece of information. Consequently, over a channel of bandwidth B we should be able to receive correctly $2B$ independent pulse amplitudes per second (the Nyquist rate). Note that because the pulses are independent, they need not be restricted to binary pulses; they can assume any amplitudes.

The transmission rate of $2B$ pulses/second over a channel of bandwidth B is the upper theoretical limit. It is not obvious how such a rate can be attained in practice. One thing is certain: because the transmission bandwidth is finite, the pulses used cannot be time limited. This means they will interfere with other pulses (intersymbol interference). If we select a pulse waveform whose amplitude is zero at the centers of all other pulses, however, we can eliminate interference at the center of each pulse and receive the data correctly. In short, the pulse used must have a bandwidth B, and it should have zero amplitude at $t = \pm n/2B$ ($n = 1, 2, 3, \ldots$) and a nonzero amplitude at $t = 0$. Only one pulse $p(t)$ satisfies these conditions (see Prob. 2.32):

$$p(t) = \operatorname{sinc}(2Bt) \qquad \textbf{(2.102)}$$

The pulse $p(t)$ is shown in Fig. 2.52a. Figure 2.52b shows a case of binary sequence **1101** at a rate of $2B$ digits/second, with a binary 1 transmitted by $p(t)$ and 0 transmitted by $-p(t)$. It is obvious that at the center of any pulse, the interference from all the remaining pulses is zero, and if we sample at the center of a pulse, we shall read the amplitude of that pulse correctly. The bandwidth of the resulting signal is B, and the pulses are transmitted at a rate of $2B$/second. This completes the description of a scheme for transmitting $2B$ pulse amplitudes/second (the Nyquist rate) over a bandwidth B. It should be stressed once again that pulse amplitudes can be chosen indepen-

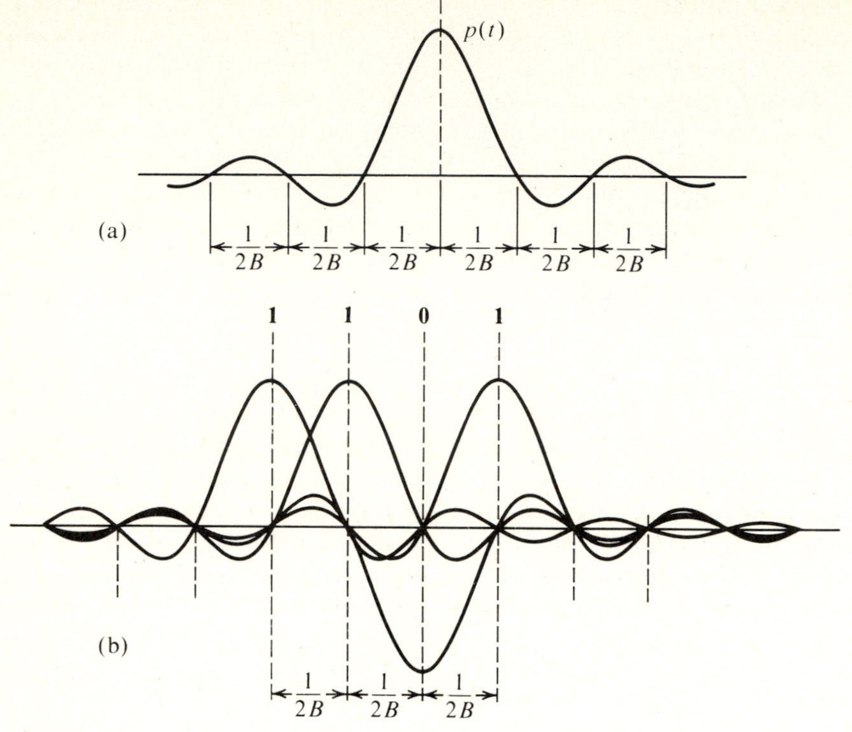

Figure 2.52 A scheme for transmitting pulses at the Nyquist rate.

dently. A binary case is a special case of this where amplitudes are restricted to only two values, 1 and -1. But, in general, this scheme is capable of transmitting numbers taking on any value.

If we use pulses of the form in Eq. (2.102) over a channel that is not distortionless, this carefully laid plan will go astray. The pulses received at the detector will not be free of interference. Hence, we must shape pulses at the transmitter such that after passing over the channel, they will arrive at the receiver in the form $p(t)$ in Eq. (2.102). If $p_1(t)$ is the transmitted pulse and $H_c(\omega)$ is the channel transfer function, then

$$\mathscr{F}^{-1}[P_1(\omega)H_c(\omega)] = \text{sinc}\,(2Bt)$$

or

$$P_1(\omega)H_c(\omega) = \frac{1}{2B}\Pi\left(\frac{\omega}{4\pi B}\right)$$

In practice, the received pulses are passed through an equalizer adjusted to ensure that the pulses at the input of the detector are such that negligible interference occurs at the pulse centers. This topic is discussed in detail in Sec. 3.5.

The rate of pulse transmission derived here is the upper theoretical limit and cannot be attained in practice. First, the pulse sinc $(2Bt)$ is an unrealizable pulse because it begins at $t = -\infty$. Any attempt to truncate it would increase the bandwidth beyond B. Second, even if the pulse sinc $(2Bt)$ were realizable, it has an undesirable feature that it decays too slowly, at a rate of $1/t$. This causes some serious practical problems. For example, suppose we are transmitting the data at a rate of $2B$ using pulses sinc $(2Bt)$. If the value of B in sinc $(2Bt)$ deviates a little, the pulse amplitude will not be zero at multiples of $1/2B$, thus causing interference with all other pulses. Because the pulses decay as $1/t$, the interference at the center of a pulse caused by all other pulses will be of the form $\Sigma\ (1/n)$. It is well known that the infinite series of this form does not converge and can add up to a large value. The same effect occurs if the sampling rate at the receiver deviates from $2B$/second. Again, the same thing happens if the synchronization is not perfect, that is, if the sampling at the receiver is not performed exactly at the center of each pulse because of time jitter, which is inevitable even in the most sophisticated systems.

The failure of this scheme occurs because sinc $(2Bt)$ does not decay fast enough to reduce interference caused by parameter deviations, which always exists in practice. If we could generate a pulse similar to sinc $(2Bt)$ (viz., $p(t) = 0$ at $t = \pm n/2B$, and $p(0) \neq 0$), but that decays faster, the scheme would function. Nyquist has investigated this problem and has shown that such pulses require a bandwidth between B and $2B$. The larger the bandwidth, the faster the rate of decay. For practical schemes, the pulses of bandwidth $2B$ are chosen in many applications. Thus, in practice we need a bandwidth* αB $(1 < \alpha \leq 2)$ to transmit $2B$ pulses/second. To put it the other way, over a channel of bandwidth B, we can transmit pulses at a rate $R = kB$ pulses/second $(1 \leq k < 2)$ without interference.

■ EXAMPLE 2.32

Determine the bandwidth of a Bell Telephone T-1 system. In this system, 24 voice channels are converted to binary PCM and then time-division multiplexed. Each telephone signal, with a nominal bandwidth of 3.5 kHz, is sampled at a rate of 8000 samples/second. The number of quantizing levels is 256.

□ **Solution:** Because $256 = 2^8$, we need eight binary pulses per sample. Because there are 8000 samples/second and a total of 24 signals to be multiplexed, we need to transmit a total of $8 \times 8000 \times 24 = 1.536 \times 10^6$ pulses/second. Hence, the transmission bandwidth is 0.768α MHz $(1 < \alpha \leq 2)$. The signaling scheme used for PCM is bipolar, which requires $\alpha = 2$ (see Sec. 3.2). Therefore, the transmission bandwidth is 1.536 MHz.† ■

---

* For binary pulses, a "duobinary scheme" that uses a bandwidth B to transmit $2B$ pulses/second is also being used. See Secs. 3.2 and 3.3.

† In practice, one framing pulse is added to each frame of $24 \times 8 = 192$ pulses. This increases the number of pulses from 1.536×10^6 to 1.544×10^6, and the corresponding bandwidth is 1.544 MHz.

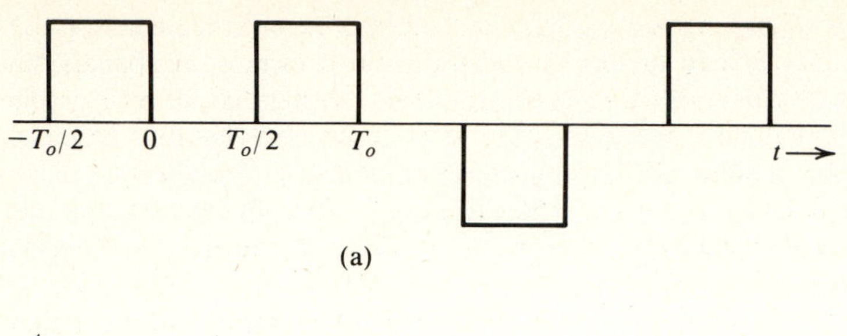

(a)

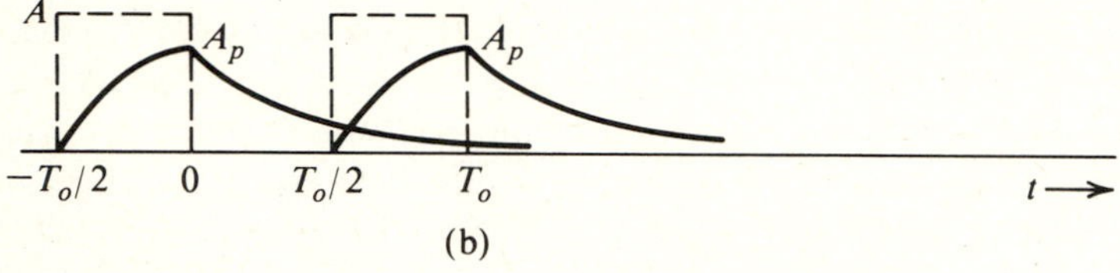

(b)

Figure 2.53 (a) Transmitted signal. (b) Received signal.

■ EXAMPLE 2.33

Rectangular pulses (Fig. 2.53*a*) are transmitted over a channel with transfer function $H_c(\omega) = a/j\omega + a$ at a rate of $2B$ pulses/second where B is the 3dB bandwidth of the channel. Determine the interference caused by a pulse at the neighboring pulse.

□ **Solution:** Figure 2.53*a* shows the transmitted pulse train and the received signal for the first two pulses only. For convenience, we assume the first pulse to start at $t = -T_o/2$. The first pulse reaches its peak value A_p at $t = 0$, and then decays as $A_p e^{-at}$. The second pulse reaches its peak A_p at $t = T_o$. This is where the second pulse is sampled at the receiver. At this instant the interference caused by the first pulse is $A_p e^{-aT_o}$. Hence, ρ, the ratio of the interfering signal to the desired signal is

$$\rho = e^{-aT_o} \tag{2.103}$$

The 3dB bandwidth of the channel is

$$B = a/2\pi$$

and the rate of pulses is $2B/s$. Hence,

$$T_o = \frac{1}{2B} = \frac{\pi}{a}$$

and

$$\rho = e^{-\pi} = 0.043$$

Thus the interference is about 4.3 percent. There will also be interference from previous pulses, but this can be neglected.

The results of this example pose an interesting question. Is it possible to communicate data at a rate higher than the Nyquist rate? In this example, it appears that communication at the Nyquist rate causes a negligible intersymbol interference. If the rate of communication is increased beyond $2B$ pulses/second (the Nyquist rate), the intersymbol interference will be more than 4.3 percent, but it will still be possible to distinguish clearly between a positive and a negative pulse. This appears to violate our earlier result that a channel bandlimited to B Hz can accurately transmit pulses at a rate not greater than $2B$ pulses/second.

There are two errors in our thinking that are causing this apparent contradiction. The first is in the arbitrariness of the 3-dB bandwidth definition. The channel under consideration has a 3-dB bandwidth of B Hz. In reality, the channel transmits frequencies up to ∞. We have arbitrarily defined 3-dB bandwidth B as that bandwidth over which the variations of $|H(\omega)|$ remain within 3 dB. Thus the channel under consideration has capabilities much more than the 3-dB bandwidth would lead us to believe.

The second error is in confusing binary data with independent data. We have shown that a channel of bandwidth B can transmit independent pulse amplitudes at a rate not greater than $2B$/s. This means pulse amplitudes can take on any arbitrary values in the range $(-\infty, \infty)$. In the case of binary pulses, on the other hand, pulse amplitudes are restricted to two fixed values only. If the pulse amplitudes are truly independent, there is absolutely no tolerance for intersymbol interference. Any interference, no matter how small, will give erroneous results. In the binary case, on the other hand, we can tolerate a great deal of interference since the objective is to distinguish only between two known amplitudes. Thus, if the objective is zero intersymbol interference (required for transmission of independent amplitude values), then the rate must not exceed $2B$/s. On the other hand, if intersymbol interference can be tolerated (as in the binary case), the rate can be increased beyond $2B$/s. Of course, higher rate causes higher intersymbol interference and leaves the system more vulnerable to channel noise. ■

2.8 THE ENERGY SPECTRAL DENSITY

The concept of the energy of a signal was discussed earlier. For a real signal* $g(t)$, the energy E_g was defined as

$$E_g = \int_{-\infty}^{\infty} g^2(t)\, dt$$

* For a complex signal $g(t)$,

$$E_g = \int_{-\infty}^{\infty} |g(t)|^2\, dt = \frac{1}{2\pi}\int_{-\infty}^{\infty} |G(\omega)|^2\, d\omega$$

See footnote on page 61.

The concept of the energy of a signal is valid only if the area under $g^2(t)$ is finite. Signals that satisfy this condition are called *energy signals* and satisfy the condition of Fourier transformability. The energy can also be expressed in terms of the spectrum $G(\omega)$ as [see Eq. (2.77c)]

$$E_g = \int_{-\infty}^{\infty} g^2(t)\, dt = \frac{1}{2\pi} \int_{-\infty}^{\infty} |G(\omega)|^2\, d\omega$$

The function $|G(\omega)|^2$ is known as the *energy spectral density* (*ESD*) and has an interesting physical interpretation.

Consider a signal $g(t)$ applied at the input of an ideal bandpass filter whose transfer function $H(\omega)$ is shown in Fig. 2.54*a*. This filter suppresses all frequencies except a narrowband $\Delta\omega$ ($\Delta\omega \to 0$) centered at frequency ω_o (Fig. 2.54*b*). If $\Delta R(\omega)$ is the Fourier transform of the response $\Delta r(t)$ of this filter, then

$$\Delta R(\omega) = G(\omega)H(\omega)$$

and the energy ΔE_r of the output signal $\Delta r(t)$ is given by [Eq. (2.77c)]

$$\Delta E_r = \frac{1}{2\pi} \int_{-\infty}^{\infty} |G(\omega)H(\omega)|^2\, d\omega$$

Because $H(\omega) = 0$ everywhere except over a narrowband $\Delta\omega$ where it is unity, we have (for $\Delta\omega \to 0$)

$$\begin{aligned}\Delta E_r &= 2\frac{1}{2\pi} |G(\omega_o)|^2\, d\omega \\ &= 2|G(\omega_o)|^2\, \Delta f\end{aligned}$$

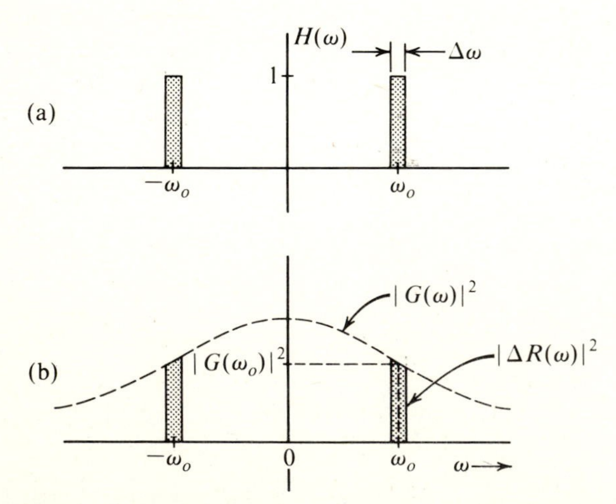

Figure 2.54 Interpretation of energy spectral density.

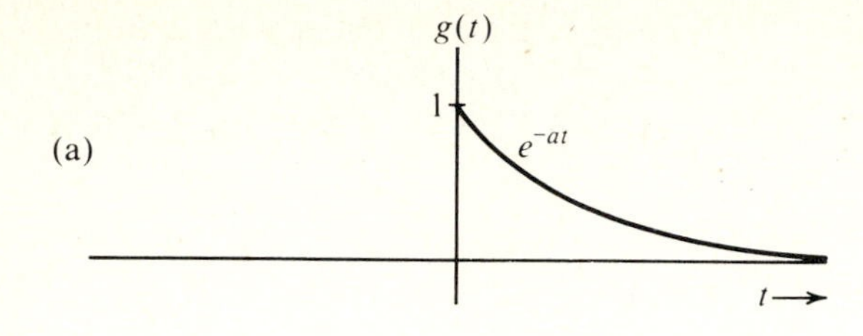

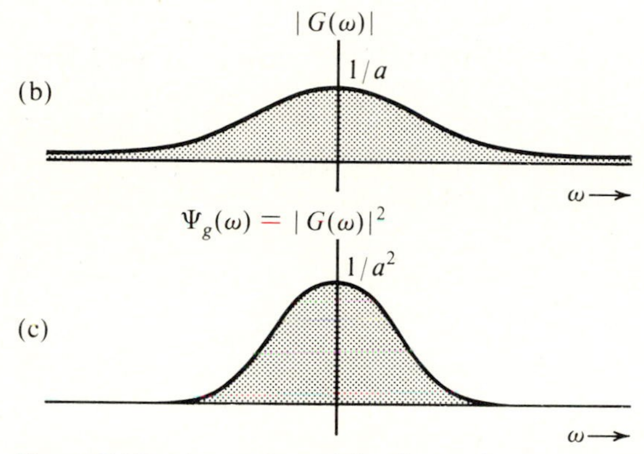

Figure 2.55 An exponential pulse and its energy spectral density.

As seen from Fig. 2.54*b*, only the frequency components of $g(t)$ that lie in the narrowband $\Delta\omega$ are transmitted intact through the filter. The remaining frequency components are completely suppressed. Therefore, $2|G(\omega_o)|^2 \Delta f$ represents the contribution to the energy of $g(t)$ by frequency components of $g(t)$ lying in the narrowband Δf centered at ω_o. Hence $|G(\omega)|^2$ is the energy per unit bandwidth (in Hz) contributed by frequency components centered at ω.

It should be noted that we have energy contribution from negative as well as positive frequency components. Moreover, the contribution by negative and positive frequency components is equal because $|G(\omega)|^2 = |G(-\omega)|^2$. Hence, we may interpret that the amount $|G(\omega)|^2$ [half of the energy $2|G(\omega)|^2$] was contributed by positive frequency components, and the remaining $|G(\omega)|^2$ was contributed by the negative frequency components.*

The function $|G(\omega)|^2$, or ESD, represents the energy per unit bandwidth (either positive or negative). The ESD $\Psi_g(\omega)$ is thus defined as

$$\Psi_g(\omega) = |G(\omega)|^2 \tag{2.104}$$

The ESD indicates the relative contribution of energy by various frequency components. Figure 2.55 shows the signal $e^{-at}u(t)$ and its ESD. The total energy E_g is

* This distinction is more for convenience than natural. Actually, a combination of negative and positive frequencies contributes the energy associated with any particular frequency band.

given by

$$E_g = \frac{1}{2\pi} \int_{-\infty}^{\infty} |G(\omega)|^2 \, d\omega \tag{2.105a}$$

In energy computations it is more convenient to use the frequency variable $f (f = \omega/2\pi)$ rather than ω. In terms of the variable f, Eq. (2.105a) becomes*

$$E_g = \int_{-\infty}^{\infty} |G(\omega)|^2 \, df \tag{2.105b}$$

This equation is also consistent with our interpretation of the ESD as the energy per unit bandwidth and shows clearly that the energy contributed by frequency components in the bandwidth Δf is the area of $|G(\omega)|^2$ over the band Δf.

Because $|G(\omega)|^2$ is an even (as well as real) function of ω, Eq. (2.105b) can be expressed as

$$E_g = 2 \int_{0}^{\infty} |G(\omega)|^2 \, df \tag{2.105c}$$

$$= 2 \int_{0}^{\infty} \Psi_g(\omega) \, df \tag{2.105d}$$

Earlier, we had shown [Eq. (2.76b)] that the ESD is a Fourier transform of the autocorrelation function $\psi_g(\tau)$

$$\psi_g(\tau) \leftrightarrow \Psi_g(\omega) \tag{2.106}$$

Energy Densities of the Input and the Output

If $g(t)$ and $r(t)$ are the input and the corresponding output of a linear system with a transfer function $H(\omega)$, then

$$R(\omega) = H(\omega)G(\omega)$$

The ESD $\Psi_g(\omega)$ and $\Psi_r(\omega)$ of the input and the output respectively are related by

$$\begin{aligned} \Psi_r(\omega) &= |R(\omega)|^2 \\ &= |H(\omega)G(\omega)|^2 \\ &= |H(\omega)|^2 \Psi_g(\omega) \end{aligned} \tag{2.107}$$

■ EXAMPLE 2.34

Determine the ESD of a square pulse $g(t) = \Pi(t/T)$ (Fig. 2.56*a*), and calculate its energy E_g. If the signal $g(t)$ is passed through an ideal lowpass filter of bandwidth f_c Hz, determine the energy E_r of the output signal $r(t)$. Sketch E_r as a function of f_c.

---

* The integrand in Eq. (2.105b and c) is $|G(2\pi f)|^2$. We retain the use of $|G(\omega)|^2$ for the convenience of brevity.

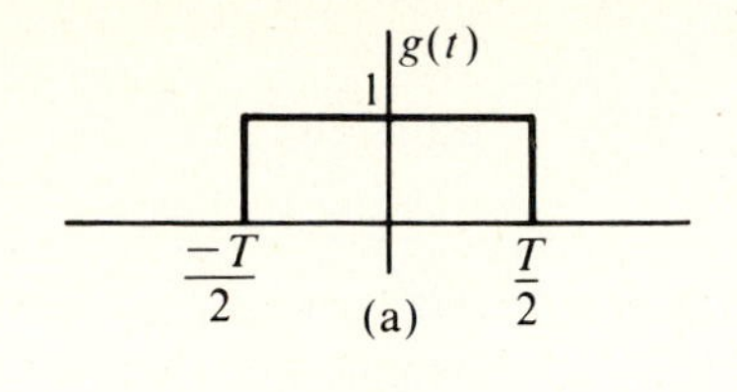

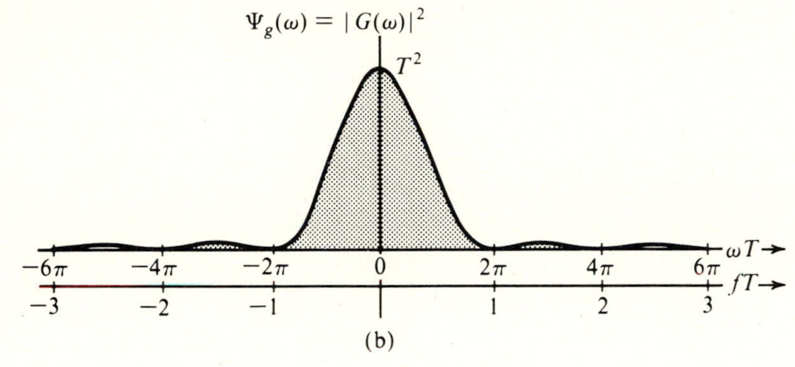

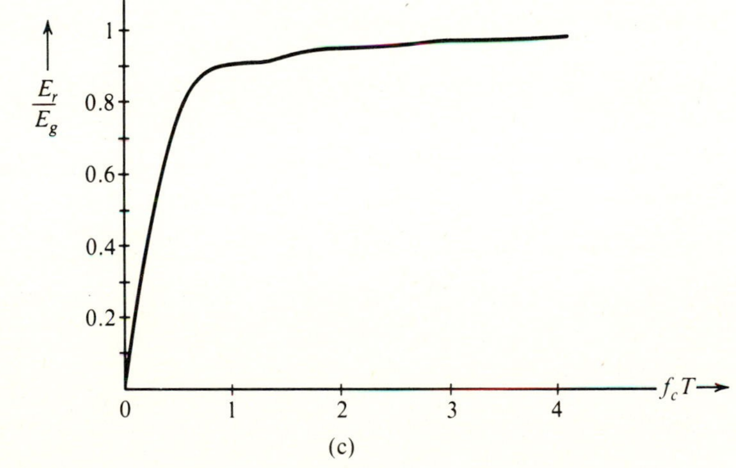

Figure 2.56 A gate function and its energy spectral density.

☐ **Solution:**

$$g(t) = \Pi\left(\frac{t}{T}\right)$$

$$G(\omega) = T \operatorname{sinc}\left(\frac{\omega T}{2\pi}\right)$$

The ESD $\Psi_g(\omega)$ is given by (Fig. 2.56*b*)

$$\Psi_g(\omega) = |G(\omega)|^2 = T^2 \operatorname{sinc}^2\left(\frac{\omega T}{2\pi}\right)$$

Thus, the energy E_g of $g(t)$ is

$$E_g = \int_{-\infty}^{\infty} g^2(t)\, dt = \int_{-T/2}^{T/2} dt = T$$

Because the ideal filter has a cutoff frequency f_c Hz, only the components below f_c Hz are transmitted. Hence E_r, the energy of the output signal is

$$E_r = 2\int_0^{f_c} T^2 \operatorname{sinc}^2\left(\frac{\omega T}{2\pi}\right) df$$

and

$$\frac{E_r}{E_g} = \frac{E_r}{T} = 2\int_0^{f_c} T \operatorname{sinc}^2(fT)\, df$$

$$= 2\int_0^{f_c T} \operatorname{sinc}^2(y)\, dy$$

The integral on the right-hand side is numerically computed. A plot of E_r/E_g vs. f_cT is shown in Fig. 2.56*c*. Note that 90.28 percent of the total energy of the pulse $g(t)$ is contained within the band $f_c = 1/T(f_cT = 1)$. ■

Energy of Modulated Signals

We shall now show that modulation causes a shift of the ESD of baseband signals and the energy of the modulated signal is half the energy of the baseband signal.

Let $g(t)$ be a baseband signal bandlimited to B Hz. The modulated signal $\varphi(t)$ is

$$\varphi(t) = g(t) \cos \omega_o t$$

Hence

$$\Phi(\omega) = \tfrac{1}{2}[G(\omega + \omega_o) + G(\omega - \omega_o)]$$

and the ESD of the modulated signal $\varphi(t)$ is

$$\Psi_\varphi(\omega) = \tfrac{1}{4}|G(\omega + \omega_o) + G(\omega - \omega_o)|^2$$

If $\omega_o \geq 2\pi B$, then $G(\omega + \omega_o)$ and $G(\omega - \omega_o)$ are nonoverlapping (Fig. 2.57), and

$$\Psi_\varphi(\omega) = \tfrac{1}{4}[|G(\omega + \omega_o)|^2 + |G(\omega - \omega_o)|^2] \tag{2.108a}$$

$$= \tfrac{1}{4}[\Psi_g(\omega + \omega_o) + \Psi_g(\omega - \omega_o)] \tag{2.108b}$$

The ESDs of $g(t)$ and $\varphi(t)$ are shown in Fig. 2.57. It is clear that modulation shifts the ESD of a baseband signal to $\pm\omega_o$.

Since the energy of a signal is proportional to the area under its ESD [Eq. (2.105)], it follows from Fig. 2.57 that the energy of $\varphi(t)$ is half that of $g(t)$, that is,

$$E_\varphi = \tfrac{1}{2}E_g \qquad (\omega_o \geq 2\pi B) \tag{2.108c}$$

In this section we have used several new functions and ideas in rapid succession. Students are generally at a loss as to why the two concepts of the autocorrelation

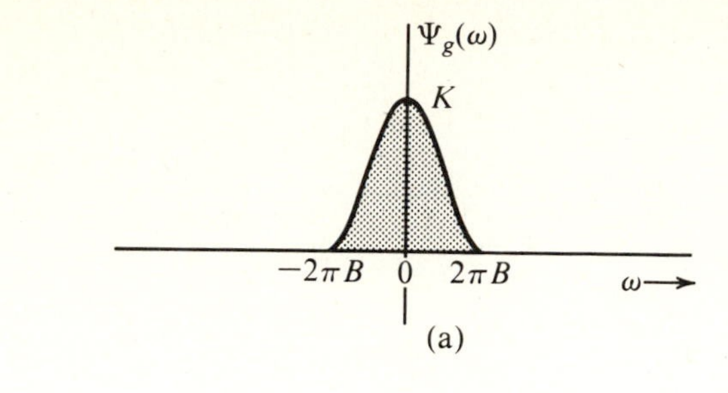

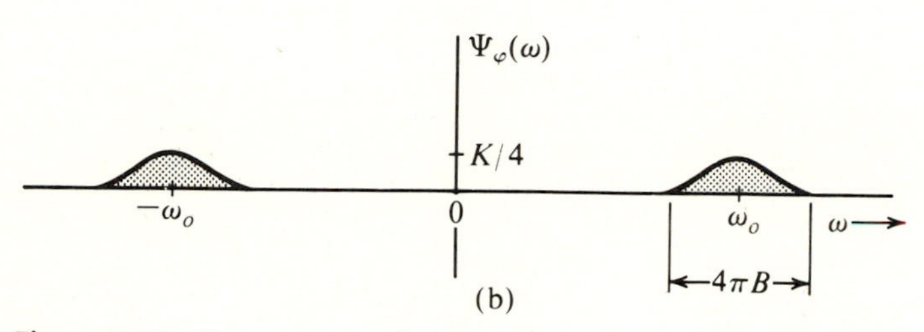

Figure 2.57 Energy spectral densities of the modulating and the modulated signals.

function $\psi_g(\tau)$ and its Fourier transform $\psi_g(\omega)$ are needed. Is not $|G(\omega)|^2$ and Eq. (2.105a) enough? Actually the ESD $|G(\omega)|^2$ can be obtained in two ways as shown below:

The signal $g(t)$ → $G(\omega)$ → $|G(\omega)|^2$, the ESD.
The signal $g(t)$ → $\psi_g(\tau)$ → $|G(\omega)|^2$, the ESD.

Strictly speaking, there is no reason to bother about $\psi_g(\tau)$ and its Fourier transform. We can obtain the ESD directly from $G(\omega)$. The only reason for introducing $\psi_g(\tau)$ is to lay the foundation for a parallel discussion about power signals in the next section. For power signals, the energy E_g is infinite and hence $G(\omega)$ does not exist, in general. In this case the only way to arrive at the power spectral density (similar to energy spectral density) is through the autocorrelation function.

2.9 THE POWER SPECTRAL DENSITY

If a signal $g(t)$ exists over the entire interval $(-\infty, \infty)$, we define the power P_g of a real $g(t)$ as the average power dissipated in a 1 ohm resistor when a voltage $g(t)$ is applied across it (or a current $g(t)$ is passed through it). Thus,*

$$P_g = \lim_{T\to\infty} \frac{1}{T} \int_{-T/2}^{T/2} g^2(t)\, dt \tag{2.109a}$$

---

* For complex signals, the power P is defined as

$$P_g = \lim_{T\to\infty} \frac{1}{T} \int_{-T/2}^{T/2} |g(t)|^2\, dt$$

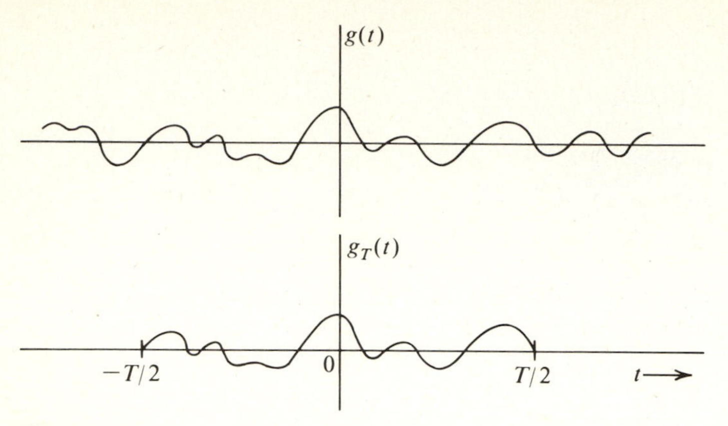

Figure 2.58

The power P_g defined above is simply the mean square value, or the time average, of the squared signal. Time averages will be denoted by a wavy bar on top. Thus,

$$P_g = \widetilde{g^2(t)} = \lim_{T\to\infty} \frac{1}{T} \int_{-T/2}^{T/2} g^2(t)\, dt \qquad \textbf{(2.109b)}$$

We observe that if E_g, the energy of $g(t)$, is finite, then its power P_g is zero, and if P_g is finite, then E_g is infinite. Signals for which E_g is finite are known as *energy signals*, and those for which P_g is nonzero and finite are known as *power signals*. It is obvious that a signal may be classified as one or the other but not both. On the other hand, there are some signal models, for example,

$$g(t) = e^{-at} \qquad -\infty < t < \infty$$

that cannot be classified as either energy or power signals, because both E_g and P_g are infinite.

In order to find the frequency-domain expression for the power P_g, we observe that power signals have infinite energy and, therefore, may not have Fourier transforms. Hence in this case we consider the truncated signal $g_T(t)$ (see Fig. 2.58). Define

$$g_T(t) = \begin{cases} g(t) & |t| < \dfrac{T}{2} \\ 0 & |t| > \dfrac{T}{2} \end{cases}$$

As long as T is finite, $g_T(t)$ has finite energy, and is Fourier transformable. Let

$$g_T(t) \leftrightarrow G_T(\omega)$$

The energy E_T of $g_T(t)$ is given by

$$E_T = \int_{-\infty}^{\infty} g_T^2(t)\, dt = \frac{1}{2\pi} \int_{-\infty}^{\infty} |G_T(\omega)|^2\, d\omega$$

But

$$\int_{-\infty}^{\infty} g_T^2(t)\, dt = \int_{-T/2}^{T/2} g^2(t)\, dt$$

Hence, the power P_g is given by [Eq. (2.109b)]

$$P_g = \lim_{T\to\infty} \frac{E_T}{T} = \lim_{T\to\infty} \frac{1}{T}\left[\frac{1}{2\pi}\int_{-\infty}^{\infty} |G_T(\omega)|^2\, d\omega\right] \tag{2.110}$$

As T increases, E_T the energy of $g_T(t)$ also increases. Thus $|G_T(\omega)|^2$ increases with T, and as $T \to \infty$, $|G_T(\omega)|^2$ also approaches ∞. However, $|G_T(\omega)|^2$ must approach ∞ at the same rate as T, because for a power signal, the integral on the right-hand side of Eq. (2.110) must converge. This convergence permits us to interchange the order of the limiting process and integration in Eq. (2.110), and we have

$$P_g = \frac{1}{2\pi}\int_{-\infty}^{\infty} \lim_{T\to\infty} \frac{|G_T(\omega)|^2}{T}\, d\omega$$

We define the *power spectral density* (*PSD*) $S_g(\omega)$ as

$$S_g(\omega) = \lim_{T\to\infty} \frac{|G_T(\omega)|^2}{T} \tag{2.111}$$

Then*

$$P_g = \frac{1}{2\pi}\int_{-\infty}^{\infty} S_g(\omega)\, d\omega \tag{2.112a}$$

$$= \frac{1}{\pi}\int_{0}^{\infty} S_g(\omega)\, d\omega \tag{2.112b}$$

As is the case with energy density, the power density $S_g(\omega)$ is also a positive, real, and even function of ω. The units of $S_g(\omega)$ are watts per hertz, and those of P_g are in watts. If we interpret P_g as the mean square value $\widetilde{g^2(t)}$, however, the units are volts^2 when $g(t)$ is a voltage.

As in the case of energy computation, it is convenient to use frequency variable f instead of ω in power computation. Equation (2.112) can be expressed as

$$P_g = \int_{-\infty}^{\infty} S_g(2\pi f)\, df = 2\int_{0}^{\infty} S_g(2\pi f)\, df \tag{2.112c}$$

Time Correlation of Power Signals

For two real power signals $g_1(t)$ and $g_2(t)$, we define the time cross-correlation function $\mathcal{R}_{g_1g_2}(\tau)$ as

$$\mathcal{R}_{g_1g_2}(\tau) = \widetilde{g_1(t)g_2(t+\tau)} = \lim_{T\to\infty} \frac{1}{T}\int_{-T/2}^{T/2} g(t)g_2(t+\tau)\, dt \tag{2.113a}$$

*One should use caution in using a unilateral expression such as $P_g = 2\int_0^\infty S_g(\omega)\, df$ when $S_g(\omega)$ contains an impulse at the origin. The impulse part should not be multiplied by the factor 2.

The time autocorrelation function $\mathcal{R}_g(\tau)$ of a real signal $g(t)$ is defined as

$$\mathcal{R}_g(\tau) = \widetilde{g(t)g(t+\tau)} = \lim_{T\to\infty} \frac{1}{T} \int_{-T/2}^{T/2} g(t)g(t+\tau)\, dt \tag{2.113b}$$

The change of variable $x = t + \tau$ allows us to write Eq. (2.113b) as*

$$\mathcal{R}_g(\tau) = \widetilde{g(t)g(t-\tau)} = \lim_{T\to\infty} \frac{1}{T} \int_{-T/2}^{T/2} g(t)g(t-\tau)\, dt \tag{2.113c}$$

From Eq. (2.113b and c) it follows that $\mathcal{R}_g(\tau)$ is an even function of τ†

$$\mathcal{R}_g(-\tau) = \mathcal{R}_g(\tau) \tag{2.113d}$$

As in the case of energy signals, the PSD $S_g(\omega)$ is the Fourier transform of the time autocorrelation function $\mathcal{R}_g(\tau)$ in Eq. (2.113c). This can be proved by considering the truncated signal $g_T(t)$ and its ESD $|G_T(\omega)|^2$. From Eq. (2.113b),

$$\mathcal{R}_g(\tau) = \lim_{T\to\infty} \frac{1}{T} \int_{-\infty}^{\infty} g_T(t)g_T(t+\tau)\, dt = \lim_{T\to\infty} \frac{\psi_{g_T}(\tau)}{T}$$

Hence [see Eq. (2.76b)]

$$\mathcal{F}[\mathcal{R}_g(\tau)] = \lim_{T\to\infty} \frac{|G_T(\omega)|^2}{T} = S_g(\omega)$$

Thus,

$$\mathcal{R}_g(\tau) \leftrightarrow S_g(\omega) \tag{2.114a}$$

and from Eq. (2.113b),

$$P_g = \widetilde{g^2(t)} = \mathcal{R}_g(0) \tag{2.114b}$$

■ EXAMPLE 2.35

Find the PSD and the power of a sinusoid (Fig. 2.59*a*)

$$g(t) = A \cos(\omega_o t + \theta)$$

□ Solution:

$$\mathcal{R}_g(\tau) = \lim_{T\to\infty} \frac{1}{T} \int_{-T/2}^{T/2} A^2 \cos(\omega_o t + \theta) \cos[\omega_o(t+\tau) + \theta]\, dt$$

$$= \frac{A^2}{2} \lim_{T\to\infty} \frac{1}{T}\left[\int_{-T/2}^{T/2} \cos \omega_o \tau\, dt + \int_{-T/2}^{T/2} \cos(2\omega_o t + \omega_o \tau + 2\theta)\, dt\right] \tag{2.115}$$

$$= \frac{A^2}{2} \cos \omega_o \tau \tag{2.116a}$$

---

*If $x = t + \tau$, Eq. (2.113*b*) becomes:

$$\mathcal{R}_g(\tau) = \lim_{T\to\infty} \frac{1}{T} \int_{-(T/2)-\tau}^{(T/2+\tau)} g(x)\, g(x-\tau)\, dx = \widetilde{g(t)\, g(t-\tau)}$$

† This is true only when $g(t)$ is a real function of τ.

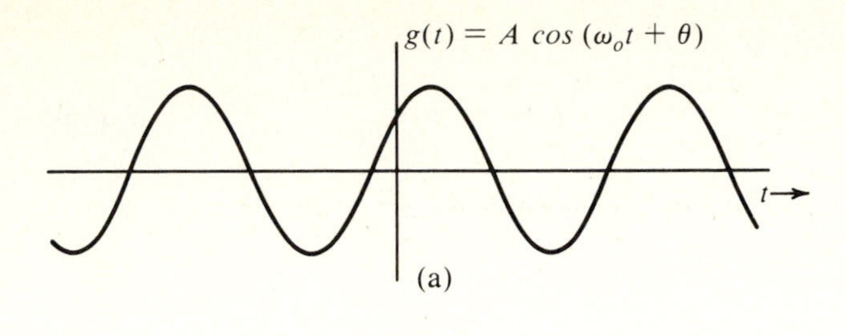

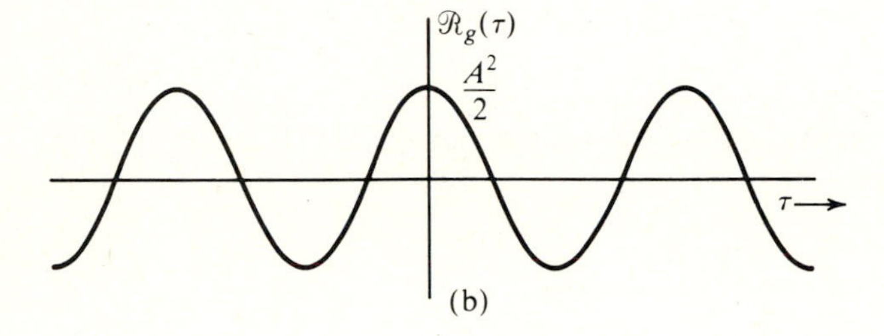

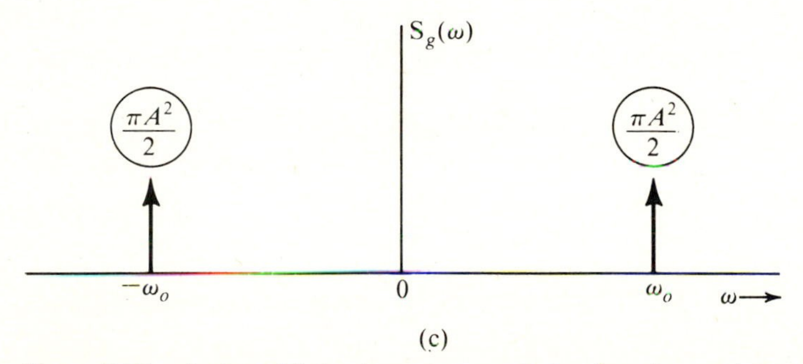

Figure 2.59 A sinusoid, its time-autocorrelation function and energy spectral density.

Observe that the autocorrelation function (Fig. 2.59*b*) is independent of phase θ. The PSD (Fig. 2.59*c*) is

$$S_g(\omega) = \mathcal{F}[\mathcal{R}_g(\tau)] = \frac{\pi A^2}{2}[\delta(\omega + \omega_o) + \delta(\omega - \omega_o)] \tag{2.116b}$$

The power, or mean square value, of $g(t)$ is (see Eq. 2.114b)

$$P_g = \mathcal{R}_g(0) = \frac{A^2}{2} \tag{2.116c}$$

We can also derive this result from $S_g(\omega)$ using Eq. (2.112c).

$$\begin{aligned} P_g &= 2\int_0^\infty S_g(\omega)\, df \\ &= 2\int_0^\infty \frac{\pi A^2}{2}\,\delta(\omega - \omega_o)\, df \end{aligned}$$

Using the fact that $d\omega = 2\pi\, df$, we have

$$P_g = \frac{A^2}{2} \tag{2.117}$$

This confirms the well-known fact that the mean square value of a sinusoid $A \cos(\omega_o t + \theta)$ is $A^2/2$. ■

PSD of Modulated Signals

Modulation causes a shift in PSD of a baseband signal. If $g(t)$ is a power signal bandlimited to B Hz, and the corresponding modulated signal $\varphi(t)$ is

$$\varphi(t) = g(t) \cos \omega_o t$$

then

$$S_\varphi(\omega) = \tfrac{1}{4}[S_g(\omega + \omega_o) + S_g(\omega - \omega_o)] \qquad \omega_o \geq 2\pi B \tag{2.118a}$$

Since the PSD is a time average of the ESD in the limit as $T \to \infty$, this result can be proved by following the argument used to derive the similar result for ESD in Eq. (2.108b).

Since the power of a signal is proportional to the area under the PSD function, it follows that

$$\widetilde{[g(t) \cos \omega_o t]^2} = \tfrac{1}{2}\widetilde{g^2(t)} \qquad \omega_o \geq 2\pi B \tag{2.118b}$$

Therefore, the mean square value (or the power) of the modulated signal is half the mean square value (or the power) of the modulating signal. The inverse Fourier transform of Eq. (2.118a) yields

$$\mathcal{R}_\varphi(\tau) = \tfrac{1}{2}\mathcal{R}_g(\tau) \cos \omega_o \tau \tag{2.118c}$$

■ EXAMPLE 2.36

Show that for

$$g(t) = A_1 \cos(\omega_1 t + \theta_1) + A_2 \cos(\omega_2 t + \theta_2) \tag{2.119a}$$

$$\mathcal{R}_g(\tau) = \frac{A_1^2}{2} \cos \omega_1 \tau + \frac{A_2^2}{2} \cos \omega_2 \tau \tag{2.119b}$$

$$S_g(\omega) = \frac{\pi}{2}(A_1^2[\delta(\omega + \omega_1) + \delta(\omega - \omega_1)] + A_2^2[\delta(\omega + \omega_2) + \delta(\omega - \omega_2)]) \tag{2.119c}$$

and

$$P_g = \widetilde{g^2(t)} = \frac{A_1^2}{2} + \frac{A_2^2}{2} \tag{2.119d}$$

☐ **Solution:** Let $g_1(t) = A_1 \cos(\omega_1 t + \theta_1)$ and $g_2(t) = A_2 \cos(\omega_2 t + \theta_2)$ then

$$\mathcal{R}_g(\tau) = \lim_{T\to\infty} \frac{1}{T} \int_{-T/2}^{T/2} [g_1(t) + g_2(t)][g_1(t+\tau) + g_2(t+\tau)]\, dt$$

$$= \mathcal{R}_{g_1}(\tau) + \mathcal{R}_{g_2}(\tau) + \mathcal{R}_{g_1 g_2}(\tau) + \mathcal{R}_{g_2 g_1}(\tau) \tag{2.120}$$

where

$$\mathcal{R}_{g_1 g_2}(\tau) = \lim_{T\to\infty} \frac{1}{T} \int_{-T/2}^{T/2} A_1 A_2 \cos(\omega_1 t + \theta_1) \cos[\omega_2(t+\tau) + \theta_2]\, dt$$

$$= \lim_{T\to\infty} \frac{A_1 A_2}{2T} \left[\int_{-T/2}^{T/2} \cos[(\omega_2 - \omega_1)t + \omega_2\tau + \theta_2 - \theta_1]\, dt \right.$$

$$\left. + \int_{-T/2}^{T/2} \cos[(\omega_2 + \omega_1)t + \omega_2\tau + \theta_1 + \theta_2]\, dt \right]$$

Each of the two integrals on the right-hand side represents an area under a sinusoid over a large time interval $(-T/2, T/2)$. This is always finite and is, at most, equal to the area under half the cycle. When this integral is divided by T $(T \to \infty)$, the result is zero. Hence, $\mathcal{R}_{g_1 g_2}(\tau)$ is zero. Similarly, $\mathcal{R}_{g_2 g_1}(\tau) = 0$. Substitution of Eq. (2.116a) in Eq. (2.120) immediately yields the desired Eq. (2.119b). The Fourier transform of Eq. (2.119b) yields Eq. (2.119c), and from Eqs. (2.114b) and (2.119b) we get Eq. (2.119d). The results in this example can be extended to a sum of any number of sinusoids. Thus, if

$$g(t) = \sum_n A_n \cos(\omega_n t + \theta_n)$$

then

$$\mathcal{R}_g(\tau) = \sum_n \frac{A_n^2}{2} \cos \omega_n \tau$$

and

$$S_g(\omega) = \frac{\pi}{2} \sum_n A_n^2 [\delta(\omega + \omega_n) + \delta(\omega - \omega_n)]$$

$$P_g = \sum_n \frac{A_n^2}{2}$$

■

Input and Output Power Spectral Densities

As in the case of the ESD, the PSDs of the input $g(t)$ and the corresponding output $r(t)$ of a linear system are related by $|H(\omega)|^2$. Thus, if $H(\omega)$ is the transfer function of a linear invariant system, then

$$S_r(\omega) = |H(\omega)|^2 S_g(\omega) \tag{2.121a}$$

and*

$$\mathcal{R}_r(\tau) = h(\tau) * h(-\tau) * \mathcal{R}_g(\tau) \tag{2.121b}$$

Equation (2.121a) can be proved along the lines used to prove Eq. (2.107) (see Prob. 2.62).

■ EXAMPLE 2.37

A power signal $g(t)$ has a PSD $S_g(\omega) = \mathcal{N}/A^2$, as shown in Fig. 2.60*a*. Determine the PSD and the mean square value of its derivative $\dot{g}(t)$.

□ **Solution:** An ideal differentiator transfer function is $j\omega$ (Fig. 2.60*b*). Hence, if $g(t)$ is applied at the input of a system with a transfer function $j\omega$, the output signal would be $\dot{g}(t)$, and

$$\begin{aligned} S_{\dot{g}}(\omega) &= |j\omega|^2 S_g(\omega) \\ &= \omega^2 S_g(\omega) \end{aligned}$$

The mean square value $\widetilde{[\dot{g}(t)]^2}$ is given by

$$\begin{aligned} \widetilde{[\dot{g}(t)]^2} &= 2\int_0^\infty \omega^2 S_g(\omega)\, df \\ &= 2\int_0^B \frac{\mathcal{N}}{A^2}(4\pi^2 f^2)\, df \\ &= \frac{8\pi^2 \mathcal{N} B^3}{3A^2} \quad \text{volt}^2 \end{aligned}$$

■

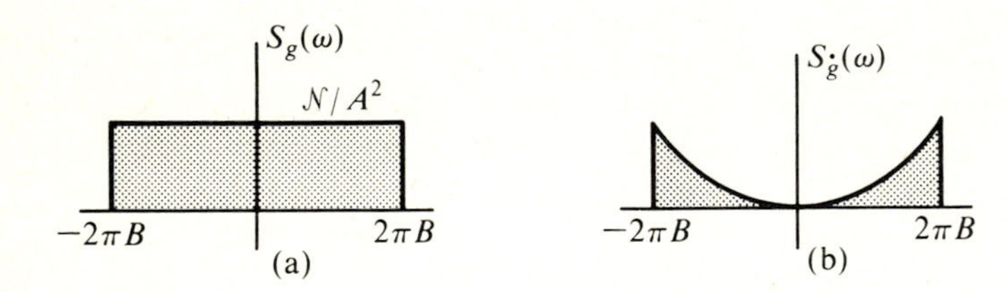

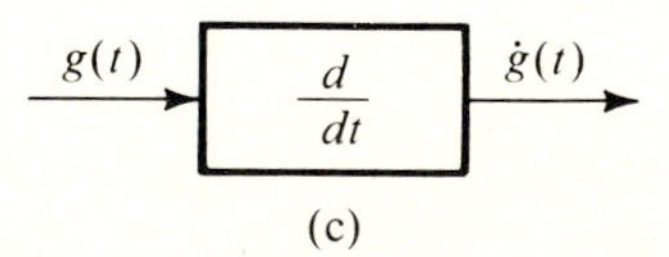

Figure 2.60 Power spectral densities at the input and the output of an ideal differentiator.

---

*If $h(t)$ is complex, Eq. (2.121b) modifies to

$$\mathcal{R}_r(\tau) = h(\tau) * h^*(\tau) * \mathcal{R}_g(\tau)$$

■ EXAMPLE 2.38

Show that if the PSD and the power of a signal $g(t)$ are $S_g(\omega)$ and P_g, respectively, then the PSD and the power of the signal $ag(t)$ are $a^2S_g(\omega)$ and a^2P_g, respectively.

□ **Solution:** When a signal $g(t)$ is passed through a linear system of transfer function $H(\omega) = a$, the output signal is $ag(t)$. Hence, from Eq. (2.121) it follows that $S_{ag}(\omega) = a^2S_g(\omega)$, and $P_{ag} = 2\int_0^\infty a^2S_g(\omega)\, df = a^2P_g$. ■

Interpretation of Power Spectral Density

The PSD has an interpretation similar to that of the energy spectral density. The PSD $S_g(\omega)$ of a signal $g(t)$ represents the relative power contributed by various frequency components, and, thus, $S_g(\omega)$ represents the power per unit bandwidth contributed by frequency components centered at ω. Thus, $2S_g(\omega_o)\,\Delta f$ is the power contribution of the frequency band Δf centered at $\omega = \omega_o$. The factor 2 is used to account for the negative as well as positive frequency components. This interpretation can be demonstrated by considering the transmission of $g(t)$ through an ideal bandpass filter of bandwidth Δf ($\Delta f \to 0$) centered at $\omega = \omega_o$, just as was done for the case of the ESD (Fig. 2.54).

2.10 THE POWER SPECTRAL DENSITY OF NOISE

Noise is a random signal, and its study requires a background in probabilities and random processes. The in-depth study of noise is presented in Chapter 5. We can, however, describe the noise PSD by measuring its power per unit bandwidth. This can be done by passing the noise signal through an ideal* narrowband filter with a variable center frequency and measuring the power, or the mean square value, of the output signal. The PSD at the filter center frequency is the filter output power divided by $2\Delta f$ where Δf is the filter bandwidth. The most commonly encountered noise in communication is the so-called *gaussian* noise discussed in Chapters 5 and 7. At this stage, rather than worrying about its statistical properties, we shall consider only its frequency-domain description, namely, its PSD. Consider a noise signal $n(t)$ (Fig. 2.61*a*) with the PSD $S_n(\omega)$ (Fig. 2.61*b*). Because a signal can be represented by a sum of sinusoids of frequencies Δf apart ($\Delta f \to 0$) over its frequency spectrum, we can express $n(t)$ as (see footnote on page 31):

$$n(t) = \sum_{\Omega} C_k \cos(\omega_k t + \theta_k) \qquad \begin{array}{l}\omega_k = k\Delta\omega \\ \Delta\omega \to 0\end{array} \tag{2.122}$$

where the summation is over Ω, the entire spectrum of $n(t)$. Amplitudes C_k's and phases θ_k's can be determined from Fourier analysis of $n(t)$. However, it is more convenient to determine C_k's from the PSD of $n(t)$. Consider the PSD of $n(t)$ over a narrowband Δf with a center frequency ω_k. As $\Delta f \to 0$, this spectrum can be approximated by two impulses centered at $\pm\omega_k$, each of strength $S_n(\omega_k)\Delta\omega = (2\pi\Delta f)S_n(\omega_k)$. The right-hand side of Eq. (2.122) has one sine component $C_k \cos(\omega_k t + \theta_k)$ over the

---

*In practice, nonideal narrowband filters can be used for this purpose (see Prob. 2.61).

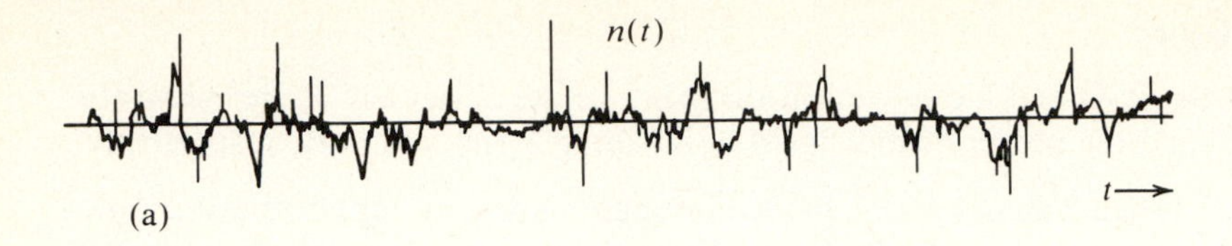

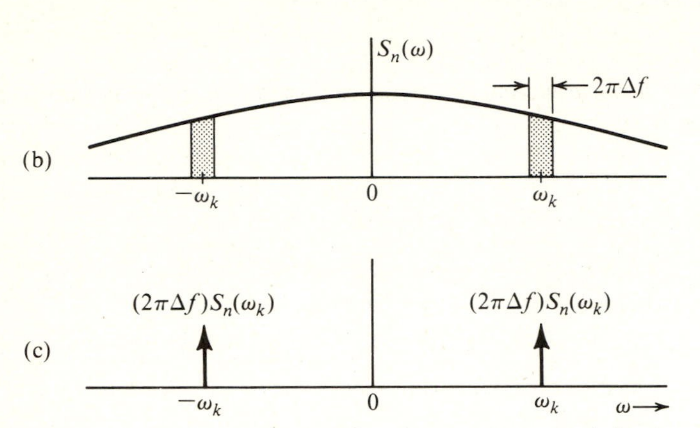

Figure 2.61 A random noise signal and its power spectral density.

narrowband, with a PSD identical to that in Fig. 2.61*c* but with impulse strengths $\pi C_k^2/2$ (see Example 2.35). Hence

$$\frac{\pi C_k^2}{2} = (2\pi \Delta f) S_n(\omega_k)$$

and

$$C_k^2 = 4 S_n(\omega_k) \Delta f \tag{2.123}$$

Thus, we can compute amplitudes C_k in Eq. (2.122) from the noise PSD $S_n(\omega)$, which in turn can be determined by passing the noise through a narrowband tuneable filter of bandwidth Δf. If the filter is tuned to ω_k, the output noise power (or mean square value) is $2S_n(\omega_k)\Delta f = C_k^2/2$.

Representing noise $n(t)$ by a sum of sinusoids as in Eq. (2.122) amounts to approximating the PSD $S_n(\omega)$ by impulses separated by $\Delta\omega$. As $\Delta\omega \to 0$, the approximation improves. Note that in this representation of noise by sinusoids, although the amplitudes C_k are determined, the phases θ_k are still undetermined. As we know, this information is not required for power computations. But it is important to remember that θ_k's are independent.* For further discussion of this sine-wave model of a noise signal in Eq. (2.122) see Rice.[7]

---

* Amplitudes C_k's and phases θ_k's are determined for a particular sample of the noise signal. Noise being a random process, every time we take a sample of $n(t)$, we shall get different values of C_k's and θ_k's. For this reason C_k's and θ_k's are called random variables implying that they are different for each noise sample. There is no need to worry about this point at this time. For more discussion, see Chapter 5.

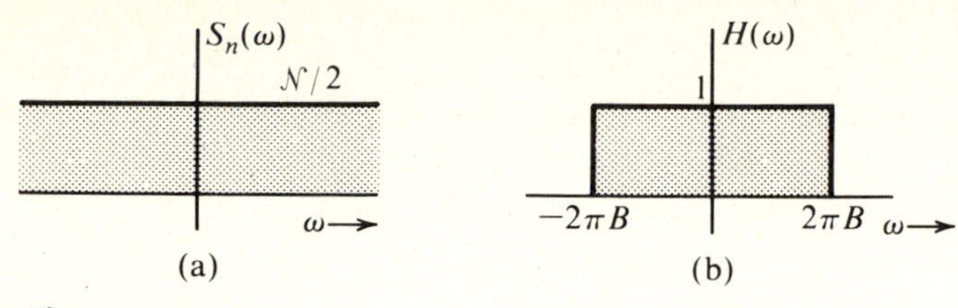

Figure 2.62

If the PSD $S_n(\omega)$ of a noise is constant with frequency, the noise is said to be *white,* a term borrowed from the concept of white light, which contains all colors (all frequency components) in the same strength. Consider a white noise with $S_n(\omega) = \mathcal{N}/2$ (Fig. 2.62*a*). It can be seen that if this noise is passed through an ideal filter of bandwidth B (Fig. 2.62*b*), the output noise power N_o is

$$N_o = 2 \int_0^B S_n(\omega)\, df = 2\left(\frac{\mathcal{N}}{2}\right)B = \mathcal{N} B \tag{2.124}$$

Hence, the white noise with $S_n(\omega) = \mathcal{N}/2$ has a power $\mathcal{N}$ per unit bandwidth.* Much of the noise encountered in practical communication systems can be modeled by a white gaussian noise, at least over the bandwidth of interest.

REFERENCES

1. B. P. Lathi, *Signals, Systems, and Communication,* Wiley, New York, 1965.
2. R. R. Goldberg, *Fourier Transforms,* Cambridge University Press, New York, 1961.
3. A. Papoulis, *The Fourier Integral and Its Application,* McGraw-Hill, New York, 1962.
4. D. A. Linden, "A Discussion of Sampling Theorem," *Proc. IRE,* vol. 47, p. 1219, July 1959.
5. A. V. Oppenheim and R. W. Schafer, *Digital Signal Processing,* Prentice-Hall, Englewood Cliffs, N.J., 1975.
6. B. P. Lathi, *Signals, Systems, and Controls,* Harper & Row, New York, 1974.
7. S. O. Rice, "Mathematical Analysis of Random Noise," *Bell Syst. Tech. J.,* vol. 23, pp. 282–332, July 1944 and vol. 24, pp. 46–156, January 1945.

PROBLEMS

2.1. If a periodic signal satisfies certain symmetry conditions, the evaluation of Fourier coefficients is somewhat simplified. Show that the following are true:

(a) If $g(t) = -g(t)$ (even symmetry), then all the sine terms in the trigonometric Fourier series vanish.

(b) If $g(t) = -g(-t)$ (odd symmetry), then all the cosine terms in the trigonometric series vanish.

(c) If $g(t) = -g(t \pm T_o/2)$ (rotation symmetry), then all even harmonics vanish.

Further, show in each case that the Fourier coefficients can be evaluated by integrating the periodic signal over the half-cycle only.

---

* Actually, $S_n(\omega) = \mathcal{N}/2$ means a power of $\mathcal{N}/2$ per unit bandwidth. This is the bandwidth of exponential components, however, which exist over negative as well as positive frequencies. Hence, the power per unit bandwidth in the sinusoidal (or conventional) sense is twice $\mathcal{N}/2$ (that is, $\mathcal{N}$).

2.2. **(a)** Show that an arbitrary function $g(t)$ can always be expressed as the sum of an even function $g_e(t)$ and an odd function $g_o(t)$.

$$g(t) = g_e(t) + g_o(t)$$

(b) Determine the odd and even components of functions **(i)** $u(t)$, **(ii)** $e^{-at}u(t)$, and **(iii)** e^{jt}.

Hint: $g(t) = \frac{1}{2}[g(t) + g(-t)] + \frac{1}{2}[g(t) - g(-t)]$.

2.3. For each of the periodic signals shown in Fig. P2.3, find the compact trigonometric Fourier series and sketch the magnitude and phase spectra.

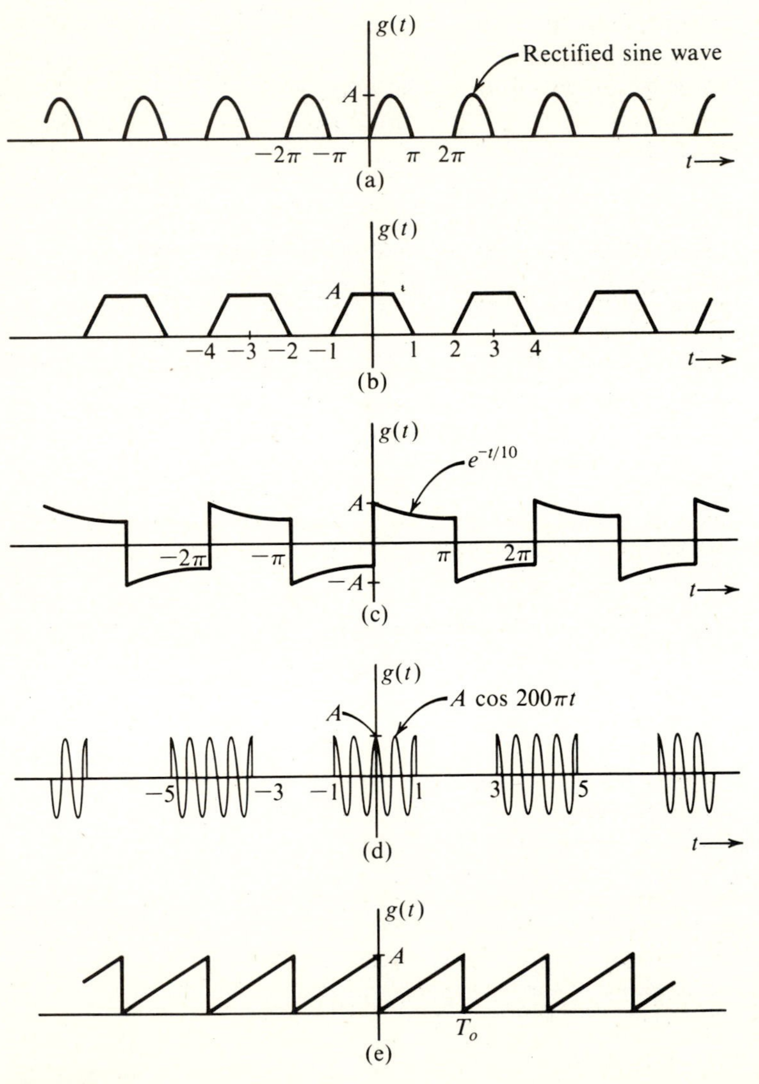

Figure P2.3

2.4. Show that the coefficients of the exponential Fourier series of an even periodic signal are real and those of an odd periodic signal are imaginary.

2.5. For each of the periodic signals in Fig. P2.3, find the exponential Fourier series and sketch the magnitude and phase spectra.

2.6. Show that the Fourier transform of $g(t)$ may also be expressed as

$$G(\omega) = \int_{-\infty}^{\infty} g(t) \cos \omega t \, dt - j \int_{-\infty}^{\infty} g(t) \sin \omega t \, dt$$

Hence, show that if $g(t)$ is an even function of t, then

$$G(\omega) = 2 \int_{0}^{\infty} g(t) \cos \omega t \, dt$$

and if $g(t)$ is an odd function of t, then

$$G(\omega) = -2j \int_{0}^{\infty} g(t) \sin \omega t \, dt$$

Hence, prove that if $g(t)$ is:	then $G(\omega)$ is:
a real and even function of t	a real and even function of ω
a real and odd function of t	an imaginary and odd function of ω
an imaginary and even function of t	an imaginary and even function of ω
a complex and even function of t	a complex and even function of ω
a complex and odd function of t	a complex and odd function of ω

2.7. Show that for real $g(t)$, the inverse transform, Eq. (2.30a), can be expressed as

$$g(t) = \frac{1}{\pi} \int_{0}^{\infty} |G(\omega)| \cos [\omega t + \theta_g(\omega)] \, d\omega$$

This is the trigonometric form of the (inverse) Fourier transform. Compare this with the compact trigonometric Fourier series.

2.8. A signal $g(t)$ can be expressed as the sum of even and odd components (see Prob. 2.2)

$$g(t) = g_e(t) + g_o(t)$$

(a) If $g(t) \leftrightarrow G(\omega)$, show that for real $g(t)$

$$g_e(t) \leftrightarrow \text{Re}\,[G(\omega)] \qquad g_o(t) \leftrightarrow j\,\text{Im}\,[G(\omega)]$$

(b) Verify these results by finding Fourier transforms of the even and odd components of the following signals: **(i)** $e^{j\omega_0 t}$, **(ii)** $u(t)$, and **(iii)** e^{-at}.

2.9. If $g(t)$ is complex with $g_r(t)$ and $g_i(t)$ as its real and imaginary components, that is,

$$g(t) = g_r(t) + jg_i(t)$$

(a) Show that

$$g_r(t) \leftrightarrow \frac{1}{2}[G(\omega) + G^*(-\omega)] \qquad g_i(t) \leftrightarrow \frac{1}{2j}[G(\omega) - G^*(-\omega)]$$

Hint: $g_r(t) - jg_i(t) = g^*(t) \leftrightarrow G^*(-\omega)$.

(b) Verify the results in part (a) for the signal $g(t) = e^{j\omega_0 t}$.

2.10. Find the Fourier transform of functions $g(t)$ shown in Fig. P2.10.

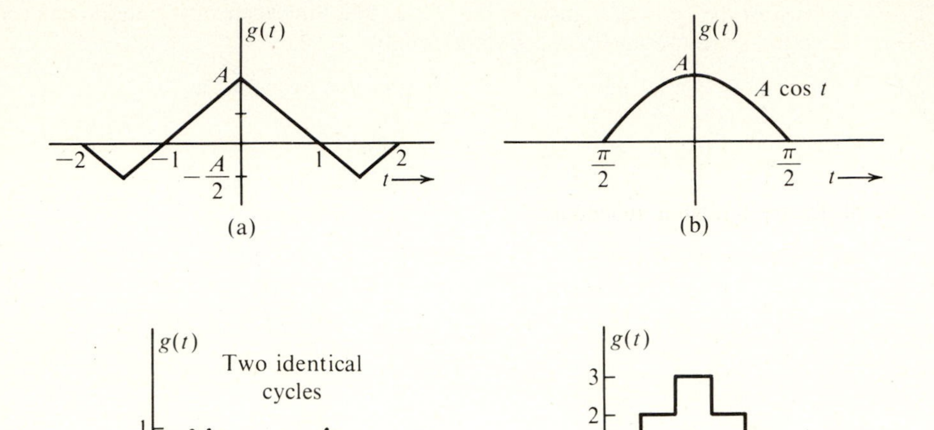

Figure P2.10

2.11. Determine signals $g(t)$ whose Fourier transforms are shown in Fig. P2.11.

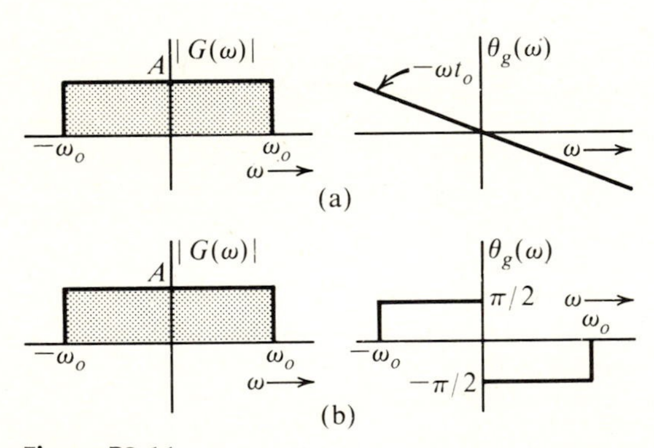

Figure P2.11

2.12. Using the sampling property of the impulse function, evaluate the following integrals:

(a) $\int_{-\infty}^{\infty} \delta(t-2) \sin \pi t \, dt$

(b) $\int_{-\infty}^{\infty} f(t_1 - t)\, \delta(t - t_2) \, dt$

(c) $\int_{-\infty}^{\infty} \delta(t+3)e^{-t} \, dt$

(d) $\int_{-\infty}^{\infty} e^{j\pi t}\, \delta(t+2) \, dt$

(e) $\int_{-\infty}^{\infty} \delta(1-t)(t^3+4) \, dt$

2.13. Show that

$$\delta(at) = \frac{1}{a}\,\delta(t)$$

Hence, show that

$$\delta(\omega) = \frac{1}{2\pi}\,\delta(f) \qquad \omega = 2\pi f$$

Hint: Show that $\int_{-\infty}^{\infty} \delta(at)\,dt = 1/a$.

2.14. Sketch the following functions:

(a) $\Pi\,(2t)$ **(b)** $\Pi\left(\frac{\omega}{100}\right)$ **(c)** $\Pi\left(\frac{t-10}{8}\right)$ **(d)** $10\,\Pi\,(2t-10)$

(e) $\Pi\left(\frac{t-10}{8}\right) + 2\,\Pi\left(\frac{t-12}{4}\right)$

2.15. Using the symmetry property, show that

(a) $\dfrac{2a}{t^2+a^2} \leftrightarrow 2\pi e^{-a|\omega|}$

(b) $\dfrac{1}{2}\left[\delta(t) - \dfrac{1}{j\pi t}\right] \leftrightarrow u(\omega)$

2.16. Using the scaling property, determine the Fourier transforms of pulses in Fig. P2.16*b*, *c* and *d* in terms of $G(\omega)$, where $G(\omega)$ is the Fourier transform of the pulse $g(t)$ in Fig. P2.16*a*.

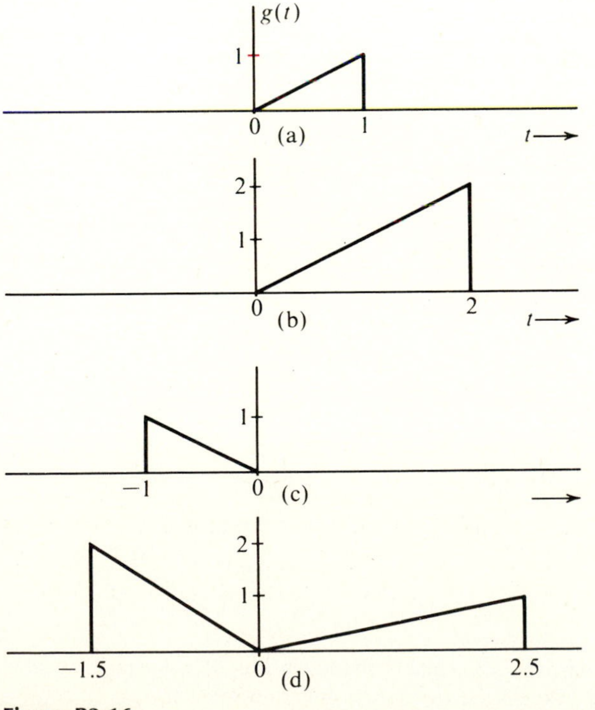

Figure P2.16

2.17. Given $\cos(\omega_o t) \leftrightarrow \pi[\delta(\omega + \omega_o) + \delta(\omega - \omega_o)]$, find the Fourier transform of $\cos n\omega_o t$ using the scaling property. *Hint:* See Prob. 2.13.

2.18. Using the time-shifting (and scaling) property, determine the Fourier transform of the signal shown in Fig. P2.18 in terms of $G(\omega)$, where $G(\omega)$ is the Fourier transform of $g(t)$ in Fig. P2.16*a*. You may also use Table 2.1 if needed.

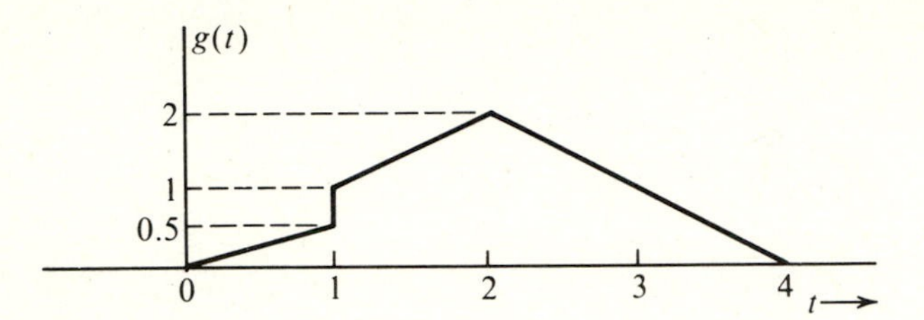

Figure P2.18

2.19. Using the time-shifting property and Table 2.1 (pair 13), determine the Fourier transform of the signal shown in Fig. P2.19.

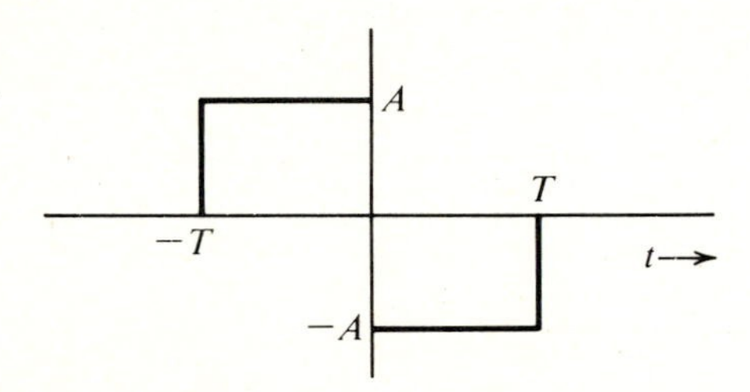

Figure P2.19

2.20. Using time-shifting property and Table 2.1 (pairs 13 and 14), determine and sketch the Fourier transforms of the signals shown in Fig. P2.20.

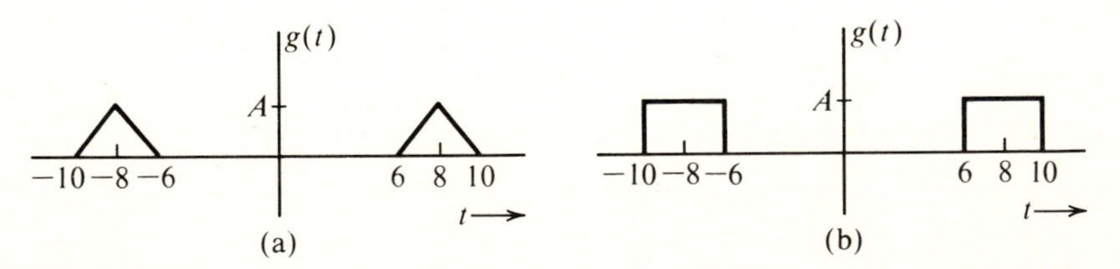

Figure P2.20

2.21. Find the Fourier transforms of the signals shown in Fig. P2.21 using the modulation property and Table 2.1. Sketch the frequency spectrum in each case.

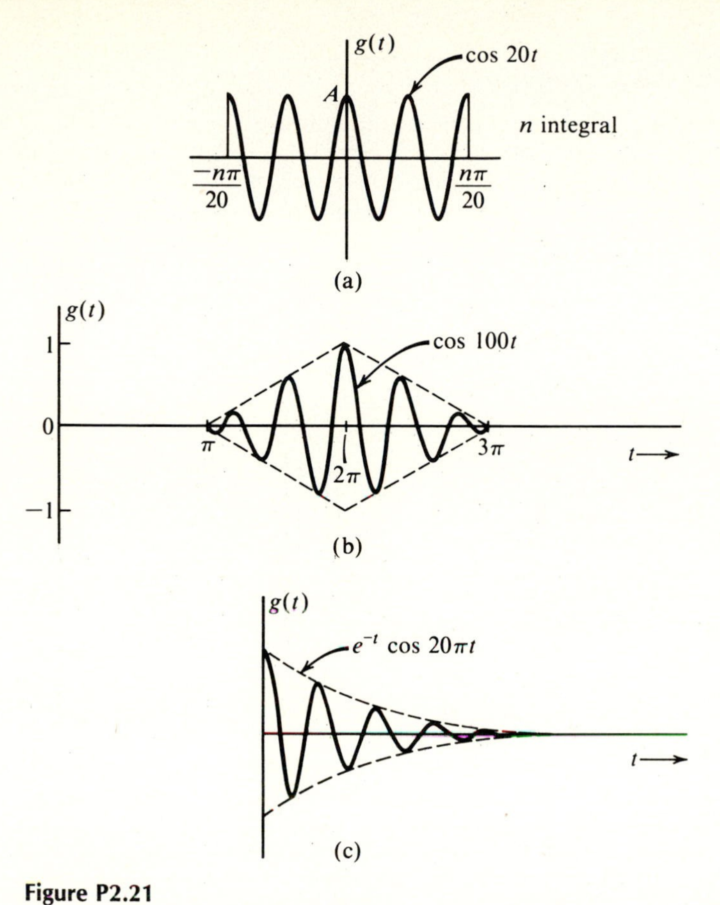

Figure P2.21

2.22. Find signals $g(t)$ whose Fourier transforms are shown in Fig. P2.22. *Hint:* Use the modulation property.

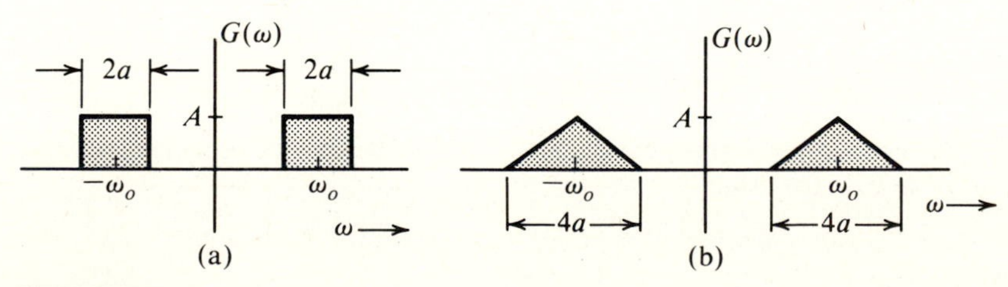

Figure P2.22

2.23. Using the time-differentiation (and time-shifting) property, find the Fourier transforms of signals shown in Fig. P2.23.

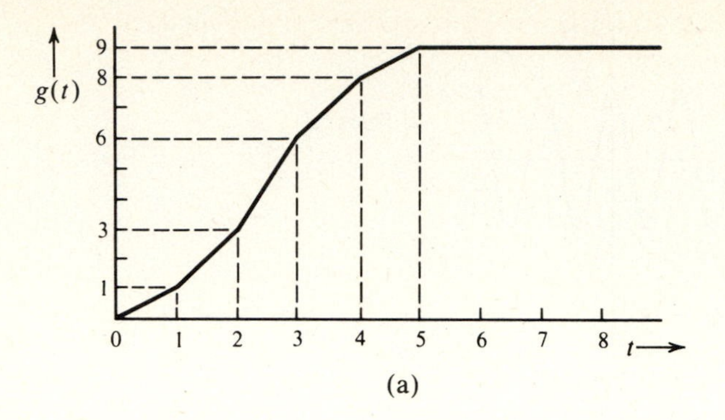

(a)

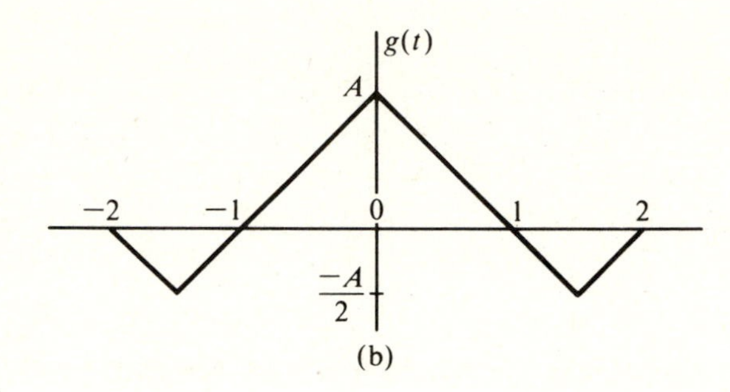

(b)

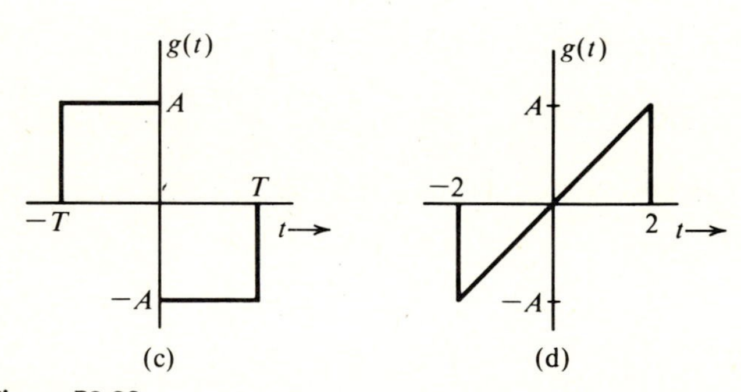

(c) (d)

Figure P2.23

2.24. Find the Fourier transform of the signal in Fig. P2.19 using the time-integration property and Table 2.1 (pair 14).

2.25. If $g(t) \leftrightarrow G(\omega)$, determine the Fourier transforms of the following:

(a) $tg(2t)$

(b) $(t - 2)g(t)$

(c) $(t - 2)g(-2t)$

(d) $t(dg/dt)$

(e) $g(1 - t)$

(f) $(1 - t)g(1 - t)$

2.26. Find the Fourier transform of a gaussian pulse

$$g(t) = \frac{1}{\sigma\sqrt{2\pi}} e^{-t^2/2\sigma^2}$$

by using the time-differentiation and frequency-differentiation properties. *Hint:* $j\omega G(\omega) = (-j/\sigma^2)(dG/d\omega)$. This gives $dG/G = -\sigma^2 \omega d\omega$. Now integrate.

2.27. The nth moment m_n of a function $g(t)$ is defined by

$$m_n = \int_{-\infty}^{\infty} t^n g(t)\, dt$$

(a) Using the frequency-differentiation property, show that

$$m_n = (j)^n \frac{d^n G(\omega)}{d\omega^n}\bigg|_{\omega = 0}$$

(b) Using this result, show that Taylor's series expansion of $G(\omega)$ can be expressed as

$$G(\omega) = m_0 - jm_1\omega - \frac{m_2\omega^2}{2!} + \frac{jm_3\omega^3}{3!} + \frac{m_4\omega^4}{4!} + \cdots$$

$$= \sum_{n=0}^{\infty} (-j)^n m_n \frac{\omega^n}{n!}$$

(c) Determine the various moments of a gate function $\Pi(t)$ and, using the above equation, find its Fourier transform.

2.28. Evaluate the following convolution integrals:

(a) $u(t) * e^{-t}u(t)$

(b) $e^{-t}u(t) * e^{-2t}u(t)$

(c) $\Pi\left(\frac{t}{T}\right) * \Pi\left(\frac{t}{T}\right)$

(d) $u(t) * u(t)$

(e) $u(t) * tu(t)$

(f) $e^{-t}u(t) * tu(t)$

Verify your results for parts (a) through (c) by using the convolution property.

2.29. Derive the time-integration property [Eq. (2.65b)] using convolution: *Hint*: Show that

$$\int_{-\infty}^{t} g(t)\, dt = g(t) * u(t)$$

2.30. Find $g_1(t) * g_2(t)$ and $g_2(t) * g_1(t)$ for the functions shown in Fig. P2.30.

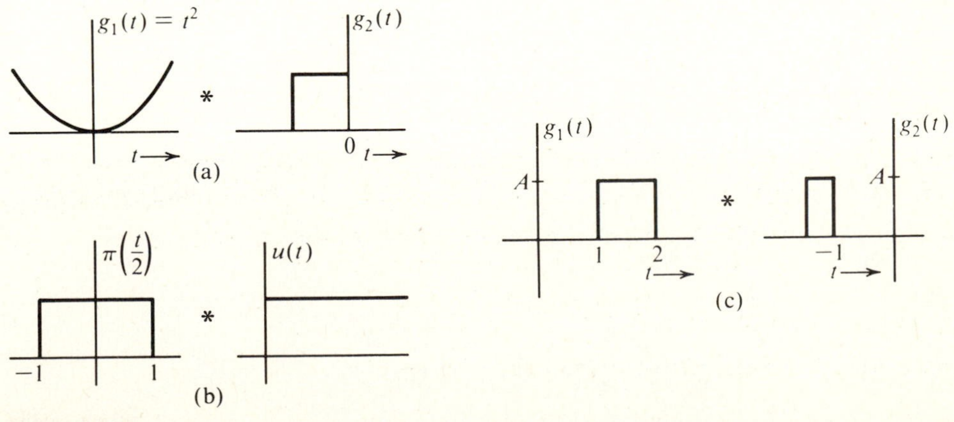

Figure P2.30

2.31. If $g_i(t) \leftrightarrow G_i(\omega)$, $i = 1, 2, 3$.

(a) Determine the Fourier transforms of the following:

(i) $[g_1(t) * g_2(t)]g_3(t)$

(ii) $[g_1(t)g_2(t)] * g_3(t)$

(b) Use the results in part (a) to determine and sketch the spectra of the signals shown in Fig. P2.31.

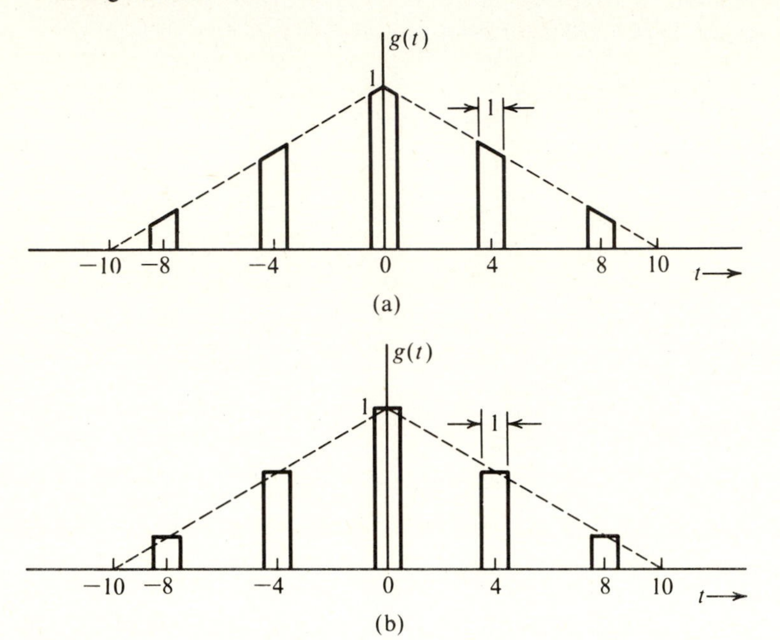

Figure P2.31

2.32. Nyquist samples of a signal $g(t)$ bandlimited to B are as follows:

$$g(0) = 1$$

$$g\left(\pm\frac{n}{2B}\right) = 0 \qquad n = 1, 2, 3, \ldots$$

Show that $g(t) = \text{sinc}\,(2Bt)$. *Hint:* Use Eq. (2.84).

2.33. A signal $g(t)$ is bandlimited to $B/2$ Hz. Nyquist samples are taken at $t = \pm n/2B$ $(n = 1, 3, 5, \ldots)$. It is found that

$$g\left(\pm\frac{1}{2B}\right) = B/2$$

$$g\left(\pm\frac{n}{2B}\right) = 0 \qquad n = 3, 5, 7, \ldots$$

Show that

$$g(t) = \frac{2B\cos \pi Bt}{\pi(1 - 4B^2t^2)} \qquad \text{and} \qquad G(\omega) = \cos\left(\frac{\omega}{2B}\right)\Pi\left(\frac{\omega}{2\pi B}\right)$$

Hint: Use Eq. (2.84).

2.34. Determine the Nyquist sampling rate and the Nyquist sampling interval for the following signals:
(a) sinc $(100t)$
(b) $\text{sinc}^2\,(100t)$
(c) sinc $(100t)$ + sinc $(50t)$
(d) sinc $(100t)$ + $\text{sinc}^2\,(60t)$

2.35. Prove the dual of the sampling theorem. Specifically, prove that if $g(t) = 0$ for $|t| > T$ (i.e., if $g(t)$ is time limited to $2T$), then $G(\omega)$ is completely specified by its samples taken uniformly at an interval not less than $1/2T$ Hz apart. Hence, show that

$$G(\omega) = \sum_{n=-\infty}^{\infty} G\left(\frac{n\pi}{T}\right) \text{sinc}\left(\frac{\omega T}{\pi} - n\right)$$

Also show that

$$\sum_{k=-\infty}^{\infty} g(t - kT) \leftrightarrow \omega_o G(\omega) \sum_{n=-\infty}^{\infty} \delta(\omega - n\omega_o) \qquad \omega_o = \frac{\pi}{T}$$

Hint: Interchange the roles of t and ω in the sampling-theorem proof.

2.36. Signal $g(t) = e^{-t}u(t)$ is not bandlimited. It is passed through an ideal filter of bandwidth B in order to bandlimit it. The output signal $\hat{g}(t)$ of this filter is an approximation of $g(t)$.
(a) Sketch spectra $|G(\omega)|$ and $|\hat{G}(\omega)|$.
(b) If the error signal $g_e(t) = g(t) - \hat{g}(t)$, sketch the spectrum $|G_e(\omega)|$.
(c) Find the energy of $g_e(t)$ in terms of B.
(d) Find the value of B that will maintain the energy of $g_e(t)$ at less than 1 percent of the energy of $g(t)$.
(e) Using the value of B found in part (d) for the filter, find the Nyquist sampling rate of $\hat{g}(t)$ and sketch the spectrum of the sampled signal.
(f) If the signal $g(t)$ is sampled directly (without passing through a bandlimiting filter) at the rate in part (e), sketch the spectrum of the sampled signal.
(g) Explain the advantage in bandlimiting the signal before sampling.

2.37. Repeat Prob. 2.36 with $g(t) = e^{-a|t|}$.

2.38. A zero-order hold circuit (Fig. P2.38) is often used to reconstruct a signal $g(t)$ from its samples.
(a) Find the unit impulse response of this circuit.
(b) Find the transfer function $H(\omega)$ and sketch $|H(\omega)|$ and $\theta_h(\omega)$.
(c) Show that when a sampled signal $g_s(t)$ is applied at the input of this circuit, the output is a staircase approximation of $g(t)$.

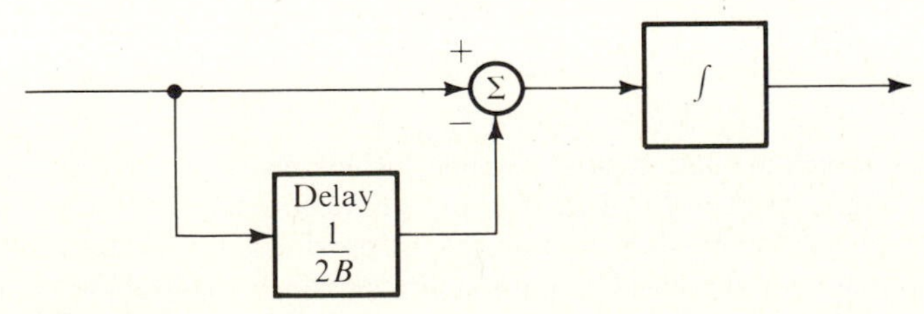

Figure P2.38

2.39. Prove that if a signal $g(t)$ bandlimited to B Hz is sampled by multiplying $g(t)$ by a pulse train $\sum_{n=-\infty}^{\infty} q(t - nT_s)$, where $q(t)$ is an arbitrary pulse not necessarily time limited, we can still recover $g(t)$ from the sampled signal $g_s(t) = g(t) \sum_{n=-\infty}^{\infty} q(t - nT_s)$, provided $T_s \leq 1/2B$. How would the sampled signal $g_s(t)$ look when $Q(\omega)$ is bandlimited to less than $2B$?

2.40. A signal $g(t)$ bandlimited to B is sampled at a rate at least equal to $2B$ samples per second. The kth sample $g(kT_s)$ is transmitted by a pulse $g(kT_s)q(t - kT_s)$, where $q(t)$ is an arbitrary pulse. The transmitted signal in this case is $\sum_k g(kT_s)q(t - kT_s)$.
(a) Show how $g(t)$ can be recovered from this signal.
(b) If $q(t) = \Pi(t/T)$, sketch the spectrum of $\sum_k g(kT_s)q(t - kT_s)$ $\quad (T < T_s)$.

2.41. Show that a minimum bandwidth needed for transmission of N time-division-multiplexed baseband signals by PAM is NB Hz, where B is the bandwidth of each of the N baseband signals.

2.42. Determine the maximum bandwidth of signals that can be transmitted through the lowpass RC filter shown in Fig. P2.42 if, over this bandwidth, the magnitude (or gain) variation is to be within 10 percent and the phase variation is to be within 7 percent of the ideal characteristics.

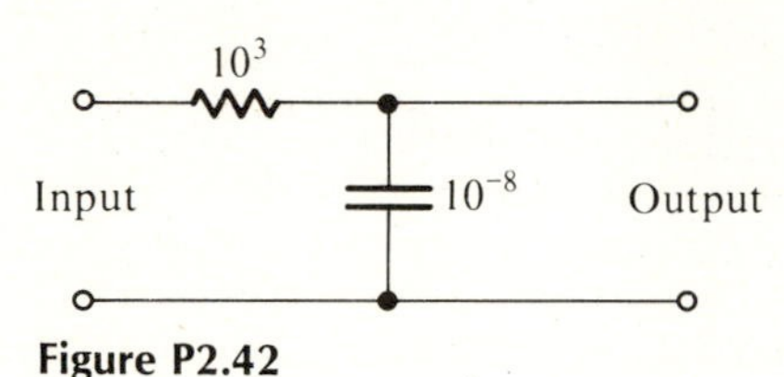

Figure P2.42

2.43. A modulated signal $g(t) \cos \omega_o t$, where $g(t) = a \cos \omega_1 t$ with $\omega_1 = 100$ and $\omega_o = 10^6$, is transmitted through the RC filter in Fig. P2.42.
(a) Is this a distortionless (or almost distortionless) transmission? Explain the reasons.
(b) What is the magnitude loss suffered by the signal?
(c) What is the envelope delay?
(d) What is the phase delay?
(e) What is the output signal?

2.44. For the RC filter in Fig. P2.42, sketch the envelope delay as a function of ω. If a narrowband signal centered at $\omega = 10^5$ is to be transmitted through this filter with the requirement that the envelope delay vary within 1 percent and the magnitude (or gain) also vary within 1 percent, what is the maximum bandwidth of the signal? What is the corresponding gain and the envelope delay?

2.45. Consider a gaussian filter

$$H(\omega) = e^{-(k\omega^2 + j\omega t_o)}$$

(a) Determine and sketch the unit impulse response of this filter.
(b) State, with reasons, whether this filter is physically realizable.
(c) Can this filter be made approximately realizable by providing a sufficient amount of delay t_o? Use your own (reasonable) criterion of approximate realizability to determine t_o.

2.46. Repeat Prob. 2.45 with

$$H(\omega) = \frac{2 \times 10^5}{\omega^2 + 10^{10}} e^{-j\omega t_o}$$

2.47. A certain channel has ideal magnitude but nonideal phase characteristics (Fig. P2.47), given by

$$|H(\omega)| = 1$$

$$\theta_h(\omega) = -\omega t_o - k \sin \omega T \qquad k \ll 1$$

(a) Show that $r(t)$, the channel response to an input pulse $g(t)$ bandlimited to B, is:

$$r(t) = g(t - t_o) + \frac{k}{2}[g(t - t_o - T) - g(t - t_o + T)]$$

Hint: Use $e^{-jk \sin \omega T} \simeq 1 - jk \sin \omega T \qquad k \ll 1$.

(b) Discuss how this channel will affect TDM and FDM signals from the point of view of interference among multiplexed signals.

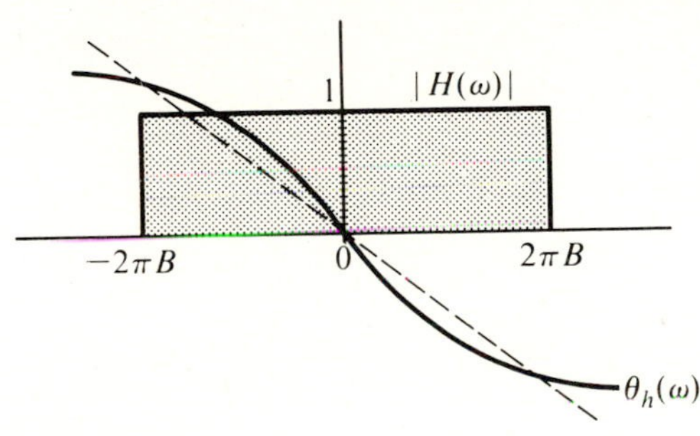

Figure P2.47

2.48. The input x and the output y of a certain nonlinear channel are related as

$$y = x + 0.22x^3$$

The input signal $x(t)$ is a sum of two modulated signals:

$$x(t) = x_1(t) \cos \omega_1 t + x_2(t) \cos \omega_2 t$$

where the spectra $X_1(\omega)$ and $X_2(\omega)$ are shown in Fig. P2.48, and ω_1 and ω_2 are $2\pi(100 \times 10^3)$ and $2\pi(110 \times 10^3)$, respectively.

(a) Sketch the spectra of the input signal $x(t)$ and the output signal $y(t)$.

(b) Can signals $x_1(t)$ and $x_2(t)$ be recovered (without distortion and interference) from the output $y(t)$?

(c) If a TDM signal consisting of two interleaved pulse trains is applied at the input, can the two trains be recovered at the output (**i**) without distortion and (**ii**) without interference?

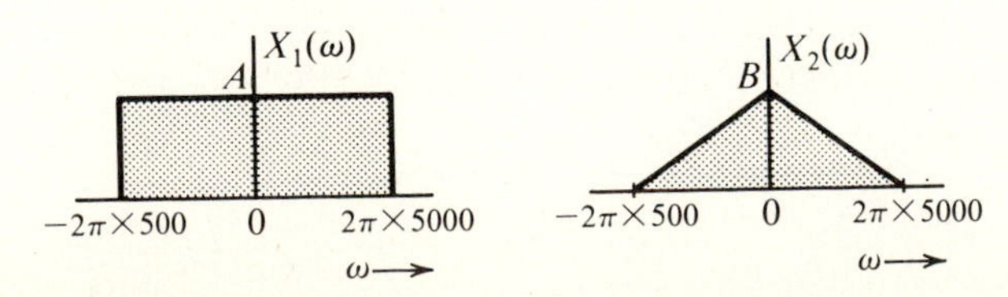

Figure P2.48

2.49. Show that $E_{g_1+g_2} = E_{g_1} + E_{g_2}$ if $g_1(t)$ and $g_2(t)$ are orthogonal signals, that is,

$$\int_{-\infty}^{\infty} g_1(t)g_2^*(t)\,dt = 0$$

Hint: See footnote on page 61 for energy of complex signals.

2.50. Show that $E_{g_1+g_2} = E_{g_1} + E_{g_2}$ if $g_1(t)$ and $g_2(t)$ have orthogonal spectra, that is,

$$\int_{-\infty}^{\infty} G_1(\omega)G_2^*(\omega)\,d\omega = 0$$

Note that nonoverlapping spectra are a special case of this. Signals with orthogonal spectra are also orthogonal.

2.51. A lowpass signal $g(t)$ is applied to a squaring device. The squarer output $g^2(t)$ is applied to a lowpass filter of bandwidth Δf (Fig. P2.51). Show that if Δf is very small, the filter output is a dc signal of amplitude $2E_g\Delta f$, where E_g is the energy of $g(t)$. *Hint:* If $g^2(t) \leftrightarrow A(\omega)$, then $A(0) = E_g$.

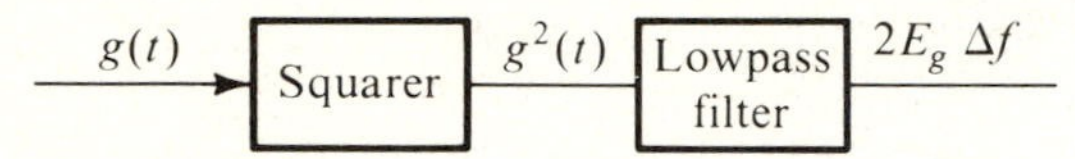

Figure P2.51

2.52. Show that for a dc signal $g(t) = A$

$$R_g(t) = A^2,\ S_g(\omega) = 2\pi A^2\,\delta(\omega),\ \text{and}\ P_g = A^2$$

2.53. Show that

$$\widetilde{[A + g(t)]^2} = A^2 + \widetilde{g^2(t)}$$

provided $g(t)$ is a zero-mean signal, that is, $\widetilde{g(t)} = 0$.

2.54. For $g(t) = A_1 \cos(\omega_1 t + \theta_1) + A_2 \cos(\omega_2 t + \theta)$, show that $\widetilde{g^2(t)} = (A_1^2 + A_2^2)/2$ by directly using Eq. (2.109b).

2.55. Find the power (mean square values) of the following signals and sketch their PSDs.

(a) $A \cos(2000\pi t) + B \sin(200\pi t)$
(b) $[A + \sin(200\pi t)] \cos(2000\pi t)$
(c) $A \cos(200\pi t) \cos(2000\pi t)$
(d) $A \sin(200\pi t) \cos(2000\pi t)$
(e) $A \sin(300\pi t) \cos(2000\pi t)$

2.56. Show that the mean square value (or the power) of a signal $A \cos(\omega_o t + \theta_1) + A \cos(\omega_o t + \theta_2)$ is $A^2 + A^2 \cos(\theta_1 - \theta_2)$.

2.57. Estimate the bandwidth B of the periodic signal $g(t)$ shown in Fig. P2.57 if the power of all components of $g(t)$ within the band B is to be at least 99.9 percent of the total power of $g(t)$.

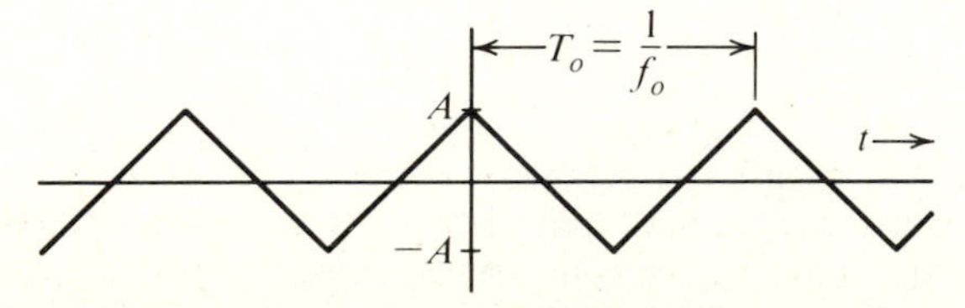

Figure P2.57

2.58. A random binary signal $x(t)$ is shown in Fig. P2.58. A binary **1** is transmitted by a pulse $p(t)$, which has an amplitude A and a width $T_o/2$, and a **0** is transmitted by the pulse $-p(t)$. **1**'s and **0**'s occur randomly, and the occurrence of **1** and **0** is equally likely. This means on the average half of the pulses are positive and half of the pulses are negative, distributed randomly. Determine $\mathcal{R}_x(\tau)$ and the PSD $S_x(\omega)$.

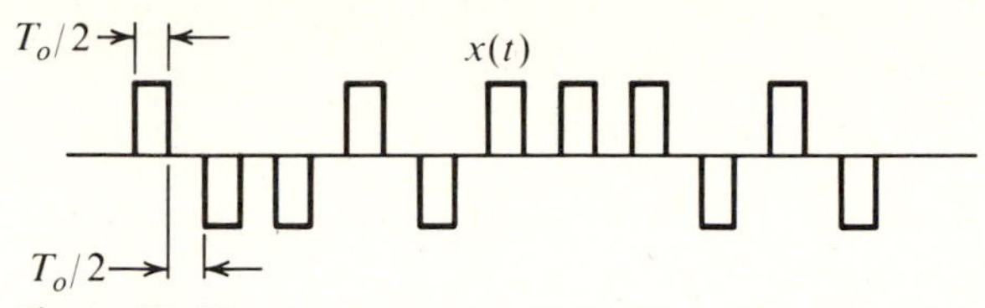

Figure P2.58

2.59. A periodic signal $g(t)$, shown in Fig. P2.59a, is transmitted through a system with transfer function $H(\omega)$ (Fig. P2.59b). For three different values of T_o ($T_o = 2\pi/3$, $\pi/3$, and $\pi/6$) find the PSD and the power (mean square value) of the output signal. Calculate the power of the input signal $g(t)$.

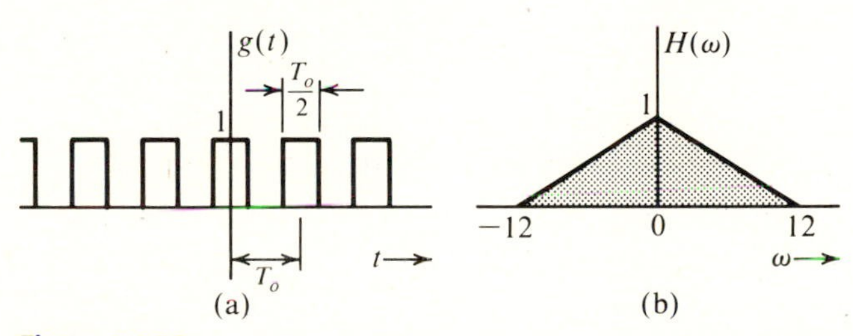

Figure P2.59

2.60. Find the mean square value of the output voltage $v_o(t)$ of the RC network shown in Fig. P2.60 if the input voltage has a PSD $S_i(\omega)$ given by:

(a) $S_i(\omega) = K$

(b) $S_i(\omega) = \Pi\,(\omega/2)$

(c) $S_i(\omega) = [\delta(\omega + 1) + \delta(\omega - 1)]$

In each case, calculate the power (the mean square value) of the input signal.

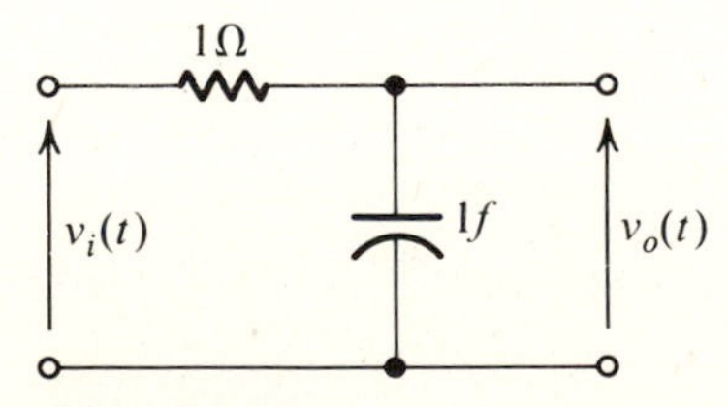

Figure P2.60

2.61. The PSD of a signal can be determined by using practical filters. A signal $g(t)$ is passed through a narrowband filter of variable center frequency ω_o. If the transfer function of this filter is $H(\omega)$ (Fig. P2.61), show that the output signal power is equal to the output signal power of an ideal filter of bandwidth B_{eq} (known as *equivalent bandwidth*) where

$$2\pi B_{eq} = \frac{1}{|H(\omega_o)|^2} \int_0^\infty |H(\omega)|^2 \, d\omega$$

Show how this result can be used to determine the PSD of a signal.

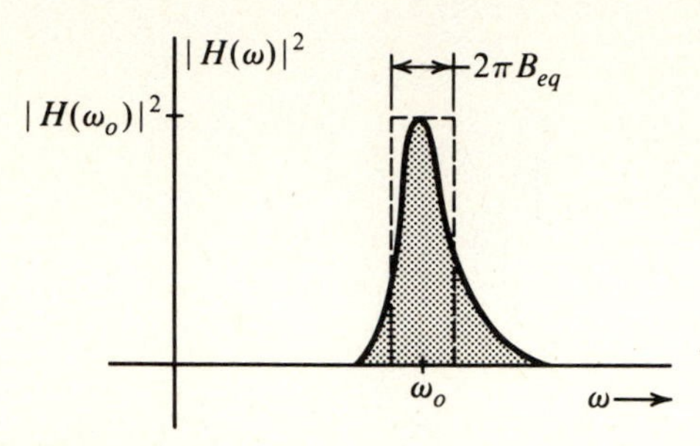

Figure P2.61

2.62. Prove Eq. (2.121a and b). *Hint:* Substitute

$$r(t) = \int_{-\infty}^{\infty} h(x) g(t - x) \, dx$$

in

$$\mathcal{R}_r(\tau) = \lim_{T \to \infty} \frac{1}{T} \int_{-T/2}^{T/2} r(t) r(t + \tau) \, dt$$

3

Digital Communication Systems

Although a significant portion of communication today is in analog form, it is being replaced rapidly by digital communication. In a period of a decade or two, most of the communication will be digital, with analog communication playing a minor role. Figure 3.1 shows the U.S. and Canadian telephone industry's terrestrial, microwave, satellite, cable, and fiber-optics transmission-system investment trend for medium-length distances[1] (500 to 2000 km).

Digital communication has several advantages over analog communication:

1. Digital communication is rugged in the sense that it is more immune to channel noise and channel distortion.
2. Regenerative repeaters along the transmission path can detect a digital signal and retransmit a new, clean (noise-free) signal. These repeaters prevent accumulation of noise along the path. This is not possible in analog communication.
3. Digital hardware implementation is flexible and permits the use of microprocessors, miniprocessors, digital switching, and large-scale integrated circuits.

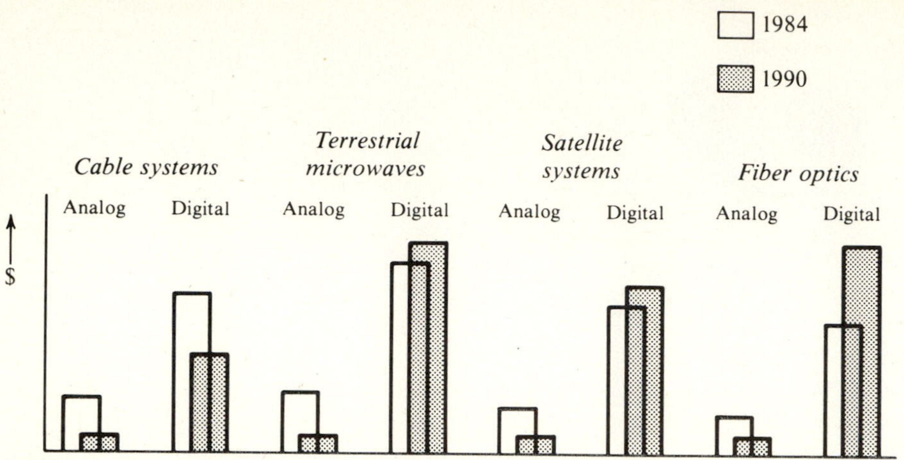

Figure 3.1 U. S. and Canadian telephone industry's medium-length (500–2000 km) wideband transmission-system investment projections for 1984 and 1990. (From K. Feher, *Digital Communications,* Prentice-Hall, Inc. 1981, by permission.)

4. Digital signals can be coded to yield extremely low error rates and high fidelity as well as privacy.
5. It is easier and more efficient to multiplex several digital signals.
6. Digital communication is inherently more efficient than analog in realizing the exchange of SNR for bandwidth.

The input to a digital system is in the form of a sequence of digits. The input could be the output from a data set, or a computer, or a digitized voice signal (PCM), digital facsimile or TV, or telemetry data, etc. Most of the discussion in this chapter is restricted to the binary case, that is, communication schemes using only two symbols, **0** and **1**. A more general case of M-ary communication which uses M symbols is briefly discussed in Sec. 3.7. In the rest of the chapter a binary digit is abbreviated as a *bit* for convenience.

Signals from several digital sources may be combined by a digital multiplexer using the process of interleaving. The output of the multiplexer is coded into electrical pulses for the purpose of transmission over a digital line. There are several possible ways of doing this. Conceptually the simplest *transmission* or *line code* is the *on-off* where **1** is transmitted by a pulse $p(t)$ and **0** is transmitted by no pulse. Another commonly used line code is the *polar* where **1** is transmitted by a pulse $p(t)$ and **0** is transmitted by the pulse $-p(t)$. This is the most efficient code, for a given transmitted power, because it is the most immune to noise. Another line code that is used in PCM is the *bipolar* or the *pseudoternary,* where **0** is transmitted by no pulse and **1** is transmitted by a pulse $p(t)$ or $-p(t)$ depending on whether the previous **1** was transmitted by $-p(t)$ or $p(t)$. In short, pulses representing consecutive **1**'s alternate as shown in Fig. 3.2. This code has the advantage that a single error in detection of pulses violates the bipolar rule of alternating pulses, and the error is readily detected (although, not corrected). A variety of line codes, each with their advantages and disadvantages, is discussed in Sec. 3.2.

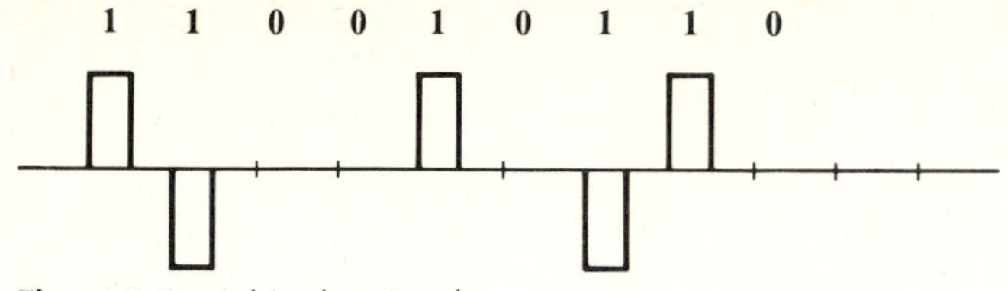

Figure 3.2 A bipolar signal.

Regenerative repeaters are used at regularly spaced intervals along a digital transmission line to detect the incoming digital signal and generate new clean pulses for further transmission along the line. This process periodically eliminates, and thereby combats accumulation of noise and signal distortion. To detect pulses, the timing information required to sample each pulse at the proper instant must be available at each repeater. This information can be extracted from the received signal itself if the line code is chosen properly. The on-off signal, for example, contains a periodic component of the desired frequency f_o as shown in Fig. 3.29. When this signal is applied to the input of a resonant circuit tuned to frequency f_o, the circuit output, which is a sinusoid of frequency f_o, can be used for timing. This timing signal (the resonant circuit output) is sensitive to the incoming bit pattern. If there are too many **0**'s in a sequence (no pulses), the sinusoidal output of the resonant circuit starts decaying thus causing error in the timing information. This problem can be overcome by using a proper line code or by using scramblers (Sec. 3.4).

3.1 DIGITAL MULTIPLEXING

Several low-bit-rate signals can be multiplexed, or combined, to form one high-bit-rate signal to be transmitted over a high-frequency medium. Because the medium is time shared by various incoming signals, this is a case of TDM. The signals from various incoming channels, or tributaries, may be of such diverse nature as a digitized voice signal (PCM), a computer output, telemetry data, a digital facsimile, and so on. The bit rate of various tributaries need not be the same.

To begin with, consider the case of all tributaries with identical bit rates. Multiplexing can be done on a bit-by-bit basis (known as bit or *digit interleaving*) as shown in Fig. 3.3*a*, or on a word-by-word basis (known as byte or *word interleaving*). Figure 3.3*b* shows the interleaving of words, or bytes, formed by four bits. The T-1 carrier, discussed in Sec. 3.9, uses eight-bit word interleaving. When the bit rate of incoming channels is not identical, the high-bit-rate channel is allocated proportionately more slots. Figure 3.3*c* shows four-channel multiplexing consisting of three channels of identical bit rate B and one channel with a bit rate of $3B$. Similar results can be attained by combining words of different lengths. It is evident that the minimum length of the multiplex frame must be a multiple of the lowest common multiple of the incoming channel bit rates, and, hence, this type of scheme is practical only when some fairly simple relationship exists among these rates. The case of completely asynchronous channels will be discussed later.

At the receiving terminal, the incoming digit stream must be divided and distrib-

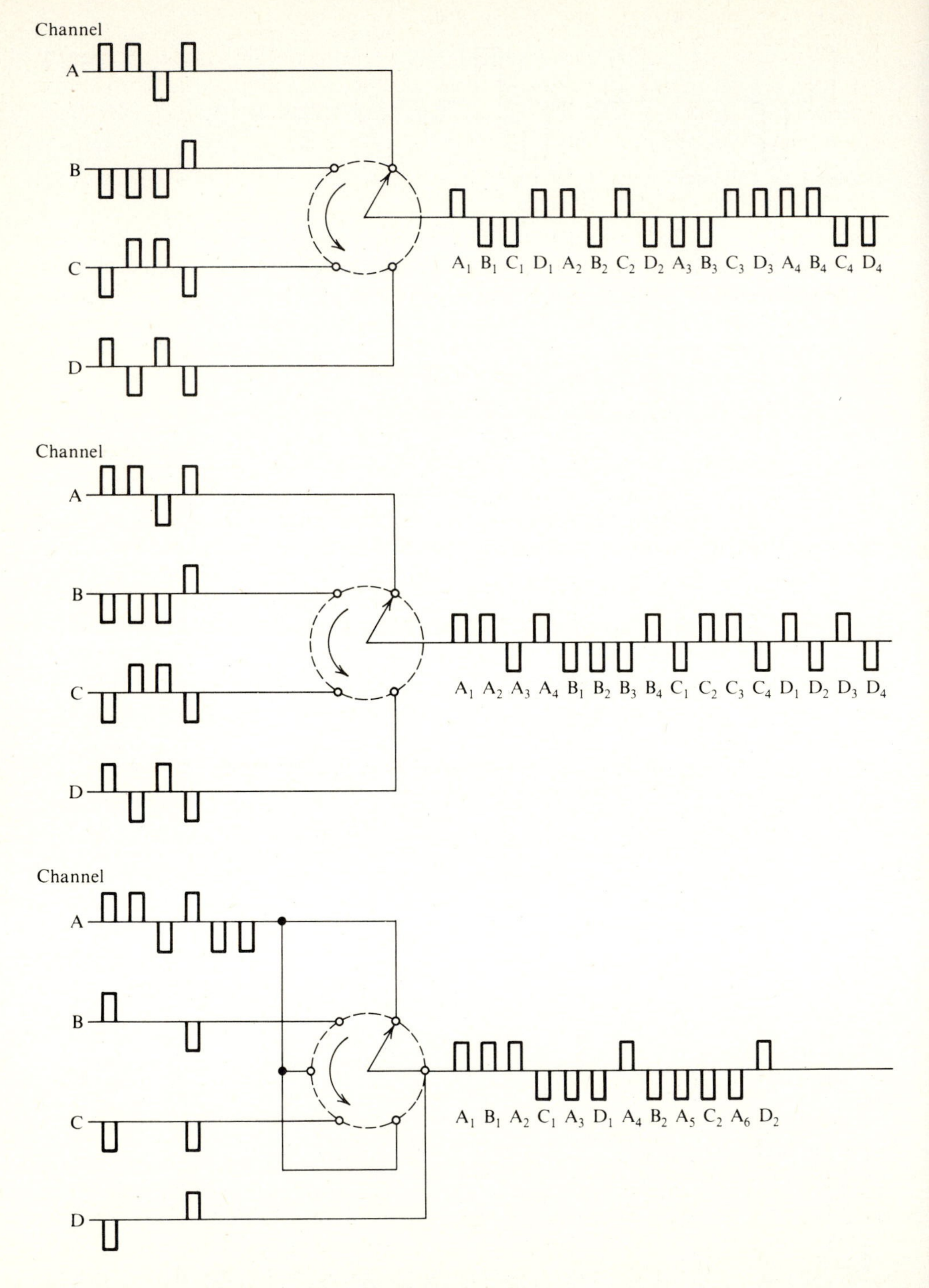

Figure 3.3 Time-division multiplexing of digital signals.

uted to the appropriate output channel. For this purpose, the receiving terminal must be able to correctly identify each bit. This requires the receiving system to uniquely synchronize in time with the beginning of each frame, with each slot in a frame, and with each bit within a slot. This is accomplished by adding framing and synchronization bits to the data bits. These bits are part of the so-called *control bits*.

Signal Format

Figure 3.4 illustrates a typical format, that of the M12 multiplexer. We have here bit-by-bit interleaving of four channels each at a rate of 1.544 Mbits/second. The main frame consists of four subframes. Each subframe has six control digits: For example

M_0 [48] C_A [48] F_0 [48] C_A [48] C_A [48] F_1 [48]

M_1 [48] C_B [48] F_0 [48] C_B [48] C_B [48] F_1 [48]

M_1 [48] C_C [48] F_0 [48] C_C [48] C_C [48] F_1 [48]

M_1 [48] C_D [48] F_0 [48] C_D [48] C_D [48] F_1 [48]

Figure 3.4 M12 multiplexer format.

the subframe 1 (first line in Fig. 3.4) has control digits M_0, C_A, F_0, C_A, C_A, and F_1. In between these control digits are 48 interleaved data bits from the four channels. (Twelve data bits from each channel.) Thus we begin with control digit M_0, followed by 48 multiplexed data bits, then add a second control bit C_A followed by the next 48 multiplexed bits, and so on. Thus, there are a total of $48 \times 6 \times 4 = 1152$ data bits and $6 \times 4 = 24$ control bits making a total 1176 bits/frame. The efficiency is $1152/1176 \simeq 98$ percent. The control bits with subscript 0 are always **0** and those with subscript 1 are always **1**. Thus, M_0, F_0 are all **0**'s and M_1 and F_1 are all **1**'s. The F digits are periodic **010101. . .** and provide the main framing pattern. The multiplexer uses this to synchronize on the frame. After locking onto this pattern, the demultiplexer searches for the **0111** pattern formed by control digits $M_0M_1M_1M_1$. This further identifies the four subframes, each corresponding to a line in Fig. 3.3. It is possible, although unlikely, that signal bits may also have a pattern **101010 . . . .** The receiver could lock onto this wrong sequence. The presence of $M_0M_1M_1M_1$ provides verification of the genuine $F_0F_1F_0F_1$ sequence. The C bits are used to transmit additional information about bit stuffing as will be discussed later.

An example of a word interleaving multiplexer (used for PCM) is discussed in Sec. 3.9.

In the majority of cases not all incoming channels are active all the time. Some of them will be transmitting data and some will be idle. This means the system is underutilized. We can, therefore, accept more input channels to take advantage of the fact that some channel will be inactive at any given time. This obviously involves much more complicated switching operations, and also requires rather careful system planning. In any random traffic situation we cannot guarantee that the number of transmission channels demanded will not exceed the number available, but by taking account of the statistics of the signal sources, it is possible to ensure that the probability of this occurring becomes acceptably low. Multiplex structures of this type have been developed for satellite systems and are known as *time-division multiple-access (TDMA) systems*.

In TDMA systems employed for telephony, the design parameters are chosen so that any overload condition only lasts for a fraction of a second, which leads to acceptable performance for speech communication. For other types of data and telegraphy transmission delays are unimportant. Hence in overload condition, the incoming data can be stored and transmitted later.

Asynchronous Channels and Bit Stuffing

In the preceding discussion, we assumed synchronization between all the incoming channels and the multiplexer. This is difficult even when all the channels are nominally at the same rate. For example, consider a 1000-km coaxial cable carrying 2×10^8 pulses/second. Assuming the nominal propagation speed in the cable to be 2×10^8 meters/second, it takes $1/200$ second of transit time and 1 million pulses will be in transit. If the cable temperature increases by 1°F, the propagation velocity will increase by about 0.01 percent. This will cause the pulses in transit to arrive sooner, thus causing a temporary increase in the rate of pulses received. Because the extra pulses cannot be accommodated in the multiplexer, they must be temporarily stored at the receiver. If the cable temperature drops, the rate of received pulses will drop, and the multiplexer will have vacant slots with no data. These slots need to be stuffed with dummy digits (*pulse stuffing*).

This shows that even in synchronously multiplexed systems, the data is rarely received at a synchronous rate. We always need a storage (known as an *elastic store*) and pulse stuffing (also known as *justification*) to accommodate such a situation. Obviously, this method of an elastic store and pulse stuffing will work even when the channels are asynchronous.

Three variants of the pulse stuffing scheme exist: (1) positive pulse stuffing, (2) negative pulse stuffing, and (3) positive/negative pulse stuffing. In positive pulse stuffing, the multiplexer rate is higher than that required to accommodate all incoming tributaries at their maximum rate. Hence, the time slots in the multiplexed signal will become available at a rate exceeding that of the incoming data so that the tributary data will tend to lag (Fig. 3.5). At some stage, the system will decide that this lag has become great enough to require pulse stuffing. The information about the stuffed-pulse

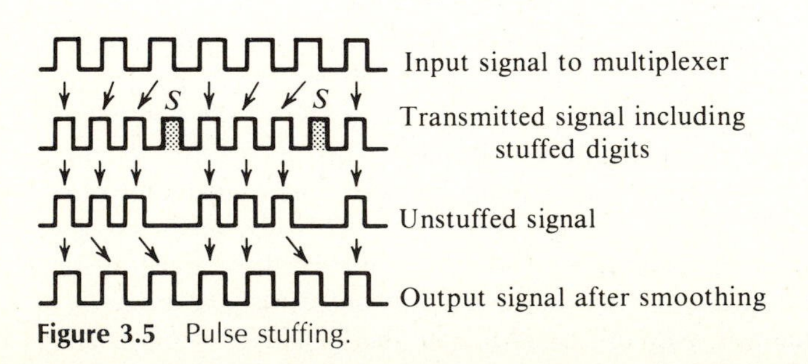

Figure 3.5 Pulse stuffing.

position is transmitted through control digits. From the control digits, the receiver knows the stuffed-pulse position and eliminates that pulse.

Negative pulse stuffing is a complement of positive pulse stuffing. The time slots in the multiplexed signal now appear at a slightly slower rate than those of the tributaries so that some of the tributary pulses cannot be accommodated in the multiplexed signal. The information about the left-out pulse and its position is transmitted through control digits. The positive/negative pulse stuffing is a combination of the above two schemes. Here the nominal rate of the multiplexer is equal to the nominal rate required to accommodate all incoming channels. Hence, we may need positive pulse stuffing at some times and negative stuffing at others. All this information is sent through control digits.

The C digits in Fig. 3.4 are used to transmit stuffing information. Only one stuffed bit/input channel is allowed per frame. This is sufficient to accommodate expected variations in the input signal rate. The bits C_A convey information about stuffing in channel A and bits C_B convey information about stuffing in channel B, and so on. The insertion of any stuffed pulse in any one subframe is denoted by setting all the three C's in that line to **1**. No stuffing is indicated by using **0**'s for all the three C's. If a bit has been stuffed, the location of the stuffed bit is the first information bit associated with the immediate channel following the F_1 bit, that is, the first bit in the last 48-bit sequence in that subframe.

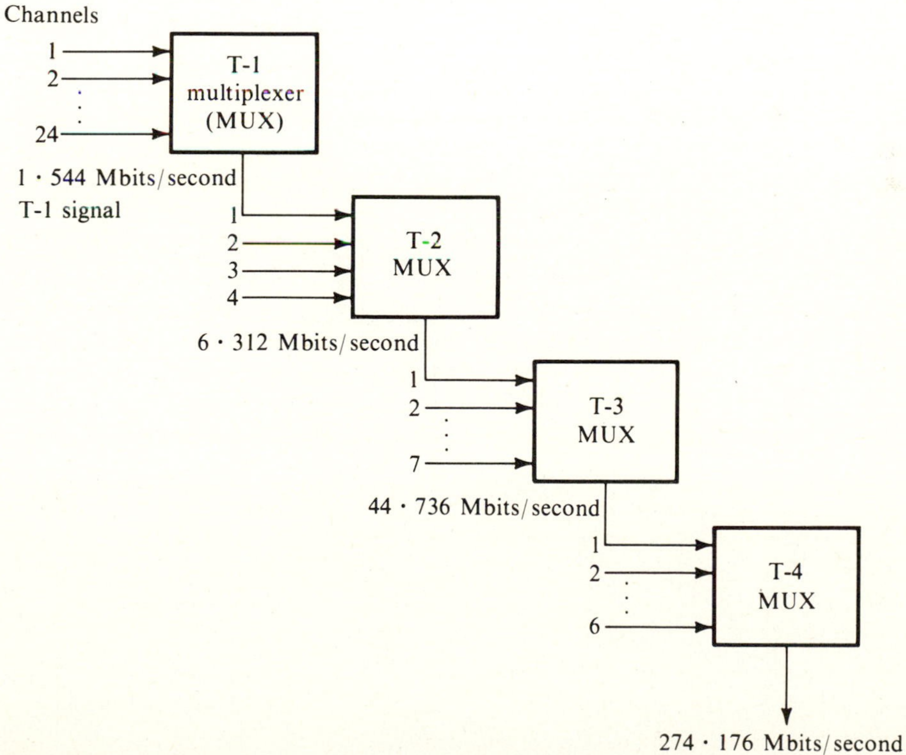

Figure 3.6 Proposed North American digital hierarchy (AT&T system).

Digital Hierarchy

Two major classes of multiplexers are used in practice. The first category is used for combining low-data-rate channels. It multiplexes channels of rates of up to 4800 bits/second into a signal of data rate of up to 9600 bits/second. The multiplexed signal is eventually transmitted over a voice-grade channel. The second class of multiplexers is at a much higher bit rate. The following is the digital hierarchy developed by the Bell System (Fig. 3.6). There are four orders, or levels, of multiplexing. The first level is the T-1 multiplexer, consisting of 24 channels of 64 kbits each. These channels need not be restricted only to digitized voice channels. Any digital signal of 64 kbits of appropriate format can be transmitted. The case of the higher levels is similar. For example, all the incoming channels of the T-2 multiplexer need not be T-1 signals obtained by multiplexing 24 channels of 64 kbits/second each. Some of them may be 1.544 Mbits/second digital signals of appropriate format. This hierarchy is proposed for North America and Japan. In Europe and the rest of the world, another hierarchy, recommended by the CCITT (Consultative Committee on International Telephony and Telegraphy) as an international standard, will be adopted. This hierarchy, based on the lowest-level PCM international standard of 2.048 Mbits/second (30 channels) is shown in Fig. 3.7.

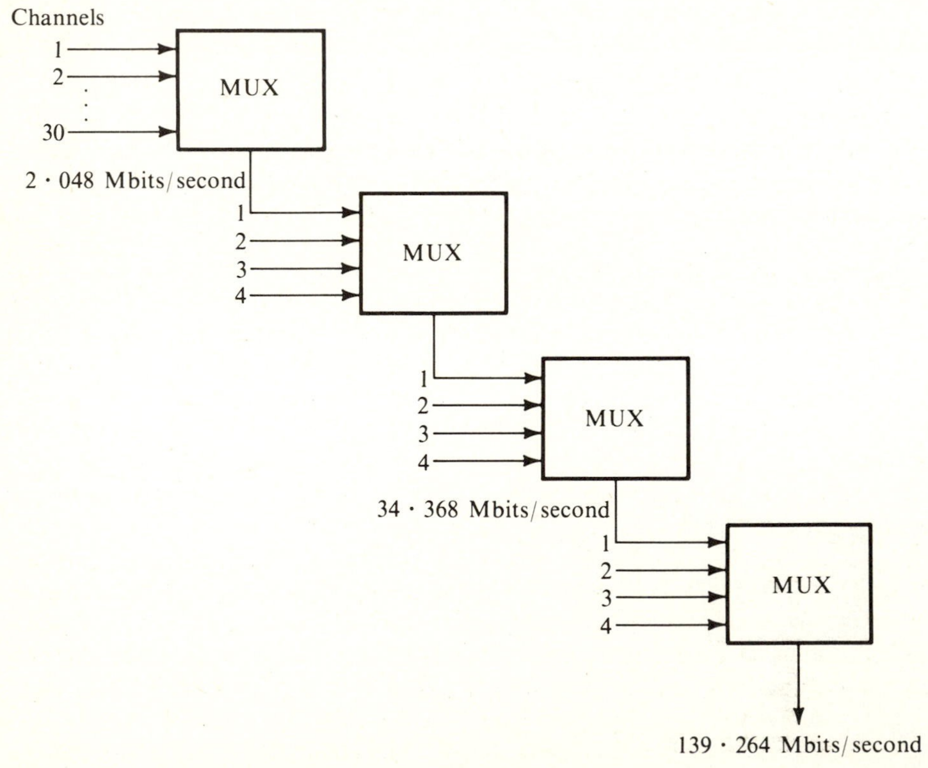

Figure 3.7 Digital hierarchy, CCITT recommendation.

3.2 LINE CODING

Digital data can be transmitted by various *transmission* or *line codes,* such as on-off, polar, bipolar, and so on. Each has its advantages and disadvantages. Among other desirable properties, a line code should have the following properties:

1. *Adequate timing content*: It should be possible to extract timing or clock information from the signal.
2. *Efficiency*: For a given bandwidth and transmitted power, the code should have the least detection error probability (i.e., the most immunity to the channel noise and ISI).
3. *Error detection and correction capability*: It should be possible to detect, and preferably correct, detection error. In the bipolar case, for example a single error will cause bipolar violation and can easily be detected. Error-correcting codes will be discussed in depth in Chapter 9.
4. *Favorable power spectral density*: The signal spectrum should be matched to the channel frequency response. For example, if a channel has high attenuation at lower frequencies, the signal spectrum should have a small PSD in this range to avoid excessive signal distortion. It is also desirable to have zero PSD at $\omega = 0$ (dc), because ac coupling is used at the repeaters. Significant power in low-frequency components causes dc wander in the pulse stream when ac coupling is used.
5. *Transparency*: It should be possible to correctly transmit a digital signal regardless of the pattern of **1**'s and **0**'s. We saw earlier that a long string of **0**'s could cause errors in timing extraction. If the data is so coded that for every possible sequence of data the coded signal is received faithfully, the code is transparent.

Before considering specific codes, we shall develop a general method of deriving the PSD of a broad class of digital signals.

Consider the pulse train in Fig. 3.8*a*, consisting of rectangular pulses of width t_o repeating every T_o interval with arbitrary amplitudes. On-off, polar, and bipolar signals are all special cases of this pulse train. Hence, we should be able to analyze many cases from the knowledge of the PSD of this pulse train. Unfortunately, it suffers from the disadvantage that it is restricted to only a rectangular pulse shape. If the basic pulse used has a different shape, we will have to derive the PSD all over again. This difficulty can be solved by a simple artifice of considering an impulse train $x(t)$ (Fig. 3.8*b*) with impulses repeating every T_o interval, the strength of the impulse at kT_o being a_k. If $x(t)$ is applied to the input of a filter that has a unit impulse response $p(t)$, then the output $y(t)$ will be a signal similar to the pulse train in Fig. 3.8*a* but with the rectangular pulses replaced by pulses of the form $p(t)$ (Fig. 3.10*b*). The pulses repeat every T_o seconds, and the pulse at kT_o is $a_k p(t)$. Also, $S_y(\omega)$, the PSD of $y(t)$, is $|P(\omega)|^2 S_x(\omega)$ [see Eq. (2.121a)]. This approach is attractive because of its generality. We now need to derive $\mathcal{R}_x(\tau)$, the time-autocorrelation function of the impulse train $x(t)$. This can be conveniently done by considering the impulses as a limiting

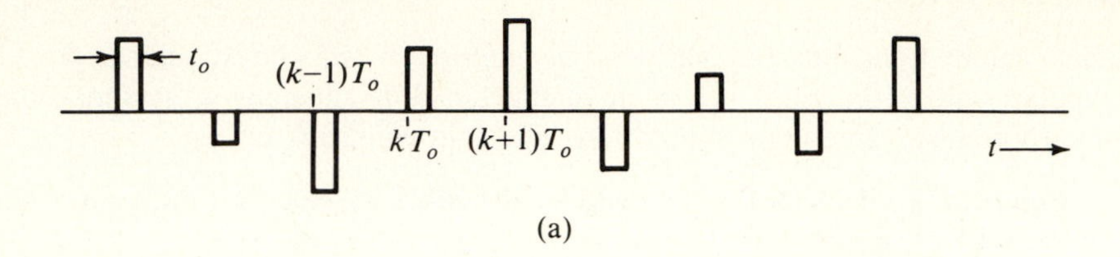

(a)

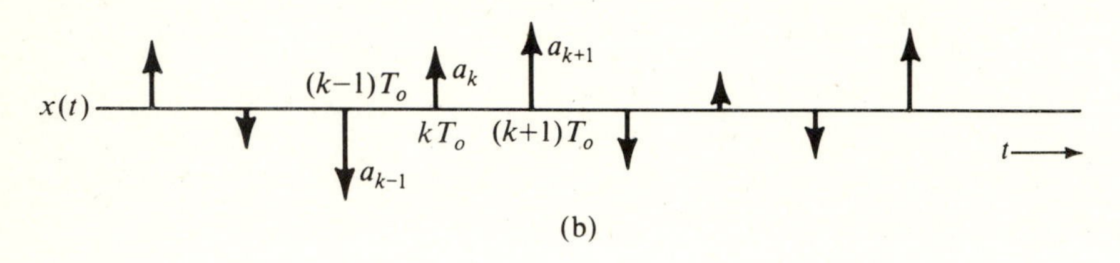

(b)

Figure 3.8

form of the rectangular pulses, as shown in Fig. 3.9*a*. Each pulse has a width $\epsilon \to 0$ and the kth pulse height is h_k. Because the strength of the kth impulse is a_k,

$$h_k \epsilon = a_k$$

If we designate the corresponding rectangular pulse train by $\hat{x}(t)$, then by definition [Eq. (2.113c)]

$$\mathcal{R}_{\hat{x}}(\tau) = \lim_{T\to\infty} \frac{1}{T} \int_{-T/2}^{T/2} x(t)x(t - \tau)\, dt \tag{3.1}$$

Because $\mathcal{R}_{\hat{x}}(\tau)$ is an even function of τ [Eq. (2.113d)], we need consider only positive τ. To begin with, consider the case of $\tau < \epsilon$. In this case the integral in Eq. (3.1) is the area under the signal $\hat{x}(t)$ multiplied by $\hat{x}(t)$ delayed by $\tau(\tau < \epsilon)$. As seen from Fig. 3.9*b*, the area associated with the kth pulse is $h_k^2(\epsilon - \tau)$, and

$$\begin{aligned}\mathcal{R}_{\hat{x}}(\tau) &= \lim_{T\to\infty} \frac{1}{T} \sum_k h_k^2(\epsilon - \tau) \\ &= \lim_{T\to\infty} \frac{1}{T} \sum_k a_k^2\left(\frac{\epsilon - \tau}{\epsilon^2}\right) \\ &= \frac{R_0}{\epsilon T_o}\left(1 - \frac{\tau}{\epsilon}\right) \end{aligned} \tag{3.2a}$$

where

$$R_0 = \lim_{T\to\infty} \frac{T_o}{T} \sum_k a_k^2 \tag{3.2b}$$

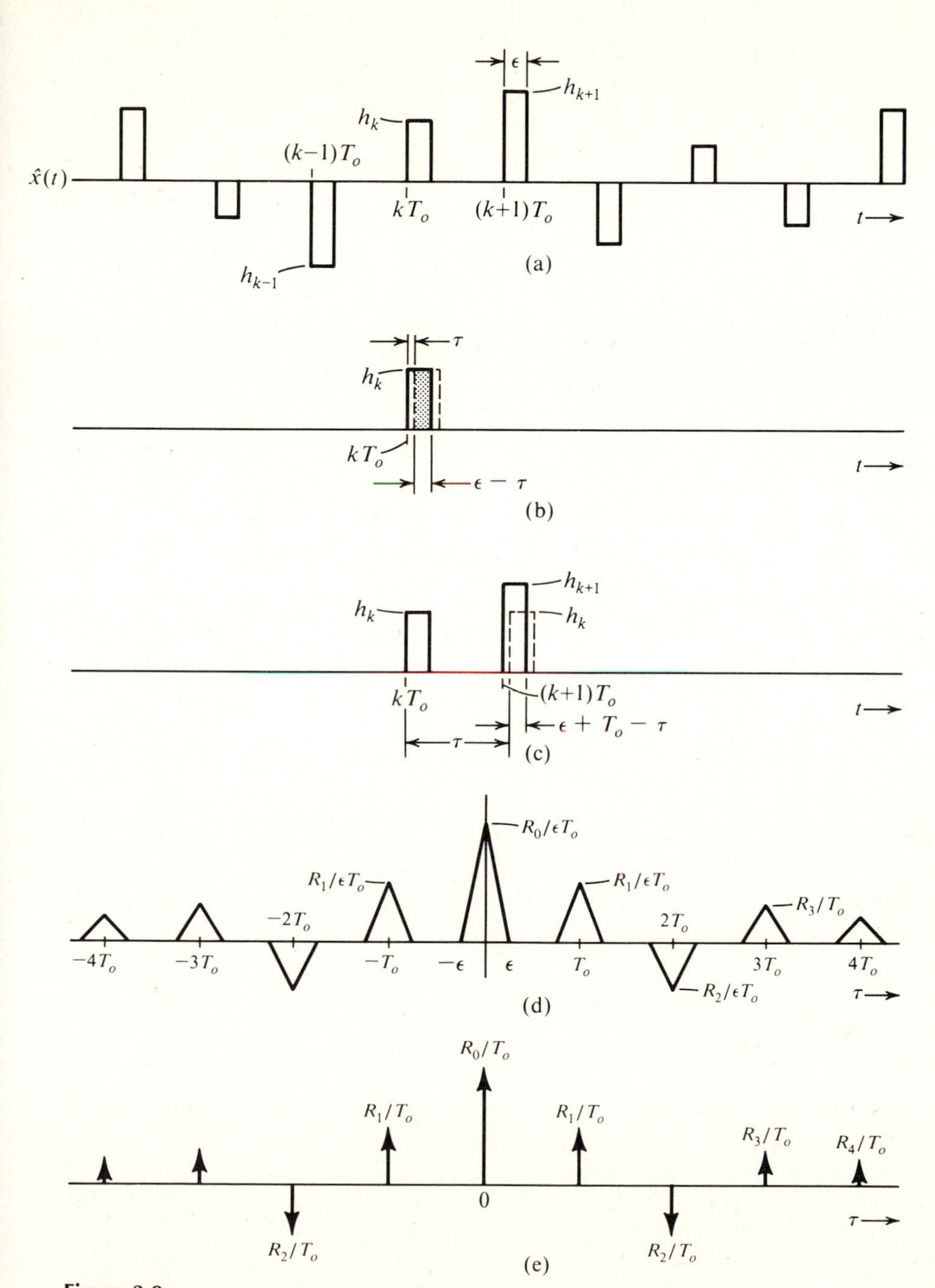

ϵ
h_{k+1}
h_k
$(k-1)T_o$
$\hat{x}(t)$
kT_o
$(k+1)T_o$
$t \longrightarrow$
h_{k-1}
(a)
τ
h_k
kT_o
$t \longrightarrow$
$\epsilon - \tau$
(b)
h_{k+1}
h_k
h_k
kT_o
$(k+1)T_o$
$t \longrightarrow$
$\epsilon + T_o - \tau$
τ
(c)
$R_0/\epsilon T_o$
$R_1/\epsilon T_o$
$R_1/\epsilon T_o$
R_3/T_o
$-2T_o$
$2T_o$
$-4T_o$
$-3T_o$
$-T_o$
$-\epsilon$
ϵ
T_o
$3T_o$
$4T_o$
$R_2/\epsilon T_o$
$\tau \longrightarrow$
(d)
R_0/T_o
R_1/T_o
R_1/T_o
R_3/T_o
R_4/T_o
0
$\tau \longrightarrow$
R_2/T_o
R_2/T_o
(e)

Figure 3.9

Because $\mathcal{R}_{\hat{x}}(\tau)$ is an even function of τ,

$$\mathcal{R}_{\hat{x}}(\tau) = \frac{R_0}{\epsilon T_o}\left(1 - \frac{|\tau|}{\epsilon}\right) \qquad |\tau| < \epsilon \tag{3.3}$$

This is a triangular pulse of height $R_0/\epsilon T_o$ and width 2ϵ centered at $\tau = 0$ (Fig. 3.9*d*). The autocorrelation $\mathcal{R}_{\hat{x}}(\tau) \to 0$ as $\tau \to \epsilon$. This is expected, because for $\tau = \epsilon$, the delayed signal $\hat{x}(t - \tau)$ does not overlap $\hat{x}(t)$ anymore, and the product $\hat{x}(t)\hat{x}(t - \tau)$ is zero. But as we increase τ further, we find that the kth pulse of $\hat{x}(t - \tau)$ will start overlapping the $(k + 1)$th pulse of $\hat{x}(t)$ as τ approaches T_o (Fig. 3.9*c*). Repeating the earlier argument, we see that $\mathcal{R}_{\hat{x}}(\tau)$ will have another triangular pulse of width 2ϵ centered at $\tau = T_o$ and of height $R_1/\epsilon T_o$ where

$$R_1 = \lim_{T\to\infty} \frac{T_o}{T} \sum_k a_k a_{k+1}$$

A similar thing happens around $\tau = 2T_o, 3T_o, \ldots$, and so on. Hence, $\mathcal{R}_{\hat{x}}(\tau)$ consists of a sequence of triangular pulses of width 2ϵ centered at $\tau = 0, \pm T_o, \pm 2T_o, \ldots$, and so on. The height of the pulses centered at $\pm nT_o$ is $R_n/\epsilon T_o$ where*

$$R_n = \lim_{T\to\infty} \frac{T_o}{T} \sum_k a_k a_{k+n}$$

In order to find $\mathcal{R}_x(\tau)$, we let $\epsilon \to 0$ in $\mathcal{R}_{\hat{x}}(\tau)$. As $\epsilon \to 0$, the width of each triangular pulse $\to 0$ and the height $\to \infty$ in such a way that the area is still finite. For the nth pulse centered at nT_o, the height is $R_n/\epsilon T_o$ and the area is R_n/T_o. Hence, (Fig. 3.9*e*)

$$\mathcal{R}_x(\tau) = \frac{1}{T_o} \sum_{n=-\infty}^{\infty} R_n\, \delta(\tau - nT_o) \tag{3.4}$$

with

$$R_n = \lim_{T\to\infty} \frac{T_o}{T} \sum_k a_k a_{k+n} \tag{3.5}$$

The PSD $S_x(\omega)$ is the Fourier transform of $\mathcal{R}_x(\tau)$. Therefore,

$$S_x(\omega) = \frac{1}{T_o} \sum_{n=-\infty}^{\infty} R_n e^{-jn\omega T_o} \tag{3.6}$$

Recognizing the fact that $R_{-n} = R_n$ (because $\mathcal{R}_x(\tau)$ is an even function of τ), we have

$$S_x(\omega) = \frac{1}{T_o}\left[R_o + 2\sum_{n=1}^{\infty} R_n \cos n\omega T_o\right] \tag{3.7}$$

Consider the pulse train $y(t)$ (Fig. 3.10*b*), made up of a basic pulse $p(t)$ (Fig. 3.10*a*)

---

* Because n, the number of pulses over T seconds is T/T_o, R_n is the time mean of the product $a_k a_{k+n}$ averaged over T seconds, that is,

$$R_n = \widetilde{a_k a_{k+n}}$$

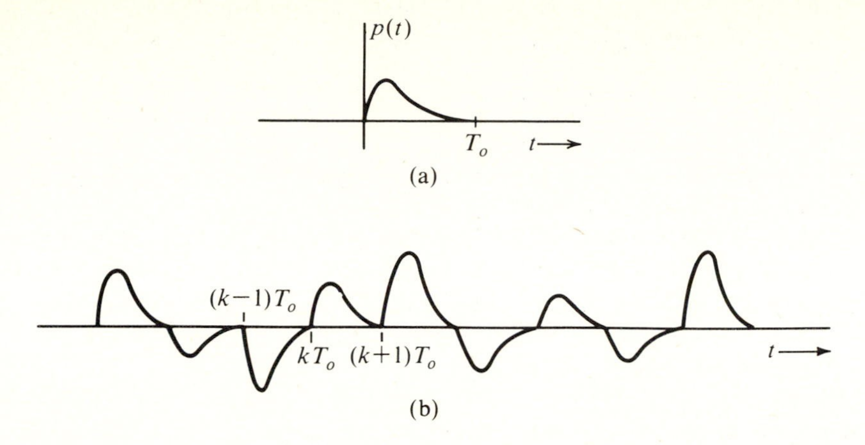

Figure 3.10

with the kth pulse $a_k p(t)$ located at kT_o. If the impulse train $x(t)$ is applied to the input of a filter with unit impulse response $p(t)$, the filter output will be the desired signal $y(t)$. Hence*

$$S_y(\omega) = |P(\omega)|^2 S_x(\omega) \tag{3.8a}$$

$$= \frac{|P(\omega)|^2}{T_o}\left[R_o + 2\sum_{n=1}^{\infty} R_n \cos n\omega T_o\right] \tag{3.8b}$$

We shall now consider several transmission codes and compare their PSDs.

On-Off Signaling

In this case the value of a_k is either 1 or 0. A total of T/T_o pulse positions exist in the interval $(-T/2, T/2)$. Assuming that **1** and **0** are equally likely, $a_k = 1$ for $T/2T_o$ pulses, and $a_k = 0$ for the remaining $T/2T_o$ pulses. Hence,

$$R_o = \lim_{T\to\infty} \frac{T_o}{T} \sum_k a_k^2$$

$$= \frac{T_o}{T}\left(\frac{T}{2T_o}\right)(1)^2 = \frac{1}{2} \tag{3.9}$$

---

*The result derived in Eq. (3.8b) may seem to be at variance with that derived by Bennett in W. R. Bennett, "Statistics of Regenerative Digital Transmission," *Bell Syst. Tech. J.*, vol. 37, pp. 1501–1543, Nov. 1958.

This is because Bennett defines

$$y(t) = \sum_n (a_n - m_1)p(t - nT_o)$$

where $m_1 = \lim_{T\to\infty} \frac{T_o}{T} \sum_k a_k$.

and

$$R_n = \lim_{T\to\infty} \frac{T_o}{T} \sum_k a_k a_{k+n}$$

The product of $a_k a_{k+n}$ is either 1 or 0, and a_k is 1 half the time and 0 half the time. The case with a_{k+n} is similar. Hence, we have four possibilities (1×1, 1×0, 0×1 or 0×0), all of them equally likely. Therefore, the product $a_k a_{k+n}$ will be 1 for a quarter of the terms, on the average, and 0 for the remaining terms. Because there are T/T_o terms over the interval $(-T/2, T/2)$,

$$R_n = \frac{T_o}{T}\left(\frac{T}{4T_o}\right)(1) = \frac{1}{4} \tag{3.10}$$

Therefore* [Eq. (3.6)]

$$S_x(\omega) = \frac{1}{4T_o} + \frac{1}{4T_o}\sum_{n=-\infty}^{\infty} e^{-jn\omega T_o} \tag{3.11}$$

Using the relationship† [see Eq. (2.24b)]

$$\sum_{n=-\infty}^{\infty} e^{-jn\omega T_o} = \frac{2\pi}{T_o}\sum_{n=-\infty}^{\infty} \delta\left(\omega - \frac{2\pi n}{T_o}\right) \tag{3.12a}$$

$$S_x(\omega) = \frac{1}{4T_o} + \frac{2\pi}{4T_o^2}\sum_{n=-\infty}^{\infty} \delta\left(\omega - \frac{2\pi n}{T_o}\right) \tag{3.12b}$$

and the desired PSD of the on-off waveform $y(t)$ is

$$S_y\omega = \frac{|P(\omega)|^2}{4T_o}\left[1 + \frac{2\pi}{T_o}\sum_{n=-\infty}^{\infty} \delta\left(\omega - \frac{2\pi n}{T_o}\right)\right] \tag{3.12c}$$

where $P(\omega)$ is the Fourier transform of the basic pulse $p(t)$ used. For the case of a half-width rectangular pulse

$$p(t) = \Pi\left(\frac{t}{T_o/2}\right) = \Pi\left(\frac{2t}{T_o}\right) \tag{3.13a}$$

$$P(\omega) = \frac{T_o}{2}\,\text{sinc}\left(\frac{\omega T_o}{4\pi}\right) \tag{3.13b}$$

and

$$S_y(\omega) = \frac{T_o}{16}\,\text{sinc}^2\left(\frac{\omega T_o}{4\pi}\right)\left[1 + \frac{2\pi}{T_o}\sum_{n=-\infty}^{\infty} \delta\left(\omega - \frac{2\pi n}{T_o}\right)\right] \tag{3.14}$$

---

* Note that the term $1/2T_o$, corresponding to R_o, is split into two—$1/4T_o$ outside the summation and $1/4T_o$ inside the summation (corresponding to $n = 0$).

† Equation (3.12a) can also be derived by recognizing that its left-hand side is the Fourier transform of the unit impulse train (taken term by term) and the right-hand side is also the Fourier transform of the unit impulse train as derived in Example 2.19.

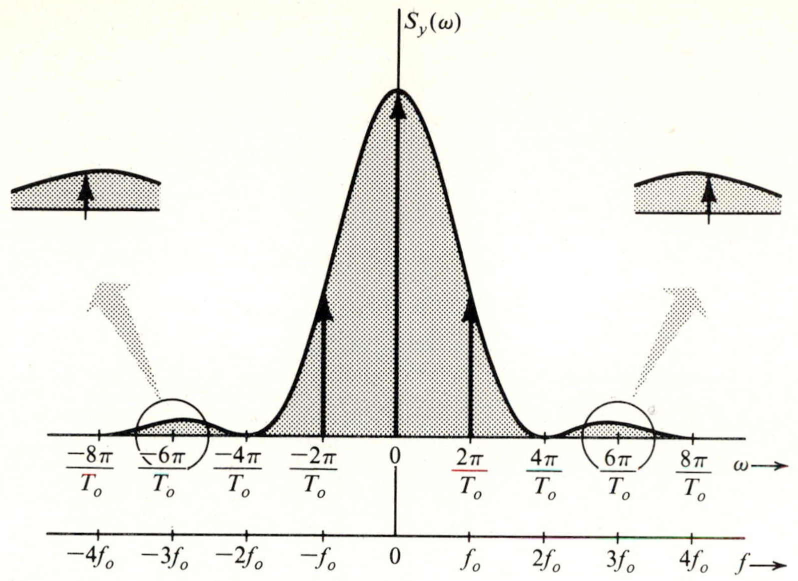

Figure 3.11 PSD of on-off signal.

This spectrum is shown in Fig. 3.11. Note that the spectrum consists of a discrete as well as a continuous part. A discrete component is present at the clock frequency $f_o = 1/T_o$. An on-off scheme using the full-width pulse $p(t) = \Pi\ (t/T_o)$ is an example of a *nonreturn to zero (NRZ)* scheme, because a pulse does not return to zero before the next pulse begins. The half-width pulse scheme, on the other hand, is an example of *return to zero (RZ)*. The reader can show that for the NRZ (full-width on-off) case, the discrete component at the clock frequency vanishes.

From Fig. 3.11 it can be seen that the essential bandwidth of the signal is $2f_o$ where f_o is the clock frequency. This is four times the theoretical bandwidth (Nyquist bandwidth) required. For a full-width pulse, the essential bandwidth reduces to f_o.

The on-off signaling is attractive from the point of view of simplicity of terminal apparatus. But it has several disadvantages. For a given transmitted power, it is less immune to noise interference than the polar scheme, which uses a positive pulse for **1** and a negative pulse for **0**. This is because the noise immunity depends on the difference of amplitudes representing **1** and **0**. Hence, for the same immunity, if on-off signaling uses pulses of amplitudes 2 and 0, polar signaling need only use pulses of amplitudes 1 and -1. It is simple to show that on-off signaling requires twice as much power as polar signaling. If a pulse of amplitude 1 or -1 has energy E, then the pulse of amplitude 2 has energy $(2)^2E = 4E$. Because $1/T_o$ digits are transmitted per second, polar signal power is $(E)(1/T_o) = E/T_o$. For the on-off case, on the other hand, each pulse energy is $4E$, but only half as many pulses are transmitted. Hence, the signal power is $(4E)\ (1/2T_o) = 2E/T_o$, which is twice that required for the polar signal.

The second disadvantage of on-off signaling is that it has a nonzero PSD at dc ($\omega = 0$). This will rule out the use of ac coupling during transmission. The ac coupling, which permits transformers and blocking capacitors to aid in impedance matching and bias removal, is very important in practice. Thirdly, the transmission bandwidth requirements are excessive. In addition, on-off signaling has no error-detection or correlation capability, and, lastly, it is not transparent. A long string of **0**'s (or offs) can create errors in timing extraction.

EXAMPLE 3.1

Derive the PSD of the on-off signal if the **1** and **0** are not equally likely. Assume that the likelihood of transmitting **1** is Q and that of transmitting **0** is $1 - Q (0 \leq Q \leq 1)$. This means if N digits are transmitted, then, on the average, NQ digits will be **1** and $N(1 - Q)$ digits will be **0** ($N \rightarrow \infty$).

In this case, there are $(T/T_o)Q$ nonzero pulses (on pulses), and

$$R_o = \frac{T_o}{T}\left(\frac{TQ}{T_o}\right)(1)^2 = Q$$

To calculate R_n, we observe that over the interval $(-T/2, T/2)$, only $(T/T_o)Q$ a_k's are 1. For each of these a_k's the likelihood of finding $a_{k+n} = 1$ is Q (i.e., out of the (TQ/T_o) a_k's that are 1, only Q fraction will find a mate a_{k+1} that is also a 1). Hence

$$\sum_k a_k a_{k+n} = Q\left(\frac{TQ}{T_o}\right) = \frac{TQ^2}{T_o}$$

and

$$R_n = \frac{T_o}{T}\left(\frac{TQ^2}{T_o}\right) = Q^2$$

From Eq. (3.8b) we have

$$\begin{aligned} S_y(\omega) &= |P(\omega)|^2\left(\frac{Q}{T_o}\right)\left[1 + 2Q\sum_{n=1}^{\infty} \cos n\omega T_o\right] \\ &= |P(\omega)|^2\left(\frac{Q}{T_o}\right)\left[(1 - Q) + Q\left(1 + 2\sum_{n=1}^{\infty} \cos n\omega T_o\right)\right] \\ &= |P(\omega)|^2\left(\frac{Q}{T_o}\right)\left[(1 - Q) + \frac{2\pi Q}{T_o}\sum_{n=-\infty}^{\infty} \delta\left(\omega - \frac{2\pi n}{T_o}\right)\right] \end{aligned}$$

The shape of the spectrum is similar to that in Fig. 3.11 except for the relative weight of the continuous and discrete components. ■

Polar Signaling

In polar signaling, **1** is transmitted by a pulse $p(t)$ and **0** is transmitted by $-p(t)$. In this case, a_k is equally likely to be 1 or -1, and a_k^2 is always 1. Hence,

$$R_o = \lim_{T\to\infty} \frac{T_o}{T} \sum_k a_k^2 = \frac{T_o}{T}\left(\frac{T}{T_o}\right)(1) = 1$$

Similarly, $a_k a_{k+n}$ can be 1 or -1. For half the combination it is 1, and for the remaining half it is -1. Therefore, $R_n = 0$ and

$$S_y(\omega) = \frac{|P(\omega)|^2}{T_o} R_o$$

$$= \frac{|P(\omega)|^2}{T_o} \tag{3.15}$$

For a half-width rectangular pulse [Eq. (3.13)]

$$S_y(\omega) = \frac{T_o}{4} \operatorname{sinc}^2\left(\frac{\omega T_o}{4\pi}\right) \tag{3.16}$$

This spectrum is identical to the continuous component of the on-off signal.

As mentioned earlier, polar signaling is more efficient than on-off. In fact, for a given transmitted power, polar signaling is the most efficient scheme. In addition, it is transparent. But, it still suffers from all the other disadvantages of on-off signaling. Note that there is no discrete clock frequency component in a polar signal. Rectification of the polar signal, however, yields a periodic signal of clock frequency and can be used to extract timing.

Bipolar (or Pseudoternary) Signaling

This is the signaling scheme used in PCM these days. A **0** is transmitted by no pulse, and **1** is transmitted by a pulse $p(t)$ or $-p(t)$, depending on whether the previous **1** was transmitted by $-p(t)$ or $p(t)$. With consecutive pulses alternating, we can avoid the dc wander and thus cause a dc null in the PSD. Bipolar signaling actually uses three symbols [$p(t)$, 0, and $-p(t)$], and, hence, it is in reality ternary rather than binary signaling.

To calculate the PSD, we have

$$R_o = \lim_{T\to\infty} \frac{T_o}{T} \sum_k a_k^2$$

On the average, half of the a_k's and 0, and the remaining half are either 1 or -1, with $a_k^2 = 1$. Because there are $T/2T_o$ pulses over the interval $(-T/2, T/2)$,

$$R_o = \frac{T_o}{T}\left(\frac{T}{2T_o}\right)(\pm 1)^2 = \frac{1}{2}$$

$$R_1 = \lim_{T\to\infty} \frac{T_o}{T} \sum_k a_k a_{k+1}$$

Because positive and negative pulses alternate, a_k and a_{k+1} cannot be 1 simultaneously. Hence, $a_k a_{k+1}$ is either 0 or -1. Two consecutive binary digits have only four possible patterns: (**11**), (**10**), (**01**), and (**00**), all four occurring in the same proportion. The product $a_k a_{k+1}$ for these combinations is $-1, 0, 0, 0$, respectively. Hence, out of T/T_o pulses in the interval $(-T/2, T/2)$, $T/4T_o$ would yield $a_k a_{k+1} = -1$, and the rest would yield $a_k a_{k+1} = 0$. Hence, $R_1 = -1/4$. In general,

$$R_n = \lim_{T\to\infty} \frac{T_o}{T} \sum_k a_k a_{k+n}$$

For $n > 1$, $a_k a_{k+n}$ is either 0, 1, or -1. The likelihood of the product $a_k a_{k+n}$ being 1 or -1 is the same (viz., 1/8). Hence, the sum $\Sigma\, a_k a_{k+n}$ will be zero.

$$R_n = 0 \qquad n > 1$$

and

$$S_y(\omega) = \frac{|P(\omega)|^2}{2T_o}\left[1 - \cos \omega T_o\right] \tag{3.17a}$$

$$= \frac{|P(\omega)|^2}{T_o} \sin^2\left(\frac{\omega T_o}{2}\right) \tag{3.17b}$$

Note that $S_y(\omega) = 0$ for $\omega = 0$ (dc), regardless of $P(\omega)$. Hence, the PSD has a dc null, which is desirable for ac coupling. For the case of half-width rectangular pulses [Eq. (3.13)]

$$S_y(\omega) = \frac{T_o}{4} \operatorname{sinc}^2\left(\frac{\omega T_o}{4\pi}\right) \sin^2\left(\frac{\omega T_o}{2}\right) \tag{3.18}$$

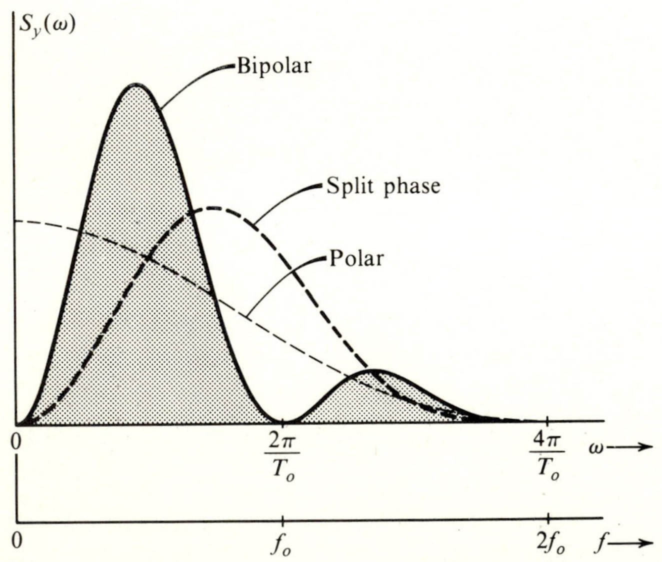

Figure 3.12 PSD of bipolar, polar, and split-phase signals normalized for equal powers.

This is shown in Fig. 3.12. The essential bandwidth of the signal is f_o ($f_o = 1/T_o$), which is half that of on-off signaling and twice the theoretical minimum bandwidth.

Bipolar signaling has several advantages: (1) Its spectrum has a dc null; (2) Its bandwidth is not excessive; (3) It has single-error-detection capability. This is because if a single detection error is made, it will cause a bipolar violation of the alternating pulse rule and will be immediately detected. If a bipolar signal is rectified, we get an on-off signal that has a discrete component at the clock frequency. Among the disadvantages, a bipolar signal requires twice as much power (3 dB) as that required for a polar signal. This is because bipolar detection is essentially equivalent to on-off signaling from the detection point of view. One has to distinguish between $+A$ or $-A$ and 0 rather than between $-A/2$ and $A/2$. Another disadvantage of bipolar signaling is that it is not transparent.

Duobinary Signaling

This scheme, proposed by A. Lender,[2] is ternary like bipolar signaling, but its bandwidth is only half that of bipolar. In this scheme, a **0** is transmitted by no pulse, and a **1** is transmitted by a pulse $p(t)$ or $-p(t)$, depending upon the polarity of the previous pulse and the number of **0**'s between them. The rule is as follows: A **1** is encoded by the same pulse as that used to encode the **1** preceding it if the two **1**'s are separated by an even number of **0**'s. Otherwise, it is encoded by the negative of the pulse used to encode the previous **1**. Figure 3.13*a* shows an example of such coding.

To derive $S_y(\omega)$, we note that

$$R_o = \lim_{T\to\infty} \frac{T_o}{T} \sum_k a_k^2$$

Because, on the average, half of the pulses are **0**, and for the remaining half, $a_k^2 = 1$, we have

$$R_o = \frac{T_o}{T}\left(\frac{T}{2T_o}\right)(1)^2 = \frac{1}{2}$$

and

$$R_1 = \lim_{T\to 0} \frac{T_o}{T} \sum_k a_k a_{k+1}$$

Half of all the a_k's are 0. The remaining half of the a_k's comprise 1 and -1 in equal proportion. Because of the duobinary coding rule, there are only four permissible sequences of $a_k a_{k+1}$ with $a_k = \pm 1$. These are: 11, 10, $-1-1$, and -10, all of which are equally likely. Half of these combinations yield $a_k a_{k+1} = 1$. The remaining half yield $a_k a_{k+1} = 0$. Hence, of all the a_k's, only one-quarter will yield the product

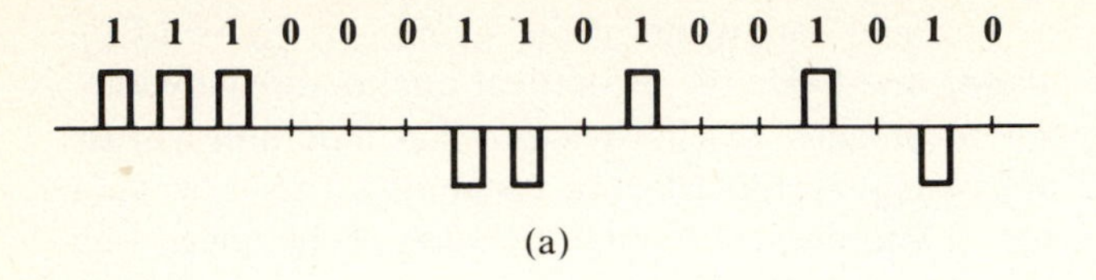

(a)

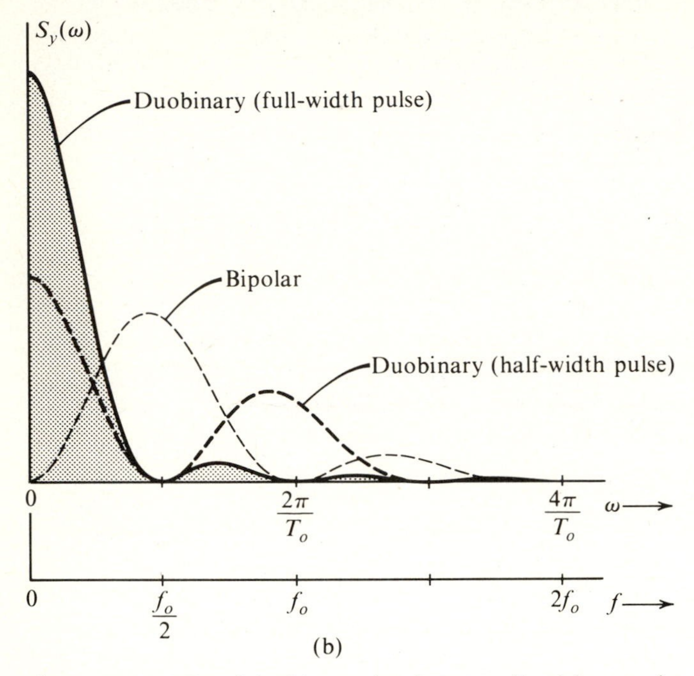

(b)

Figure 3.13 PSD of duobinary signals (normalized for equal powers).

$a_k a_{k+1} = 1$, and

$$R_1 = \frac{T_o}{T}\left(\frac{T}{4T_o}\right)(1) = \frac{1}{4}$$

Also,

$$R_2 = \lim_{T\to\infty} \frac{T_o}{T} \sum_k a_k a_{k+2}$$

Of all possible combinations of a_k and a_{k+2}, only those with $a_k = \pm 1$ and $a_{k+2} = \pm 1$ will contribute to the product $a_k a_{k+2}$. Consider all permissible sequences of $a_k a_{k+1} a_{k+2}$ that contribute to the product $a_k a_{k+2}$. These are 111, $10-1$, $-1-1-1$, and -101. These are all equally likely sequences, with half of the sequences yielding $a_k a_{k+2} = 1$ and the remaining half yielding $a_k a_{k+2} = -1$. Hence, $R_2 = 0$. Similarly, we can show that

$$R_n = 0 \qquad n > 1$$

and

$$S_y(\omega) = \frac{|P(\omega)|^2}{2T_o}(1 + \cos \omega T_o) \quad \textbf{(3.19a)}$$

$$= \frac{|P(\omega)|^2}{T_o} \cos^2\left(\frac{\omega T_o}{2}\right) \quad \textbf{(3.19b)}$$

For a half-width rectangular pulse $p(t)$ [Eq. (3.13)],

$$S_y(\omega) = \frac{T_o}{4} \operatorname{sinc}^2\left(\frac{\omega T_o}{4\pi}\right) \cos^2\left(\frac{\omega T_o}{2}\right) \quad \textbf{(3.20)}$$

This PSD is shown in Fig. 3.13*b*. Note that $S_y\omega = 0$ at $\omega = \pi/T_o$ since cos $(\omega T_o/2) = 0$ at this frequency.

As a first approximation, we may say that the essential bandwidth of duobinary is π/T_o or $f_o/2$ Hz, which is half that of the bipolar. Note, however, that there is sizable power in components above the frequency $f_o/2$. By choosing the proper shape for $p(t)$, it is possible to achieve negligible PSD at frequencies above $f_o/2$. For a full-width rectangular pulse $p(t)$, for example, most of the power is concentrated in components below $f_o/2$, as seen from Fig. 3.13b. Thus by shaping $p(t)$ properly, in duobinary it is possible to achieve the Nyquist signaling rate, that is, f_o digits/second over a bandwidth $f_o/2$. In the next section, we shall discuss the topic of pulse shaping in greater detail.

Like a bipolar signal, a duobinary signal is not transparent. Timing can be extracted by rectifying the signal. Like a bipolar signal, it suffers a 3 dB power disadvantage in comparison to polar. It also has error-detection capability, because correct reception implies that between successive pulses of the same polarity an even number of zeros must occur and between successive pulses of opposite polarity an odd number of zeros must occur. Hence, this signaling is similar to bipolar except that it requires only half the bandwidth. But it suffers from the nonzero PSD at ω=0. Lender proposed a modified duobinary technique[7] that eliminates the problem of the nonzero PSD at ω=0 (see Sec. 3.3). Using this promising technique, GTE Lenkurt, Inc., came out in 1976 with a duobinary repeater that is fully compatible with the AT&T T-1 carrier system. This duobinary repeater accepts two T-1 systems (each with 24 channels) and converts them into duobinary. It is thus able to transmit 48 PCM channels over a cable where only 24 channels were transmitted earlier. Modified duobinary signaling is discussed in Sec. 3.3.

Split-Phase (or Manchester) Signaling

Because the PSD $S_y(\omega)$ is the product of $|P(\omega)|^2$ and $S_x(\omega)$, the spectrum can be shaped by controlling $P(\omega)$ or $S_x(\omega)$. The dc null was obtained in the bipolar case by making $S_x(\omega) = 0$ at $\omega = 0$. The same effect is obtained by choosing $p(t)$ such that $P(\omega) = 0$ at $\omega = 0$. Because

$$P(\omega) = \int_{-\infty}^{\infty} p(t)e^{-j\omega t}\, dt$$

then

$$P(0) = \int_{-\infty}^{\infty} p(t)\, dt \tag{3.21}$$

Hence, if the area under $p(t)$ is made zero, $P(0)$ is zero, and we have a dc null in the PSD. There are several ways of doing this. For a rectangular pulse, one possible shape of $p(t)$ is shown in Fig. 3.14*a*. Here each bit is represented by two successive pulses

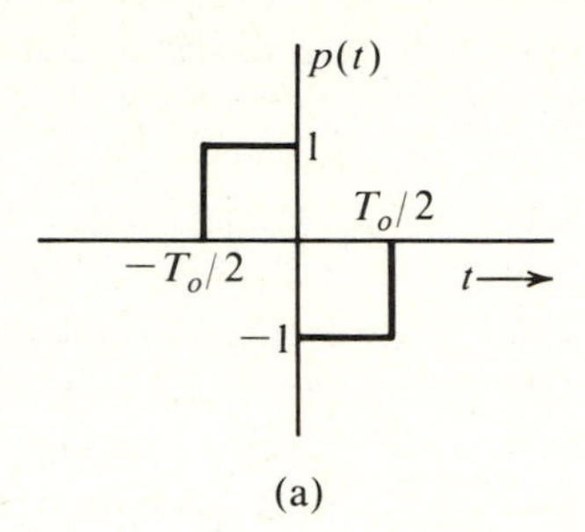

(a)

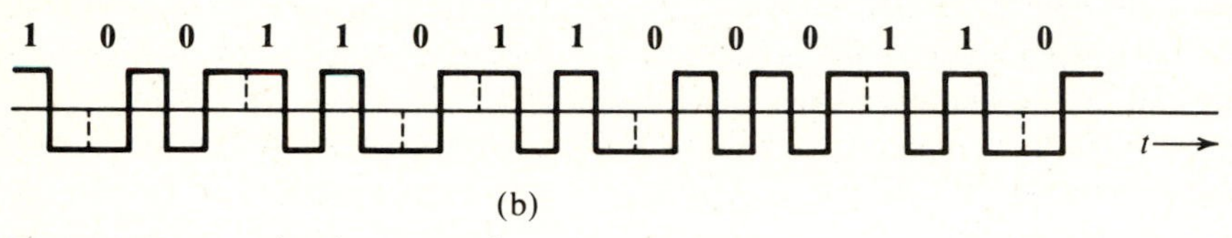

(b)

Figure 3.14 A split-phase signal.

of opposite polarity. The binary **0** is transmitted by $-p(t)$, as shown in Fig. 3.14*b*. This signaling is known as *Manchester,* or *split-phase* (also *twinned-binary*), signaling. Because this is a polar signal, the PSD $S_y(\omega)$ is given by [Eq. (3.15)]

$$S_y(\omega) = \frac{|P(\omega)|^2}{T_o}$$

For the pulse $p(t)$ in Fig. 3.14*a*,

$$p(t) = \Pi\left(\frac{t + T_o/4}{T_o/2}\right) - \Pi\left(\frac{t - T_o/4}{T_o/2}\right)$$

and

$$\begin{aligned} P(\omega) &= \frac{T_o}{2}\operatorname{sinc}\left(\frac{\omega T_o}{4\pi}\right) e^{j\omega T_o/4} - \frac{T_o}{2}\operatorname{sinc}\left(\frac{\omega T_o}{4\pi}\right) e^{-j\omega T_o/4} \\ &= jT_o \operatorname{sinc}\left(\frac{\omega T_o}{4\pi}\right)\sin\left(\frac{\omega T_o}{4}\right) \end{aligned} \tag{3.22}$$

and

$$S_y(\omega) = T_o \operatorname{sinc}^2\left(\frac{\omega T_o}{4\pi}\right)\sin^2\left(\frac{\omega T_o}{4}\right) \tag{3.23}$$

Comparison of this with the bipolar PSD in Eq. (3.17) shows that the bandwidth for split-phase signaling is twice that for bipolar signaling (Fig. 3.12). Split-phase, however, has one advantage over bipolar in that it is transparent, because every pulse position is occupied, and a long string of **0**'s will not cause any difficulty in timing extraction.

High-Density Bipolar (HDB) Signaling

The problem of nontransparency in bipolar signaling is eliminated by adding pulses when the number of consecutive **0**'s exceeds n. Such a modified coding is designated as *high-density bipolar coding,* HDBn, where n can take on any value 1, 2, 3, . . . , and so on. The most important of the HDB codes is HDB3.

The basic idea of the HDBn code is that when a run of more than n binary **0**'s occurs, the $n + 1$ zeros are replaced by one of the special binary digit sequences. The sequences are chosen to include some binary **1**'s in order to increase the timing content of the signal. The **1**'s included deliberately violate the bipolar rule for easy identification of the substituted sequence. In HDB3 coding, for example, the special sequences used are **000V** and **100V** where **V** is **1**. The **V** bit is encoded by a pulse of such a polarity as to violate the bipolar rule, whereas the **1** bit in **100V** is encoded by a pulse of polarity in conformity with the bipolar rule. The choice of sequence **000V** or **100V** is made in such a way that consecutive **V** pulses alternate signs in order to avoid dc wander and to maintain the dc null in the PSD. The sequence **100V** is used when there has been an even number of **1**'s following the last special sequence. Figure 3.15*a* shows an example of this coding. Note that in the sequence **100V**, **1** and **V** are both encoded by the same pulse. The decoder has to check two things—the bipolar violations and the number of **0**'s preceding each violation to determine if the previous **1** is also a substitution.

Despite deliberate bipolar violations, HDB signaling retains error-detecting capability. Any single error will insert a spurious bipolar violation (or will delete one of the deliberate violations). This will become apparent when, at the next violation, the alternation of violations does not appear. This also shows that deliberate violations can be detected despite single errors.

The derivation of the PSD for HDB signaling is tedious, requiring calculations of R_n for large values of n. For example, in the case of HDB3, it has been found necessary to calculate as far as R_{63} to produce a satisfactory PSD near the origin.[3] Figure 3.15*b* shows the PSD of HDB3 as well as that of a bipolar signal to facilitate comparison.

There are many other transmission codes, too numerous to list here. A list of codes and appropriate references can be found in Bylanski and Ingram.[4]

Comments on Spectral Shaping. Superficially, it may appear that because a pulse train is nothing but a succession of the basic pulse $p(t)$, the frequency components (the spectrum) of the pulse train $y(t)$ should be the same as those of the pulse $p(t)$. From the discussion in this section, it is clear that this is not so. If pulses are transmitted in a certain pattern, some components are cancelled completely or partially, and the spectrum of $y(t)$ is altered. Consider, for example, transmission of periodic data **10101010. . .** by on-off and polar signaling. In the former case, we have a pulse $p(t)$ followed by no pulse, and this pattern repeats itself. If $p(t)$ is a rectangular pulse, the

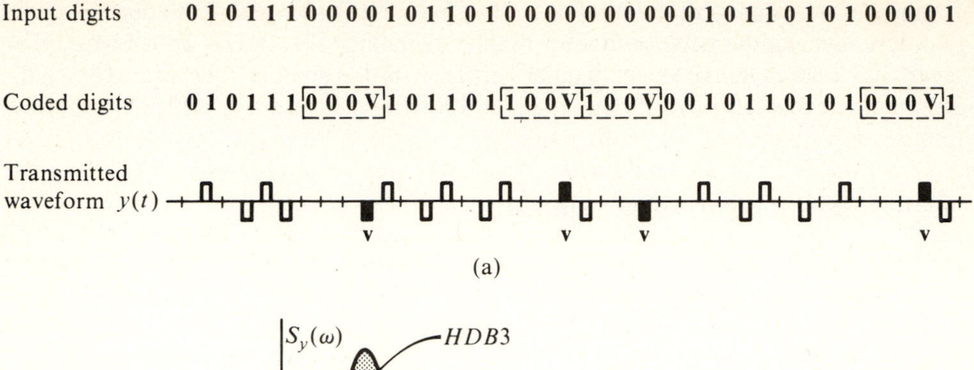

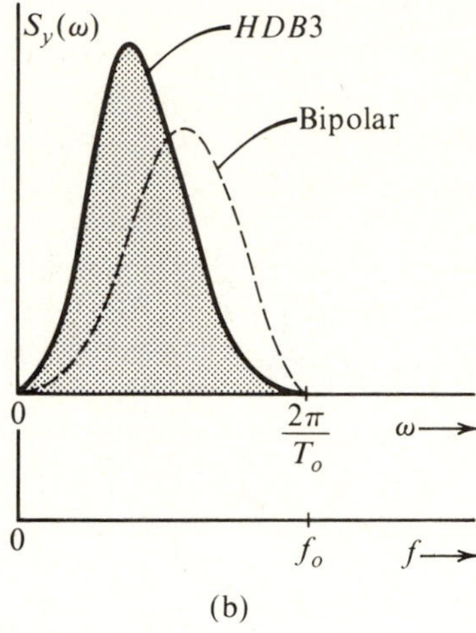

(b)

Figure 3.15 HDB3 signal and its PSD.

train $y(t)$ consists of a repetitive pattern of $p(t)$ followed by $-p(t)$. In this case the dc components of the pulses cancel out, and $y(t)$ has no dc component. Thus, the PSD of $y(t)$ is decided not only by the pulse spectrum $P(\omega)$ but also by the pulse pattern. This fact is clearly brought out by Fig. (3.8*a*).

$$S_y(\omega) = |P(\omega)|^2 S_x(\omega)$$

Note that $S_x(\omega)$ depends on R_n ($n = 1, 2, 3, \ldots$), where R_n is the time correlation of pulse amplitudes separated by an n-pulse interval.

$$R_n = \widetilde{a_k a_{k+n}}$$

In Sec. 3.3, we shall discuss methods of shaping $p(t)$.

3.3 PULSE SHAPING

The PSD $S_y(\omega)$ of a digital signal $y(t)$ can be controlled by $S_x(\omega)$ (the pulse pattern) or by $P(\omega)$ (the pulse shape). In the last section we discussed how to control PSD by the pulse pattern. The control of PSD by pulse shaping will now be discussed.

In Sec. 3.2, the basic pulse $p(t)$ was a simple rectangular pulse. Strictly speaking, in this case the bandwidth of $S_y(\omega)$ is infinite since $P(\omega)$ has infinite bandwidth. But we found that the essential bandwidth of $S_y(\omega)$ was finite. For example, most of the power of a bipolar signal is contained within the essential band 0 to f_o. Note, however, that the PSD is small but is still nonzero in the range $f > f_o$. Therefore, when such a signal is transmitted over a channel of bandwidth f_o, a significant portion of its spectrum is transmitted, but a small portion of the spectrum is suppressed. This will cause distortion in the received signal. But since the power in the suppressed components is small, the distortion is not severe. Nevertheless, there will be intersymbol interference (ISI) caused by the distortion which manifests itself as pulse spreading.

To resolve this difficulty of ISI, let us review our problem in brief. We need to transmit a pulse every T_o interval, the kth pulse being $a_k p(t)$. The channel has a finite bandwidth and we are required to detect the pulse amplitude a_k correctly (that is, without ISI). In our discussion so far we are considering time-limited pulses. Since such pulses cannot be bandlimited, part of their spectra is suppressed by a bandlimited channel. This causes pulse distortion and the consequent ISI. We can try to resolve this difficulty by using pulses which are bandlimited to begin with so that they can be transmitted intact over a bandlimited channel. But bandlimited pulses cannot be time limited. Obviously, various pulses will overlap and cause ISI. Thus, whether we begin with time-limited pulses or bandlimited pulses, it appears that ISI cannot be avoided. It is inherent in the finite transmission bandwidth. Fortunately, there is an escape from this blind alley. As noted in Chapter 2, pulse amplitudes can be detected correctly despite pulse spreading (or overlapping), if there is no ISI at the decision-making instants. This can be accomplished by a properly shaped bandlimited pulse. To eliminate ISI, Nyquist proposed three different criteria for pulse shaping.[5]

Nyquist's First Criterion for Zero ISI

In the first method, Nyquist achieves zero ISI by choosing the pulse shape such that it has a nonzero amplitude at its center (say $t = 0$) and zero amplitudes at $t = \pm nT_o$ ($n = 1, 2, 3, \ldots$), where T_o is the separation between successive transmitted pulses. It can be seen that each transmitted pulse causes zero ISI at all the remaining pulse centers, or signaling instants. If we restrict the pulse bandwidth to $f_o/2$, then only one pulse, sinc $(f_o t)$, (Fig. 3.16*a*) has this property (see Prob. 2.32).

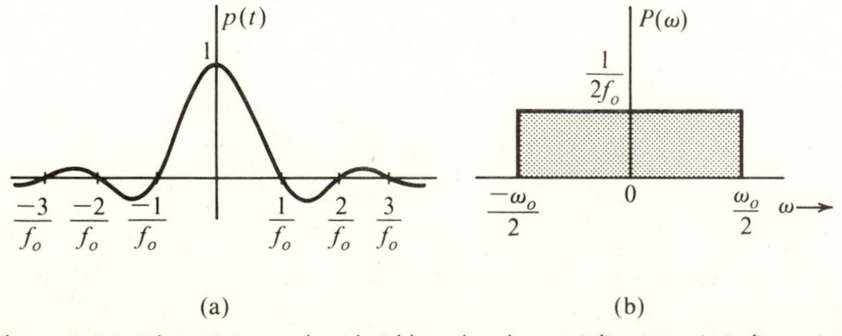

Figure 3.16 The minimum bandwidth pulse that satisfies Nyquist's first criterion and its spectrum.

$$\text{sinc}\,(f_o t) = \begin{cases} 1 & t = 0 \\ 0 & t = \pm nT_o \quad \left(T_o = \dfrac{1}{f_o}\right) \end{cases} \tag{3.24}$$

Using this pulse, we can transmit at a rate of f_o pulses/second without ISI, over a bandwidth $f_o/2$ (the Nyquist rate).

Because

$$\text{sinc}\,(f_o t) \leftrightarrow \frac{1}{f_o}\,\Pi\left(\frac{\omega}{2\pi f_o}\right) \tag{3.25}$$

the sinc $(f_o t)$ can be generated as an impulse response of an ideal filter of bandwidth $f_o/2$ (Fig. 3.16*b*).

It was shown in Sec. 2.7 that the sinc pulse has serious practical problems. A slight error in either the transmission rate or the sampling rate at the receiver, or a small time jitter in sampling instants, can cause sufficiently large ISI for the scheme to fail. This is because sinc $(f_o t)$ decays too slowly (as $1/t$). The solution is to find a pulse $p(t)$ that satisfies Eq. (3.24) but that decays faster than $1/t$. Nyquist has shown that such a pulse requires a bandwidth $kf_o/2$, with $1 < k < 2$.

This can be proved as follows. The desired pulse sinc $p(t)$ satisfies [Eq. (3.24)]

$$p(t) = \begin{cases} 1 & t = 0 \\ 0 & t = \pm nT_o \end{cases} \tag{3.26}$$

Let $p(t) \leftrightarrow P(\omega)$, where the bandwidth of $P(\omega)$ is in the range $(f_o/2,\ f_o)$ (Fig. 3.17*a*). If we sample $p(t)$ every T_o seconds by multiplying $p(t)$ by a train of impulses, then because of Eq. (3.26), all the samples except the one at the origin are zero.

$$p(t) \sum_{n=-\infty}^{\infty} \delta(t - nT_o) = \delta(t)$$

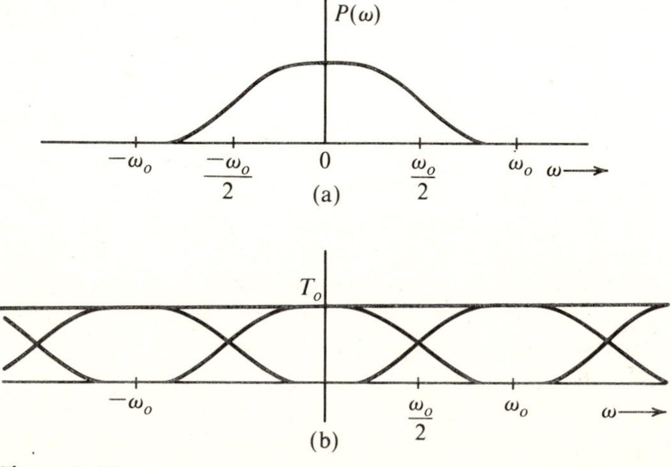

Figure 3.17

We now take the Fourier transform of both sides of the above equation. Using the results of Eq. (2.80) to obtain the transform of the left-hand side, we have

$$\frac{1}{T_o} \sum_{n=-\infty}^{\infty} P(\omega - n\omega_o) = 1 \tag{3.27}$$

or

$$\sum_{n=-\infty}^{\infty} P(\omega - n\omega_o) = T_o \tag{3.28}$$

Thus, the sum of the spectra formed by repeating $P(\omega)$ every ω_o is a constant T_o, as shown in Fig. 3.17*b*. Consider the spectrum in Fig. 3.17*b* over the range $0 < \omega < \omega_o$. Over this range only two terms $P(\omega)$ and $P(\omega - \omega_o)$ in the summation in Eq. (3.28) are involved. Hence

$$P(\omega) + P(\omega - \omega_o) = T_o \qquad 0 < \omega < \omega_o$$

Letting $\omega = x + \omega_o/2$

$$P\left(x + \frac{\omega_o}{2}\right) + P\left(x - \frac{\omega_o}{2}\right) = T_o \qquad |x| < \frac{\omega_o}{2} \tag{3.29}$$

Use of the result in Eq. (2.32) in Eq. (3.29) yields

$$P\left(\frac{\omega_o}{2} + x\right) + P^*\left(\frac{\omega_o}{2} - x\right) = T_o \qquad |x| < \frac{\omega_o}{2} \tag{3.30}$$

If we assume $P(\omega)$ of the form

$$P(\omega) = |P(\omega)|\, e^{-j\omega t_d}$$

then the term $e^{-j\omega t_d}$ is a time delay, and only $|P(\omega)|$ need satisfy Eq. (3.30). Because $|P(\omega)|$ is real, Eq. (3.30) implies

$$\left|P\left(\frac{\omega_o}{2} + x\right)\right| + \left|P\left(\frac{\omega_o}{2} - x\right)\right| = T_o \qquad |x| < \frac{\omega_o}{2} \tag{3.31}$$

Hence $|P(\omega)|$ should be of the form shown in Fig. 3.18. The bandwidth of $P(\omega)$ is $(\omega_o/2) + \omega_x$. If we define the roll-off factor $r = \omega_x/(\omega_o/2)$, then $0 \leq r \leq 1$, and the bandwidth of $P(\omega)$ is $(1 + r)\, f_o/2$.

Although the phase of $P(\omega)$ must be linear up to the frequency where $|P(\omega)|$ goes to 0, for most practical applications it is sufficient to equalize the phase characteristics up to the 10- to 15-dB attenuation point (i.e., the point where $P(\omega)$ is 10 to 15 dB below its peak). A spectrum with the same characteristics is required in what is known as vestigial sideband modulation (see Sec. 4.6). For this reason, we shall refer to the spectrum $P(\omega)$ in Eqs. (3.30) and (3.31) as a *vestigial spectrum*.

The pulse $p(t)$ in Eq. (3.26) has zero ISI at the centers of all other pulses transmitted at a rate of f_o pulses/second. Shaping the pulse $p(t)$ so that it causes zero ISI at the centers of all the remaining pulses (or signaling instants) is the first of the three methods proposed by Nyquist for zero ISI. This criterion for zero ISI is known

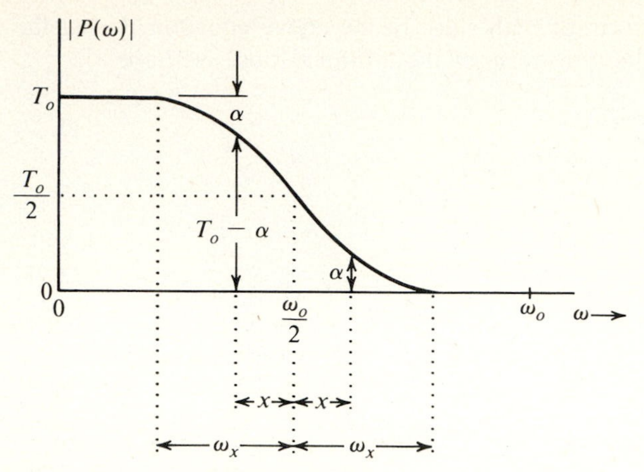

Figure 3.18 A vestigial spectrum.

as Nyquist's first criterion. We have shown that a pulse with a vestigial spectrum [Eq. (3.30) or Eq. (3.31)] satisfies Nyquist's first criterion for zero ISI.

Because $0 \leq r < 1$, the bandwidth of $P(\omega)$ is restricted to the range $f_o/2$ to f_o. The pulse $p(t)$ can be generated as a unit impulse response of a filter with transfer function $P(\omega)$. But because $P(\omega) = 0$ over a band, it violates the Paley-Wiener criterion and is therefore unrealizable. Because the vestigial roll-off characteristic is gradual, however, it can be more closely approximated by a practical filter. One family of spectra that satisfies Nyquist's first criterion is

$$P(\omega) = \begin{cases} \dfrac{1}{2}\left[1 - \sin\left(\dfrac{\pi\left(\omega - \dfrac{\omega_o}{2}\right)}{2\omega_x}\right)\right] & \left|\omega - \dfrac{\omega_o}{2}\right| < \omega_x \\ 0 & |\omega| > \dfrac{\omega_o}{2} + \omega_x \\ 1 & |\omega| < \dfrac{\omega_o}{2} - \omega_x \end{cases} \tag{3.32}$$

Figure 3.19*a* shows three curves, corresponding to $\omega_x = 0$ $(r = 0)$, $\omega_x = \omega_o/4$ $(r = 0.5)$ and $\omega_x = \omega_o/2$ $(r = 1)$. The respective impulse responses are shown in Fig. 3.19*b*. It can be seen that increasing ω_x (or r) improves $p(t)$; that is, more gradual cutoff reduces the oscillatory nature of $p(t)$ and causes it to decay more rapidly. For this reason let us consider the maximum value of ω_x, namely, $\omega_x = \omega_o/2$ $(r = 1)$. In this case Eq. (3.32) becomes

$$P(\omega) = \frac{1}{2}\left(1 + \cos\frac{\omega}{2f_o}\right)\Pi\left(\frac{\omega}{4\pi f_o}\right) \tag{3.33a}$$

$$= \cos^2\left(\frac{\omega}{4f_o}\right)\Pi\left(\frac{\omega}{4\pi f_o}\right) \tag{3.33b}$$

This characteristic is known in the literature as the *raised-cosine* characteristic, be-

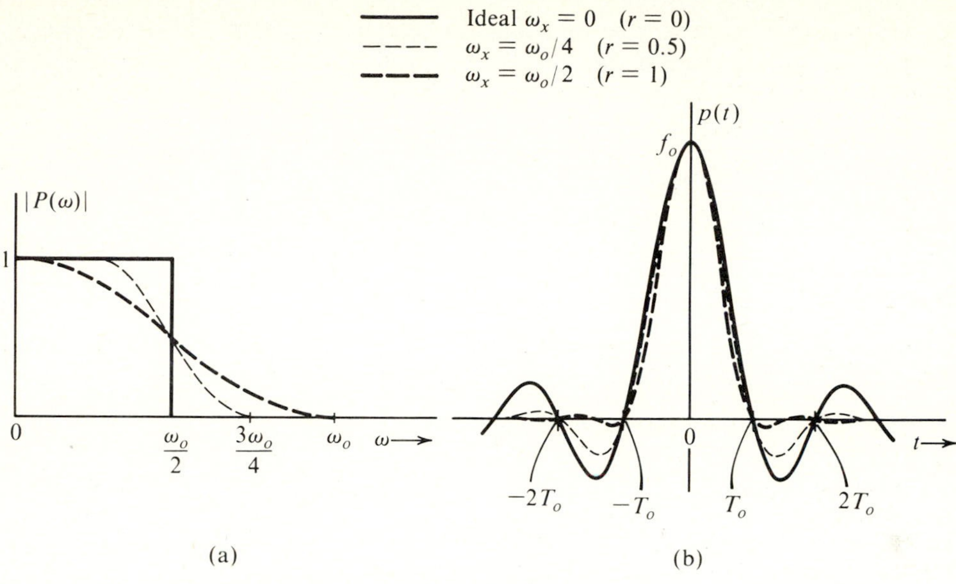

Figure 3.19 Pulses satisfying Nyquist's first criterion.

cause it represents a cosine raised by its peak amplitude. It is also known as the *full-cosine roll-off* characteristic. The unit impulse response of this spectrum can be found from Eq. (3.33a), using the result in Example 2.16 [Eq. (2.58)]

$$p(t) = f_o\left[\text{sinc}\,(2f_ot) + \frac{1}{2}\,\text{sinc}\,(2f_ot - 1) + \frac{1}{2}\,\text{sinc}\,(2f_ot + 1)\right] \tag{3.34a}$$

Expressing sinc $(x) = \sin(\pi x)/\pi x$ in this equation, we get

$$p(t) = \frac{f_o}{1 - 4f_o^2t^2}\,\text{sinc}\,(2f_ot) \tag{3.34b}$$

$$= f_o\,\frac{\cos \pi f_ot}{1 - 4f_o^2t^2}\,\text{sinc}\,(f_ot) \tag{3.34c}$$

This pulse is shown in Fig. 3.19*b* ($\omega_x = \omega_o/2$). We can make several important observations about the raised-cosine pulse. First, the bandwidth of this pulse is f_o and has a value f_o at $t = 0$ and is zero not only at all the remaining signaling instants but also at points midway between all the signaling instants. Secondly, it decays rapidly, as $1/t^3$. As a result, the raised-cosine pulse is relatively insensitive to deviations of f_o, sampling rate, timing jitter, and so on. Furthermore, the pulse-generating filter with transfer function $P(\omega)$ [Eq. (3.33b)] is closely realizable. The phase characteristics that go along with this filter are very nearly linear, so that no additional phase equalization is needed. Lastly, we shall see that the raised-cosine pulse also satisfies Nyquist's second criterion for zero ISI.

It should be remembered that it is the pulses received at the detector input that should have the Nyquist form for zero ISI. In practice, because the channel is not ideal (distortionless), in general, transmitted pulses should be shaped so that after passing through the channel with transfer function $H_c(\omega)$, they will be received in the proper shape (such as raised-cosine pulses) at the receiver. Hence, the transmitted pulse $p_i(t)$ should satisfy

$$P_i(\omega)H_c(\omega) = P(\omega)$$

where $P(\omega)$ is the vestigial spectrum in Eq. (3.31).

■ EXAMPLE 3.2

Determine the pulse transmission rate in terms of the transmission bandwidth B_T and the roll-off factor r. Assume a scheme using Nyquist's first criterion.

☐ **Solution:** The transmission bandwidth B_T is given by

$$2\pi B_T = \frac{\omega_o}{2} + \omega_x$$

$$= \frac{(1 + r)\,\omega_o}{2}$$

Therefore

$$B_T = \frac{(1 + r)\,\omega_o}{4\pi} = \frac{(1 + r)}{2} f_o$$

The basic pulse is of the form in Eq. (3.26), with zero values at $t = \pm nT_o = \pm n/f_o$ ($n = 1, 2, 3, \ldots$). Hence, in order to use Nyquist's first criterion, the pulses must be separated by a $1/f_o$ interval. The rate of transmission is f_o pulses/second. Hence,

$$f_o = \frac{2}{1 + r} B_T$$

Because $0 < r < 1$, the pulse transmission rate varies from $2B_T$ to B_T, depending on the choice of r. A smaller r gives a higher signaling rate. But the pulse $p(t)$ decays slowly, creating the same problems as those discussed for the sinc pulse. For the raised-cosine pulse $r = 1$ and $f_o = B_T$, half the Nyquist rate. ■

Nyquist's Second Criterion for Zero ISI

This scheme doubles the transmission rate (or reduces the bandwidth to half) by using a simple trick ("doubling the dotting speed") known to telegraphers more than 80 years ago.[6] Early telegraphers found that a telegraph system designed to transmit data at a rate f_o could also receive data at a rate $2f_o$ unambiguously, provided the received signal was interpreted correctly. To explain this behavior, consider a system designed to transmit data at a rate of f_o bits/second using polar signaling. When **1** is transmitted by a full-width rectangular pulse $p(t)$, the bandwidth is large enough so that the received pulse will rise to positive amplitude K, and when **0** is transmitted by $-p(t)$, the received pulse will rise to negative amplitude $-K$. If the transmitted pulse rate is doubled, transmitted pulse width is halved, and the received pulses cannot reach their full values. But if **1** is followed by **1**, we have two half-width pulses in succession—making one full-width pulse. This causes the received pulse to reach full positive value K. Similarly, if **0** is followed by **0**, the received pulse reaches full negative value $-K$. But if **1** is followed by **0** or vice versa, the received pulse will stay close to 0. Thus, the received signal can be interpreted as shown in the table below.

Received amplitude	Transmitted digit
K	**1** (previous digit also **1**)
$-K$	**0** (previous digit also **0**)
0	Complement of the previous digit

Thus, the data can be received unambiguously even when the rate is doubled. As we shall see, this trick really uses Nyquist's second criterion.

For this scheme to function, the pulse $p(t)$ should satisfy

$$p\left(\pm\frac{T_o}{2}\right) = C$$

and

$$p\left(\pm\frac{nT_o}{2}\right) = 0 \qquad n = 3, 5, 7, \ldots \tag{3.35}$$

This means the pulse $p(t)$ causes zero interference at points midway between all the signaling instants ($t = \pm 3T_o/2, \pm 5T_o/2, \ldots$) with the sole exceptions of the points midway between itself and its immediate neighbors ($t = \pm T_o/2$ (Fig. 3.20*a*). If we restrict the pulse bandwidth to $f_o/2$, then only the following pulse $p(t)$ satisfies this property in Eq. (3.35) (see Prob. 2.33):

$$p(t) = \frac{2f_o \cos \pi f_o t}{\pi(1 - 4f_o^2 t^2)} \tag{3.36a}$$

The Fourier transform* $P(\omega)$ of $p(t)$ is

$$P(\omega) = \cos\left(\frac{\omega}{2f_o}\right) \Pi\left(\frac{\omega}{2\pi f_o}\right) \tag{3.36b}$$

The pulse $p(t)$ and its spectrum $P(\omega)$ are shown in Fig. 3.20.

---

* Recognize that

$$p(t) = \frac{f_o}{2}\left[\operatorname{sinc} f_o\left(t + \frac{T_o}{2}\right) + \operatorname{sinc} f_o\left(t - \frac{T_o}{2}\right)\right]$$

Use of Eq. (2.58) yields $P(\omega)$ in Eq. (3.36b).

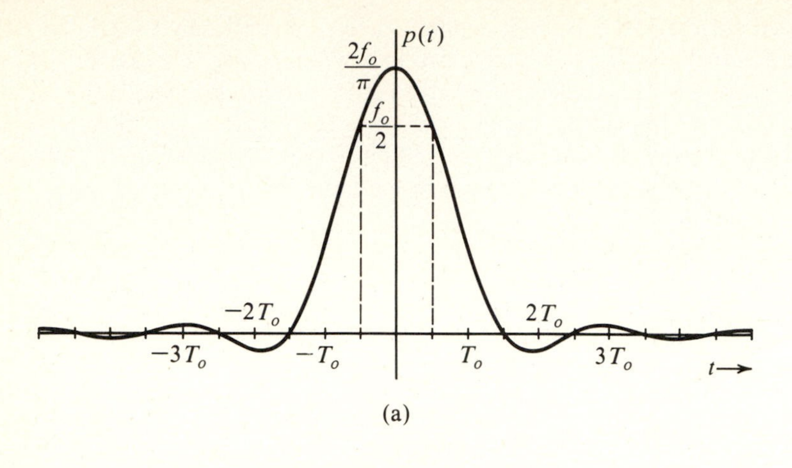

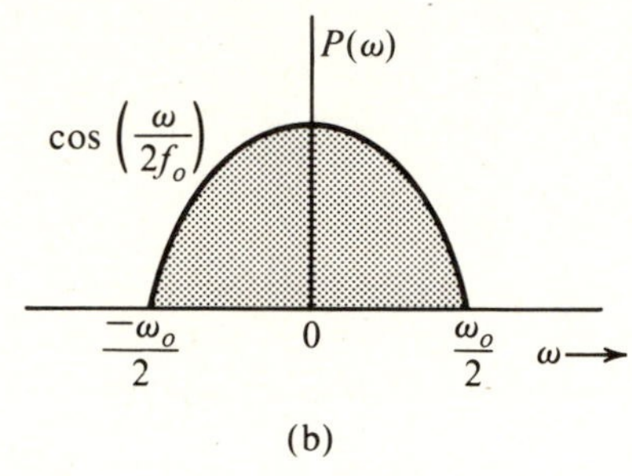

Figure 3.20 The minimum bandwidth pulse that satisfies Nyquist's second criterion and its spectrum.

The functioning of this scheme can be explained as follows. A **1** is transmitted by $p(t)$ and a **0** is transmitted by $-p(t)$. Note that

$$p\left(\pm\frac{T_o}{2}\right) = \frac{f_o}{2}$$

and

$$p\left(\pm\frac{nT_o}{2}\right) = 0 \qquad n = 3, 5, 7, \ldots$$

When a **1** is followed by **0** or vice versa, we have two pulses of opposite polarity in succession ($p(t)$ followed by $-p(t)$ or vice versa). Hence, at the midpoint of the two pulses, the pulse amplitudes are equal and of opposite signs, and the sum is zero (Fig. 3.21*a*). Thus, a sequence of two digits of opposite polarity is recognized by a zero amplitude at the midpoint of the two signaling instants. On the other hand, if **1** is followed by **1**, the amplitudes add to $2(f_o/2) = f_o$ at the midpoints of the two signaling instants (Fig. 3.21*b*). Similarly, **0** followed by **0** yields a negative value $-f_o$.

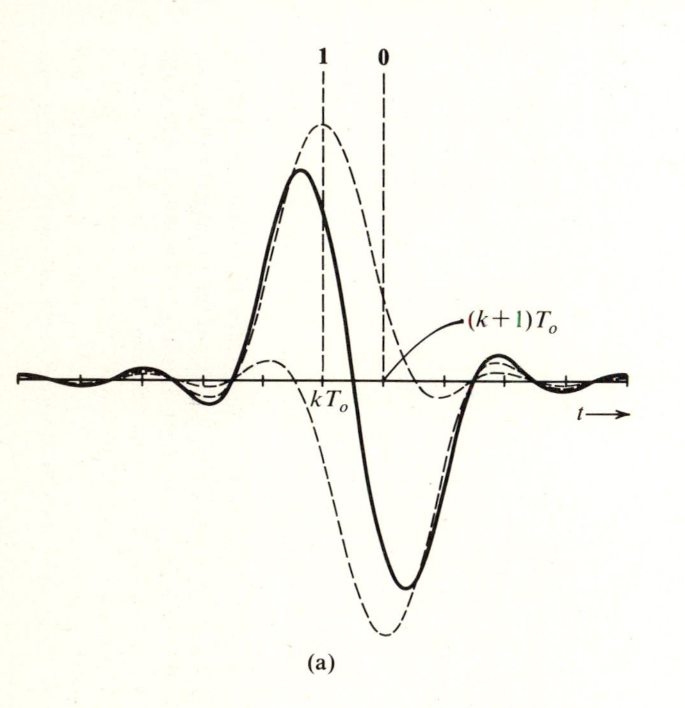

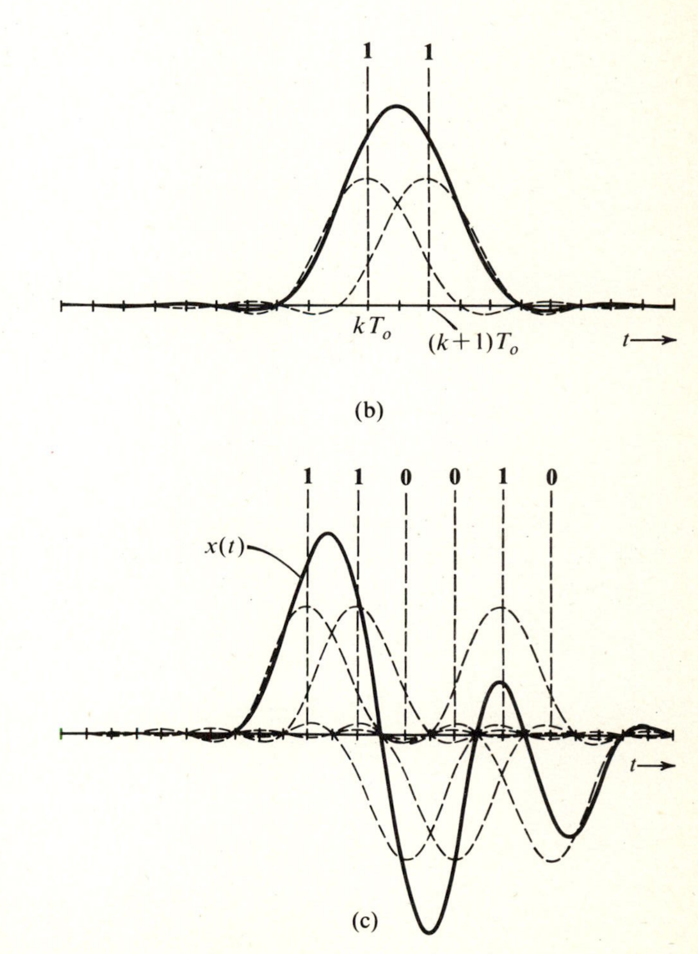

Figure 3.21 Communication using Nyquist's second criterion pulses.

A typical sequence **110010** and the corresponding waveform $x(t)$ are shown in Fig. 3.21*c*. The rate of transmission is f_o digits/second using a bandwidth $f_o/2$ Hz. To detect the received sequence, we sample the waveform $x(t)$ (Fig. 3.21*c*) at the midpoints of the signaling instants. Three possible sample values exist: f_o, 0, and $-f_o$. If the sample value is f_o, the digit detected is **1** (the preceding digit is also **1**), and if the sample value is $-f_o$, the digit detected is **0** (the preceding digit is also **0**). But the sample value of **0** means a transition; that is, the digit detected is **0** if the preceding digit is **1** and vice versa. This is easily verified from Fig. 3.21*c*. An example of detection is given in Fig. 3.22 (assuming no errors caused by channel noise). This example also indicates the error-detecting property of this scheme. Examination of samples of the waveform $x(t)$ in Fig. 3.22 shows that there are always an even number of zero-valued samples between two full-valued samples of the same polarity and an odd number of zero-valued samples between two full-valued samples of opposite polarity. Thus, the first sample of $x(t)$ is f_o, and the next full-valued sample (the fourth sample) is f_o. Between these full-valued samples of the same polarity, there are an even number (i.e., 2) of zero-valued samples. If one of the sample values is detected wrong, this rule is violated, and the error is detected.

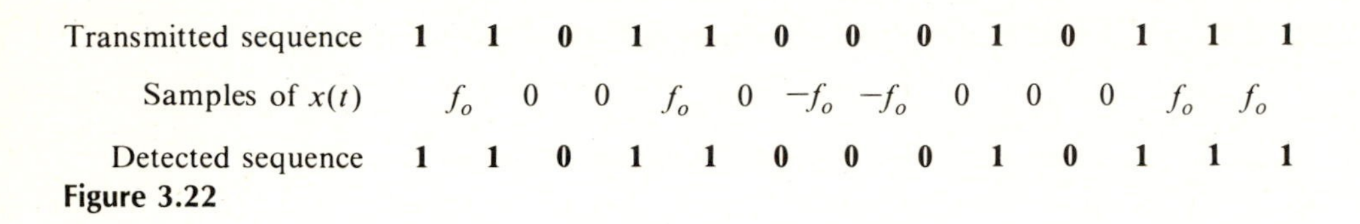

Transmitted sequence	**1**	**1**	**0**	**1**	**1**	**0**	**0**	**0**	**1**	**0**	**1**	**1**	**1**
Samples of $x(t)$		f_o	0	0	f_o	0	$-f_o$	$-f_o$	0	0	0	f_o	f_o
Detected sequence	**1**	**1**	**0**	**1**	**1**	**0**	**0**	**0**	**1**	**0**	**1**	**1**	**1**

Figure 3.22

The observant reader cannot fail to see the connection between the duobinary signaling scheme in Sec. 3.2 and the present scheme. In the duobinary case, binary data digits are transmitted by ternary (3-valued) pulses. Pulses of the same polarity are separated by an even number of zero pulses, and those of opposite polarity are separated by an odd number of zero pulses. In other words, when the received waveform is sampled at signaling instants, we obtain exactly the same pattern of sample values as that in Fig. 3.22 [sample of $x(t)$]. The two schemes are similar in nature (see Prob. 3.13). Hence, schemes using Nyquist's second criterion are duobinary.*

Precoding. In the detection scheme here, a zero-valued sample implies transition, that is, the digit is detected as **1** if the previous digit is **0** and vice versa. This means the digit interpretation is based on the preceding digit. If a digit were detected wrong,

* Also known as *correlative* or *partial-response* schemes.

the error would tend to propagate. Lender proposed precoding of data digits to eliminate the error propagation. If $a_1 a_2 a_3 \ldots a_k \ldots$ is the data digit sequence, we generate a new sequence $b_1 b_2 b_3 \ldots b_k \ldots$ by the rule

$$b_k = a_k \oplus b_{k-1}$$

where $\oplus$ is modulo 2 addition defined by the rule:

$$\mathbf{1} \oplus \mathbf{1} = \mathbf{0} \oplus \mathbf{0} = \mathbf{0}$$

$$\mathbf{1} \oplus \mathbf{0} = \mathbf{0} \oplus \mathbf{1} = \mathbf{1}$$

The sequence $b_1 b_2 b_3 \ldots$ is now transmitted using pulses $p(t)$ and $-p(t)$ in Fig. 3.20*a* for **1** and **0**, respectively. If the *k*th data digit is **0** ($a_k = 0$), $b_k = b_{k-1}$, and the *k*th sample of the waveform $x(t)$ will be either f_o or $-f_o$. On the other hand, if $a_k = \mathbf{1}$, b_k will be the complement of b_{k-1}, and the value of the *k*th sample of $x(t)$ will be zero.* Hence, the decoding rule is

$$a_k = \mathbf{0} \qquad \text{if } x(kT_o) = \pm f_o$$

$$a_k = \mathbf{1} \qquad \text{if } x(kT_o) = 0$$

In this case, the *k*th digit is decoded from the *k*th sample of $x(t)$ and does not depend upon the previous digit, thus eliminating the error propagation.

Nyquist's second criterion functions by achieving zero ISI at the midpoints of the signaling instants. The pulse $p(t)$ in Eq. (3.36a) decays rapidly as $1/t^2$, and, hence, the scheme is not sensitive to perturbations in the transmission rate or the sampling rate, jitter, and so on. This pulse can be generated as an impulse response of a filter with transfer function $P(\omega)$ in Eq. (3.36b). This filter can be easily approximated in practice.† In addition to the fact that the transmission rate is doubled by using Nyquist's second criterion, this scheme is capable of detecting single errors.

A careful examination of the raised-cosine pulse (Fig. 3.19 for $\omega_x = \omega_o/2$) shows that it satisfies Nyquist's second criterion [Eq. (3.35)] also. Thus, data transmitted by a raised-cosine pulse can be detected using Nyquist's first criterion (sampling $x(t)$ at the signaling instants) or by the second criterion (sampling $x(t)$ at the midpoints of the sampling instants). But it uses twice the bandwidth required for $p(t)$ in Eq. (3.36a).

The Modified Duobinary Scheme. The above scheme suffers from the fact that its PSD is nonzero at $\omega = 0$. This drawback is corrected in the *modified duobinary scheme* proposed by Lender.[7] This scheme also permits data transmission at a rate of f_o bits/second over a bandwidth $f_o/2$. The transmitted signal's PSD has a dc null. The basic pulse $p(t)$ used in the modified duobinary scheme is (Fig. 3.23*a*)

* This precoding scheme is known by the more familiar name of differential coding in the literature (see Sec. 6.6 for further discussion).

† It can be closely realized with a delay.

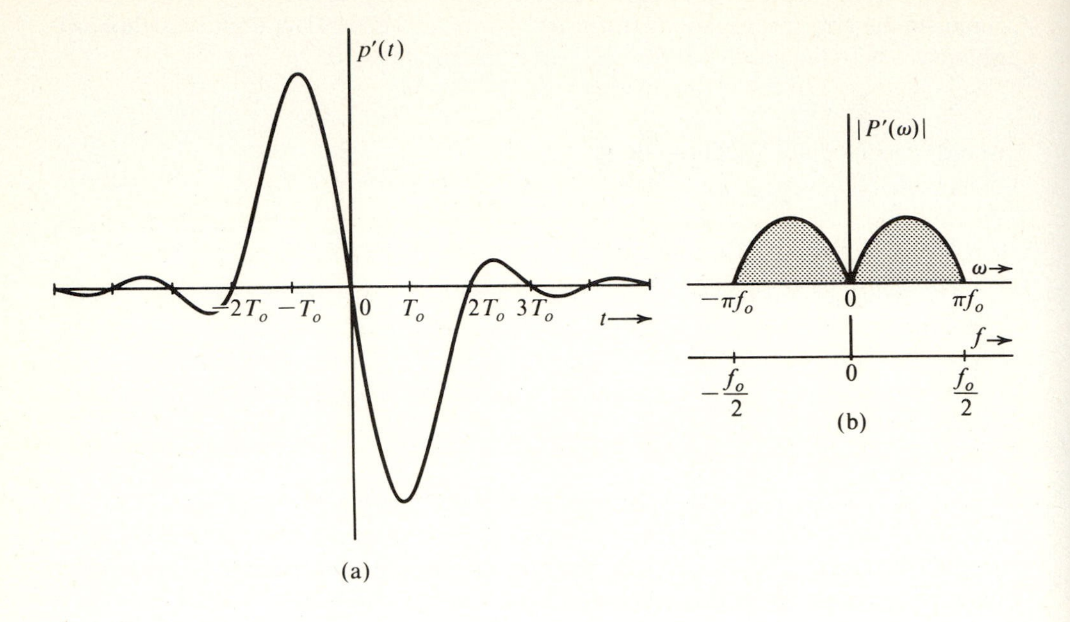

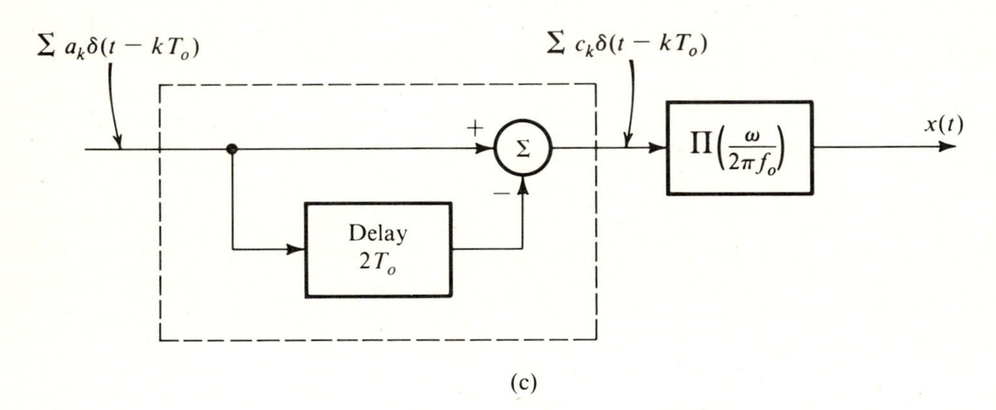

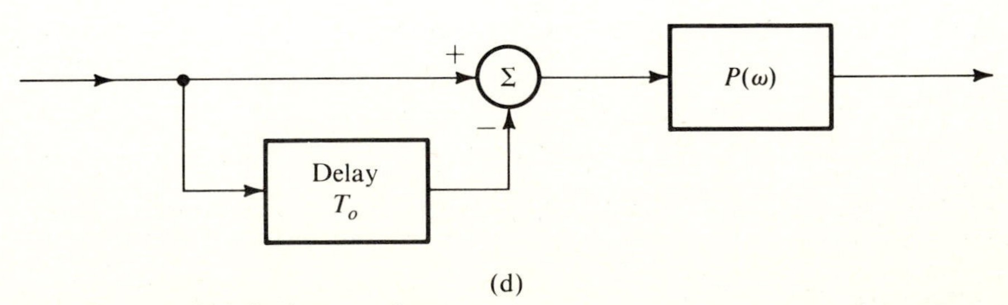

Figure 3.23 Modified duobinary scheme.

$$p'(t) = \frac{2f_o \sin \pi f_o t}{\pi(f_o^2 t^2 - 1)} \tag{3.37a}$$

The Fourier transform* $P'(\omega)$ of this pulse is (Fig. 3.23*b*)

$$P'(\omega) = 2j \sin \omega T_o \, \Pi\left(\frac{\omega}{2\pi f_o}\right) \tag{3.37b}$$

We shall soon see that this filter is closely realizable. All we need to show is that using $p'(t)$, it is possible to transmit f_o bits/second with zero ISI. Figure 3.23*c* shows a system with transfer function

$$\hat{P}(\omega) = (1 - e^{-2j\omega T_o}) \, \Pi\left(\frac{\omega}{2\pi f_o}\right)$$

$$= e^{-j\omega T_o}(2j \sin \omega T_o) \, \Pi\left(\frac{\omega}{2\pi f_o}\right)$$

Hence, Fig. 3.23*c* is the realization of $P'(\omega)$ in Eq. (3.37b) with a delay T_o. Let the data to be transmitted be $a_0 a_1 a_2 \ldots a_k \ldots$ where a_k is either 0 or 1. We apply an impulse train $\Sigma\, a_k\, \delta(t - kT_o)$ at the input of $\hat{P}(\omega)$. At the adder output, we have another train $\Sigma\, c_k\, \delta(t - kT_o)$, where

$$c_k = a_k - a_{k-2}$$

Because a_k can be 0 or 1, c_k has three possible values: 0, ± 1. From the above equation

$$a_k = c_k + a_{k-2}$$

Hence,

if $c_k = 1$, a_k must be 1

if $c_k = -1$, a_k must be 0

if $c_k = 0$, $a_k = a_{k-2}$

Thus, knowing c_k, we can uniquely determine a_k.

The impulse train $\Sigma\, c_k\, \delta(t - kT_o)$ is applied to an ideal filter whose impulse response is f_o sinc $(f_o t)$. Hence, data c_k is transmitted by the pulse sinc $(f_o t)$, which satisfies Nyquist's first criterion. The filter output $x(t) = f_o \, \Sigma\, c_k \operatorname{sinc} f_o(t - kT_o)$ can be sampled, and the data c_k can be recovered with zero ISI. Knowing c_k we can uniquely determine a_k. This proves that the modified duobinary scheme can transmit

---

*Recognize that $p'(t)$ in Eq. (3.37a) is

$$p'(t) = f_o \operatorname{sinc}[f_o(t + T_o)] - f_o \operatorname{sinc}[f_o(t - T_o)]$$

Hence

$$P'(\omega) = \Pi\left(\frac{\omega}{2\pi f_o}\right)[e^{j\omega T_o} - e^{-j\omega T_o}]$$

f_o digits/second over a bandwidth $f_o/2$ with zero ISI. We shall now show that $P'(\omega)$ is readily realizable. For this purpose, we express $P'(\omega)$ as [Eq. (3.37b)]

$$P'(\omega) = 4j \sin\left(\frac{\omega T_o}{2}\right) \cos\left(\frac{\omega T_o}{2}\right) \Pi\left(\frac{\omega}{2\pi f_o}\right)$$

$$= 2e^{-j\omega T_o/2}[1 - e^{-j\omega T_o}] \cos\left(\frac{\omega}{2f_o}\right) \Pi\left(\frac{\omega}{2\pi f_o}\right)$$

$$= 2e^{-j\omega T_o/2}[1 - e^{-j\omega T_o}] P(\omega)$$

where $P(\omega)$ is the filter [Eq. (3.36b)] satisfying Nyquist's second criterion. Figure 3.23*d* shows a realization of $P'(\omega)$ according to the above equation (within a delay of $T_o/2$). As mentioned earlier, $P(\omega)$ can easily be approximated in practice. Hence $P'(\omega)$ is readily realized.

Precoding. We have seen that the digit interpretation in the modified duobinary scheme is based on the preceding digit. Thus, in Fig. 3.23*c*, if $c_k = 0$, $a_k = a_{k-2}$. This tends to propagate error. In duobinary signaling, precoding was used to eliminate this problem. Lender proposed a similar precoding for the modified duobinary scheme. If $a_1 a_2 \ldots, a_k \ldots$ is the data digit sequence, we generate a new sequence $b_1 b_2 \ldots b_k \ldots$ by the rule

$$b_k = a_k \oplus b_{k-2}$$

This precoded sequence $b_1 b_2 \ldots b_k \ldots$ is now transmitted using the modified duobinary scheme. This is equivalent to applying the precoded sequence at the input of the system in Fig. 3.23*c*. The output sequence $c_1 c_2 \ldots c_k$ in Fig. 3.23*c* is given by

$$c_k = b_k - b_{k-2}$$

$$= (a_k \oplus b_{k-2}) - b_{k-2}$$

Because a_k's and b_k's can be either 0 or 1, c_k can be 0 or ± 1. From the above equation it follows that

if $c_k = 1$ or -1, a_k must be 1

if $c_k = 0$, a_k must be 0

Thus, knowing c_k we can determine a_k uniquely without depending on the previous digit, thereby eliminating the error propagation.

Nyquist's Third Criterion for Zero ISI

In this scheme, ISI is eliminated by shaping the pulse so that the total area under the pulse in the signaling interval is nonzero but is zero in any other signaling interval. The pulse detection is accomplished by measuring the area in each signaling interval. One pulse $p(t)$ that satisfies this criterion has a spectrum

$$P(\omega) = \frac{1}{\text{sinc}\left(\dfrac{\omega}{2\pi f_o}\right)} \Pi\left(\frac{\omega}{2\pi f_o}\right)$$

It can be shown that $p(t)$, the inverse transform of this $P(\omega)$, satisfies the third criterion. This scheme is inferior to the first and second from the point of view of noise immunity.[8]

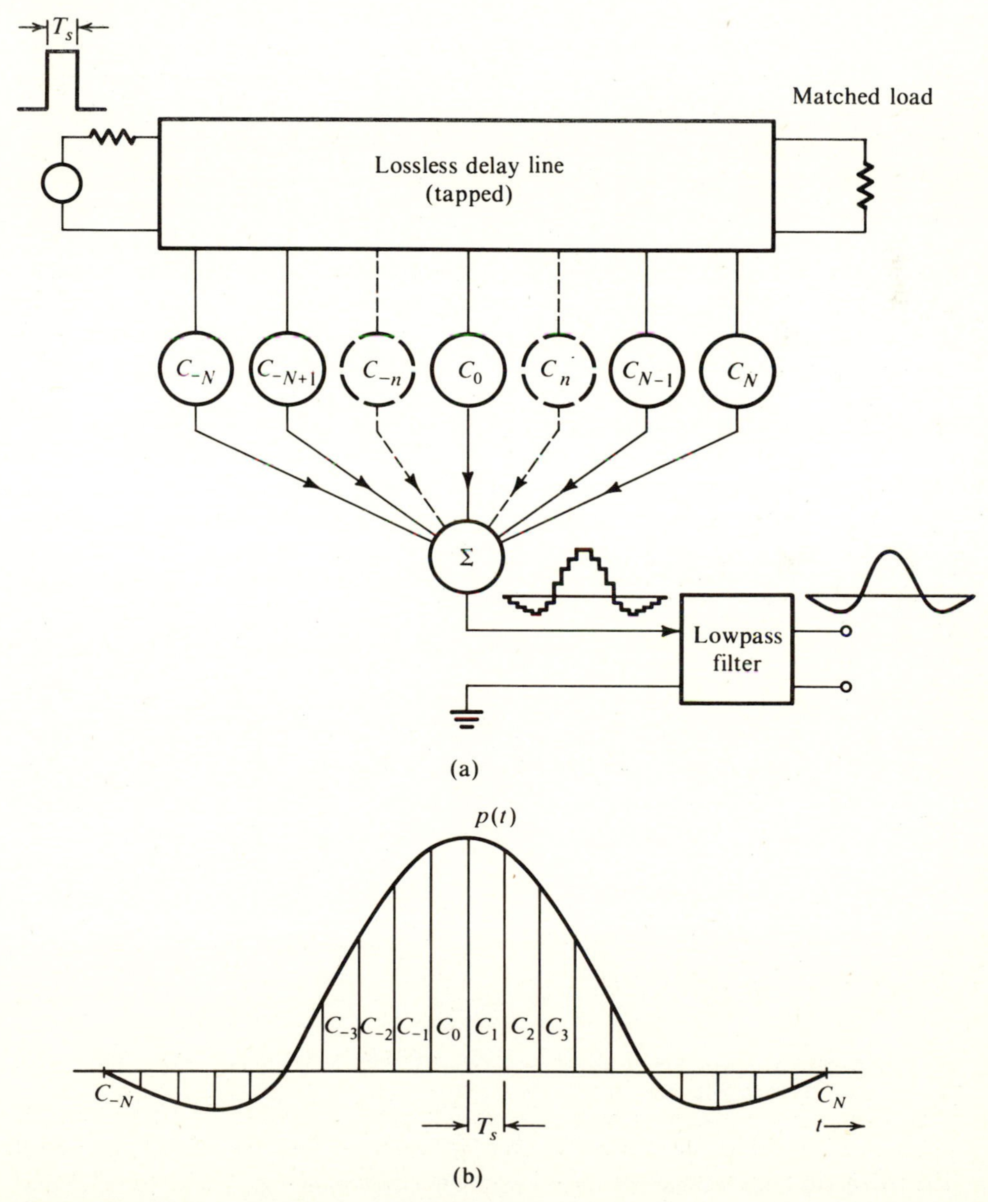

Figure 3.24 Pulse generation by transversal filter.

Pulse Generation

A pulse $p(t)$ satisfying a Nyquist criterion can be generated as the unit impulse response of a filter with transfer function $P(\omega)$. This may not always be easy. A better method is to generate the waveform directly, using the transversal filter in Fig. 3.24*a*. Sample values are obtained from the graph of the pulse $p(t)$ to be generated (Fig. 3.24*b*), and the constants of the filter are set in proportion to these sample values in sequence, as shown in Fig. 3.24*a*. When a narrow rectangular pulse with the width of the sampling interval is applied at the input of the transversal filter, the output will be a staircase approximation of $p(t)$. When the output is passed through a lowpass filter, the output will be a smoother version. The approximation can be improved by reducing the pulse sampling interval.

It should be stressed once again that the pulses arriving at the detector input of the receiver need to have the Nyquist form. Hence, the transmitted pulses should be so shaped that after passing through the channel, they are received in the form of Nyquist pulses. In practice, however, pulses need not be shaped rigidly at the transmitter. The final shaping can be carried out by an equalizer at the receiver, as discussed later (Sec. 3.5).

3.4 SCRAMBLING

Scramblers can remove long strings of **0**'s in binary data. In general, a scrambler tends to make the data more random by removing long strings of **1**'s or **0**'s. Figure 3.25 shows a typical scrambler and unscrambler.

The scrambler consists of a feedback shift register, and the matching unscrambler has a feedforward shift register, as shown in Fig. 3.25. Each stage in the shift register delays a bit by one unit. To show that the matched unscrambler does indeed unscramble, consider the output sequence T of the scrambler (Fig. 3.25*a*). If S is the input sequence, then

$$S \oplus D^3T \oplus D^5T = T \tag{3.38}$$

where D represents the delay operator; that is, D^nT is the sequence T delayed by n units.

Remember that the modulo 2 sum of any sequence with itself gives a sequence of all **0**'s. Adding $(D^3 \oplus D^5)T$ to both sides of the above equation, we get

$$\begin{aligned} S &= T \oplus (D^3 \oplus D^5)T \\ &= [1 + (D^3 \oplus D^5)]T \\ &= (1 \oplus F)T \text{ with } F = D^3 \oplus D^5 \end{aligned} \tag{3.39}$$

To design the unscrambler at the receiver, we start with the received sequence T. From Eq. (3.39), it follows that

$$S = T \oplus FT = T \oplus (D^3 + D^5)T$$

This is readily implemented by the arrangement shown in Fig. 3.25*b*.

Note that a single detection error in the received sequence T will affect three

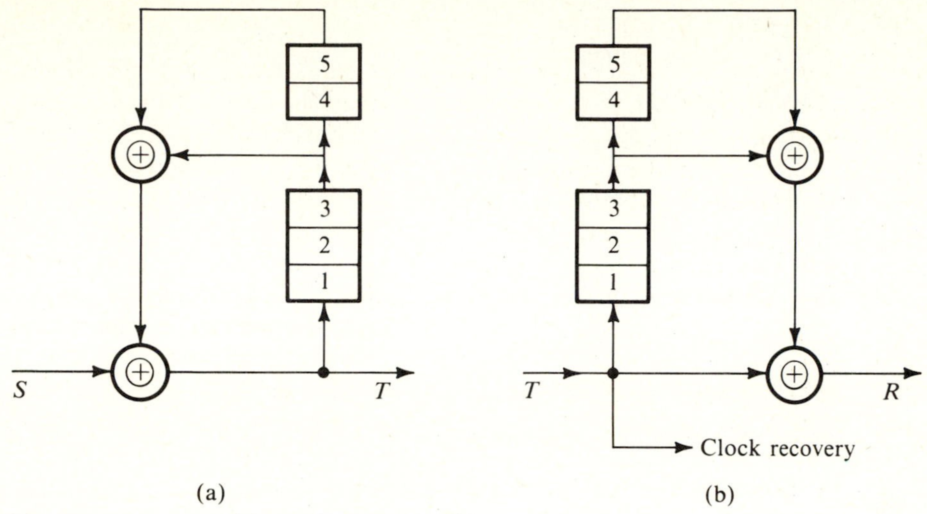

Figure 3.25 A scrambler and an unscrambler.

output bits in R. Hence, scrambling has the disadvantage of causing multiple errors for a single received bit error.

■ EXAMPLE 3.3

The data stream **101010100000111** is fed to the scrambler in Fig. 3.25*a*. Find the scrambler output T, assuming the initial content of the registers to be zero.

☐ **Solution:** From Fig. 3.25*a* we observe that the sequence S enters the register and is returned as $(D^3 \oplus D^5)S = FS$ through the feedback path. This new sequence FS again enters the register and is returned as F^2S, and so on. Hence,

$$T = S \oplus FS \oplus F^2S \oplus F^3S \oplus \cdots$$

$$= (1 \oplus F \oplus F^2 \oplus F^3 \oplus \cdots)S \qquad \textbf{(3.40)}$$

Recognizing that

$$F = D^3 \oplus D^5$$

$$F^2 = (D^3 \oplus D^5)(D^3 \oplus D^5) = D^6 \oplus D^{10} \oplus D^8 \oplus D^8$$

Because modulo 2 addition of any sequence with itself is zero, $D^8 \oplus D^8 = 0$, and

$$F^2 = D^6 \oplus D^{10}$$

and

$$F^3 = (D^6 \oplus D^{10})(D^3 \oplus D^5) = D^9 \oplus D^{11} \oplus D^{13} \oplus D^{15}$$

and so on. Hence [see Eq. (3.40)],

$$T = (1 \oplus D^3 \oplus D^5 \oplus D^6 \oplus D^9 \oplus D^{10} \oplus D^{11} \oplus D^{12} \oplus D^{13} \oplus D^{15} \cdots)S$$

Because D^nS is simply the sequence S delayed by n bits, various terms in the above equation correspond to the sequences shown below.

$$\begin{aligned}
S &= \mathbf{101010100000111} \\
D^3S &= \mathbf{000101010100000111} \\
D^5S &= \mathbf{00000101010100000111} \\
D^6S &= \mathbf{000000101010100000111} \\
D^9S &= \mathbf{000000000101010100000111} \\
D^{10}S &= \mathbf{0000000000101010100000111} \\
D^{11}S &= \mathbf{00000000000101010100000111} \\
D^{12}S &= \mathbf{000000000000101010100000111} \\
D^{13}S &= \mathbf{0000000000000101010100000111} \\
D^{15}S &= \mathbf{000000000000000101010100000111} \\
T &= \mathbf{101110001101001}
\end{aligned}$$

Note that the input sequence contains the periodic sequence **10101010 . . .** , as well as a long string of **0**'s. The scrambler output effectively removes the periodic component as well as the long string of **0**'s. The input sequence has 15 digits. The scrambler output up to the 15th digit only is shown, because all the output digits beyond 15 depend on input digits beyond 15, which are not given.

It is easy to verify that the unscrambler output is indeed S when the above sequence T is applied at its input (see Prob. 3.16). ■

3.5 THE REGENERATIVE REPEATER

One of the significant advantages of digital systems over analog systems is their ability to reconstruct the transmitted pulse train at intervals along the channel in order to combat the accumulation of signal distortion (or dispersion) and noise. The process of regenerating the pulse train is performed by regenerative repeaters.

Basically, a regenerative repeater performs three functions: (1) reshaping incoming pulses by means of an equalizer, (2) the extraction of timing information required to sample incoming pulses at optimum instants, and (3) decision making based on the pulse samples. The schematic of a repeater is shown in Fig. 3.26.

The Preamplifier and Equalizer

A pulse train is attenuated and distorted by the transmission medium. The distortion is in the form of dispersion, which is caused by an attenuation of high-frequency components of the pulse train. Theoretically, an equalizer should have a frequency characteristic that is the inverse of that of the transmission medium. This will restore higher-frequency components and eliminate pulse dispersion. Unfortunately, this also increases the received noise by increasing its bandwidth and boosting its high-frequency components. For digital signals, however, complete equalization is really

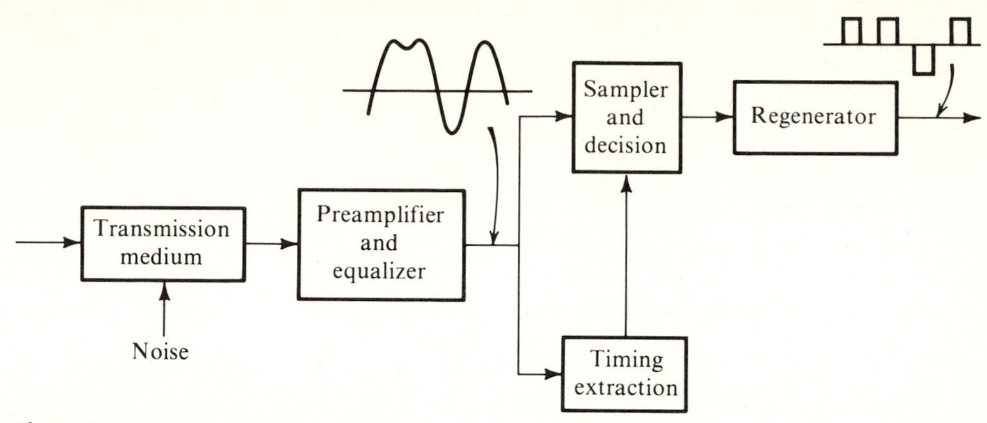

Figure 3.26 A regenerative repeater.

not necessary, because a detector has to make relatively simple decisions—such as whether the pulse is positive or negative (or whether the pulse is present or absent). Therefore, considerable pulse dispersion can be tolerated. Pulse dispersion results in pulse spreading and consequent interference with neighboring pulses (intersymbol interference, or ISI). This causes errors in pulse detection. The interfering noise also causes errors in pulse detection. For this reason, design of an optimum equalizer involves an inevitable compromise between reducing ISI and reducing interfering noise. A judicious choice of the equalization characteristics is a central feature of all digital communication systems.

It is really not necessary to eliminate or minimize ISI; that is, the interference with neighboring pulses for all t. All that is needed is to eliminate or minimize interference with neighboring pulses at their respective sampling instants only, because the decision is based only on sample values. This can be accomplished by the transversal-filter equalizer encountered earlier (Fig. 3.24*a*). The time delay τ between successive taps is chosen to be T_o, the pulse sampling interval.

To begin with, set the multiplier constants $C_0 = 1$ and $C_k = 0$ for all other values of k in the transversal filter in Fig. 3.24*a*. Thus the output of the filter will be the same as the input delayed by NT_o. Suppose a single pulse is transmitted, and the corresponding received pulse at the input of the transversal filter is $p_r(t)$, as shown in Fig. 3.27*a*. With the above tap-gain setting, the filter output will be exactly $p_r(t - NT_o)$, that is, $p_r(t)$ delayed by NT_o. This delay is not relevant to our discussion. Hence, for convenience, we shall assume that $p_r(t)$ in Fig. 3.27*a* also represents the filter output. To avoid interference with other pulses, we only need to ensure that the pulse amplitude is negligible at all the multiples of T_o. From Fig. 3.27*a*, we see that the pulse amplitudes a_1, a_{-1}, and a_2 at T_o, $-T_o$, and $2T_o$, respectively, are not negligible. By adjusting various multiplier constants (C_k's), we can accomplish the desired goal. Suppose a_1 is the largest of the three values (a_1, a_{-1}, and a_2) to be corrected, with $a_1/a_0 = r$. Set the constant $C_1 = -r$. This will cause an additional pulse $-r$ times the pulse in Fig. 3.27*a* to be generated with a delay of T_o. It has an amplitude $-ra_0 = -a_1$ at T_o and will cause the filter output to be zero at $t = T_o$, as desired. But this pulse

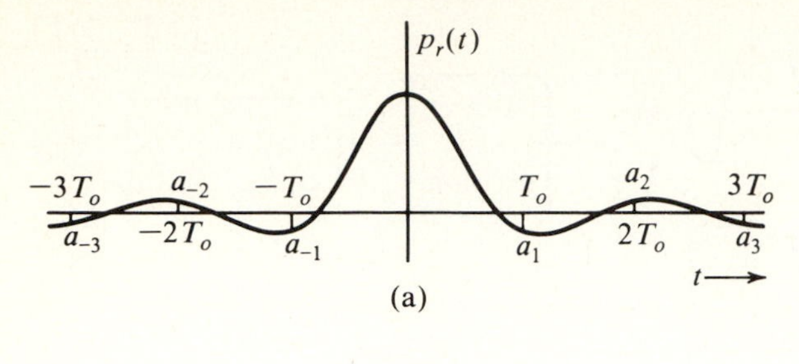

(a)

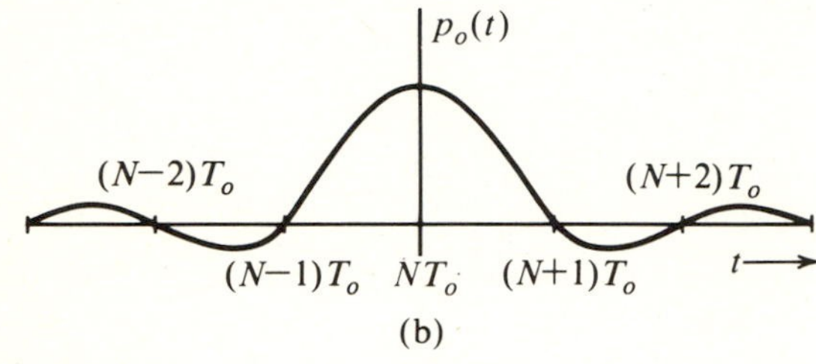

(b)

Figure 3.27

causes minor modifications (perhaps undesirable) at other instants. For example, at $t = 0$ it will cause a change by $-ra_{-1}$. Because $a_{-1} < a_1$, this change is less than r^2a_0. Hence, this new setting will cause an undesirable change of at most r^2a_0 at other instants. If the next biggest error is found to be at $-T_o$, the constant C_{-1} should be adjusted in the same manner. It will probably be necessary at some stage to readjust C_1. If r is small, however, it is clear that this procedure will converge.

This problem can also be tackled analytically to obtain exact values of tap settings as discussed in Appendix 3.1.

Automatic and Adaptive Equalization. The setting of the tap gains of an equalizer can be done automatically by using a special sequence of pulses prior to the data transmission and by using an iterative technique to obtain optimum tap gains. In adaptive equalizers, the tap gains are adjusted continuously during transmission. In practice both methods are used.[9,10]

The Eye Diagram

The ISI can be conveniently studied on an oscilloscope through what is known as *the eye diagram*. A random binary pulse sequence is sent over the channel. The channel output is applied to the vertical input of an oscilloscope. The time base of the scope is triggered at the same rate as that of the incoming pulses, and it yields a sweep lasting exactly T_o, the interval of one pulse. The oscilloscope shows the superposition of several traces, which is nothing but the input signal (vertical input) cut up every T_o and then superimposed (Fig. 3.28). The oscilloscope pattern thus formed looks like a human eye, and, hence, the name eye diagram.

As an example, consider the transmission of a binary signal by polar rectangular pulses. If the channel is ideal with infinite bandwidth, pulses will be received without distortion. When this signal is cut up, every pulse interval or each piece will be either

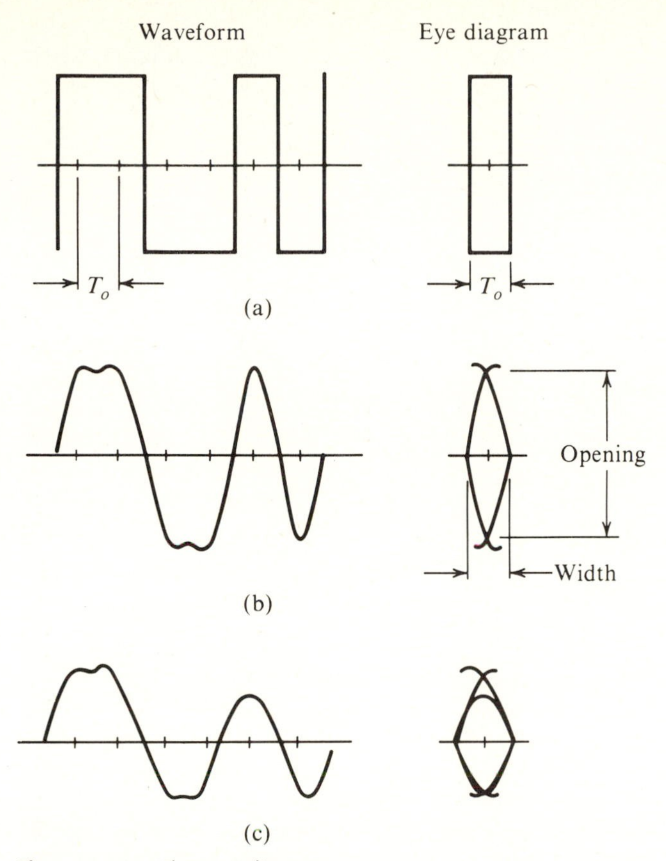

Figure 3.28 The eye diagram.

a positive or a negative rectangular pulse. When those are superimposed, the resulting eye diagram will be as shown in Fig. 3.28*a*. If the channel is not distortionless or has finite bandwidth, or both, received pulses will no longer be rectangular but will be rounded and spread out. If the equalizer is adjusted properly to eliminate ISI at the pulse sampling instants, the resulting eye diagram will be rounded (Fig. 3.28*b*) but will still have full opening at the midpoint of the eye. This is because the midpoint of the eye represents the sampling instant of each pulse, where the pulse amplitude is maximum; also, at this point there is no interference from other pulses (because of zero ISI). If ISI is not zero, pulse values at their respective sampling instants will deviate from full-scale values by a varying amount in each trace, causing a blur and thus closing the eye partially at the midpoint, as shown in Fig. 3.28*c*.

In the presence of channel noise, the eye will tend to close in all cases. Smaller noise will cause proportionately less closing. The decision threshold as to which symbol (**1** or **0**) is transmitted is the midpoint of the eye.* Observe that for zero ISI,

---

*This is true for a two-level decision, e.g., when $p(t)$ and $-p(t)$ are used for **1** and **0**, respectively. For a three-level decision (bipolar signaling, for example), there will be two thresholds.

the system can tolerate noise up to half the eye opening at the midpoint. Because the ISI reduces the eye opening, it clearly reduces noise tolerance. The eye diagram is also used to determine optimum tap settings of the equalizer. Taps are adjusted to obtain the maximum eye opening.

The eye diagram is useful in deciding the optimum sampling or decision-making instant (the instant when the eye opening is maximum), as well as the amount of noise that can be tolerated. The width of the eye indicates the time interval over which the decision can be made. If the decision-making instant deviates from the instant when the eye has a maximum opening, the margin to noise tolerance is reduced. This causes higher error probability in pulse detection. Because in any system the sampling instants deviate from the ideal (because of the presence of jitter), the eye diagram allows one to study the effects of jitter.

Timing Extraction

The received digital signal needs to be sampled at precise instants. This requires a clock signal at the receiver in synchronism with the clock signal at the transmitter (*symbol* or *bit synchronization*). Three general methods of synchronization exist:

1. Derivation from a primary or a secondary standard (e.g., transmitter and receiver slaved to a master timing source).
2. Transmitting a separate synchronizing signal (pilot clock).
3. Self-synchronization, where the timing information is extracted from the received signal itself.

The first method is suitable for large volumes of data and high-speed communication systems because of its high cost. In the second method, part of the channel capacity is used to transmit timing information and is suitable when the available capacity is large compared to the data rate. The third method is a very efficient method of timing extraction or clock recovery because the timing is derived from the digital signal itself. An example of the self-synchronization method will be discussed here.

We have already shown [Eq. (3.14)] that a digital signal, such as an on-off binary signal (Fig. 3.29*a*), contains a discrete component of the clock frequency itself. This can also be seen from the fact that such a waveform can be expressed as a sum of two waveforms: (1) a random component (Fig. 3.29*b*), and (2) a periodic component (Fig. 3.29*c*) with the same fundamental frequency as the clock frequency. Hence, when the on-off binary signal is applied to a resonant circuit tuned to the clock frequency, the output signal is the desired clock signal.

Not all the binary signals contain a discrete component of the clock frequency. For example, a bipolar signal has no discrete component of any frequency [see Eq. (3.18)]. In such cases, timing can be extracted by using a nonlinear operation. In the bipolar case, for instance, a simple rectification converts a bipolar signal to an on-off signal, which can readily be used to extract timing.

Small random deviations of the incoming pulses from their ideal location (known as *timing jitter*) are always present, even in the most sophisticated systems. Although the source emits pulses at the right instants, subsequent operations during transmission (e.g., at repeaters) tend to deviate pulses from these original positions. The Q of the

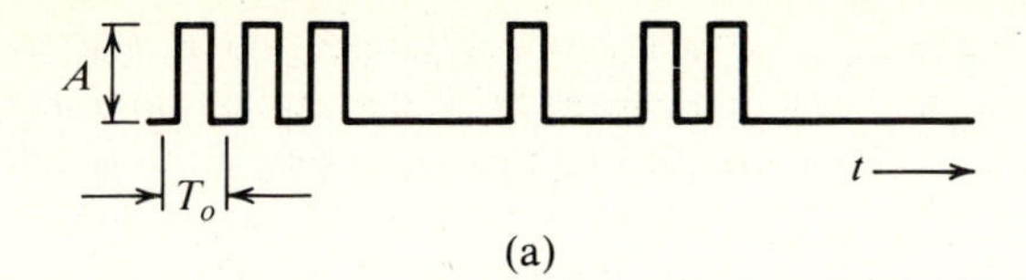

(a)

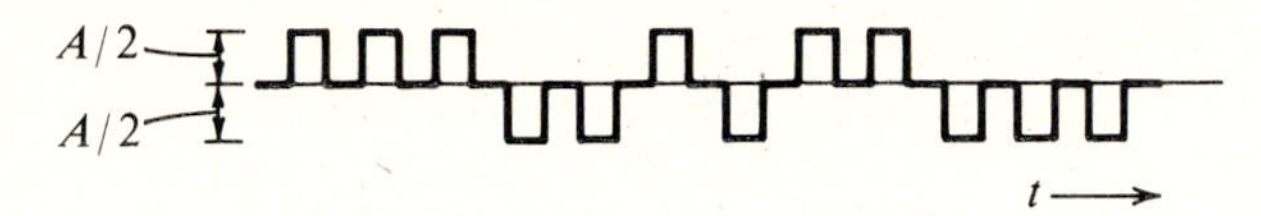

(b) Random component

(c) Periodic component

Figure 3.29 The on-off signal and its components.

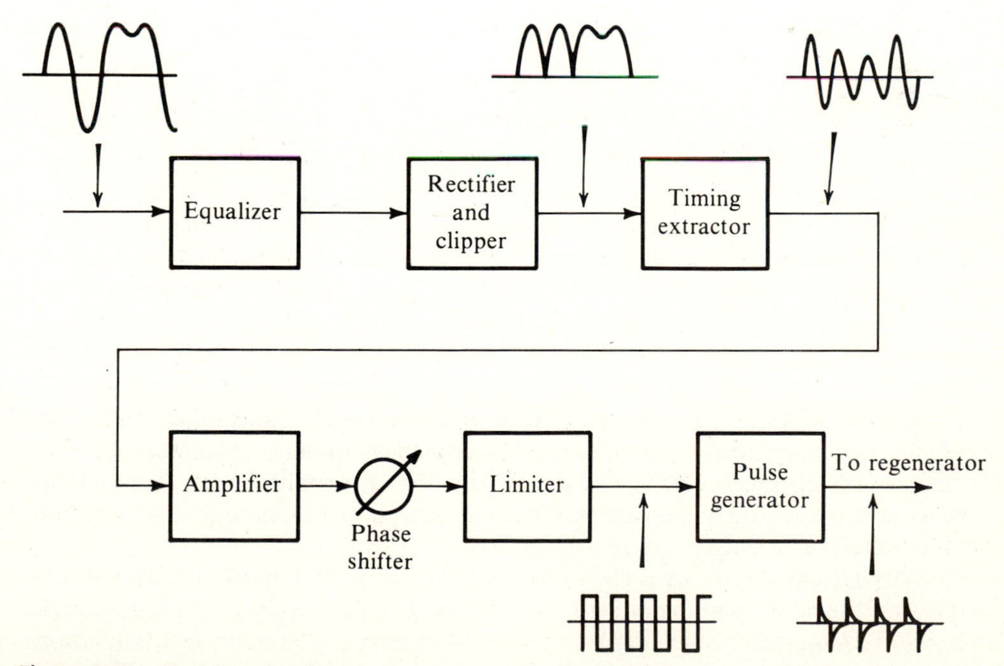

Figure 3.30 Timing extraction.

tuned circuit used for timing extraction must be large enough to provide an adequate suppression of timing jitter, yet small enough to meet the stability requirements. During the intervals where there are no pulses in the input, the oscillation continues because of the flywheel effect of the high-Q circuit. But still the oscillator output is sensitive to the pulse pattern; for example, during a long string of **1**'s the output amplitude will increase, whereas during a long string of **0**'s it will decrease. This introduces additional jitter in the timing signal extracted.

The complete timing extractor and time-pulse generator for a bipolar case is shown in Fig. 3.30. The sinusoidal output of the oscillator is passed through a phase shifter that adjusts the phase of the timing signal so that the timing pulses occur at the maximum eye opening. This method is used to recover the clock at each of the regenerators in a PCM system. The jitter introduced by successive regenerators adds up, and after a certain number of regenerators it is necessary to use a regenerator with a more sophisticated clock recovery system.

■ EXAMPLE 3.4

Show that an on-off signal using full-width pulses (Fig. 3.31*b*) does not contain a discrete component of the clock frequency. Find a scheme to extract timing from such a signal.

If we decompose this signal into random and periodic components as in Fig. 3.29, we find that the periodic component degenerates to a dc signal because of the full-width pulses. Hence, this signal has no periodic component. Timing can still be extracted as shown in Fig. 3.31*a*, which includes the nonlinear operation of rectification. The signal is differentiated, then rectified, and then threshold-detected to yield on-off pulses of smaller width. When these pulses are applied to a circuit tuned to the clock frequency, each pulse excites the circuit and generates the sinusoid of the clock frequency. During the intervals where there are no pulses, the oscillation continues because of the flywheel effect of the high-Q circuit. ■

Timing Jitter. Variations of the pulse positions or sampling instants cause timing jitter. This results from several causes, some of which are dependent on the pulse pattern being transmitted whereas others are not. The former are cumulative along the chain of regenerative repeaters because all the repeaters are affected in the same way, whereas the other forms of jitter are random from regenerator to regenerator and therefore tend to partially cancel out their mutual effects over a long-haul link. Random forms of jitter are caused by noise, interference, and mistuning of the clock circuits. The pattern-dependent jitter results from clock mistuning, amplitude-to-phase conversion in the clock circuit, and ISI, which alters the position of the peaks of the input signal according to the pattern. The rms value of the jitter over a long chain of N repeaters can be shown to increase as $\sqrt{N}$.

Jitter accumulation over a digital link may be reduced by buffering the link with an elastic store and clocking out the digit stream under the control of a highly stable phase-lock loop. Jitter reduction is necessary about every 200 miles in a long digital link to keep the maximum jitter within reasonable limits.

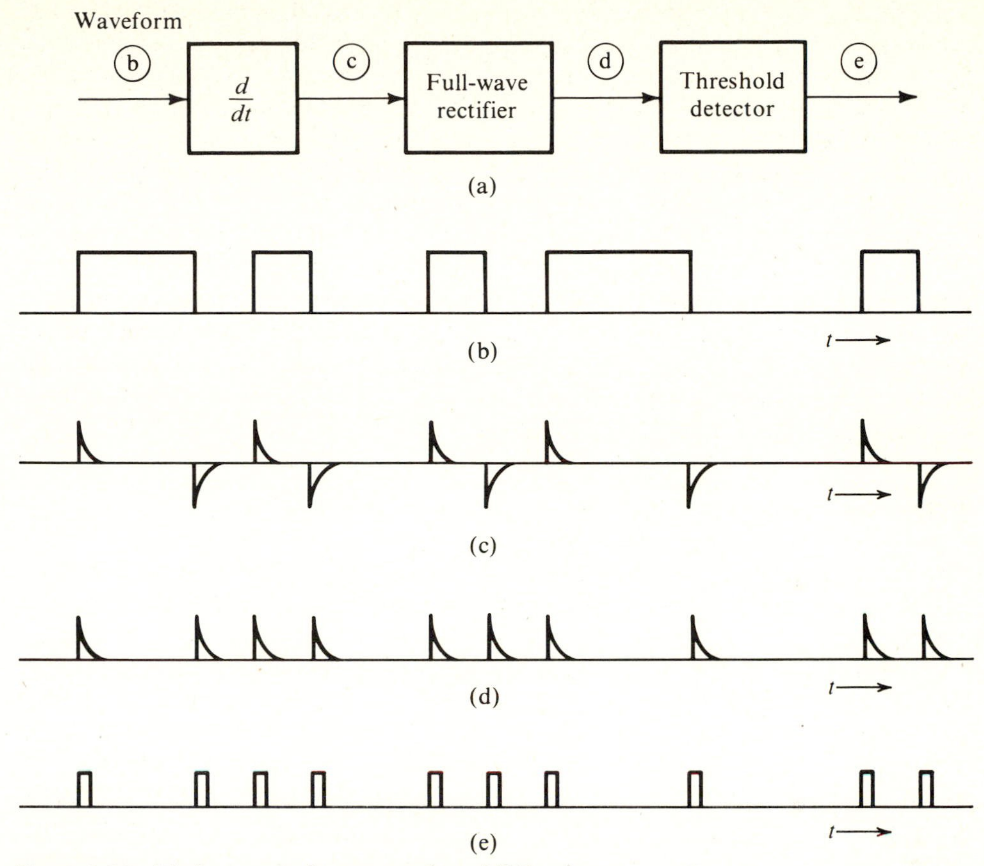

Figure 3.31 Timing or clock recovery from NRZ pulses.

3.6 DETECTION ERROR PROBABILITY

The signal received at the detector consists of the desired pulse train plus a random channel noise. This can cause error in pulse detection. Consider, for example, the case of polar transmission using a basic pulse $p(t)$ (Fig. 3.32*a*). This pulse has a peak amplitude A_p. A typical received pulse train is shown in Fig. 3.32*b*. Pulses are sampled at their peak values. If noise were absent, the sample of the positive pulse (corresponding to **1**) will be A_p and that of the negative pulse (corresponding to **0**) would be $-A_p$.* Because of noise, these samples would be $\pm A_p + n$ where n is the random noise amplitude (see Fig. 3.32*b*). From the symmetry of the situation, the detection threshold is zero, i.e., if the pulse sample value is positive, the digit is detected as **1**; if the sample value is negative, the digit is detected as **0**.

The decision whether **1** or **0** is transmitted could be made readily from the pulse sample, except that n is random, meaning its exact value is unpredictable. It may have a large or a small value and can be negative as well as positive. It is possible that **1**

* This assumes zero ISI.

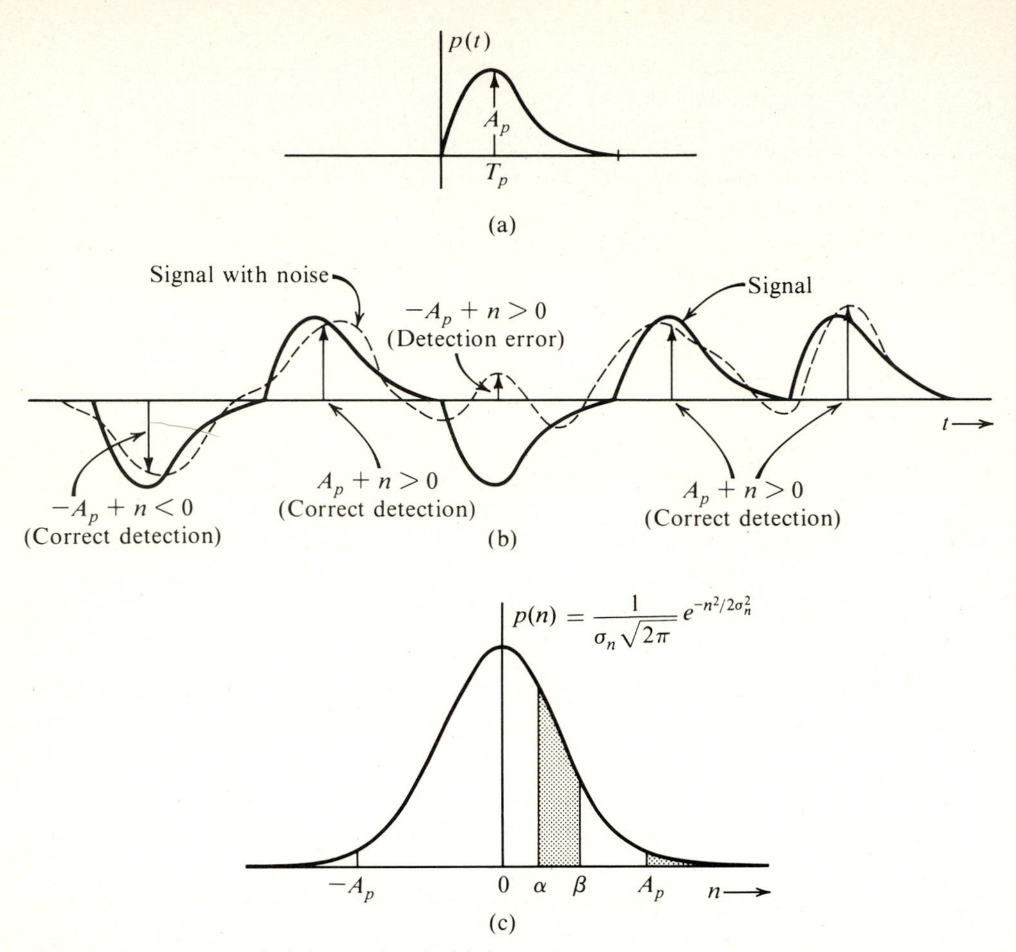

Figure 3.32 Error probability in threshold detection.

is transmitted, but n at the sampling instant may have a large negative value. This will make the sample value $A_p + n$ small or even negative. On the other hand, if **0** is transmitted, and n has a large positive value at the sampling instant, the sample value $-A_p + n$ can be positive and the digit will be detected wrongly as **1**. This is clear from Fig. 3.32*b*.

The amplitude n of the so-called *gaussian* noise ranges from $-\infty$ to ∞ although the likelihood (or the probability) that n will take on very large values decreases rapidly as $e^{-n^2/2\sigma_n^2}$, where σ_n is the rms value of the noise. Still, occasionally, n can take on large positive or negative values causing detection errors as discussed earlier. When **0** is transmitted, the sample value of the received pulse is $-A_p + n$. If $n > A_p$, the sample value is positive and the digit will be detected wrongly as **1**. If $P(\varepsilon|\mathbf{0})$ is the likelihood or the probability of error, given that **0** is transmitted, then

$$P(\varepsilon|\mathbf{0}) = \text{likelihood or the probability that } n > A_p \tag{3.41a}$$

similarly

$$P(\varepsilon\,|\,\mathbf{1}) = \text{probability that } n < -A_p \tag{3.41b}$$

The detailed discussion of random signals and error probabilities is deferred to Chapters 5 and 6. The following brief discussion is just to give a broad understanding.

The probabilities in Eq. (3.41) can be computed if we know the relative distribution of amplitudes of the random noise. Such distribution is specified by the *probability density function* (*PDF*) of n. For the case of gaussian noise, the PDF $p(n)$ is given by

$$p(n) = \frac{1}{\sigma_n \sqrt{2\pi}} e^{-n^2/2\sigma_n^2} \tag{3.42}$$

where σ_n is the rms value of the noise signal. This PDF in Fig. 3.32c, shows the relative distribution of values of n. The likelihood (or the probability) that an amplitude of n lies in a range (α, β) is given by the area under PDF over the range (α, β), that is

$$\text{Likelihood } (\alpha < n < \beta) = \frac{1}{\sigma_n \sqrt{2\pi}} \int_{\alpha}^{\beta} e^{-n^2/2\sigma_n^2}\, dn \tag{3.43}$$

It follows from Eqs. (3.41) and (3.43) that:

$$p(\varepsilon\,|\,\mathbf{0}) = \frac{1}{\sigma_n \sqrt{2\pi}} \int_{A_p}^{\infty} e^{-n^2/2\sigma_n^2}\, dn \tag{3.44a}$$

$$= \frac{1}{\sqrt{2\pi}} \int_{A_p/\sigma_n}^{\infty} e^{-x^2/2}\, dx \tag{3.44b}$$

This integral cannot be obtained in a closed form. It has been computed numerically and can be found in standard mathematical tables. This integral, being a function of A_p/σ_n, can be denoted by $Q(A_p/\sigma_n)$ for convenience. Thus,

$$P(\varepsilon\,|\,\mathbf{0}) = Q\left(\frac{A_p}{\sigma_n}\right) \tag{3.45}$$

Note that since PDF of n is symmetrical about the origin, the probability that $n > A_p$ is the same as the probability that $n < -A_p$. Hence $P(\varepsilon\,|\,\mathbf{1})$ is the same as $P(\varepsilon\,|\,\mathbf{0})$. The function $Q(x)$ is plotted in Fig. 5.11d and is tabulated on p.

A very good approximation to the Q function is

$$Q(x) \simeq \frac{1}{x\sqrt{2\pi}} \left(1 - \frac{0.7}{x^2}\right) e^{-x^2/2} \qquad x > 2 \tag{3.46}$$

The error in this approximation is just about 1 percent for $x > 2.15$. The error decreases as x increases.

To get a rough idea of the orders of magnitude, let us consider the peak pulse amplitude A_p to be k times the noise rms value, that is, $A_p = k\sigma_n$. In this case

$$P(\varepsilon\,|\,\mathbf{0}) = P(\varepsilon\,|\,\mathbf{1}) = Q(k)$$

The following table shows error probabilities for various values of k.

k	1	2	3	4	5	6
$P(\varepsilon \mid \mathbf{0})$	0.1587	0.0227	0.00135	3.16×10^{-5}	2.87×10^{-7}	9.9×10^{-10}

The error probability of 10^{-6} means, on the average, only one out of a million pulses will be detected wrongly. Thus when A_p is five times the noise rms amplitude, the error probability is 2.87×10^{-7}. This means, on the average, only 1 out of 3,484,320 pulses will be detected wrongly.

In general, if the separation between pulse amplitudes to be distinguished is $2A_p$ (as in the polar case discussed here), the error probability is $Q(A_p/\sigma_n)$. For on-off or bipolar (or duobinary), the separation between pulses to be distinguished is only A_p and, consequently, the error probability is

$$P(\varepsilon \mid \mathbf{0}) = P(\varepsilon \mid \mathbf{1}) = Q\left(\frac{A_p}{2\sigma_n}\right) \tag{3.47}$$

3.7 *M*-ARY COMMUNICATION

Digital communication uses only a finite number of symbols for communication, the minimum number being two (the binary case). Thus far we have restricted ourselves to only the binary case. We shall now briefly discuss some aspects of M-ary communication (communication using M symbols). This subject is discussed in depth in Chapters 6 and 7.

It is easy to show that the information transmitted by each symbol increases with M. For example, when $M = 4$ (4-ary, or quaternary, case), we have four basic symbols, or pulses, available for communication (Fig. 3.33*a*). A sequence of two binary digits can be transmitted by just one 4-ary symbol. This is because only four possible sequences of two binary digits exist (viz., **11, 10, 01,** and **00**). Because we have four distinct symbols available, we can assign one of the four symbols to each of these combinations (Fig. 3.33*a*). This signaling (*multiamplitude signaling*) allows us to transmit each pair of binary digits by one 4-ary pulse (Fig. 3.33*b*). Hence, to transmit n binary digits, we need only $n/2$ 4-ary pulses. This means one 4-ary symbol can transmit the information of two binary digits. Similarly, one 8-ary symbol can transmit the information of three binary digits, and one 16-ary symbol can transmit the information of four binary digits. In general, the information I_M transmitted by an M-ary symbol is

$$I_M = \log_2 M \text{ binary digits, or bits} \tag{3.48}$$

This means we can increase the rate of information transmission by increasing M. But the transmitted power increases as M, because to have the same noise immunity, the minimum separation between pulse amplitudes should be comparable to that of binary pulses. Therefore pulse amplitudes increase with M (see Fig. 3.33). It will be shown

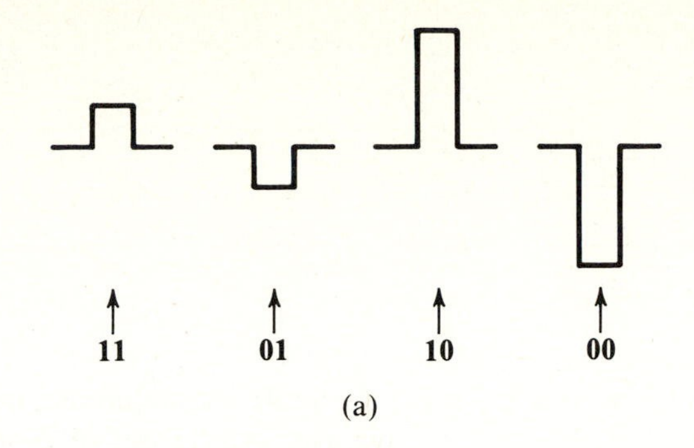

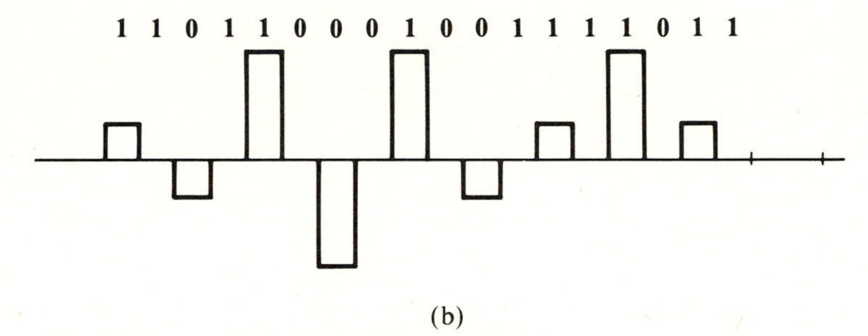

Figure 3.33 4-ary multiamplitude signal.

in Chapter 6 that the transmitted power increases as M^2 (see Prob. 3.21). Thus, to increase the rate of communication by a factor of $\log_2 M$, the power required increases as M^2. Because the transmission bandwidth depends only on the pulse rate and not on pulse amplitudes, the bandwidth is independent of M.

Figure 3.33 shows just one possible M-ary scheme (multiamplitude signaling). There are infinite possible ways of structuring M waveforms. For example, we may use M orthogonal pulses $\varphi_1(t)$, $\varphi_2(t)$, . . . , $\varphi_M(t)$ with the property

$$\int_0^{T_o} \varphi_i(t)\varphi_j(t)\, dt = \begin{cases} C & i = j \\ 0 & i \neq j \end{cases}$$

Figure 3.34 shows one possible set of M orthogonal signals

$$\varphi_k(t) = \begin{cases} \sin 2\pi k f_o t & 0 < t < T_o \qquad k = 1, 2, \ldots M \\ 0 & \text{otherwise} \end{cases}$$

It can readily be shown that all of these M pulses are mutually orthogonal. Since the highest pulse frequency is Mf_o, the transmission bandwidth in this case is Mf_o. In general it can be shown that the bandwidth of an orthognal M-ary scheme is M times that of the binary scheme [see Sec. 7.3, Eq. (7.72)]. Therefore in an M-ary orthogonal scheme, the rate of communication is increased by a factor $\log_2 M$ at the cost of an increase in transmission bandwidth by a factor M. It will be shown in Chapter 6 that,

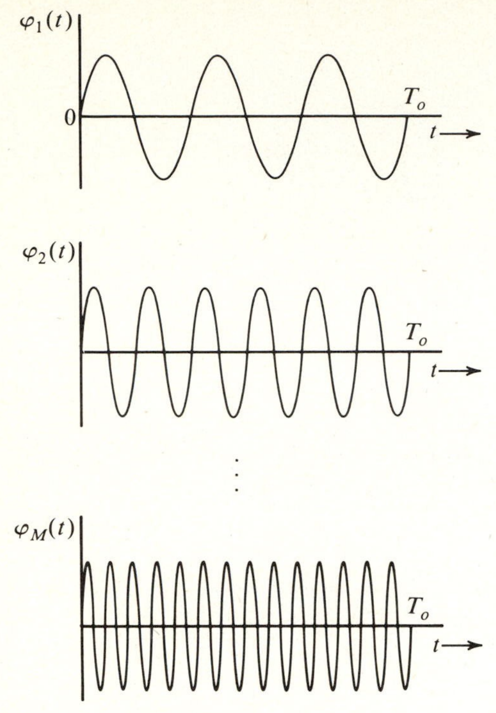

Figure 3.34 *M*-ary orthogonal pulses.

for a comparable noise immunity, the transmitted power is practically independent of M in the orthogonal scheme.

In Chapters 6 and 7, we shall discuss several other types of M-ary signaling. The nature of the exchange between the transmission bandwidth and the transmitted power (or SNR) depends upon the choice of pulses. For example, in orthogonal signaling, the transmitted power is practically independent of M but the transmission bandwidth increases with M. Contrast this to the multiamplitude case, where the transmitted power increases roughly with M^2 and the bandwidth remains constant. Thus, M-ary signaling allows us great flexibility in exchanging signal power (or SNR) with the transmission bandwidth. The choice of the appropriate system will depend upon the particular circumstances. For example, it will be appropriate to use multiamplitude signaling if the bandwidth is at a premium (as in telephone lines) and to use orthogonal signaling when power is at a premium (as in space communication). Because of its simplicity, however, binary communication is perhaps the single most important mode of communication in practice today.

■ EXAMPLE 3.5

Find the PSD of the quaternary (4-ary) signal in Fig. 3.33*b*, assuming that each of the four pulses is equally likely.

☐ **Solution:** In this case, the a_k's can take on four values, ± 1 and ± 2, all of which are equally likely. Also,

$$R_0 = \lim_{T\to\infty} \frac{T_o}{T} \sum_k a_k^2$$

If pulses occur every T_o interval, there are $T/4T_o$ pulses of values 1, −1, 2, and −2 each. Thus

$$R_0 = \frac{T_o}{T}\left[\frac{T}{4T_o}(1^2 + (-1)^2 + (2)^2 + (-2)^2\right] = 2.5$$

and

$$R_1 = \lim_{T\to\infty} \frac{T_o}{T} \sum_k a_k a_{k+1}$$

The product $a_k a_{k+1}$ can take on any of the following six values: 1, 2, 4, −1, −2, or −4. All these values are equally likely. Hence, $R_1 = 0$.

In a similar way, we can show that

$$R_n = 0 \qquad n > 0$$

and

$$S_x(\omega) = \frac{2.5}{T_o}$$

and

$$S_y(\omega) = \frac{2.5|P(\omega)|^2}{T_o}$$

where $P(\omega)$ is the Fourier transform of the basic pulse of amplitude $A/2$.

Hence, the PSD is identical (except for a multiplying constant) to that of polar binary signaling. Therefore, they have an identical transmission bandwidth for a given T_o. For a given T_o, however, the quaternary scheme transmits twice as much information as does the binary. ■

3.8 DIGITAL CARRIER SYSTEMS

Thus far, we have discussed baseband digital systems, where signals are transmitted directly without any shift in the frequencies of the signal. Because baseband signals have sizable power at low frequencies, they are suitable for transmission over a pair of wires or coaxial cables. Local telephone communication as well as short-haul PCM (between telephone exchanges) are examples of this. Baseband signals cannot be transmitted over a radio link because this would necessitate impracticably large antennas to efficiently radiate the low-frequency spectrum of the signal. Hence for such a purpose, the signal spectrum must be shifted to a high-frequency range. A spectrum shift to higher frequencies is also required to transmit several messages simultaneously by sharing the large bandwidth of the transmission medium (FDM). As seen in Chapter 2, the spectrum of a signal can be shifted to a higher frequency by modulating a high-frequency sinusoid (carrier) by the baseband signal. Two basic forms of modulation exist: amplitude modulation and angle modulation. In amplitude modulation, the carrier amplitude is varied in proportion to the modulating signal (i.e., the baseband signal). This is shown in Fig. 3.35. An unmodulated carrier cos $\omega_c t$ is shown

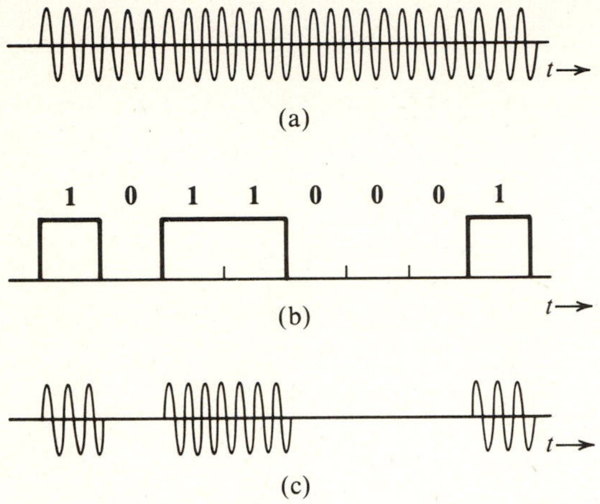

Figure 3.35 (a) The carrier cos $\omega_c t$. (b) The modulating signal $s(t)$. (c) ASK: the modulated signal $s(t)$ cos $\omega_c t$.

in Fig. 3.35*a*. The on-off baseband signal $y(t)$ (the modulating signal) is shown in Fig. 3.35*b*. When the carrier amplitude is varied in proportion to $y(t)$, we have the modulated carrier $y(t)$ cos $\omega_c t$, shown in Fig. 3.35*c*. Note that the modulated signal is still an on-off signal. This modulation scheme of transmitting binary data is known as *on-off keying (OOK)* or *amplitude-shift keying (ASK)*.

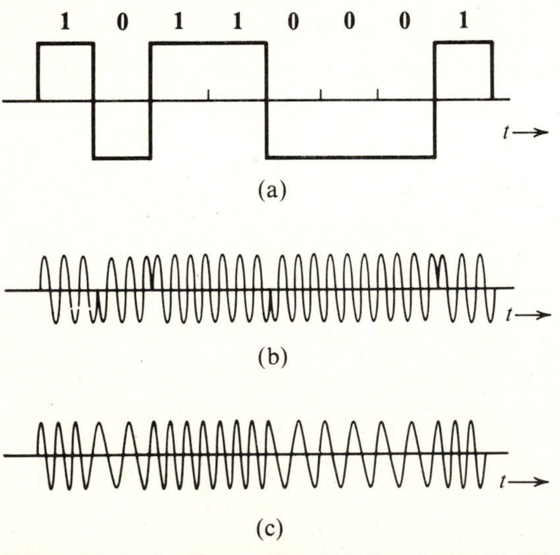

Figure 3.36 (a) The modulating signal $s(t)$. (b) PSK: the modulated signal $s(t)$ cos $\omega_c t$. (c) FSK: the modulated signal.

If the baseband signal $y(t)$ were polar (Fig. 3.36*a*), the corresponding modulated signal $y(t) \cos \omega_c t$ would appear as shown in Fig. 3.36*b*. In this case, if $p(t)$ is the basic pulse, we are transmitting **1** by a pulse $p(t) \cos \omega_c t$ and **0** by $-p(t) \cos \omega_c t = p(t) \cos (\omega_c t + \pi)$. Hence, the two pulses are π radians apart in phase. The information resides in the phase of the pulse. For this reason this scheme is known as *phase-shift keying (PSK)*. Note that the transmission is still polar.

When the data is transmitted by varying the frequency, we have the case of *frequency-shift keying (FSK)*, as shown in Fig. 3.36*c*. A **0** is transmitted by a pulse of frequency ω_{c_0}, and **1** is transmitted by a pulse of frequency ω_{c_1}. The information about the transmitted data resides in the carrier frequency.

Modulation causes a shift in the baseband signal spectrum [Eq. (2.118a)]. The ASK signal in Fig. 3.35*c*, for example, is $y(t) \cos \omega_c t$ where $y(t)$ is an on-off signal (using a full-width or NRZ pulse). Hence the PSD of the ASK signal is the same as that of an on-off signal (Fig. 3.11) shifted to $\pm\omega_c$ as shown in Fig. 3.37*a*.* The PSK

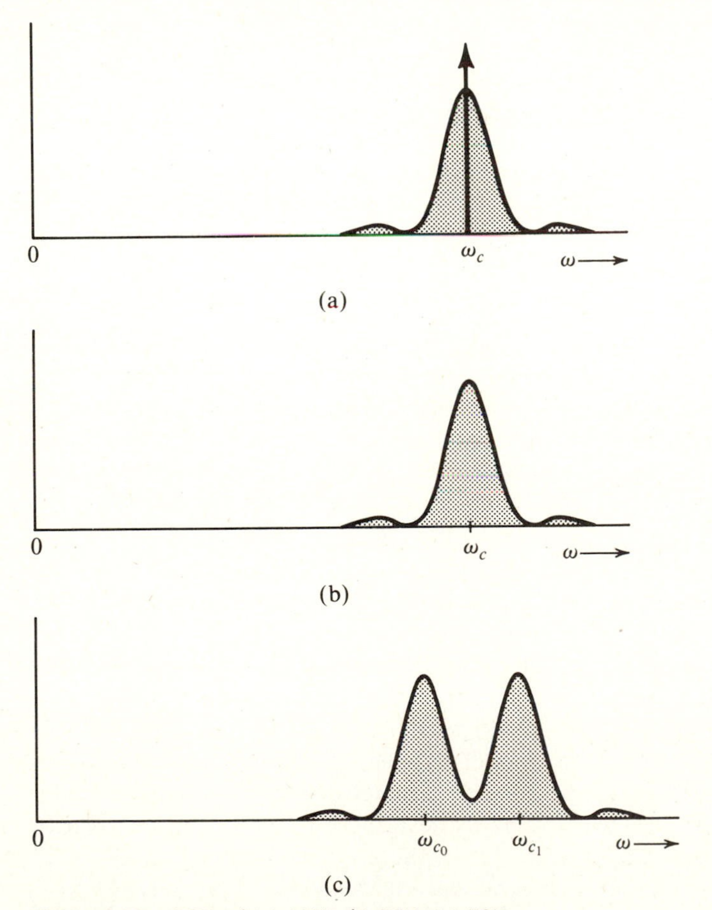

Figure 3.37 PSD of (a) ASK, (b) PSK, (c) FSK.

*Note that an on-off signal in Fig. 3.35*b* uses a full-width rectangular pulse which has no discrete components except at dc. Therefore, the ASK spectrum has discrete component only at $\omega = \omega_c$.

signal, on the other hand, is $y(t) \cos \omega_c t$ where $y(t)$ is a polar signal. Therefore, the PSD of a PSK signal is the same as that of the polar baseband signal shifted to $\pm\omega_c$ as shown in Fig. 3.37*b*. Note that this is the same as the PSD of the ASK minus its discrete components.

The FSK signal may be viewed as a sum of two interleaved ASK signals, one with a modulating frequency ω_{c_0}, and the other with a modulating frequency ω_{c_1}. Hence, the spectrum of FSK is the sum of two ASK spectra at frequencies ω_{c_0} and ω_{c_1} as shown in Fig. 3.37*c*. No discrete components appear in this spectrum. It can be shown that by properly choosing ω_{c_0} and ω_{c_1}, discrete components can be eliminated. Note that the bandwidth of FSK is higher than that of ASK or PSK.

We can also modulate bipolar, duobinary, or any other scheme discussed earlier. Use of the basic rectangular pulse in Figs. 3.35 or 3.36 is for the sake of convenience. In practice, baseband pulses may be specially shaped to eliminate ISI.

As observed earlier, polar signaling is the most efficient scheme from the point of view of noise immunity. The PSK, being polar, requires 3 dB less power than ASK (or FSK) for the same noise immunity, that is, for the same error probability in pulse detection.

3.9 DIGITAL TRANSMISSION OF ANALOG SIGNALS

As seen in Chapter 1, it is possible to convert an analog signal into a digital form by approximating the signal. We shall discuss here two important techniques of transmission of analog signals by digital means: (1) pulse code modulation (PCM) and (2) delta modulation (DM).

Pulse-Code Modulation

As mentioned in Chapter 1, a gap of more than 20 years occurred between the invention of PCM and its implementation, because of the unavailability of suitable switching devices. Vacuum tubes, the devices available before the invention of the transistor, were not only bulky but were poor switches and dissipated a lot of power as heat. Systems using vacuum tubes as switches were large, rather unreliable, and tended to overheat. PCM was just waiting for the invention of the transistor, which happens to be small, consumes little power, and is a nearly ideal switch.

Coincidentally, at about the same time, the demand in telephone service had grown to the point where the old system was overloaded, particularly in large cities. It was not easy to install new underground cables because in many cities the available space under the streets was already occupied by other services (such as water, gas, sewer, etc.). Moreover, digging up streets and causing many dislocations was not very attractive. An attempt was made on a limited scale to increase the capacity by frequency-division multiplexing several voice channels through amplitude modulation. Unfortunately, the cables were primarily designed for the audio voice range (0 to 4 kHz) and suffered severely from noise; furthermore, cross talk at high frequencies between pairs of channels on the same cable was unacceptable. Ironically, PCM—requiring a bandwidth several times larger than that required for FDM signals—offered the solution. This is because PCM with closely spaced regenerative repeaters can work satisfactorily on noisy, poor-high-frequency-performance lines. The repeaters, spaced 6000 feet apart, clean up the signal and regenerate new pulses before the pulses get too distorted and noisy. This is the history of the Bell System's T-1 carrier system.[11] A pair of wires that used to transmit one audio signal of bandwidth 4 kHz is now used to transmit 24 time-division-multiplexed PCM telephone signals with a total bandwidth of 1.544 MHz.

A schematic of a T-1 carrier system is shown in Fig. 3.38*a*. All 24 channels are sampled in a sequence. The sampler output represents a time-division-multiplexed PAM signal. The multiplexed PAM signal is now applied to the input of an encoder

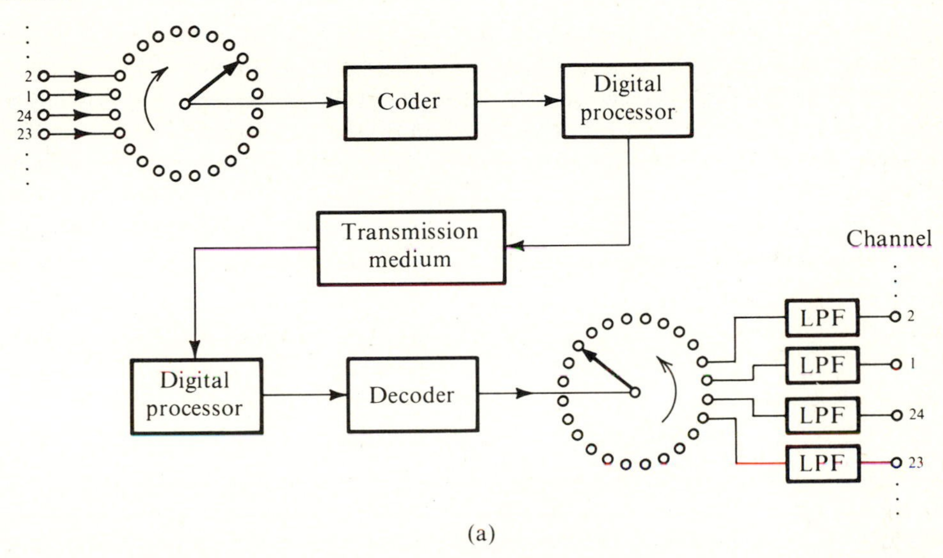

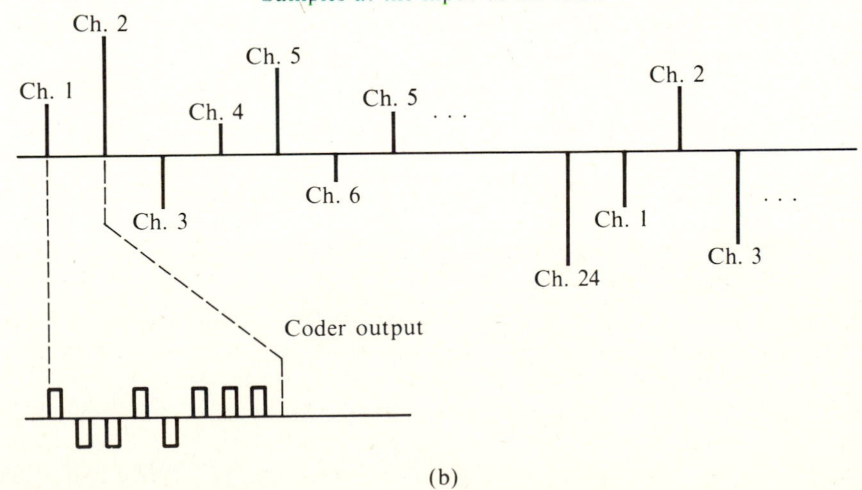

Figure 3.38 T-1 carrier system.

that quantizes each sample and encodes it into eight binary pulses—a binary code word* (see Fig. 3.38*b*). The signal, now converted to a digital form, is sent over the transmission medium using bipolar code. Regenerative repeaters spaced 6000 feet apart, detect the pulses and transmit new pulses. At the receiver, the decoder converts the binary pulses into samples (decoding). The samples are then demultiplexed (i.e., distributed to each of the 24 channels). The desired audio signal is reconstructed by passing the samples through a lowpass filter in each channel.

The commutators in Fig. 3.38 are not mechanical but are high-speed electronic switching circuits. Several schemes are available for this purpose.[12] Sampling is done by electronic gates (such as a series bridge diode circuit, shown in Fig. 4.5*b*) opened periodically by narrow pulses of 2 μs duration.

After the Bell System introduced the T-1 carrier system in the United States, dozens of variations were proposed or adopted elsewhere before the CCITT standardized its 30-channel PCM system with a pulse rate of 2.048 Mbits/second (in contrast to T-1, with 24 channels and 1.544 Mbits/second). The 30-channel system is used all over the world, except in North America and Japan. Because of the widespread adoption of the T-1 carrier system in the United States and Japan before the CCITT standardization, it now seems likely that the two standards will continue to be used in different parts of the world, with appropriate interfaces used in international communication.

Quantizing

Let m_p be the peak amplitude of a message $m(t)$. The amplitude range $(-m_p, m_p)$ is divided into L uniformly spaced intervals, each of width $2m_p/L$ (Fig. 3.39*a*).† A sample amplitude value is approximated by the midpoint of the interval in which it lies (Fig. 3.39*a*). The input-output characteristics of a quantizer are shown in Fig. 3.39*b*. The error caused by quantization has a range $(-m_p/L, m_p/L)$. The signal reconstructed from the quantized samples will be a distorted version of the desired message signal. If $m(kT_s)$ is the kth sample of the signal $m(t)$, and if $\hat{m}(kT_s)$ is the corresponding quantized sample, then from Eq. (2.84),

$$m(t) = \sum_k m(kT_s) \operatorname{sinc}(2Bt - k)$$

and

$$\hat{m}(t) = \sum_k \hat{m}(kT_s) \operatorname{sinc}(2Bt - k)$$

where $\hat{m}(t)$ is the signal reconstructed from quantized samples. The distortion component in the reconstructed signal is $q(t) = \hat{m}(t) - m(t)$.

---

* In an earlier version, each sample was encoded by seven bits. An additional bit was added for signaling.

† We are assuming that $[m(t)]_{\max} = |m(t)_{\min}| = m_p$.

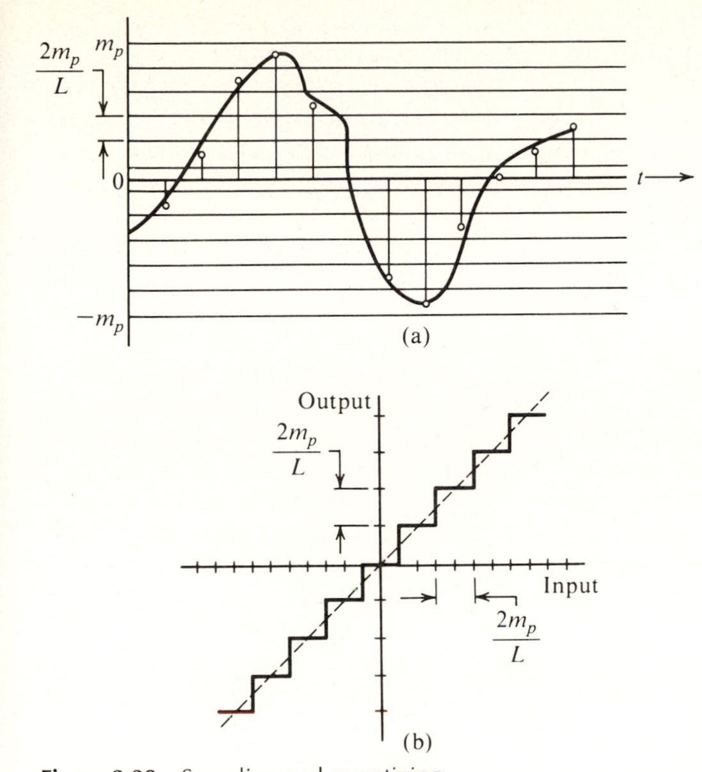

Figure 3.39 Sampling and quantizing.

$$q(t) = \sum_k [\hat{m}(kT_s) - m(kT_s)] \text{ sinc } (2Bt - k)$$

$$= \sum_k q(kT_s) \text{ sinc } (2Bt - k)$$

where $q(kT_s)$ is the quantization error in the kth sample. The signal $q(t)$ is the undesired signal, and, hence, acts as noise, known as *quantization noise*. To calculate the power, or the mean square value of $q(t)$, we have

$$\widetilde{q^2(t)} = \lim_{T\to\infty} \frac{1}{T} \int_{-T/2}^{T/2} q^2(t)\, dt$$

$$= \lim_{T\to\infty} \frac{1}{T} \int_{-T/2}^{T/2} \left(\sum_k q(kT_s) \text{ sinc } (2Bt - k) \right)^2 dt \tag{3.49}$$

Use of the result of Example 2.25 in Eq. (3.49) yields

$$\widetilde{q^2(t)} = \lim_{T\to\infty} \frac{1}{2BT} \sum_k q^2(kT_s) \tag{3.50}$$

Because the sampling rate is $2B$, the total number of samples over the averaging interval T is $2BT$. Hence, the right-hand side of Eq. (3.50) represents the average, or mean, of the square of the quantization error $q(kT_s)$. In other words, $\widetilde{q^2(t)}$ is the sum of the squares of the quantization errors of each sample divided by the total number of samples. The quantization error (Fig. 3.39*a*) lies in the range $(-m_p/L, m_p/L)$. Let us divide this range into N uniform intervals, each of value Δ ($\Delta \to 0$). Assuming that a quantization error is equally likely to lie anywhere in the range $(-m_p/L, m_p/L)$, the number of samples with a quantization error of $k\Delta$ (k integer, $-N/2 \leq k \leq N/2$) will be the same for all values of k. Let this number be M. Because there are N intervals, the total number of samples is MN, and the mean square of the quantization error is

$$\begin{aligned}\widetilde{q^2(t)} &= \frac{1}{MN}\sum_{k=-N/2}^{N/2} M(k\Delta)^2 \\ &= \frac{2\Delta^2}{N}\sum_{k=1}^{N/2} k^2 \\ &= \frac{2\Delta^2}{N}\left[\frac{\left(\frac{N}{2}\right)\left(\frac{N}{2}+1\right)(N+1)}{6}\right]\end{aligned}$$

Because the averaging is done over $T \to \infty$ [Eq. (3.50)], the total number of samples $MN \to \infty$. Also, because $N\Delta = 2m_p/L$ and $\Delta \to 0$, $N \to \infty$. Hence,

$$\widetilde{q^2(t)} = \frac{2\Delta^2}{N}\frac{N^3}{24} = \frac{(N\Delta)^2}{12} = \frac{m_p^2}{3L^2}$$

Because $\widetilde{q^2(t)}$ is the mean square value, or power, of the quantization noise,* we shall denote it by N_q.

$$N_q = \widetilde{q^2(t)} = \frac{m_p^2}{3L^2} \qquad \textbf{(3.51)}$$

Assuming that the pulse detection error at the receiver is negligible, the reconstructed signal $\hat{m}(t)$ at the receiver output is

$$\hat{m}(t) = m(t) + q(t)$$

The desired signal at the output is $m(t)$, and the (quantization) noise is $q(t)$. If the power of the message signal $m(t)$ is $\widetilde{m^2(t)}$, then

$$S_o = \widetilde{m^2(t)} \qquad \textbf{(3.52a)}$$

---

* Those who are familiar with the theory of probability can derive this result directly by noting that the probability density of the quantization error q is $1/(2m_p/L) = L/2m_p$ over the range $|q| \leq m_p/L$ and is zero elsewhere. Hence,

$$\overline{q^2} = \int_{-m_p/L}^{m_p/L} q^2 p(q)\,dq = \int_{-m_p/L}^{m_p/L} \frac{L}{2m_p} q^2\,dq = \frac{m_p^2}{3L^2}$$

$$N_o = N_q = \frac{m_p^2}{3L^2} \tag{3.52b}$$

and

$$\frac{S_o}{N_o} = 3L^2 \frac{\widetilde{m^2(t)}}{m_p^2} \tag{3.53}$$

Because the quantization noise is $m_p^2/3L^2$, once a quantizer with L levels is designed for a certain peak message value m_p and implemented in a system, the quantization noise is fixed. The signal power $\widetilde{m^2(t)}$, however, varies from talker to talker by as much as 40 dB (power ratio 10^4). The signal power can also vary because of the different lengths of the connecting circuits. This means the SNR in Eq. (3.53) can vary widely, depending on the talker and the length of the circuit. Even for the same talker, the quality of the received signal will deteriorate markedly when he speaks quietly. Statistically, it is found that smaller amplitudes predominate in speech and larger amplitudes are much less frequent. This means the SNR will be low most of the time. Moreover, transmission quality will vary with the talker and the circuit length.

The root of this difficulty lies in the fact that the quantizing steps are of uniform value $\Delta v = 2m_p/L$. The quantization noise $N_q = (\Delta v)^2/12$ is directly proportional to the square of the step size. The problem can be solved by using smaller steps for smaller amplitudes (nonuniform quantizing), as shown in Fig. 3.40*a*. The same result is obtained by first compressing signal samples and then using a uniform quantization. The input-output characteristics of a compressor are shown in Fig. 3.40*b*. The horizontal axis is the normalized input signal (i.e., the input signal amplitude m divided by the signal peak value m_p). The vertical axis is the output signal y. The compressor maps input signal increments Δm into larger increments Δy for small input signals and vice versa for large input signals. Hence, a given interval Δm contains a larger number of steps (or smaller step size) when m is small. The quantization noise is smaller for smaller input signal power. An approximately logarithmic compression characteristic yields a quantization noise nearly proportional to the signal power $\widetilde{m^2(t)}$, thus making the SNR practically independent of the input signal power over a large dynamic range,[13,14] as shown in Fig. 3.43.

Among several choices, two compression laws have been accepted as desirable standards by the CCITT: the μ-law used in North America and Japan, and the A-law used in Europe and the rest of the world and international routes. The μ-law is given by

$$y = \frac{\operatorname{sgn}(m)}{\ln(1+\mu)} \ln\left(1 + \mu\left|\frac{m}{m_p}\right|\right) \qquad \left|\frac{m}{m_p}\right| \leq 1 \tag{3.54a}$$

The A-law is

$$y = \begin{cases} \dfrac{A}{1+\ln A}\left(\dfrac{m}{m_p}\right) & \left|\dfrac{m}{m_p}\right| \leq \dfrac{1}{A} \\[2ex] \dfrac{\operatorname{sgn}(m)}{1+\ln A}\left[1 + \ln A\left|\dfrac{m}{m_p}\right|\right] & \dfrac{1}{A} \leq \left|\dfrac{m}{m_p}\right| \leq 1 \end{cases} \tag{3.54b}$$

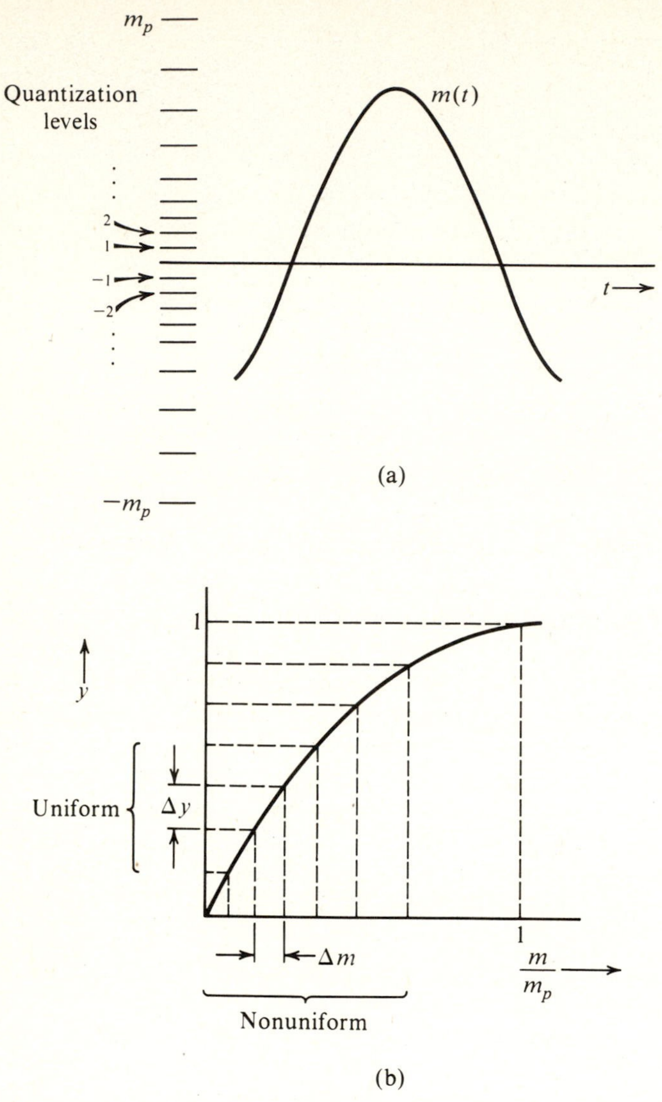

Figure 3.40 Nonuniform quantization.

These characteristics are shown in Figs. 3.41 and 3.42.

The compression parameter μ (or A) determines the degree of compression. To obtain a nearly constant S_o/N_o over an input-signal-power dynamic range of 40 dB, μ should be greater than 100. Values of $\mu = 100$ and 255 are standardized in American systems. For the A-law, a value of $A = 87.6$ gives comparable results and has been standardized by the CCITT.

The compressed samples must be restored to their original value at the receiver by using an expandor with a characteristic complementary to that of the compressor. The compressor and the expandor together are called the *compandor*. It was men-

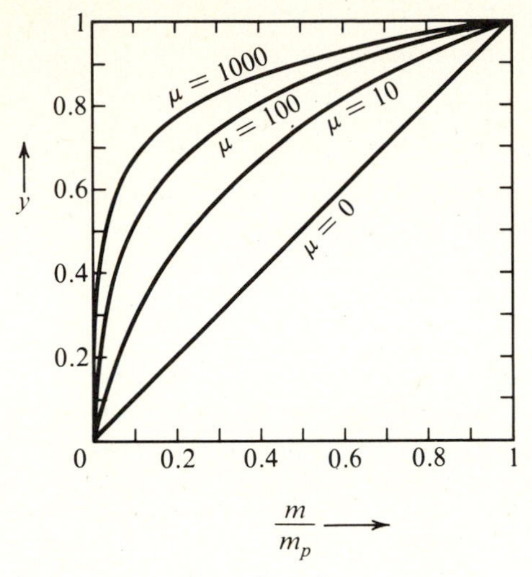

Figure 3.41 The μ-law characteristics.

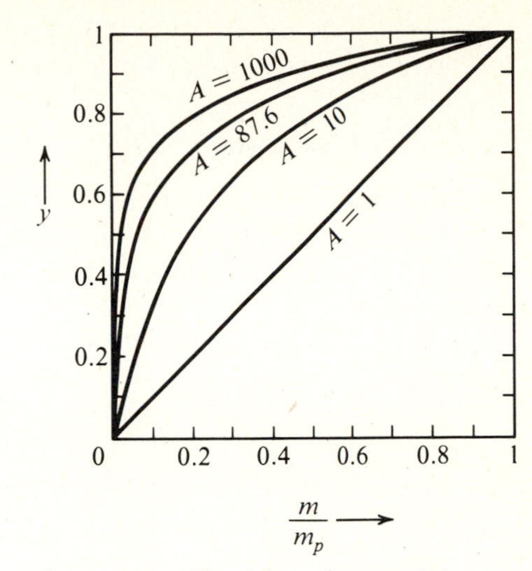

Figure 3.42 The A-law characteristics.

tioned in Sec. 2.7 that instantaneous compression of a signal increases its bandwidth. But in PCM, we are compressing not the signal $m(t)$ but its samples. Because the number of samples does not change, the problem of bandwidth increase does not arise here.

It can be shown (see Sec. 6.4) that when a μ-law compandor is used, the output SNR is

$$\frac{S_o}{N_o} \simeq \frac{3L^2}{[\ln(1+\mu)]^2} \qquad \mu^2 \gg \frac{m_p^2}{\widetilde{m^2(t)}} \tag{3.55}$$

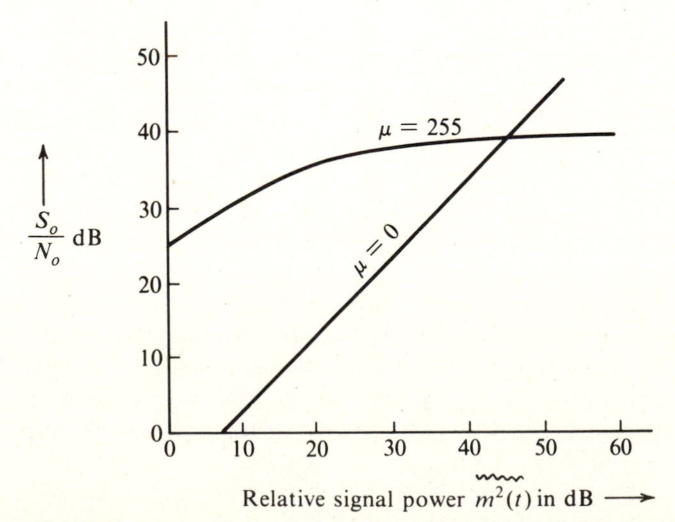

Figure 3.43 Signal-to-quantization-noise ratio in PCM with and without compression.

The output SNR for the case of $\mu = 255$ and $\mu = 0$ (uniform quantization) as a function of $\widetilde{m^2}$ (the message signal power) is shown in Fig. 3.43.

Transmission Bandwidth

Because a group of n binary pulses yields 2^n distinct patterns, $L = 2^n$, or $n = \log_2 L$. For a message signal $m(t)$ bandlimited to B, there are $2B$ samples/second, and each sample is coded into n binary pulses, requiring a total of $2nB$ pulses/second. Since bipolar signaling is used, the transmission bandwidth is $2nB$ Hz. For example, when $L = 256$, $n = 8$; a 4-kHz audio signal requires a 64-kHz bandwidth. To transmit 24 time-division-multiplexed PCM signals, we need a $64{,}000 \times 24 = 1.536$ MHz bandwidth.* Because of repeaters, however, PCM can satisfactorily utilize a poor-high-frequency channel. This is why the T-1 carrier system can use the same pair of wires that were used earlier to transmit a single audio signal of 4 kHz.

The Output SNR

The output SNR in Eq. (3.53) or Eq. (3.55) can be expressed as

$$\frac{S_o}{N_o} \simeq 3k(2)^{2n} \tag{3.56a}$$

where $k = \widetilde{m^2(t)}/m_p^2$ [uncompressed case, in Eq. (3.53)]

or

$$k = 1/[\ln(1+\mu)]^2 \qquad \text{[compressed case, in Eq. (3.55)]}$$

Hence,

$$\left(\frac{S_o}{N_o}\right) \text{dB} = 10 \log \left(\frac{S_o}{N_o}\right)$$

$$= (\alpha + 6n) \text{ dB} \tag{3.56b}$$

where $\alpha = 10 \log (3k)$. From Eq. (3.56a), we observe that the SNR increases exponentially with the bandwidth. This trade of SNR with bandwidth is attractive and comes close to the upper theoretical limit. A small increase in bandwidth yields a large benefit in terms of SNR. From Eq. (3.56b) it can be seen that increasing n by one (increasing one bit in the code word) quadruples the output SNR (6 dB increase). Thus, if we increase n from 8 to 9, the SNR quadruples, but the bandwidth increases only from 64 kHz to 72 kHz (an increase of only 12.5 percent). We shall see in Chapter 4 that frequency and phase modulation also exchange SNR for bandwidth. But it requires a doubling of the bandwidth to quadruple the SNR. In this respect PCM is strikingly superior to FM or PM.

---

* Actually, the required bandwidth is 1.544 MHz because of additional pulses required for synchronization.

■ EXAMPLE 3.6

Compare the case of $L = 64$ with the case of $L = 256$ from the point of view of transmission bandwidth and the output SNR. Assume $\mu = 100$.

For $L = 64$, $n = 6$, and the transmission bandwidth is 48 kHz.

$$\frac{S_o}{N_o} = \alpha + 36 \text{ dB}$$

$$\alpha = 10 \log \frac{3}{[\ln (101)]^2} = -8.51$$

Hence,

$$\frac{S_o}{N_o} = 27.49 \text{ dB}$$

For $L = 256$, $n = 8$, and the transmission bandwidth is 64 kHz.*

$$\frac{S_o}{N_o} = \alpha + 6n = 39.49 \text{ dB}$$

The SNR for $L = 256$ is 12 dB (ratio of 16) superior to the SNR of $L = 64$. The former requires 33 percent more bandwidth, as compared to the latter. ■

The Compandor

A logarithmic compressor can be realized by a semiconductor diode, because the v-i characteristic of such a diode is of the desired form in the first quadrant:

$$V = \frac{KT}{q} \ln \left[1 + \frac{I}{I_s} \right]$$

Two matched diodes in parallel with opposite polarity provide the approximate characteristic in the first and the third quadrants (ignoring the saturation current). In practice, adjustable resistors are placed in series with each diode and a third variable resistor is added in parallel. By adjusting various resistors, the resulting characteristic is made to fit a finite number of points (usually seven) on the ideal characteristics.

An alternative approach is to use a piecewise linear approximation to the logarithmic characteristics. A 15-segmented approximation (Fig. 3.44) to the eight bit ($L = 256$) with $\mu = 255$ law is widely used in the D-2 channel bank that is used in conjunction with the T-1 carrier system. The segmented approximation is only marginally inferior in terms of SNR.[15]

The Encoder

The multiplexed PAM output is applied at the input of the encoder, which quantizes and encodes each sample into a group of n binary digits. A variety of encoders is

---

* Some recent systems that use duobinary signaling (rather than bipolar) require only half this bandwidth. See the discussion on duobinary signaling in Sec. 3.2.

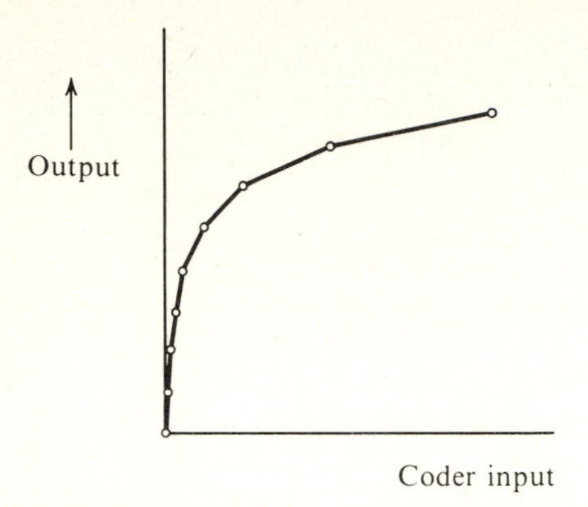

Figure 3.44 Piecewise linear compressor characteristic.

available.[16] We shall discuss here the "*digit-at-a-time*" encoder, which makes n sequential comparisons to generate an n-bit code word. The sample is compared with a voltage obtained by a combination of reference voltages proportional to 2^7, 2^6, 2^5, . . . , 2^0. The reference voltages are conveniently generated by a bank of resistors R, $2R$, 2^2R, . . . , 2^7R.

The encoding involves answering successive questions, beginning with whether or not the sample is in the upper or lower half of the allowed range. The first code digit **1** or **0** is generated, depending on whether the sample is in the upper or the lower half of the range. In the second step, another digit **1** or **0** is generated, depending on whether the sample is in the upper or the lower half of the subinterval in which it has been located. This process continues until the last binary digit in the code is generated.

Decoding is the inverse of encoding. In this case, each of the n digits is applied to a resistor of different value. The kth digit is applied to a resistor 2^kR. The currents in all the resistors are added. The sum is proportional to the quantized sample value. For example, a binary code word **10010110** will give a current proportional to $2^7 + 0 + 0 + 2^4 + 0 + 2^2 + 2^1 + 0 = 150$. This completes the D/A conversion.

Synchronizing and Signaling

Binary code words corresponding to samples of each of the 24 channels are multiplexed in a sequence, as shown in Fig. 3.45. A segment containing one code word (corresponding to one sample) from each of the 24 channels is called a *frame*. Each frame has $24 \times 8 = 192$ information bits. Because the sampling rate is 8000 samples/second, each frame takes 125 μs. At the receiver, it is necessary to be sure where each frame begins in order to separate information bits correctly. For this purpose a *framing bit* is added at the end of each frame. This makes a total of 193 bits/frame. Framing bits are chosen so that a sequence of framing bits, one at the end of each frame, forms a special pattern that is unlikely to be formed in a speech signal. For example, the sequence **101010. . .** is impossible in a speech signal because it implies a frequency component of 4 kHz in the signal that has been cut off at 3400 Hz before sampling.

A sequence formed by the 193rd bit from each frame is examined by the logic of

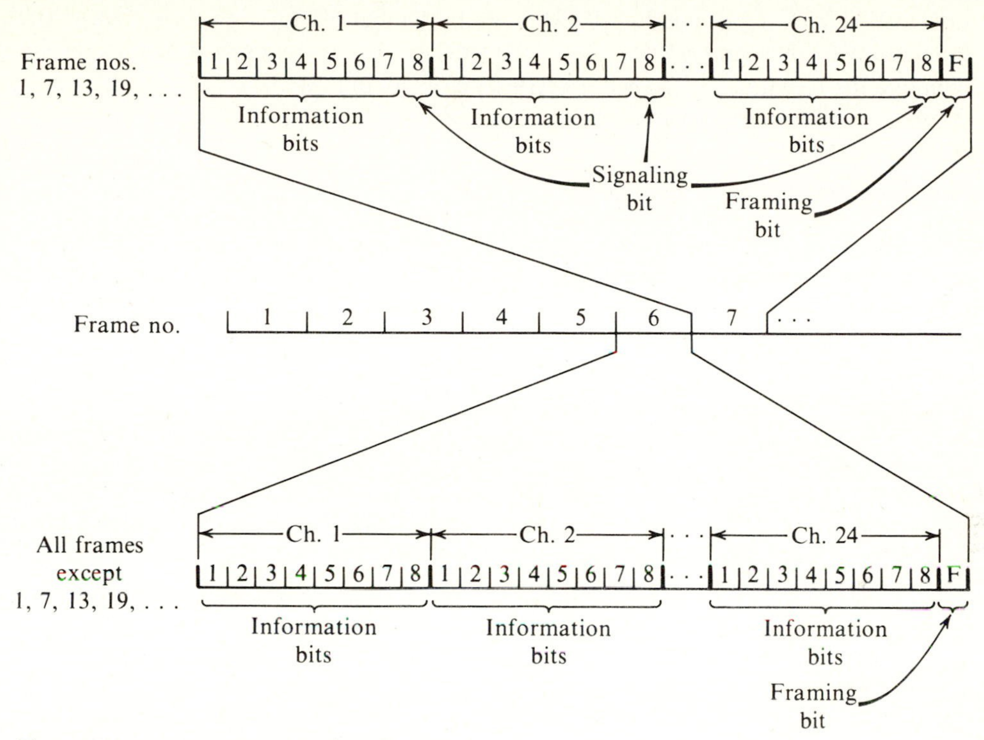

Figure 3.45 T-1 system signaling format.

the receiving terminal. If this sequence does not follow the given coded pattern, then a synchronization loss is detected, and the next position is examined to determine if it is actually the framing bit. It takes about 0.4 to 6 ms to detect and about 50 ms (in the worst possible case) to reframe.

In addition to information and framing bits, we need to transmit signaling bits corresponding to dialing pulses, as well as telephone on-hook/off-hook signals. Rather than create extra time slots for this information,* we use one information bit (the least significant bit) of every sixth sample of a signal to transmit this information. This means every sixth sample of each voice signal will have a possible error corresponding to the least significant digit. Every sixth frame, therefore, has $7 \times 24 = 168$ information bits, 24 signaling bits, and 1 framing bit. In all the remaining frames, there are 192 information bits and 1 framing bit. The signaling bits for each signal occur at a rate of $8000/6 = 1333$ bits/second. The frame format is shown in Fig. 3.45. Note that frame numbers 1, 7, 13, 19, . . . (etc.) include signaling bits, whereas the remaining frames do not.

The Conference on European Postal and Telegraph Administration (CEPT) has standardized a PCM with 256 time slots per frame. Each frame has $30 \times 8 = 240$

---

*In the earlier version of T-1, quantizing levels $L = 128$ required only seven information bits. The eighth bit was used for signaling.

information bits, corresponding to 30 speech channels (with eight bits each). The remaining 16 bits/frame are used for frame synchronization and signaling. Therefore, although the bit rate is 2.048 Mbits/second corresponding to 32 voice channels, only 30 voice channels are transmitted.

DELTA MODULATION

Basically, *delta modulation* (*DM*) can be viewed as a simple method of converting analog signals into digital signals. It is distinguished from PCM by its simplicity and efficiency.

A basic delta modulator (Fig. 3.46) consists of a comparator and a sampler in the direct path and an integrator-amplifier in the feedback path. The analog signal $m(t)$ is compared with the feedback signal $\hat{m}(t)$. The error signal $\epsilon(t) = \hat{m}(t) - m(t)$ is applied to a comparator. If ϵ is positive, the comparator output is a constant signal of amplitude E, and if ϵ is negative, the comparator output is $-E$. Thus the comparator output $m_c(t)$ is given by

$$m_c(t) = E \operatorname{sgn}[\epsilon(t)] \tag{3.57}$$

The comparator output is sampled by a sampler at a rate of f_s samples/second where f_s is typically much higher than the Nyquist rate. The sampler thus produces a pulse train $d(t)$ that consists of positive pulses when $m(t) > \hat{m}(t)$ and negative pulses when $m(t) < \hat{m}(t)$. The pulse train $d(t)$ is the delta-modulated pulse train (Fig. 3.46*d*). The modulated signal $d(t)$ is amplified and integrated in the feedback path to generate $\hat{m}(t)$ [Fig. 3.46*c*], which tries to follow $m(t)$. Each pulse in $d(t)$ gives rise to a step function (positive or negative depending on the pulse polarity) in $\hat{m}(t)$. If, for example, $m(t) > \hat{m}(t)$, a positive pulse is generated in $d(t)$, which gives rise to a positive step in $\hat{m}(t)$ trying to equalize $\hat{m}(t)$ to $m(t)$ in small steps at every sampling instant. The waveforms $m(t)$ and $\hat{m}(t)$ are shown in Fig. 3.46*c*. It can be seen that $\hat{m}(t)$ is a kind of staircase approximation of $m(t)$. When $\hat{m}(t)$ is passed through a lowpass filter, the coarseness of the staircase in $\hat{m}(t)$ is eliminated, and we get a smoother and better approximation to $m(t)$. The demodulator at the receiver consists of an amplifier-integrator (identical to that in the feedback path of the modulator) followed by a lowpass filter (Fig. 3.46*b*).

DM, PCM and DPCM

In PCM, the analog signal samples are quantized in L levels, and this information is transmitted by n pulses per sample ($n = \log_2 L$). A little reflection shows that in DM, the modulated signal carries information not about the signal samples but about the difference between successive samples. If the difference is positive or negative, a positive or a negative pulse (respectively) is generated in the modulated signal $d(t)$. Basically, therefore, DM carries the information about the derivative of $m(t)$, and, hence, the name delta modulation. This can also be seen from the fact that integration of the modulated signal yields $\hat{m}(t)$, which is an approximation of $m(t)$.

In PCM, the information of each quantized sample is transmitted by a code word of n pulses, whereas in DM the information of the difference between successive samples is transmitted by a code word of one pulse. We may therefore expect a large

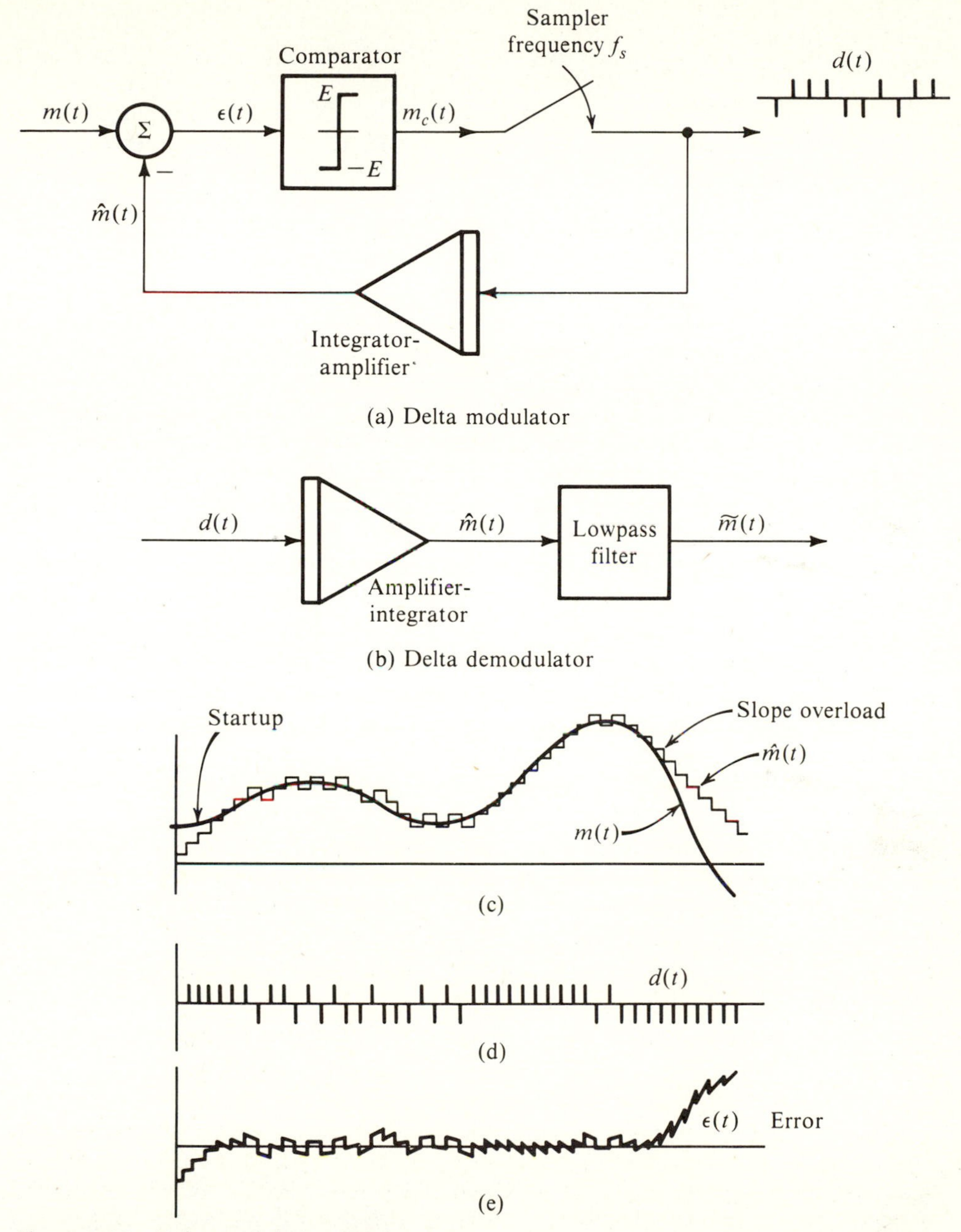

Figure 3.46 Delta modulation (DM).

quantization error in DM. Fortunately, this is not true, because in DM a compensating factor is that the sampling rate is much higher, typically four to eight times higher than that in PCM.

A variation of DM uses a code word of k pulses instead of one to transmit the information about the difference between the successive samples. In this scheme the error $\epsilon(t)$ [Fig. 3.46*e*] is quantized in 2^k levels, and the quantized value of $\epsilon(t)$ is then transmitted by a binary code word of k pulses (as in PCM). This scheme is known as *delta modulation by PCM (DPCM)*. It has been shown[17] that for speech signals, DPCM performance is superior to that of both PCM and DM. Unfortunately, the process of quantization and codification makes DPCM as complex as PCM. The simplicity of DM as well as the simplicity of multiplexing in PCM are both lost in DPCM.

DPCM has been used in digitized visual telephone (Picturephone) signals to conserve the bandwidth. The pulse rate required is only 6.312 Mbits/second, in contrast to the pulse rate of about 92 Mbits/second for digitized (PCM) TV video signals.

Threshold of Coding and Overloading

Threshold and overloading effects can be clearly seen in Fig. 3.46*c*. Variations in $m(t)$ smaller than the step value (threshold of coding) are lost in DM. At the same time, if $m(t)$ changes too fast (that is, $\dot{m}(t)$ is too high), $\hat{m}(t)$ cannot follow $m(t)$, and overloading occurs. This is the so-called *slope overload*, and it is one of the basic limiting factors in the performance of DM. We should expect slope overload rather than amplitude overload in DM, because DM basically carries the information about $\dot{m}(t)$.

The slope overload occurs when $\hat{m}(t)$ cannot follow $m(t)$. During the sampling interval T_s, $\hat{m}(t)$ is capable of changing by σ, where σ is the height of the step. Hence, the maximum slope that $\hat{m}(t)$ can follow is σ/T_s, or σf_s, where f_s is the sampling frequency. Hence, no overload occurs if

$$|\dot{m}(t)| < \sigma f_s \tag{3.58}$$

Consider the case of tone modulation:

$$m(t) = A \cos \omega t \tag{3.59}$$

The condition for no overload is

$$|\dot{m}(t)|_{\max} = \omega A < \sigma f_s \tag{3.60}$$

Hence, the maximum amplitude $A_{\max}$ of this signal that can be tolerated without overload is given by

$$A_{\max} = \frac{\sigma f_s}{\omega} \tag{3.61}$$

The overload amplitude of the modulating signal is inversely proportional to the frequency ω. For higher modulating frequencies, the overload occurs for smaller amplitudes. For voice signals, which contain all frequency components up to (say) 4 kHz, calculating $A_{\max}$ by using $\omega = 2\pi \times 4000$ in Eq. (3.61) will give an overly conservative value of $A_{\max}$. It has been shown by de Jager[18] that $A_{\max}$ for voice signals can be calculated by using $\omega_r \simeq 2\pi \times 800$ in Eq. (3.61).

$$[A_{\max}]_{\text{voice}} \simeq \frac{\sigma f_s}{\omega_r} \tag{3.62}$$

Thus, the maximum voice signal amplitude, A_{max}, that can be used without causing slope overload in DM is the same as the maximum amplitude of a sinusoidal signal of reference frequency f_r ($f_r \simeq 800$) that can be used without causing slope overload in the same system.

Fortunately, the voice spectrum (as well as TV video signal) also decays with frequency and closely follows the overload characteristics (curve c, Fig. 3.47). For

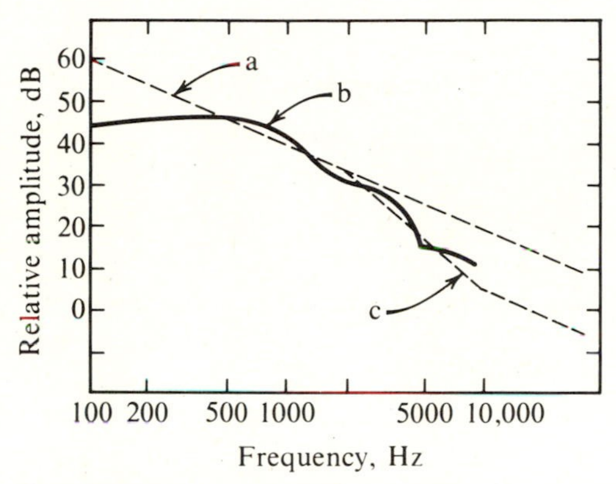

Figure 3.47 Voice signal spectrum.

this reason DM is well suited for voice (and TV) signals. Actually, the voice signal spectrum (curve b) decreases as $1/\omega$ up to 2000 Hz, and beyond this frequency, it decreases as $1/\omega^2$. If we had used a double integration in the feedback circuit instead of a single integration, A_{max} in Eq. (3.61) would be proportional to $1/\omega^2$. Hence, a better match between the voice spectrum and the overload characteristics is achieved by using a single integration up to 2000 Hz and a double integration beyond 2000 Hz.

Double Integration

The effect of double integration may also be seen from other points of view. When a narrow pulse is integrated once, it gives a step function, and $\hat{m}(t)$ in the case of single integration is made up of a coarse step function. If a narrow pulse is integrated twice, the output is a ramp, and we expect $\hat{m}(t)$ in this case to be a smoother function (see Fig. 3.48*b*) that follows $m(t)$ more closely. The double integration is provided by two *RC* circuits in cascade (Fig. 3.48), where $R_2C_2 \gg R_1C_1$ so that the second *RC* section does not overload the first. By adjusting the values R_1C_1 and R_2C_2 properly, the overload characteristic is made to closely match the voice spectrum, as shown in Fig. 3.47 (curve c). Here $1/R_1C_1$ is chosen to be 200π ($f_1 = 100$ Hz), and $1/R_2C_2$ is chosen to be 4000π ($f_2 = 2000$ Hz). This circuit provides single integration up to 2000 Hz and double integration beyond 2000 Hz to closely match the voice spectrum (see Fig. 3.47).

The double integration, unfortunately, causes a problem of hunting. In the case of single integration, each successive pulse causes a step change of σ in $\hat{m}(t)$. In the double integration, on the other hand, each successive pulse causes a fixed change in

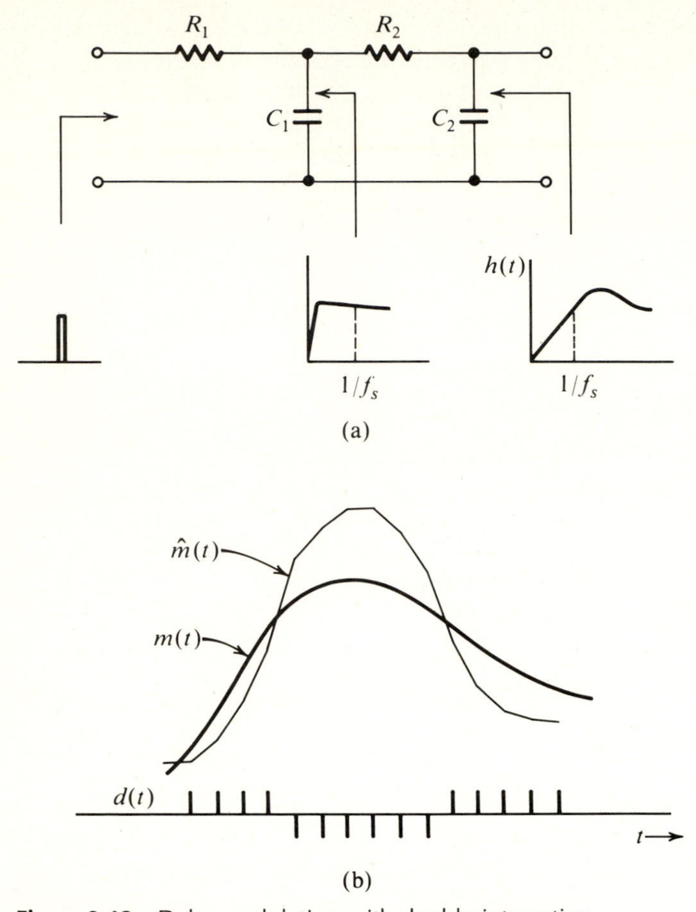

Figure 3.48 Delta modulation with double integration.

the slope of $\hat{m}(t)$. When $m(t)$ is increasing, the modulator output $d(t)$ generally is a sequence of positive pulses. Each pulse increases the slope of $\hat{m}(t)$. Now when $m(t)$ starts decreasing, $\hat{m}(t)$ could have at that moment a large slope, and it could take a long time before the slope decreases sufficiently to catch up with $m(t)$. This is what causes overshoot in $\hat{m}(t)$ at the point where $m(t)$ has maxima (see Fig. 3.48). The same process repeats when $m(t)$ has minima, with a consequent undershoot in $\hat{m}(t)$. This type of hunting can eventually cause some sort of oscillation and create instability.

The hunting described above appears because the circuit is not capable of recognizing quickly enough the rapid changes in $m(t)$ in order to adjust itself. This defect could be corrected if one could predict the future value of $\hat{m}(t)$ and compare it to $m(t)$ for generating pulses in $d(t)$.

The circuit in Fig. 3.49*a* gives double integration as well as prediction. We can show that $E_2(t)$ approximately represents the future value of $E_1(t)$, that is,

$$E_2(t) \simeq E_1(t + \tau) \tag{3.63}$$

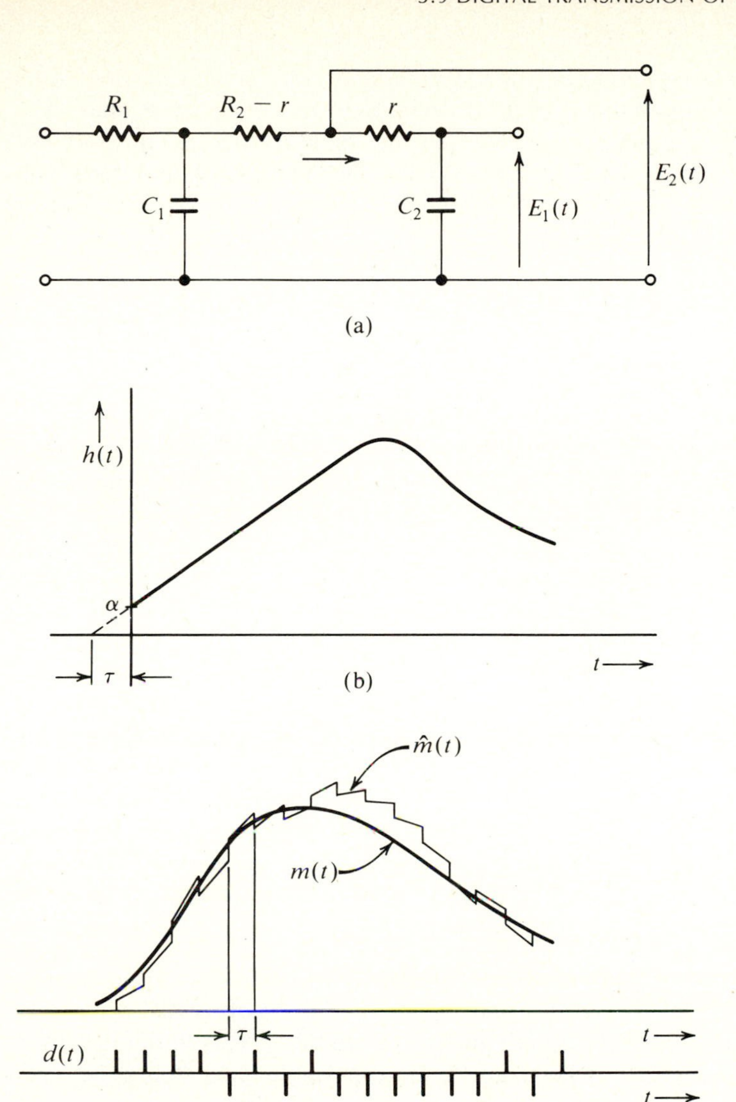

Figure 3.49 Double integrator with prediction.

To show this, we observe that

$$E_1(t + \tau) \simeq E_1(t) + \tau \frac{dE_1}{dt} \tag{3.64}$$

$$= E_1(t) + \tau \frac{i}{C_2}$$

Letting $r = \tau / C_2$, we have

$$E_1(t + \tau) \simeq E_1(t) + ri = E_2(t)$$

The prediction time τ is chosen to be one sampling interval. The waveform of $E_2(t)$ is approximately the same as that of $E_1(t)$ but shifted to the left by τ seconds. The impulse response $h(t)$ of this circuit is therefore the same as that in Fig. 3.48 but shifted to the left by τ seconds (see Fig. 3.49*b*). Thus, the impulse response starts with a value α at $t = 0$, and its slope eventually decreases to zero after a few intervals of τ seconds.

The voltage $E_2(t)$ is fed back to the comparator. The signals $m(t)$ and $\hat{m}(t)$ in this case are shown in Fig. 3.49*c*.

Adaptive Delta Modulation

The DM discussed so far suffers from one serious disadvantage. The dynamic range of amplitudes is too small, because of the threshold and overload effects discussed earlier. To correct this problem, some type of signal compression is necessary. In DM, a suitable method appears to be the adaptation of the step value σ (or the slope in the case of double integration) according to the input signal level. For example, in Fig. 3.46, when the signal $m(t)$ is falling rapidly, slope overload occurs. If we can increase the step size during this period, the overload could be avoided. On the other hand, if the slope of $m(t)$ is small, reduction of step size will reduce the threshold level as well as the quantization noise. In adaptive delta modulation, a circuit detects the overload as well as threshold conditions and adjusts the step size accordingly. Hence, the adaptive DM has a much larger dynamic range.

In the literature, various adaptive procedures have been proposed. These methods fall into two broad categories: the continuous and the discrete. In the continuous method, the step value varies continuously, whereas in the discrete case, the step value can vary discretely. We shall discuss here one system of continuous adaptation.

Continuous Adaptation. From the several systems proposed in the literature, we choose here the system of Tomozawa and Kaneko[19] (Fig. 3.50). This scheme is identical to ordinary DM except that the modulator output $d(t)$ is multiplied by a signal-level-detector output before being applied to the double integrator in the feedback path. The level-detector output indicates overload and threshold conditions. A long sequence of positive or negative pulses in $d(t)$ indicates overload, and alternating positive and negative pulses indicate threshold. The level detector consists of an integrator, a rectifier, and a smoothing filter. When overload occurs, $d(t)$ has a long string of positive or negative pulses. Hence, the integrator output will be high (positive or negative). The rectifier is used because we are interested in knowing whether overload has occurred and not its sign. Thus for overload conditions, the level detector output is high, and the step size is increased. On the other hand, for threshold conditions the level detector output is low, and the step size is reduced. The ratio of the largest to the smallest step size of 20 dB was obtained by Tomozawa and Kaneko.

Another notable method of continuous adaptation, that of Greefkes and Riemens[20] (see also Jayant[21]), uses digital control for companding; the control signal is derived from the pattern of transmitted pulses.

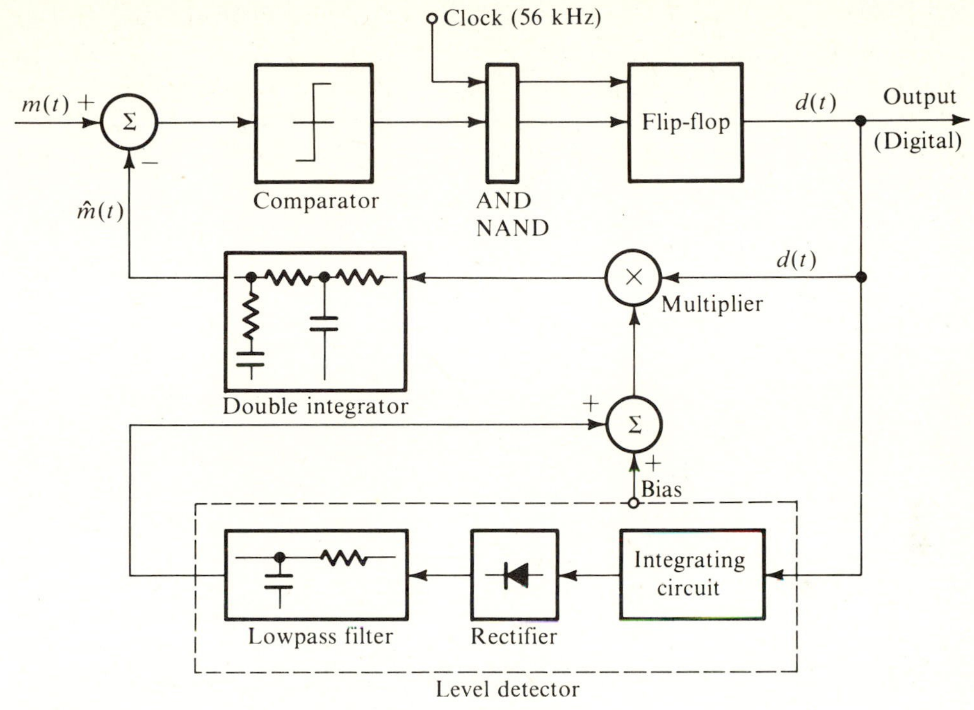

Figure 3.50 Continuous adaptive delta modulation.

Output SNR

The quantization noise in DM is shown in Fig. 3.46*e*. It can be seen that the quantization-error amplitude lies in the range $(-\sigma, \sigma)$, where σ is the step height of $\hat{m}(t)$. The error waveform, $\epsilon(t)$, is quite irregular, and we can assume the quantization-error amplitude to lie anywhere in the range $(-\sigma, \sigma)$. The situation is similar to that encountered in PCM, where the quantization-error amplitude was in the range $(-m_p/L, m_p/L)$. Hence from Eq. (3.51), the mean square error $\overline{\epsilon^2}$ of the quantization error is

$$\overline{\epsilon^2} = \frac{\sigma^2}{3} \tag{3.65}$$

From Fig. 3.46*e*, it is apparent that the quantization-error PSD will be a continuous spectrum, with most of the power in the frequency range extending well beyond the sampling frequency f_s. At the output, most of this will be suppressed by the baseband filter of bandwidth B. Hence, the output quantization-noise power N_q will be well below that indicated in Eq. (3.65). To compute N_q, we shall make an assumption that the PSD of the quantization noise is uniform and concentrated in the band 0—f_s. This assumption has been experimentally verified. Because the total power $\sigma^2/3$ is uniformly spread over the bandwidth f_s, the power within the baseband B is

$$N_q = \left(\frac{\sigma^2}{3}\right)\frac{B}{f_s}$$

$$= \frac{\sigma^2 B}{3f_s} \tag{3.66}$$

The output signal power is $S_o = \widetilde{m^2(t)}$. Assuming no overload distortion, the output noise is $N_o = N_q$, and

$$\frac{S_o}{N_o} = \frac{3f_s\widetilde{m^2(t)}}{\sigma^2 B} \tag{3.67}$$

If m_p is the peak signal amplitude, then from Eq. (3.62),

$$m_p = \frac{\sigma f_s}{\omega_r} \tag{3.68}$$

and

$$\frac{S_o}{N_o} = \frac{3f_s^3\widetilde{m^2(t)}}{\omega_r^2 B m_p^2} \tag{3.69a}$$

Because we need to transmit f_s pulses/second, the transmission bandwidth $B_T = kf_s/2 (1 < k < 2)$. Hence,

$$\frac{S_o}{N_o} = 24\left(\frac{B}{\omega_r}\right)^2\left(\frac{B_T}{kB}\right)^3\left(\frac{\widetilde{m^2(t)}}{m_p^2}\right) \tag{3.69b}$$

For voice signals, $B = 4000$ and $\omega_r = 2\pi(800)$. Hence,

$$\frac{S_o}{N_o} = \frac{150}{\pi^2}\left(\frac{B_T}{kB}\right)^3\left(\frac{\widetilde{m^2(t)}}{m_p^2}\right) \tag{3.70}$$

Thus the output SNR varies as the cube of the bandwidth expansion ratio. This result is derived for the single integration case. For double-integration DM, Greefkes and de Jager have shown that [22]

$$N_o = \frac{\sigma^2}{8\pi^2 c}\frac{Bf_2^2}{f_s^3} \tag{3.71}$$

where $c = (0.026)^2$ and f_2 is the frequency at which double integration starts. Typically $f_2 = 1800$ to 2000 Hz. Hence,

$$\frac{S_o}{N_o} = 8\pi^2 c \frac{f_s^3}{Bf_2^2\sigma^2}\widetilde{m^2(t)}$$

Substituting Eq. (3.68) in the above equation we get

$$\frac{S_o}{N_o} = \frac{64cB^4}{f_r^2 f_2^2}\left(\frac{B_T}{kB}\right)^5\left(\frac{\widetilde{m^2(t)}}{m_p^2}\right) \tag{3.72}$$

Substituting $B = 4000$, $f_r = 800$, $f_2 = 1800$, and $k = 1$, we get

$$\frac{S_o}{N_o} = 5.34\left(\frac{B_T}{B}\right)^5 \frac{\widetilde{m^2(t)}}{m_p^2} \tag{3.73}$$

It should be remembered that these results are valid only for voice signals. In all above developments, we have ignored the pulse detection error at the receiver. It is assumed that the transmitted pulse amplitudes are several times the channel noise rms value, and, hence, the pulse detection error is negligible. In the case of large channel noise, we shall have an additional source of noise caused by pulse detection errors. This noise component will cause signal degradation (threshold effect) for signals of small transmitted power.

Comparison with PCM

To compare DM with PCM, the output SNR for voice signals as a function of the bandwidth expansion ratio B_T/B is plotted in Fig. 3.51 for tone modulation for which

$$\frac{\widetilde{m^2}}{m_p^2} = \frac{1}{2}$$

The transmission bandwidth is assumed to be the theoretical minimum bandwidth for DM as well as PCM. From this figure it is clear that DM with double integration has a performance superior to PCM with compression (which is the practical case) for the

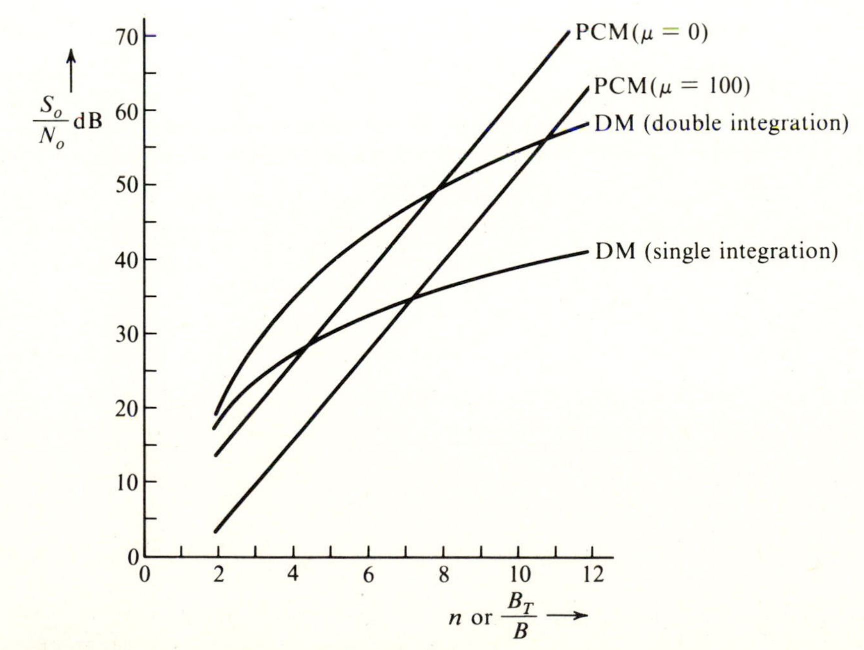

Figure 3.51 Comparison of DM and PCM.

range up to* $B_T/B = 10$. For larger values, PCM surpasses DM. Remember that this is true only of voice and TV signals, for which DM is ideally suited. For other types of signals, DM does not compare as well with PCM. Because the delta-modulated signal is a digital signal, it has all the advantages of digital systems, such as the use of regenerative repeaters and other advantages mentioned earlier. As far as detection errors are concerned, DM is more immune to this kind of error than is PCM, where the weight of the detection error depends on the digit location; thus for $n = 8$, the error in the first digit is 128 times as large as the error in the last digit. For DM, on the other hand, each digit has equal importance. Experiments have shown that an error probability P_e of the order of 10^{-1} does not affect the intelligibility of voice signals in DM, whereas P_e as low as 10^{-4} can cause serious error leading to threshold in PCM. For multiplexing several channels, however, DM suffers from the fact that each channel requires its own coder and decoder, whereas for PCM, one coder and one decoder is shared by all channels. But this very fact of an individual coder and decoder for each channel also permits more flexibility in DM. On the route between terminals, it is easy to drop one or more channels and insert other incoming channels. For PCM, such operations can be performed only at the terminals. This is particularly attractive for rural areas with low population density and where the population grows progressively. The individual coder-decoder also avoids cross talk, thus, alleviating the stringent design requirements in the multiplexing circuits in PCM. Considering all these factors, DM could well compete against PCM, particularly over short or medium-length links. Indeed, the French company Telecommunications Radioelectriques et Telephoniques (TRT) has developed a 60-channel multiplexed system using DM (continuous adaptation). The interesting thing is that it uses the bandwidth of 2.048 MHz—the same as that used by the PCM 30-channel system. This system has been used in France on an experimental basis since 1968, between Poitiers and Neuville. Apparently it has functioned satisfactorily, without failure, and the French PTT has ordered a number of similar terminals. Because DM offers better protection against noise, French authorities have also selected DM equipment for the airborne radio communications and position-location equipment proposed by TRT for the DIOSCUROS project (air traffic control over the Atlantic via satellites).

Despite several advantages, DM is unlikely to be widely adopted for voice-signal transmission for the simple reason that PCM was there first.

---

*Recent investigations using real speech signals indicate the crossover point for B_T/B to be between 7 and 8 rather than at 10.

APPENDIX 3.1 Equalizer Tap Setting Computations

The transversal filter output $p_o(t)$ (Fig. 3.27) is given by

$$p_o(t) = \sum_{n=-N}^{N} C_n p_r[t - (n + N)T_o] \tag{A3.1}$$

The samples of $p_o(t)$ at $t = (k + N)T_o$ are

$$p_o[(k + N)T_o] = \sum_{n=-N}^{N} C_n p_r[(k - n)T_o] \qquad k = 0, \pm 1, \pm 2, \pm 3, \ldots \tag{A3.2}$$

These samples are required to be zero for all $k \neq 0$, and the sample for $k = 0$ is required to be unity. Abbreviating $p_r(kT_o)$ by $p_r(k)$ for convenience, we have from Eq. (A3.2) a set of infinite simultaneous equations in terms of $2N + 1$ variables. Hence, it is not possible to realize this condition. However, if we specify the values of $p_o(t)$ only at $2N + 1$ points as

$$p_o[(k + N)T_o] = \begin{cases} 1 & k = 0 \\ 0 & k = \pm 1, \pm 2, \ldots, \pm N \end{cases} \tag{A3.3}$$

then a unique solution exists. This assures that a pulse will have zero interference at sampling instances of N preceding and N succeeding pulses. Because the pulse amplitude decays rapidly, interference beyond the Nth pulse is not significant, in general, for $N > 2$. Substitution of the condition Eq. (A3.3) in Eq. (A3.2) yields a set of $2N + 1$ simultaneous equations in $2N + 1$ variables:

$$\begin{bmatrix} 0 \\ 0 \\ \cdots \\ 0 \\ 0 \\ 1 \\ 0 \\ \cdots \\ 0 \\ 0 \end{bmatrix} = \begin{bmatrix} p_r(0) & p_r(-1) & \cdots & p_r(-2N) \\ p_r(1) & p_r(0) & \cdots & p_r(-2N+1) \\ \cdots & \cdots & \cdots & \cdots \\ & & & \\ p_r(N-1) & p_r(N-2) & \cdots & p_r(-N-1) \\ p_r(N) & p_r(N-1) & \cdots & p_r(-N) \\ p_r(N+1) & p_r(N) & \cdots & p_r(-N+1) \\ \cdots & \cdots & \cdots & \cdots \\ p_r(2N-1) & p_r(2N-2) & \cdots & p_r(1) \\ p_r(2N) & p_r(2N-1) & & p_r(0) \end{bmatrix} \begin{bmatrix} C_{-N} \\ C_{-N+1} \\ \cdots \\ \\ C_{-1} \\ C_0 \\ C_1 \\ \cdots \\ C_{N-1} \\ C_N \end{bmatrix} \tag{A3.4}$$

The tap-gain C_k's can be obtained by solving this set of equations.

■ **EXAMPLE 3.7**

For the received pulse $p_r(t)$ in Fig. 3.27*a*, let

$$a_0 = p_r(0) = 1$$

$$a_1 = p_r(1) = -0.3,\ a_2 = p_r(2) = 0.1,\ a_3 = p_r(3) = -0.03$$

$$a_{-1} = p_r(-1) = -0.2,\ a_{-2} = p_r(-2) = 0.05,\ a_{-3} = p_r(-3) = -0.01$$

Design a three-tap equalizer. Assume $p_r(\pm k) \simeq 0$ for $k \geq 4$. In this case $N = 1$. Substituting the above values in Eq. (A3.4), we get

$$\begin{bmatrix} 0 \\ 1 \\ 0 \end{bmatrix} = \begin{bmatrix} 1 & -0.2 & 0.05 \\ -0.3 & 1 & -0.2 \\ 0.1 & -0.3 & 1 \end{bmatrix} \begin{bmatrix} C_{-1} \\ C_0 \\ C_1 \end{bmatrix}$$

This yields $C_{-1} = 0.210$, $C_0 = 1.13$, and $C_1 = 0.318$.

Substituting these tap values and the data $p_r(k)$ in Eq. (A3.2), we obtain

$$p_o(-2T_o) = 0.0008 \qquad p_o(2T_o) = 0$$

$$p_o(-T_o) = 0.0113 \qquad p_o(3T_o) = 0.011$$

$$p_o(0) = 0 \qquad p_o(4T_o) = -0.0021$$

$$p_o(T_o) = 1$$

The output $p_o(t)$ is sketched in Fig. 3.27*b*.

We could have estimated the tap-gain C_k's by using the trial and error method discussed in Sec. 3.5. For example, because $a_1 = p_r(1) = -0.3$, $C_1 = 0.3$, which is close to C_1 (=0.318) calculated from Eq. (A3.4). Similarly, because $a_{-1} = p_r(-1) = -0.2$, $C_{-1} = 0.2$, which is close to C_2 (=0.21) calculated above. ■

REFERENCES

1. K. Feher, *Digital Communication, Microwave Applications,* Prentice-Hall, Englewood Cliffs, N.J., 1981, Chap. 7.
2. A. Lender, "Duobinary Technique for High Speed Data Transmission," *IEEE Trans. Commun. and Electron.*, vol. CE-82, pp. 214–218, May 1963.
3. D. W. Davis and D. L. A. Barber, *Communication Networks for Computers,* Wiley, New York, 1973.
4. P. Bylanski and D. G. W. Ingram, *Digital Transmission Systems,* Peter Peregrinus Ltd., Herts., England, 1976.
5. H. Nyquist, "Certain Topics in Telegraph Transmission Theory," *AIEE Trans.*, vol. 47, p. 817, April 1928.
6. W. R. Bennett and J. R. Davey, *Data Transmission,* McGraw-Hill, New York, 1965.
7. A. Lender, "Correlative Level Coding for Binary-Data Transmission," *IEEE Spectrum,* vol. 3, no. 2, pp. 104–115, February 1966.
8. E. D. Sunde, *Communication Systems Engineering Technology,* Wiley, New York, 1969.
9. R. W. Lucky and H. R. Rudin, "Generalized Automatic Equalization for Communication Channels," *IEEE Int. Comm. Conf.*, vol. 22, 1966.
10. R. W. Lucky, J. Salz, and E. J. Weldon, Jr., *Principles of Data Communication,* McGraw-Hill, New York, 1968.

11. W. R. Bennett, *Introduction to Signal Transmission,* McGraw-Hill, New York, 1970.
12. E. L. Gruenberg, *Handbook of Telemetry and Remote Control,* McGraw-Hill, New York, 1967.
13. B. Smith, "Instantaneous Companding of Quantized Signals," *Bell Syst. Tech. J.*, vol. 36, pp. 653–709, May 1957.
14. K. W. Cattermole, *Principles of Pulse-Code Modulation,* Ilife, England, 1969.
15. C. L. Dammann, L. D. McDaniel, and C. L. Maddox, "D-2 Channel Bank Multiplexing and Coding," *Bell Syst. Tech. J.*, vol. 51, pp. 1675–1700, October 1972.
16. Bell Telephone Laboratories, *Transmission Systems for Communication,* 4th ed., 1970.
17. H. Van De Weg, "Quantizing Noise of a Single Integration Delta Modulation System with an N-Digit Code," *Philips Res. Rep.*, no. 8, pp. 367–385, 1953.
18. F. de Jager, "Delta Modulation, A Method of PCM Transmission Using the 1-unit Code," *Philips Res. Rep.*, no. 7, pp. 442–466, 1952.
19. A. Tomozawa and H. Kaneko, "Companded Delta Modulation for Telephone Transmission," *IEEE Trans. Commun. Technol.*, vol. CT-14, pp. 9–157, February 1968.
20. J. A. Greefkes and K. Riemens, "Code Modulation with Digitally Controlled Companding for Speech Transmission," *Philips Tech. Rev.*, vol. 31, pp. 335–353, 1970.
21. N. S. Jayant, "Digital Coding of Speech Waveform: PCM, DPCM, and DM Quantizers," *Proc. IEEE,* vol. 62, pp. 611–632, May 1974.
22. J. A. Greefkes and F. de Jager, "Continuous Delta Modulation," *Philips Res. Rep.*, no. 23, pp. 233–246, 1968.

PROBLEMS

3.1. A signal $m_1(t)$ is bandlimited to 3 kHz, and three other signals, $m_2(t)$, $m_3(t)$, and $m_4(t)$, are bandlimited to 1 kHz each. These signals are to be sampled at the Nyquist rate. Suggest a suitable multiplexing arrangement (as in Fig. 3.3*a*). What must be the speed of the commutator (in samples/second)? If the commutator output is quantized ($L = 1024$) and binary coded, what is the output bit rate?

3.2. In Prob. 3.1, if the bandwidth of $m_1(t)$ is 3 kHz but those of $m_2(t)$, $m_3(t)$, and $m_4(t)$, are 1.2 kHz each, what must be the minimum speed of the commutator (in samples/second) in order to sample each signal at least at the Nyquist rate?

3.3. Four signals, $m_1(t)$, $m_2(t)$, $m_3(t)$, and $m_4(t)$, are to be sampled at a rate at least equal to the Nyquist rate of each, and then time-division multiplexed. The bandwidths of the signals are 2 kHz, 1 kHz, 600 Hz, and 500 Hz, respectively. Suggest a suitable TDM scheme and determine the commutator speed in samples/second. If the commutator output is quantized ($L = 512$) and binary coded, what is the output bit rate?

3.4. Derive the time-autocorrelation function and the PSD of an on-off signal (Fig. P3.4) directly, without using the results in Eqs. (3.6) or (3.8). Assume that a positive or a zero pulse is equally likely, and $t_o \leq T_o/2$.

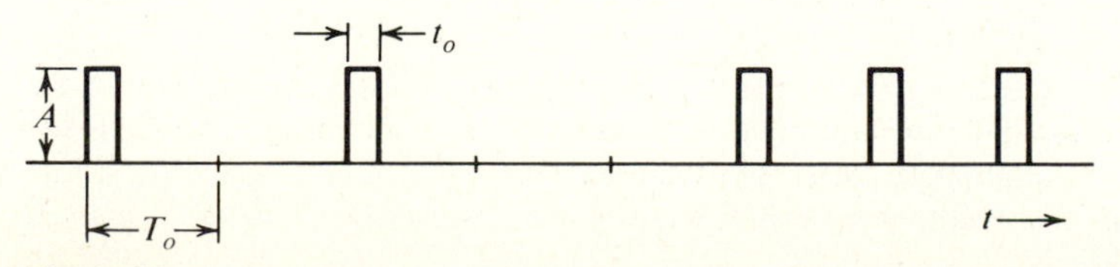

Figure P3.4

3.5. Repeat Prob. 3.4 for the bipolar signal.

3.6. It is possible to force the PSD to vanish at some finite values of ω in the band by using Nth-order bipolar signaling. In this case we separate the incoming digits into N digit streams; the first stream consists of digit numbers $1, N + 1, 2N + 1, \ldots$; the second stream consists of digit numbers $2, N + 2, \ldots$; and the Nth stream consisting of digits $N, 2N, 3N, \ldots$. Each stream is now independently coded using bipolar coding. All N streams are time multiplexed to get the final signal. Show that the PSD will be zero at frequencies $f_o/N, 2f_o/N, \ldots, f_o$, where f_o is the bit rate of the input digits (Fig. P3.6).

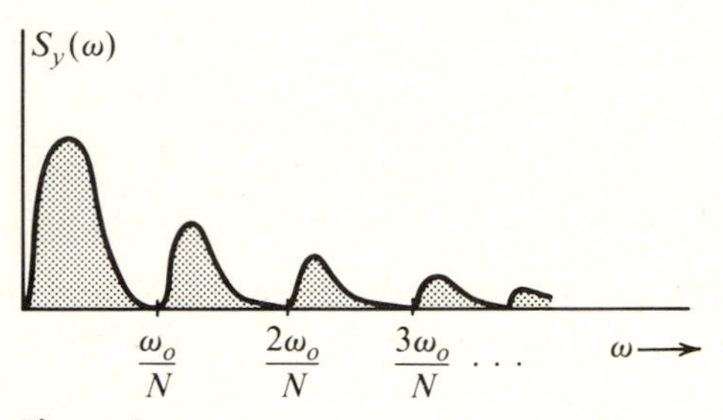

Figure P3.6

3.7. Determine the transmission bandwidth required for the multiplexed signals in Probs. 3.1, 3.2, and 3.3. Assume duobinary signaling.

3.8. Information digits a_k are processed as shown in Fig. P3.8. The output digits b_k are ternary, with values $+2$, 0, and -2. These digits are transmitted by using no pulse for

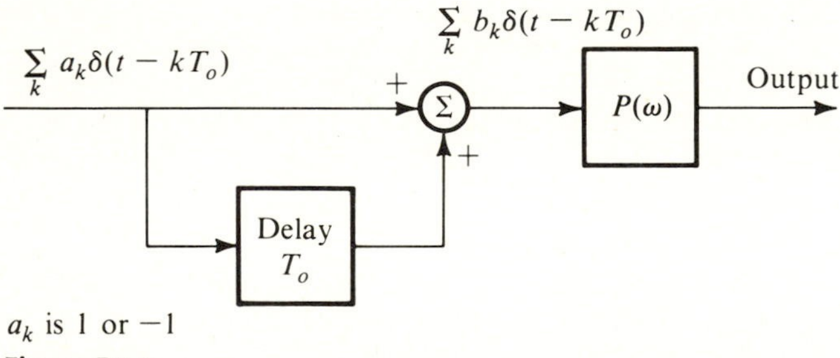

Figure P3.8

0 and a pulse $p(t)$ or $-p(t)$ for digits $+2$ and -2, respectively. Show that the resulting signal has the same PSD as that of the duobinary signal [Eq. (3.19)]. Because $a_k = b_k - a_{k-1}$, show that the received data can be decoded provided the first information digit a_1 is known.

3.9. Show that in bipolar signaling, the transmission bandwidth is independent of the pulse width, whereas in on-off and polar signaling, the transmission bandwidth is inversely proportional to the pulse width. For example, the bandwidth required for a full-width pulse is half that required for a half-width pulse.

3.10. A *top-hat* pulse (Fig. P3.10*a*) and a dipulse (Fig. P3.10*b*) are used in binary signaling to shape the PSD of the transmitted signal. Because the area under $p(t)$ is zero, the PSD

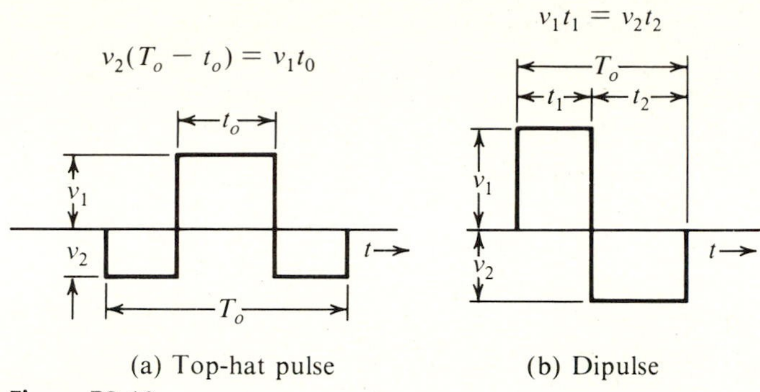

(a) Top-hat pulse (b) Dipulse

Figure P3.10

has a dc null [see Eq. (3.32)]. Show that when polar signaling is used, the transmission bandwidth in the case of either pulse is twice that of bipolar signaling. Show also that the power tends to concentrate at the lower-frequency end in the case of the dipulse and at the higher-frequency end in the case of the top-hat pulse. Assume $v_1 = v_2$ for both pulses.

3.11. If $P_r(\omega)$ and $P_i(\omega)$ are real and imaginary parts of the vestigial filter $P(\omega)$ in Eq. (3.30), show that

$$P_r\left(\frac{\omega_o}{2} + x\right) + P_r\left(\frac{\omega_o}{2} - x\right) = C$$

$$x < \frac{\omega_o}{2}$$

$$P_i\left(\frac{\omega_o}{2} + x\right) = P_i\left(\frac{\omega_o}{2} - x\right)$$

3.12. Binary data is transmitted over a telephone line of bandwidth 3.6 kHz. Pulses are shaped according to Nyquist's first criterion using the roll-off factor $r = 0.2$.

(a) Give a rough sketch of the pulse spectrum $P(\omega)$.

(b) Determine the rate of pulse transmission in bits/second.

3.13. Figure P3.13*a* shows a scheme using Nyquist's second criterion.Given

$$P(\omega) = \cos\left(\frac{\omega}{2f_o}\right) \Pi\left(\frac{\omega}{2\pi f_o}\right)$$

and $a_1 a_2 \ldots a_k, \ldots$ is the transmitted pulse amplitude sequence, where a_k can be -1 or 1 (polar signaling).

(a) Show that the scheme in Fig. P3.13*a* is equivalent to that in Fig. P3.13*b* with $P(\omega)$ in Eq. (3.36b).

(b) Assume a typical sequence for a_k and determine the sequence b_k in Fig. P3.13*b*.

(c) Show that the sequence b_k is three-valued, with values -2, 0, and 2 and it has the property of the duobinary sequence in Sec. 3.2; that is, an even number of 0's occur between two successive nonzero digits of the same polarity, and an odd number of 0's occur between two successive nonzero digits of opposite polarity.

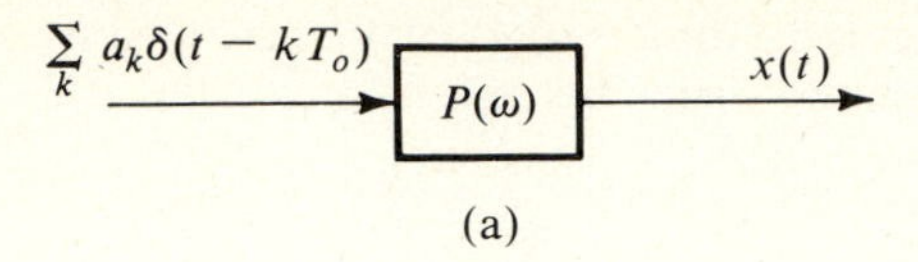

(a)

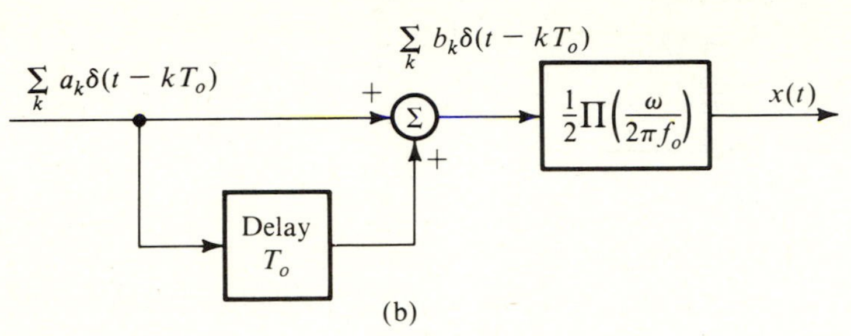

(b)

Figure P3.13

3.14. **(a)** Show that if $P_r(\omega) + jP_i(\omega)$ [where $P_r(\omega)$ is real and symmetrical about $\omega_o/2$ and $P_i(\omega)$ is real with odd symmetry about $\omega_o/2$] is added to $P(\omega)$ in Eq. (3.36b), the resulting spectrum's inverse transform still satisfies Nyquist's second criterion.

(b) Show that the raised-cosine spectrum is obtained by adding to $P(\omega)$ in Eq. (3.36b) a real $P_r(\omega)$ that is symmetrical about $\omega_o/2$.

3.15. In a binary data transmission using Nyquist's second criterion, sample values at the midpoint of signaling pulses were read as follows:

$$f_o \; f_o \; 0 \; 0 \; 0 \; -f_o \; 0 \; 0 \; -f_o \; 0 \; f_o \; 0 \; 0 \; -f_o \; 0 \; f_o \; f_o \; 0 \; -f_o$$

(a) Explain if there is any error in detection.

(b) Can you guess the correct transmitted digit sequence? There is more than one possible correct sequence. Give as many as possible correct sequences assuming that more than one detection error is extremely unlikely.

3.16. In Example 3.3, when the sequence S = **101010100000111** was applied to the input of the scrambler in Fig. 3.25*a*, the output T was found to be **101110001101001**. Verify that when this sequence T is applied to the input of the unscrambler in Fig. 3.25*b*, the output is the original sequence, S = **101010100000111**.

3.17. A scrambler is shown in Fig. P3.17. Design the corresponding unscrambler. If a sequence S = **101010100000111** is applied to the input of this scrambler, determine the output sequence T. Verify that if this T is applied to the input of the unscrambler, the output is the sequence S.

3.18. In a certain binary communication system, the pulse $p_r(t)$ received (see Fig. 3.27*a*) has the following values at the sampling instants:

$$p_r(0) = 1$$

$$p_r(T_o) = 0.1 \qquad p_r(-T_o) = 0.3$$

$$p_r(2T_o) = -0.02 \qquad p_r(-2T_o) = -0.07$$

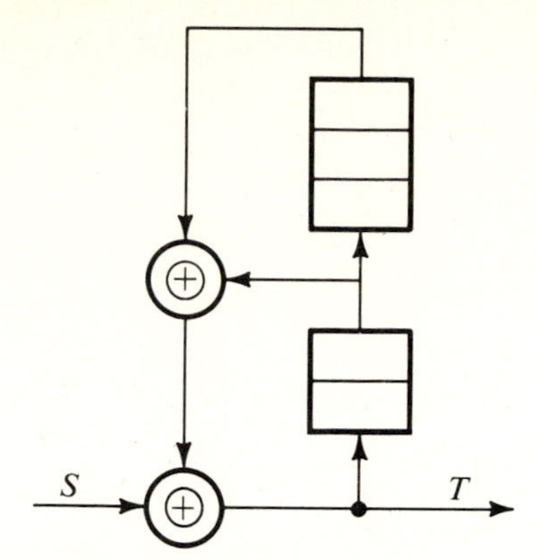

Figure P3.17

Determine approximate values for the tap settings of a three-tap transversal equalizer using the trial-and-error method discussed in the text.

3.19. Show that the noise immunity of bipolar and duobinary signaling is identical to that of on-off signaling.

3.20. For bipolar signaling with the received peak amplitude $A_p = 0.001$ determine the detection error probability if the channel noise is gaussian with rms value 0.0002. What is the error probability for: **(a)** on-off signaling, **(b)** bipolar signaling, **(c)** duobinary signaling? Assume zero ISI in pulse detection.

3.21. Consider a case of binary transmission using polar signaling that uses half-width rectangular pulses of amplitudes $A/2$ and $-A/2$. The data rate is f_o digits/second.

(a) What is the transmission bandwidth and the transmitted power.

(b) This data is to be transmitted by M-ary rectangular half-width pulses of amplitudes $\pm A/2$, $\pm 3A/2$, $\pm 5A/2$, . . . , $\pm[(M-1)/2]A$. The minimum pulse amplitude separation is A in order to maintain about the same noise immunity. If each of the M-ary pulses is equally likely to occur, determine the transmitted power. Also, determine the transmission bandwidth.

(c) If M-ary pulses, described in part (b), are transmitted at a rate of f_o pulses/second, determine the transmission rate of binary digits and the transmission bandwidth. Show that the transmitted power is $(M^2-1)A^2/24 \simeq M^2A^2/24$.

3.22. Binary data is transmitted over a certain channel at a certain rate f_o bits/second. To reduce the transmission bandwidth, it is decided to transmit this data using M-ary multiamplitude signaling.

(a) By what factor is the bandwidth reduced?

(b) By what factor is the transmitted power increased, assuming minimum separation between pulse amplitudes to be the same in both cases?

3.23. A TV signal's bandwidth is 4.2 MHz. If the number of quantizing levels is to be 512, determine the number of binary pulses/second of the corresponding PCM signal and the transmission bandwidth.

3.24. A message signal $m(t)$ is transmitted by binary PCM without compression. If the signal-to-quantization-noise ratio is required to be at least 50 dB, determine the minimum value of L required. Assume that $m(t)$ is sinusoidal.

3.25. A signal $m(t)$ is such that its amplitude is uniformly distributed in the range $(-m_p, m_p)$, that is, the amplitude is equally likely to take on any value in the range $(-m_p, m_p)$.

(a) Show that

$$\widetilde{m^2(t)} = \frac{m_p^2}{3}$$

Hint: $\widetilde{m^2}$ can be derived by the method used to derive N_q in Eq. (3.51).

(b) The signal $m(t)$ is sampled and quantized into L uniform levels and then coded into binary pulses (a PCM signal). Show that the signal-to-quantization-noise ratio at the receiver is

$$\frac{S_o}{N_q} = L^2$$

3.26. For a TV signal transmitted by binary PCM, the signal-to-quantization-noise ratio is required to be at least 50 dB (assuming no compression). Determine the number of quantizing levels L and the resulting SNR, assuming uniform amplitude distribution of the video signal. *Hint:* Use the results in Prob. 3.25, and remember that L is of the form 2^n.

3.27. For the PCM signal in Prob. 3.23, determine L if the compression parameter $\mu = 255$ and the minimum SNR required is 50 dB. Determine the output SNR with this value of L. *Hint:* Remember that L is of the form 2^n.

3.28. In multiamplitude scheme with $M = 16$,

(a) Determine the minimum transmission bandwidth required to transmit data at a rate of 12,000 bits/second.

(b) If Nyquist's first criterion with a roll-off factor $r = 0.2$ is used to shape pulses, determine the transmission bandwidth.

3.29. In M-ary PCM, multiamplitude M-ary pulses are used to encode quantized samples. Show that

$$\frac{S_o}{N_o} = 3\frac{\widetilde{m^2(t)}}{m_p^2}(M)^{2n}$$

where n is the number of M-ary pulses used to encode each signal sample.

3.30. In a single-integration DM system, the voice signal is sampled at a rate of 64,000 samples/second. The maximum signal amplitude $A_{\max} = 1$.

(a) Determine the minimum value of the step size σ to avoid slope overload.

(b) Determine the quantization-noise power N_q if the voice signal bandwidth is 3.5 kHz.

(c) Assuming that the voice signal is sinusoidal, determine S_o and the SNR.

(d) Assuming that the voice signal amplitude is uniformly distributed in the range $(-1, 1)$, determine S_o and the SNR.

(e) Determine the transmission bandwidth.

3.31. Repeat parts (c) and (d) in Prob. 3.30 if double-integration DM is used, with $f_2 = 1800$ Hz.

4

Modulation

PART I Amplitude (Linear) Modulation

Modulation is a process that causes a shift of the range of frequencies in a signal. It is used to gain certain advantages, as mentioned in Chapter 1. Before discussing modulation, it is important to distinguish between communication that does not use modulation (*baseband communication*) and communication that uses modulation (*carrier communication*).

4.1 BASEBAND AND CARRIER COMMUNICATION

The term baseband is used to designate the band of frequencies of the signal delivered by the source or the input transducer (see Fig. 1.2). In telephony, the baseband is the audio band (band of voice signals) of 0 to 3.5 kHz. In television, the baseband is the video band occupying 0 to 4.3 MHz. For digital data or PCM using bipolar signaling at a rate of f_o pulses/second, the baseband is 0 to f_o Hz.

In baseband communication, baseband signals are transmitted without modulation, that is, without any shift in the range of frequencies of the signal. Because the baseband signals have sizable power at low frequencies, they cannot be transmitted over a radio link but are suitable for transmission over a pair of wires or coaxial cables. Local telephone communication and short-haul PCM (between two exchanges) use baseband communication. Because baseband communication uses only baseband frequencies, its uses are rather restricted. Also, because the transmission of signals at lower frequencies is in general more difficult, it is desirable to shift the signal spectrum to a higher-frequency range by modulation. Moreover, the vast spectrum of frequencies available because of technological advances cannot be utilized by a baseband scheme. By modulating several baseband signals and shifting their spectra to nonoverlapping bands, one can use all the available bandwidth more efficiently. Long-haul communication over a radio link also requires modulation to shift the signal spectrum to higher frequencies to enable efficient power radiation using antennas of reasonable dimensions. Yet another use of modulation is to exchange transmission bandwidth for the SNR.

Communication that uses modulation to shift the frequency spectrum of a signal is known as *carrier communication*. In this mode, one of the basic parameters (amplitude, frequency, or phase) of a *sinusoidal carrier* of high frequency ω_c is varied in proportion to the baseband signal $m(t)$. This results in amplitude modulation (AM), frequency modulation (FM), or phase modulation (PM), respectively. The latter two types of modulation are similar, in essence, and are grouped under the name *angle modulation*. Modulation is used to transmit analog as well as digital baseband signals.

A comment about pulse-modulated signals (PAM, PWM, PPM, PCM, and DM) is in order here. Despite the term modulation, these signals are baseband signals. The term modulation is used here in another sense. Pulse-modulation schemes are really baseband coding schemes, and they yield baseband signals. These signals must still modulate a carrier in order to shift their spectra.

4.2 AMPLITUDE MODULATION: DOUBLE SIDEBAND (DSB)

In amplitude modulation, the amplitude A_c of the unmodulated carrier $A_c \cos(\omega_c t + \theta_c)$ is varied in proportion to the baseband signal (known as the *modulating signal*). The frequency ω_c and the phase θ_c are constant. We can assume $\theta_c = 0$ without a loss of generality. If the carrier amplitude A_c is made directly proportional to the modulating signal $m(t)$, the modulated carrier is $m(t) \cos \omega_c t$ (Fig. 4.1*c*). As seen earlier [Eq. (2.60a)], this type of modulation simply shifts the spectrum of $m(t)$ to the carrier frequency (Fig. 4.1*c*); that is, if

$$m(t) \leftrightarrow M(\omega)$$

$$m(t) \cos \omega_c t \leftrightarrow \tfrac{1}{2}[M(\omega + \omega_c) + M(\omega - \omega_c)] \tag{4.1}$$

The bandwidth of the modulated signal is $2B$, which is twice the bandwidth of the modulating signal $m(t)$. From Fig. 4.1*c*, we observe that the modulated carrier spectrum centered at ω_c is composed of two parts: a portion that lies above ω_c, known as the *upper sideband (USB)*, and a portion that lies below ω_c, known as the *lower sideband (LSB)*. Similarly, the spectrum centered at $-\omega_c$ has upper and lower side-

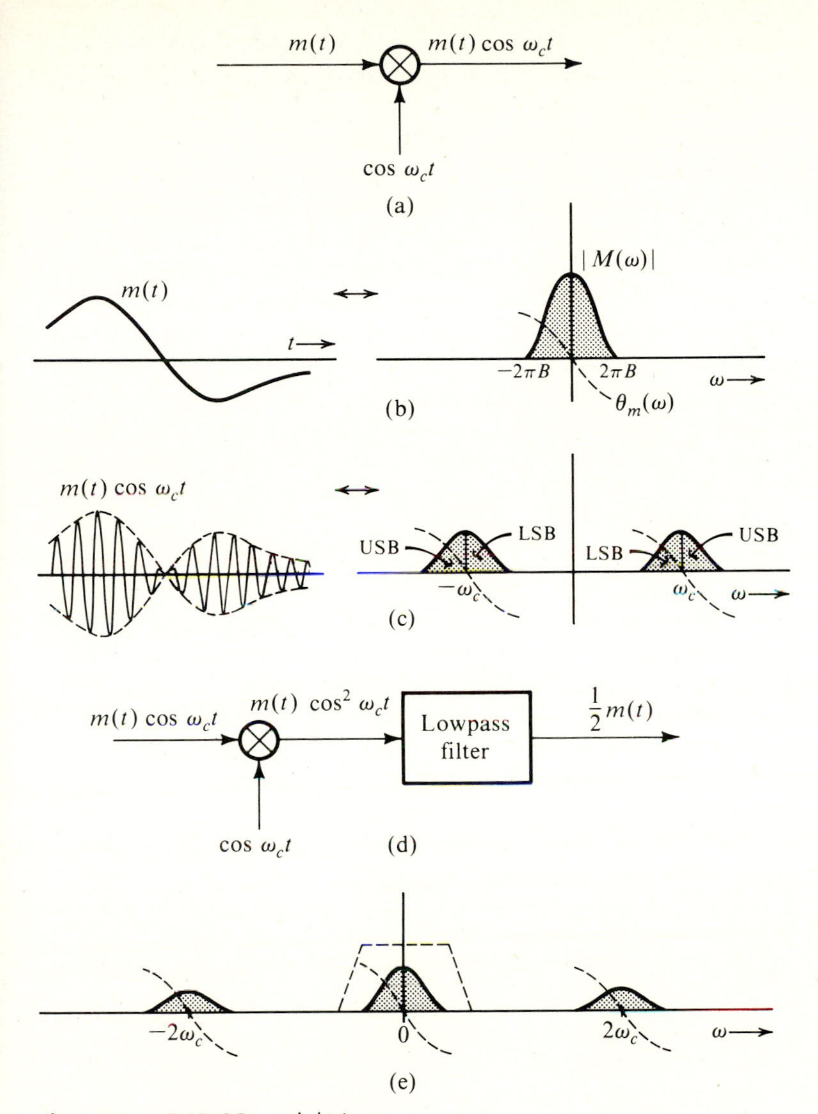

Figure 4.1 DSB-SC modulation.

bands. For instance, if $m(t) = \cos \omega_m t$, then the modulated signal

$$\begin{aligned} m(t) \cos \omega_c t &= \cos \omega_m t \cos \omega_c t \\ &= \tfrac{1}{2}[\cos (\omega_c + \omega_m)t + \cos (\omega_c - \omega_m)t] \end{aligned}$$

The component of frequency $\omega_c + \omega_m$ is the upper sideband, and that of frequency $\omega_c - \omega_m$ is the lower sideband, corresponding to the modulating signal of frequency ω_m. Thus, each component of frequency ω_m in the modulating signal gets translated into two components, of frequencies $\omega_c + \omega_m$ and $\omega_c - \omega_m$, in the modulated signal.

Note that the modulated signal $m(t) \cos \omega_c t$, as seen from the above equation, has components of frequencies $\omega_c \pm \omega_m$ but does not have a component of the carrier frequency ω_c. For this reason, this scheme is referred to as *double-sideband-suppressed-carrier (DSB-SC) modulation.*

The DSB-SC modulation translates the frequency spectrum by $\pm\omega_c$ (that is, $+\omega_c$ and $-\omega_c$), as seen from Eq. (4.1). To recover the original signal $m(t)$ from the modulated signals, it is necessary to retranslate the spectrum to its original position. The process of retranslating the spectrum to its original position is referred to as *demodulation,* or *detection*. Observe that if the modulated carrier spectrum (Fig. 4.1*c*) is shifted again by $\pm\omega_c$, we get back the desired baseband spectrum plus an unwanted spectrum at $\pm 2\omega_c$, which can be suppressed by a lowpass filter (Fig. 4.1*e*). This means that in order to demodulate, we should multiply the incoming modulated carrier by $\cos \omega_c t$ and pass the product through a lowpass filter (Fig. 4.1*d*). This conclusion can be directly verified from the identity

$$(m(t) \cos \omega_c t)(\cos \omega_c t) = \tfrac{1}{2}[m(t) + m(t) \cos 2\omega_c t] \tag{4.2a}$$

and

$$(m(t) \cos \omega_c t)(\cos \omega_c t) \leftrightarrow \tfrac{1}{2}M(\omega) + \tfrac{1}{4}[M(\omega + 2\omega_c) + M(\omega - 2\omega_c)] \tag{4.2b}$$

It can be seen from Fig. 4.1*e* that a lowpass filter allows the desired spectrum $M(\omega)$ to pass and suppresses the unwanted high-frequency spectrum centered at $\pm 2\omega_c$.

A possible form of lowpass-filter characteristics is shown (dotted) in Fig. 4.1*e*. The demodulator is shown in Fig. 4.1*d*. It is interesting to observe that the process at the receiver is similar to that required at the transmitter. This method of recovering the baseband signal is called *synchronous detection,* or *coherent detection,* where we use a carrier of exactly the same frequency (and phase) as was used for modulation. Thus, for demodulation, we need to generate a local carrier at the receiver in synchronism with the carrier that was used at the modulator.

The relationship of B to ω_c is of interest. From Fig. 4.1 it is obvious that $\omega_c \geq 2\pi B$ in order to avoid the overlap of $M(\omega + \omega_c)$ and $M(\omega - \omega_c)$. If $\omega_c < 2\pi B$, the information of $m(t)$ is lost in the process of modulation, and it is impossible to retrieve $m(t)$ from the modulated signal $m(t) \cos \omega_c t$. Theoretically, therefore, the only requirement is that $\omega_c \geq 2\pi B$. The practical factors, however, impose additional restrictions. A radiating antenna can radiate only a narrowband without distortion. This means that to avoid distortion caused by the radiating antenna, $\omega_c/2\pi B \gg 1$. The broadcast band AM radio uses is the band 550 kHz to 1600 kHz, or a ratio of $\omega_c/2\pi B$ roughly in the range of 100 to 300.

■ EXAMPLE 4.1

Baseband signals shown in Fig. 4.2*a* and *b* modulate a carrier of frequency ω_c. Assuming DSB-SC modulation, sketch the modulated waveforms.

The DSB-SC waveform for the signal in Fig. 4.2*a* is shown in Fig. 4.2*b*. The signal in Fig. 4.2*c* is a digital signal (polar signaling). The modulated waveform is shown in Fig. 4.2*d*. The modulated signal is also polar. This is a PSK scheme (see Sec. 3.8). This example is given here to stress that modulation is used to process analog as well as digital signals. ■

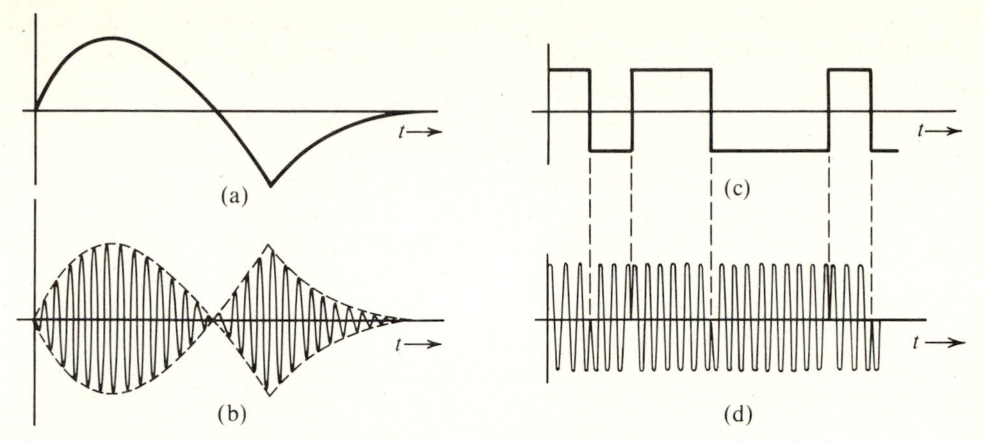

Figure 4.2 DSB-SC modulation: (a) and (c) modulating waveforms; (b) and (d) modulated waveforms.

Modulators

Modulation can be achieved in several ways. We shall discuss here some important categories of modulators.

Multiplier Modulators. Modulation is performed directly by multiplying $m(t)$ by $\cos \omega_c t$ using an analog multiplier whose output is proportional to the product of two input signals (Fig. 4.3a). In a variable-gain amplifier, the gain parameter (such as the

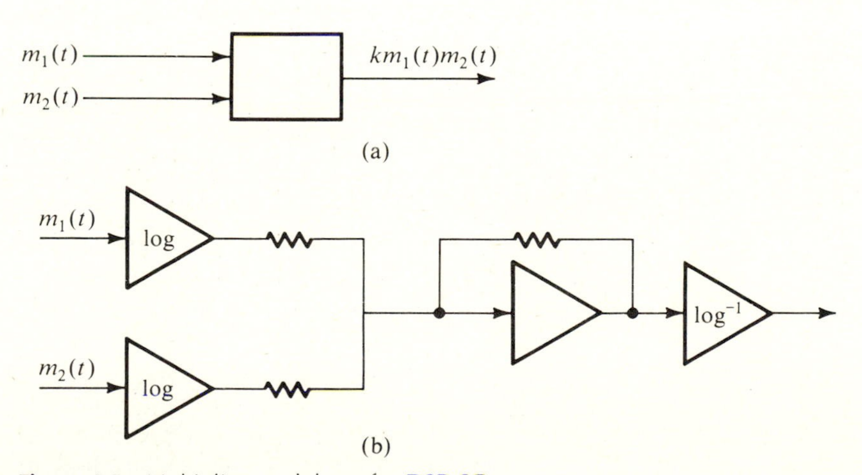

Figure 4.3 Multiplier modulator for DSB-SC.

β of a transistor) is controlled by one of the signals, say, $m_1(t)$. The amplifier gain is no longer constant but is $km_1(t)$ (varying with time). The output is the gain times the input signal $m_2(t)$, that is, $km_1(t)m_2(t)$. Note that this type of modulator is a linear time-varying system.

Another way to multiply two signals is through logarithmic amplifiers, an adder, and an inverse logarithmic amplifier, as shown in Fig. 4.3*b*. The figure is self-explanatory. For practical implementation of both of the above modulators, see Sheingold.[1]

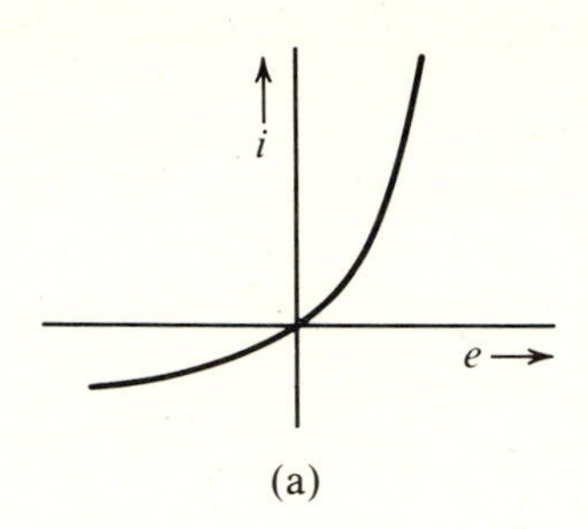

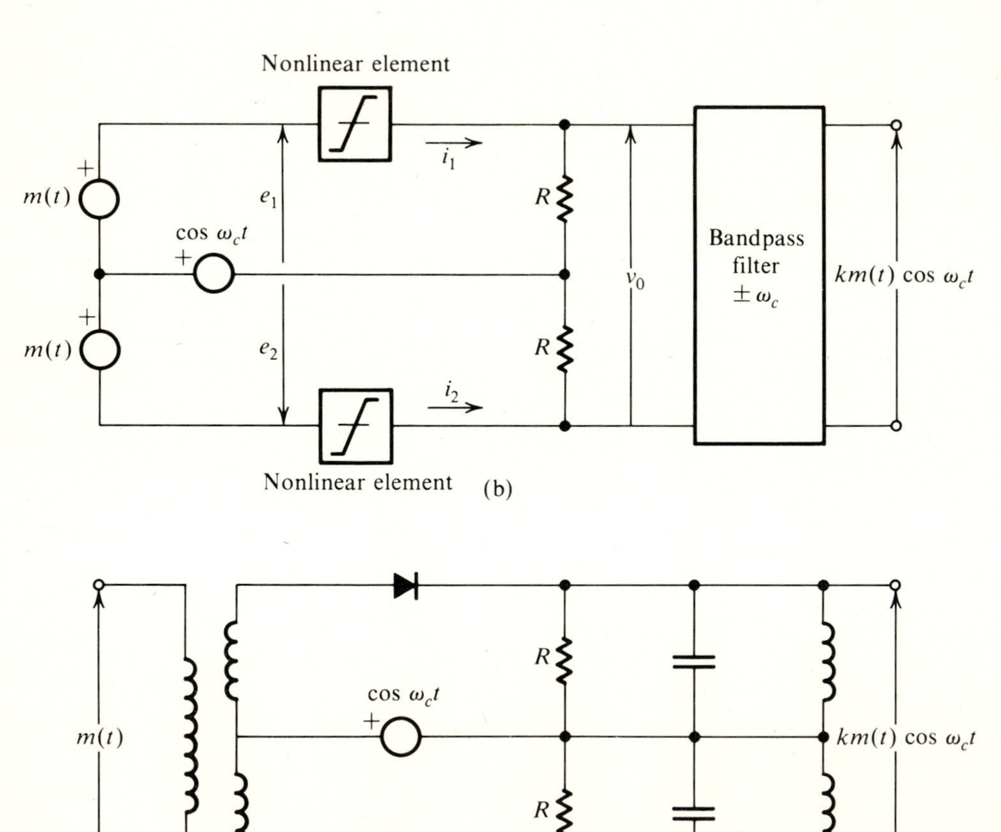

Figure 4.4 Nonlinear DSB-SC modulator.

Nonlinear Modulators. Modulation can also be achieved by using nonlinear devices. The characteristic of a typical nonlinear device is shown in Fig. 4.4*a*. A semiconductor diode or a transistor is an example of such a device.

Nonlinear characteristics such as these may be approximated by a power series:

$$i = ae + be^2 \tag{4.3}$$

One possible scheme using nonlinear elements for producing modulation is shown in Fig. 4.4*b*.

To analyze this circuit, consider the nonlinear element in series with the resistor R as a composite nonlinear element whose terminal voltage e and the current i are related by the power series in Eq. (4.3). The voltages e_1 and e_2 (Fig. 4.4*b*) are given by

$$e_1 = \cos \omega_c t + m(t) \quad \text{and} \quad e_2 = \cos \omega_c t - m(t)$$

Hence, currents i_1 and i_2 are given by

$$\begin{aligned} i_1 &= ae_1 + be_1^2 \\ &= a[\cos \omega_c t + m(t)] + b[\cos \omega_c t + m(t)]^2 \end{aligned} \tag{4.4a}$$

and

$$i_2 = a[\cos \omega_c t - m(t)] + b[\cos \omega_c t - m(t)]^2 \tag{4.4b}$$

The output voltage v_o is given by

$$v_o = i_1 R - i_2 R = 2R[2bm(t) \cos \omega_c t + am(t)] \tag{4.4c}$$

The signal $am(t)$ in this equation can be filtered out by using a bandpass filter tuned to ω_c at the output terminals. Implementation of this scheme using diodes is shown in Fig. 4.4*c*.

Switching Modulators. The multiplication operation required for modulation can be replaced by a simpler switching operation if we realize that a modulated signal can be obtained by multiplying $m(t)$ not only by a pure sinusoid but by any periodic signal $\varphi(t)$ of the fundamental radian frequency ω_c. Such a periodic signal can be expressed by a trigonometric Fourier series as

$$\varphi(t) = \sum_{n=0}^{\infty} C_n \cos (n\omega_c t + \theta_n) \tag{4.5a}$$

Hence,

$$m(t)\, \varphi(t) = \sum_{n=0}^{\infty} C_n m(t) \cos (n\omega_c t + \theta_n) \tag{4.5b}$$

This shows that the spectrum of the product $m(t)\, \varphi(t)$ is the spectrum $M(\omega)$ shifted to $\pm\omega_c, \pm 2\omega_c, \ldots, \pm n\omega_c, \ldots$. If this signal is passed through a bandpass filter of bandwidth $2B$ and tuned to ω_c, then we get the desired modulated signal $c_1 m(t) \cos (\omega_c t + \theta_1)$.

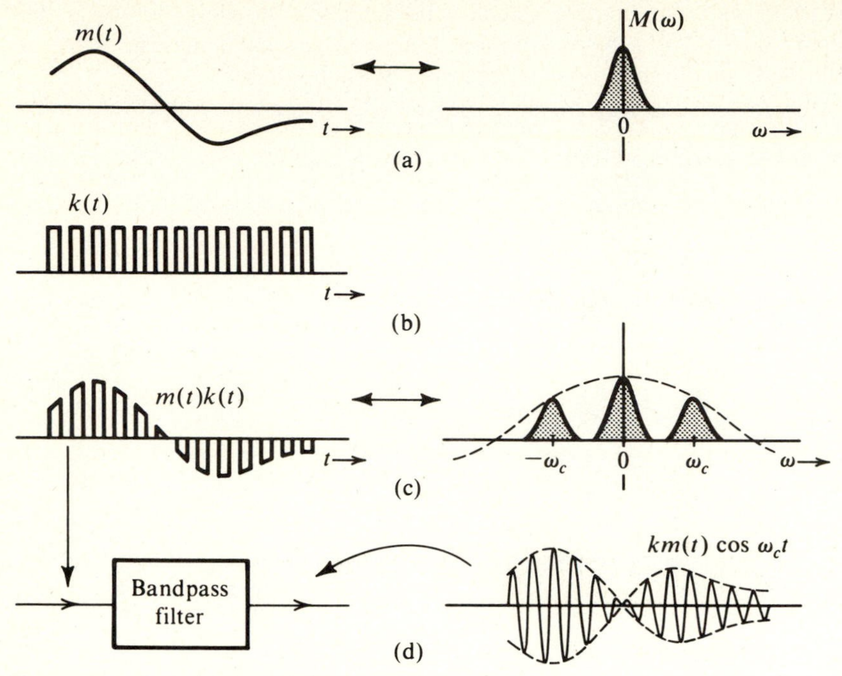

Figure 4.5 Switching modulator for DSB-SC.

The square pulse train $k(t)$ in Fig. 4.5*b* is a periodic signal whose Fourier series was found earlier [Eq. (2.13b) with $A = 1$] as

$$k(t) = \frac{1}{2} + \frac{2}{\pi} \sum_{n=1,3,5,\cdots} \frac{(-1)^{(n-1)/2}}{n} \cos n\omega_c t \tag{4.6}$$

$$= \frac{1}{2} + \frac{2}{\pi}\left(\cos \omega_c t - \frac{1}{3} \cos 3\omega_c t + \frac{1}{5} \cos 5\omega_c t - \cdots\right)$$

The signal $m(t)k(t)$ is given by

$$m(t)k(t) = \frac{1}{2}m(t) + \frac{2}{\pi}\left(m(t) \cos \omega_c t - \frac{1}{3}m(t) \cos 3\omega_c t + \frac{1}{5}m(t) \cos 5\omega_c t - \cdots\right) \tag{4.7a}$$

and

$$m(t)k(t) \leftrightarrow \frac{1}{2}M(\omega) + \frac{1}{\pi} \sum_{n=1,3,5,\cdots} \frac{(-1)^{(n-1)/2}}{n} [M(\omega + n\omega_c) + M(\omega - n\omega_c)] \tag{4.7b}$$

The product $m(t)k(t)$ and its spectrum are shown in Fig. 4.5*c*. When the signal

$m(t)k(t)$ is passed through a bandpass filter tuned to ω_c, the output is the desired modulated signal $(2/\pi)m(t)\cos\omega_c t$ (Fig. 4.5d).

Now here is the payoff. Multiplication of a signal by a square pulse train is in reality a switching operation. It involves switching the signal $m(t)$ on and off periodically and can be accomplished by simple switching elements controlled by $k(t)$. Figure 4.6a shows a *shunt-bridge diode modulator,* which will implement this

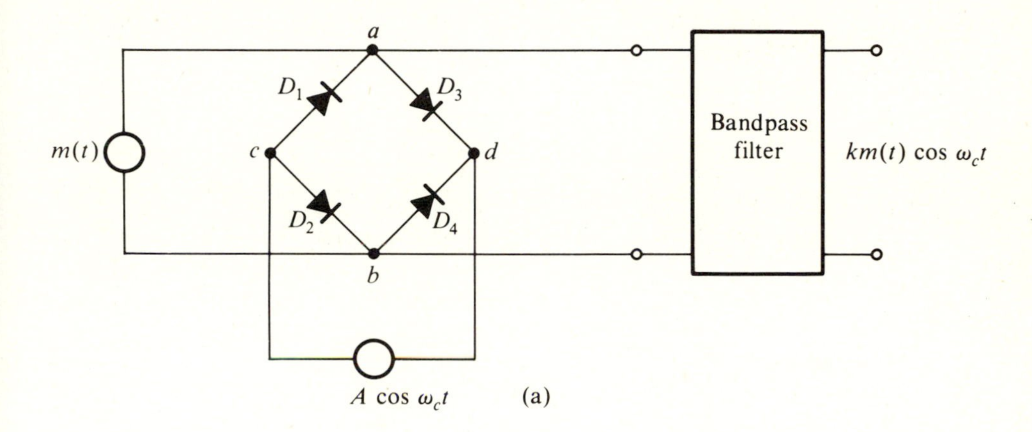

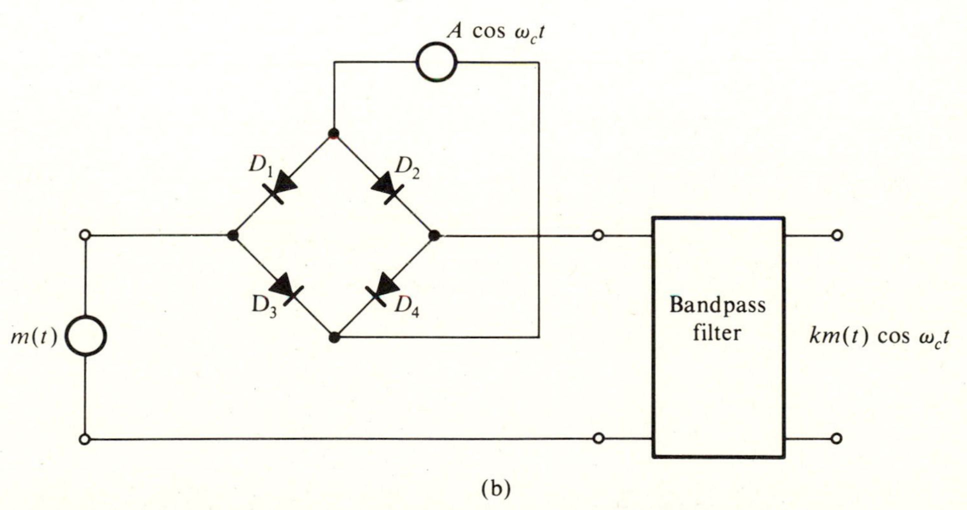

Figure 4.6 Switching modulators for DSB-SC. (a) Shunt-bridge diode modulator. (b) Series-bridge diode modulator.

scheme. Diodes D_1, D_2 and D_3, D_4 are matched pairs. When the signal $\cos\omega_c t$ is of a polarity that will make terminal c positive with respect to d, all the diodes conduct, assuming that the amplitude $A \gg m(t)$. Because diodes D_1 and D_2 are matched, terminals a and b have the same potential, and the input to the bandpass filter is shorted during this half-cycle. During the next half-cycle, terminal d is positive with respect to c, and all four diodes open, thus connecting $m(t)$ to the input of the bandpass filter. This switching on and off of $m(t)$ repeats for each cycle of the carrier. The effective input to the bandpass filter is $m(t)k(t)$, and the output is the desired modulated signal $c_1 m(t) \cos \omega_c t$.

In Fig. 4.6*a*, the diode bridge is in parallel with $m(t)$; hence, the name shunt-bridge diode modulator. If we place the diode bridge in series with $m(t)$ (Fig. 4.6*b*), it will accomplish the same purpose by periodically interrupting the input to the bandpass filter. This arrangement is called a *series-bridge diode modulator*.

Another switching modulator, known as the *ring modulator,* is shown in Fig. 4.7*a*. During the positive half-cycles of the carrier, diodes D_1 and D_3 conduct, and D_2

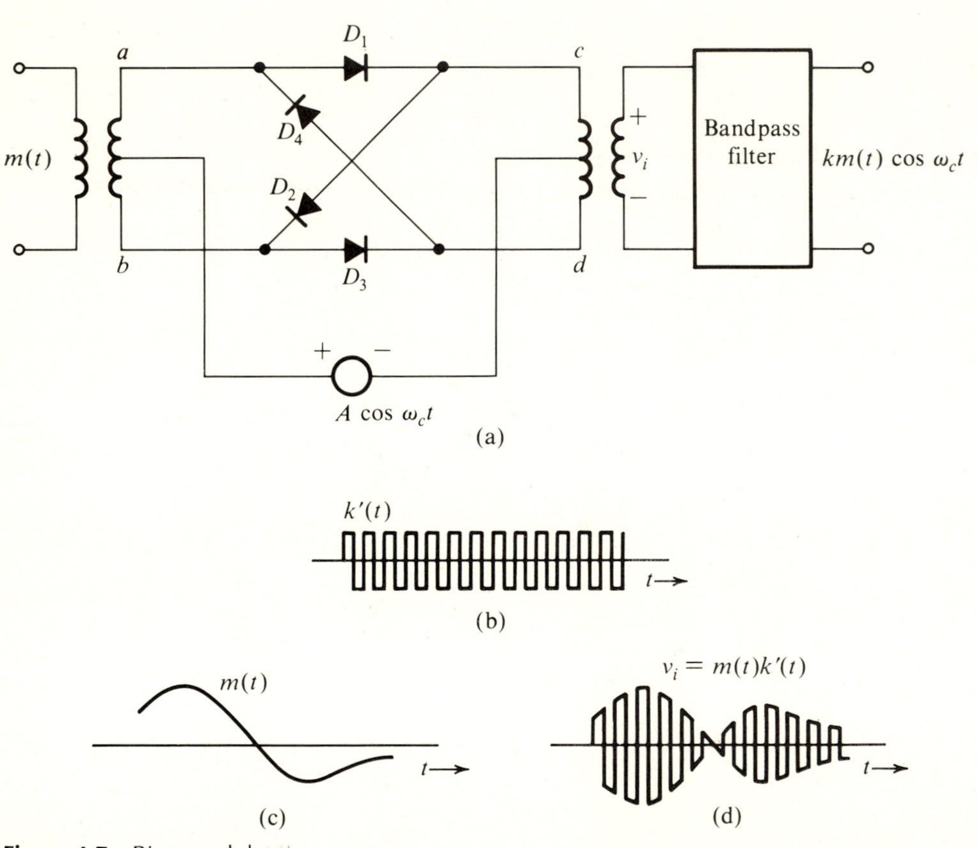

Figure 4.7 Ring modulator.

and D_4 are open. Hence, terminal a is connected to c, and terminal b is connected to d. During the negative half-cycles of the carrier, diodes D_1 and D_3 are open, and D_2 and D_4 are shorted, thus connecting terminal a to d and terminal b to c. Hence, the output is proportional to $m(t)$ during the positive half-cycle and to $-m(t)$ during the negative half-cycle. In effect, $m(t)$ is multiplied by a square pulse train $k'(t)$, shown in Fig. 4.7*b*. Note that $k'(t)$ is the same as $k(t)$ with $A = 2$ and its dc term eliminated

[Eq. (2.13b)]. Hence

$$k'(t) = \frac{4}{\pi} \sum_{n=1,3,5,\cdots} \frac{(-1)^{(n-1)/2}}{n} \cos n\omega_c t \tag{4.8a}$$

and

$$m(t)k'(t) = \frac{4}{\pi} \sum_{n=1,3,5,\cdots} \frac{(-1)^{(n-1)/2}}{n} m(t) \cos n\omega_c t \tag{4.8b}$$

The signal $m(t)k'(t)$ is shown in Fig. 4.7*d*. When this waveform is passed through a bandpass filter tuned to ω_c (Fig. 4.7*a*), the filter output will be the desired signal $(4/\pi)m(t) \cos \omega_c t$.

Balanced Modulators. We shall see in the next section that it is much easier to generate a signal of the form $m(t) \cos \omega_c t + A \cos \omega_c t$ than the DSB-SC signal ($m(t) \cos \omega_c t$). We can generate the DSB-SC signal using two such generators in a balanced configuration (Fig. 4.8) that will suppress the carrier term. The outputs of

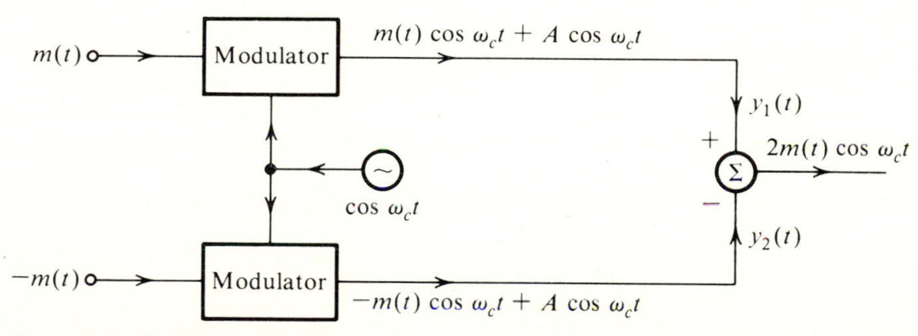

Figure 4.8 Balance modulator.

the two generators are

$$y_1(t) = m(t) \cos \omega_c t + A \cos \omega_c t$$

$$y_2(t) = -m(t) \cos \omega_c t + A \cos \omega_c t$$

The balanced modulator output $y(t)$ is

$$y(t) = y_1(t) - y_2(t) = 2m(t) \cos \omega_c t$$

For a perfect suppression of the carrier, both modulators should be matched as closely as possible.

The nonlinear modulator in Fig. 4.4 is an example of a balanced modulator. If we consider only the upper (or the lower) half of this modulator, it generates $y_1(t)$ [see Eq. (4.4a)]. By matching the upper and the lower modulator, the carrier is suppressed. Another example of a balanced modulator appears in Prob. 4.6.

■ EXAMPLE 4.2 Frequency Mixer or Converter

A frequency mixer, or frequency converter, is used to change the carrier frequency of a modulated signal $m(t) \cos \omega_c t$ from ω_c to some other frequency ω_I. This can be done by multiplying $m(t) \cos \omega_c t$ by $2 \cos (\omega_c + \omega_I)t$ or $2 \cos (\omega_c - \omega_I)t$ and then bandpass-filtering the product, as shown in Fig. 4.9*a*.

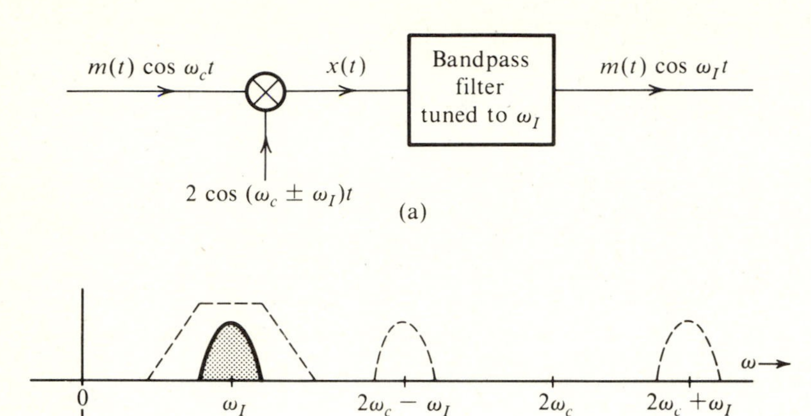

Figure 4.9 Frequency mixer (or converter).

The product $x(t)$ is

$$x(t) = 2m(t) \cos \omega_c t \cos (\omega_c \pm \omega_I)t$$
$$= m(t)[\cos \omega_I t + \cos (2\omega_c \pm \omega_I)t]$$

The spectrum of $m(t) \cos (2\omega_c \pm \omega_I)$ is centered at $2\omega_c \pm \omega_I$ and is eliminated by a bandpass filter tuned to ω_I (assuming $\omega_I < \omega_c - 2\pi B$) as shown in Fig. 4.9*b*. The filter output is $m(t) \cos \omega_I t$.

The operation of frequency mixing, or frequency conversion (also known as heterodyning), is identical to the operation of modulation with a modulating carrier frequency that differs from the incoming carrier frequency by ω_I. Any one of the modulators discussed earlier can be used for frequency mixing. When the local carrier (modulating carrier) is $\omega_c + \omega_I$, the operation is called *up-conversion*, and when it is $\omega_c - \omega_I$, the operation is *down-conversion*. ■

Demodulation of DSB-SC Signals

As discussed earlier, demodulation of a DSB-SC signal is identical to modulation (see Fig. 4.1*d*). At the receiver, we multiply the incoming signal by a local carrier of frequency and phase in synchronism with the carrier used at the modulator. The product is then passed through a lowpass filter. The only difference between the modulator and demodulator is the output filter. In the modulator, the multiplier output is passed through a bandpass filter tuned to ω_c, whereas in the demodulator, the multiplier output is passed through a lowpass filter. Therefore, all four types of modulators discussed earlier can also be used as demodulators, provided the bandpass filters at the output are replaced by lowpass filters of bandwidth B.

For demodulation, the receiver must generate a carrier in phase and frequency synchronism with the incoming carrier. These demodulators are called synchronous, or coherent (also homodyne) demodulators.*

■ EXAMPLE 4.3

Analyze the switching modulator in Fig. 4.6*a* when it is used as a synchronous demodulator.

The input signal is $m(t) \cos \omega_c t$. The carrier causes the periodic switching on and off of the input signal. The output is $m(t) \cos \omega_c t \, k(t)$.

$$m(t) \cos \omega_c t \, k(t) = m(t) \cos \omega_c t \left[\frac{1}{2} + \frac{2}{\pi} \sum_{n=1,3,5,\cdots} \frac{(-1)^{(n-1)/2}}{n} \cos n\omega_c t \right]$$

$$= \frac{1}{\pi} m(t) + \text{other terms centered at } \omega_c, 2\omega_c, \ldots$$

When this signal is passed through a lowpass filter, the output is the desired signal $(1/\pi)m(t)$. ■

4.3 AMPLITUDE MODULATION (AM)

Generally speaking, suppressed-carrier systems need sophisticated circuitry at the receiver for the purpose of generating a local carrier of exactly the right frequency and phase for synchronous demodulation. But such systems are very efficient from the point of view of power requirements at the transmitter. In point-to-point communications, where there is one transmitter for each receiver, substantial complexity in the receiver system can be justified, provided it results in a large enough saving in expensive high-power transmitting equipment. On the other hand, for a broadcast system with a multitude of receivers for each transmitter, it is more economical to have one expensive high-power transmitter and simpler, less expensive receivers. For such applications, a large carrier signal is transmitted along with the suppressed-carrier-modulated signal $m(t) \cos \omega_c t$, thus eliminating the need to generate a local carrier signal at the receiver. This is the so-called AM (amplitude modulation), in which the transmitted signal $\varphi_{\text{AM}}(t)$ is given by

$$\varphi_{\text{AM}}(t) = m(t) \cos \omega_c t + A \cos \omega_c t \tag{4.9a}$$

$$= [A + m(t)] \cos \omega_c t \tag{4.9b}$$

The spectrum of $\varphi_{\text{AM}}(t)$ is the same as that of $m(t) \cos \omega_c t$ plus two additional impulses

---

* The terms synchronous, coherent, and homodyne mean the same thing. The term homodyne is used to contrast with heterodyne, where a different carrier frequency is used for the purpose of translating the spectrum (see Example 4.2).

at $\pm\omega_c$

$$\varphi_{AM}(t) \leftrightarrow \tfrac{1}{2}[M(\omega + \omega_c) + M(\omega - \omega_c)] + \pi A[\delta(\omega + \omega_c) + \delta(\omega - \omega_c)] \tag{4.9c}$$

The modulated signal $\varphi_{AM}(t)$ is shown in Fig. 4.10d. Because $E(t)$ is the envelope of

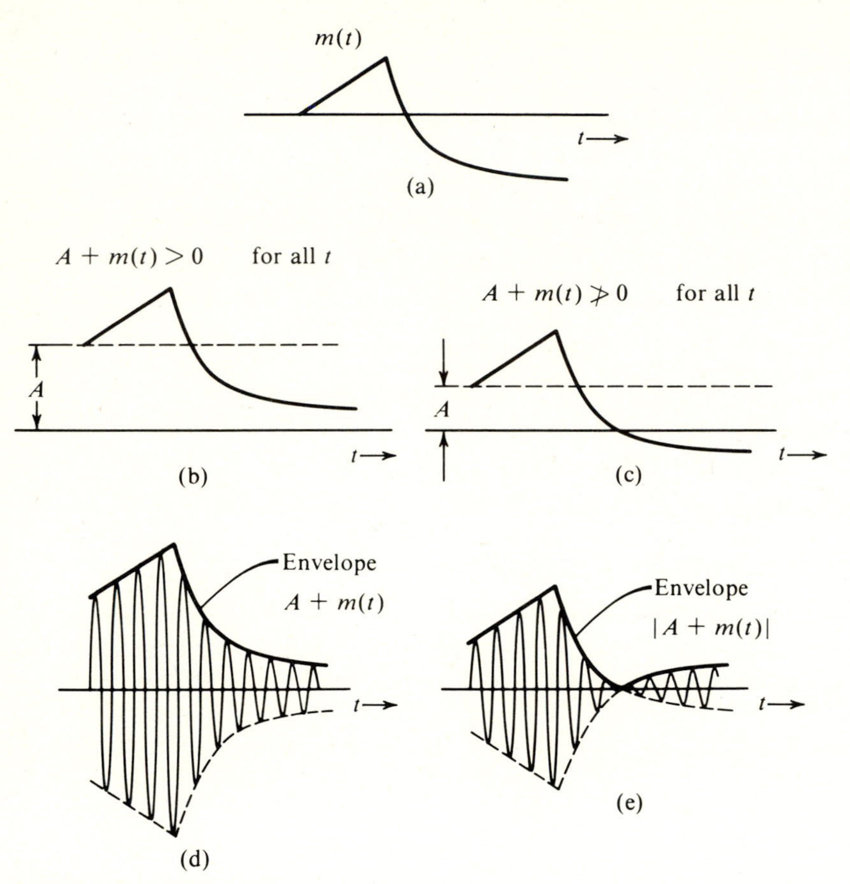

Figure 4.10 AM signal and its envelope.

the signal $E(t) \cos \omega_c t$ (provided* $E(t) > 0$ for all t), the envelope of $\varphi_{AM}(t)$ in Eq. (4.9b) is $A + m(t)$ (provided $A + m(t) > 0$ for all t). This fact is also evident from Fig. 4.10. If A is large enough to make $A + m(t)$ positive for all t, the recovery of $m(t)$ from $\varphi_{AM}(t)$ simply reduces to envelope detection.

The condition for demodulation by an envelope detector is

$$A + m(t) > 0 \qquad \text{for all } t \tag{4.10a}$$

This is the same as

$$A \geq -m(t)_{\min} \tag{4.10b}$$

---

* $E(t)$ must also be a slowly varying signal as compared to $\cos \omega_c t$.

We define the modulation index μ as

$$\mu = \frac{-m(t)_{\min}}{A} \tag{4.11a}$$

Equation (4.10a and b) yields

$$\mu \leq 1 \tag{4.11b}$$

as the required condition for proper demodulation of AM by an envelope detector.

For $\mu > 1$ (overmodulation), the option of envelope detection is no longer available. We then need to use synchronous demodulation. Note that synchronous demodulation can be used for any value of μ (see Prob. 4.11). The envelope detector, which is considerably simpler and less expensive than the synchronous detector, can be used only for $\mu \leq 1$. For $\mu \geq 1$, it is possible to extract the required local carrier from the received signal by using a narrowband filter* tuned to ω_c.

■ EXAMPLE 4.4

Sketch $\varphi_{\text{AM}}(t)$ for modulation indices of $\mu = 0.5$ and $\mu = 1$, when $m(t) = \alpha \cos \omega_m t$. This case is referred to as *tone modulation* because the modulating signal is a pure sinusoid (or tone).

In this case, the modulation index is

$$\mu = \frac{-m(t)_{\min}}{A} = \frac{\alpha}{A}$$

Hence

$$m(t) = \alpha \cos \omega_m t = \mu A \cos \omega_m t$$

and

$$\varphi_{\text{AM}}(t) = [A + m(t)] \cos \omega_c t = A[1 + \mu \cos \omega_m t] \cos \omega_c t \tag{4.12}$$

Figure 4.11*a* and *b* shows the modulated signals corresponding to $\mu = 0.5$ and $\mu = 1$, respectively. ■

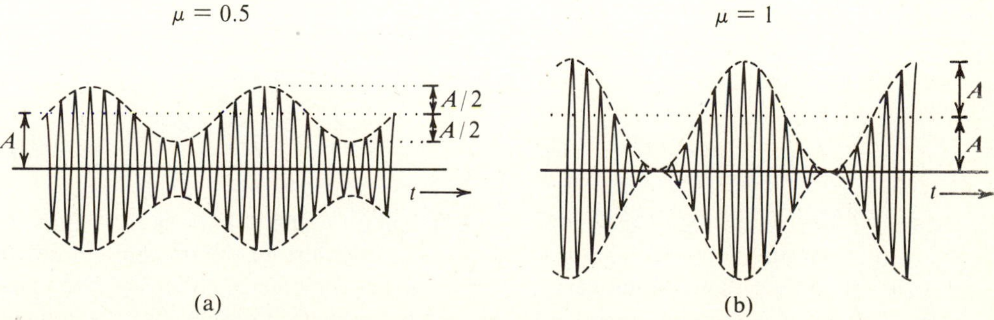

Figure 4.11 Tone modulated AM (a) $\mu = 0.5$. (b) $\mu = 1$.

* The carrier can be readily extracted by using a phase-lock loop, which will be discussed in Sec. 4.15.

Sideband and Carrier Power

The advantage of envelope detection in AM has its price. In AM, the carrier term does not carry any information, and, hence, the carrier power is wasted.

$$\varphi_{\text{AM}}(t) = \underbrace{A \cos \omega_c t}_{\text{carrier}} + \underbrace{m(t) \cos \omega_c t}_{\text{sidebands}}$$

The carrier power P_c is the mean square value of $A \cos \omega_c t$, which is $A^2/2$. Hence

$$P_c = \frac{A^2}{2}$$

The sideband power P_s is the mean square value of $m(t) \cos \omega_c t$. Therefore, P_s is half the mean square value of $m(t)$ [see Eq. (2.118)]:

$$P_s = \tfrac{1}{2}\,\widetilde{m^2(t)}$$

The total power P_t is $P_c + P_s$:

$$P_t = P_c + P_s = \tfrac{1}{2}[A^2 + \widetilde{m^2(t)}]$$

The percentage of the total power carried by the sidebands is η, given by

$$\eta = \frac{P_s}{P_t} \times 100\% = \frac{\widetilde{m^2(t)}}{A^2 + \widetilde{m^2(t)}} \times 100\% \tag{4.13}$$

For the special case of tone modulation

$$m(t) = \mu A \cos \omega_m t \text{ and } \widetilde{m^2(t)} = \frac{(\mu A)^2}{2}$$

and

$$\eta = \frac{\mu^2}{2 + \mu^2} \times 100\% \tag{4.14a}$$

with the condition that $\mu \leq 1$. It can be seen that η_{max} occurs at $\mu = 1$ and is given by

$$\eta_{\text{max}} = 33\% \tag{4.14b}$$

Thus for tone modulation, the maximum efficiency (which occurs at $\mu = 1$) is 33 percent. This means under the best conditions, only a third of the transmitted power carries the information. Moreover, this efficiency is for tone modulation. For voice signals it is even worse (of the order of 25 percent or lower). Volume compression and peak limiting is commonly used in AM to ensure that full modulation ($\mu = 1$) is maintained most of the time.

EXAMPLE 4.5

Determine η and the percentage of the total power carried by the sidebands of the AM wave for tone modulation when (1) $\mu = 0.5$, and (2) $\mu = 0.3$. From Eq. (4.14a)

$$\eta = \frac{\mu^2}{2 + \mu^2} \times 100\%$$

For $\mu = 0.5$,

$$\eta = \frac{0.25}{2.25} 100\% = 11.11\%$$

Hence, only about 11 percent of the total power is in the sidebands. For $\mu = 0.3$,

$$\eta = \frac{0.09}{2.09} 100\% = 4.3\%$$

Hence, only 4.3% of the total power is the useful power (power in sidebands). ■

Generation of AM Signals

AM signals can be generated by any DSB-SC generator if the modulating signal is $A + m(t)$ instead of just $m(t)$. AM can be generated in simpler ways, however. For example, the nonlinear modulator used for DSB-SC is a balanced modulator. We can show that to generate AM, we need use only one of the two branches. Referring to Fig. 4.4, the output of the upper modulator is i_1R, given by [Eq. (4.4a)]

$$\begin{aligned} i_1R &= R[a[\cos \omega_c t + m(t)] + b[\cos \omega_c t + m(t)]^2] \\ &= \underbrace{aR \cos \omega_c t + 2bm(t) \cos \omega_c t}_{\text{AM}} + \underbrace{aRm(t) + bm^2(t) + b \cos^2 \omega_c t}_{\text{suppressed by bandpass filter}} \end{aligned}$$

When this signal is passed through a bandpass filter tuned to ω_c, the last three terms are suppressed, leaving only the first two terms, which in fact represent an AM signal.

It is the same story with switching modulators. There is no need to use a diode bridge, as in Fig. 4.6 or Fig. 4.7. Figure 4.12 shows an AM modulator with a single

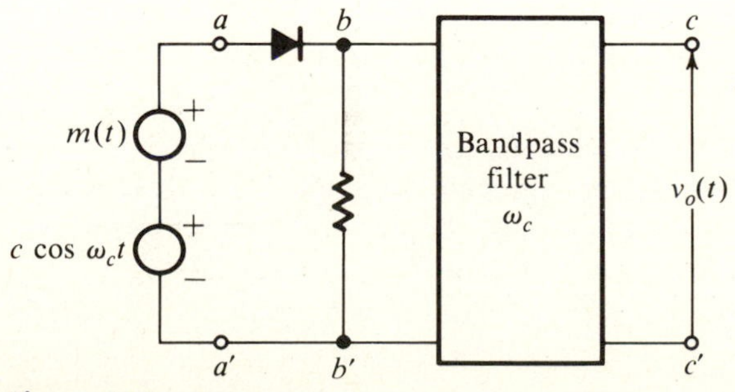

Figure 4.12 AM generator.

diode acting as a switch. The input is $c \cos \omega_c t + m(t)$ with $c \gg m(t)$, so that the switching action of the diode is controlled by $c \cos \omega_c t$. The diode opens and shorts periodically with $\cos \omega_c t$, in effect multiplying the input signal $[c \cos \omega_c t + m(t)]$ by $k(t)$. The voltage across terminals bb', is

$$\begin{aligned}
v_{bb'}(t) &= [c \cos \omega_c t + m(t)]k(t) \\
&= [c \cos \omega_c t + m(t)] \\
&\qquad \left[\frac{1}{2} + \frac{2}{\pi}\left(\cos \omega_c t - \frac{1}{3}\cos 3\omega_c t + \frac{1}{5}\cos 5\omega_c t - \cdots\right)\right] \\
&= \underbrace{\frac{c}{2}\cos \omega_c t + \frac{2}{\pi} m(t) \cos \omega_c t}_{\text{AM}} + \underbrace{\text{other terms}}_{\substack{\text{suppressed by} \\ \text{bandpass filter}}}
\end{aligned}$$

The bandpass filter tuned to ω_c suppresses all the other terms, yielding the desired AM signal at the output.

Demodulation of AM Signals

The AM signal can be demodulated coherently by a locally generated carrier (see Prob. 4.11). Coherent, or synchronous, demodulation of AM,* however, will defeat the very purpose of AM and, hence, is rarely used in practice. We shall consider here three noncoherent methods of AM demodulation: 1) rectifier detection, 2) envelope detection, and 3) square-law detection.

Rectifier Detector. If AM is applied to a diode and a resistor circuit (Fig. 4.13), the negative part of the AM wave will be suppressed. The output across the resistor is a rectified version of the AM signal. In essence, the AM signal is multiplied by $k(t)$. Hence, the rectified output v_R is

$$\begin{aligned}
v_R &= \{[A + m(t)] \cos \omega_c t\} k(t) \\
&= [A + m(t)] \cos \omega_c t \\
&\qquad \left[\frac{1}{2} + \frac{2}{\pi}\left(\cos \omega_c t - \frac{1}{3}\cos 3\omega_c t + \frac{1}{5}\cos 5\omega_c t - \cdots\right)\right] \\
&= \frac{1}{\pi}[A + m(t)] + \text{other terms of higher frequencies}
\end{aligned}$$

When v_R is applied to a lowpass filter of cutoff B, the output is $[A + m(t)]/\pi$, and all the other terms in v_R of frequencies higher than B are suppressed. The dc term A/π may be blocked by a capacitor (Fig. 4.13) to give the desired output $m(t)/\pi$. The output can be doubled by using a full-wave rectifier.

It is interesting to note that rectifier detection is in effect synchronous detection

---

*By AM, we mean the case $\mu \leq 1$.

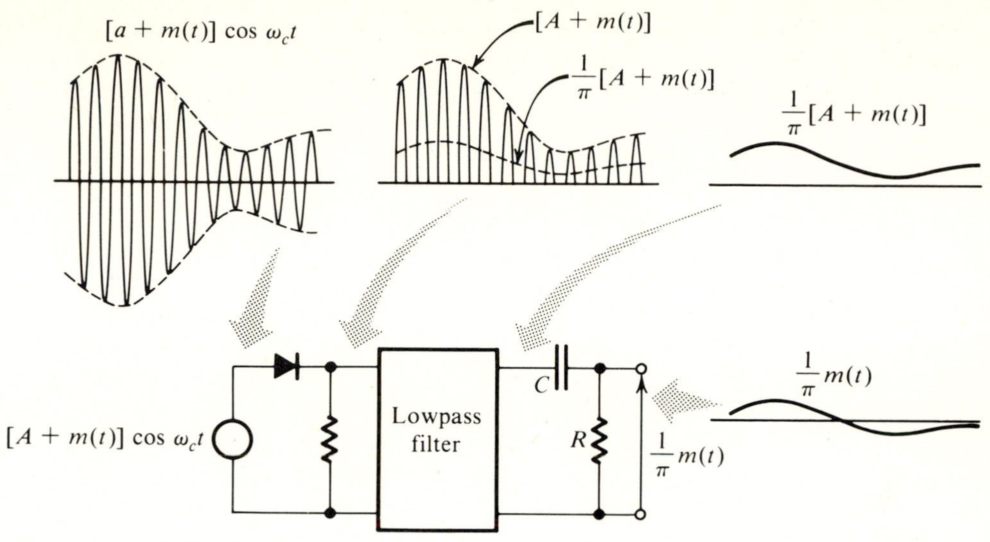

Figure 4.13 Rectifier detector for AM.

performed without using a local carrier. The high carrier content in the received signal makes this possible.

Envelope Detector. In an envelope detector, the output of the detector follows the envelope of the modulated signal. The circuit shown in Fig. 4.14 functions as an

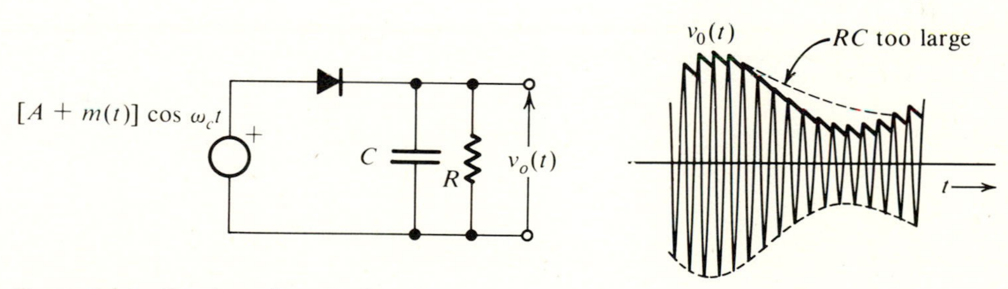

Figure 4.14 Envelope detector for AM.

envelope detector. On the positive cycle of the input signal, the capacitor C charges up to the peak voltage of the input signal. As the input signal falls below this peak value, the diode is cut off, because the capacitor voltage (which is very nearly the peak voltage) is greater than the input signal voltage, thus causing the diode to open. The capacitor now discharges through the resistor R at a slow rate. During the next positive cycle, when the input signal becomes greater than the capacitor voltage, the diode conducts again. The capacitor again charges to the peak value of this (new) cycle. The capacitor discharges slowly during the cutoff period, thus changing the capacitor voltage very slightly.

During each positive cycle, the capacitor charges up to the peak voltage of the input signal and then decays slowly until the next positive cycle. The output voltage thus follows the envelope of the input. A ripple signal of frequency ω_c, however, is caused by capacitor discharge between positive peaks. This ripple is reduced by increasing the time constant RC so that the capacitor discharges very little between the positive peaks ($RC \gg 1/\omega_c$). Making RC too large, however, would make it impossible for the capacitor voltage to follow the envelope (see Fig. 4.14). Thus, RC should be large compared to $1/\omega_c$ but should be small compared to $1/2\pi B$, where B is the highest frequency in $m(t)$ (see Example 4.6). This, incidentally, also requires that $\omega_c \gg 2\pi B$, a condition that is necessary for a well-defined envelope.

The envelope-detector output is $A + m(t)$ with a ripple of frequency ω_c. The dc term A can be blocked out by a capacitor or a simple RC highpass filter. The ripple may be reduced further by another (lowpass) RC filter.

■ EXAMPLE 4.6

For tone modulation (Example 4.4), determine the upper limit on RC to ensure that the capacitor voltage follows the envelope.

□ **Solution:** Figure 4.15 shows the envelope and the voltage across the capacitor.

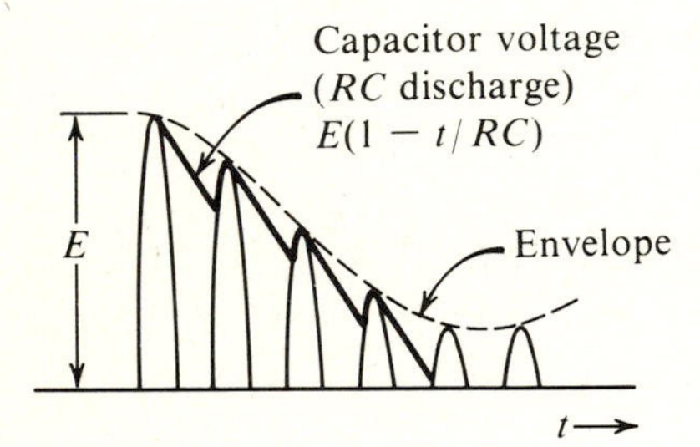

Figure 4.15

The capacitor discharges from the peak value E at some arbitrary instant $t = 0$. The voltage v_C across the capacitor is given by

$$v_C = Ee^{-t/RC}$$

Because the time constant is much larger than the interval between the two successive cycles of the carrier ($RC \gg 1/\omega_c$), the capacitor voltage v_C discharges exponentially for a short time compared to its time constant. Hence, the exponential can be approximated by a straight line obtained from the first two terms in Taylor's series of $Ee^{-t/RC}$.

$$v_C \simeq E\left(1 - \frac{t}{RC}\right)$$

The slope of the discharge is $-E/RC$. In order for the capacitor to follow the envelope

$E(t)$, the magnitude of the slope of the RC discharge must be greater than the magnitude of the slope of the envelope $E(t)$. Hence

$$\left|\frac{dv_C}{dt}\right| = \frac{E}{RC} \geq \left|\frac{dE}{dt}\right| \tag{4.15}$$

But the envelope $E(t)$ of a tone-modulated carrier is [Eq. (4.12)]

$$E(t) = A[1 + \mu \cos \omega_m t]$$

$$\frac{dE}{dt} = -\mu A \omega_m \sin \omega_m t$$

Hence, Eq. (4.15) becomes

$$\frac{A(1 + \mu \cos \omega_m t)}{RC} \geq \mu A \omega_m \sin \omega_m t \qquad \text{for all } t$$

or

$$RC \leq \frac{1 + \mu \cos \omega_m t}{\mu \omega_m \sin \omega_m t} \qquad \text{for all } t$$

The worst possible case occurs when the right-hand side is the minimum. This is found (as usual, by taking the derivative and setting it to zero) to be when $\cos \omega_m t = -\mu$. For this case, the right-hand side is $\sqrt{(1 - \mu 2)}/\mu\omega_m$. Hence,

$$RC \leq \frac{1}{\omega_m}\left(\frac{\sqrt{1 - \mu^2}}{\mu}\right)$$ ■

Square-Law Detector. An AM signal can be demodulated by squaring it and then passing the squared signal through a lowpass filter. This can be readily seen from the fact that

$$\varphi_{\text{AM}}(t) = [A + m(t)] \cos \omega_c t$$

and

$$\varphi_{\text{AM}}^2(t) = \frac{[A^2 + 2Am(t) + m^2(t)]}{2}(1 + \cos 2\omega_c t)$$

The lowpass filter output $y_o(t)$ is

$$y_o(t) = \frac{A^2}{2}\left[1 + 2\frac{m(t)}{A} + \left(\frac{m(t)}{A}\right)^2\right]$$

Usually, $m(t)/A \ll 1$ for most of the time. Only when $m(t)$ is near its peak is this violated. Hence,

$$y_o(t) \simeq \frac{A^2}{2} + A\,m(t)$$

A blocking capacitor will suppress the dc term, yielding the output $A\,m(t)$. Note that the square-law detector causes signal distortion. The distortion, however, is negligible when μ is small. This type of detection can be performed by any nonlinear device that does not have odd symmetry. Whenever a modulated signal passes through any nonlinearity of this type, it will create a demodulation component, whether intended or not.

4.4 AMPLITUDE MODULATION: SINGLE SIDEBAND (SSB)

The DSB spectrum has two sidebands: the upper sideband (USB) and the lower sideband (LSB), either of which contains the complete information of the baseband signal (Fig. 4.16). A scheme in which only one sideband is transmitted is known as

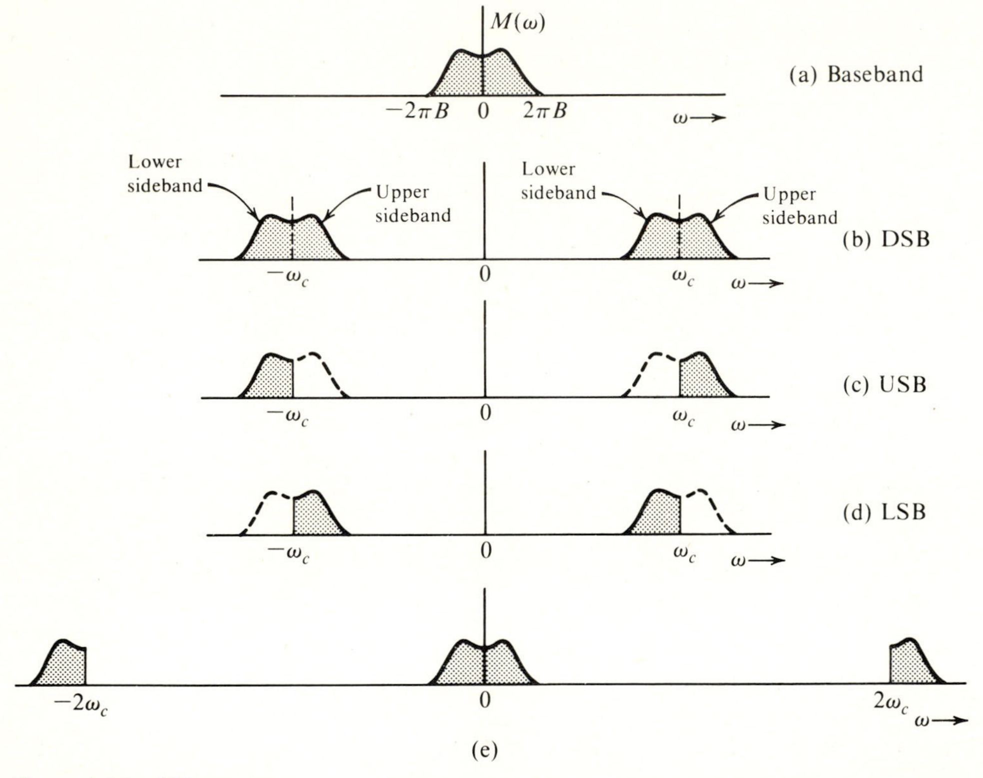

Figure 4.16 SSB spectra.

single sideband (SSB) transmission, and it requires only half the bandwidth of a DSB signal.

An SSB signal can be coherently (synchronously) demodulated. For example, multiplication of a USB signal (Fig. 4.16*c*) by cos $\omega_c t$ shifts its spectrum by $\pm\omega_c$, yielding the spectrum in Fig. 4.16*e*. Lowpass filtering of this signal yields the desired baseband signal. The case is similar with LSB signals. Hence, demodulation of SSB signals is identical to that of DSB-SC signals. Note that we are talking of SSB signals without an additional carrier, and, hence, they are suppressed-carrier signals (SSB-SC).

To determine the time-domain expression of an SSB signal, we introduce the concept of the *preenvelope* of a signal $m(t)$. The spectra $M_+(\omega) = M(\omega)u(\omega)$ and $M_-(\omega) = M(\omega)u(-\omega)$ are shown in Fig. 4.17*b* and *c*. If $m_+(t)$ and $m_-(t)$ are the

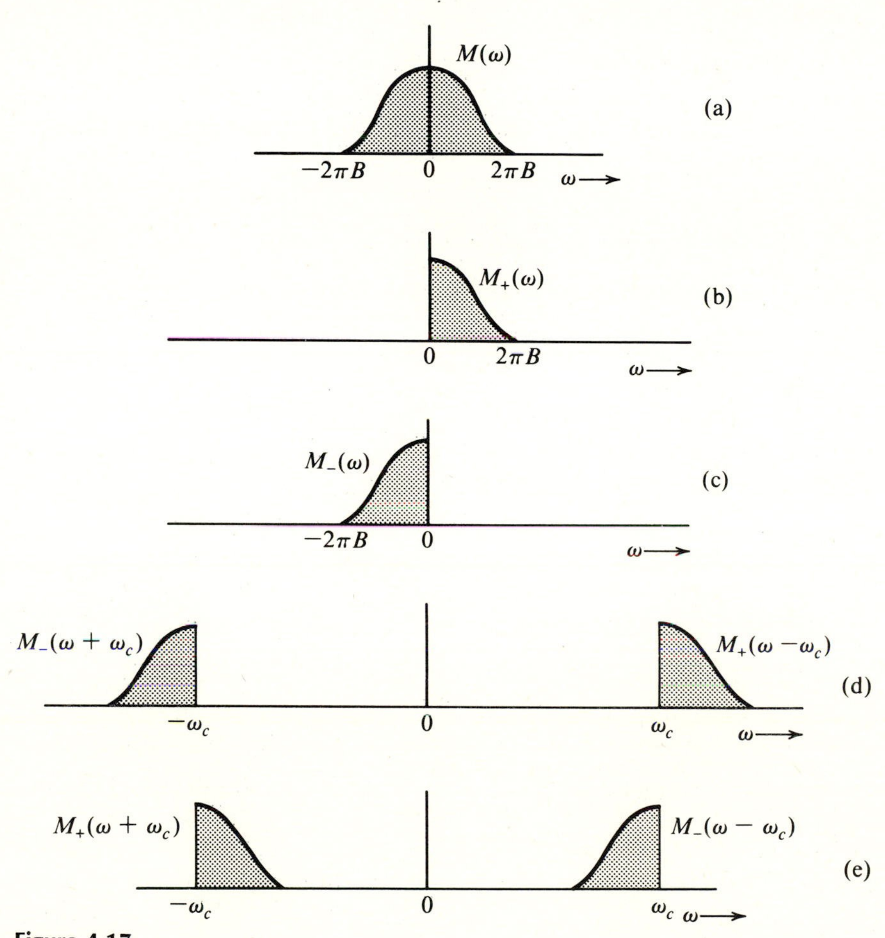

Figure 4.17

inverse Fourier transforms of $M_+(\omega)$ and $M_-(\omega)$, then we call $2m_+(t)$ the preenvelope* of $m(t)$. Because $|M_+(\omega)|$ and $|M_-(\omega)|$ are not even functions of ω, $m_+(t)$ and $m_-(t)$ are complex signals. Also, from Eq. (2.34) it follows that $M_+(-\omega)$ and $M_-(\omega)$ are conjugates and, hence, $m_+(t)$ and $m_-(t)$ are conjugates (see Prob. 2.9). Also, because $m_+(t) + m_-(t) = m(t)$,

$$m_+(t) = \tfrac{1}{2}[m(t) + jm_h(t)] \qquad \textbf{(4.16a)}$$

*In the literature, $2m_+(t)$ is also known as the *analytic signal*.

and

$$m_-(t) = \tfrac{1}{2}[m(t) - jm_h(t)] \tag{4.16b}$$

To determine $m_h(t)$, we note that

$$\begin{aligned} M_+(\omega) &= M(\omega)u(\omega) \\ &= \tfrac{1}{2}M(\omega)[1 + \operatorname{sgn}(\omega)] \\ &= \tfrac{1}{2}M(\omega) + \tfrac{1}{2}M(\omega)\operatorname{sgn}(\omega) \end{aligned} \tag{4.17a}$$

From Eqs. (4.16a) and (4.17a), it follows that

$$jm_h(t) \leftrightarrow M(\omega)\operatorname{sgn}(\omega) \tag{4.17b}$$

or

$$M_h(\omega) = -jM(\omega)\operatorname{sgn}(\omega) \tag{4.17c}$$

From Eq. (2.52) and the time-convolution property, it follows that

$$m_h(t) = \frac{1}{\pi}\int_{-\infty}^{\infty} \frac{m(\alpha)}{t - \alpha}\, d\alpha \tag{4.17d}$$

The signal $m_h(t)$ is the *Hilbert transform** of $m(t)$. From Eq. (4.17c), it follows that if $m(t)$ is passed through a transfer function $H(\omega) = -j \operatorname{sgn}(\omega)$, then the output is $m_h(t)$, the Hilbert transform of $m(t)$. Because

$$H(\omega) = -j\operatorname{sgn}(\omega) \tag{4.18a}$$

$$\begin{aligned} &= -j = 1e^{-j\pi/2} \qquad \omega > 0 \\ &= j = 1e^{j\pi/2} \qquad \omega < 0 \end{aligned} \tag{4.18b}$$

it follows that $|H(\omega)| = 1$, and $\theta_h(\omega) = -\pi/2$ for $\omega > 0$ and $\pi/2$ for $\omega < 0$, as shown in Fig. 4.18. Thus, if we delay the phase of every component of $m(t)$ by $-\pi/2$,

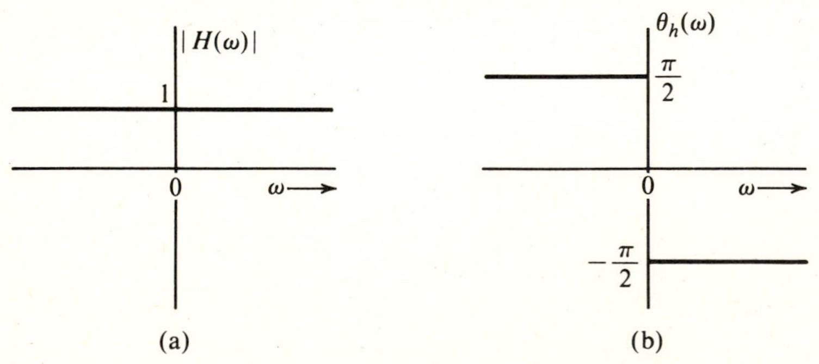

Figure 4.18 Transfer function of an ideal $\pi/2$ phase shifter.

---

* The integral in Eq. (4.17d) is an improper integral because the integrand generally goes to infinity at $\alpha = t$. The Hilbert transform is therefore defined as the Cauchy-principle value, that is,

$$m_h(t) = \lim_{\epsilon \to 0}\left[\int_{-\infty}^{t-\epsilon} \frac{m(\alpha)}{t - \alpha}\, d\alpha + \int_{t+\epsilon}^{\infty} \frac{m(\alpha)}{t - \alpha}\, d\alpha\right]$$

the resulting signal is $m_h(t)$, the Hilbert transform of $m(t)$. A Hilbert transformer is just a phase shifter (that shifts by $-\pi/2$). We can now express the SSB signal in terms of $m(t)$ and $m_h(t)$. Because the USB signal $\varphi_{\text{USB}}(\omega)$ is (Fig. 4.17*d*)

$$\Phi_{\text{USB}}(\omega) = M_+(\omega - \omega_c) + M_-(\omega + \omega_c)$$

$$\varphi_{\text{USB}}(t) = m_+(t)\, e^{j\omega_c t} + m_-(t)\, e^{-j\omega_c t}$$

Substituting Eq. (4.16a and b) in the above equation yields

$$\varphi_{\text{USB}}(t) = m(t) \cos \omega_c t - m_h(t) \sin \omega_c t \tag{4.19a}$$

Similarly, it can be shown that

$$\varphi_{\text{LSB}}(t) = m(t) \cos \omega_c t + m_h(t) \sin \omega_c t \tag{4.19b}$$

Hence, a general SSB signal $\varphi_{\text{SSB}}(t)$ can be expressed as

$$\varphi_{\text{SSB}}(t) = m(t) \cos \omega_c t \mp m_h(t) \sin \omega_c t \tag{4.19c}$$

where the minus sign applies to the USB and the plus sign applies to the LSB.

■ EXAMPLE 4.7

Determine the Hilbert transform of (1) $m(t) = \cos(\omega_o t + \theta)$, and (2) $m(t) = 2a/(t^2 + a^2)$.

□ **Solution:**

(1) $M(\omega) = \pi[\delta(\omega - \omega_o)e^{j\theta} + \delta(\omega + \omega_o)e^{-j\theta}]$

Hence

$$M_h(\omega) = -j \operatorname{sgn}(\omega) M(\omega) = -j\pi[\delta(\omega - \omega_o)e^{j\theta} - \delta(\omega + \omega_o)e^{-j\theta}]$$

and [see Eq. (2.62b), with $\varphi = \theta - \pi/2$]

$$m_h(t) = \sin(\omega_o t + \theta)$$

This represents a phase shift of $-\pi/2$, as expected.

(2) From Table 2.1, pair 15, and the symmetry property, we have

$$m(t) = \frac{2a}{t^2 + a^2} \leftrightarrow 2\pi e^{-a|\omega|}$$

Hence,

$$M_h(\omega) = -j \operatorname{sgn}(\omega) M(\omega) = -j2\pi[e^{-a\omega}u(\omega) - e^{a\omega}u(-\omega)]$$

From Eq. (2.51), it follows that

$$m_h(t) = -j\left[\frac{1}{a - jt} - \frac{1}{a + jt}\right] = \frac{2t}{t^2 + a^2}$$

■

■ EXAMPLE 4.8

Find $\varphi_{\text{SSB}}(t)$ for the sinusoidal modulating signal $\cos \omega_m t$ (tone modulation), and verify that the spectrum of $\varphi_{\text{SSB}}(t)$ is indeed an SSB spectrum.

☐ Solution:

$$m(t) = \cos \omega_m t$$

In this case (see Example 4.7)

$$m_h(t) = \cos\left(\omega_m t - \frac{\pi}{2}\right) = \sin \omega_m t$$

Hence, from Eq. (4.19c),

$$\varphi_{\text{SSB}}(t) = \cos \omega_m t \cos \omega_c t \mp \sin \omega_m t \sin \omega_c t$$
$$= \cos(\omega_c \pm \omega_m)t$$

Thus,

$$\varphi_{\text{USB}}(t) = \cos(\omega_c + \omega_m)t \qquad \varphi_{\text{LSB}}(t) = \cos(\omega_c - \omega_m)t$$

To verify these results, consider the spectrum of $m(t)$ (Fig. 4.19*a*) and its DSB-SC (Fig. 4.19*b*), USB (Fig. 4.19*c*), and LSB (Fig. 4.19*d*) spectra. It is evident that the

(a)

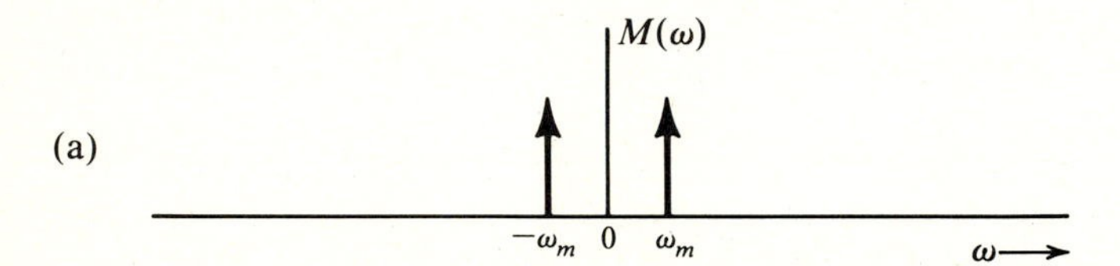

(b)

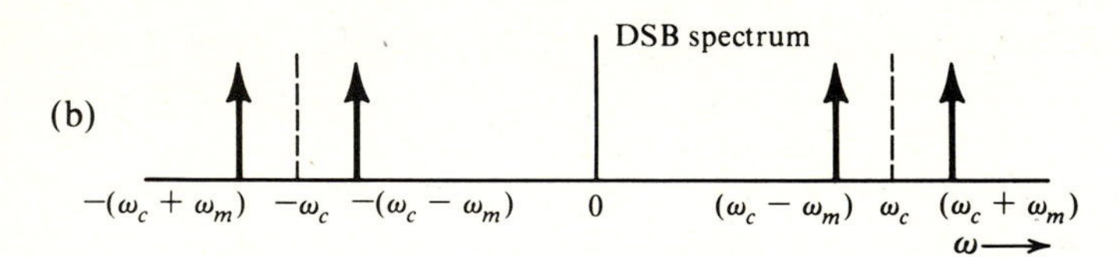

(c)

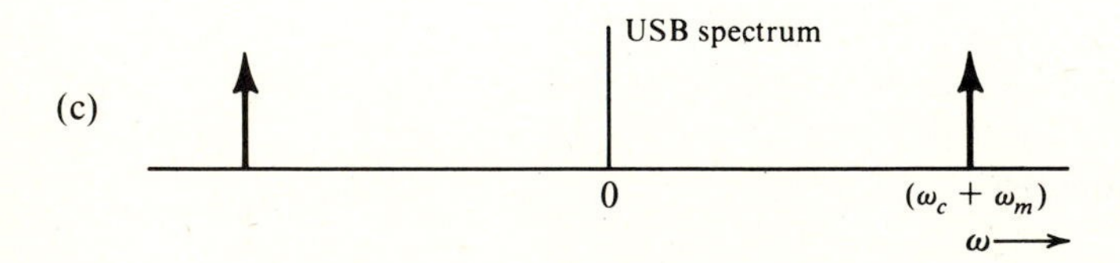

(d)

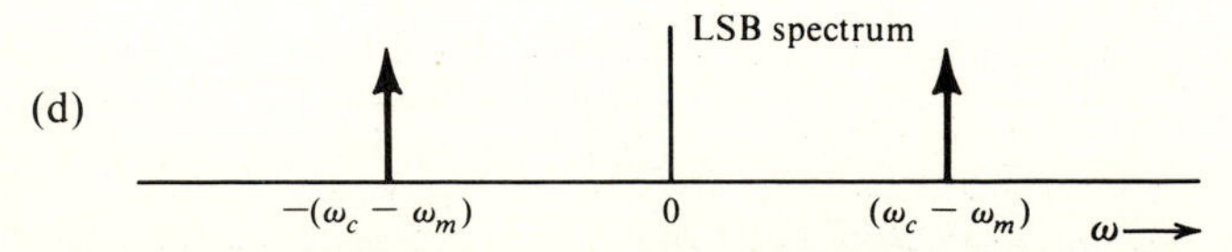

Figure 4.19 SSB spectra for tone modulation.

spectra in Fig. 4.19*c* and *d* do indeed correspond to the $\varphi_{\mathrm{USB}}(t)$ and $\varphi_{\mathrm{LSB}}(t)$ derived above. ■

Generation of SSB Signals[2]

Two methods are commonly used to generate SSB signals. The first method uses sharp cutoff filters to eliminate the undesired sideband, and the second method uses phase-shifting networks to achieve the same goal. Yet another method, known as Weaver's method,[3] can also be used to generate SSB, provided the baseband signal spectrum has little power near the origin (see Prob. 4.27).

The Selective-Filtering Method. This is the most commonly used method of generating SSB signals. In this method, a DSB-SC signal is passed through a sharp cutoff filter to eliminate the undesired sideband.

To obtain the USB, the filter should pass all components above ω_c unattenuated and completely suppress all components below ω_c. This, as we know, is impossible. There must be some separation between the passband and the stopband. Fortunately, the voice signal provides this condition, because its spectrum shows little power content at the origin (Fig. 4.20*a*). In addition, articulation tests have shown that for speech signals, frequency components below 300 Hz are not important. In other words, we may suppress all speech components below 300 Hz without affecting the intelligibility appreciably.* Thus, filtering of the unwanted sideband becomes relatively easy for speech signals because we have a 600-Hz transition region around the cutoff frequency ω_c (Fig. 4.20*b*).

To minimize adjacent channel interference, the undesired sideband should be attenuated at least 40 dB. If the carrier frequency is too high (for example, f_c = 10 MHz), the ratio of the gap band (600 Hz) to the carrier frequency is small, and, thus, a transition of 40 dB in amplitude over 600 Hz could still pose a problem. In such a case, the modulation is carried out in more than one step. First, the baseband signal DSB-modulates a low-frequency carrier ω_{c_1}. The unwanted sideband is easily suppressed, because the ratio of the gap band to the carrier is reasonably high. After suppression of the unwanted sideband (Fig. 4.20*c*), the resulting spectrum is identical to the baseband spectrum except that the gap band is now $2\omega_{c_1}$ rather than the original 600 Hz. The first step, then, amounts to increasing the gap band width to $2\omega_{c_1}$. In the second step, this signal (with the large gap band) DSB-modulates a carrier of high frequency ω_{c_2}. The unwanted sideband (shown dotted in Fig. 4.20*d*) is now easily suppressed because of the large gap band. If the carrier frequency is too high, the process may have to be repeated.

The Phase-Shift Method. Equation (4.19) is the basis of this method. Figure 4.21 shows the implementation of Eq. (4.19). The box marked "$-\pi/2$" is a $\pi/2$ phase shifter, which delays the phase of every frequency component by $\pi/2$. Hence, it is

* Similarly, suppression of speech-signal components above 3500 Hz cause no appreciable change in intelligibility.

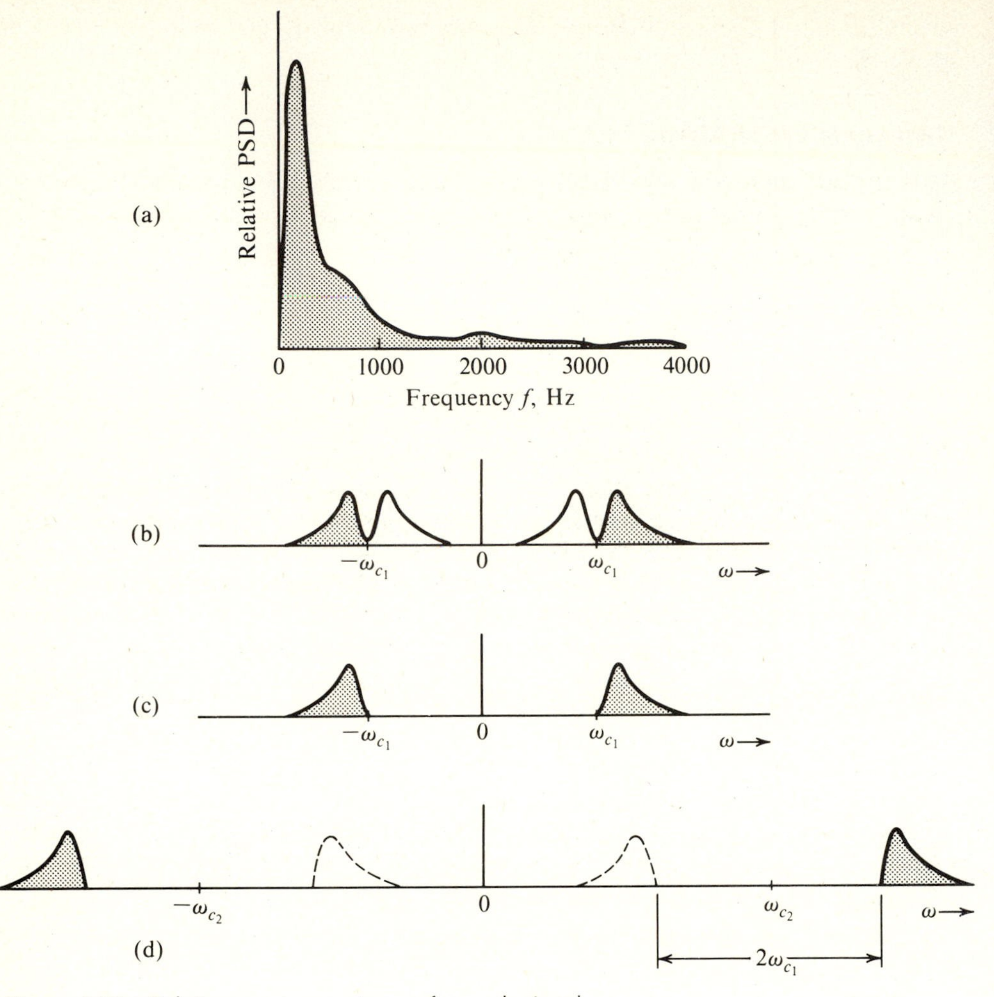

Figure 4.20 Relative power spectrum of speech signal.

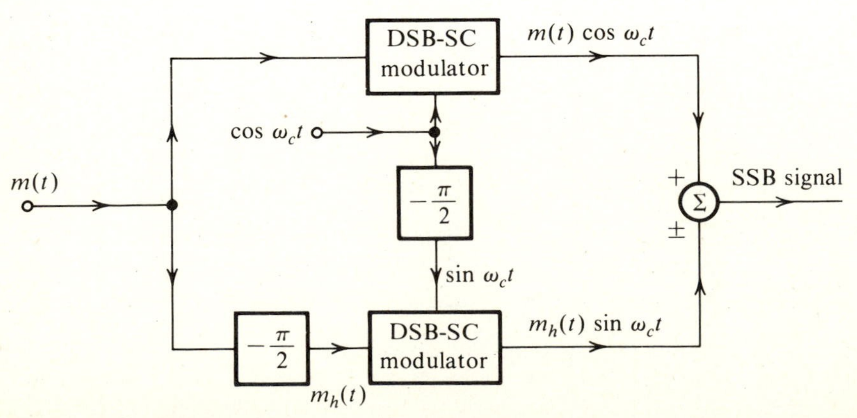

Figure 4.21 SSB generator.

a Hilbert transformer. Note that an ideal phase shifter is also unrealizable. We can, at most, approximate it over a finite band.

Demodulation of SSB-SC Signals

It was shown earlier that SSB-SC signals can be coherently demodulated. We can readily verify this in another way:

$$\varphi_{\text{SSB}}(t) = m(t) \cos \omega_c t \mp m_h(t) \sin \omega_c t$$

Hence,

$$\begin{aligned}\varphi_{\text{SSB}}(t) \cos \omega_c t &= \tfrac{1}{2}m(t)[1 + \cos 2\omega_c t] \mp \tfrac{1}{2}m_h(t) \sin 2\omega_c t \\ &= \tfrac{1}{2}m(t) + \tfrac{1}{2}[m(t) \cos 2\omega_c t \mp m_h(t) \sin 2\omega_c t] \end{aligned} \tag{4.20}$$

Thus, the product $\varphi_{\text{SSB}}(t) \cos \omega_c t$ yields the baseband signal and another SSB signal with a carrier $2\omega_c$. The spectrum in Fig. 4.16*e* shows precisely this result. A lowpass filter will suppress the unwanted SSB term, giving the desired baseband signal $m(t)/2$. Hence, the demodulator is identical to the synchronous demodulator used for DSB-SC. Thus, any one of the synchronous DSB-SC demodulators discussed in Sec. 4.2 can be used to demodulate an SSB-SC signal.

Envelope Detection of SSB Signals with a Carrier. We now consider SSB signals with an additional carrier (SSB + C). Such a signal can be expressed as

$$\varphi(t) = A \cos \omega_c t + [m(t) \cos \omega_c t + m_h(t) \sin \omega_c t] \tag{4.21}$$

Although $m(t)$ can be recovered by synchronous detection (multiplying $\varphi(t)$ by $\cos \omega_c t$) if A, the carrier amplitude, is large enough, $m(t)$ can also be recovered from $\varphi(t)$ by envelope or rectifier detection. This can be shown by rewriting $\varphi(t)$ as

$$\begin{aligned}\varphi(t) &= [A + m(t)] \cos \omega_c t + m_h(t) \sin \omega_c t \\ &= E(t) \cos (\omega_c t + \theta)\end{aligned}$$

where $E(t)$, the envelope of $\varphi(t)$, is given by

$$\begin{aligned}E(t) &= \{[A + m(t)]^2 + m_h^2(t)\}^{1/2} \\ &= A\left[1 + \frac{2m(t)}{A} + \frac{m^2(t)}{A^2} + \frac{m_h^2(t)}{A^2}\right]^{1/2}\end{aligned}$$

If $A \gg |m(t)|$, then in general* $A \gg |m_h(t)|$, and the terms $m^2(t)/A^2$ and $m_h^2(t)/A^2$ can be ignored; thus,

$$E(t) = A\left[1 + \frac{2m(t)}{A}\right]^{1/2}$$

Using binomial expansion and discarding higher-order terms (because $m(t)/A \ll 1$),

---

*This may not be true for all t, but it is true for most t.

we get

$$E(t) \simeq A\left[1 + \frac{m(t)}{A}\right]$$
$$= A + m(t)$$

It is evident that for a large carrier, the envelope of $\varphi(t)$ has the form of $m(t)$, and the signal can be demodulated by an envelope detector.

In AM, envelope detection requires the condition $A \geq -m(t)_{\text{min}}$, whereas for SSB + C, the condition is $A \gg |m(t)|$. Hence, in SSB, the required carrier amplitude is much larger than that in AM, and, consequently, the efficiency of SSB + C is much less than that of AM.

Quadrature Amplitude Modulation (QAM)

DSB signals occupy twice the bandwidth required for SSB signals. This disadvantage can be overcome by transmitting two DSB signals using carriers of the same frequency but in phase quadrature (Fig. 4.22). Both modulated signals occupy the same bandwidth. The two baseband signals can be separated at the receiver by synchronous detection using two local carriers in phase quadrature. This can be shown by considering the multiplier output $x_1(t)$ of channel 1 (Fig. 4.22):

$$x_1(t) = 2[m_1(t) \cos \omega_c t + m_2(t) \sin \omega_c t] \cos \omega_c t$$
$$= m_1(t) + m_1(t) \cos 2\omega_c t + m_2(t) \sin 2\omega_c t$$

The last two terms are suppressed by the lowpass filter, yielding the desired output $m_1(t)$. Similarly, the output of channel 2 can be shown to be $m_2(t)$. This scheme is known as *quadrature amplitude modulation (QAM),* or quadrature multiplexing.

Thus, two signals can be transmitted over a bandwidth $2B$ either by using SSB transmission and frequency-division multiplexing or by using DSB transmission and quadrature multiplexing. The former is preferable to the latter for practical reasons. A slight error in the phase of the carriers at the demodulator in QAM will not only result in loss of signal but will also lead to interference between the two channels. Similar difficulties arise when the local frequency is in error (see Prob. 4.29). In addition, unequal attenuation of the upper and lower sidebands during transmission also leads to crosstalk or cochannel interference.* This too can be easily verified. Unequal attenuation of the channel will destroy the symmetry of the magnitude spectrum about $\pm\omega_c$, and the received signal component $\varphi_1(t)$ from $m_1(t)$ will have a form [see Eq. (2.61a)]

$$\varphi_1(t) = m_{c_1}(t) \cos \omega_c t + m_{s_1}(t) \sin \omega_c t$$

instead of $m_1(t) \cos \omega_c t$. Demodulation of this signal will result in $m_{c_1}(t)$ in channel 1 and $m_{s_1}(t)$ in channel 2. Both $m_{c_1}(t)$ and $m_{s_1}(t)$ are derived from $m_1(t)$. Hence, channel 1 will receive the desired signal in a distorted form, and channel 2 will receive

*Cochannel refers to channels having the same carrier frequency.

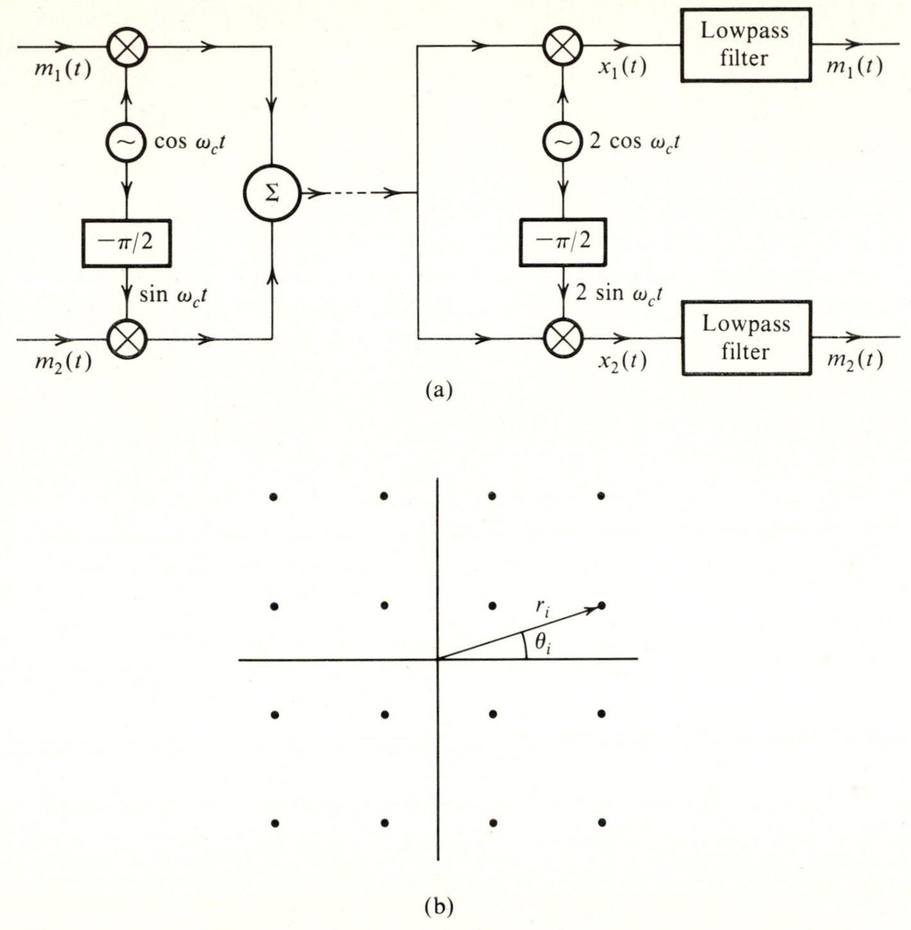

Figure 4.22 (a) QAM or quadrature multiplexing. (b) 16-point QAM ($M = 16$).

an interference signal. Similarly, the component $\varphi_2(t)$ from $m_2(t)$ will yield the distorted version of desired signal in channel 2 and an interference signal in channel 1.

If we were to multiplex two SSB signals on adjacent channels, a phase error in the carriers would cause distortion in each channel but not interference. Similarly, the channel nonidealities would cause distortion in each channel without causing interference.* The only interference in SSB multiplexing is caused by incomplete suppression of the unwanted sideband. For this reason, SSB multiplexing is preferred to quadrature multiplexing. Quadrature multiplexing is used in color television to multiplex the so-called chrominance signals, which carry the information about colors. There synchronization pulses are transmitted to keep the local oscillator at the right frequency and phase. QAM is also used in digital signal transmission. We shall

---

* We are assuming linear distortion here. Nonlinearities in the channel will cause interference in two adjacent SSB channels.

present here one application of QAM in digital signal transmission over telephone lines.

To transmit baseband pulses at a rate of 2400 bits/second, we need a minimum bandwidth of 1200 Hz. In practice, however, using Nyquist's first criterion pulses, the required bandwidth is $(1 + r)$ 1200, where r is a roll-off factor (see Example 3.2). In this particular application, 12.5 percent roll-off ($r = 0.125$) is used. This gives a bandwidth of 1350 Hz. Modulation doubles the required bandwidth to 2700 Hz. Thus, we need a bandwidth of 2700 Hz to transmit data at a rate of 2400 bits/second. Use of quadrature multiplexing will double the rate over the same bandwidth. Note that in the above method, we are transmitting 2 PSK signals ($m_1(t)$ and $m_2(t)$ in Fig. 4.22*a*) using carriers in phase quadrature. For this reason, it is also known as *quadrature PSK (QPSK)*.

We can increase the transmission rate further by using M-ary QAM.* One practical case with $M = 16$ uses the following 16 pulses (16 symbols).

$$\begin{aligned} p_i(t) &= a_i p(t) \cos \omega_c t + b_i p(t) \sin \omega_c t \\ &= r_i p(t) \cos (\omega_c t + \theta_i) \qquad\qquad i = 1, 2, \ldots, 16 \end{aligned}$$

where $p(t)$ is a properly shaped baseband pulse. The choices of r_i and θ_i for 16 pulses is shown graphically in Fig. 4.22*b*.

Since $M = 16$, each pulse can transmit the information of $\log_2 16 = 4$ binary digits (see Sec. 3.7). This can be done as follows: there are 16 possible sequences of four binary digits and there are 16 combinations (a_i, b_i) in Fig. 4.22*b*. Thus, every possible four-bit sequence is transmitted by a particular (a_i, b_i) or (r_i, θ_i). Therefore, one single pulse $r_i p(t) \cos (\omega_c t + \theta_i)$ transmits four bits. Thus a normal transmission line that can transmit 2400 bits/second can now transmit 9600 bits/second.

Modulation as well as demodulation can be performed by using the system in Fig. 4.22*a*. The inputs are $m_1(t) = a_i p(t)$ and $m_2(t) = b_i p(t)$. The two outputs at the demodulator are $a_i p(t)$ and $b_i p(t)$. From the knowledge of (a_i, b_i), we can determine the four transmitted bits. Complete analysis of 16-ary QAM can be found in Example 7.3.

At each end of the telephone line, we need a modulator and demodulator to transmit as well as to receive data. The two devices, *mo*dulator and *dem*odulator are usually packaged in one unit called a *modem*.

4.5 EFFECTS OF FREQUENCY AND PHASE ERRORS IN SYNCHRONOUS DEMODULATION

In the suppressed-carrier amplitude-modulated systems (DSB-SC and SSB-SC), one must generate a local carrier at the receiver for the purpose of synchronous demodulation. Ideally, the local carrier must be in frequency and phase synchronism with the incoming carrier. Any discrepancy in the frequency or phase of the local carrier gives rise to distortion in the detector output. We shall now discuss this topic in more detail for DSB and SSB signals.

---

*The M-ary QAM discussed here is also called *amplitude phase shift keying (APK)*.

DSB-SC Signals

Let the received signal be $m(t) \cos \omega_c t$ and the local carrier be $\cos [(\omega_c + \Delta\omega)t + \delta]$. The local-carrier frequency and phase errors in this case are $\Delta\omega$ and δ, respectively. The product of the received signal and the local carrier is $e_d(t)$, given by

$$\begin{aligned} e_d(t) &= m(t) \cos \omega_c t \cos [(\omega_c + \Delta\omega)t + \delta] \\ &= \tfrac{1}{2} m(t)\{\cos [(\Delta\omega)t + \delta] + \cos [(2\omega_c + \Delta\omega)t + \delta]\} \end{aligned} \tag{4.22}$$

The second term on the right-hand side is filtered out by the lowpass filter, leaving the output $e_o(t)$ as

$$e_o(t) = \tfrac{1}{2} m(t) \cos [(\Delta\omega)t + \delta] \tag{4.23}$$

If $\Delta\omega$ and δ are both zero (no frequency or phase error), then

$$e_o(t) = \tfrac{1}{2} m(t)$$

as expected. Let us consider two special cases. If $\Delta\omega = 0$, Eq. (4.23) reduces to

$$e_o(t) = \tfrac{1}{2} m(t) \cos \delta \tag{4.24a}$$

This output is proportional to $m(t)$ when δ is a constant. The output is maximum when $\delta = 0$ and is minimum (zero) when $\delta = \pm\pi/2$. Thus, the phase error in the local carrier causes the attenuation of the output signal without causing any distortion, as long as δ is constant. Unfortunately, the phase error δ may vary randomly with time. This may occur, for example, because of variations in the propagation path. This causes the gain factor $\cos \delta$ at the receiver to vary randomly and is undesirable.

Next we consider the case where $\delta = 0$ and $\Delta\omega \neq 0$. In this case, Eq. (4.23) becomes

$$e_o(t) = \tfrac{1}{2} m(t) \cos (\Delta\omega)t \tag{4.24b}$$

The output here is not merely an attenuated replica of the original signal but is also distorted. Because $\Delta\omega$ is usually small, the output is the signal $m(t)$ multiplied by a low-frequency sinusoid. This is a "beating" effect and is a rather serious type of distortion.

Because $\omega_c \gg 2\pi B$, $\Delta\omega$ (which is a small fraction of ω_c) can still be appreciable in comparison to $2\pi B$. A human ear can tolerate a drift of up to 30 Hz. One may use quartz crystal oscillators, which generally are very stable. Crystals can be cut for the same frequency at the transmitter and the receiver. For a very high carrier frequency, however, even quartz-crystal performance may not be adequate. In such a case, a carrier, or *pilot*, is transmitted at a reduced level (usually about -20 dB) along with the sidebands. The pilot is separated at the receiver by a very narrowband filter tuned to the pilot frequency. It is amplified and used to synchronize the local oscillator.

Another way of obtaining a correct carrier at the receiver is shown in Fig. 4.23*a*. The incoming signal is squared and then passed through a narrow bandpass filter tuned to $2\omega_c$. The output of this filter is the sinusoid $k \cos 2\omega_c t$, which is in phase synchronism with the incoming carrier but has exactly twice the desired frequency. The filter output is passed through a 2-to-1 frequency divider to obtain a local carrier in phase and frequency synchronism with the incoming carrier. The analysis is straight-

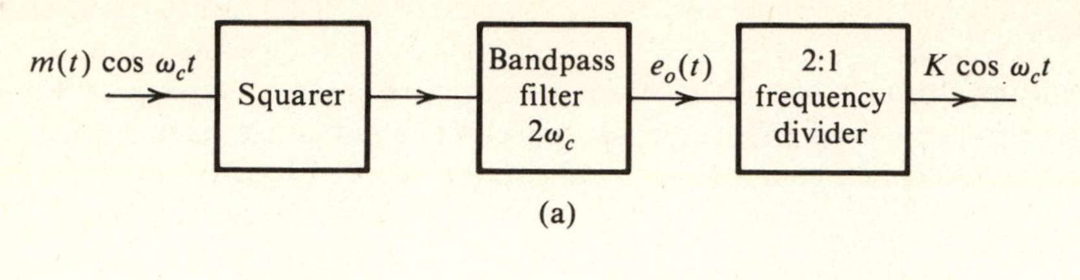

(a)

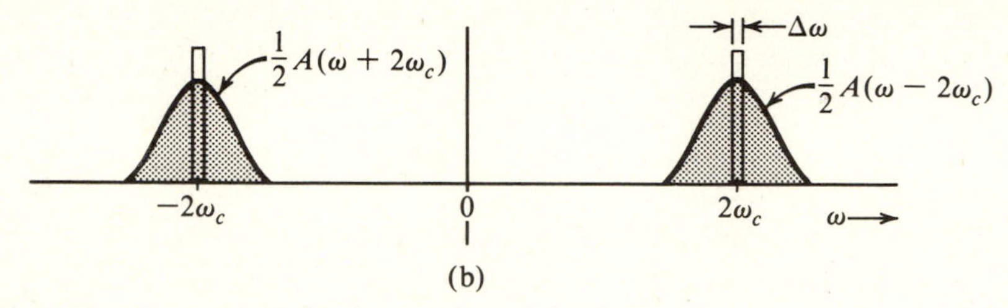

(b)

Figure 4.23 Generation of coherent demodulation carrier using signal squaring.

forward. The squarer output is

$$[m(t) \cos \omega_c t]^2 = \tfrac{1}{2}m^2(t) + \tfrac{1}{2}m^2(t) \cos 2\omega_c t$$

The bandpass filter suppresses the $m^2(t)$ term. However, $m^2(t) \cos 2\omega_c t$ does not pass entirely, because the filter has a narrowband. The bandwidth of $m^2(t)$ is $2B$, and $m^2(t) \cos 2\omega_c t$ has a bandwidth of $4B$ centered at $2\omega_c$. If

$$m^2(t) \leftrightarrow A(\omega)$$

$$m^2(t) \cos 2\omega_c t \leftrightarrow \tfrac{1}{2}[A(\omega + 2\omega_c) + A(\omega - 2\omega_c)]$$

If the bandpass filter's bandwidth $\Delta f \ll 4B$, then the filter will accept only the part of the spectrum centered at $\pm 2\omega_c$ (Fig. 4.23*b*). The output spectrum $E_o(\omega)$ is essentially two narrow pulses, each of area $A(0)\Delta\omega/2$. These pulses can be approximated by impulses. Hence,

$$E_o(\omega) \simeq \frac{A(0)\Delta\omega}{2}[\delta(\omega + 2\omega_c) + \delta(\omega - 2\omega_c)]$$

and

$$e_o(t) = A(0)\Delta f \cos 2\omega_c t$$

Also, because

$$A(\omega) = \mathcal{F}[m^2(t)] = \int_{-\infty}^{\infty} m^2(t)e^{-j\omega t}\, dt$$

$$A(0) = \int_{-\infty}^{\infty} m^2(t)\, dt = E_m$$

Hence

$$e_o(t) = E_m\Delta f \cos 2\omega_c t \qquad \textbf{(4.25)}$$

The signal $e_o(t)$ is in phase synchronism with the incoming carrier but has twice the desired frequency. This is corrected by the 2-to-1 frequency divider. The output is now in frequency and phase synchronism with the incoming carrier. One qualification is in order. Because the incoming signal sign is lost in the squarer, we have a sign ambiguity (or phase ambiguity of π) in the carrier generated. This is immaterial for analog signals. For a digital baseband signal, however, the carrier sign is essential, and this method, therefore, cannot be used directly without further modifications.*

Yet another scheme, proposed by Costas,[4] for generating a local carrier is shown in Fig. 4.24. It is assumed that the nominal carrier frequency is known a priori. A local oscillator generates a carrier $\cos(\omega_c t + \theta)$. Various signals are indicated in Fig. 4.24.

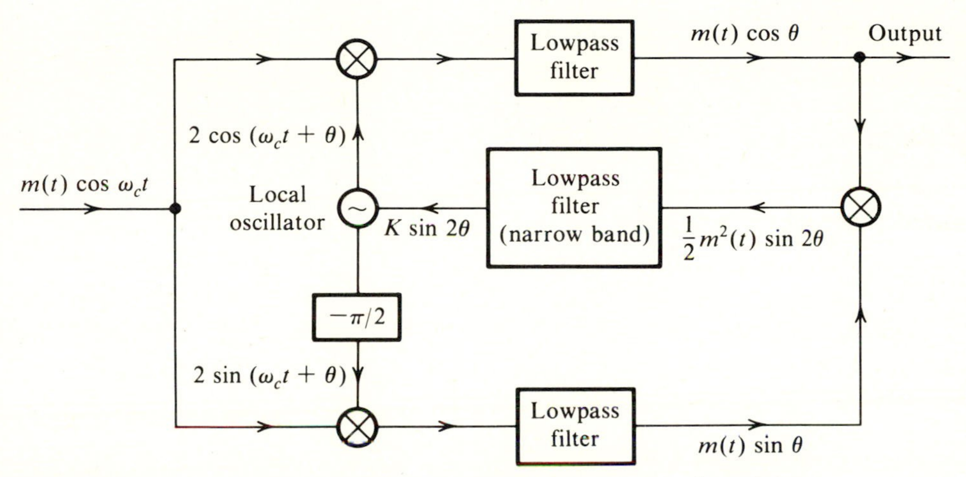

Figure 4.24 Costas phase-lock loop for generation of coherent demodulation carrier.

The two lowpass filters suppress high-frequency terms to yield $m(t)\cos\theta$ and $m(t)\sin\theta$, respectively. These outputs are further multiplied to give $m^2(t)\sin 2\theta$. When this is passed through a narrowband lowpass filter, the output is $k\sin 2\theta$, where k is $E_m\Delta f$ [see Eq. (4.25)].When θ is small, $k\sin 2\theta \simeq 2k\theta$. This signal is used to adjust the phase of the local oscillator. When the local oscillator is in exact synchronism with the incoming signal, the output of the upper arm is the desired demodulated signal.

SSB-SC Signals

The incoming SSB signal at the receiver is given by† [Eq. (4.19)]

$$\varphi_{\text{SSB}}(t) = m(t)\cos\omega_c t + m_h(t)\sin\omega_c t$$

Let the local carrier be $\cos[(\omega_c + \Delta\omega)t + \delta]$. The product of the incoming signal and

*This involves differential coding (see Sec. 4.7) of binary digits. Another method is to establish the correct polarity of the carrier initially and then monitor for π radian change.

†Here we are considering the LSB. The discussion, however, applies equally to the USB.

the local carrier is $e_d(t)$, given by

$$e_d(t) = \varphi_{\text{SSB}}(t) \cos [(\omega_c + \Delta\omega)t + \delta]$$
$$= [m(t) \cos \omega_c t + m_h(t) \sin \omega_c t] \cos [(\omega_c + \Delta\omega)t + \delta]$$

The lowpass filter suppresses the sum frequency components centered at $(2\omega_c + \Delta\omega)$ leaving the difference frequency components centered $\Delta\omega$. Hence the output $e_o(t)$ is

$$e_o(t) = \tfrac{1}{2}\{m(t) \cos [(\Delta\omega)t + \delta] - m_h(t) \sin [(\Delta\omega)t + \delta]\} \tag{4.26}$$

Observe that if $\Delta\omega$ and δ are both zero, the output is

$$e_o(t) = \tfrac{1}{2}m(t)$$

as expected. It is interesting to compare the effects of phase and frequency errors for DSB and SSB systems. If $\Delta\omega = 0$, we observed that for DSB, the signal remains undistorted, although it is attenuated by a factor $\cos \delta$. If δ is close to $\pm\pi/2$, however, the signal attenuation can be very high. For SSB signals, on the other hand, when $\Delta\omega = 0$, the output is given by

$$e_o(t) = \tfrac{1}{2}[m(t) \cos \delta - m_h(t) \sin \delta] \tag{4.27a}$$

We shall now show that the distortion in Eq. (4.27a) is a phase distortion where each frequency component of $M(\omega)$ acquires a phase shift δ. From Eq. (4.27a), we have

$$E_o(\omega) = \tfrac{1}{2}[M(\omega) \cos \delta - M_h(\omega) \sin \delta]$$

But [Eq. (4.18)]

$$M_h(\omega) = -j \operatorname{sgn}(\omega) M(\omega) = \begin{cases} -jM(\omega) & \omega > 0 \\ jM(\omega) & \omega < 0 \end{cases}$$

and

$$E_o(\omega) = \begin{cases} \tfrac{1}{2}M(\omega)e^{j\delta} & \omega > 0 \\ \tfrac{1}{2}M(\omega)e^{-j\delta} & \omega < 0 \end{cases} \tag{4.27b}$$

It is apparent that $e_o(t)$ has the same magnitude spectrum as that of $m(t)$, but the phase of each component is shifted by δ. Thus, the phase error in the local carrier gives rise to a phase distortion in the detector output. Phase distortion, however, is usually not a serious problem with voice signals, because the human ear is somewhat insensitive to phase distortion. Such distortion may change the quality of speech, but the voice is still intelligible. In video signals and data transmission, however, phase distortion may be intolerable.

It can be seen by setting $\delta = 0$ in Eq. (4.26) that the effect of a frequency error in SSB transmission is equivalent to generating another SSB signal with a carrier frequency $\Delta\omega$. This means each component of $m(t)$ is shifted by $\Delta\omega$ (Fig. 4.25*b*). Again, this may make the voice sound only slightly different, as long as $\Delta\omega$ is within limits. For voice signals, a frequency shift of ± 20 Hz is tolerable. Most U.S. systems, however, restrict the shift to ± 2 Hz.

For the case of DSB transmission, on the other hand, the output signal is $m(t) \cos \Delta\omega t$ [Eq. (4.24b)], the spectrum of which is shown in Fig. 4.25*c*. Whereas

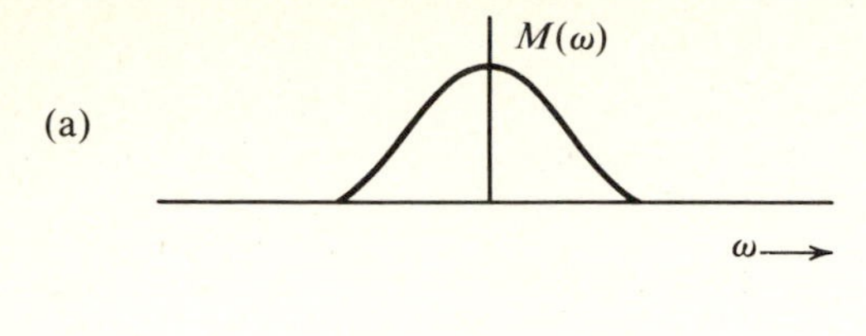

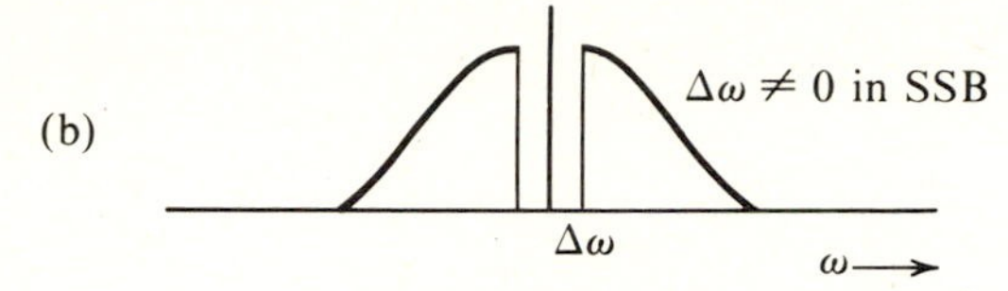

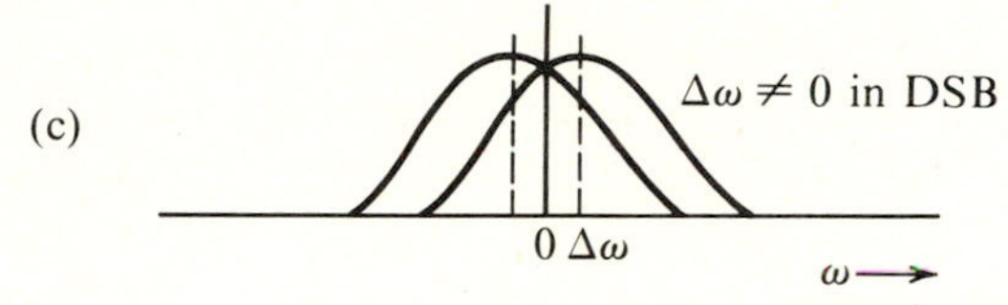

Figure 4.25 Effect of frequency drift in a locally generated carrier.

in SSB each frequency component is shifted by $\Delta\omega$, in DSB each component is shifted by $\Delta\omega$ and $-\Delta\omega$. There is a "beating" effect of multiplying $m(t)$ by a time-varying signal $\cos \Delta\omega t$. This distortion is more serious than that in SSB. Thus, for voice signals, a frequency or phase error in the local carrier is more serious in DSB-SC transmission than in SSB-SC transmission.

For the purpose of synchronization at the SSB receiver, one may use highly stable crystal oscillators, with crystals cut for the same frequency at the transmitter and the receiver. At very high frequencies, where even quartz crystals may have inadequate performance, a pilot carrier may be transmitted. These are the same methods used for DSB-SC. The received-signal squaring technique, however, as well as the Costas loop used in DSB-SC cannot be used for SSB-SC. This can be seen by expressing the SSB signal as

$$\varphi_{\text{SSB}}(t) = E(t) \cos [\omega_c t + \theta(t)]$$

where $E(t) = \sqrt{m^2(t) + m_h^2(t)}$

$$\theta(t) = -\tan^{-1}\left(\frac{m_h(t)}{m(t)}\right)$$

Squaring this signal yields

$$\begin{aligned}\varphi^2_{\text{SSB}}(t) &= E^2(t) \cos^2 [\omega_c t + \theta(t)] \\ &= \frac{E^2(t)}{2}(1 + \cos [2\omega_c t + 2\theta(t)])\end{aligned}$$

The signal $E^2(t)$ is eliminated by a bandpass filter. Unfortunately, the remaining signal is not a pure sinusoid of frequency $2\omega_c$ (as was the case for DSB). There is nothing we can do to remove the time-varying phase $2\theta(t)$ from this sinusoid. Hence, for SSB the squaring technique does not work. The same argument can be used to show that the Costas loop will not work either.

■ EXAMPLE 4.9

Verify Eq. (4.26) (frequency and phase-error distortion in SSB) for a tone modulation.

□ **Solution:**

$$m(t) = \cos \omega_m t$$

and

$$\varphi_{\text{LSB}}(t) = \cos (\omega_c + \omega_m)t$$

Let the local carrier be $2 \cos [(\omega_c + \Delta\omega)t + \delta]$. When $\varphi_{\text{LSB}}(t)$ is multiplied by the local carrier and then lowpass filtered, the sum frequency terms are suppressed, and the output $e_o(t)$ is

$$e_o(t) = \cos [(\omega_m + \Delta\omega)t + \delta]$$

When $\delta = 0$

$$e_o(t) = \cos (\omega_m + \Delta\omega)t$$

Thus, the output has a frequency shift of $\Delta\omega$. When $\Delta\omega = 0$

$$e_o(t) = \cos (\omega_m t + \delta)$$

The output is the desired signal with a phase shift δ. ■

Carrier Reinsertion Techniques of Detecting Suppressed-Carrier Signals

Suppressed-carrier signals can be detected by reinserting a sufficient amount of the carrier at the receiver. After a sufficient amount of the carrier has been reinserted, one may use either rectifier detection or envelope detection. The phase and the frequency of the reinserted carrier should be properly synchronized in order to avoid distortion. It can be shown that frequency and phase errors in the added carrier cause exactly the same kinds of distortion (in DSB or SSB) as those discussed earlier in this section (see Probs. 4.30, 4.31, and 4.32).

Comparison of Various AM Systems

We have discussed various aspects of AM and AM-SC (DSB-SC and SSB-SC) systems. It is interesting to compare these systems from various points of view.

The AM system has an advantage over the AM-SC systems (that is, DSB-SC and SSB-SC) at the receiver. The detectors required for AM are relatively simpler (rectifier or envelope detectors) than those required for suppressed-carrier systems. For this reason, all AM broadcast systems use AM. In addition, AM signals are easier to generate at high power levels, as compared to suppressed-carrier signals. The balanced modulators required in the latter are somewhat difficult to design.

Suppressed-carrier systems have an advantage over AM in that they require less power to transmit the same information. Under normal conditions, the carrier takes up to 75 percent (or even more) of the total transmitted power. This necessitates a rather expensive transmitter for AM. For suppressed-carrier systems, however, the receiver is much more complex and consequently more expensive. For a point-to-point communication system, where there are only a few receivers for one transmitter, the complexity in a receiver is justified, whereas for public broadcast systems, where there are millions of receivers for each transmitter, AM is the obvious choice.

The AM signal also suffers from the phenomenon of fading, as mentioned earlier. Fading is strongly frequency dependent; that is, various frequency components suffer different attenuation and nonlinear phase shifts. This is known as *selective* fading. The effect of fading is more serious on AM signals than on AM-SC signals, because in AM the carrier must maintain a certain strength in relation to the sidebands [Eq. (4.10)]. Because of selective fading, the carrier may be attenuated to the point where Eq. (4.10) is no longer satisfied. In such a case, the received signal detected by an envelope detector will show severe distortion. Even if the carrier is not badly attenuated and Eq. (4.10) is satisfied, selective fading can still badly distort the AM signal because of unequal attenuation and nonlinear phase shifts of the two sidebands and the carrier. The effect of selective fading becomes pronounced at higher frequencies. Therefore, suppressed-carrier systems are preferred at higher frequencies. The AM system is generally used for medium-frequency broadcast transmission.

Next we compare the DSB-SC system with the SSB-SC system. Here we find that the balance is mostly in favor of SSB-SC. The following are the advantages of SSB-SC over DSB-SC.

1. SSB-SC needs only half the bandwidth needed for DSB-SC. Although this difference can be balanced out by quadrature multiplexing two DSB-SC signals, practical difficulties of crosstalk are much more serious in quadrature multiplexing.
2. Frequency and phase errors in the local carrier used for demodulation have more serious effects in DSB-SC than in SSB-SC, particularly for voice signals.
3. Selective fading disturbs the relationship of the two sidebands in DSB-SC and causes more serious distortion than in the case of SSB-SC, where only one sideband exists.

For these reasons, DSB-SC is rarely used in audio communication. Long-haul telephone systems use SSB-SC multiplexed systems with a pilot carrier. For short-haul systems, DSB is sometimes used. PCM, however, is gradually replacing both.

SSB compares poorly to DSB in one respect: the generation of high-level SSB signals is more difficult than that of DSB signals. This disadvantage is overcome in what is called *vestigial sideband transmission*.

4.6 AMPLITUDE MODULATION: VESTIGIAL SIDEBAND (VSB)

A vestigial-sideband system is a compromise between DSB and SSB. It inherits the advantages of DSB and SSB but avoids their disadvantages. VSB signals are relatively easy to generate, and, at the same time, their bandwidth is only slightly (typically 25 percent) greater than that of SSB signals.

In VSB, instead of rejecting one sideband completely (as in SSB), a gradual cutoff of one sideband, as shown in Fig. 4.26*d*, is accepted. The roll-off characteristic of the

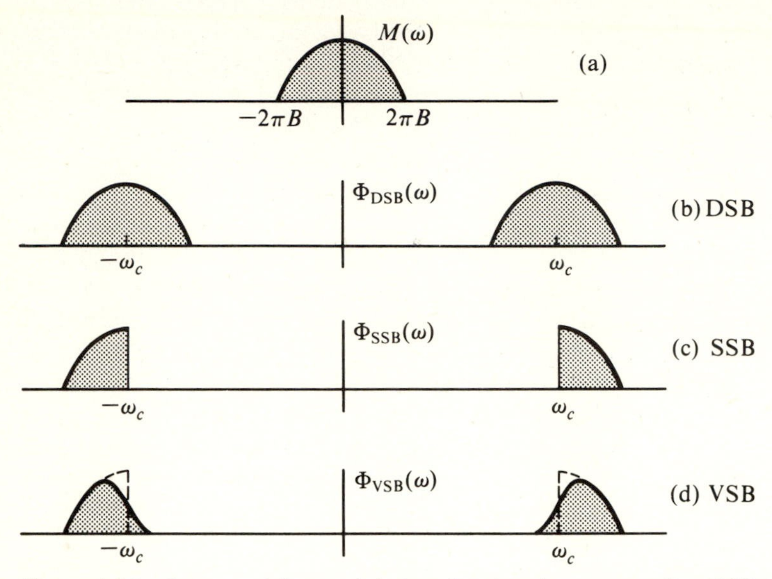

Figure 4.26 Spectra of the modulating signal and corresponding DSB, SSB, and VSB signals.

filter is such that the partial suppression of the transmitted sideband (the upper sideband in Fig. 4.26*d*) in the neighborhood of the carrier is exactly compensated for by the partial transmission of the corresponding part of the suppressed sideband (the lower sideband in Fig. 4.26*d*). Because of this spectral shaping, the baseband signal can be recovered exactly by a synchronous detector (Fig. 4.27*a*). If a large carrier is transmitted along with the VSB signal, the baseband signal can be recovered by an envelope (or a rectifier) detector.

To determine $H(\omega)$, the vestigial shaping filter to produce VSB from DSB (Fig. 4.27*a*), we have

$$\Phi_{VSB}(\omega) = [M(\omega + \omega_c) + M(\omega - \omega_c)]H(\omega) \tag{4.28}$$

We require that $m(t)$ be recoverable from $\varphi_{VSB}(t)$ using synchronous demodulation of the latter. This is done by multiplying the incoming VSB signal $\varphi_{VSB}(t)$ by $2 \cos \omega_c t$. The product $e_d(t)$ is given by

$$e_d(t) = 2\varphi_{VSB}(t) \cos \omega_c t \leftrightarrow [\Phi_{VSB}(\omega + \omega_c) + \Phi_{VSB}(\omega - \omega_c)]$$

Substitution of Eq. (4.28) in the above equation and eliminating the spectra at $\pm 2\omega_c$ (suppressed by a lowpass filter), the output $e_o(t)$ at the lowpass filter (Fig. 4.27*a*) is given by

$$e_o(t) \leftrightarrow M(\omega)[H(\omega + \omega_c) + H(\omega - \omega_c)]$$

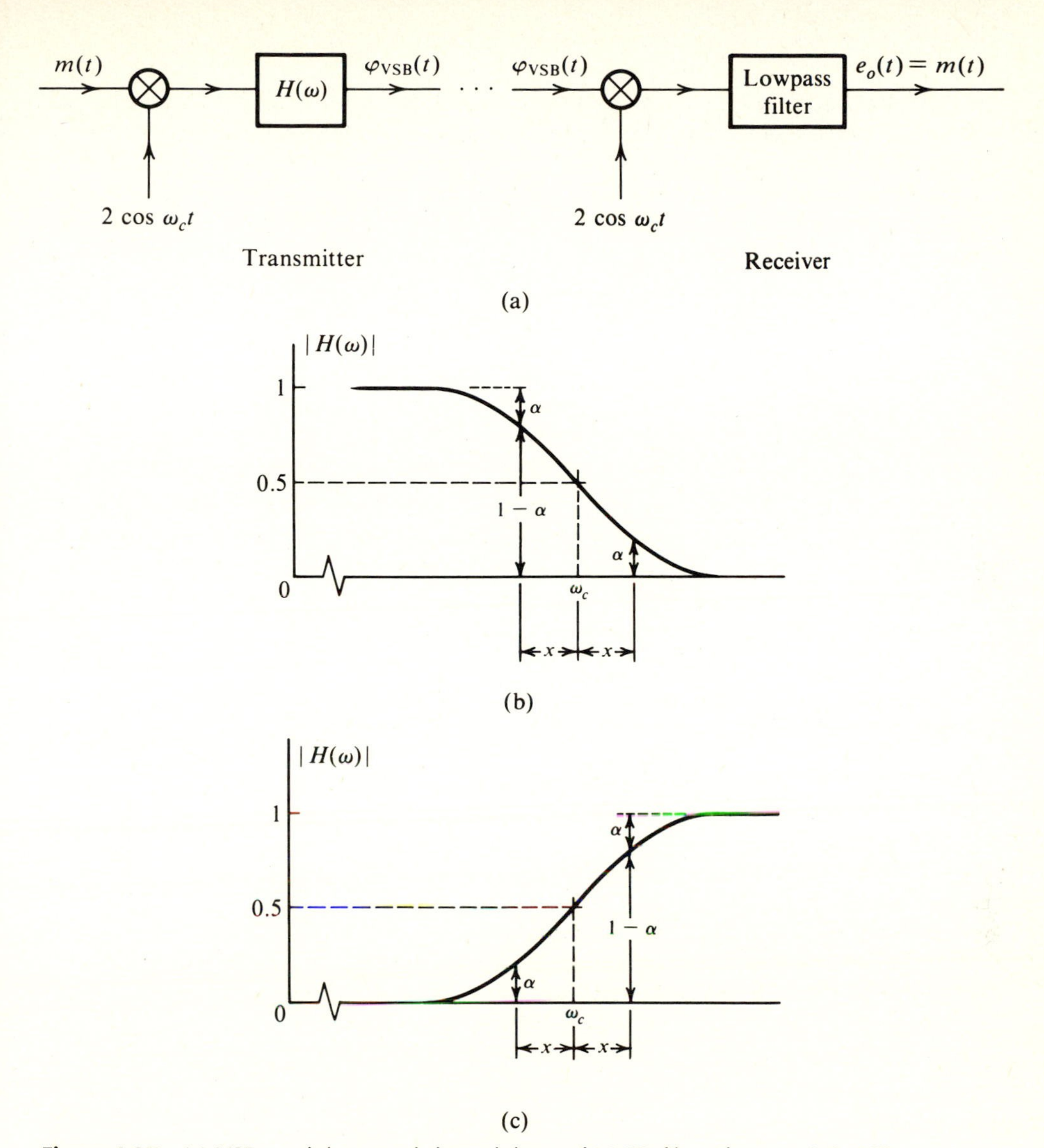

Figure 4.27 (a) VSB modulator and demodulator. (b) VSB filter characteristic LSB. (c) VSB filter characteristic (USB).

For distortionless reception, we must have

$$e_o(t) \leftrightarrow CM(\omega) \qquad (C \text{ is a constant})$$

Choosing $C = 1$, we have

$$H(\omega + \omega_c) + H(\omega - \omega_c) = 1 \qquad |\omega| \leq 2\pi B \tag{4.29}$$

For any real filter, $H(-\omega) = H^*(\omega)$. Hence, Eq. (4.29) can be expressed as

$$H(\omega_c + \omega) + H^*(\omega_c - \omega) = 1 \qquad |\omega| \leq 2\pi B \tag{4.30a}$$

or

$$H(\omega_c + x) + H^*(\omega_c - x) = 1 \qquad |x| \leq 2\pi B \tag{4.30b}$$

This is precisely the vestigial filter discussed in Sec. 3.3 [Eq. (3.30)].

If we consider a filter with a transfer function of the form $|H(\omega)|e^{-j\omega t_d}$, the term $e^{-j\omega t_d}$ represents a pure delay. Hence, only $|H(\omega)|$ need satisfy Eq. (4.30b). Because $|H(\omega)|$ is real, Eq. (4.30b) implies that

$$|H(\omega_c + x)| + |H(\omega_c - x)| = 1 \qquad |x| \leq 2\pi B \tag{4.31}$$

Figure 4.27*b* and *c* shows two possible forms of $|H(\omega)|$ that satisfy Eq. (4.31). The filters in Fig. 4.27*b* and *c* correspond to VSB filters that retain the LSB and the USB, respectively.

It is not necessary to realize the desired VSB spectral shaping in one filter $H(\omega)$. It can be done in two stages: one filter in the transmitter ($H(\omega)$ in Fig. 4.27*a*), and the remaining filter at the input of the receiver. This is precisely what is done in broadcast television systems. If the two filters still do not achieve the desired vestigial characteristics, the remaining equalization can be achieved in the lowpass filter of the receiver (Fig. 4.27*a*).

To find $\varphi_{\text{VSB}}(t)$, we note that $\Phi_{\text{VSB}}(\omega)$ is a bandpass spectrum not symmetrical about its center frequency ω_c. Hence, it can be expressed in terms of quadrature components [Eq. (2.61a)]:

$$\varphi_{\text{VSB}}(t) = m_c(t) \cos \omega_c(t) + m_s(t) \sin \omega_c t \tag{4.32}$$

We can show that $m_c(t) = m(t)$ by referring to Fig. 4.27*a*. When $\varphi_{\text{VSB}}(t)$ is multiplied by $2 \cos \omega_c t$ and then lowpass filtered, the output is $m(t)$. In our case

$$2\varphi_{\text{VSB}}(t) \cos \omega_c t = m_c(t) + m_c(t) \cos 2\omega_c t + m_s(t) \sin 2\omega_c t$$

When this signal is lowpass filtered, the output is $m_c(t)$. Hence, it follows that

$$m_c(t) = m(t) \tag{4.33a}$$

To determine $m_s(t)$, we observe that if $\varphi_{\text{VSB}}(t)$ in Eq. (4.32) is multiplied by $2 \sin \omega_c t$ and then lowpass filtered, the output is $m_s(t)$. Because

$$2\varphi_{\text{VSB}}(t) \sin \omega_c t \leftrightarrow j[\Phi_{\text{VSB}}(\omega + \omega_c) - \Phi_{\text{VSB}}(\omega - \omega_c)]$$

substituting Eq. (4.28) in the above equation and eliminating the high-frequency terms we get

$$m_s(t) \leftrightarrow M(\omega)[H(\omega + \omega_c) - H(\omega - \omega_c)]$$

The use of Eq. (4.29) in the above equation yields

$$m_s(t) \leftrightarrow M(\omega)[1 - 2H(\omega - \omega_c)] \tag{4.33b}$$

In conclusion, the vestigial-modulated signal $\varphi_{\text{VSB}}(t)$ is

$$\varphi_{VSB}(t) = m(t) \cos \omega_c t + m_s(t) \sin \omega_c t \tag{4.34a}$$

If we change the sign of $m_s(t)$ in Eq. (4.34a), it amounts to turning the filter $H(\omega)$ about ω_c. This gives the VSB signal, which retains the upper sideband. Hence, $\varphi_{VSB}(t)$ can be expressed as

$$\varphi_{VSB}(t) = m(t) \cos \omega_c t \pm m_s(t) \sin \omega_c t \tag{4.34b}$$

This expression is similar to that of an SSB signal [Eq. (4.19c)] with $m_s(t)$ replacing $m_h(t)$.

It can be seen that VSB does the trick by partially suppressing the transmitted sideband and compensating for this with a gradual roll-off filter. Thus, VSB is a clever compromise between SSB and DSB, with advantages of both at a small cost in increased bandwidth, which is slightly larger than that of SSB (typically 25 to 30 percent larger). It can be generated from DSB signals by using relatively simple filters with gradual roll-off characteristics. Its immunity to selective fading is comparable to that of SSB. Also, if a sufficiently large carrier $A \cos \omega_c t$ is added, the resulting signal (VSB + C) can be demodulated by an envelope detector with relatively small distortion. This mode (VSB + C) combines the advantages of AM, SSB, and DSB. All these properties make VSB attractive for commercial television broadcasting. The baseband video signal of television occupies an enormous bandwidth of 4.5 MHz, and a DSB signal needs a bandwidth of 9 MHz. It would seem desirable to use SSB in order to conserve the bandwidth. Unfortunately, this creates two problems. First, the baseband video signal has sizable power in the low-frequency region, and consequently it is very difficult to suppress one sideband completely. Second, for a broadcast receiver, an envelope detector is preferred over a synchronous one in order to reduce the cost. An envelope detector gives much less distortion for VSB signals than for SSB signals.

The VSB characteristics in television signals are achieved in two steps (Fig. 4.28). At the transmitter, the spectrum is shaped without rigidly controlling the transition region (Fig. 4.28*b*). The receiver also has an input filter, which further shapes the incoming signal to have the desired complimentary symmetry required of a vestigial signal (Fig. 4.28*c*). The reason for the two-step shaping is that it is much easier to shape the spectrum at the receiver, where power levels are much lower. The envelope detector in a VSB system does cause some distortion. Because the eye is relatively insensitive to amplitude distortion, this problem is not too serious.

Envelope Detection of VSB + C Signals

That VSB + C signals can be envelope detected may be proved by using exactly the same argument used in proving the case for SSB + C signals. This is because the modulated signals in both cases have the same form, with $m_h(t)$ in SSB replaced by $m_s(t)$ in VSB.

We have shown that SSB + C requires a much larger carrier than DSB + C (AM) for envelope detection. Because VSB + C is an in-between case, it is to be expected that the added carrier in VSB will be larger than that in AM but smaller than that in SSB + C.

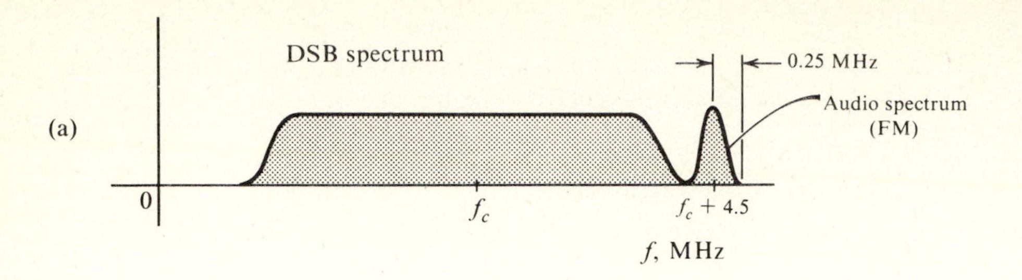

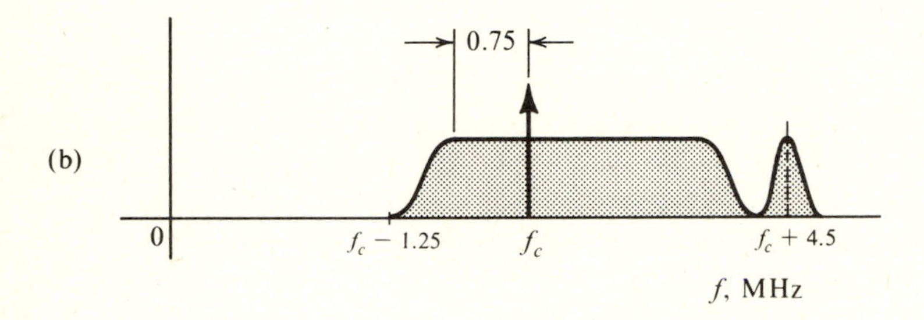

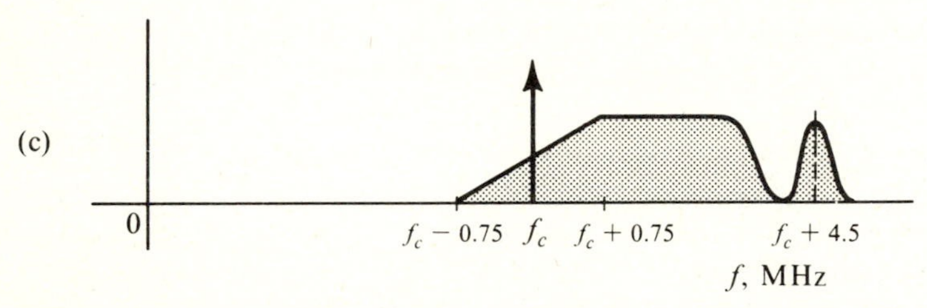

Figure 4.28 TV signal spectra. (a) DSB signal. (b) Signal transmitted. (c) Output signal of the vestigial filter in the receiver.

■ EXAMPLE 4.10

A vestigial-filter transfer function $H(\omega)$ is shown in Fig. 4.29, with $\omega_c = 10^5$. Find the VSB-modulated signal $\varphi_{\text{VSB}}(t)$ when $m(t) = \cos \omega_m t (\omega_m = 1000)$. This is the case of tone modulation. Referring to Fig. 4.27*a*, the DSB signal is

$$2 \cos \omega_m t \cos \omega_c t = \cos (\omega_c - \omega_m)t + \cos (\omega_c + \omega_m)t$$

There are two sinusoids, of frequencies $\omega_c + 1000$ and $\omega_c - 1000$, respectively. These sinusoids are transmitted through $H(\omega)$ in Fig. 4.29, which has a gain of 0.75 and 0.25 at $\omega_c - 1000$ and $\omega_c + 1000$, respectively. Hence $\varphi_{\text{VSB}}(t)$, the filter output, is

$$\begin{aligned}\varphi_{\text{VSB}}(t) &= 0.75 \cos (\omega_c - \omega_m)t + 0.25 \cos (\omega_c + \omega_m)t \\ &= \underbrace{\cos \omega_m t \cos \omega_c t}_{m(t)} + \underbrace{0.5 \sin \omega_m t \sin \omega_c t}_{m_s(t)}\end{aligned}$$

■

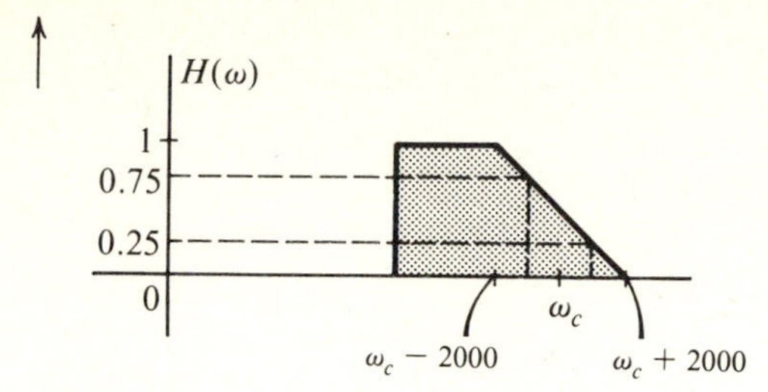

Figure 4.29

Linearity of Amplitude Modulation

In all the types of modulation discussed thus far, the modulated signal (excluding the carrier term) satisfies the principles of superposition. For example, if modulating signals $m_1(t)$ and $m_2(t)$ produce modulated signals* $\varphi_1(t)$ and $\varphi_2(t)$, respectively, then the modulating signal $k_1m_1(t) + k_2m_2(t)$ produces the modulated signal $k_1\varphi_1(t) + k_2\varphi_2(t)$. The reader can verify linearity for all types of amplitude modulation (DSB, SSB, AM, and VSB). This property is valuable in analysis. Because any signal can be expressed as a sum (discrete or in continuum) of sinusoids, the complete description of the modulation system can be expressed in terms of tone modulation. For example, if $m(t) = \cos \omega_m t$ (tone modulation), the DSB-SC signal is

$$\cos \omega_m t \cos \omega_c t = \tfrac{1}{2}[\cos (\omega_c - \omega_m)t + \cos (\omega_c + \omega_m)t]$$

This shows that DSB-SC translates a frequency ω_m to two frequencies, $\omega_c - \omega_m$ (LSB) and $\omega_c + \omega_m$ (USB). We can generalize this result to any nonsinusoidal modulating signal $m(t)$. This is precisely the result obtained earlier by using a more general analysis.

4.7 DIGITAL CARRIER SYSTEMS

As seen earlier, digital signals can be modulated by several schemes such as ASK, PSK, FSK, etc. Demodulation of digital-modulated signals is similar to that of analog-modulated signals. For example, ASK (see Fig. 3.35) can be demodulated coherently (synchronous) or noncoherently (envelope detection). The noncoherent scheme performance is close to the performance of the coherent scheme when the noise is small. The difference in the two schemes is pronounced when the noise is large. This behavior is similar to that observed in analog signals.

In PSK, a **1** is transmitted by a pulse $A \cos \omega_c t$ and a **0** is transmitted by a pulse $-A \cos \omega_c t$ (see Fig. 4.2*d*). The information in PSK signals therefore resides in the carrier phase. These signals cannot be demodulated noncoherently (envelope detection) because the envelope is the same for both **1** and **0**. The coherent detection is similar to that used for analog signals. Methods of carrier synchronization are also the same as those used for analog signals. A small pilot can be transmitted along with the

---

*Note that we are excluding the carrier term from $\varphi_1(t)$ and $\varphi_2(t)$. In short, superposition applies to the suppressed-carrier portion only. For more discussion, see Van Trees.[5]

modulated signal. In the absence of a pilot, one of the self-synchronization methods such as the Costas loop or the signal squaring technique discussed in Sec. 4.5 can be used. Because these techniques yield a carrier with sign ambiguity (or phase ambiguity of π), they cannot be used directly to demodulate PSK. This is because a sign ambiguity in the demodulating carrier can detect a negative pulse as a positive pulse (detect **0** as **1**) and vice versa. This problem can be solved by encoding the data by *differential code* before modulation.

In this case, a **1** is encoded by the same pulse used to encode the previous data bit (no transition) and a **0** is encoded by the negative of the pulse used to encode the previous data bit (transition). This is shown in Fig. 4.30*a*. Thus a transition in the

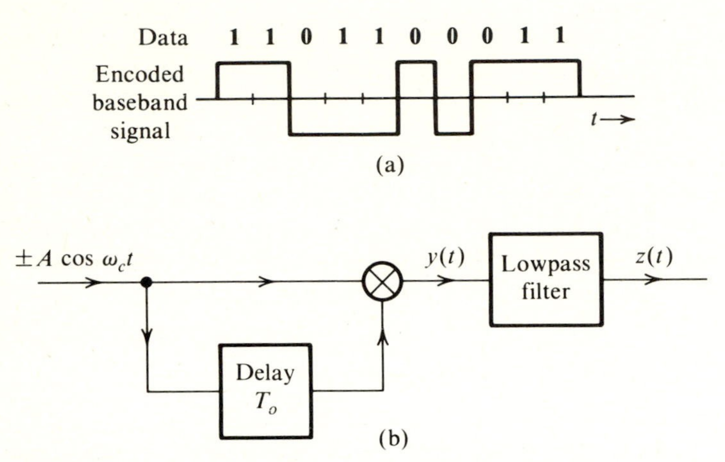

Figure 4.30 (a) Differential coding. (b) Differential PSK receiver.

received pulse sequence indicates **0** and no transition indicates **1**.* Therefore, the absolute signs of the received pulses are not important for detection. What is important is the change in signs of successive pulses. These sign changes are correctly detected even if the demodulating carrier has a sign ambiguity.

Differential coding also facilitates noncoherent detection of PSK. In this scheme, known as *differential PSK* or *DPSK* (Fig. 4.30*b*), we avoid generation of a local carrier by observing that the received modulated signal itself is a carrier ($\pm A \cos \omega_c t$) with a possible sign ambiguity. For demodulation, in place of the carrier, we use the received signal delayed by T_o (one bit interval). If the received pulse is identical to the previous pulse, the product $y(t) = A^2 \cos^2 \omega_c t$, and the lowpass filter output $z(t) = A^2/2$. If the received pulse is of opposite sign, $y(t) = -A^2 \cos^2 \omega_c t$ and $z(t) = -A^2/2$. In differential coding, two pulses of the same polarity in succession (no transition) indicates a **1** and two pulses of opposite polarity in succession (transition) indicates a **0**. Hence, the positive value of $z(t)$ is immediately detected as a **1** and the negative $z(t)$ is detected as a **0**.

---

*Precoding discussed in connection with duobinary is actually differential coding where a **0** is transmitted by no transition and a **1** is transmitted by a transition.

The FSK can be viewed as two interleaved ASK signals with carrier frequencies ω_0 and ω_1 respectively (Fig. 3.36*c*). Therefore, FSK can be detected coherently or noncoherently. In noncoherent detection, the incoming signal is applied to a bank of two filters tuned to ω_0 and ω_1. Each filter is followed by an envelope detector (see Fig. 6.35). The outputs of the two envelope detectors are sampled and compared. A **0** is transmitted by a pulse of frequency ω_0 and this pulse will appear at the output of the filter tuned to ω_0. Practically no signal appears at the output of the filter tuned to ω_1. Hence the sample of the envelope detector output following the ω_0 filter will be greater than the sample of the envelope detector output following the ω_1 filter and the receiver decides that a **0** was transmitted. In the case of a **1**, the situation is reversed.

FSK can also be detected coherently by generating two references of frequencies ω_0 and ω_1, and demodulating the received signal by two demodulators using the two carriers and then comparing the outputs of the two demodulators.

From the point of view of noise immunity, coherent PSK is superior to all other schemes. PSK also requires smaller bandwidth than FSK (see Fig. 3.37). Quantitative discussion of this topic can be found in Chapter 6, Part II.

Digital signals can also be modulated using SSB (or VSB) to conserve bandwidth. For the purpose of generating SSB signals, however, it is desirable to have a dc null in the PSD of the baseband signal (Sec. 4.4). Therefore, bipolar, duobinary, or split phase signals are suitable for SSB modulation. Although SSB conserves the bandwidth, the same result is achieved by QPSK (or QAM) discussed earlier.

4.8 INTERFERENCE AND NOISE IN AM SYSTEMS

Signals transmitted over the same medium will interfere with each other, even if their nominal spectra do not overlap. This is because the real spectrum of a signal may be much wider than the nominal spectrum. For example, in an SSB spectrum, the unwanted spectrum is never completely suppressed, and it will interfere with the adjacent channel. In an AM broadcast, carrier frequencies are spaced 10 kHz apart, allowing a 10-kHz bandwidth to each station. But it is impossible to suppress completely the components beyond the 10-kHz bandwidth. Hence, some interference will occur among adjacent channels. If one of the stations is very powerful, it could be unfortunate for its weak neighbor.

To get some qualitative notion of the effect of interference in AM systems, let us consider the DSB-SC signal $m(t) \cos \omega_c t$ and an interfering sinusoid $I \cos [(\omega_c + \omega_d)t + \theta]$ whose spectra are shown in Fig. 4.31*a*. The signal $r(t)$ received at the input of the receiver is

$$r(t) = m(t) \cos \omega_c t + I \cos [(\omega_c + \omega_d)t + \theta] \qquad \textbf{(4.35a)}$$

Note that the ratio of the peak amplitudes of the desired signal and the interfering signal is m_p/I. The received signal $r(t)$ is demodulated by multiplying it by $2 \cos \omega_c t$ (synchronous demodulation) and then lowpass filtering (Fig. 4.31*b*). The filter output $y_d(t)$ is

$$y_d(t) = m(t) + I \cos (\omega_d t + \theta) \qquad \textbf{(4.35b)}$$

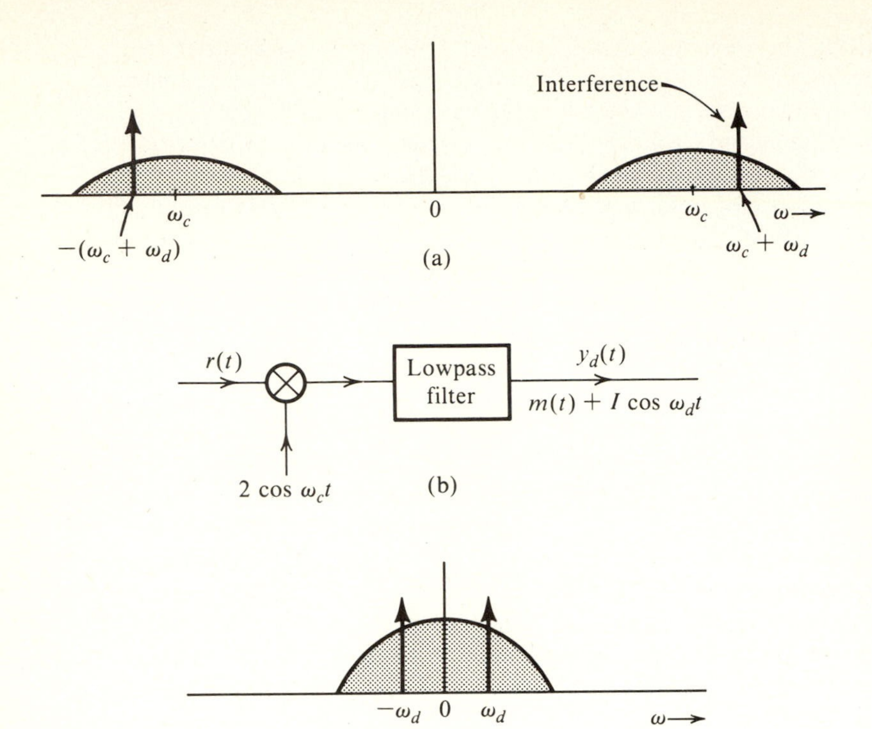

Figure 4.31 Effect of interference in DSB systems.

The spectrum of $y_d(t)$ is shown in Fig. 4.31*c*. Observe that, after demodulation, the ratio of the peak signal amplitude to that of the interfering signal is still m_p/I. Thus, the demodulation neither increases nor decreases the effect of interference in DSB-SC. Demodulation merely translates the sinusoidal interference from frequency $\omega_c + \omega_d$ to ω_d without changing its amplitude (or power). If there are two interfering sinusoids

$$r(t) = m(t) \cos \omega_c t + I_1 \cos [(\omega_c + \omega_{d_1})t + \theta_1] + I_2 \cos [(\omega_c + \omega_{d_2})t + \theta_2]$$

then it can readily be seen that

$$y_d(t) = m(t) + I_1 \cos (\omega_{d_1} t + \theta_1) + I_2 \cos (\omega_{d_2} t + \theta_2) \tag{4.35c}$$

This is the direct consequence of the linearity of DSB-SC.

Let us now consider the case of AM with envelope detection. The received signal in this case is

$$\begin{aligned} r(t) &= [A + m(t)] \cos \omega_c t + I \cos (\omega_c + \omega_d)t \qquad \textbf{(4.36a)} \\ &= \{[A + m(t)] + I \cos \omega_d t\} \cos \omega_c t - I \sin \omega_d t \sin \omega_c t \end{aligned}$$

and the envelope is

$$E(t) = [[A + m(t) + I \cos \omega_d t]^2 + I^2 \sin^2 \omega_d t]^{1/2}$$

If the interference is small ($I \ll A$)

$$E(t) \simeq A + m(t) + I \cos \omega_d t \tag{4.36b}$$

The constant A is blocked, and the remaining output is identical to that obtained for DSB-SC with coherent demodulation [Eq. (4.35b)].

Next, consider the case of a large interference, where $I \gg A$. Rewriting $r(t)$ in Eq. (4.36a) as

$$\begin{aligned} r(t) &= [A + m(t)] \cos (\omega_c + \omega_d - \omega_d)t + I \cos (\omega_c + \omega_d)t \\ &= \{[A + m(t)] \cos \omega_d t + I\} \cos (\omega_c + \omega_d)t \\ &\quad + [A + m(t)] \sin \omega_d t \sin (\omega_c + \omega_d)t \end{aligned}$$

if $I \gg A$, the second term on the right-hand side can be ignored. The envelope $E(t)$ of the first term is

$$E(t) = [A + m(t)] \cos \omega_d t + I \tag{4.36c}$$

The constant I is blocked by a capacitor to yield $[A + m(t)] \cos \omega_d t$ as the output. Note that the output in this case is the desired output multiplied by the interference signal. The interference in the output is not additive as in Eq. (4.36b), but is multiplicative. This results in mutilation of the desired signal. The degradation of the desired signal is caused by the so-called *threshold effect* and is a consequence of the nonlinearity of the envelope detector.

The Output Signal-to-Noise Ratio

A rigorous treatment of noise analysis can be undertaken only after the study of random processes (Chapter 5). It is possible, however, to conduct a simple analysis of this problem using the sinusoidal model of noise developed in Chapter 2. Because channel noise is a kind of interference, we can use the results of interference analysis directly.

The SNR at the receiver output (output SNR) is an important parameter because it indicates the quality of the output signal. A large SNR indicates a better signal. Hence, the output SNR serves as a measure of the merit of a system. For a fair comparison of various systems, we should compare their output SNRs for a given transmitted power. Because the received message-signal power (at the receiver input) is directly proportional to the transmitted power, we can also compare the output SNRs for a given received message-signal power S_i. In this discussion, the message signal $m(t)$ is assumed to be a power signal.

Let us compare the output SNRs of DSB-SC, SSB-SC, VSB-SC, and AM systems for a given power S_i received at the input of the receiver. The channel noise will be assumed to be white with a PSD $\mathcal{N}/2$ (Fig. 2.62).

$$S_n(\omega) = \frac{\mathcal{N}}{2}$$

This noise can be represented by a sum of sinusoids of frequencies Δf apart ($\Delta f \rightarrow 0$), as in Eq. (2.122). The modulated signal spectrum and the noise spectrum [using the sinusoidal model in Eq. (2.122)] over the modulated signal band* are shown in Fig. 4.32.

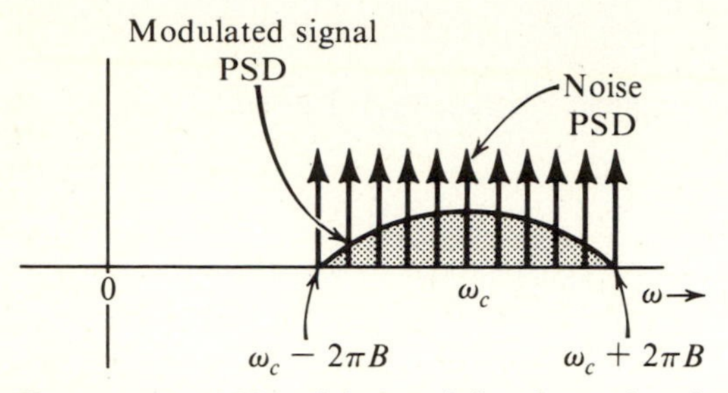

Figure 4.32 PSDs of the modulated signal and interfering noise in a DSB system.

1. DSB-SC. For DSB-SC, from Eq. (4.35b or c), it can be seen that the demodulator translates each interfering sinusoid of frequency $\omega_c + \omega_d$ to frequency ω_d. The amplitude and, hence, the power of each interfering sinusoid remains unchanged at the demodulator output. Therefore, the total interference (or noise) power N_o at the demodulator output is the same as the channel noise power over the band $2B$ that is occupied by the received signal. This is $2\mathcal{N}B$ [see Eq. (2.124)]. Hence†

$$N_o = 2\mathcal{N}B \tag{4.37a}$$

The output signal is $m(t)$ [Eq. (4.35b or c)]. Hence S_o, the output signal power is

$$S_o = \widetilde{m^2(t)} \tag{4.37b}$$

The received message signal is $m(t) \cos \omega_c t$. Hence, the received-message-signal power S_i is

$$S_i = \tfrac{1}{2}\widetilde{m^2(t)} \tag{4.37c}$$

Therefore

$$\frac{S_o}{N_o} = \frac{\widetilde{m^2(t)}}{2\mathcal{N}B} = \frac{S_i}{\mathcal{N}B} \tag{4.37d}$$

---

* Components lying outside of this band will be suppressed by the input filter of the receiver.

† Here we see that the noise sinusoid at frequencies $\omega_c + \omega_d$ as well as $\omega_c - \omega_d$ in the received signal are translated to the same baseband frequency ω_d. Hence, the power of the output noise at frequency ω_d is double that of the power of the received noise component at frequency $\omega_c + \omega_d$ (or $\omega_c - \omega_d$). One may argue that if two components add, their power should quadruple rather than double (see Example 2.38). Fortunately this does not happen, because the two components have random phases and they add incoherently with the consequence that their powers add rather than quadruple. This can be seen as follows: We can readily show that the power of a pair of signals $A \cos(\omega_d t + \theta_1) + A \cos(\omega_d t + \theta_2)$ is $A^2 + A^2 \cos(\theta_1 - \theta_2)$ (see Prob. 2.56). Because θ_1 and θ_2 are independent and random, the second term $A^2 \cos(\theta_1 - \theta_2)$ can take on any value in the range $(-A^2, A^2)$ depending on θ_1 and θ_2. When we consider the sum of a large number of such pairs (Fig. 4.32) the second term contribution averages to zero, and the effective power of the signal $A \cos(\omega_d t + \theta_1) + A \cos(\omega_d t + \theta_2)$ is just A^2 (the sum of the powers of two signals).

Thus

$$\frac{S_o}{N_o} = \gamma \tag{4.37e}$$

where

$$\gamma = \frac{S_i}{\mathcal{N}B} \tag{4.38}$$

Because $\mathcal{N}$ and B are constants, γ is directly proportional to S_i. Hence, comparing various systems for the output SNR for a given S_i is the same as comparing these systems for the output SNR for a given γ. If for a given γ, one system has a higher output SNR than the other system, the former is superior to the latter, at least from the SNR point of view. Equation (4.38) gives the output SNR for DSB-SC. The output SNR is plotted as a function of γ (both in dB) in Fig. 4.33.

2. SSB-SC. Consider an SSB-SC signal

$$\varphi_{\text{SSB}}(t) = m(t) \cos \omega_c t + m_h(t) \sin \omega_c t \tag{4.39}$$

This signal can be generated from $2m(t) \cos \omega_c t$ by suppressing the unwanted sideband. The power of $2\, m(t) \cos \omega_c t = 4[\widetilde{m^2(t)}/2] = 2\widetilde{m^2(t)}$. Suppression of one sideband reduces the power by half. Hence, the power* of $\varphi_{\text{SSB}}(t)$ in Eq. (4.39) is $\widetilde{m^2(t)}$. The received signal $r(t)$ is $\varphi_{\text{SSB}}(t)$ in Eq. (4.39) plus the channel noise. If we multiply $r(t)$ by $2 \cos \omega_c t$ and pass the product through a lowpass filter, the output is $m(t)$ plus noise. Hence S_o, the output signal power, is

$$S_o = \widetilde{m^2(t)} \tag{4.40a}$$

The output noise power can be computed in exactly the same way as in DSB-SC. The N_o would have been the same in SSB-SC as in DSB-SC except for the fact that the noise is only over half the bandwidth (viz, B). Hence, N_o in this case will be half that in Eq. (4.37a).

$$N_o = \mathcal{N}B \tag{4.40b}$$

Also, S_i, the power of the received message signal $\varphi_{\text{SSB}}(t)$, is

$$S_i = \widetilde{m^2(t)} \tag{4.40c}$$

and

$$\frac{S_o}{N_o} = \frac{\widetilde{m^2(t)}}{\mathcal{N}B} = \frac{S_i}{\mathcal{N}B} = \gamma \tag{4.40d}$$

Hence, the output SNR in SSB-SC is identical to the output SNR in DSB-SC for a given power transmitted. Also, because SSB-SC and DSB-SC have identical trans-

---

* This result may also be derived by observing that the powers $m_h(t)$ and $m(t)$ are identical because $m_h(t)$ is obtained by passing $m(t)$ through a Hilbert transformer ($|H(\omega)| = 1$). Hence, $m(t) \cos \omega_c t$ and $m_h(t) \sin \omega_c t$ each have a power $\widetilde{m^2(t)}/2$.

mission bandwidths,* theoretically SSB-SC and DSB-SC have identical capabilities in every respect. The output SNR vs. γ plot is identical to that of DSB-SC (Fig. 4.33).

3. VSB-SC. The performance of VSB-SC is almost the same as that of SSB-SC. The signal power S_i in VSB is somewhat smaller than that in SSB. The output power S_o in VSB is identical to that in SSB. The output noise power N_o in VSB is slightly larger than that in SSB, however, because the transmission bandwidth of VSB is slightly larger than that of SSB. The reduced value of S_i compensates partly for the increased value of N_o and the SNR in VSB is close to that of SSB. Therefore,

$$\frac{S_o}{N_o} \simeq \gamma \tag{4.41}$$

4. AM. The received signal is envelope detected. We shall consider here the small-noise as well as the large-noise case.

The Small-Noise Case. The received message signal is $[A + m(t)] \cos \omega_c t$. Hence S_i, the received message-signal power, is (see Prob. 2.53)

$$S_i = \frac{\widetilde{[A + m(t)]^2}}{2} = \frac{A^2 + \widetilde{m^2(t)}}{2} \tag{4.42a}$$

The envelope-detector output is exactly the same as that for DSB-SC except for a constant A, which is blocked by a capacitor [see Eqs. (4.35b) and (4.36b)]. Hence, S_o and N_o will be the same as those for DSB-SC [Eqs. (4.37a and b)]

$$S_o = \widetilde{m^2(t)} \tag{4.42b}$$

$$N_o = 2\mathcal{N}B \tag{4.42c}$$

Hence,

$$\begin{aligned} \frac{S_o}{N_o} &= \frac{\widetilde{m^2(t)}}{2\mathcal{N}B} \\ &= \frac{\widetilde{m^2(t)}}{A^2 + \widetilde{m^2(t)}}\left(\frac{S_i}{\mathcal{N}B}\right) \\ &= \frac{\widetilde{m^2(t)}}{A^2 + \widetilde{m^2(t)}}\gamma \end{aligned} \tag{4.42d}$$

Thus,

$$\frac{S_o}{N_o} < \gamma$$

For a tone modulation, for example under the best possible conditions ($\mu = 1$),

---

* We need to use quadrature multiplexing in DSB-SC in order to utilize its bandwidth capabilities fully.

$m(t) = A \cos \omega_m t$ and $\widetilde{m^2(t)} = A^2/2$. This gives

$$\frac{S_o}{N_o} = \frac{\gamma}{3}$$

The Large-Noise Case. As seen in Eq. (4.36c), when the interference is large, it is no longer possible to extract the desired signal at the output. The output noise is no longer additive as in the earlier cases but is multiplicative (i.e., the desired signal is multiplied by the noise signal). This is the so-called threshold phenomenon. It can be shown that the onset of the threshold occurs when the noise power is about one-tenth of the message-signal power, that is, $\gamma \simeq 10$ dB (see Example 6.12). Hence, S_o/N_o as a function of γ starts deteriorating rapidly below $\gamma = 10$ (Fig. 4.33).

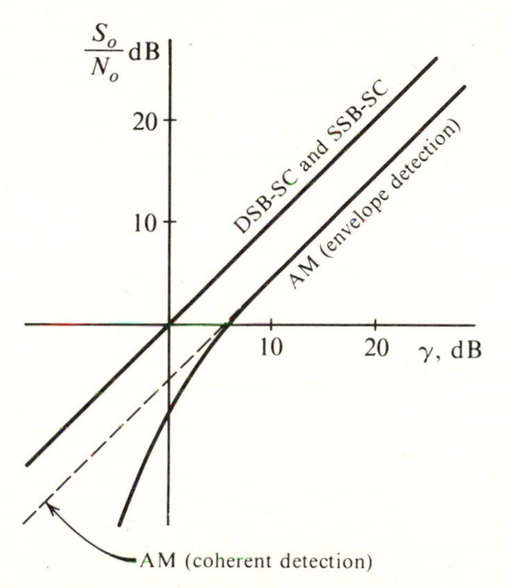

Figure 4.33 Performance of DSB-SC, SSB-SC, and AM systems.

4.9 THE SUPERHETERODYNE AM RECEIVER

The radio receiver used in the AM system is the so-called *superheterodyne* AM receiver (Fig. 4.34). It consists of a radio-frequency (RF) section, a frequency converter (see Example 4.2), an intermediate-frequency (IF) amplifier, an envelope detector, and an audio amplifier.

The RF section is basically a tuneable filter and an amplifier that picks up the desired station by tuning the filter to the right frequency band. The next section, the frequency converter, translates the carrier from ω_c to a fixed IF frequency of 455 kHz. For this purpose, it uses a local oscillator whose frequency f_{Lo} is exactly 455 kHz above the incoming carrier frequency f_c; that is, $f_{Lo} = f_c + f_{IF}(f_{IF} = 455$ kHz). The tuning of the local oscillator and the RF tuneable filter is done by one knob. Tuning capacitors in both circuits are ganged together and are designed so that the tuning frequency of the local oscillator is always 455 kHz above the tuning frequency of the

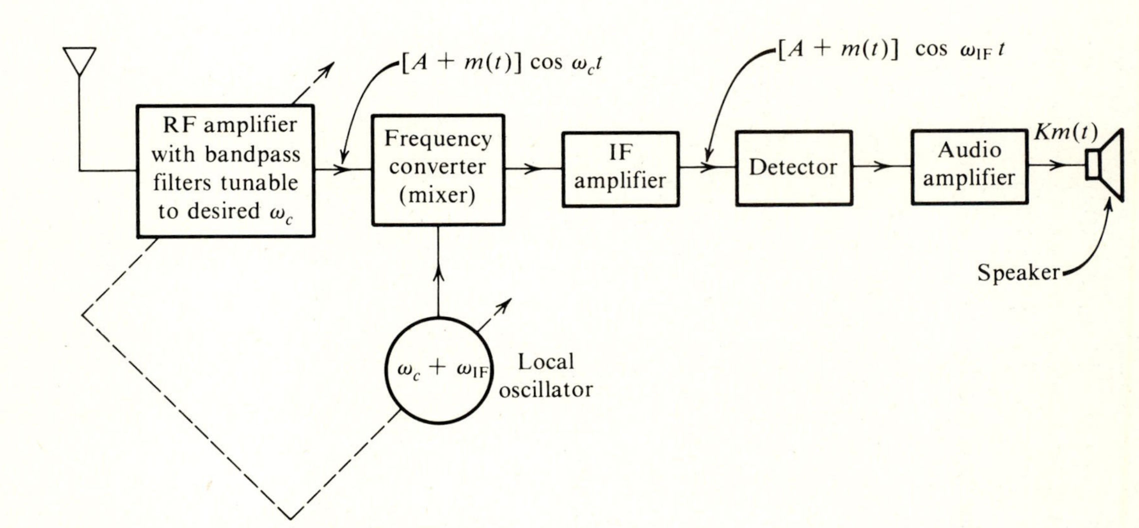

Figure 4.34 A superheterodyne receiver.

RF filter. This means every station that is tuned in is translated to a fixed carrier frequency of 455 kHz by the frequency converter.

The reason for translating all the stations to a fixed carrier of 455 kHz is to obtain adequate selectivity. It is difficult to design sharp bandpass filters of bandwidth 10 kHz (the modulated audio spectrum) if the center frequency f_c is very high. This is particularly true if this filter is tuneable. Hence, the RF filter cannot provide adequate selectivity, and a lot of adjacent channel interference will occur. But when this signal is translated to an IF frequency by a converter, it is further amplified by an IF amplifier (usually a 3-stage amplifier), which does have good selectivity. This is because the IF frequency is reasonably low, and, secondly, its center frequency is fixed. Hence, even though the IF amplifier input contains a lot of adjacent-channel components, because of its high selectivity, the IF section satisfactorily suppresses all interference and amplifies the signal for envelope detection.

In reality, practically all of the selectivity is realized in the IF section; the RF section plays a negligible role. The main function of the RF section is image frequency suppression. As observed in Example 4.2, the mixer, or converter, output consists of components of the difference between the incoming (f_c) and the local-oscillator (f_{Lo}) frequencies (that is, $f_{\text{IF}} = f_{Lo} - f_c$). Now, if the incoming carrier frequency $f_c = 1000$ kHz, then $f_{Lo} = f_c + f_{\text{IF}} = 1000 + 455 = 1455$ kHz. But another carrier, with $f_c' = 1455 + 455 = 1910$ kHz, will also be picked up because the difference $f_c' - f_{Lo}$ is also 455 kHz. The station at 1910 kHz is said to be the *image* of the station at 1000 kHz. Stations that are $2f_{\text{IF}} = 910$ kHz apart are called *image stations* and would both appear simultaneously at the IF output if it were not for the RF filter at receiver input. The RF filter may provide poor selectivity against adjacent stations separated by 10 kHz, but it can provide reasonable selectivity against a station separated from another by 910 kHz.

The receiver (Fig. 4.34) converts the carrier frequency to IF by using a local oscillator of frequency f_{Lo} higher than the incoming carrier frequency (up-conversion) and, hence, is called a superheterodyne receiver. The principle of superheterodyning, first introduced by E. H. Armstrong, is used in AM and FM as well as television receivers. The reason for up-conversion rather than down-conversion is that the former leads to a smaller tuning range for the local oscillator than does the latter. The broadcast-band frequencies range from 550 to 1600 kHz. The up-conversion f_{Lo} ranges from 1005 to 2055 kHz, whereas the down-conversion range of f_{Lo} would be 95 to 1145 kHz. It is much easier to design an oscillator that is tuneable over a smaller frequency ratio.

The importance of the superheterodyne principle cannot be overstressed in radio broadcasting. In the early days (before 1919), the entire selectivity against adjacent stations was realized in the RF filter. Because this filter has poor selectivity, it was necessary to have several stages (several resonant circuits) in cascade for adequate selectivity. Each filter was individually tuned in the earliest receivers. It was very time-consuming and cumbersome to tune in a station by bringing all resonant circuits into synchronism. This was improved upon by ganging variable capacitors together by mounting them on the same shaft rotated by one knob. But variable capacitors are bulky, and there is a limit to the number than can be ganged together. This limited the selectivity available from receivers. Consequently, adjacent carrier frequencies had to be separated widely, resulting in fewer frequency bands. It was the superheterodyne receiver that made it possible to accommodate many more radio stations.

4.10 TELEVISION

In television, the central problem is transmission of visual images by electrical signals. The image, or picture, can be thought of as a frame subdivided into several small squares, known as picture elements. A large number of picture elements in a given image means cleaner reproduction (better resolution) at the receiver (see Fig. 4.35).

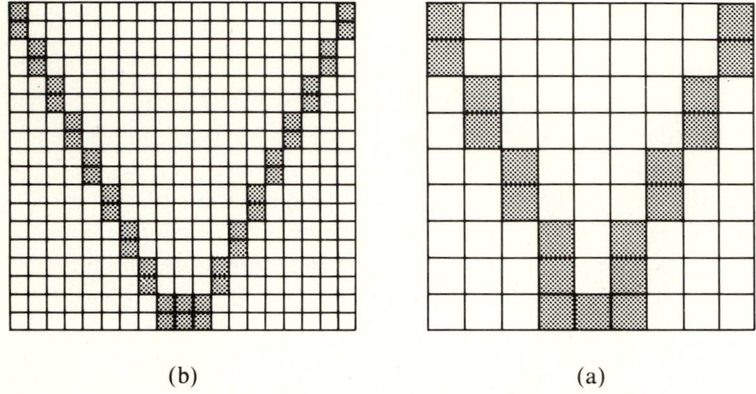

Figure 4.35 Effect of the number of picture elements on resolution.

The information of the entire picture is transmitted by transmitting an electrical signal proportional to the brightness level of the picture elements taken in a certain sequence. We start from the upper left-hand corner with element number 1 and scan the first row of elements (Fig. 4.36); then we come back to the start of the second row, scan the

Figure 4.36 Scanning pattern (raster).

second row, and continue this way until we finish the last row. The electrical signal thus generated during the entire scanning interval has the information of the picture.

The image is furnished by the television camera tube. There exist a variety of camera tubes. The image orthicon is one example. In this tube, the optical system generates a focused image on a photo cathode, which eventually produces an electrically *charged image* on another surface, known as the *target mosaic*. What this means is that every point on the target-mosaic surface acquires a positive electric charge proportional to the brightness of the image. Thus, instead of a light image, we have a charge image. An electron gun now scans the target-mosaic surface with an electron beam in the manner shown in Fig. 4.36. The beam is controlled by a set of voltages across horizontal and vertical deflection plates. Periodic saw-tooth signals (Fig. 4.37) are applied to these plates. The beam scans the horizontal line 1–2 in 53.5

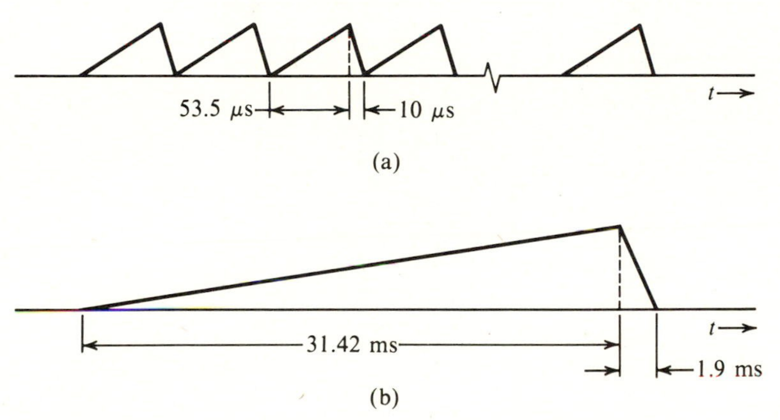

Figure 4.37 (a) Horizontal deflection signal. (b) Vertical deflection signal.

μs and quickly flies back in 10 μs to the left to point 3 and scans line 3–4, and so on. On the target mosaic, where there is a high positive charge (corresponding to a higher brightness level), more electrons from the beam will be absorbed, and the return beam will have fewer electrons, giving a smaller current. Areas corresponding to darker elements (less positive charge) will return a large current. The scanning lines are not perfectly horizontal but have a small downward slope, because during the horizontal deflection the beam is also continuously deflected downwards because of a slower vertical deflection signal (Fig. 4.37*b*). When all the horizontal lines are scanned in 31.42 msec, the vertical deflection signal goes to zero, which means the beam goes back to point 1 again and is ready to start the next frame.

Scanning is continuous at a rate of 60 picture frames per second. The electrical signal thus generated is a video signal corresponding to the visual image. This signal with some modifications (to be discussed later) VSB modulates the video carrier of frequency f_c (see Fig. 4.29). This carrier is transmitted along with the frequency-modulated audio carrier of frequency f_α that is 4.5 MHz higher than the video carrier frequency f_c, that is, $f_\alpha = f_c + 4.5$ MHz.

The receiver is similar to an oscilloscope. An electron gun with horizontal and vertical deflection plates generates an electron beam that scans the screen exactly in the same pattern and in synchronism with the scanning at the transmitter. When the electron beam flies back horizontally after completing each horizontal line, it will leave an unwanted flyback trace on the screen. To avoid this, a blanking pulse (known as the horizontal blanking pulse) is added during the flyback interval (which occurs at the end of each horizontal sweep). Similarly, a vertical blanking pulse is added at the end of each vertical sweep to eliminate the unwanted vertical retrace. These blanking pulses are added at the transmitter itself. We also need to add scan-synchronization information at the transmitter. This is done by adding a large pulse to each blanking pulse. A typical video signal is shown in Fig. 4.38. It is evident that

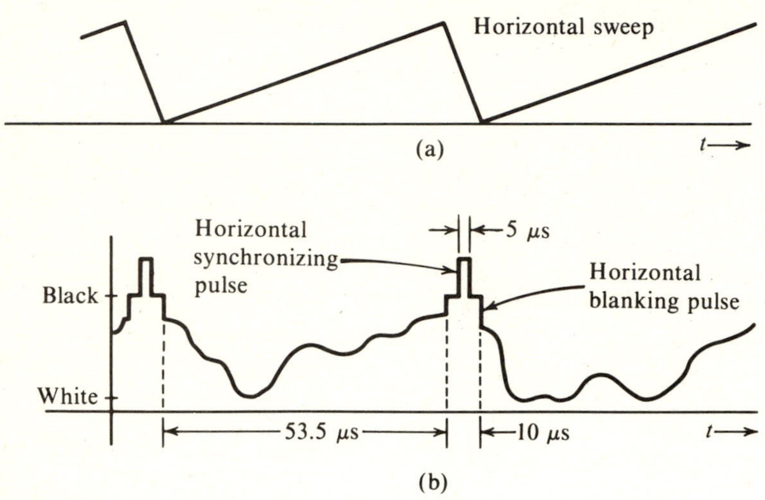

Figure 4.38 TV video signal.

over the entire flyback interval, the blanking pulse (at black level) will eliminate the trace. Similarly, vertical blanking and synchronizing pulses (which are much wider than the corresponding horizontal pulses) are added to the video signal at the end of each vertical sweep. The video signal is now VSB + C modulated (see Fig. 4.28) and transmitted along with the frequency-modulated audio signal. The transmitter block diagram is shown in Fig. 4.39*a*. The receiver block diagram is shown in Fig. 4.39*b*. This is a superheterodyne receiver and is used for the same reasons we use superheterodyne radio receivers (see Fig. 4.34). The converter shifts the entire spectrum (video as well as frequency-modulated audio) to the IF frequency. This signal is now amplified and envelope detected. The audio signal is still of frequency-modulated form with a carrier of 4.5 MHz. It is separated and demodulated. The video signal is amplified. Synchronizing pulses are separated and applied to the vertical and horizontal sweep generators. The video signal is clamped to the blanking pulses (dc restoration) and then applied to the picture tube.

Bandwidth Considerations

The number of horizontal lines used in the United States is 495 per frame. The time required for vertical retrace at the end of the scan is equivalent to that required for 30 horizontal lines. Hence, each frame is considered to have a total of 525 lines,* out of

* In Europe, a total of 625 horizontal lines are used.

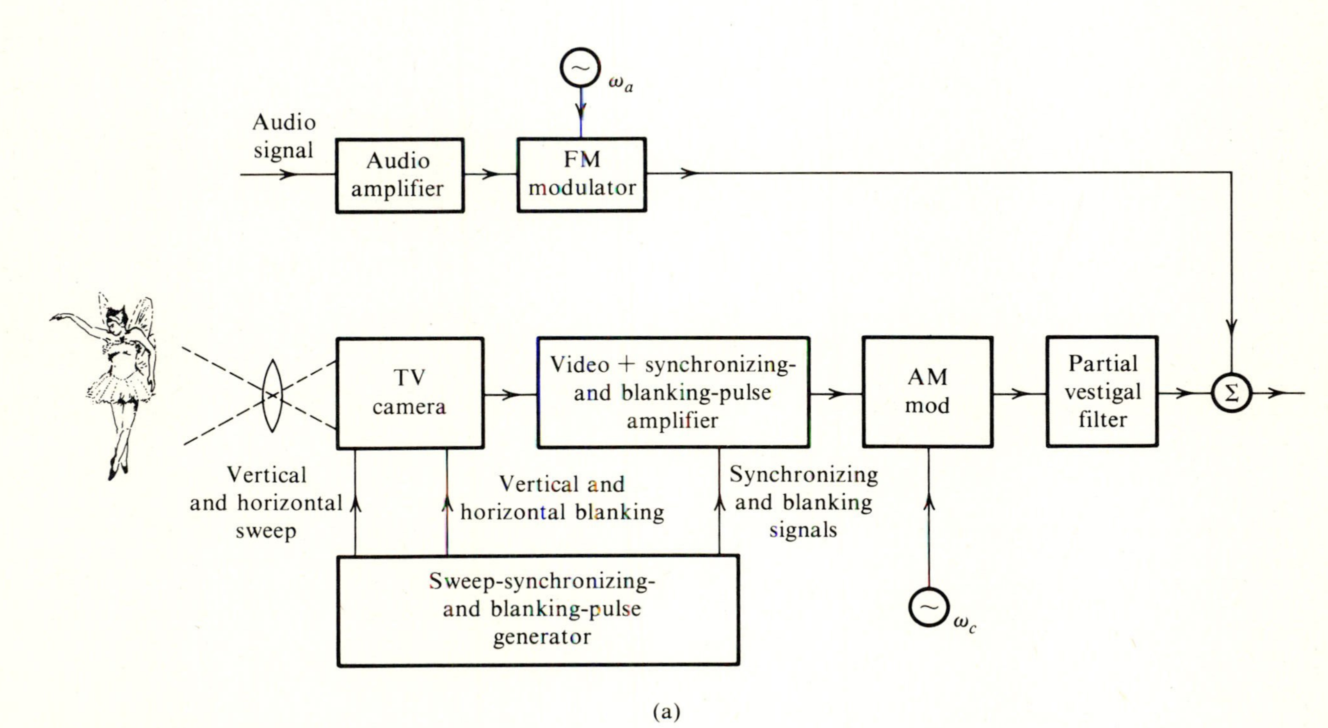

Figure 4.39 (a) TV transmitter.

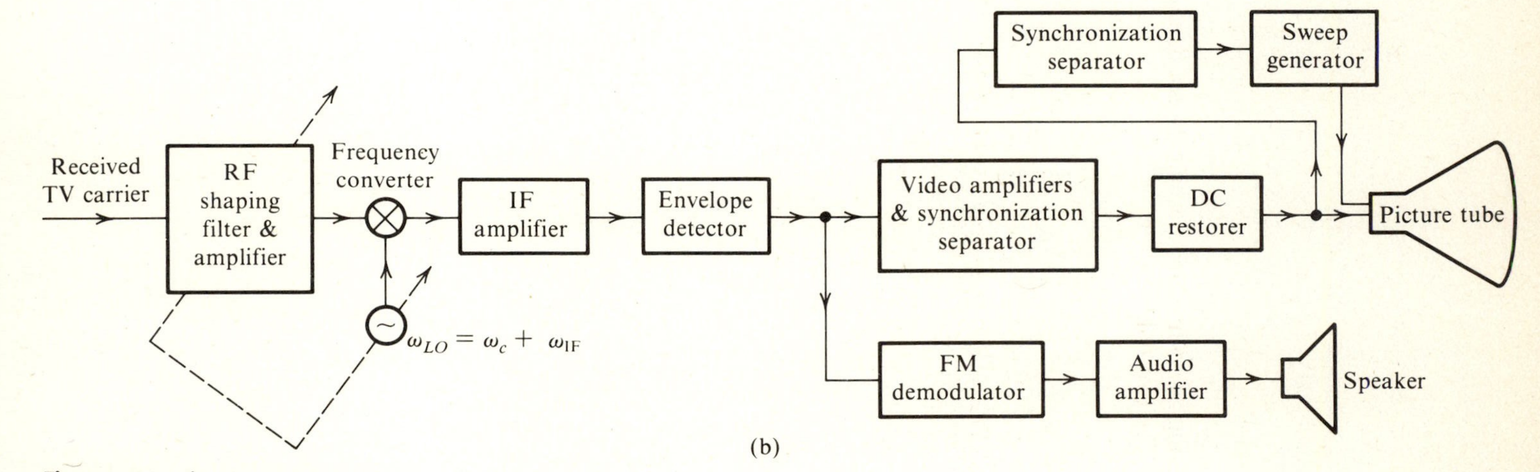

Figure 4.39 (b) TV receiver.

which only 495 are active. Images must be transmitted in a rapid succession of frames in order to create the illusion of continuity and avoid the flicker and jerky motion seen in old Charlie Chaplin movies. Because of the retinal property of retaining an image for a brief period even after the object is removed, it is necessary to transmit about 40 images, or frames, per second. In television we transmit only 30 frames/second in order to conserve bandwidth. To eliminate the flicker effect caused by the low frame rate, scanning the 495 lines is done in two successive patterns. In the first scanning pattern (called the first field), the entire image is scanned using only 247.5 lines (solid lines shown in Fig. 4.36). In the second scanning pattern (or second field), the image is scanned again by using 247.5 lines interlaced between lines of the first field (shown dotted in Fig. 4.36). The two fields together constitute a complete image, or frame. Thus, in reality there are only 30 complete frames per second, and a total equivalent of $525 \times 525 \times 30 = 8.26 \times 10^6$ picture elements per second.* We can estimate the transmission bandwidth of a video signal by observing that transmitting a video signal amounts to transmitting 8.26×10^6 pieces of information (or pulses) per second. Hence, the theoretical bandwidth required is half this, namely, 4.13 MHz (see Sec. 2.7).

The Video Spectrum

To begin with, consider a simple case of transmission of a still image. The scanning procedure discussed earlier is equivalent to scanning an array of the same image repeating itself in both dimensions, as shown in Fig. 4.40*a*. The brightness level b for this figure is a function of x (horizontal) and y (vertical) and can be expressed as $b(x, y)$. Because the picture repeats in the x as well as the y dimension, $b(x, y)$ is a periodic function of x as well as y, with periods of α and β, respectively. Hence, $b(x, y)$ can be represented by a two-dimensional Fourier series with fundamental frequencies $2\pi/\alpha$ and $2\pi/\beta$, respectively.

$$b(x, y) = \sum_{m=-\infty}^{\infty} \sum_{n=-\infty}^{\infty} B_{mn} \exp\left[j2\pi\left(\frac{mx}{\alpha} + \frac{ny}{\beta}\right)\right] \tag{4.43a}$$

If the scanning beam moves with a velocity v_x and v_y in the x and y direction, respectively, then $x = v_x t$ and $y = v_y t$, and the video signal $e(t)$ is

$$e(t) = \sum_{m=-\infty}^{\infty} \sum_{n=-\infty}^{\infty} B_{mn} \exp\left[j2\pi\left(m\frac{v_x}{\alpha}t + n\frac{v_y}{\beta}t\right)\right] \tag{4.43b}$$

But α/v_x is the time required to scan one horizontal line, and β/v_y is the time required to scan the complete image.

$$\frac{\alpha}{v_x} = \frac{1}{(30)(525)} \quad \text{and} \quad \frac{\beta}{v_y} = \frac{1}{30}$$

---

* Actually, the ratio of the image width to the image height (aspect ratio) is 4/3. Hence, the number of picture elements will increase by a factor 4/3. But this factor is almost cancelled out because the scanning pattern does not align perfectly with the checkerboard pattern in Fig. 4.35, thus reducing the resolution by a factor of 0.70 (the Kerr factor).

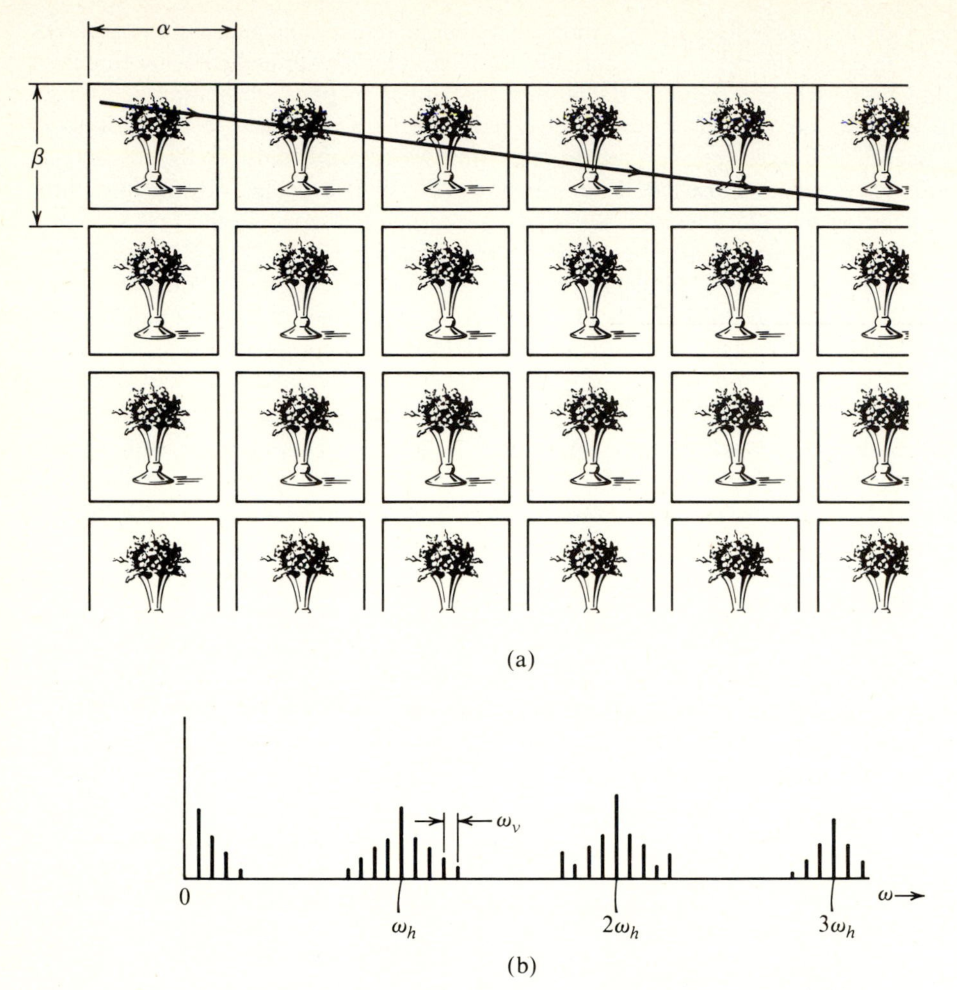

Figure 4.40 (a) Model for scanning process using doubly periodic image fields. (b) Spectrum of the monochrome video signal.

and

$$e(t) = \sum_{m=-\infty}^{\infty} \sum_{n=-\infty}^{\infty} B_{mn} \exp[j2\pi(15{,}750m + 30n)t] \tag{4.43c}$$

The video signal is periodic with fundamentals f_h = 15.75 kHz (horizontal-sweep frequency) and f_v = 30 Hz. The harmonics are spaced at 15.75 kHz intervals, and around each harmonic is clustered a satellite of harmonics 30 Hz apart, as shown in Fig. 4.40*b*.

This spectrum was derived for still-picture transmission. When motion or change occurs from frame to frame, $b(x, y)$ will not be periodic, and the spectrum will not be a line spectrum but will have spreading or smearing. But empty spaces still exist between harmonics of f_h (15.75 kHz). We take advantage of these gaps to transmit the additional information of a color TV signal over the same bandwidth.

The FCC allows a 6-MHz bandwidth for television broadcasting, with the frequency allocations as shown below.

Channel Number	Frequency Band, MHz
VHF 2, 3, 4	54–72
VHF 5, 6	76–88
VHF 7–13	174–216
VHF 14–83	470–890

Compatible Color Television (CCTV)

All colors can be synthesized by mixing the three primary colors—blue, yellow, and red—in the right amounts. In television, blue, green (the combination of blue and yellow), and red are used instead, for the practical reason of the availability of phosphers that glow with these colors when excited by an electron beam.

In color TV cameras, the optical system resolves the image into three primary colors (red, green, and blue) images. A set of three camera tubes produces three video signals $m_r(t)$, $m_g(t)$, and $m_b(t)$ from these images. We could transmit the three video signals and synthesize the color image at the receiver from the three signals. This causes two difficulties, however. It requires three times as much bandwidth as that of monochrome (black-and-white) television, and, secondly, it is not compatible with the existing monochrome system because a monochrome television will receive only one of the primary colors.

These problems are solved by using signal matrixing. The information about $m_r(t)$, $m_g(t)$, and $m_b(t)$ can be transmitted by three signals, each of which is a linear combination of $m_r(t)$, $m_g(t)$, and $m_b(t)$, provided the three combinations are linearly independent. Thus, we can transmit the signals $m_Y(t)$, $m_I(t)$, and $M_Q(t)$ given by

$$m_Y(t) = 0.30m_r(t) + 0.59m_g(t) + 0.11m_b(t)$$

$$M_I(t) = 0.60m_r(t) - 0.28m_g(t) - 0.32m_b(t)$$

$$m_Q(t) = 0.21m_r(t) - 0.52m_g(t) + 0.31m_b(t)$$

Signals $m_r(t)$, $m_g(t)$, and $m_b(t)$ are normalized to a maximum value of 1 so that each of these signal's amplitudes lies in the range of 0 to 1. Hence, $m_Y(t)$ is always positive, whereas $m_I(t)$ and $m_Q(t)$ are bipolar. The signal $m_Y(t)$ is known as the *luminance* signal because it has been found that this particular combination of the three primary-color signals closely matches the luminance of the conventional monochrome video signal. Hence, a black and white set need use only this signal for its operation.

The signals $m_I(t)$ and $m_Q(t)$ are known as the *chrominance* signals.* We could

---

* These signals have an interesting interpretation in terms of the *hue* and *saturation* of colors. Hue refers to the color, such as red, yellow, green, blue, or any color in between. Saturation, or color intensity, refers to the purity of the color. For example, a deep red has 100 percent saturation, but pink—which is a dilution of red with white—will have a lesser amount of saturation. Saturation is given by $\sqrt{m_I^2(t) + m_Q^2(t)}$, and hue is given by an angle $\tan^{-1}[m_Q(t)/m_I(t)]$. Each color has a certain hue, or angle. For example, red, blue, and green are at angles 19°, 136°, and 242°, respectively.

have chosen some other combinations instead of $m_I(t)$ and $m_Q(t)$. But these particular combinations are chosen because they efficiently use certain features of human color vision,[6] as explained below.

Multiplexing Luminance and Chrominance Signals. The luminance signal (the Y signal) is transmitted as a monochrome video signal occupying a bandwidth of 4.2 MHz. The chrominance signals (I and Q signals) also have the same bandwidth (viz, 4.2 MHz each). Subjective tests have shown, however, that the human eye is not perceptive to changes in chrominance (hue and saturation) over smaller areas. This means we can cut out high-frequency components without affecting the quality of the picture, because the eye would not have perceived them anyway. This enables us to limit the bandwidths of the I and Q signals to 1.6 and 0.6 MHz, respectively. The Q signal and the 0 to 0.6 MHz portion of the I signal are sent by QAM, whereas $m_{IH}(t)$, the 0.6 to 1.6 MHz portion of the I signal is sent by LSB (see Figs. 4.41 and 4.42). The subcarrier has frequency* $f_{cc} = 3.583125$ MHz. Thus, the modulated Q signal is $x_Q(t) = m_Q \sin \omega_{cc}t$. The modulated I signal is

$$x_I(t) = \underbrace{[m_I(t) - m_{IH}(t)] \cos \omega_{cc}t}_{\text{DSB(QAM)}} + \underbrace{m_{IH}(t) \cos \omega_{cc}t + m_{IH_h}(t) \sin \omega_{cc}t}_{\text{LSB}}$$

$$= m_I(t) \cos \omega_{cc}t + m_{IH_h}(t) \sin \omega_{cc}t$$

The composite multiplexed signal $m_v(t)$ is

$$m_v(t) = m_Y(t) + m_Q \sin \omega_{cc}t + m_I(t) \cos \omega_{cc}t + m_{IH_h}(t) \sin \omega_{cc}t$$

In addition, a sample of the subcarrier (color burst) is added to the above multiplexed signal for frequency and phase synchronization of the locally generated subcarrier at the receiver. The color burst is added on the trailing edge of the horizontal blanking pulse. The composite video signal $m_v(t)$ is now sent by VSB + C, as discussed in Sec. 4.6.

The Receiver. Because the CCTV system is compatible with monochrome receivers, let us see what happens if we apply the signal $m_v(t)$ to a monochrome picture tube. It may seem necessary to remove the chrominance signals from $m_v(t)$ before applying it to the picture tube. Fortunately, this is not necessary, because the interference of the chrominance signals with the luminance signal, although present on the screen, is practically invisible to the human eye. This happens because of the way chrominance

* $f_{cc} = 227.5 f_h = 227.5 \times 15.75$ kHz $= 3.583125$ MHz. Thus, f_{cc} lies midway between $227 f_h$ and $228 f_h$. This causes the chrominance signals' spectra to be shifted to gaps midway between harmonics of f_h (Fig. 4.42d). In practice, f_{cc} is made slightly smaller than $227.5 f_h$($f_{cc} = 3.579545$ MHz) to avoid an objectionable beat frequency with the audio carrier,[6] which lies 4.5 MHz above the picture carrier. Because $f_h = f_{cc}/227.5$, $f_h = 15.7326$ kHz, and the field repetition frequency is actually 59.94 rather than 60.

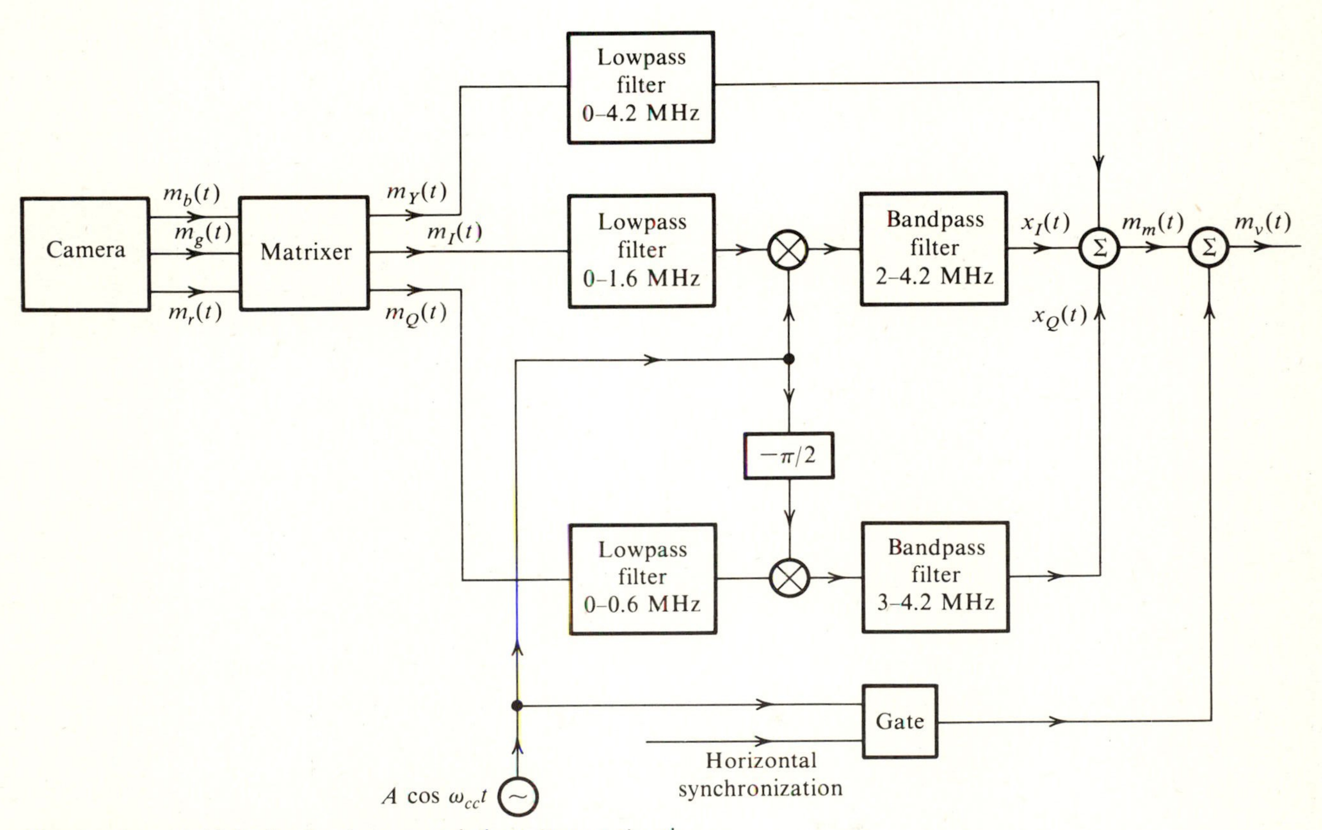

Figure 4.41 Multiplexing luminance and chrominance signals.

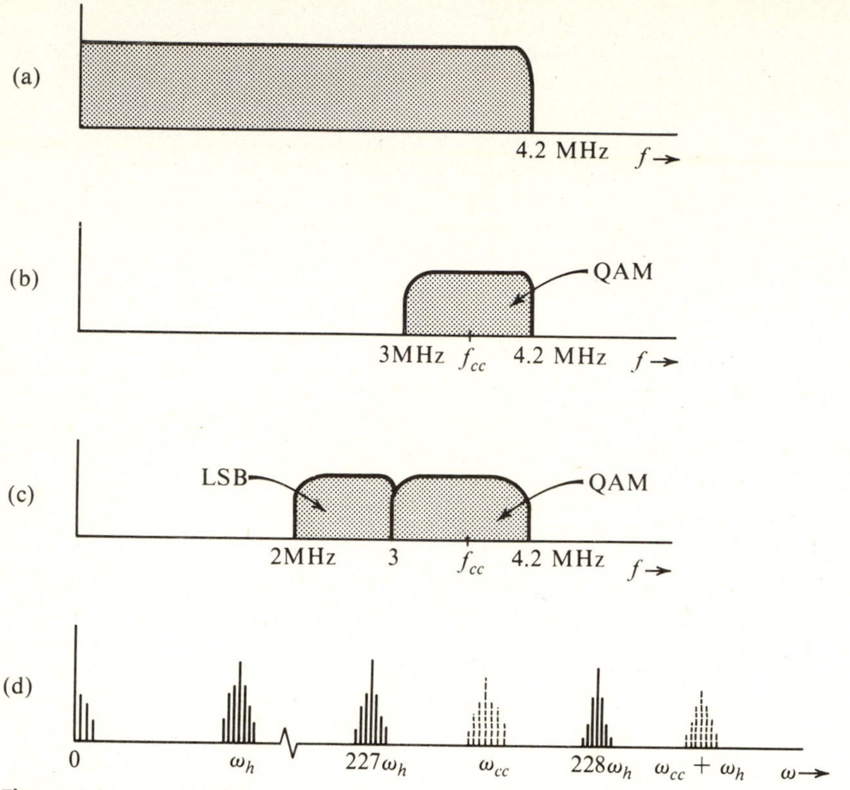

Figure 4.42 (a) Band occupied by $m_Y(t)$. (b) Band occupied by $m_Q(t)$. (c) Band occupied by $m_I(t)$. (d) Interleaving of the chrominance and luminance signal spectra.

signals are interlaced in the frequency domain and because of the persistence of human vision that tends to average out brightness over time as well as space.

The chrominance signal is superimposed on the luminance signal. Figure 4.43 shows the nature of the chrominance signals. Recall that $\omega_{cc} = 227.5\ \omega_h$. During one horizontal line, then, there will be 227.5 chrominance signal cycles. Hence, the chrominance signal continuously changes from positive to negative and vice versa in the horizontal direction. In addition, because there are 227.5 cycles in one line, if a chrominance signal begins with a positive cycle in the beginning of a line, it will end with a positive cycle at the end of the line. The next horizontal line will begin with a negative cycle of the chrominance signal (Fig. 4.43). Hence, in any given frame, the chrominance signal not only reverses its phase along the horizontal (x) direction but also reverses its phase along the vertical (y) direction (on the next horizontal line). But this is not all. During one field, the chrominance signal completes (227.5)(525) cycles, and it returns to a given spot during the next field with opposite polarity. Hence, the chrominance signals reverse phase spatially (in the vertical and horizontal directions) as well as temporally at any given spot. Because the human eye is not sensitive to rapid time variations or rapid space variations, it can notice only space and

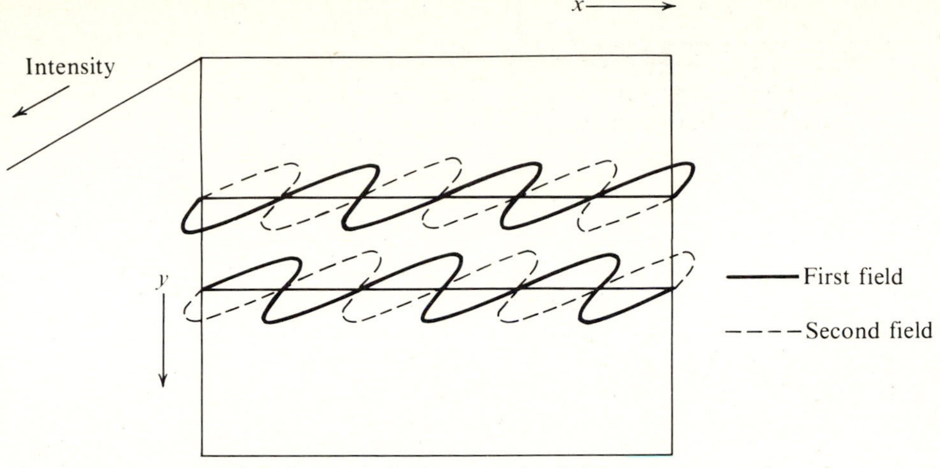

Figure 4.43 Temporal and spatial phase reversals of chrominance signals.

time averages. This makes the chrominance signals practically invisible to the human eye. Thus the color signal is compatible with an unmodified monochrome receiver.

Demultiplexing. In a color receiver the received signal is demodulated exactly as in the monochrome case. This yields $m_v(t)$. This signal must now be demultiplexed to separate $m_Y(t)$, $m_I(t)$, and $m_Q(t)$. The demultiplexing scheme is shown in Fig. 4.44. The output of the 4.2-MHz filter contains $m_Y(t)$, as well as modulated $m_I(t)$ and $m_Q(t)$ [Fig. 4.42*b* and *c*]. Because of the frequency interlacing discussed earlier, however, these signals are practically invisible. Hence, the output of the 4.2-MHz filter serves the function of $m_Y(t)$. Next we demodulate $m_v(t)$ using carriers in phase quadrature. To determine the various signals in Fig. 4.44, we observe that the signal $z(t)$ in Fig. 4.44 consists of modulated $m_I(t)$ and $m_Q(t)$, plus the part of $m_Y(t)$ in the band 2 to 4.2 MHz. Let us denote this high-frequency component of $m_Y(t)$ by $m_{YH}(t)$. Then

$$z(t) = m_{YH}(t) + m_Q(t) \sin \omega_{cc}t + m_I(t) \cos \omega_{cc}t + m_{IH_h}(t) \sin \omega_{cc}t$$

Hence,

$$x_1(t) = 2m_{YH}(t) \cos \omega_{cc}t + m_Q(t) \sin 2\omega_{cc}t + m_I(t)[1 + \cos 2\omega_c t] + m_{IH_h}(t) \sin 2\omega_{cc}t$$

The double-frequency terms will be suppressed by the bandpass filter. In addition, the signal $2m_{YH}(t) \cos \omega_{cc}t$ will be invisible because of the frequency-interlacing effect. This is because the spectrum of this signal is the spectrum of $m_Y(t)$ shifted to $\omega_{cc} = 227.5\ \omega_h$, and it will become invisible because of the frequency-interlacing discussed earlier. Hence, the filter output of the 0 to 1.6 MHz filter yields $m_I(t)$. Similarly, the output of the 0 to 0.6 MHz filter* yields $m_Q(t)$. The three signals $m_Y(t)$, $m_I(t)$, and $m_Q(t)$ are then matrixed to obtain $m_r(t)$, $m_g(t)$, and $m_b(t)$.

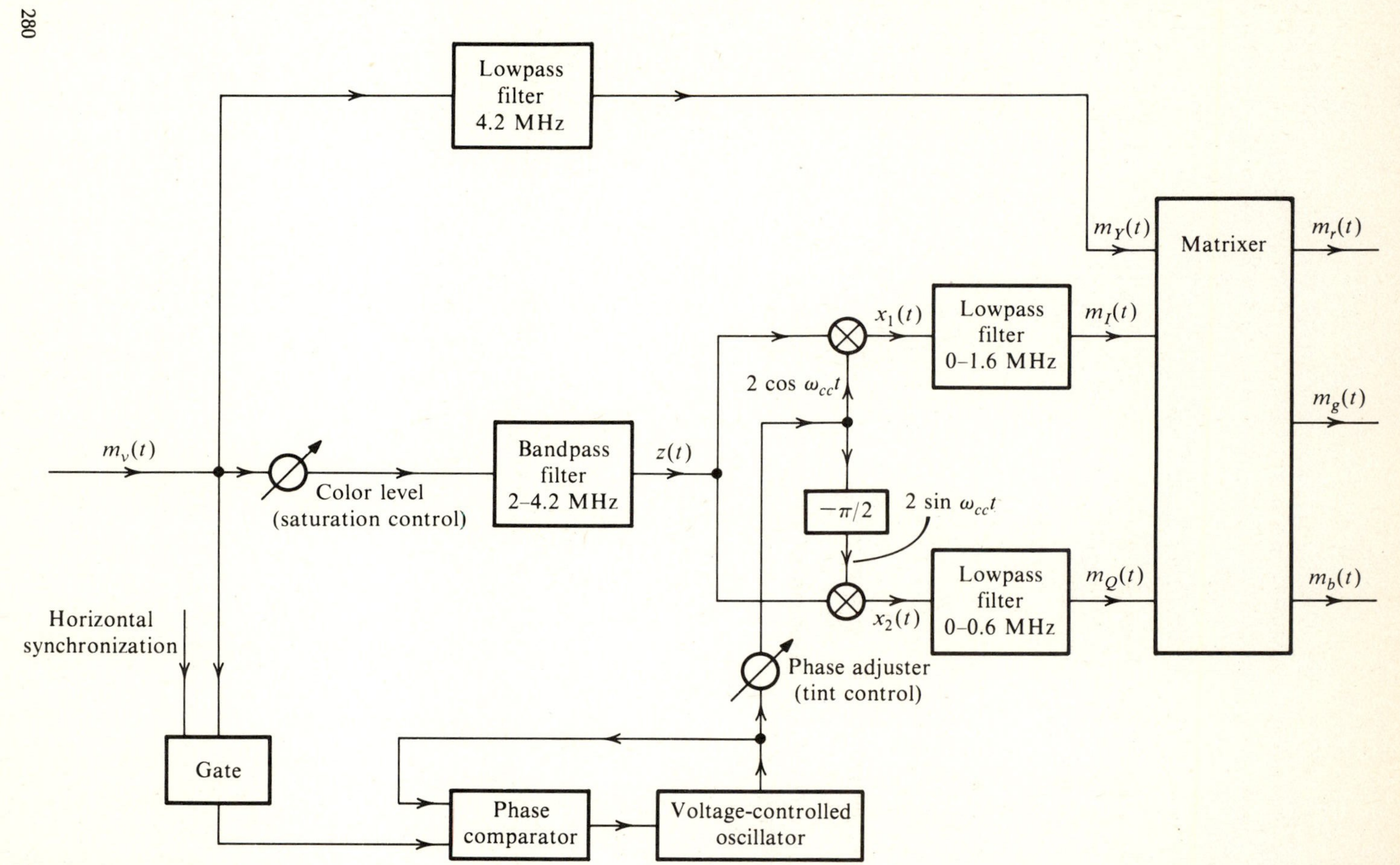

Figure 4.44 Color TV receiver.

For synchronizing the locally generated color carrier ω_{cc}, we separate the color burst (Fig. 4.44) and compare it with the locally generated carrier. The difference is applied to the voltage-controlled oscillator (VCO) that generates the local carrier. The phase of the locally generated carrier is adjustable. This is the so-called tint control.

4.11 FREQUENCY-DIVISION MULTIPLEXING

Signal multiplexing allows transmission of several signals on the same channel. In Chapter 3, time-division multiplexing (TDM), where several signals time share the same channel, was discussed. In frequency-division multiplexing (FDM), several signals share the band of a channel. Each signal is modulated by a different carrier frequency. The various carriers are adequately separated to avoid overlap (or interference) between the spectra of various modulated signals (Fig. 4.45). These carriers are referred to as *subcarriers*. Each signal may use a different kind of modulation (for example, DSB-SC, AM, SSB, VSB, or even FM or PM). The modulated-signal spectra may be separated by a small guard band to avoid interference and facilitate signal separation at the receiver.

When all of the modulated spectra are added, we have a composite signal that may be considered as a baseband signal to further modulate a high-frequency (radio frequency, or RF) carrier for the purpose of transmission.

At the receiver, the incoming signal is first demodulated by the RF carrier to retrieve the composite baseband, which is then bandpass filtered to separate each modulated signal. Then each modulated signal is individually demodulated by an appropriate subcarrier to obtain all the basic baseband signals.

Telephone-Channel Multiplexing

Until recently, almost all long-haul telephone channels were multiplexed by FDM using SSB signals. This multiplexing technique, standardized by the CCITT, provides considerable flexibility in branching, dropping off, or inserting blocks of channels at points en route. The typical arrangement in North American FDM telephone hierarchy is shown[7] in Fig. 4.46*a*. A basic *group* consists of 12 frequency-division multiplexed SSB voice channels, each of bandwidth 4 kHz (first-level multiplexing). A basic group uses LSB spectra and occupies a band 60 to 108 kHz. An alternate group configuration of 12 USB voice signals, occupying a band of 148 to 196 kHz is also used (Fig. 4.46*c*).

A basic *supergroup* of 60 channels is formed by multiplexing five basic groups, and it occupies a band 312 to 552 kHz. An alternate supergroup configuration that uses USB spectra is also shown (Fig. 4.46*d*).

A basic *mastergroup* of 600 channels is formed by multiplexing 10 supergroups.†

*This filter will suppress $m_{IH}(t)$, whose components lie in the range 0.6 to 1.6 MHz.

† This is true for the North American hierarchy. In the CCITT hierarchy, a basic mastergroup is formed by multiplexing five supergroups (300 voice channels).

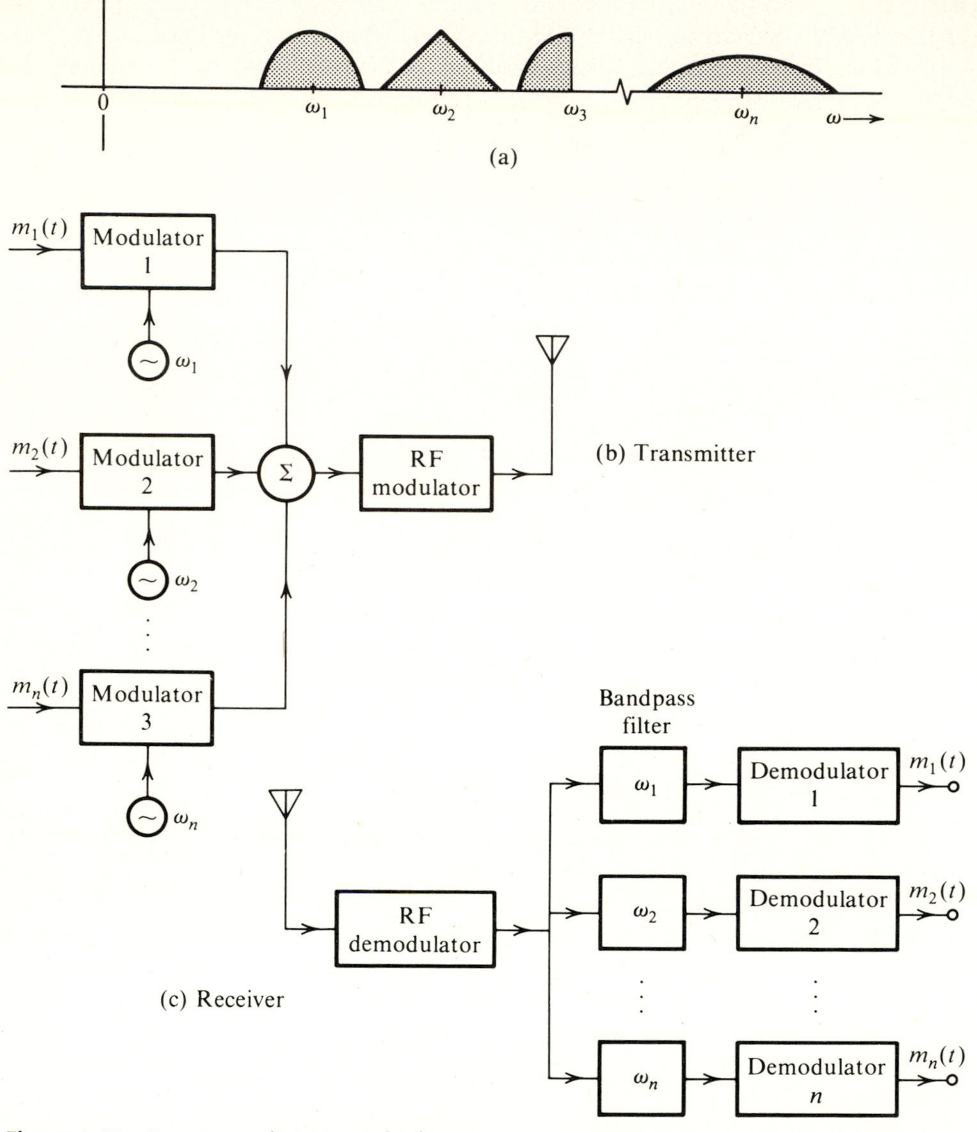

Figure 4.45 Frequency division multiplexing.

There are two standard mastergroup configurations: the L600 and the U600, as shown in Fig. 4.46*e*.

Modern broadband transmission systems can transmit even larger groupings than mastergroups. For the L3 carrier and TH microwave, three mastergroups and one supergroup comprising 1860 message channels are combined. The L4 system utilizes six U600 mastergroups multiplexed to form 3600 channels. The multiplexed signal is

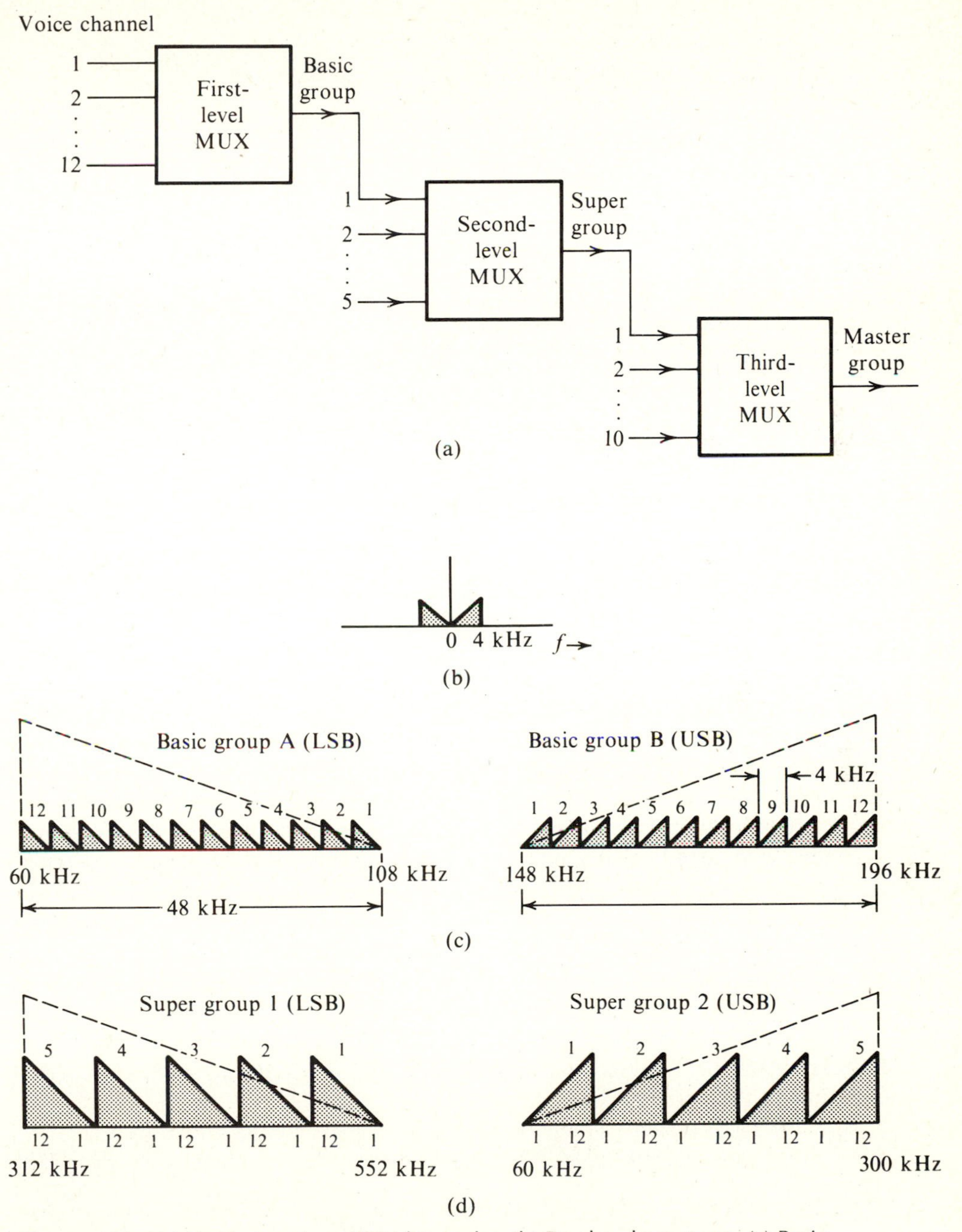

Figure 4.46 (a) North American FDM hierarchy. (b) Baseband spectrum. (c) Basic groups, 12 channels. (d) Supergroups, 60 channels.

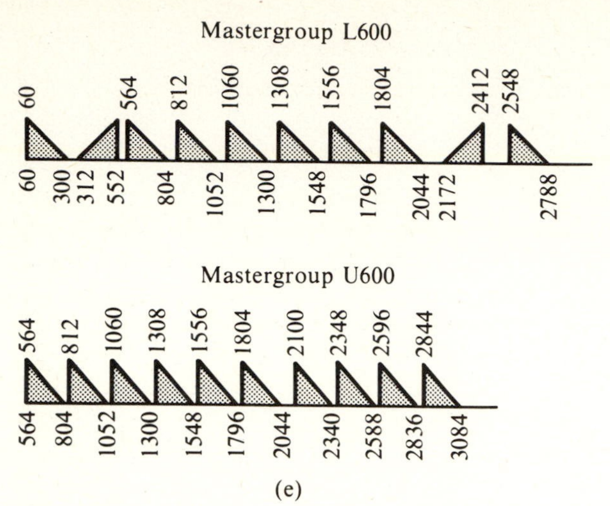

Figure 4.46 (e) Mastergroups, 600 channels.

fed into the baseband input of a microwave radio channel or directly into a coaxial transmission system.

Time-Assignment System Interpolation. Long-distance speech communication uses separate paths (two one-way circuits) for the two directions (send and receive) of transmission. Because on the average, a talker talks half the time and listens half the time, each of the circuits will be used half the time only. Measurements on working transatlantic lines show that speech activity is present only about 40 percent of the time on each circuit. This means each circuit is free about 60 percent of the time, on the average. This fact is used to interpolate additional talkers onto communication facilities using a *time-assignment-system-interpolation (TASI)* concept.[8] TASI systems have mainly been used with submarine-cable voice-channel facilities. New systems may operate with mixed satellite and cable voice channels. TASI is a voice-operated switching and channel-assignment system. A new incoming speech signal is assigned to a channel that is temporarily unused by another talker. The new talker will keep the channel until he is silent and his channel is needed for another talker. During the low-traffic period, the system may require no switching, and the talker may keep the same channel throughout his conversation. During high-traffic periods, however, successive portions of the same talker's speech signal may be assigned and switched to different channels. The assignment and switching is accomplished in milliseconds, so that little of the initial syllable is lost.

Using the TASI system, a large number of customers can be served by a smaller number of channels. For example, the TASI B system uses 96 channels to serve 235 customers.

PART II Angle (Exponential) Modulation

4.12 THE CONCEPT OF THE GENERALIZED ANGLE AND ANGLE MODULATION

In AM signals, the amplitude of a carrier is modulated by a signal $m(t)$, and, hence, the information content of $m(t)$ is in the amplitude variations of the carrier. Because a sinusoidal signal is described by three parameters—amplitude, frequency, and phase—there exists a possibility of carrying the same information by varying either the frequency or the phase of the carrier. By definition, however, a sinusoidal signal has a constant frequency and phase, and, hence, the variation of either of these parameters appears to be contradictory to the definition of a sinusoidal signal. We now have to extend the concept of a sinusoid to a generalized function whose frequency and phase may vary with time.

In frequency modulation, we wish to vary the carrier frequency in proportion to the modulating signal $m(t)$. This means the carrier frequency is changing continuously every instant. Prima facie, this does not make much sense because to define a frequency, we must have a sinusoidal signal at least over one cycle with the same frequency. This problem reminds us of the introduction to the concept of *instantaneous velocity* in our first course in mechanics. We are used to thinking of velocity as being constant over an interval and cannot even imagine that it can vary at each instant. But gradually the idea sinks in. We never forget, however, the wonder and amazement that was caused by the idea when it was first introduced. A similar experience awaits the student with the concept of *instantaneous frequency*.

Let us consider a generalized sinusoidal signal $\varphi(t)$ given by

$$\varphi(t) = A \cos \theta(t) \tag{4.44}$$

where $\theta(t)$ is the *generalized angle* and is a function of t. Figure 4.47 shows a

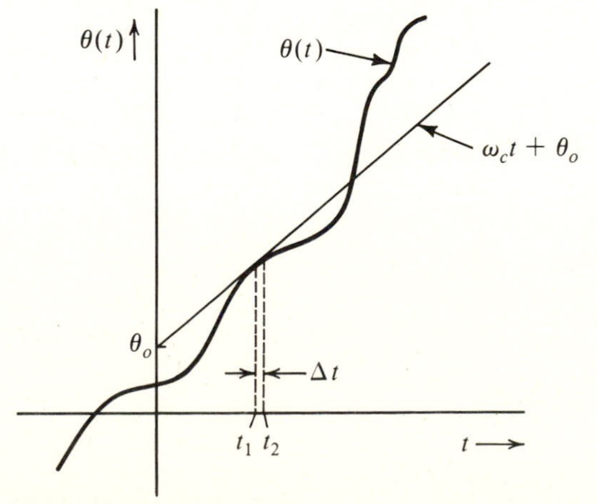

Figure 4.47 Concept of instantaneous frequency.

hypothetical case of $\theta(t)$. For a sinusoid $A \cos(\omega_c t + \theta_o)$, the generalized angle is $\omega_c t + \theta_o$. This is a straight line with a slope ω_c and intercept θ_o, as shown in Fig. 4.47. For the hypothetical case, $\theta(t)$ is tangential to the angle $(\omega_c t + \theta_o)$ over a small interval Δt. The crucial point is that over this small interval, the signal $\varphi(t) = A \cos \theta(t)$ and the sinusoid $A \cos(\omega_c t + \theta_o)$ are identical; that is,

$$\varphi(t) = A \cos(\omega_c t + \theta_o) \qquad t_1 < t < t_2$$

We are certainly justified in saying that over this small interval Δt, the frequency of $\varphi(t)$, is ω_c. Because $(\omega_c t + \theta_o)$ is tangential to $\theta(t)$ over this small interval, the frequency of $\varphi(t)$ is the slope of its angle $\theta(t)$ over this small interval. We can generalize this concept at every instant and say that the instantaneous frequency ω_i at any instant t is the slope of $\theta(t)$ at t. Thus, for $\varphi(t)$ in Eq. (4.44)

$$\omega_i(t) = \frac{d\theta}{dt} \tag{4.45a}$$

$$\theta(t) = \int_{-\infty}^{t} \omega_i(\alpha)\, d\alpha \tag{4.45b}$$

Now we can see the possibility of transmitting the information of $m(t)$ by varying the angle θ of a carrier. Such techniques of modulation, where the angle of the carrier is varied in some manner with a modulating signal $m(t)$, are known as *angle modulation,* or *exponential modulation*. Two simple possibilities are: *phase modulation (PM)* and *frequency modulation (FM)*. In PM, the angle $\theta(t)$ is varied linearly with $m(t)$:

$$\theta(t) = \omega_c t + \theta_o + k_p m(t)$$

where k_p is a constant and ω_c is the carrier frequency. Assuming $\theta_o = 0$ without loss of generality

$$\theta(t) = \omega_c t + k_p m(t) \tag{4.46a}$$

The resulting PM wave is

$$\varphi_{\text{PM}}(t) = A \cos[\omega_c t + k_p m(t)] \tag{4.46b}$$

The instantaneous frequency $\omega_i(t)$ is given by

$$\omega_i(t) = \frac{d\theta}{dt} = \omega_c + k_p \dot{m}(t) \tag{4.46c}$$

Hence in phase modulation, the instantaneous frequency ω_i varies linearly with the derivative of the modulating signal. If the instantaneous frequency ω_i is varied linearly with the modulating signal, we have frequency modulation. Thus in FM, the instantaneous frequency ω_i is

$$\omega_i(t) = \omega_c + k_f m(t) \tag{4.47a}$$

where k_f is a constant. The angle $\theta(t)$ is now

$$\begin{aligned} \theta(t) &= \int_{-\infty}^{t} [\omega_c + k_f m(\alpha)]\, d\alpha \\ &= \omega_c t + k_f \int_{-\infty}^{t} m(\alpha)\, d\alpha \end{aligned} \tag{4.47b}$$

Here we have assumed the constant term in $\theta(t)$ to be zero without loss of generality. The FM wave is

$$\varphi_{\text{FM}}(t) = A \cos \left[\omega_c t + k_f \int_{-\infty}^{t} m(\alpha)\, d\alpha \right] \tag{4.47c}$$

From Eqs. (4.46b) and (4.47c), it is apparent that PM and FM are not only very similar but are inseparable. In fact, by looking at an angle-modulated carrier, there is no way of telling if it is FM or PM, because a PM wave corresponding to $m(t)$ is the FM wave corresponding to $\dot{m}(t)$ (Fig. 4.48*a*) and an FM wave corresponding to $m(t)$ is the PM

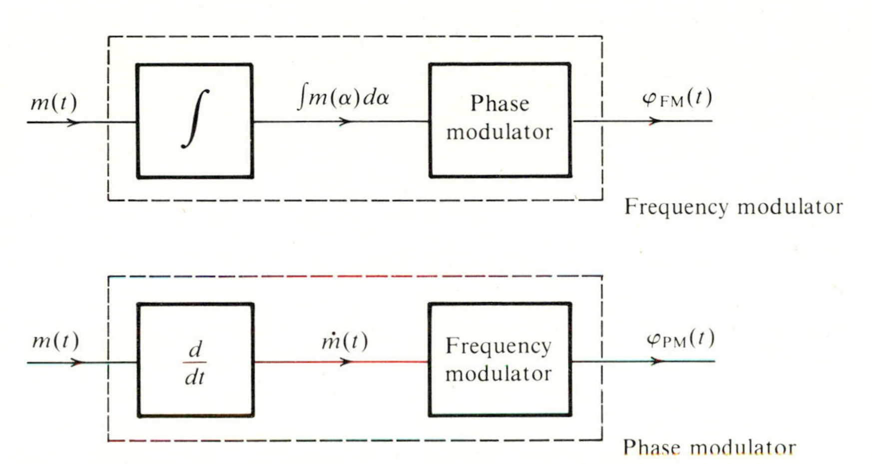

Figure 4.48 Phase and frequency modulation are inseparable.

wave corresponding to $\int_{-\infty}^{t} m(\alpha)\, d\alpha$ (Fig. 4.48*b*). One of the methods of generating FM in practice (the Armstrong indirect-FM system) actually integrates $m(t)$ and uses it to phase-modulate a carrier (see Fig. 4.51).

In reality, FM and PM can be considered as special cases of exponential modulation for which the modulated wave $\varphi_{\text{EM}}(t)$ is

$$\varphi_{\text{EM}}(t) = A \cos \left[\omega_c t + k \int_{-\infty}^{t} m(\alpha)\, h(t - \alpha)\, d\alpha \right] \tag{4.48}$$

where k is a constant and $h(t)$ is the unit impulse response of a linear time-invariant system. If $h(t) = \delta(t)$, we have PM, and if $h(t) = u(t)$, we have FM. There is no reason to restrict ourselves to these two cases only. We shall see later, in Sec. 4.16, that in general, for optimum performance, $h(t)$ is neither $\delta(t)$ (PM) nor $u(t)$ (FM) but is something else, depending on the modulating signal spectrum and the channel characteristics.

■ EXAMPLE 4.11

Sketch FM and PM waves for the modulating signal $m(t)$ shown in Fig. 4.49*a*. The constants k_f and k_p are $2\pi(10^5)$ and 10π, respectively, and the carrier frequency f_c is 100 MHz.

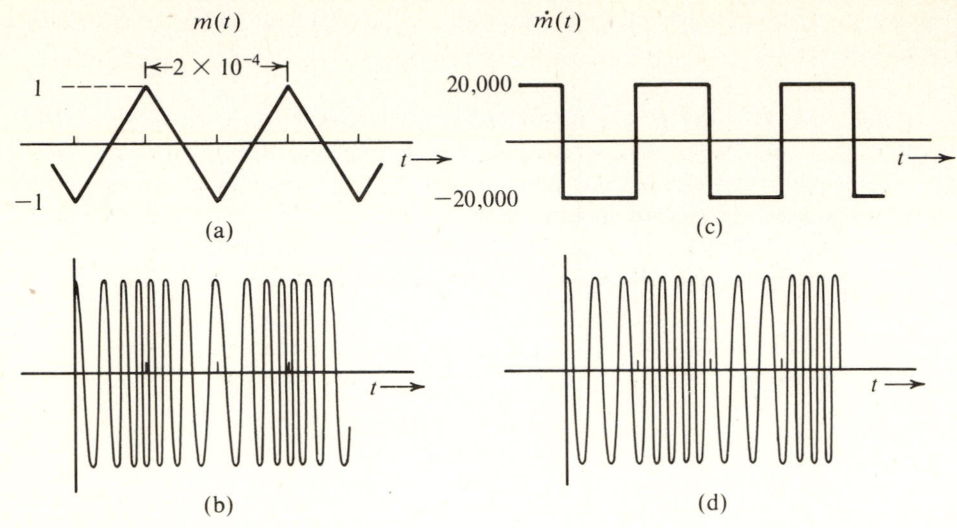

Figure 4.49 FM and PM waveforms.

☐ **Solution:**

For FM

$$\omega_i = \omega_c + k_f m(t)$$

Dividing throughout by 2π, we have the equation in terms of the variable f (frequency). The instantaneous frequency f_i is

$$f_i = f_c + \frac{k_f}{2\pi} m(t)$$

$$= 10^8 + 10^5\, m(t)$$

$$(f_i)_{\min} = 10^8 - 10^5\, [m(t)]_{\min} = 99.9 \text{ MHz}$$

$$(f_i)_{\max} = 10^8 + 10^5\, [m(t)]_{\max} = 100.1 \text{ MHz}$$

Because $m(t)$ increases and decreases linearly with time, the instantaneous frequency increases linearly from 99.9 to 100.1 MHz over a half-cycle and decreases linearly from 100.1 to 99.9 MHz over the remaining half-cycle of the modulating signal (Fig. 4.49*b*).

For PM

PM for $m(t)$ is FM for $\dot{m}(t)$. This also follows from Eq. (4.46c).

$$f_i = f_c + \frac{k_p}{2\pi} \dot{m}(t)$$

$$= 10^8 + 5\, \dot{m}(t)$$

$$(f_i)_{\min} = 10^8 - 5\, [\dot{m}(t)]_{\min} = 10^8 - 10^5 = 99.9 \text{ MHz}$$

$$(f_i)_{\max} = 10^8 + 5\, [\dot{m}(t)]_{\max} = 10^8 + 10^5 = 100.1 \text{ MHz}$$

Because $\dot{m}(t)$ switches back and forth from a value of $-20{,}000$ to $20{,}000$, the carrier frequency switches back and forth from 99.9 to 100.1 MHz every half-cycle of $\dot{m}(t)$, as shown in Fig. 4.49*d*.

This indirect method of sketching PM (using $\dot{m}(t)$ to frequency-modulate a carrier) works as long as $m(t)$ is a continuous signal. If $m(t)$ is discontinuous, $\dot{m}(t)$ contains impulses, and this method fails. In such a case, a direct approach should be used. This is demonstrated in the next example. ■

■ EXAMPLE 4.12

Sketch FM and PM waves for the digital modulating signal $m(t)$ shown in Fig. 4.50*a*. The constants k_f and k_p are $2\pi(10^5)$ and $\pi/2$, respectively, and $f_c = 100$ MHz.

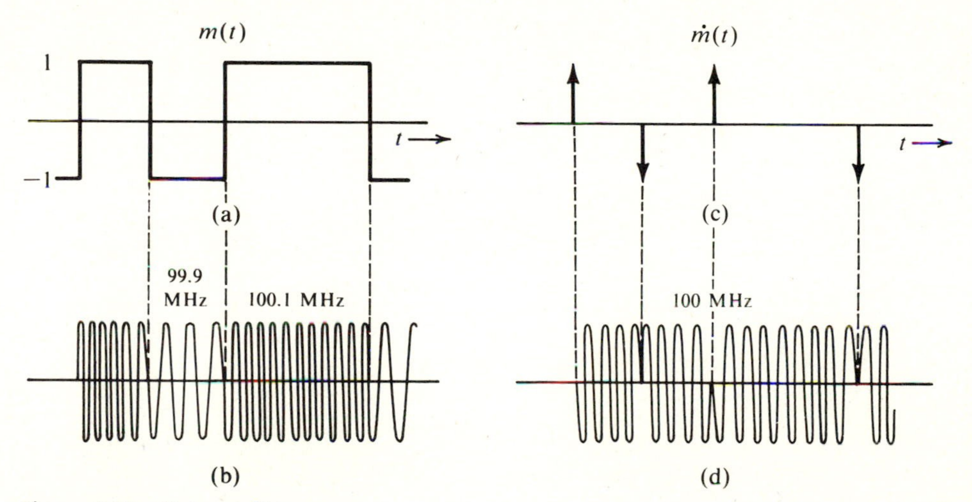

Figure 4.50 FM and PM waves.

□ Solution:

For FM

$$f_i = f_c + \frac{k_f}{2\pi} m(t) = 10^8 + 10^5\, m(t)$$

Because $m(t)$ switches from 1 to -1 and vice versa, the FM wave frequency switches back and forth between 99.9 MHz and 100.1 MHz, as shown in Fig. 4.50*b*. This scheme of carrier frequency modulation by a digital signal (Fig. 4.50*b*) is called *frequency-shift keying (FSK)*, because information digits are transmitted by shifting the carrier frequency (see Sec. 3.8).

For PM

$$f_i = f_c + \frac{k_p}{2\pi} \dot{m}(t) = 10^8 + \frac{1}{4}\, \dot{m}(t)$$

The derivative $\dot{m}(t)$ (Fig. 4.50*c*) contains impulses, and it is not immediately apparent how an instantaneous frequency can be changed by an infinite amount and then changed back to the original frequency in zero time. Let us consider the direct approach

$$\begin{aligned}\varphi_{\text{PM}}(t) &= A \cos\left[\omega_c t + k_p m(t)\right] \\ &= A \cos\left[\omega_c t + \frac{\pi}{2} m(t)\right] \\ &= \begin{cases} A \sin \omega_c t & \text{when } m(t) = -1 \\ -A \sin \omega_c t & \text{when } m(t) = 1 \end{cases}\end{aligned}$$

This scheme of carrier phase modulation by a digital signal (Fig. 4.50*d*) is called *phase-shift keying (PSK),* because information digits are transmitted by shifting the carrier phase. Note that PSK may also be viewed as a DSB-SC modulation by $m(t)$ (see Example 4.1).

The PM wave $\varphi_{\text{PM}}(t)$ in this case has phase discontinuities at instants where impulses of $\dot{m}(t)$ are located. At these instants, the carrier phase shifts by π instantaneously. A finite phase shift in zero time implies infinite instantaneous frequency at these instants. This agrees with our observation about $\dot{m}(t)$.

When $m(t)$ is a digital signal (as in Fig. 4.50*a*), $\varphi_{\text{PM}}(t)$ shows a phase discontinuity where $m(t)$ has a jump discontinuity. We shall now show that in such a case the phase deviation $k_p m(t)$ must be restricted to a range $(-\pi, \pi)$ in order to avoid ambiguity in demodulation. For example, if k_p were $3\pi/2$ in the present example, then

$$\varphi_{\text{PM}}(t) = A \cos\left[\omega_c t + \frac{3\pi}{2} m(t)\right]$$

In this case $\varphi_{\text{PM}}(t) = A \sin \omega_c t$ when $m(t) = 1$ or $-1/3$. This will certainly cause ambiguity at the receiver when $A \sin \omega_c t$ is received. Such ambiguity never arises if $k_p m(t)$ is restricted to the range $(-\pi, \pi)$.

What causes this ambiguity? When $m(t)$ has jump discontinuities, the phase of $\varphi_{\text{PM}}(t)$ changes instantaneously, and it can be demodulated only by detecting the phase. Such a phase demodulator output is directly proportional to the phase of the input wave $\varphi_{\text{PM}}(t)$. Because a phase $\varphi_o + 2n\pi$ is indistinguishable from the phase φ_o, ambiguities will be inherent in the demodulator unless the phase variations are limited to the range $(-\pi, \pi)$. This means k_p should be small enough to restrict the phase change $k_p m(t)$ to the range $(-\pi, \pi)$.

No such restriction on k_p is required if $m(t)$ is continuous. In this case the phase $\varphi_o + 2n\pi$ will exhibit n additional carrier cycles over the case of phase of only φ_o. We can detect the PM wave by using an FM demodulator followed by an integrator (see Prob. 4.55). The additional n cycles will be detected by the FM demodulator, and the subsequent integration will yield a phase $2n\pi$. Hence, the phase φ_o and $\varphi_o + 2n\pi$ can be detected without ambiguity. This conclusion can also be verified from Example 4.11, where the maximum phase change $\Delta\varphi = 10\pi$.

Because a bandlimited signal can have no jump discontinuities, we can say that when $m(t)$ is bandlimited, k_p has no restrictions. ■

The Power of an Angle-Modulated Wave

Although the instantaneous frequency and phase of an angle-modulated wave can vary with time, the amplitude A always remains constant. Hence, the power of an angle-modulated wave (PM or FM) is always $A^2/2$, regardless of the value of k_p or k_f.

4.13 BANDWIDTH OF ANGLE-MODULATED WAVES

We shall first consider the bandwidth of an FM wave and then extend the discussion to PM.

Let us define:

$$a(t) = \int_{-\infty}^{t} m(\alpha)\, d\alpha \tag{4.49}$$

and

$$\hat{\varphi}_{\text{FM}}(t) = A \exp[j(\omega_c t + k_f a(t))] \tag{4.50a}$$

Then

$$\varphi_{\text{FM}}(t) = \text{Re } \hat{\varphi}_{\text{FM}}(t) \tag{4.50b}$$

Now

$$\hat{\varphi}_{\text{FM}}(t) = A\left[1 + jk_f a(t) - \frac{k_f^2}{2!} a^2(t) + \cdots + j^n \frac{k_f^n}{n!} a^n(t) + \cdots\right] e^{j\omega_c t} \tag{4.51a}$$

and

$$\begin{aligned}\varphi_{\text{FM}}(t) &= \text{Re } [\hat{\varphi}_{\text{FM}}(t)] \\ &= A\left[\cos \omega_c t - k_f a(t) \sin \omega_c t - \frac{k_f^2}{2!} a^2(t) \cos \omega_c t \right. \\ &\qquad \left. + \frac{k_f^3}{3!} a^3(t) \sin \omega_c t + \cdots \right]\end{aligned} \tag{4.51b}$$

The modulated wave consists of an unmodulated carrier plus various amplitude-modulated terms, such as $a(t) \sin \omega_c t$, $a^2(t) \cos \omega_c t$, $a^3(t) \sin \omega_c t$, . . . , and so on. The signal $a(t)$ is an integral of $m(t)$. If $M(\omega)$ is bandlimited to B, $A(\omega)$ is also bandlimited* to B. The spectrum of $a^2(t)$ is simply $A(\omega) * A(\omega)/2\pi$ and is band-limited to $2B$. Similarly, the spectrum of $a^n(t)$ is bandlimited to nB. Hence, the spectrum consists of an unmodulated carrier plus spectra of $a(t)$, $a^2(t)$, . . . , $a^n(t)$, . . . , and so on, centered at ω_c. Clearly, the modulated wave is not band-limited. It has an infinite bandwidth and is not related to the modulating-signal spectrum in any simple way, as was the case in AM.

---

*This is because integration is a linear operation equivalent to passing a signal through a transfer function $1/j\omega$ [see Eq. (2.65d)]. Hence, if $M(\omega)$ is bandlimited to B, $A(\omega)$ must also be bandlimited to B.

Although the theoretical bandwidth of an FM wave is infinite, we shall see that most of the modulated-signal power resides in a finite bandwidth. There are two distinct possibilities in terms of bandwidths—the narrowband FM and the wideband FM.

Unlike AM, angle modulation is nonlinear. The principle of superposition does not apply. This may be verified from the fact that

$$A \cos [\omega_c t + k_f a_1(t)] + A \cos [\omega_c t + k_f a_2(t)] \neq A \cos [\omega_c t + k_f(a_1(t) + a_2(t))]$$

To give a simple example, consider the modulating signal

$$m(t) = k_1 \cos \omega_1 t + k_2 \cos \omega_2 t$$

The amplitude modulation will simply shift frequencies ω_1 and ω_2 of the modulating signal to $\omega_c \pm \omega_1$ and $\omega_c \pm \omega_2$. In FM [Eq. (4.51)], however, we have not only the frequencies $\omega_c \pm \omega_1$ and $\omega_c \pm \omega_2$ but also frequencies $\omega_c \pm (n\omega_1 \pm m\omega_2)$, where n and m are integers taking on all possible values. It is clear that FM generates intermodulation frequencies $\omega_c \pm (n\omega_1 \pm m\omega_2)$. The principle of superposition does not hold. If, however, k_f is very small (that is, if $|k_f a(t)| \ll 1$), then all but the first two terms in Eq. (4.51) are negligible, and we have

$$\varphi_{FM}(t) \simeq A[\cos \omega_c t - k_f a(t) \sin \omega_c t] \tag{4.52}$$

This is a linear modulation. This expression is similar to that of the AM wave. Because the bandwidth of $a(t)$ is B, the bandwidth of $\varphi_{FM}(t)$ in Eq. (4.52) is only $2B$. For this reason, the case ($|k_f a(t)| \ll 1$) is called *narrowband FM (NBFM)*. The *narrowband PM (NBPM)* case is similarly given by

$$\varphi_{PM}(t) = A[\cos \omega_c t - k_p m(t) \sin \omega_c t] \tag{4.53}$$

A comparison of NBFM [Eq. (4.52)] with AM [Eq. (4.9a)] brings out clearly the similarities and differences between the two types of modulation. Both cases have a carrier term and sidebands centered at $\pm\ \omega_c$. The modulated-signal bandwidths are identical (viz., $2B$). The sideband spectrum for FM has a phase shift of $\pi/2$ with respect to the carrier, whereas that of AM is in phase with the carrier. It must be remembered, however, that despite apparent similarities, the AM and FM signals have very different waveforms. In an AM signal, the frequency is constant and the amplitude varies with time, whereas in an FM signal the amplitude is constant and the frequency varies with time.

Equations (4.52) and (4.53) suggest a possible method of generating narrowband FM and PM signals by using DSB-SC modulators. The block-diagram representation of such systems is shown in Fig. 4.51.

Wideband FM (WBFM)

If the deviation in the carrier frequency is large enough (i.e., if the constant k_f is chosen large enough so that the condition $|k_f a(t)| \ll 1$ is not satisfied), the analysis of FM signals becomes very involved for a general modulating signal $m(t)$.

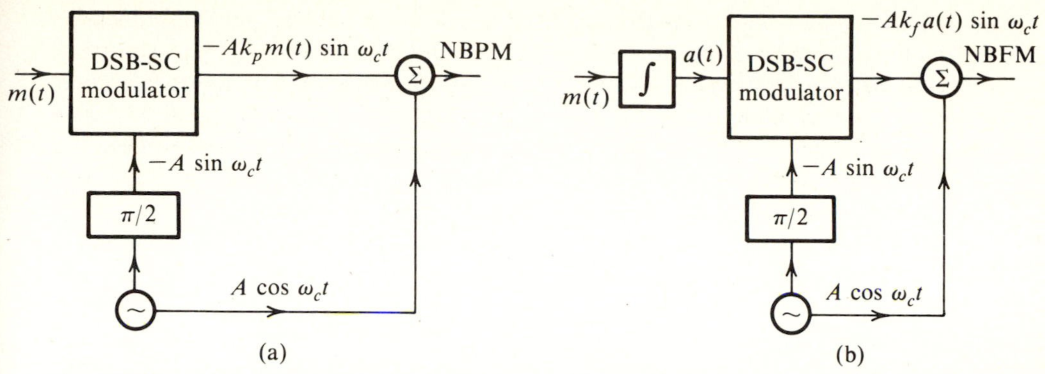

Figure 4.51 Narrowband FM and PM wave generation.

We know from the earlier discussion that the theoretical bandwidth of the FM wave is infinite. It will be seen, however, that most of the power of the FM wave resides in a finite band. In order to estimate this bandwidth, we observe that the instantaneous frequency of the modulated wave is

$$\omega_i = \omega_c + k_f m(t)$$

Thus, the instantaneous frequency varies in the range $\omega_c - k_f m_p$ to $\omega_c + k_f m_p$ if we assume that $|m(t)_{\min}| = m(t)_{\max} = m_p$. Because the instantaneous frequency ω_i varies in the range $\omega_c - k_f m_p$ to $\omega_c + k_f m_p$, we may be justified in assuming that the modulated-wave spectrum lies more or less in this range. If this is the case, the estimated bandwidth B_{FM} of the FM wave is given by

$$2\pi B_{\text{FM}} \simeq 2k_f m_p$$

Note that $k_f m_p$ is the *maximum deviation* of the carrier frequency ω_c. Let us define

$$\Delta\omega = k_f m_p \tag{4.54}$$

or

$$\Delta f = \frac{k_f}{2\pi} m_p \tag{4.55}$$

Hence,

$$B_{\text{FM}} \simeq 2\Delta f \tag{4.56}$$

where Δf is the maximum deviation of the carrier frequency f_c.

We shall soon see that this expression is valid only when $\Delta f \gg B$. In the case $\Delta f < B$, we have a case of NBFM where $B_{\text{FM}} \simeq 2B$ and not $2\Delta f$. This was indeed the fallacy that gave birth to FM in the first place. In the twenties, pioneers in the radio field were trying to find a modulation system that would occupy a narrower bandwidth than that required for AM. Using the above argument, it was concluded that the FM bandwidth was $2\Delta f$, and by choosing Δf to be small, the FM bandwidth could be made as small as one wished!

The fallacy here is equating the instantaneous frequency to the spectral frequency. In the case $\Delta f \gg B$, this works out all right. But when $\Delta f < B$, the spectrum spreading effect, which could be ignored in WBFM because Δf was so large, cannot be ignored. For example, if $m(t)$ is a slowly varying signal, ω_i changes slowly, and over several cycles the carrier frequency appears to be a constant, ω_i. Hence, the carrier spectrum will have a component of frequency ω_i. Actually, this is an approximation of the reality. When we have a sinusoidal signal of frequency ω_i over the entire time interval $(-\infty, \infty)$, it gives rise to a spectral component of frequency ω_i (impulse at ω_i). But when the sinusoidal signal exists only over a finite time interval, the spectrum is spread out. This is exactly what happens in FM. Because of variations in ω_i, the carrier appears to have a constant frequency ω_i only over a finite time interval and then changes to a new value of ω_i. Hence, each interval over which ω_i can be assumed constant gives rise to a spectral component ω_i plus the frequency spread mentioned earlier. We shall soon show that this frequency spread is of the order of B. Strictly speaking, in WBFM, the carrier bandwidth will be $2\Delta f$ plus this spread (of the order of B) on both ends of the spectrum. But because in WBFM $\Delta f \gg B$, we can ignore this spread. In NBFM, on the other hand, $\Delta f < B$, and the spread can no longer be ignored. In fact, this spreading is what determines the bandwidth.

To demonstrate this result, let us approximate the baseband signal $m(t)$ by a staircase signal, as shown in Fig. 4.52*a*. Because the signal $m(t)$ is bandlimited to B, to a first approximation, we can assume the signal to be constant over the Nyquist interval $1/2B$. The FM wave for a staircase-approximated signal will consist of a sequence of sinusoidal pulses, each of constant frequency and a duration of $1/2B$ seconds.

Note that an abrupt change of carrier frequency occurs at every sampling instant. One such pulse is shown in Fig. 4.52*b*. The spectrum of each pulse can be obtained by using pair 13 from Table 2.1 and the modulation property. The spectrum of a typical pulse in Fig. 4.52*b* is shown in Fig. 4.52*c*. This spectrum occupies the band $\omega_i - 4\pi B$ to $\omega_i + 4\pi B$. Clearly, the spectrum of the entire FM wave will lie in the frequency range $\omega_c - k_f m_p - 4\pi B$ to $\omega_c + k_f m_p + 4\pi B$, and the bandwidth B_{FM} is given by

$$2\pi B_{FM} = 2k_f m_p + 8\pi B \tag{4.57}$$

and

$$B_{FM} = 2(\Delta f + 2B) \tag{4.58}$$

For wideband FM, $\Delta f \gg B$, and the bandwidth of an FM carrier is approximately twice the carrier frequency deviation Δf.

In the literature, other rules of thumb for the FM bandwidth are found. The most commonly used rule, known as *Carson's rule,* is

$$B_{FM} = 2(\Delta f + B) \tag{4.59}$$

Carson's rule gives a better bandwidth estimate than does Eq. (4.58) in the narrowband case, where $\Delta f \ll B$, and $B_{FM} \simeq 2B$. This agrees with our earlier results.

In other cases, where $\Delta f \ll B$ is not satisfied (wideband and intermediate cases), Eq. (4.58) gives a better estimate than does Carson's rule. Carson's estimate is a bit

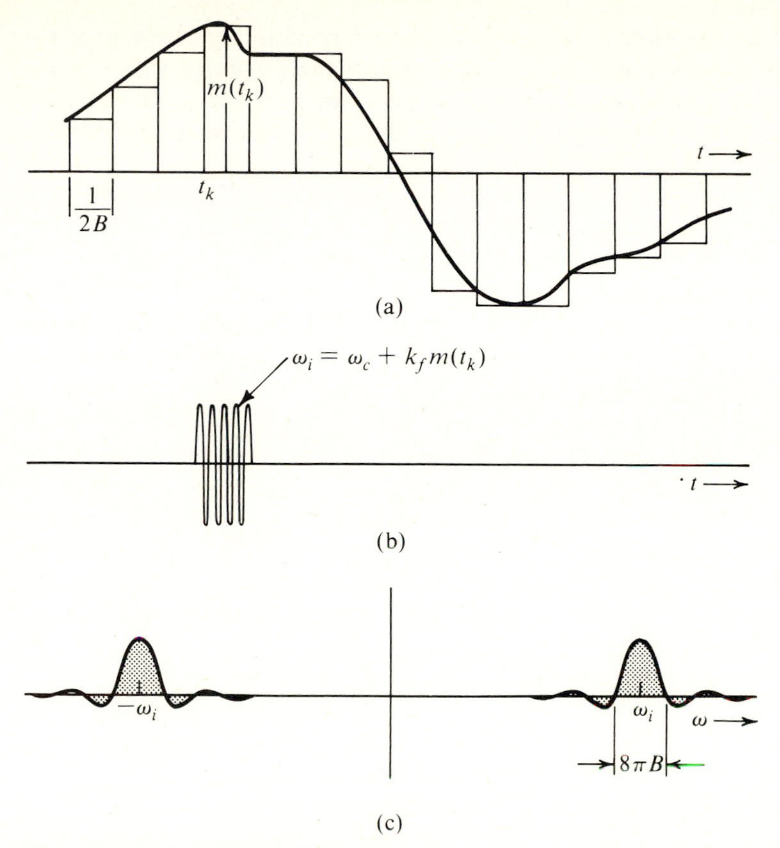

Figure 4.52 Estimation of FM wave bandwidth.

too low, whereas the estimate of Eq. (4.58) is slightly higher than the actual bandwidth. The truth lies somewhere between these two estimates, although for equipment design, one should be more conservative and use Eq. (4.58).

For convenience, we define a *deviation ratio* β as

$$\beta = \frac{\Delta f}{B} \tag{4.60}$$

Then Eqs. (4.58) and (4.59) can be combined as

$$B_{\text{FM}} = 2B(\beta + k) \tag{4.61}$$

where k varies between 1 and 2 depending on the value of β. In terms of β, we define the modulation as narrowband if $\beta \ll 1$ and wideband if $\beta \gg 1$.* It will be shown in Chapter 6 that the border between narrowband and wideband is roughly $\beta = 0.5$.

* We can readily show that the condition $\beta \gg$ or $\ll 1$ is equivalent to $|k_f a(t)| \gg$ or $\ll 1$ where $a(t)$ is the integral of $m(t)$ (see the case of tone modulation on page 296).

The deviation ratio controls the amount of modulation and, consequently, plays a role similar to the modulation index in AM. Indeed, for the special case of tone-modulated FM, the deviation ratio β is called the *modulation index*.

We can verify the bandwidth relations for a specific case of tone modulation; that is,

$$m(t) = \alpha \cos \omega_m t$$

From Eq. (4.49),*

$$a(t) = \frac{\alpha}{\omega_m} \sin \omega_m t \tag{4.62}$$

Thus, from Eq. (4.50a), we have

$$\hat{\varphi}_{\text{FM}}(t) = A \exp\left[j\omega_c t + \frac{k_f \alpha}{\omega_m} \sin \omega_m t \right]$$

Because

$$\Delta\omega = k_f m_p = \alpha k_f$$

The deviation ratio (or the modulation index, in this case) is

$$\beta = \frac{\Delta f}{f_m} = \frac{\Delta\omega}{\omega_m} = \frac{\alpha k_f}{\omega_m} \tag{4.63}$$

Hence,

$$\begin{aligned} \varphi_{\text{FM}}(t) &= A e^{j[\omega_c t + \beta \sin \omega_m t]} \\ &= A e^{j\omega_c t}\left[e^{j\beta \sin \omega_m t}\right] \end{aligned} \tag{4.64}$$

The exponential term in the brackets is a periodic signal with period $2\pi/\omega_m$ and can be expanded by the exponential Fourier series, as usual.

$$e^{j\beta \sin \omega_m t} = \sum_{n=-\infty}^{\infty} C_n e^{jn\omega_m t}$$

where

$$C_n = \frac{\omega_m}{2\pi} \int_{-\pi/\omega_m}^{\pi/\omega_m} e^{j\beta \sin \omega_m t} e^{-jn\omega_m t}\, dt$$

Letting $\omega_m t = x$, we get

$$C_n = \frac{1}{2\pi} \int_{-\pi}^{\pi} e^{j(\beta \sin x - nx)}\, dx \tag{4.65}$$

The integral on the right-hand side cannot be evaluated in a closed form but must be integrated by expanding the integrand in infinite series. This integral has been extensively tabulated and is denoted by $J_n(\beta)$, the Bessel function of the first kind and nth order.[9] These functions are plotted in Fig. 4.53*a* as a function of n for various values

* Here we are assuming that the constant $a(-\infty) = 0$.

of β. Thus,

$$e^{j\beta \sin \omega_m t} = \sum_{n=-\infty}^{\infty} J_n(\beta)\, e^{jn\omega_m t} \tag{4.66}$$

Substituting Eq. (4.66) in Eq. (4.64), we get

$$\hat{\varphi}_{\text{FM}}(t) = A \sum_{n=-\infty}^{\infty} J_n(\beta)\, e^{j(\omega_c t + n\omega_m t)}$$

and

$$\varphi_{\text{FM}}(t) = A \sum_{n=-\infty}^{\infty} J_n(\beta) \cos\,(\omega_c + n\omega_m)t \tag{4.67}$$

The modulated signal has a carrier component and an infinite number of sidebands of frequencies $\omega_c \pm \omega_m$, $\omega_c \pm 2\omega_m$, . . . , $\omega_c \pm n\omega_m$, . . . , as shown in Fig. 4.53*b*. The strength of the nth sideband at $\omega = \omega_c + n\omega_m$ is* $J_n(\beta)$. From the plots of $J_n(\beta)$ in Fig. 4.53*a*, it can be seen that for a given β, $J_n(\beta)$ decreases with n. For a

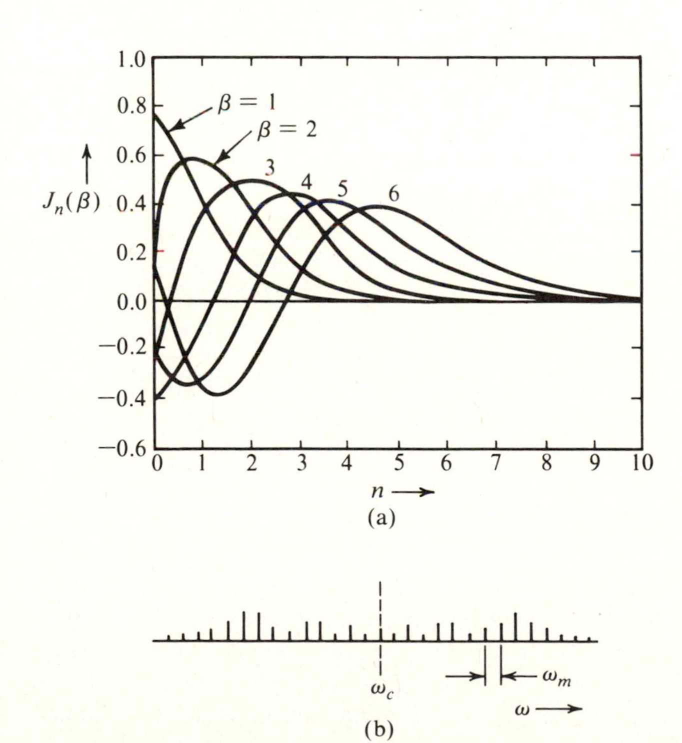

Figure 4.53 (a) Variations of $J_n(\beta)$ as a function of n for various values of β. (b) Tone-modulated FM wave spectrum.

---

* Also $J_{-n}(\beta) = (-1)^n J_n(\beta)$. Hence, the magnitude of the lower sideband at $\omega = \omega_c - n\omega_m$ is the same as that of the upper sideband at $\omega = \omega_c + n\omega_m$.

sufficiently large n, $J_n(\beta)$ is negligible, and there are only a finite number of significant sidebands. It can be seen from Fig. 4.53a that $J_n(\beta)$ is negligible for $n > \beta + 2$. Hence, the number of significant sidebands is $\beta + 2$. The bandwidth of the FM carrier is given by

$$B_{\text{FM}} = 2nf_m = 2(\beta + 2)f_m \tag{4.68a}$$

$$= 2(\Delta f + 2B) \tag{4.68b}$$

which verifies our previous result [Eq. (4.58)]. When $\beta \ll 1$ (NBFM), there is only one significant sideband and the bandwidth $B_{\text{FM}} = 2f_m = 2B$.

In the literature, tone modulation is discussed in great detail. Because FM is a nonlinear modulation, however, the principle of superposition does not apply, and the results derived for tone modulation may not be true for other modulating signals. Indeed, as we shall see in Sec. 4.16, conclusions drawn from tone modulation can be totally misleading. It is important in broadcasting, however, because many specifications and restrictions by the FCC are based on tone modulation.

The method for finding the spectrum of a tone-modulated FM wave can be used for finding the spectrum of an FM wave when $m(t)$ is a general periodic signal. In this case

$$\hat{\varphi}_{\text{FM}}(t) = Ae^{j\omega_c t}\left[e^{jk_f a(t)}\right]$$

Because $a(t)$ is a periodic signal, $e^{jk_f a(t)}$ is also a periodic signal, which can be expressed as an exponential Fourier series in the above expression. After this, it is relatively straightforward to write $\varphi_{\text{FM}}(t)$ in terms of the carrier and the sidebands.

Phase Modulation

All the results derived for FM can be directly applied to PM. Thus, for PM

$$\Delta\omega = k_p m'_p \tag{4.69a}$$

where*

$$m'_p = [\dot{m}(t)]_{\max} \tag{4.69b}$$

$$B_{\text{PM}} = 2[\Delta f + kB] \qquad 1 < k < 2 \tag{4.70a}$$

$$= 2\left[\frac{k_p m'_p}{2\pi} + kB\right] \tag{4.70b}$$

One important difference exists between FM and PM with regard to Δf. In FM, $\Delta\omega = k_f m_p$ depends only on the peak value of $m(t)$. It is independent of the spectrum of $m(t)$. On the other hand, in PM, $\Delta\omega = k_p m'_p$ depends on the peak value of $\dot{m}(t)$. But $\dot{m}(t)$ depends strongly on the frequency spectrum of $m(t)$. The presence of higher-frequency components in $m(t)$ causes rapid variations, resulting in a higher value of m'_p. Similarly, predominance of lower-frequency components will result in a lower value of m'_p. Hence, whereas the WBFM carrier bandwidth [Eq. (4.58)] is

* We are assuming that $|\dot{m}(t)_{\min}| = m'_p$.

practically independent* of the spectrum of $m(t)$, the WBPM carrier bandwidth [Eq. (4.70)] strongly depends on the spectrum of $m(t)$. For $m(t)$ with a spectrum concentrated at lower frequencies, B_{PM} will be smaller than that when the spectrum of $m(t)$ is concentrated at higher frequencies.

These conclusions can be verified for tone modulation, where $m(t) = \alpha \cos \omega_m t$ and $\dot{m}(t) = -\alpha\omega_m \sin \omega_m t$. Hence,

$$(\Delta\omega)_{FM} = k_f m_p = \alpha k_f$$

$$(\Delta\omega)_{PM} = k_p m'_p = \alpha\omega_m k_p$$

For a wideband case, the carrier bandwidth $\simeq 2\Delta f$, which is practically independent of the spectrum of $m(t)$ in the case of FM. But it is strongly dependent on the spectrum of $m(t)$ in the case of PM.

■ EXAMPLE 4.13

Estimate B_{FM} and B_{PM} for the modulating signal $m(t)$ in Fig. 4.49a for $k_f = \pi \times 10^4$ and $k_p = \pi/4$.

☐ **Solution:** First we must determine the essential bandwidth B of $m(t)$. Using the methods in Chapter 2, $m(t)$ can be expressed by the Fourier series

$$m(t) = \sum_n C_n \cos n\omega_o t \qquad \omega_o = \frac{2\pi}{2 \times 10^{-4}} = 10^4\,\pi$$

where

$$C_n = \begin{cases} \dfrac{8}{\pi^2 n^2} & n = 1, 3, 5, \ldots \\ 0 & n \text{ is even} \end{cases}$$

It can be seen that the harmonic amplitudes decrease rapidly with n. The third harmonic is only 11 percent of the fundamental, and the fifth harmonic is only 4 percent of the fundamental. This means the third and the fifth harmonic powers are 1.21 percent and 0.16 percent, respectively, of the fundamental component power. Hence, we are justified in assuming the essential bandwidth of $m(t)$ as the frequency of the third harmonic, that is, $3(10^4/2)$. Thus,

$$B = 1.5 \times 10^4$$

For FM

$$\Delta f = \frac{1}{2\pi} k_f m_p = 0.5 \times 10^4$$

Hence, $\Delta f < B$ (narrowband case), and

$$B_{FM} = 2(\Delta f + B) = 2(0.5 + 1.5)\,10^4 = 4 \times 10^4$$

*Except for its weak dependence on B [Eq. (4.58)].

For PM

$$\Delta f = \frac{1}{2\pi} k_p m'_p = 2500$$

Hence, $\Delta f \ll B$ (narrowband case), and

$$B_{PM} = 2(\Delta f + B) = 3.5 \times 10^4$$ ■

■ EXAMPLE 4.14

Estimate B_{FM} and B_{PM} for the modulating signal $m(t)$ in Fig. 4.49*a* for $k_f = 2\pi \times 10^5$ and $k_p = 10\pi$.

For FM

$$\Delta f = \frac{1}{2\pi} k_f m_p = 10^5$$

Because $B = 1.5 \times 10^4$, $\Delta f \gg B$ (wideband case), and

$$B_{FM} = 2(\Delta f + 2B) = 2.6 \times 10^5$$

For PM

$$\Delta f = \frac{1}{2\pi} k_p m'_p = 10^5 \gg B \text{ (wideband case), hence,}$$

$$B_{PM} = 2(\Delta f + 2B) = 2.6 \times 10^5$$ ■

■ EXAMPLE 4.15

Repeat Example 4.14 if $m(t)$ is time-expanded by a factor of 2; that is, if the period of $m(t)$ is 4×10^{-4}.

□ **Solution:** In this case, $B = 7.5$ kHz, $m_p = 1$ is unchanged, but $m'_p = 10^4$.

For FM

$$\Delta f = \frac{1}{2\pi} k_f m_p = 10^5$$

$$B_{FM} = 2(\Delta f + 2B) = 2(10^5 + 15 \times 10^3) = 2.3 \times 10^5$$

For PM

$$\Delta f = \frac{1}{2\pi} k_p m'_p = 5 \times 10^4$$

$$B_{PM} = 2(\Delta f + 2B) = 1.3 \times 10^5$$

Note that time expansion of $m(t)$ has very little effect on the FM bandwidth, but it halves the PM bandwidth. This verifies our observation that the PM spectrum is strongly dependent on the spectrum of $m(t)$. ■

A Historical Note

Today, nobody doubts that FM has a place in broadcasting and communication. As recently as the late sixties, the future of FM broadcasting seemed doomed because of uneconomical operations.

The history of FM is full of strange ironies. The impetus behind the development of FM was the necessity to reduce the transmission bandwidth. Superficial reasoning showed that it was feasible to reduce the transmission bandwidth by using FM. But the experimental results showed otherwise. The transmission bandwidth of FM was actually larger than that of AM! Careful mathematical analysis by Carson showed that FM indeed required a larger bandwidth than that of AM. Unfortunately, Carson did not recognize the compensating advantage of FM in its ability to suppress noise. Without much basis, he concluded that FM introduced inherent distortion and had no compensating advantages whatsoever.[10] The opinion of one of the ablest mathematicians of the day in the communication field thus set back the development of FM. The noise-suppressing advantage of FM was later proved by Major Edwin H. Armstrong,[11] a brilliant engineer whose contributions to the field of radio systems are comparable with those of Hertz and Marconi. It was largely the work of Armstrong that was responsible for rekindling the interest in FM.

Although Armstrong did not invent the concept of FM, he must be considered the father of modern FM. To quote from the early British text *Frequency Modulation Engineering* by Christopher E. Tibbs: "The subject of frequency modulation as we understand it today may be considered to date from Armstrong's paper of 1936. It is true that a good deal of the knowledge of the subject existed prior to that date, but Armstrong was the first to point out in a truly remarkable paper those peculiar characteristics to which modern technique owes its value."[12]

Armstrong was one of the leading architects who laid the groundwork for the mass-communication system. His work on FM came toward the close of his career. Before that, he was well known for several breakthrough contributions to the radio field. *Fortune* magazine says[13]: "Wideband frequency modulation is the fourth, and perhaps the greatest, in a line of Armstrong inventions that have made most of modern broadcasting what it is. Major Armstrong is the acknowledged inventor of the regenerative 'feedback' circuit, which brought radio art out of the crystal-detector headphone stage and made the amplification of broadcasting possible; the superheterodyne circuit, which is the basis of practically all modern radio; and the super-regenerative circuit now in wide use in . . . shortwave systems."

Armstrong was the last of the breed of the lone attic inventors. For the sake of establishing FM broadcasting, he fought a long and a costly battle with the radio broadcast establishment, which, abetted by the FCC (Federal Communications Commission), fought tooth and nail to resist FM. In 1944, the FCC, on the basis of erroneous testimony of a technical expert, abruptly shifted the allocated bandwidth of FM from 42 to 50 MHz to 88 to 108 MHz. This dealt a crippling blow to FM by making obsolete all the equipment (transmitters, receivers, antennas, etc.) that had been built and sold for the old FM bands. Armstrong continued to fight the decision, and in 1947 he succeeded in getting the technical expert to admit his error. In spite of all this, the FCC allocations remained unchanged. Armstrong spent a sizable fortune that he made from previous inventions in legal struggles. The broadcast industry, which so strongly resisted FM, turned around and used his inventions without paying him royalties. Armstrong spent nearly half of his life in the law courts in some of the longest, most notable, and acrimonious patent suits of the era.[12] In the end, with his funds depleted, his energy drained, and his family-life shattered, a despondent Armstrong committed suicide (in 1954) by walking out of a window 13 stories above the street.

Applications of FM

Frequency modulation (and angle modulation in general) has a number of unique features that recommend it for various radio systems.

The constant amplitude of FM makes it less susceptible to nonlinearities. Consider, for example, a nonlinear channel with input x and output y related by

$$y = a_1 x + a_2 x^2 + a_3 x^3$$

For an FM wave, the input $x(t) = A \cos [\omega_c t + \psi(t)]$, where

$$\psi(t) = k_f \int m(\alpha)\, d\alpha.$$

Hence,

$$\begin{aligned} y(t) &= A a_1 \cos [\omega_c t + \psi(t)] + a_2 A^2 \cos^2 [\omega_c t + \psi(t)] \\ &\qquad + a_3 A^3 \cos^3 [\omega_c t + \psi(t)] \\ &= \frac{a_2 A^2}{2} + \left(A a_1 + \frac{3}{4} a_3 A^3\right) \cos [\omega_c t + \psi(t)] + \frac{a_2 A^2}{2} \cos [2\omega_c t + 2\psi(t)] \\ &\qquad + \frac{a_3 A^3}{4} \cos [3\omega_c t + 3\psi(t)] \end{aligned}$$

The term $\cos [2\omega_c t + 2\psi(t)] = \cos [2\omega_c t + 2k_f \int m(\alpha)\, d\alpha]$ represents an FM wave with carrier $2\omega_c$ and with twice the frequency deviation ($\Delta\omega = 2k_f m_p$). Similarly, the term $\cos [3\omega_c t + 3\psi(t)]$ is an FM wave with carrier $3\omega_c$ and triple the frequency deviation. Hence, nonlinearity generates components at unwanted frequencies. But the desired term $\cos [\omega_c t + \psi(t)]$ is undistorted, and by using a bandpass filter centered at ω_c, we can suppress all unwanted terms in $y(t)$ and obtain the desired signal component without distortion. Recall that a similar nonlinearity in AM not only caused unwanted modulation with carrier frequencies $2\omega_c$ and $3\omega_c$ but also caused distortion of the desired signal (see Example 2.31). This is the primary reason why angle modulation is used in microwave radio relay systems, where nonlinear operation of amplifiers and other devices has thus far been unavoidable at the required power levels. In addition, the constant amplitude of FM gives it a kind of immunity against rapid fading. The effect of amplitude variations caused by rapid fading can be eliminated by using automatic gain control and bandpass limiting (discussed in Sec. 4.14). These features make FM attractive for microwave radio relay systems. Angle modulation is also less vulnerable than AM to small signal interference from adjacent channels (see Sec. 4.15). Finally, FM is capable of exchanging SNR for the transmission bandwidth.

In telephone systems, several channels are multiplexed using SSB signals (see Sec. 4.11, Fig. 4.46). The multiplexed signal is frequency modulated and transmitted over a microwave radio relay system with many links in tandem. In this application, however, FM is used not to realize the noise reduction but to realize other advantages of constant amplitude, and, hence, NBFM rather than WBFM is used.

Wideband FM is widely used in space and satellite communication systems. The large bandwidth expansion reduces the required SNR and thus reduces the transmitter power requirement—which is very important because of weight considerations in space. Wideband FM is also used for high-fidelity radio transmission over rather limited areas.

4.14 GENERATION OF FM WAVES

Basically, there are two ways of generating FM waves: *indirect generation* and *direct generation*.

The Indirect Method of Armstrong

In this method, NBFM is generated by integrating $m(t)$ and using it to phase modulate a carrier, as shown in Fig. 4.51*b* [or Eq. (4.52)]. The NBFM is then converted to WBFM by using frequency multipliers (Fig. 4.54). A frequency multiplier is just a

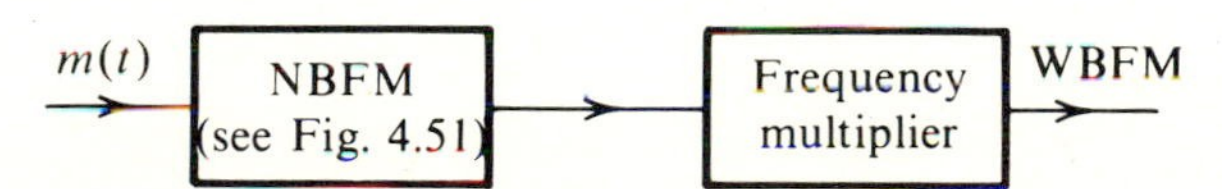

Figure 4.54 Simplified block diagram of Armstrong indirect FM wave generator.

nonlinear device. A simple square-law device, for example, can multiply the frequency of a factor of 2. For a square-law device, the input $e_i(t)$ and the output $e_o(t)$ are related by

$$e_o(t) = [e_i(t)]^2$$

If

$$e_i(t) = \varphi_{\text{FM}}(t) = \cos\left(\omega_c t + k_f \int_{-\infty}^{t} m(\alpha)\, d\alpha\right)$$

then

$$e_o(t) = \cos^2\left(\omega_c t + k_f \int_{-\infty}^{t} m(\alpha)\, d\alpha\right)$$

$$= \frac{1}{2} + \frac{1}{2}\cos\left(2\omega_c t + 2k_f \int_{-\infty}^{t} m(\alpha)\, d\alpha\right)$$

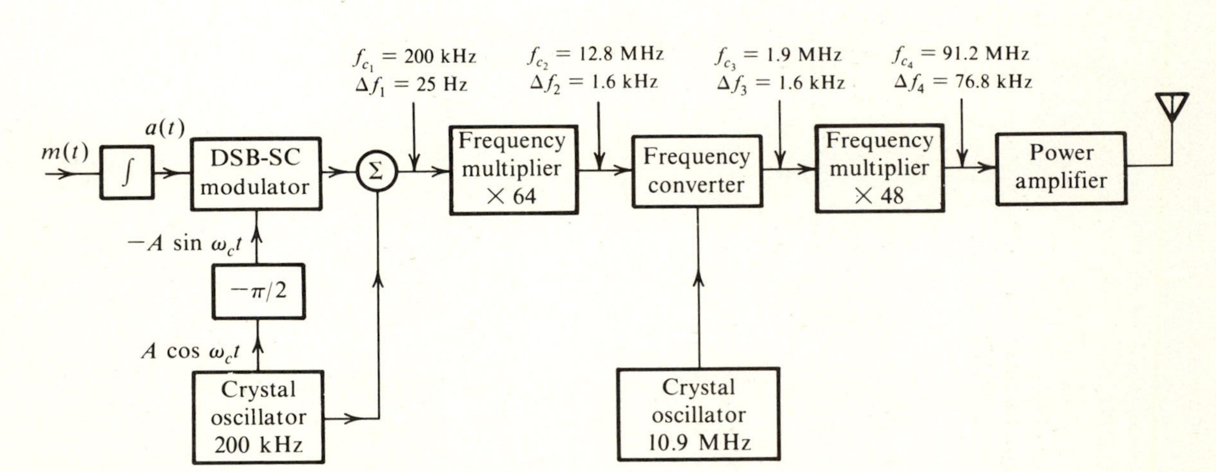

Figure 4.55 Armstrong indirect FM transmitter.

The dc term is filtered out to give the output, whose carrier frequency as well as frequency deviation are multiplied by two. Any nonlinear device, such as a diode or a transistor, can be used for this purpose. These devices have the characteristic

$$e_o(t) = a_0 + a_1 e_i(t) + a_2 e_i^2(t) + \cdots + a_n e_i^n(t)$$

Hence, the output will have spectra at $\omega_c, 2\omega_c, \ldots, n\omega_c$, with frequency deviations $\Delta f, 2\Delta f, \ldots, n\Delta f$, respectively. We then use the appropriate filter to choose the desired multiplier value.

It should be remembered that the NBFM generated by Armstrong's method (Fig. 4.51*b*) has some distortion because of the approximation of Eq. (4.51) by Eq. (4.52). The output of the Armstrong NBFM modulator, as a result, also has some amplitude modulation. Amplitude limiting in the frequency multipliers removes most of this distortion.

A simplified diagram of a commercial FM transmitter using Armstrong's method is shown in Fig. 4.55. The final output is required to have a carrier frequency of 91.2 MHz and $\Delta f = 75$ kHz. We begin with NBFM with a carrier frequency $f_{c_1} = 200$ kHz generated by a crystal oscillator. This frequency is chosen because it is easy to construct stable crystal oscillators as well as balanced modulators at this frequency. The deviation Δf is chosen to be 25 Hz in order to maintain $\beta \ll 1$, as required in NBPM. For tone modulation $\beta = \Delta f/f_m$. The baseband spectrum (required for hi-fidelity purposes) ranges from 50 Hz to 15 kHz. The choice of $\Delta f = 25$ is reasonable because it gives $\beta = 0.5$ for the worst possible case ($f_m = 50$).

In order to achieve $\Delta f = 75$ kHz, we need a multiplication of $75{,}000/25 = 3000$. This can be done by two multiplier stages, of 64 and 48, as shown in Fig. 4.55. This gives a total multiplication of $64 \times 48 = 3072$, and $\Delta f = 76.8$ kHz. The multiplication is effected by using frequency doublers and triplers in cascade, as needed. Thus, a multiplication of 64 can be obtained by six doublers in cascade, and a multiplication of 48 can be obtained by four doublers and a tripler in cascade. Multiplication of $f_c = 200$ kHz by 3072, however, would yield a final carrier of about 600 MHz. This difficulty is avoided by using a frequency translation, or conversion, after the first multiplier (Fig. 4.55). The frequency converter shifts the entire spectrum without altering Δf. Hence, we have $f_{c_3} = 1.9$ MHz and $\Delta f_3 = 1.6$ kHz. Further multiplication, by 48, yields $f_{c_4} = 91.4$ MHz and $\Delta f = 76.8$ kHz.

This scheme has an advantage of frequency stability, but it suffers from inherent noise caused by excessive multiplication and distortion at lower modulating frequencies, where $\Delta f/f_m$ is not small enough.

■ EXAMPLE 4.16

Discuss the nature of distortion inherent in the Armstrong indirect FM generator.

☐ **Solution:** Two kinds of distortions arise in this scheme: amplitude distortion and frequency distortion. The NBFM wave is given by [Eq. (4.52)]

$$\varphi_{\text{FM}}(t) = A[\cos \omega_c t - k_f a(t) \sin \omega_c t] \tag{4.71a}$$

$$= AE(t) \cos [\omega_c t + \theta(t)] \tag{4.71b}$$

where

$$E(t) = \sqrt{1 + k_f^2 a^2(t)} \tag{4.72a}$$

$$\theta(t) = \tan^{-1} k_f a(t) \tag{4.72b}$$

Amplitude distortion occurs because the amplitude $AE(t)$ of the modulated waveform is not constant. This is not a serious problem, because amplitude variations can be eliminated by a bandpass limiter, which consists of an amplitude limiter followed by a bandpass filter. This is discussed in detail in the next section (see Fig. 4.58).

The phase $\theta(t)$ [Eq. (4.72b)] is

$$\theta(t) = \tan^{-1} k_f a(t)$$

Ideally, $\theta(t)$ should be $k_f a(t)$. The instantaneous frequency $\omega_i(t)$ is

$$\begin{aligned}\omega_i = \dot{\theta}(t) &= \frac{k_f \dot{a}(t)}{1 + k_f^2 a^2(t)} \\ &= \frac{k_f m(t)}{1 + k_f^2 a^2(t)} \\ &= k_f m(t)[1 - k_f^2 a^2(t) + k_f^4 a^4(t) - \cdots]\end{aligned} \tag{4.73a}$$

For tone modulation, $m(t) = \alpha \cos \omega_m t$, $a(t) = \alpha \sin \omega_m t/\omega_m$, and the modulation index $\beta = \alpha k_f/\omega_m$. Hence,

$$\omega_i(t) = \beta\omega_m \cos \omega_m t[1 - \beta^2 \sin^2 \omega_m t + \beta^4 \sin^4 \omega_m t - \cdots] \tag{4.73b}$$

It is evident from Eq. (4.73b) that this scheme has odd-harmonic distortion, the most important term being the third harmonic. Ignoring the remaining terms, Eq. (4.73b) becomes

$$\begin{aligned}\omega_i(t) &\simeq \beta\omega_m \cos \omega_m t[1 - \beta^2 \sin^2 \omega_m t] \\ &= \beta\omega_m\left(1 - \frac{\beta^2}{4}\right) \cos \omega_m t + \frac{\beta^3 \omega_m}{4} \cos 3\omega_m t \\ &\simeq \underbrace{\beta\omega_m \cos \omega_m t}_{\text{desired}} + \underbrace{\frac{\beta^3 \omega_m}{4} \cos 3\omega_m t}_{\text{distortion}} \qquad \text{for } \beta \ll 1\end{aligned}$$

The ratio of the third harmonic distortion to the desired signal is $\beta^2/4$. For the generator in Fig. 4.55, the worst possible case occurs at the lower modulation frequency of 50 Hz, where $\beta = 0.5$. In this case the third harmonic distortion is 1/16th, or 6.25 percent. ■

Direct Generation

An oscillator whose frequency can be controlled by an external voltage is a *voltage-controlled oscillator (VCO)*. In a VCO, the oscillation frequency varies linearly with the control voltage. We can generate an FM wave by using the modulating signal $m(t)$

as a control signal. This gives

$$\omega_i(t) = \omega_c + k_f m(t)$$

One can construct a VCO using an operational amplifier and an hysteric comparator[1] (such as a Schmitt trigger circuit). Another way of accomplishing the same goal is to vary one of the reactive parameters (C or L) of the resonant circuit of an oscillator. For example, in a Harltey oscillator (Fig. 4.56), the oscillation frequency is given by

$$\omega_o = \frac{1}{\sqrt{LC}} \qquad L = L_1 + L_2 \tag{4.74a}$$

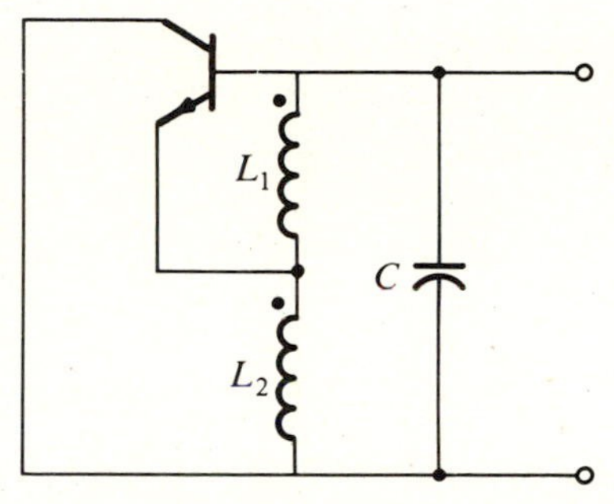

Figure 4.56 Direct generation of FM by varying a reactance parameter.

If the capacitance C is varied by the modulating signal $m(t)$, the oscillator output will be the desired FM wave. Thus, if

$$C = C_0 - km(t) \tag{4.74b}$$

$$\omega_o = \frac{1}{\sqrt{LC_0\left[1 - \dfrac{km(t)}{C_0}\right]}}$$

$$= \frac{1}{\sqrt{LC_0}\left[1 - \dfrac{km(t)}{C_0}\right]^{1/2}} \tag{4.75a}$$

$$\simeq \frac{1}{\sqrt{LC_0}}\left[1 + \frac{km(t)}{2C_0}\right] \qquad \frac{km(t)}{C_0} \ll 1 \tag{4.75b}$$

$$= \omega_c\left[1 + \frac{km(t)}{2C_0}\right] \qquad \omega_c = 1/\sqrt{LC_0}$$

$$= \omega_c + k_f m(t) \qquad k_f = \frac{k\omega_c}{2C_0}$$

A reverse-biased semiconductor diode acts as a capacitor whose capacitance varies with the bias voltage. The capacitance of these diodes, known under several trade names (such as varicaps, varactors, or voltacaps), can be approximated as a linear function of the bias voltage $m(t)$ [Eq. (4.75b) over a limited range]. Note that

$$\Delta C = km_p = \frac{2k_f C_0 m_p}{\omega_c}$$

Hence,

$$\frac{\Delta C}{C_0} = \frac{2k_f m_p}{\omega_c} = \frac{2\Delta f}{f_c}$$

In practice $\Delta f / f_c$ is usually small, and, hence, ΔC is a small fraction of C_0, which helps limit the harmonic distortion that arises because of the approximation in Eq. (4.75b).

Direct FM generation generally produces sufficient frequency deviation and requires little frequency multiplication. But this method has poor frequency stability. In practice, feedback is used to stabilize the frequency. The output frequency is compared with a constant frequency generated by a stable crystal oscillator. An error signal (error in frequency) is detected and fed back to the oscillator to correct the error.

4.15 DEMODULATION OF FM

In an FM wave, the information resides in the instantaneous frequency $\omega_i = \omega_c + k_f m(t)$. Hence, a frequency-selective network with a transfer function of the form $|H(\omega)| = a\omega + b$ over the FM band would yield an output proportional to the instantaneous frequency (Fig. 4.57*a*).* There are several possible networks with such characteristics. The simplest among them is an ideal differentiator with the transfer function $j\omega$.

If we apply $\varphi_{FM}(t)$ to an ideal differentiator, the output is

$$\dot{\varphi}_{FM}(t) = \frac{d}{dt}\left\{A \cos\left[\omega_c t + k_f \int_{-\infty}^{t} m(\alpha)\, d\alpha\right]\right\}$$

$$= A[\omega_c + k_f m(t)] \sin\left[\omega_c t + k_f \int_{-\infty}^{t} m(\alpha)\, d\alpha\right] \tag{4.76}$$

The signal $\dot{\varphi}_{FM}(t)$ is both amplitude and frequency modulated (Fig. 4.57*b*), the envelope being $A[\omega_c + k_f m(t)]$. Because $\Delta\omega = k_f m_p < \omega_c$, $\omega_c + k_f m(t) > 0$ for all t, and $m(t)$ can be obtained by envelope detection of $\dot{\varphi}_{FM}(t)$ (Fig. 4.57*c*).

The amplitude A of the incoming FM carrier is assumed to be constant. If the amplitude A were not constant, but a function of time, there would be an additional term containing dA/dt on the right-hand side of Eq. (4.76). Even if this term were neglected, the envelope of $\dot{\varphi}_{FM}(t)$ would be $A(t)[\omega_c + k_f m(t)]$, and the envelope-detector output would be proportional to $m(t)A(t)$. Hence, it is essential to maintain A as a constant. Several factors, such as channel noise, fading, and so on, cause A to vary. This variation in A should be removed before applying the signal to the FM detector.

---

* Provided the variations of ω_i are slow in comparison to the time constant of the network.

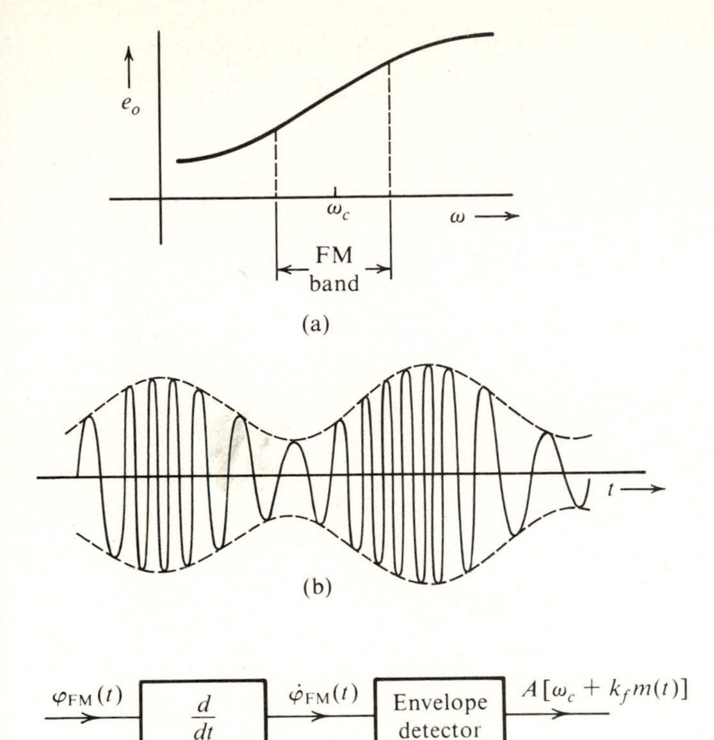

Figure 4.57 (a) FM demodulator frequency response. (b) Output of a differentiator to the input FM wave. (c) FM demodulation by direct differentiation.

Bandpass Limiter

The amplitude variations of an angle-modulated carrier can be eliminated by what is known as a *bandpass limiter*, which consists of a hard limiter followed by a bandpass filter (Fig. 4.58*a*). The input-output characteristic of a hard limiter is shown in Fig. 4.58*b*. The incoming angle-modulated wave can be expressed as

$$v_i(t) = A(t) \cos \theta(t) \tag{4.77}$$

where

$$\theta(t) = \omega_c t + k_f \int_{-\infty}^{t} m(\alpha)\, d\alpha \tag{4.78}$$

The output $v_o(t)$ of the hard limiter can be expressed as a function of θ (because $A(t) \geq 0$).

$$v_o(\theta) = \begin{cases} 1 & \cos \theta > 0 \\ -1 & \cos \theta < 0 \end{cases} \tag{4.79}$$

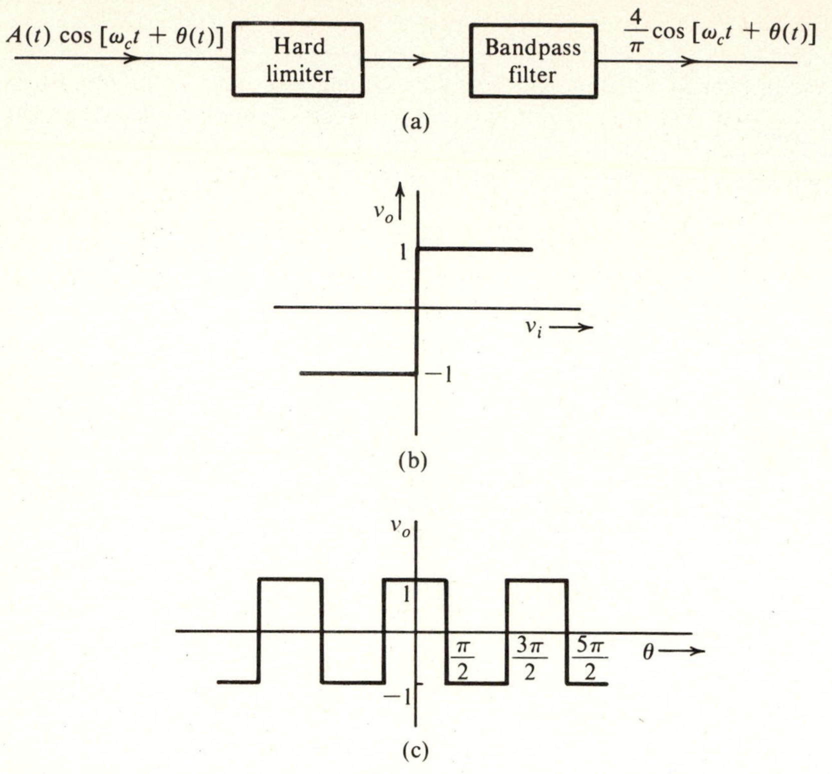

Figure 4.58 (a) Hard limiter and bandpass filter used to remove amplitude variations in FM wave. (b) Hard limiter input-output characteristic. (c) Hard limiter output vs. input signal angle.

Hence, v_o as a function of θ is a periodic square-wave function with a period 2π (Fig. 4.58c), which can be expanded by a Fourier series [see Eq. (4.8a)].

$$v_o(\theta) = \frac{4}{\pi}\left[\cos\theta - \frac{1}{3}\cos 3\theta + \frac{1}{5}\cos 5\theta + \cdots\right] \tag{4.80}$$

This is valid for any real variable θ. At any instant t, $\theta = \omega_c t + k_f \int m(\alpha)\,d\alpha$, and the output is $v_o[\omega_c t + k_f \int m(\alpha)\,d\alpha]$. Hence, the output v_o as a function of time is given by

$$\begin{aligned} v_o[\theta(t)] = v_o[\omega_c t + k_f \textstyle\int m(\alpha)\,d\alpha] = \frac{4}{\pi}\Big[&\cos[\omega_c t + k_f \textstyle\int m(\alpha)\,d\alpha] \\ &- \frac{1}{3}\cos 3[\omega_c t + k_f \textstyle\int m(\alpha)\,d\alpha] \\ &+ \frac{1}{5}\cos 5[\omega_c t + k_f \textstyle\int m(\alpha)\,d\alpha]\cdots\Big] \end{aligned}$$

The output, therefore, has the original FM wave plus a frequency-multiplied FM wave with multiplication factors of 3, 5, 7, . . . , and so on. We can pass the output of the hard limiter through a bandpass filter with a center frequency ω_c and a bandwidth B_{FM}, as shown in Fig. 4.58*a*. The filter output, $e_o(t)$, is the desired angle-modulated carrier with a constant amplitude.

$$e_o(t) = \frac{4}{\pi} \cos \left[\omega_c(t) + k_f \int m(\alpha)\, d\alpha\right] \tag{4.81}$$

Although we derived these results for FM, it applies to PM (angle modulation in general) as well. The bandpass filter not only maintains the constant amplitude of the angle-modulated carrier but also partially suppresses the channel noise when the noise is small.[14]

Practical Frequency Demodulators

A simple realization of the scheme in Fig. 4.57*b* is shown in Fig. 4.59*a*. A balanced version of this scheme is discussed in Clarke and Hess.[15]

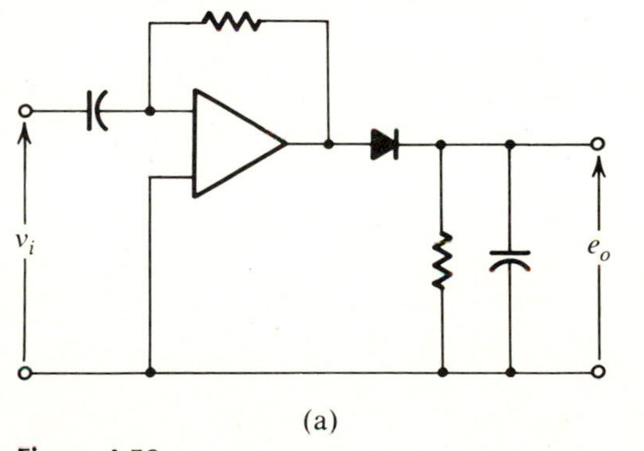

(a)

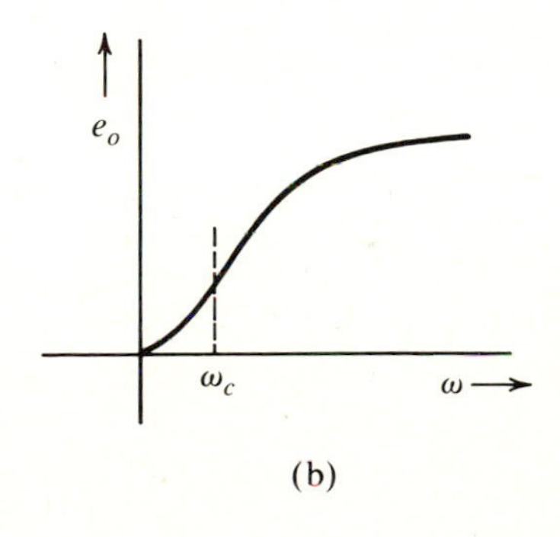

(b)

Figure 4.59

A simple tuned circuit followed by an envelope detector (Fig. 4.60*a*) also can serve as a frequency detector, because below (or above) the resonance frequency, $|H(\omega)|$ is approximately linear (Fig. 4.60*b*). Because the operation is on the slope of $|H(\omega)|$, this method is also called *slope detection*. It suffers from the fact that the slope of $|H(\omega)|$ is linear over only a small band and, hence, causes considerable distortion in the output. This fault can be partially corrected by a balanced configuration (Fig. 4.57*c*) that uses two resonant circuits, one tuned above and the other tuned below ω_c. The inputs to the two circuits are equal but of opposite sign. The outputs are envelope detected and subtracted to give the desired signal proportional to $m(t)$. The composite output vs. the frequency characteristics of this demodulator can be obtained by subtracting the two individual characteristics, as shown in Fig. 4.60*d*. The resulting composite characteristics are linear over a wider bandwidth, and the output is zero for an unmodulated carrier input.

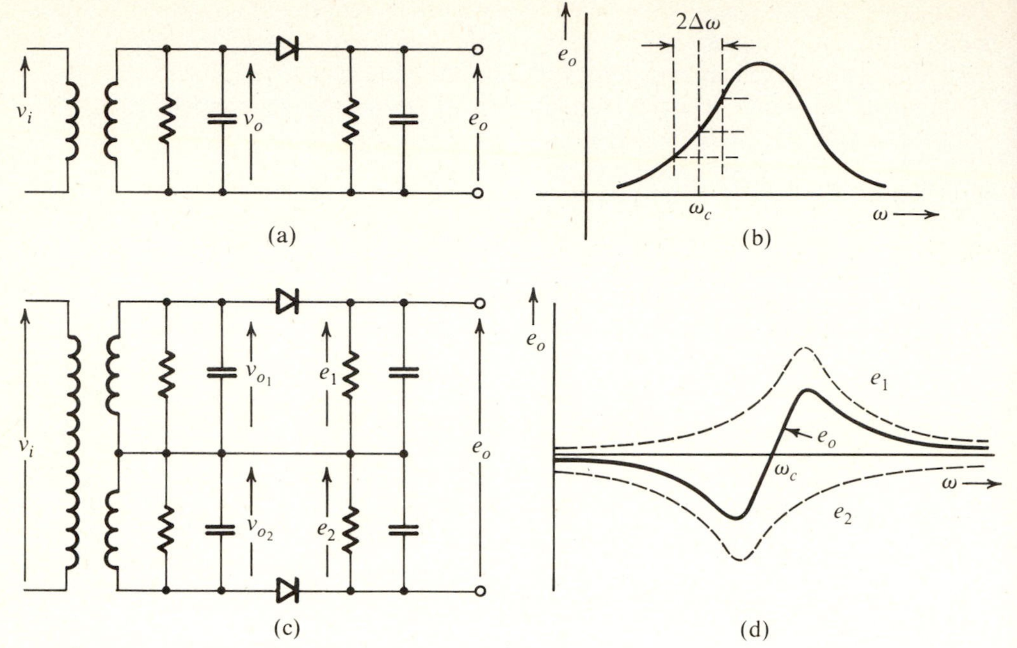

Figure 4.60 FM demodulators (discriminators).

Another balanced demodulator that is widely used is the *phase-shift detector* (Fig. 4.61*a*). It uses a doubly tuned circuit with an ac connection between the primary and the center of the secondary. Capacitors C_c and C have negligible impedance at $\omega = \omega_c$. The entire primary voltage appears across the *RF* choke *L*. Figure 4.61*b* shows the simplified equivalent circuit. The voltage across the upper envelope detector is $E_{d_1} = E_p + E_s/2$, and that across the lower is $E_{d_2} = E_p - E_s/2$.

The parameters are adjusted so that the circuit resonates at $\omega = \omega_c$. For the sake of analysis we can ignore the envelope detector and the RF choke, because they draw very little current. At resonance ($\omega = \omega_c$), the current I is in phase with the voltage E_p. Hence, the voltage across the capacitor C_2, which is caused solely by I, lags E_p by 90°. But the voltage across C_2 is E_s. Hence, E_s lags E_p by 90° at $\omega = \omega_c$. For $\omega > \omega_c$, the circuit is inductive, and I lags E_p. Hence, E_s lags E_p by more than 90°. Similarly, E_s lags E_p by less than 90° for $\omega < \omega_c$ (Fig. 4.61*c*). The output v_o is $|E_{d_1}| - |E_{d_2}|$, the difference of two envelope-detector outputs. From Fig. 4.61*c*, it is clear that $v_o = 0$ at $\omega = \omega_c$, with v_o increasing for $\omega > \omega_c$ and decreasing for $\omega < \omega_c$, as shown in Fig. 4.61*d*.

A variant of the phase-shift detector is the *ratio detector* (Fig. 4.62). This circuit is identical to the phase-shift detector except in two respects: one diode is reversed, and there is a voltage-stabilizing capacitor C_s across the load resistors. Also, the output v_o is taken from the centers of the *R*'s and *C*'s of the envelope-detector circuits. The capacitor C_s is chosen large enough to maintain a constant voltage E_c that is equal to the peak voltage across the diode input. This feature eliminates variations in the amplitude of the FM wave (amplitude limiting). Any sudden changes in the amplitude

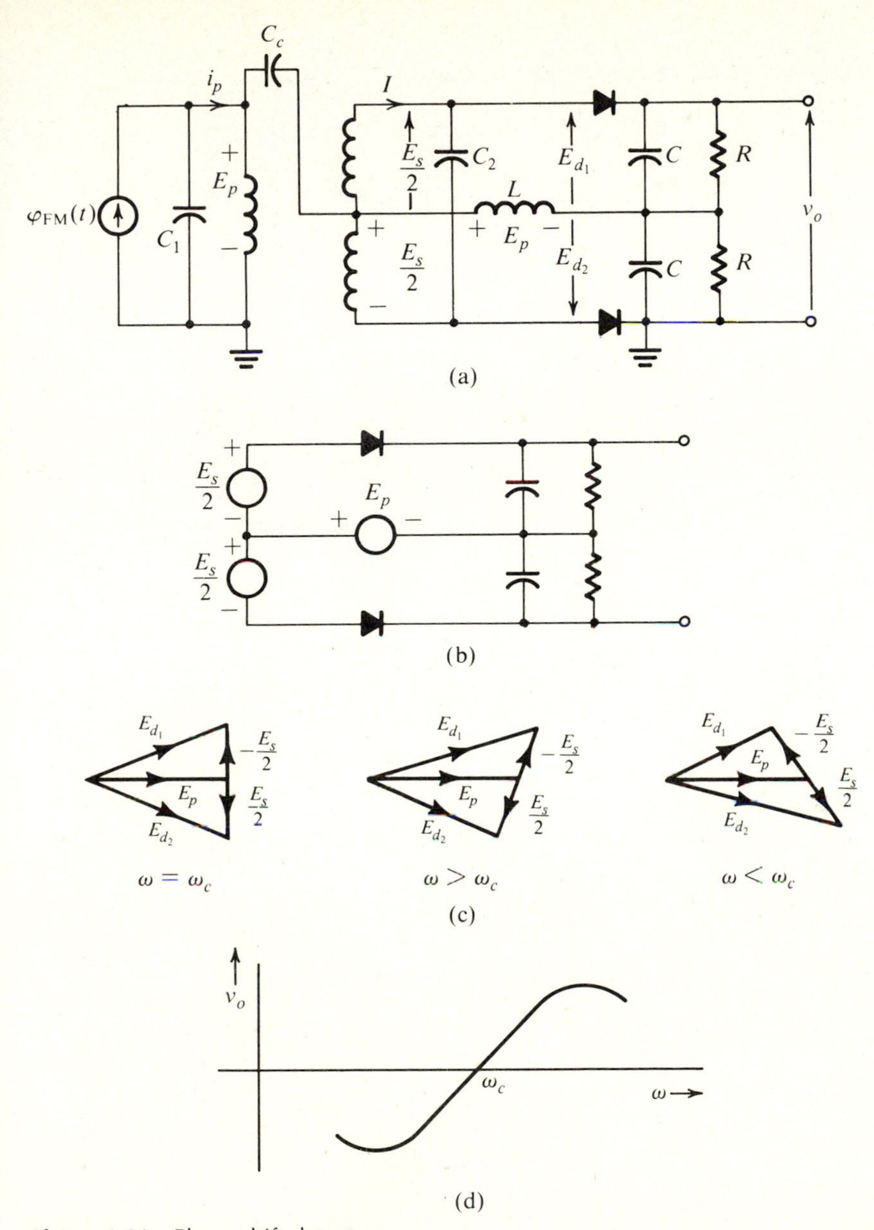

Figure 4.61 Phase-shift detector.

of the incoming wave will be suppressed by the large capacitor. Also,

$$E_c = e_1 + e_2$$

and

$$v_o = e_1 - \frac{E_c}{2} = e_1 - \frac{e_1 + e_2}{2} = \frac{e_1 - e_2}{2}$$
$$= k[|E_{d_1}| - |E_{d_2}|]$$

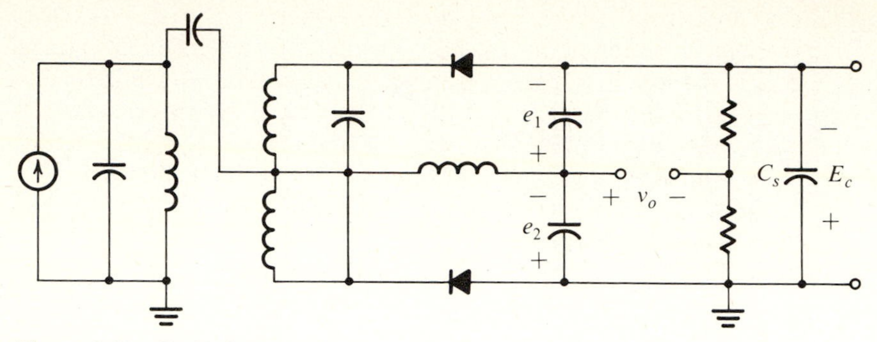

Figure 4.62 Ratio detector.

As seen earlier, $e_1 = e_2$ at $\omega = \omega_c$, $e_1 > e_2$ for $\omega > \omega_c$, and $e_1 < e_2$ for $\omega < \omega_c$. Hence, v_o vs. ω characteristic is similar to those shown in Fig. 4.61*d*.

The chief advantage of this circuit over the phase-shift detector is the feature of amplitude limiting. Both of these circuits are widely used in practice. All the frequency-sensitive detectors discussed above are also known as *frequency discriminators*.

An FM signal can also be demodulated by measuring the number of zero crossings/second. Because the number of zero crossings/second is twice the instantaneous frequency, the desired modulating signal is proportional to the number of zero crossings/second. The interval T over which the number of zero crossings is counted is small compared to $1/B$ but large compared to $1/f_c$. This ensures that the zero-crossing rate will be nearly constant over T, and at the same time there will be sufficient zero crossings over the interval to ensure accuracy. An example of such a detector is described in Clarke and Hess.[15]

Yet another demodulation scheme, using a delay element, is shown in Fig. 4.63.

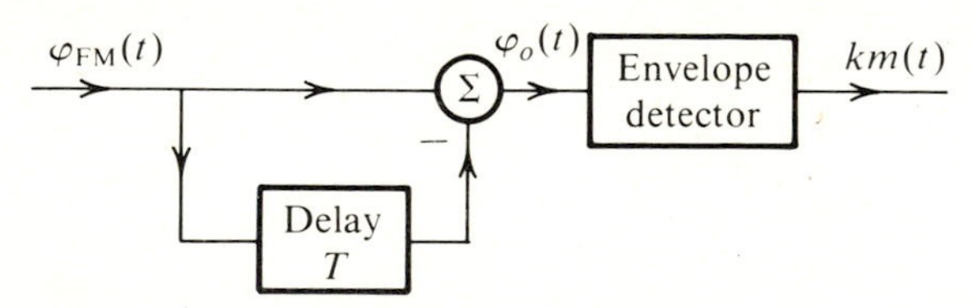

Figure 4.63 Differentiator type detector.

For a small delay T, the output $\varphi_o(t)$ (Fig. 4.63) is proportional to $\dot{\varphi}_{FM}(t)$, because

$$\varphi_o(t) = \varphi_{FM}(t) - \varphi_{FM}(t - T) \simeq T\,\dot{\varphi}_{FM}(t) \qquad T \ll 1/f_c$$

The signal $\varphi_o(t)$ is now envelope-detected to obtain the desired signal. The delay T can be implemented by a section of a transmission line.

Phase-Lock Loop. The phase-lock loop (PLL) is primarily used in tracking the phase and frequency of the carrier component of an incoming signal. It is, therefore, a useful device for synchronous demodulation of AM signals with suppressed carrier or with a little carrier (the pilot). It can also be used for the demodulation of angle-modulated signals. In the presence of strong noise, the PLL is more effective than conventional techniques (discussed earlier). For the low-noise case, conventional techniques perform as well as the PLL. The PLL has the advantage when the signal power is too low and the noise power is too high (low SNR). For this reason, the PLL is used in such applications as space-vehicle-to-earth data links, where there is a premium on transmitter weight; or where the loss along the transmission path is very large; and, more recently, in commercial FM receivers.

The operation of the PLL is similar to that of a feedback system (Fig. 4.64*a*). In

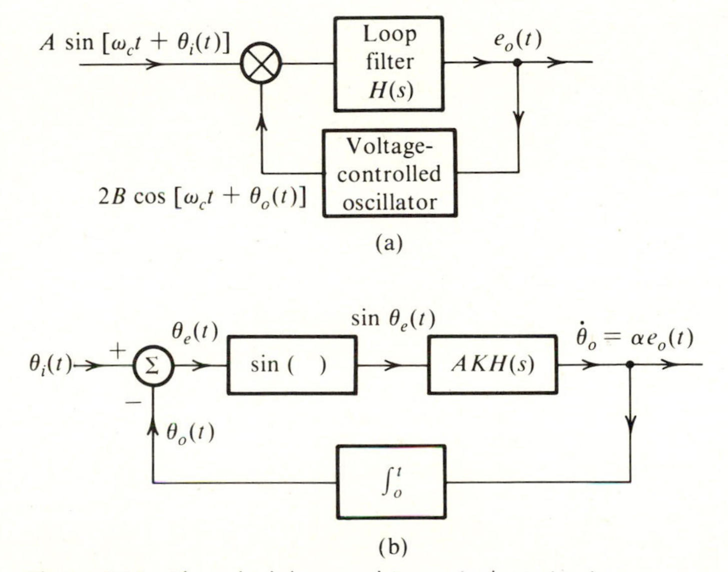

Figure 4.64 Phase-lock loop and its equivalent circuit.

a typical feedback system, the signal fed back tends to follow the input signal. If the signal fed back is not equal to the input signal, the difference (known as the error) will change the signal fed back until it is close to the input signal. A PLL operates on a similar principle, except that the quantity fed back and compared is not the amplitude, but a generalized phase $\theta(t)$. Hence, the VCO adjusts its frequency until its generalized angle comes close to the angle of the incoming signal. At this point, the frequency and phase of the two signals are in synchronism (except for a difference of a constant phase).

PLL Operation. A PLL consists of (1) a multiplier that serves as a phase comparator, (2) a loop filter $H(s)$, and (3) a VCO. The output of the loop filter $H(s)$ is $e_o(t)$, which acts as an input to the VCO (Fig. 4.64*a*). Let the quiescent frequency of the VCO (the VCO frequency when $e_o(t) = 0$) be ω_c. The instantaneous VCO frequency is

$$\omega_i = \omega_c + \alpha e_o(t)$$

Hence, the VCO output is $2C \cos [\omega_c t + \theta_o(t)]$, where

$$\dot{\theta}_o(t) = \alpha e_o(t) \tag{4.82}$$

and α and C are constants of the PLL.

Let the incoming signal (input to the PLL) be $A \sin [\omega_c t + \theta_i(t)]$. It is not necessary for the VCO quiescent frequency equal to the incoming signal frequency.* If the incoming signal is $A \sin [\omega_{in} t + \psi(t)]$, then it can be expressed as $A \sin [\omega_c t + \theta_i(t)]$, where $\theta_i(t) = (\omega_{in} - \omega_c)t + \psi(t)$. Hence, the analysis that follows is general and not restricted to equal frequencies of the incoming signal and the quiescent VCO signal.

The multiplier output $2AC \sin [\omega_c t + \theta_i(t)] \cos [\omega_c t + \theta_o(t)]$ yields the sum and difference frequency terms. The sum frequency term is suppressed by the loop filter. Hence, the effective input to the loop filter is $AC \sin [\theta_i(t) - \theta_o(t)]$. If $h(t)$ is the unit impulse response of the loop filter,

$$\begin{aligned} e_o(t) &= h(t) * AC \sin [\theta_i(t) - \theta_o(t)] \\ &= AC \int_0^t h(t - x) \sin [\theta_i(x) - \theta_o(x)]\, dx \end{aligned} \tag{4.83}$$

Substituting Eq. (4.82) in Eq. (4.83),

$$\dot{\theta}_o(t) = AK \int_0^t h(t - x) \sin \theta_e(x)\, dx \tag{4.84}$$

where $K = \alpha C$ and $\theta_e(t)$ is the phase error, defined as

$$\theta_e(t) = \theta_i(t) - \theta_o(t) \tag{4.85}$$

This equation immediately suggests a model for the PLL, as shown in Fig. 4.64*b*.

This is an exact model of the PLL operation. It is a nonlinear feedback system, because it contains a nonlinear multiplier sin (·) in its path. The VCO angle $\theta_o(t)$ tries to track the incoming angle $\theta_i(t)$. The error $\theta_e(t)$ is used to correct $\theta_o(t)$ and to bring it as close to $\theta_i(t)$ as possible. In proper operation, the error $\theta_e \rightarrow 0$. Under this condition, the VCO output and the incoming signal are synchronized in frequency, and their phase angles differ by exactly $\pi/2$. The two signals are said to be mutually *phase coherent,* or in *phase-lock*. The VCO thus tracks the frequency and the phase of the incoming signal. In the Costas scheme of demodulating DSB-SC signals, the PLL is used to generate a local carrier that is phase coherent with the incoming signal carrier (see Sec. 4.5). To show that the PLL can also demodulate FM waves, we note that in proper operation θ_e approaches a small constant. Hence,

$$\theta_o(t) = \theta_i(t) - \theta_e$$

---

*All that is needed is to set the VCO quiescent frequency as close as possible to the incoming frequency.

For the incoming FM carrier* $A \sin [\omega_c t + \theta_i(t)]$,

$$\theta_i(t) = k_f \int_{-\infty}^{t} m(\alpha)\, d\alpha \tag{4.86}$$

Hence,

$$\theta_o(t) = k_f \int_{-\infty}^{t} m(\alpha)\, d\alpha - \theta_e$$

and

$$e_o(t) = \frac{1}{\alpha}\, \dot{\theta}_o(t) \simeq \frac{k_f}{\alpha}\, m(t) \tag{4.87}$$

Thus, the PLL acts as an FM demodulator. If the incoming signal is a PM wave, $\theta_o(t) = \theta_i(t) = k_p m(t)$, and $e_o(t) = k_p \dot{m}(t)/\alpha$. In this case we need to integrate $e_o(t)$ to obtain the desired signal.

Being a nonlinear system, the detailed analysis of the PLL is rather involved and beyond our scope. Complete analysis of two special cases is carried out in Appendix 4.1. The first one is that of a small error; that is, $\theta_e(t) \ll \pi/2$, so that $\sin \theta_e(t) \simeq \theta_e(t)$ and the operation becomes linear. The second case is that of an ideal lowpass filter, with $H(s) = 1$ (first-order PLL).

4.16 INTERFERENCE AND NOISE IN ANGLE-MODULATED SYSTEMS

Let us consider the simple case of the interference of an unmodulated carrier $A \cos \omega_c t$ with another sinusoid $I \cos (\omega_c + \omega_d)t$. The received signal $r(t)$ is

$$r(t) = A \cos \omega_c t + I \cos (\omega_c + \omega_d)t \tag{4.88}$$

$$= (A + I \cos \omega_d t) \cos \omega_c t - I \sin \omega_d t \sin \omega_c t$$

$$= E_r(t) \cos [\omega_c t + \varphi_d(t)] \tag{4.89}$$

where

$$\varphi_d(t) = \tan^{-1} \frac{I \sin \omega_d t}{A + I \cos \omega_d t} \tag{4.90}$$

When the interfering signal is small in comparison to the carrier ($I \ll A$),

$$\varphi_d(t) \simeq \frac{I}{A} \sin \omega_d t \tag{4.91}$$

The phase demodulator output is $\varphi_d(t)$, and the frequency demodulator output is $\dot{\varphi}_d(t)$.

---

* Here we are using $\sin [\omega_c t + \theta_i(t)]$ rather than the usual $\cos [\omega_c t + \theta_i(t)]$. This is really immaterial, because a cosine can be expressed as a sine with a $\pi/2$ phase addition. Because the final step [Eq. (4.87)] involves differentiation of the angle, the constant phase vanishes.

Hence, the demodulator output $y_d(t)$ is given by

$$y_d(t) = \frac{I}{A} \sin \omega_d t \qquad \text{for PM} \tag{4.92}$$

$$y_d(t) = \frac{I\omega_d}{A} \cos \omega_d t \qquad \text{for FM} \tag{4.93}$$

Observe that in either case, the interference output is inversely proportional to the carrier amplitude A. Thus, the larger the carrier amplitude A, the smaller the interference effect. This behavior is very different from that observed in AM signals (Sec. 4.8), where the interference output is independent of the carrier amplitude. Hence, angle-modulated systems suppress weak interference much better than do AM systems.

Because of the suppression of weak interference in FM, we observe what is known as the "capture effect." For two transmitters with carrier-frequency separation less than the audio range, instead of getting interference, we observe that the stronger carrier effectively suppresses (captures) the weaker carrier.

Subjective tests show that an interference level as low as 35 dB in the audio signals can cause objectionable effects. Hence in AM, the interference level should be kept below 35 dB. On the other hand, for FM, because of the capture effect, the interference level need only be below 6 dB. A more extensive discussion of interference in angle-modulated systems can be found in Panter.[16]

When we compare PM with FM, we observe that the interference amplitude is constant for all ω_d in PM but increases linearly with ω_d in FM (Fig. 4.65).

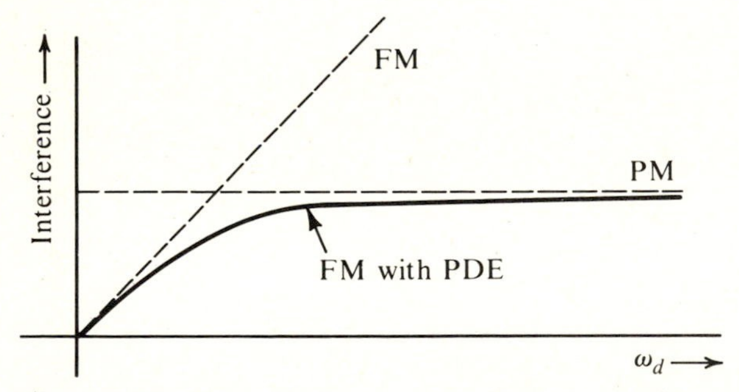

Figure 4.65 Effect of interference in PM, FM, and FM with PDE.

The results in Eqs. (4.92) and (4.93) can be extended for more than one interfering sinusoid. For example, if there are two interfering sinusoids $I_1 \cos (\omega_c + \omega_{d_1})t$ and $I_2 \cos (\omega_c + \omega_{d_2})t$, then the received signal is

$$r(t) = A \cos \omega_c t + I_1 \cos (\omega_c + \omega_{d_1})t + I_2 \cos (\omega_c + \omega_d)t \tag{4.94}$$

$$= E_r(t) \cos [\omega_c t + \varphi_d(t)] \tag{4.95}$$

where

$$\varphi_d(t) = \frac{I_1 \sin \omega_{d_1} + I_2 \sin \omega_{d_2} t}{A + I_1 \cos \omega_{d_1} t + I_2 \cos \omega_{d_2} t} \tag{4.96}$$

For small interference, I_1 and $I_2 \ll A$, and

$$\varphi_d(t) \simeq \frac{1}{A}[I_1 \sin \omega_{d_1}t + I_2 \sin \omega_{d_2}t] \tag{4.97}$$

Hence, the demodulator output $y_d(t)$ is

$$y_d(t) = \frac{1}{A}[I_1 \sin \omega_{d_1}t + I_2 \sin \omega_{d_2}t] \qquad \text{for PM} \tag{4.98}$$

$$= \frac{1}{A}[I_1 \omega_{d_1} \cos \omega_{d_1}t + I_2 \omega_{d_2} \cos \omega_{d_2}t] \qquad \text{for FM} \tag{4.99}$$

Hence, the system behaves linearly, at least for small interfering signals; that is, the combined effect of several independent interfering signals can be computed by finding the effect caused by each interference (assuming all the remaining interfering signals to be zero) and then adding all individual effects.

The Output Signal-to-Noise Ratio

As in the case of AM, we can consider noise as an interference. The results in Eqs. (4.92) and (4.93) or Eqs. (4.98) and (4.99) can be directly used to find the output noise power in PM and FM.

PM. From Eq. (4.92), it can be seen that if a sinusoid $I \cos (\omega_c + \omega_d)t$ interferes with the desired signal on the transmission path, its effect is to generate an interfering (noise) sinusoid $(I/A) \sin \omega_d t$ in the demodulator output. This means that the demodulator translates each interfering sinusoid of frequency $\omega_c + \omega_d$ to a frequency ω_d in the output signal. The amplitude of each sinusoid is multiplied by $1/A$, and the phase is shifted by $\pi/2$. The phase shift is of no consequence for power computation. Multiplication by $1/A$ implies multiplication of the PSD by $1/A^2$ (Example 2.38). This means the channel-noise power density around ω_c is translated to the origin at the demodulator output. Because the channel noise is assumed to be white and a PSD $\mathcal{N}/2$, the noise PSD at the demodulator output would be $(\mathcal{N}/2)/A^2 = \mathcal{N}/2A^2$. This is only half the PSD, however, which arises from channel-noise components above ω_c. Components below ω_c contribute an equal amount of power, because the interference $I \cos (\omega_c - \omega_d)$ yields $(-I/A) \sin \omega_d t$ at the demodulator output* and generates the remaining half of the PSD $(\mathcal{N}/2)/A^2 = \mathcal{N}/2A^2$. Hence $S_{n_o}(\omega)$, the PSD at the demodulator output, is $\mathcal{N}/A^2$.

$$S_{n_o}(\omega) = \frac{\mathcal{N}}{A^2} \tag{4.100}$$

The demodulator noise output is also white (Fig. 4.66). This noise passes through a

*Note that the two sinusoids $I \cos (\omega_c + \omega_d)t$ and $I \cos (\omega_c - \omega_d)t$ have independent phases (not shown here) and cause incoherent addition as discussed in the footnote on p. 262. Therefore, powers resulting from two interferences add.

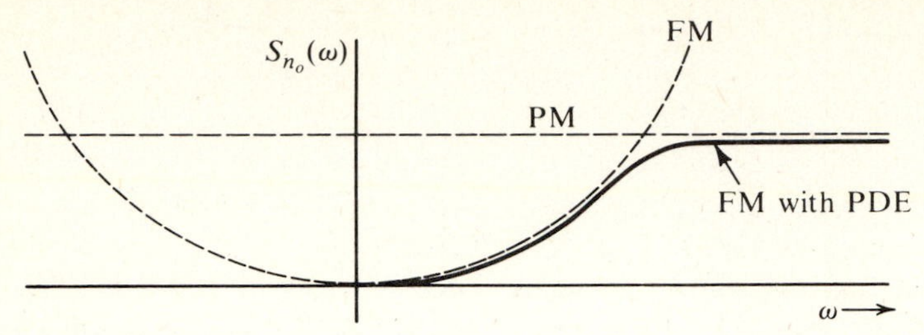

Figure 4.66 The output noise PSD in PM, FM, and FM with PDE.

baseband filter of bandwidth B. Hence, the noise output N_o is

$$N_o = \frac{\mathcal{N}}{A^2}(2B) = \frac{2\mathcal{N}B}{A^2} \tag{4.101}$$

The message signal at the demodulator input is $A \cos [\omega_c t + k_p m(t)]$. Hence,

$$S_i = \frac{A^2}{2} \tag{4.102}$$

The PM demodulator detects the phase $k_p m(t)$. Hence, the output signal* is $k_p m(t)$, and its power S_o is

$$S_o = k_p^2 \widetilde{m^2(t)} \tag{4.103}$$

Hence

$$\begin{aligned} \frac{S_o}{N_o} &= k_p^2 \widetilde{m^2(t)} \left(\frac{A^2/2}{\mathcal{N}B}\right) \\ &= k_p^2 \widetilde{m^2(t)}\, \gamma \end{aligned} \tag{4.104}$$

Also, the frequency deviation $\Delta\omega$ is [Eq. (4.69)]

$$\Delta\omega = k_p m_p'$$

Hence,

$$\frac{S_o}{N_o} = (\Delta\omega)^2 \frac{\widetilde{m^2(t)}}{m_p'^2}\, \gamma \tag{4.105}$$

For tone modulation,

$$m(t) = \alpha \cos \omega_m t \qquad \dot{m}(t) = -\alpha\omega_m \sin \omega_m t$$

Hence,

$$\widetilde{m^2(t)} = \alpha^2/2 \qquad m_p'^2 = \alpha^2\omega_m^2$$

---

* Here we are assuming that the noise power can be computed by assuming zero message signal and that the output message-signal power can be computed by assuming zero noise. This is true for linear systems. Fortunately, it is also valid for angle modulation. Full justification of this is given in Chapter 6.

and

$$\frac{S_o}{N_o} = \frac{1}{2}\left(\frac{\Delta\omega}{\omega_m}\right)^2 \gamma \tag{4.106}$$

FM. We can use a similar technique to compute $S_{n_o}(\omega)$ for FM. From Eq. (4.93) it can be seen that a noise sinusoid $I \cos(\omega_c + \omega)t$ in the received signal is translated as a noise sinusoid $(I\omega/A)\cos \omega t$ at the demodulator output. This means that in FM, the demodulator output noise sinusoid at frequency ω is multiplied by ω/A rather than $1/A$, as was the case in PM. Hence, $S_{n_o}(\omega)$ for FM will be $(\mathcal{N})(\omega^2/A^2)$; that is,

$$S_{n_o}(\omega) = \frac{\mathcal{N}\omega^2}{A^2} \tag{4.107}$$

This PSD is parabolic with ω (Fig. 4.66).

We come to the same conclusion by considering an FM demodulator as a PM demodulator followed by a differentiator (Prob. 4.55, Fig. P4.55*b*). The noise PSD at the PM demodulator output is $\mathcal{N}/A^2$ [Eq. (4.100)]. This passes through an ideal differentiator with $H(\omega) = j\omega$. Hence, the PSD of the output noise will be $|j\omega|^2(\mathcal{N}/A^2) = \mathcal{N}\omega^2/A^2$. This noise is passed through a baseband filter of bandwidth B. Hence, the demodulator noise power output N_o is

$$\begin{aligned} N_o &= 2\int_0^B S_{n_o}(\omega)\, df \\ &= \frac{2\mathcal{N}}{A^2}\int_0^B \omega^2\, df = \frac{2\mathcal{N}}{A^2}\int_0^B (2\pi f)^2\, df \\ &= \frac{8\pi^2\mathcal{N}B^3}{3A^2} \end{aligned} \tag{4.108a}$$

The message signal at the demodulator input is $A\cos[\omega_c t + k_f \int m(\alpha)\, d\alpha]$. Hence,

$$S_i = A^2/2 \tag{4.108b}$$

The FM demodulator detects the instantaneous frequency variation. Hence, the message-signal output of the demodulator is $k_f m(t)$,* and

$$S_o = k_f^2 \widetilde{m^2(t)} \tag{4.108c}$$

Thus,

$$\begin{aligned} \frac{S_o}{N_o} &= \frac{k_f^2 \widetilde{m^2(t)}}{8\pi^2\mathcal{N}B^3/3A^2} \\ &= \frac{3k_f^2 \widetilde{m^2(t)}}{(2\pi B)^2}\left(\frac{S_i}{\mathcal{N}B}\right) \\ &= 3\frac{k_f^2 \widetilde{m^2(t)}}{(2\pi B)^2}\gamma \end{aligned} \tag{4.109a}$$

---

* In a practical demodulator, the output will be $\alpha k_f m(t)$, where α is a constant of the demodulator. Because the signal and the noise at the output are both multiplied by α, the output SNR is unaffected.

Because $\Delta\omega = k_f m_p$

$$\frac{S_o}{N_o} = 3\left(\frac{\Delta\omega}{2\pi B}\right)^2 \left(\frac{\widetilde{m^2(t)}}{m_p^2}\right)\gamma \tag{4.109b}$$

$$= 3\beta^2\gamma\left(\frac{\widetilde{m^2(t)}}{m_p^2}\right) \tag{4.109c}$$

where* β is the deviation ratio $(\Delta f/B)$.

It is clear that

$$\frac{S_o}{N_o} \leq 3\beta^2\gamma$$

For tone modulation,

$$m(t) = \alpha \cos \omega_m t, \qquad m_p = \alpha$$

and

$$\frac{S_o}{N_o} = \frac{3}{2}\left(\frac{\Delta\omega}{\omega_m}\right)^2 \gamma \tag{4.110}$$

We can draw some useful conclusions from Eqs. (4.105) and (4.109). First, in both PM and FM, the output SNR is directly proportional to $(\Delta\omega)^2$. In the wideband case, the transmission bandwidth is roughly twice the frequency deviation. Hence, in wideband PM and FM, the output SNR increases as the square of the transmission bandwidth. Therefore, doubling of the transmission bandwidth quadruples the output SNR (increases it by 6 dB). This is a far cry from PCM case, where increasing just one bit in a code word quadruples the SNR. Increasing one bit in an n-bit code word increases the bandwidth by a fraction $1/n$. Hence, although angle modulation can effect an exchange between the SNR and the transmission bandwidth, it is not an efficient transaction.

Second, between PM and FM, for tone modulation FM yields three times as much SNR as does PM. Does this mean FM is superior to PM? Not at all! All it says is that FM is superior to PM for tone modulation. But nobody has ever used tone modulation to transmit real-life information. Angle modulation is nonlinear modulation, so conclusions for tone modulation cannot be applied blindly to other baseband signals. In fact, these conclusions in many instances are downright misleading. For example, in the literature, PM gets short shrift as being inferior to FM; a conclusion based purely on tone modulation analysis.† For this reason PM is benignly neglected in the texts. PM does not deserve this disrepute. We shall now show that in reality, PM is superior to FM in all practical cases. First let us give an example to show that PM can be superior to FM.

---

* If m_{rms} is the rms amplitude of $m(t)$, the crest, or peak factor, K_m of $m(t)$ is defined as $K_m = m_p/m_{\text{rms}}$. It can be seen that $\widetilde{m^2(t)}/m_p^2 = 1/K_m^2$.

† An additional reason given for the alleged inferiority of PM is that the phase deviation in PM has to be restricted to π for unambiguous demodulation. As seen in Example 4-12, this is simply not true.

■ EXAMPLE 4.17

When $m(t) = 8 \cos \omega_o t + \cos 5\omega_o t$, show that for a given transmission bandwidth, the output SNR in PM is four times that of FM.

□ Solution:

$$m(t) = 8 \cos \omega_o t + \cos 5\omega_o t, \text{ and } m_p = 9$$

$$\dot{m}(t) = -(8\omega_o \sin \omega_o t + 5\omega_o \sin 5\omega_o t), \text{ and* } m_p' = 13\omega_o$$

The bandwidth $B = 5\omega_o/2\pi$.

From Eqs. (4.105) and (4.109b),

$$\frac{(S_o/N_o)_{\text{PM}}}{(S_o/N_o)_{\text{FM}}} = \frac{(2\pi B)^2 m_p^2}{3m_p'^2} = \frac{(5\omega_o)^2 81}{(13\omega_o)^2 3} = 3.994$$ ■

Which Is Superior: PM or FM?

We have seen that for some signals, PM is superior to FM, whereas the opposite may be true for other signals. From Eqs. (4.105) and (4.109b), we have

$$\frac{(S_o/N_o)_{\text{PM}}}{(S_o/N_o)_{\text{FM}}} = \frac{(2\pi B)^2 m_p^2}{3m_p'^2} \tag{4.111}$$

If $m(t)$ has a large peak amplitude, and its derivative $\dot{m}(t)$ has a relatively smaller peak amplitude, PM tends to be superior to FM. For opposite conditions, FM tends to be superior to PM. A general conclusion is rather difficult at this time. But let us consider a class of signals with the same peak amplitude m_p. For this class of signals, it is obvious that PM performance improves in relation to FM as m_p' is reduced. Because m_p' depends upon how rapidly the signal $m(t)$ varies, it is a function of how the power of $m(t)$ is distributed in various frequency components. If the PSD of $m(t)$ is predominantly concentrated at lower frequencies, the signal is a low-frequency signal, m_p' will be smaller, and PM will perform better than FM. On the other hand, if the PSD of $m(t)$ is predominantly concentrated at higher frequencies, m_p' will tend to be larger, and FM will perform better than PM (see Fig. 4.67). This explains why FM is superior for tone

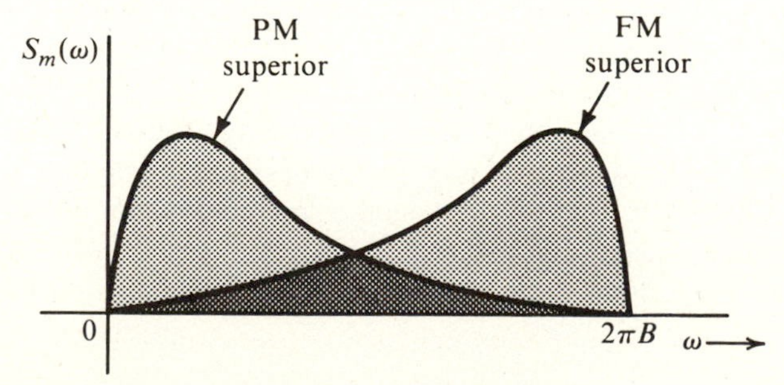

Figure 4.67 Comparison of FM and PM.

---

* m_p' can be found from $(d/dt)\dot{m}(t) = -(8\omega_o^2 \cos \omega_o t + 25\omega_o^2 \cos 5\omega_o t) = 0$. This gives $t_{\max} = 3\pi/2\omega_o$, and $\dot{m}(t_{\max}) = m_p' = 13\omega_o$.

modulation (all power concentrated at the highest frequency) and why PM is superior for $m(t)$ in Example 4.17 (power concentrated at lower frequencies). To determine if PM or FM is superior, a more convenient criterion, in terms of the PSD (rather than m_p and m'_p) of the modulating signal, is derived in Chapter 6.

Most signals encountered in real life, including voice and music, have PSDs that are concentrated at lower frequencies, with relatively little power in the higher frequencies (see Fig. 4.20*a*). In such a case PM is expected to be superior to FM. It may seem logical for radio broadcast stations to use PM rather than FM. Why then is FM rather than PM used in radio broadcasting? The reason is that the so-called broadcast FM is not really FM but is FM modified by preemphasis-deemphasis (PDE). In this mode, it is neither FM nor PM but a general form of angle modulation that is superior to PM and FM.

Preemphasis and Deemphasis

It was mentioned earlier that PM and FM are special cases of angle modulation in Eq. (4.48).

$$\varphi_{\text{EM}}(t) = A \cos\left[\omega_c t + k \int_{-\infty}^{t} m(\alpha)h(t - \alpha)\, d\alpha\right]$$

This can be realized by passing $m(t)$ through a filter $H(\omega)$ [impulse response $h(t)$] and then phase modulating a carrier using the resulting signal (Fig. 4.68). The filter $H(\omega)$

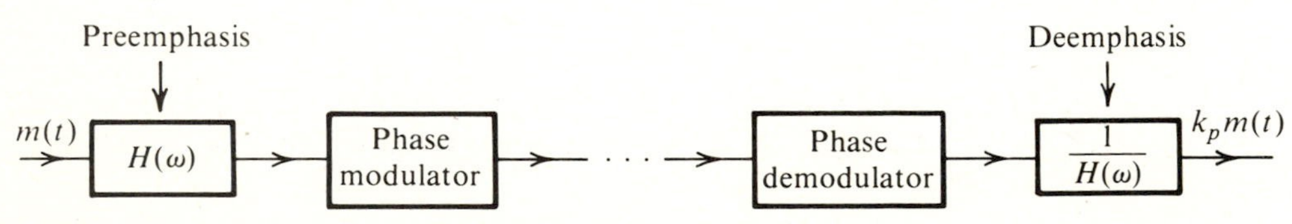

Figure 4.68 PDE in a phase-modulated system.

is known as the preemphasis filter, which redistributes the PSD of the baseband signal. At the receiver, the incoming signal is phase demodulated and then deemphasized through the complementary filter $1/H(\omega)$ in order to get back the original signal. The optimum $H(\omega)$ depends on the nature of $m(t)$ and the channel noise. PM and FM are special cases, for which $H(\omega) = 1$ and $1/j\omega$ (integrator), respectively. Neither PM nor FM may be optimum for audio signals. One must use the appropriate preemphasis filter for optimum performance. When PDE is used, the system is neither PM nor FM.

Figure 4.69 shows the PDE system used in practice. Note that the system in Fig. 4.69 is equivalent to that in Fig. 4.68 with $H_p(\omega) = j\omega H(\omega)$.

Figure 4.69 PDE in a frequency-modulated system.

To get a feel for PDE filters, let us examine the PSDs of $m(t)$ and the output noise. For a radio broadcast, the baseband signal $m(t)$ has a bandwidth of 15 kHz, even though the PSD of $m(t)$ is concentrated within 2 kHz and is small beyond 2 kHz (Fig. 4.20*a*). The output noise, on the other hand, is parabolic (Fig. 4.66). Hence, the noise is strongest in the frequency range where the signal is the weakest. If we boost the high-frequency components of the signal at the transmitter (preemphasis), and then attenuate them correspondingly at the FM receiver (deemphasis), we get back $m(t)$ undistorted. But the noise will be considerably weakened. This is because unlike $m(t)$, the noise enters after the transmitter and is not boosted. It undergoes only deemphasis, or attenuation of high-frequency components, at the receiver. Because the noise PSD is parabolic, attenuation of high-frequency components cuts down the noise significantly.

Optimum PDE filters will be discussed in Chapter 6. For historical and practical reasons, optimum PDE filters are not used in practice. Figure 4.69 shows a system with preemphasis (before modulation) and deemphasis (after demodulation) as used in commercial FM broadcasting. Filters $H_p(\omega)$ and $H_d(\omega)$ are shown in Fig. 4.70. The

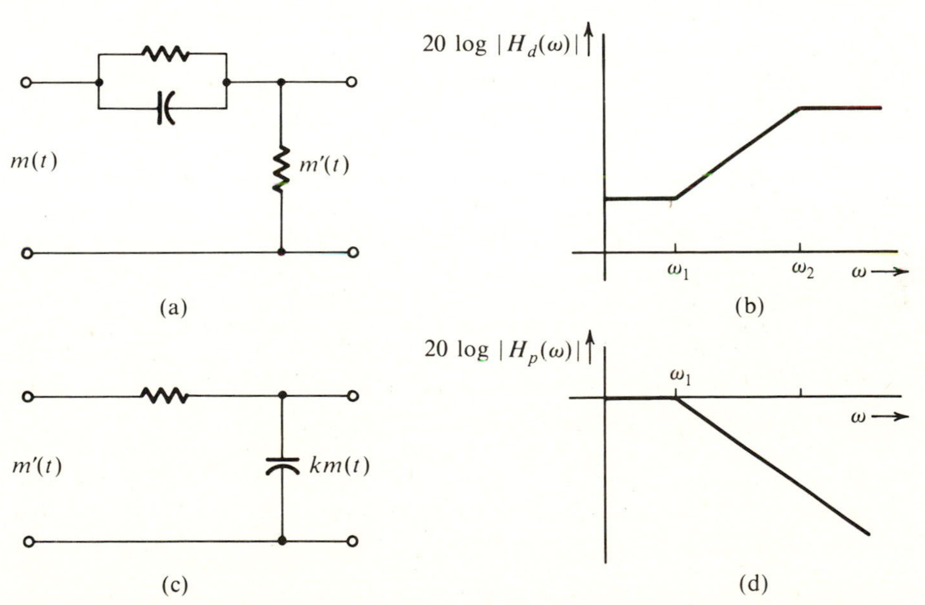

Figure 4.70 (a) Preemphasis filter. (b) Its frequency response. (c) Deemphasis filter. (d) Its frequency response.

frequency f_1 is 2.1 kHz, and f_2 is typically 30 kHz or more (well beyond audio range), so that f_2 does not even enter into the picture. These filters can be realized by simple *RC* circuits (Fig. 4.70). The choice of $f_1 = 2.1$ kHz was apparently made on an experimental basis. It was found that this choice of f_1 maintained the same peak amplitude m_p with or without preemphasis.[17] This satisfied the constraint of a fixed transmission bandwidth.

The preemphasis transfer function is

$$H_p(\omega) = K\frac{j\omega + \omega_1}{j\omega + \omega_2} \tag{4.112a}$$

where K, the gain, is set at a value of ω_2/ω_1. Thus,

$$H_p(\omega) = \left(\frac{\omega_2}{\omega_1}\right)\frac{j\omega + \omega_1}{j\omega + \omega_2} \tag{4.112b}$$

For $\omega \ll \omega_1$,

$$H_p(\omega) \simeq 1 \tag{4.112c}$$

For frequencies $\omega_1 \ll \omega \ll \omega_2$,

$$H_p(\omega) \simeq \frac{j\omega}{\omega_1} \tag{4.112d}$$

Thus, the preemphasizer acts as a differentiator at intermediate frequencies (2.1 to 15 kHz). This means that FM with PDE is FM over the modulating-signal frequency range 0 to 2.1 kHz and is nearly PM over the range 2.1 to 15 kHz.

The deemphasis filter $H_d(\omega)$ is given by

$$H_d(\omega) = \frac{\omega_1}{j\omega + \omega_1}$$

Note that for $\omega \ll \omega_2$, $H_p(\omega) \simeq (j\omega + \omega_1)/\omega_1$. Hence, $H_p(\omega)H_d(\omega) \simeq 1$ over the baseband 0 to 15 kHz.

To compute the improvement in SNR resulting from PDE, we observe that the parabolic PSD of the output noise [Eq. (4.107)] passes through a deemphasis filter

$$H_d(\omega) = \frac{\omega_1}{j\omega + \omega_1}$$

Thus N_o', the noise power at the deemphasis filter output, is

$$\begin{aligned}
N_o' &= 2\int_0^B S_{n_o}(\omega)|H_d(\omega)|^2\,df \\
&= 2\int_0^B \frac{\mathcal{N}\omega^2}{A^2}\frac{\omega_1^2}{\omega^2 + \omega_1^2}\,df \\
&= \frac{2\mathcal{N}\omega_1^2}{A^2}\int_0^B \frac{4\pi^2 f^2}{4\pi^2 f^2 + \omega_1^2}\,df \\
&= \frac{\mathcal{N}\omega_1^2}{A^2}\left[2B - 2f_1 \tan^{-1}\frac{B}{f_1}\right] \qquad f_1 = \omega_1/2\pi
\end{aligned}$$

The noise power N_o without PDE is found in Eq. (4.108a). Hence, the improvement factor N_o/N_o' is

$$\frac{N_o}{N_o'} = \frac{1}{3\left[1 - \frac{f_1}{B}\tan^{-1}\left(\frac{B}{f_1}\right)\right]}\left(\frac{B}{f_1}\right)^2 \tag{4.113a}$$

Substituting $B = 15$ kHz and $f_1 = 2.1$ kHz, we get

$$\frac{N_o}{N_o'} = 21.25 \tag{4.113b}$$

$$= 13.27 \text{ dB} \tag{4.113c}$$

Because the output signal power itself remains unchanged by PDE, the output SNR increases by a factor of 21 (13 dB).

The side benefit of PDE is improvement in the interference characteristics. Because the interference enters after the transmitter stage, it undergoes only the deemphasis operation and not the boosting, or preemphasis. Hence, the interference amplitudes for frequencies beyond 2.1 kHz undergo attenuation that is roughly linear with frequency. The resulting interference characteristic (with PDE) is shown in Fig. 4.65. Note that with PDE, we have the best of both PM and FM—that is, low interference at low frequencies (as in FM) and a constant interference at high frequencies (as in PM).

We could also use preemphasis-deemphasis in AM broadcasting to improve the output SNR. In practice, however, this is not done for several reasons. First, the output noise PSD in AM is flat, and not parabolic as in FM. Hence, the deemphasis does not yield such a dramatic improvement in AM as it does in FM. Second, introduction of PDE would necessitate modifications in receivers already in use. Third, increasing high-frequency component amplitudes (preemphasis) would increase interference with adjacent stations (no such problem arises in FM). Moreover, an increase in the modulation index at high frequencies would make detector design more difficult.

Threshold

The interference analysis in angle modulation was based on the small-interference assumption. Hence, all the SNR relationships derived here are valid for the small-noise case. For large noise, we encounter the familiar threshold phenomenon. This can be seen from the capture effect. It was shown that for angle modulation, if two signals interfere with each other, the stronger one captures (i.e., effectively suppresses) the weaker one. Hence, when the noise is small, the desired signal captures the noise, yielding a higher SNR. But this is a double-edged sword, which also works the other way. If the noise is strong, it can effectively capture (or suppress) the signal. The onset of the threshold begins when the channel noise power is about one-tenth the signal power. The signal power is $A^2/2$, and for white noise with a PSD $\mathcal{N}/2$, the channel noise power is $\mathcal{N}B_{FM}$, where B_{FM}, the transmission bandwidth, is [Eq. (4.61)]

$$B_{FM} = 2B(\beta + 2)$$

Hence, the threshold occurs when

$$2\mathcal{N}B(\beta + 2) = \frac{1}{10}\frac{A^2}{2}$$

$$20(\beta + 2) = \frac{A^2/2}{\mathcal{N}B}$$

By definition, the right-hand side of the above equation is γ. Hence, the threshold occurs for values of γ below $20(\beta + 2)$; or

$$\gamma_{\text{thresh}} = 20(\beta + 2) \tag{4.114}$$

The effect of the threshold is shown in Fig. 4.71 for tone modulation, where $\widetilde{m^2(t)}/m_p^2$ is $\frac{1}{2}$, and [see Eq. (4.109c)]

$$\frac{S_o}{N_o} = \frac{3}{2}\beta^2\gamma \tag{4.115}$$

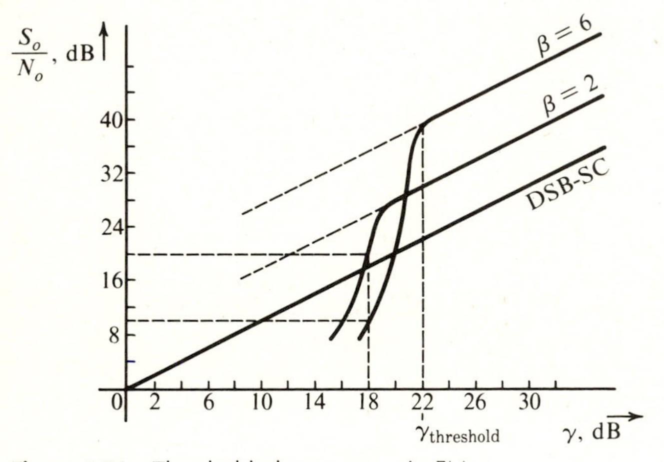

Figure 4.71 Threshold phenomenon in FM.

Figure 4.71 shows S_o/N_o in decibels as a function of γ for two values of β. For $\beta = 2$, the threshold occurs at $\gamma_{\text{thresh}} = 20(\beta + 2) = 80$, which is 19 dB. For $\beta = 6$, $\gamma_{\text{thresh}} = 20(6 + 2) = 160$, which is 22 dB. Observe the rapid deterioration of the SNR below the threshold. To understand the threshold properly, consider the operation on the curve $\beta = 2$ (bandwidth expansion ratio $B_{\text{FM}}/B = 8$). If the transmitted power is such that $\gamma = 18$ dB, we are in the threshold region ($\gamma_{\text{thresh}} = 19$ dB), and the output SNR is low (about 20 dB; see Fig. 4.71). If we try to increase the SNR by the usual method of increasing the transmission bandwidth (remember: doubling the bandwidth increases the SNR by 6 dB), we are in for a surprise. Suppose we go to $\beta = 6$ (transmission bandwidth $B_{\text{FM}} = 16B$); we see from Fig. 4.71 that at $\gamma = 18$ dB, the S_o/N_o is about 10 dB. That is, increasing the transmission bandwidth has actually reduced the SNR and put us deeper into the threshold. Why? This is because for $\beta = 2$, we are already in the threshold. By increasing the transmission bandwidth further, we are increasing the channel noise proportionately. Hence, we go even deeper into threshold, and instead of improving, the SNR deteriorates even further. It can be seen that the SNR for $\beta = 6$ at this point is inferior even to the DSB or SSB curve. This shows that in angle modulation, we cannot realize the SNR improvement indefinitely merely by increasing the transmission bandwidth. Beyond a point, bandwidth increases prove to be counterproductive.

Using Eq. (4.115), the output SNR is plotted as a function of γ for various values of β in Fig. 4.72. The dotted line shows the limit of operation above the threshold.

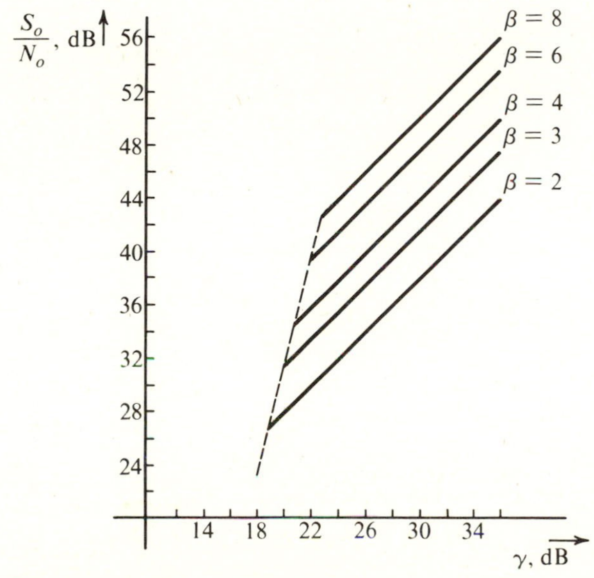

Figure 4.72 Performance of FM system.

Although this plot is valid for tone modulation only, it can be used for any general case with $\widetilde{m^2(t)}/m_p^2 = \lambda$. This is because the SNR for a general case [Eq. (4.109c)] and for tone modulation [Eq. (4.115)] differ by a factor $\lambda/0.5 = 2\lambda$. Hence for a general case, we simply shift the curves in Fig. 4.72 upwards by 10 log (2λ) dB.

4.17 THE FM RECEIVER

The FCC has assigned a frequency range of 88 to 108 MHz for FM broadcasting, with a separation of 200 kHz between adjacent stations and a peak frequency deviation $\Delta f = 75$ kHz.

A monophonic FM receiver is identical to the superheterodyne AM receiver in Fig. 4.34 except that the IF frequency is 10.7 MHz and the envelope detector is replaced by an amplitude limiter and a frequency discriminator followed by a deemphasizer.

Earlier FM broadcasts were monophonic. Stereophonic FM broadcasting, in which two audio signals L (left microphone) and R (right microphone) are used for a more natural effect, was proposed later. The FCC ruled that the stereophonic system had to be compatible with the original monophonic system. This meant that the older monophonic receivers should be able to receive the signal $L + R$, and the total transmission bandwidth for the two signals (L and R) should still be 200 kHz, with a Δf of 75 kHz for the two combined signals. This would ensure that the older receivers could continue to receive monophonic as well as stereophonic broadcasts, although in the latter case the stereo effect would be absent.

A transmitter and a receiver for a stereo broadcast are shown in Fig. 4.73*a* and *c*, respectively. At the transmitter, the two signals L and R are added and subtracted to obtain $L + R$ and $L - R$. These signals are preemphasized. The preemphasized signal $(L - R)'$ DSB-SC modulates a carrier of 38 kHz obtained by doubling the frequency of a 19-kHz signal that is used as a pilot. The signal $(L + R)'$ is used directly. All three signals (the third being the pilot) form a composite baseband signal $m(t)$ (Fig. 4.73*b*).

$$m(t) = (L + R)' + (L - R)' \cos \omega_c t + \alpha \cos \frac{\omega_c t}{2} \tag{4.116}$$

The reason for using a pilot of 19 kHz rather than 38 kHz is that it is easier to extract the pilot at 19 kHz, because there are no signal components within 4 kHz of that frequency.

The receiver operation (Fig. 4.73*c*) is self-explanatory. A monophonic receiver consists of only the upper branch of the receiver and, hence, receives only $L + R$. This is of course the complete audio signal without the stereo effect. Hence, the system is compatible. The pilot is extracted, and (after doubling its frequency) it is used to demodulate coherently the signal $(L - R)' \cos \omega_c t$.

An interesting aspect of stereo transmission is that the peak amplitude of the composite signal $m(t)$ in Eq. (4.116) is practically the same as that of the monophonic signal (if we ignore the pilot), and, hence, Δf—which is proportional to the peak signal amplitude for stereophonic transmission—remains practically the same as that for the monophonic case. This can be explained by the so-called *interleaving* effect, as follows.

The L' and R' signals are very similar in general. Hence, we can assume their peak amplitudes to be equal to A_p. Under the worst possible conditions, L' and R' will reach their peaks at the same time, yielding [Eq. (4.116)]

$$|m(t)|_{\max} = 2A_p + \alpha$$

In the monophonic case, the peak amplitude of the baseband signal $(L + R)'$ is $2A_p$. Hence, the peak amplitudes in the two cases differ only by α, the pilot amplitude. To account for this, the peak sound amplitude in the stereo case is reduced to 90 percent of its full value. This amounts to a reduction in the signal power by a ratio of $(0.9)^2 = 0.81$, or 1 dB. Thus, the effective SNR is reduced by 1 dB because of the inclusion of the pilot.

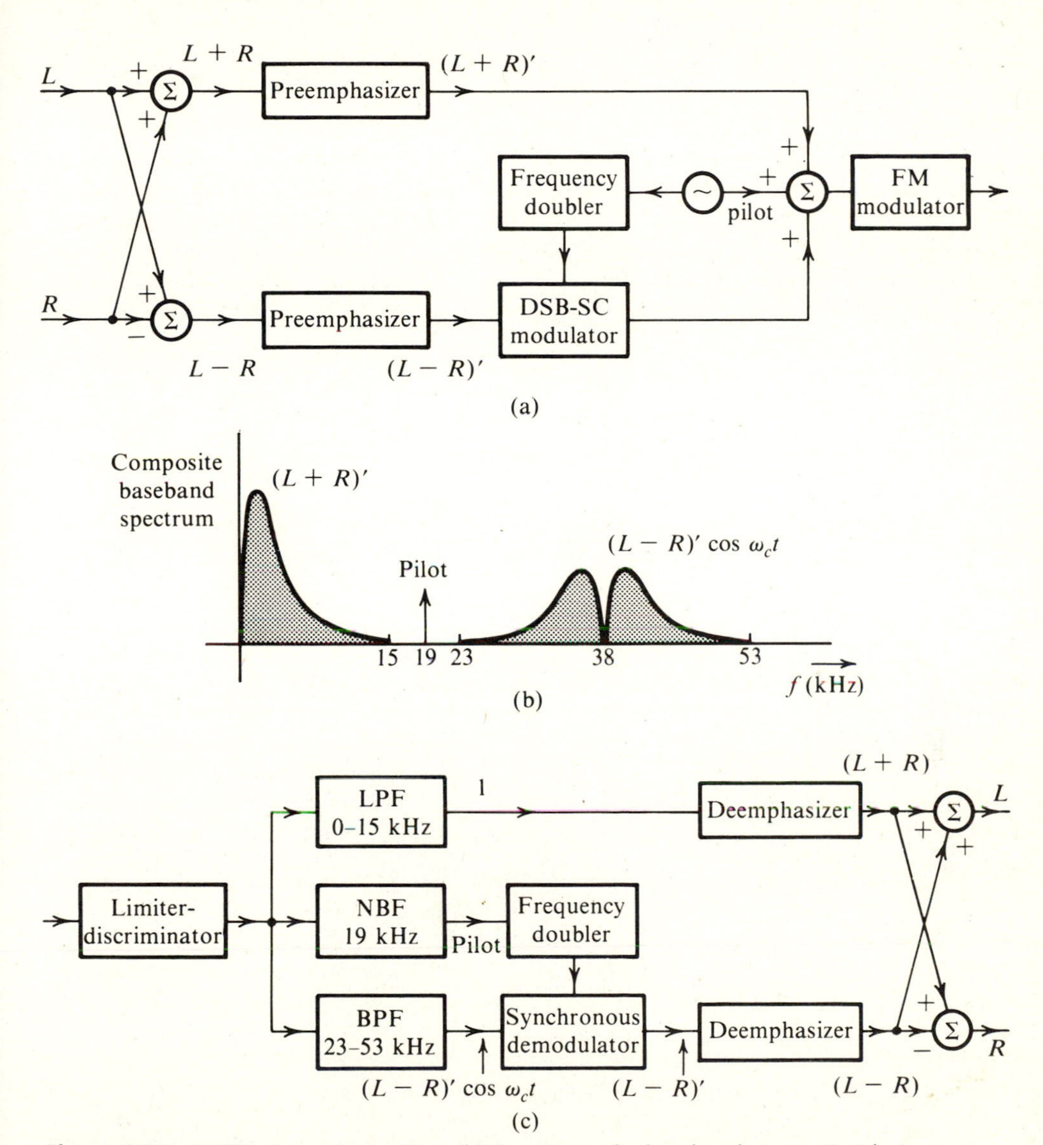

Figure 4.73 (a) FM stereo transmitter. (b) Spectrum of a baseband stereo signal. (c) FM stereo receiver.

Transmission media	Wavelength	Designation	Frequency	Applications
Optical fibers	10^{-7}	Ultraviolet	10^{15}	Optical communication
		Visible light		
	10^{-6} (1 micron)	Infrared	10^{14}	
Waveguides, line-of-sight.	1 cm	30—300 GHz extremely high frequency (EHF)	10^{11}	Research, radio astronomy, radar landing systems
Line-of-sight relaying, line-of-sight ionosphere penetration, waveguides	10 cm	3—30 GHz superhigh frequency (SHF)	10^{10}	Satellite and space communication, microwave relay, radar (airborne, approach, surveillance, and weather)
Tropospheric scatter, line-of-sight relaying	1 M	0.3—3 GHz ultrahigh frequency (UHF)	10^{9}	TV (UHF), space telemetry radar, military satellite communication
Coaxial cables, skywave (ionospheric and tropospheric scatter)	10 M	30—300 MHz very high frequency (VHF)	10^{8}	TV (VHF) and FM, land transportation (taxis, buses, railroads), air traffic control
Coaxial cables, ionospheric reflection (sky wave)	100 M	3—30 MHz high frequency (HF)	10^{7}	Business, amateur and citizens band, military communication, mobile radio telephone
	1 km	0.3—3 MHz medium frequency (MF)	10^{6}	AM broadcasting, amateur, mobile, public safety
Wire pairs, surface ducting (ground wave)	10 km	30—300 kHz low frequency (LF)	10^{5}	Navigational aids, radio beacons, industrial (power line) communication
	100 km	3—30 kHz very low frequency (VLF)	10^{4}	Navigation, telephony, telegraphy, frequency and timing standards
	1000 km	0.3—3 kHz voice frequency (VF)	10^{3}	Telephony, data terminals
		30—300 Hz extremely low frequency (ELF)	10^{2}	Macrowave submarine communication

Figure 4.74 Various frequency bands with typical uses and transmission media.

4.18 TRANSMISSION MEDIA

Message-bearing electrical signals are transmitted over a distance through a variety of transmission media, ranging from a pair of wires to optical fibers, depending on the nature of the electrical signals.

An electrical signal is a form of an electromagnetic wave of a certain frequency and a wavelength. Thus, a telegraph signal, a radio broadcast signal, and a light beam from the sun, a star, or a laser are all forms of electromagnetic energy of different frequencies. Various frequency bands of the electromagnetic spectrum are assigned to specific types of communication, as shown in Fig. 4.74. For each band, we need to use an appropriate transmission medium suitable for the frequency range (as indicated in Fig. 4.74). The following is a brief discussion of various transmission media encountered in practice.

1. Wire-Pair Cables

Several pairs of wires, each pair forming one transmission path, are packed in one cable. The crosstalk between pairs is reduced by twisting the pairs. These are used mainly for telephone signals and low-rate data communication (see Fig. 4.74).

2. Coaxial Cables

At higher frequencies, wire pairs are unsuitable, because they have higher electrical resistance due to the skin effect, and they suffer an increased loss of energy due to radiation from the wires. For higher frequencies, a coaxial cable is appropriate, because it eliminates both of the above problems. In addition, there is virtually no crosstalk between various coaxial cables that are bound together in one large cable. This is because the current in each coaxial cable is concentrated on the inside of the outer shell and the outside of the inner conductor, creating a shielding effect. For this same reason, coaxial cables are much more immune to noise and crosstalk. Hence, a signal can withstand much more attenuation before requiring amplification. Whereas a single pair of wires can carry 12 or 24 voice channels, a single coaxial cable can carry 1800 to 3600 (or even more) voice channels.

3. Radio Systems

At still higher frequencies, radiation of electromagnetic waves through free space becomes attractive because of reduced antenna dimensions. The energy radiated from a transmitting antenna may reach the receiving antenna over any of several possible propagation paths, as shown in Fig. 4.75. The wave that reaches the receiving antenna after being reflected from the ionosphere is the *sky wave*. The waves that are reflected at abrupt changes in the effective dielectric constant of the troposphere (the region within 10 km of the earth's surface) are the *tropospheric waves*. Energy propagated over all other paths is considered to be the *ground wave,* which can be divided into a *space wave* and a *surface wave*. The space wave is made up of the *direct wave* (the signal that travels the direct path from transmitter to receiver) and the *ground reflected wave,* which is the signal that arrives at the receiver after being reflected from the surface of the earth. The surface wave is a wave that is guided along the earth's surface, much as an electromagnetic wave is guided by a transmission line.

A received signal can be the resultant of a number of the above waves. All these mechanisms of propagation may be present over a radio link. Some of them are negligible in certain frequency ranges, however. In the VLF band, for example, the wavelengths are so long that they are comparable to the heights of the lowest iono-

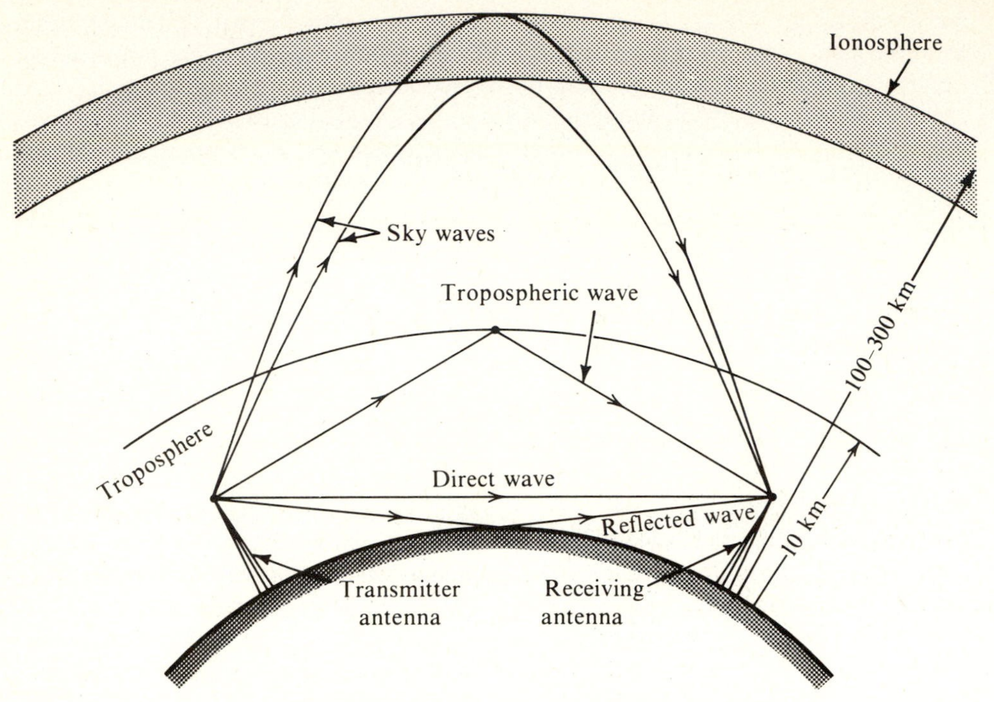

Figure 4.75 Several possible propagation paths in a radio system.

spheric layers (about 100 km). The ionosphere and the earth's surface act as conducting planes to form a waveguide. Thus, VLF signals can have worldwide coverage. This band is used for telegraph transmission, for navigational aids, and for distributing standard frequencies.

In the LF band, propagation is mainly from the ground wave, which provides stable transmission over distances up to about 1500 km. This band is used for long-wave sound broadcasting. In the MF and HF bands it is the sky wave that predominates. These bands are used for sound broadcasting (AM and amateur radio, etc.) and long-distance communication to ships and aircraft.

At frequencies above 30 MHz, the radio waves pass through the ionosphere instead of being reflected by it. Hence, radio communication in the VHF and UHF bands depends on the direct-wave mechanism. The range of the direct wave is the "line-of-sight" distance. Because of the tropospheric waves, however, the range increases beyond the optical horizon.

Microwave transmission (SHF) over distances beyond the optical horizon can be obtained by means of the mechanism of *tropospheric scattering* (tropospheric reflection and refraction). Microwave (along with coaxial cable) is used for bulk transmission. Of the two, microwave is the main contender. Microwave has been used more extensively than coaxial in recent years for the building of long-haul trunks. Like coaxial cables, microwave links today carry thousands of voice channels and are in widespread use for the transmission of television signals. They require line-of-sight transmission by a chain of relaying antennas throughout the region. Relay towers are usually spaced 30 miles apart. Thus, long-distance telephone and television signals are picked up every 30 miles, amplified, and retransmitted. A long-distance microwave circuit has fewer repeaters as compared to a coaxial cable circuit, because the repeaters are spaced about 30 miles apart in the former and 2 to 4 miles apart in the latter. This results in a superior quality signal in the microwave system in comparision to that in the coaxial system.

During the past 20 years, extensive use has been made of the frequency bands of 2 to 10 GHz for analog-signal transmission. For this reason, the development of digital microwave systems has been concentrated at higher frequencies. At present, systems operating in the 11-GHz and the 19-GHz region are being considered. These frequencies present a severe problem of atmospheric loss of signal caused by water-vapor absorption and oxygen absorption.

Satellite Communication Systems. A communication satellite provides a form of microwave relay. Because it is high in the sky, its line-of-sight range is much longer, and, therefore, it relays signals over longer distances. The early communication satellites were in relatively low-altitude orbits and, consequently, sped around the earth in a few hours. This required the ground antennas to move constantly in order to beam signals to them. The satellites were overhead for only a brief period. Hence, early transatlantic TV transmission was confined to a 5-minute session of Walter Cronkite. In 1965, the Early Bird satellite was launched into a much higher equatorial orbit of 36,000 km (or 22,400 miles), so that it traveled around the earth once in 24 hours. Because the earth also rotates once in 24 hours, the satellite appears to be stationary over the earth. Small thrusters make adjustments to keep it as nearly stationary as possible. A growing number of stationary satellites are now being used for military and civilian purposes. Such satellites are known as *synchronous satellites*. Satellites are powered by solar batteries. The number and type of satellites to be used

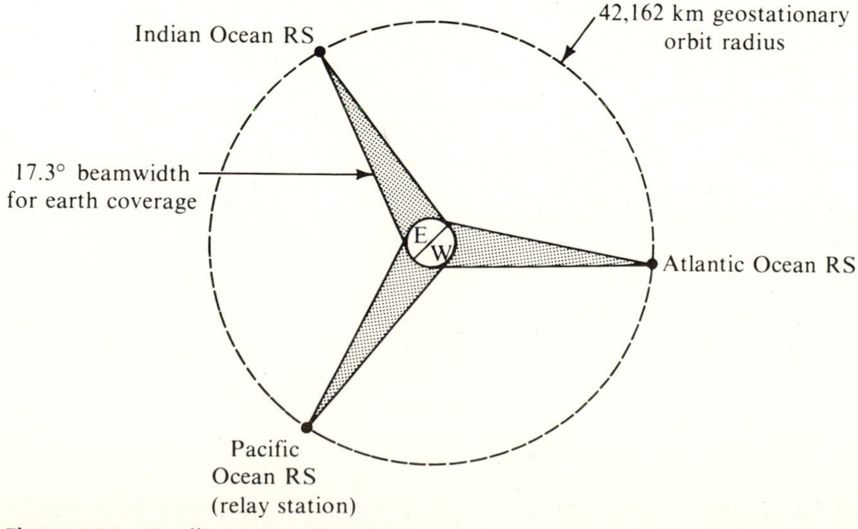

Figure 4.76 Satellite communication.

in a satellite relay network depends on the network coverage desired. Although three satellites in geostationary orbits (Fig. 4.76) can provide global coverage, the desire to satisfy an increased communications demand and other reasons make four or more satellites with closer spacing an obvious consideration for a global system. A smaller number of satellites are required for domestic systems that must provide regional coverage for a nation or group of nations.

A disadvantage of a satellite link is a long propagation delay caused by great distances between the earth stations and the geostationary satellite. For a one-way channel, this delay is about 270 ms. Thus, in a two-way telephone conversation, an interval of more than half a second occurs between speaking and receiving an answer. Because of the delay, it is necessary to use echo suppressors. In practice, the delay is not objectionable to telephone users connected by a single satellite link. Connections made between two links in tandem, however, would be unsatisfactory. Consequently, the CCITT recommends that very long intercontinental connections be made over a tandem connection of a satellite link one way and a submarine cable the other way.

4. Waveguides

Waveguide transmission can be considered as radio transmission confined within a controlled environment. Hence, it is possible to eliminate the atmospheric absorption and other instabilities that are inherent in free-space transmission. Waveguides are capable of transmitting much higher frequencies, and signal losses decrease with frequency. Theoretically, the losses should continue to decrease indefinitely as frequency increases, although an upper limit is set by today's engineering. Waveguides constructed by Bell Laboratories can carry up to 200,000 voice signals in one direction. Waveguide systems are not in commercial use today, although technical problems have now been largely solved. Their introduction is restricted by economic factors and traffic requirements, because these systems are applicable only to very high-capacity routes.

5. Optical Fibers

These are discussed in the next section.

4.19 OPTICAL COMMUNICATION

In a little over 10 years, lightwave communication using optical fibers has progressed from a laboratory proposal to a commercial reality. We can anticipate the widespread use of optical fibers on a routine basis beginning in the early eighties. Along with this will come reduced cost of existing services and the introduction of new services made more economical by this new transmission medium.

The optical band is just an extension of the radio and microwave spectrum. The laser made available a coherent optical frequency of the order of 10^{15}. Even if we use only a 0.1 percent bandwidth, it still corresponds to a 1000-GHz bandwidth—a transmission capacity previously undreamed of. It is, however, difficult to achieve this capacity in practice, at least at this time. The practical optical-communication systems today operate at a rate of 2 Mbits/second to 1 Gbits/second. But the technology is progressing rapidly, and researchers expect to achieve rates of several Gbits/second.

The present-day transmission system is more akin to the spark-gap transmitter and receiver used at the turn of the century. Yet these crude systems represent an effective solution to real system needs. Perhaps the ultimate systems will follow in the course of time, in one or two decades, but a dramatic increase in the need for ground communication will have to emerge before they can be utilized.

Two technological developments made optical communication a reality: the development of light sources that can be modulated at high digit rates, and the production of low-loss glass fibers that act as optical waveguides. These light sources fall into two categories: *light emitting diodes* (LEDs), which produce noncoherent light, and *lasers*, which produce coherent light. The former seem to be much more practical today. LEDs can be directly modulated by varying the drive current at a rate up to a few hundred Mbits/second. The laser is a threshold device which turns on at about 100 mA of drive current, although recent reports indicate much lower drive currents. The laser can be modulated by varying the drive current up to a rate of the order of Gbits/second.

Silicon pin and avalanche photodiodes are used as detectors. For longer wavelengths, germanium and indium gallium arsenide detectors are being considered. These detectors convert light into electrical currents with excellent efficiency and low noise.

The optical fibers produced today have low losses, of less than 1 dB/km. This means systems using repeater spacing of the order of 30 km are practicable.[18]

The advantages of optical communication are low noise, absence of crosstalk, fewer repeaters, and a very large capacity. The Bell System is making a standard fiber-optic system available to its operating companies with the first one going to its Atlanta telephone network. Systems are also installed in Bernal Heights, California; Smyrna, Georgia; Pittsburgh, Pennsylvania; San Francisco, California; White Plains, New York; and Newark, New Jersey. Although most of these links are short, one installation will serve the 40-mile route between Pittsburgh and Greensburg, Pennsylvania. In the years to come, the Bell System expects to install fiber links between telephone centrals at a faster rate.

Recently, Pacific Telephone Company announced plans[19]

> to construct a "lightwave network" in California geared to meet increasing demands for telecommunications services including high-speed computer linkups. . . . The 633-mile-long, . . . laser communications project will be the largest in the world when it is completed in 1985. . . . The project, which will eventually link San Francisco, Sacramento, Los Angeles, San Diego and cities in between, is needed to meet the needs of California's burgeoning telecommunications market.
>
> The network's first leg will go into service in February 1983, connecting Sacramento to Stockton, Oakland and San Jose. The completed system will increase Pacific Telephone's present long-distance telecommunications capacity by roughly two-thirds. . . . Increased demand in the West Coast telecommunications market [is attributed in large part to] the growing dependence of many businesses on telephone lines to tie together computer systems and tie in to remote computer data banks.
>
> The system, which uses digital rather than analog technology, will connect 19 super-capacity electronic switching stations in Northern and Southern California. . . . Each of the 19 centers will be capable of processing up to 550,000 calls an hour.
>
> [All this will be accomplished by using a cable (no wider than a human finger) consisting of 144 hair thin strands of optical fiber.]

Longer fiber-optic links, including a transatlantic cable, are under serious consideration. Bell researchers are developing a multifiber cable for undersea telephone transmission that will increase repeater distance to 35 km while increasing two-way voice-channel capacity to 36,000. Transatlantic service is expected to start in 1988, with transmission of streams at 274 Mbits/second. The cable will be 6500-km long and reach ocean depths of 6.5 km.

Submarine cables also figure prominantly in Japan's telecommunication picture. Nippon Telegraph and Telephone is developing a fiber-optic system that is to be operational in 1985. It is to transmit from 5000 to 18,000 voice channels at 400 Mbits/second, with repeaters spaced 25 km apart.

Despite all the advantages of optical communication, in countries with sophisticated communication systems (such as the United States), fiber optics will not suddenly take over the earlier modes, simply because of the large investments already made in existing systems. As *Fortune* magazine says,[20] "A. T. & T. alone has $111 billion of plant in place, most of it copper, and the $25 million it is spending on fiber optics this year is an insignificant fraction of the $16 billion the company will put into expansion and modernization." In the United States, fiber optics will gradually creep into use. In the short term, fibers have a much brighter future in places such as Canada's western provinces, where telecommunication services are not as common. Saskatchewan, for instance, is considering plans for a province-wide 3200-km fiber-cable system, in which the broadband network will link the telephone exchange to the province's cities and major towns and carry voice data and video. At the same time, Bell Canada has been in the forefront of techniques to expand the use of fiber optics beyond telephone applications. Already, close to 3600 km of fiber cables (44,000 km of fibers) has been installed throughout the Canadian telephone trunk network, and these lines are being expanded, as needs demand, to transmit data and video signals.[18] China has also expressed interest, with the security angle one of fiber optics' selling points. The United States may thus find itself in the peculiar position of being bypassed by less-developed countries in the application of a technology that it has been instrumental in perfecting and in whose development it has until now pretty much led the world.

APPENDIX 4.1 Phase-Lock Loop Analysis

Small-Error Analysis

In this case, $\sin \theta_e \simeq \theta_e$, and the block diagram in Fig. 4.64*b* reduces to that in Fig. A4.1*a*. Straightforward calculation gives

$$\frac{\Theta_o(s)}{\Theta_i(s)} = \frac{AKH(s)/s}{1 + [AKH(s)/s]} = \frac{AKH(s)}{s + AKH(s)} \tag{A4.1}$$

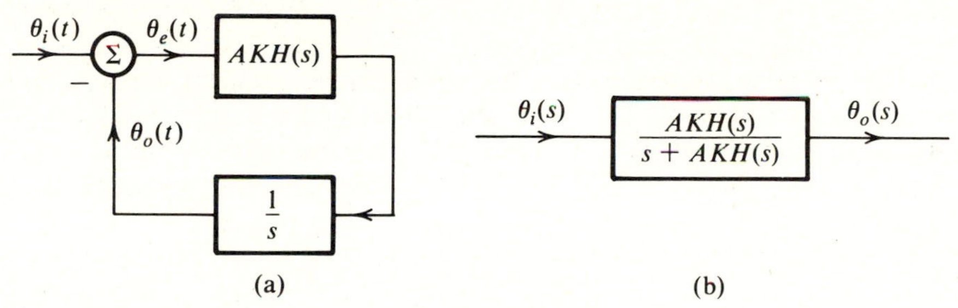

Figure A4.1 Equivalent circuits of a linearized PLL.

Therefore, the PLL acts as a filter with transfer function $AKH(s)/[s + AKH(s)]$, as shown in Fig. A4.1*b*. The error $\Theta_e(s)$ is given by

$$\Theta_e(s) = \Theta_i(s) - \Theta_o(s) = \left[1 - \frac{\Theta_o(s)}{\Theta_i(s)}\right]\Theta_i(s)$$

$$= \frac{s}{s + AKH(s)}\Theta_i(s) \tag{A4.2}$$

One of the important applications of the PLL is in acquisition of the frequency and the phase for the purpose of synchronization. Let the incoming signal be $A \sin (\omega_o t + \varphi_o)$. We wish to generate a local signal of frequency ω_o and phase* φ_o. Assuming the quiescent frequency of the VCO to be ω_c ($\omega_c \neq \omega_o$),

$$\theta_i(t) = (\omega_o - \omega_c)t + \varphi_o$$

and

$$\Theta_i(s) = \frac{(\omega_o - \omega_c)}{s^2} + \frac{\varphi_o}{s}$$

* With a difference $\pi/2$.

Consider the special case of $H(s) = 1$. Substituting the above equation in Eq. (A4.2),

$$\Theta_e(s) = \frac{s}{s + AK}\left[\frac{\omega_o - \omega_c}{s^2} + \frac{\varphi_o}{s}\right]$$

$$= \frac{(\omega_o - \omega_c)/AK}{s} - \frac{(\omega_o - \omega_c)/AK}{s + AK} + \frac{\varphi_o}{s + AK}$$

Hence,

$$\theta_e(t) = \frac{(\omega_o - \omega_c)}{AK}[1 - e^{-AKt}] + \varphi_o{}^{-AKt} \tag{A4.3a}$$

Observe that

$$\lim_{t\to\infty} \theta_e(t) = \frac{\omega_o - \omega_c}{AK} \tag{A4.3b}$$

Hence, after the transient dies (in about $4/AK$ seconds), the phase error maintains a constant value of $(\omega_o - \omega_c)/AK$. This means the PLL frequency eventually equals the incoming frequency ω_o. There is, however, a constant phase error. The PLL output is

$$2C \cos\left[\omega_o t + \varphi_o - \frac{\omega_o - \omega_c}{AK}\right]$$

For a second-order PLL using

$$H(s) = \frac{s + a}{s} \tag{A4.4a}$$

$$\Theta_e(s) = \frac{s}{s + AKH(s)}\Theta_i(s) \tag{A4.4b}$$

$$= \frac{s^2}{s^2 + AK(s + a)}\left[\frac{\omega_o - \omega_c}{s^2} + \frac{\theta_o}{s}\right]$$

The final-value theorem directly yields[21]

$$\lim_{t\to\infty} \theta_e(t) = \lim_{s\to 0} s\Theta_e(s) = 0 \tag{A4.5}$$

In this case, the PLL eventually acquires both the frequency and the phase of the incoming signal.

It must be remembered that the above analysis assumes a linear model, which is valid only when $\theta_e(t) \ll \pi/2$. This means the frequency ω_o and ω_c must be very close for this analysis to be valid. For a general case, one must use the nonlinear model in Fig. 4.64*b*. For such an analysis, the reader is referred to Viterbi[22] or Lindsey.[23]

To analyze PLL behavior as an FM demodulator, we consider the case of a small-error (linear model of the PLL) with $H(s) = 1$. For this case, Eq. (A4.1) becomes

$$\Theta_o(s) = \frac{AK}{s + AK}\Theta_i(s)$$

If $E_o(s)$ and $M(s)$ are Fourier transforms of $e_o(t)$ and $m(t)$, respectively, then from Eqs. (4.86 and 4.87) we have

$$\Theta_i(s) = \frac{k_f M(s)}{s} \qquad \text{and} \qquad s\Theta_o(s) = \alpha E_o(s)$$

Hence,

$$E_o(s) = \left(\frac{k_f}{\alpha}\right)\frac{AK}{s + AK}M(s)$$

Thus, the PLL output $e_o(t)$ is a distorted version of $m(t)$ and is equivalent to the output of a single-pole circuit (such as a simple RC circuit) with transfer function $k_f AK/\alpha(s + AK)$ with $m(t)$ as the input. To reduce distortion, we must choose AK well above the radian bandwidth of $m(t)$, so that $e_o(t) \simeq k_f m(t)/\alpha$.

In the presence of small noise, the behavior of the PLL is comparable to that of a frequency discriminator. The advantage of the PLL over a frequency discriminator appears only when the noise is large. The PLL is generally used as an FM demodulator for a very narrowband modulating signal $m(t)$, such as those arising in space applications.

FM signals can also be demodulated using feedback in a circuit similar to the PLL. This method, known as the *frequency-compressive feedback method*, will be discussed in Chapter 6.

Using small-error analysis, it can be shown that a first-order loop cannot track an incoming signal whose instantaneous frequency is varying linearly with time. Moreover, such a signal can be tracked within a constant phase (constant phase error) by using a second-order loop [Eq. (A4.4a)], and it can be tracked with zero phase error using a third-order loop.[22]

First-Order-Loop Analysis

Here we shall use the nonlinear model in Fig. 4.64*b*, but for the simple case of $H(s) = 1$. For this case $h(t) = \delta(t)$ and Eq. (4.84) gives

$$\dot{\theta}_o(t) = AK \sin \theta_e(t)$$

and

$$\dot{\theta}_e = \dot{\theta}_i - AK \sin \theta_e(t) \tag{A4.6}$$

Let us here consider the problem of frequency and phase acquisition. The incoming signal is $A \sin (\omega_o t + \varphi_o)$ and the VCO has a quiescent frequency ω_c. Hence,

$$\theta_i(t) = (\omega_o - \omega_c)t + \varphi_o$$

and

$$\dot{\theta}_e = (\omega_o - \omega_c) - AK \sin \theta_e(t) \tag{A4.7}$$

For a better understanding of the PLL behavior, let us sketch $\dot{\theta}_e$ vs. θ_e (Fig. A4.2). To satisfy Eq. (A4.7), the loop operation must stay along the sinusoidal trajectory shown in Fig. A4.2. When $\dot{\theta}_e = 0$, the system is in equilibrium, because at these points, θ_e stops varying with time. Thus $\theta_e = \theta_1$, θ_2, θ_3, or θ_4 are all equilibrium points.

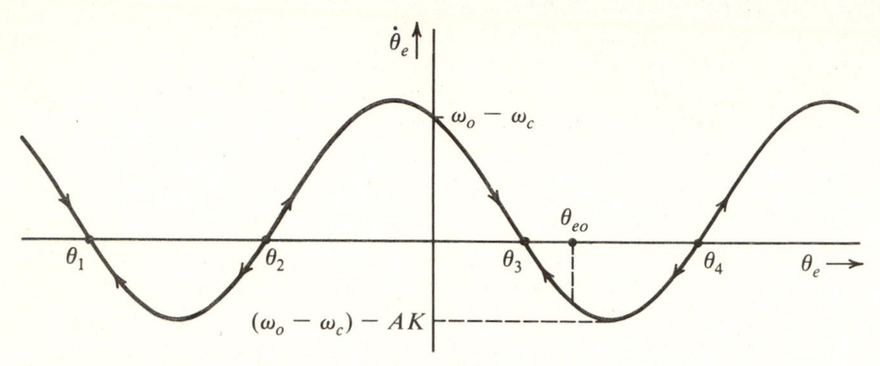

Figure A4.2 Trajectory of a first-order PLL.

If the initial phase error $\theta_e(0) = \theta_{eo}$ (Fig. A4.2), we observe that $\dot{\theta}_e$ corresponding to this value of θ_e is negative. Hence, the phase error will start decreasing along the sinusoidal trajectory until it reaches the value θ_3, where equilibrium is attained. Hence, in steady state, the phase error is a constant θ_3. This means the loop is in frequency lock; that is, the VCO frequency is now ω_o but there is a phase error of θ_3. Note, however, that if $|\omega_o - \omega_c| > AK$, there are no equilibrium points in Fig. A4.2, the loop never achieves lock, and θ_e continues to move along the trajectory forever. Hence, this simple loop can achieve phase lock provided the incoming frequency ω_o does not differ from the quiescent VCO frequency ω_c by more than AK.

In Fig. A4.2, several equilibrium points exist. Half of these points, however, are unstable equilibrium points, meaning that a slight perturbation in the system state will move the operating point farther away from these equilibrium points. Points θ_1 and θ_3 are stable points, because any small perturbation in the system state will tend to bring it back to these points. Consider, for example, the point θ_3. If the state is perturbed along the trajectory toward the right, $\dot{\theta}_e$ is negative, which tends to reduce θ_e and bring it back to θ_3. If the operating point is perturbed from θ_3 toward the left, $\dot{\theta}_e$ is positive, θ_e will tend to increase, and the operating point will return to θ_3. On the other hand, at point θ_2 if the point is perturbed toward the right, $\dot{\theta}_e$ is positive, and θ_e will increase until it reaches θ_3. Similarly, if at θ_2 the operating point is perturbed toward the left, $\dot{\theta}_e$ is negative, and θ_e will decrease until it reaches θ_1. Hence, θ_2 is an unstable equilibrium point. The slightest disturbance, such as noise, will dislocate it either to θ_1 or θ_3. In a similar way, we can show that θ_4 is an unstable point and that θ_1 is a stable equilibrium point.

The equilibrium point θ_3 occurs where $\dot{\theta}_e = 0$. Hence, from Eq. (A4.7),

$$\theta_3 = \sin^{-1} \frac{\omega_o - \omega_c}{AK}$$

If $\theta_3 \ll \pi/2$, then

$$\theta_3 \simeq \frac{\omega_o - \omega_c}{AK}$$

which agrees with our previous result of the small-error analysis [Eq. (A4.3b)].

The first-order loop suffers from the fact that it has a constant phase error. Moreover, it can acquire frequency lock only if the incoming frequency and the VCO quiescent frequency differ by not more than AK rps. Higher-order loops overcome these disadvantages, but they create a new problem of stability.[22]

In space vehicles, because of the Doppler shift and the oscillator drift, the frequency of the received signal has a lot of uncertainty. The Doppler shift itself could be as high as ±75 kHz, whereas the desired modulating signal may be just 10 Hz. To receive such a signal by conventional receivers would require a filter of bandwidth 150 kHz, when the desired signal has a bandwidth of only 10 Hz. This would cause an undesirable increase in the noise received (by a factor of 15,000), because the noise power is proportional to the bandwidth. The PLL proves convenient here because it tracks the received frequency continuously, and the filter bandwidth required is only 10 Hz.

REFERENCES

1. D. H. Sheingold, ed., *Nonlinear Circuits Handbook,* Analog Devices Inc., Norwood, Mass., 1974.
2. Single Sideband Issue, *Proc. IRE,* vol. 44, no. 12, Dec. 1956.
3. D. K. Weaver, Jr., "A Third Method of Generation and Detection of Single Sideband Signals," *Proc. IRE,* vol. 44, pp. 1703–1705, Dec. 1956.
4. J. P. Costas, "Synchronous Communication," *Proc. IRE,* vol. 44, pp. 1713–1718, Dec. 1956.
5. H. L. VanTrees, *Detection, Estimation, and Modulation Theory* (Part 1), Wiley, New York, 1968, Chapter 6.
6. L. H. Hansen, *Introduction to Solid-State Television Systems,* Prentice-Hall, Englewood Cliffs, N.J., 1969.
7. Bell Telephone Laboratories, *Transmission Systems for Communication,* 4th ed., 1970.
8. J. M. Fraser, D. B. Bullock, and N. G. Long, "Overall Characteristics of a TASI System," *Bell Syst. Tech. J.,* vol. 51, pp. 1439–1454, July 1962.
9. E. Jahnke and F. Emde, *Table of Functions,* Dover Publications, New York, 1945.
10. J. Carson, "Notes on the Theory of Modulation," *Proc. IRE,* vol. 24, pp. 57–64, Feb. 1922.
11. Armstrong, E. H., "A Method of Reducing Disturbances in Radio Signaling by a System of Frequency Modulation," *Proc. IRE,* vol. 24, pp. 689–740, May 1936.
12. Lessing, L., *Man of High Fidelity: Edwin Howard Armstrong,* J. B. Lippincott, Philadelphia, 1956.
13. "A Revolution in Radio," *Fortune,* vol. 20, p. 116, Oct. 1939.
14. W. B. Davenport, Jr., "Signal-to-Noise Ratios in Bandpass Limiters," *J. Appl. Phys.,* vol. 24, pp. 720–727, June 1953.
15. K. K. Clarke and D. T. Hess, *Communication Circuits: Analysis and Design,* Addison-Wesley, Reading, Mass., 1971.

16. P. F. Panter, *Modulation, Noise, and Spectral Analysis,* McGraw-Hill, New York, 1965.
17. L. B. Arguimbau, *Vacuum Tube Circuits and Transistors,* Wiley, New York, 1964, p. 466.
18. N. Mokoff, "Technology '82: Fiber Optics," *IEEE Spectrum,* vol. 19, pp. 39–40, Jan. 1982.
19. Sacramento *Bee*, February 16, 1982.
20. G. Bilinsky, "Fiber Optics Finally Sees the Light of Day," *Fortune,* vol. 101, pp. 111–120, March 24, 1980.
21. B. P. Lathi, *Signals, Systems, and Controls,* Harper & Row, New York, 1974.
22. A. J. Viterbi, *Principles of Coherent Communication,* McGraw-Hill, New York, 1966.
23. W. C. Lindsey, *Synchronization Systems in Communication and Control,* Prentice-Hall, Englewood Cliffs, N.J., 1972.

PROBLEMS

4.1. Find the DSB-SC signal and sketch its spectrum for:

(a) $m(t) = \cos 100t$
(b) $m(t) = \cos 100t + \cos 300t$
(c) $m(t) = \cos 200t \cos 300t$
if the carrier frequency $\omega_c = 2000$. In each case, identify the USB and LSB spectra.

4.2. It is desired to generate a modulated signal $km(t) \cos \omega_c t$ by multiplying $m(t)$ by a sinusoid $\cos \omega_c t$. Unfortunately, the generator output is highly distorted and is given by $\cos^3 \omega_c t$ rather than $\cos \omega_c t$. Show how the desired modulated signal can be generated using this distorted carrier and a multiplier. Calculate the value of k. What is the restriction on the value of ω_c if $m(t)$ is bandlimited to B Hz.

4.3. It is desired to generate a modulated signal $km(t) \cos \omega_c t$ with $f_c = 300$ kHz using the ring modulator shown in Fig. 4.7. The sinusoid generator that is available has a fixed frequency of 100 kHz. Show how the desired signal can be obtained using a ring modulator. Calculate the value of k.

4.4. Fig. P4.4 shows a DSB-SC modulator. The carrier available at the multiplier is distorted

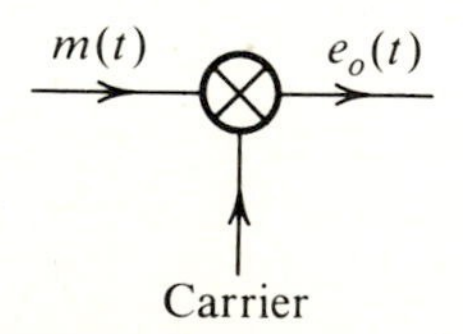

Figure P4.4

and is given by $a_1 \cos \omega_c t + a_2 \cos^2 \omega_c t$.
(a) Find $e_o(t)$ and sketch its spectrum.
(b) Show how you can obtain the desired DSB-SC signal from $e_o(t)$.
(c) If $m(t)$ is bandlimited to B, determine the minimum value of ω_c that can be used to obtain a distortionless DSB-SC signal in part (b).

4.5. A modulated signal $m(t)\cos\omega_c(t)$ is amplified by a nonlinear amplifier whose output $e_o(t)$ and input $e_i(t)$ are related as

$$e_o(t) = a_1 e_i(t) + a_2 e_i^2(t)$$

(a) Find and sketch the spectrum of the output $e_o(t)$.
(b) Show how you can obtain the desired DSB-SC signal from $e_o(t)$.
(c) If $m(t)$ is bandlimited to B, determine the minimum value of ω_c in order to recover $m(t)$ from the modulated signal $e_o(t)$.

4.6. The two diodes used in Fig. P4.6*a* are identical, with voltage-current characteristics as

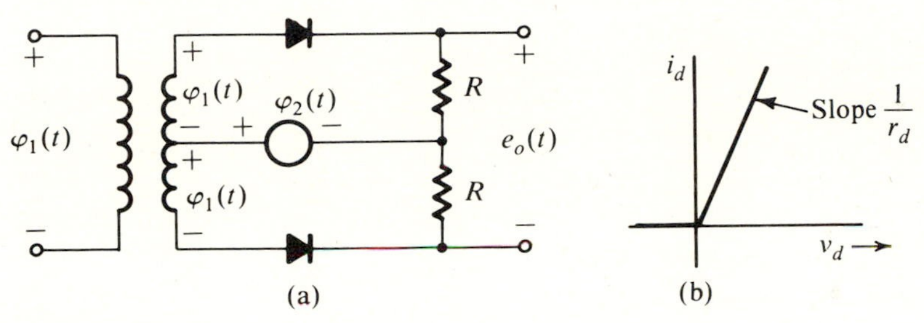

Figure P4.6

shown in Fig. P4.6*b*.
(a) Find the output $e_o(t)$ if $\varphi_2(t) = A\cos\omega_c t$, with

$$A \gg |\varphi_1(t)| \quad \text{for all } t.$$

(b) Show that this circuit can be used as a DSB-SC modulator or a synchronous demodulator by suitably filtering the output $e_o(t)$.
Hint: The diode switching is controlled by $A\cos\omega_c t$. Hence, the diodes open and close periodically with frequency f_c.

4.7. In Prob. 4.6, if $\varphi_1(t) = \sin(\omega_c t + \theta)$ and $\varphi_2(t) = A\cos\omega_c t$ $(A \gg 1)$, show that the output contains a *dc* term proportional to $\sin\theta$. Hence, show that this circuit can be used as a phase discriminator (to measure the phase difference between two sinusoids) if the output is terminated by a large capacitor. *Hint:* See the hint in Prob. 4.6. The output $e_o(t)$ is periodic. Find its dc component.

4.8. The bridge circuit is commonly employed in measurements. A resistance bridge is often used to measure quantities that can vary a resistance linearly. A strain gage, for example, is a device in which a strain-sensitive element is attached to the body to be strained. The resistance of the element then varies in proportion to the strain. Similarly, some elements are temperature sensitive, and their resistance varies linearly with the temperature (a thermistor, for example). The resistance bridge shown in Fig. P4.8 has three fixed resistors, as shown. The fourth resistor is a variable resistor, which varies in proportion to the quantity to be measured. Let this resistor have a quiescent value of KR. The value R_s of this resistor is given by $R_s = KR[1 + \alpha m(t)]$ where $m(t)$ is the quantity to be measured (for example, strain or temperature) and α is the constant of proportionality. A sinusoidal source $A\cos\omega_c t$ is applied at terminals aa'. Show that the output at terminals bb' is a DSB-SC signal if $|\alpha m(t)| \ll 1$.

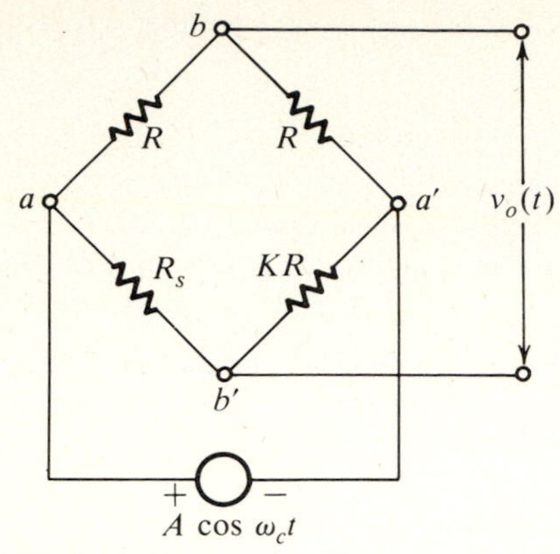

Figure P4.8

4.9. In one of the scrambling schemes used to discourage wiretapping, the message-signal spectrum is inverted using the procedure shown in Fig. P4.9*b*. The scrambled signal

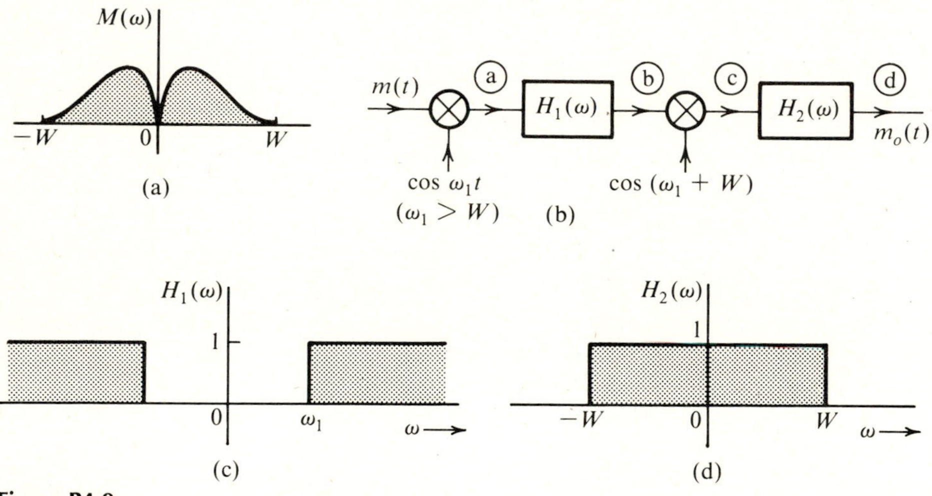

Figure P4.9

waveform $m_o(t)$ is entirely different from the original waveform $m(t)$.

(a) Sketch the spectra of the signals at points a, b, c, and d, assuming $m(t)$ to be bandlimited to B.

(b) Show how you can descramble the signal $m_o(t)$, that is, how can $m(t)$ be recovered from $m_o(t)$.

Hint: Because the spectrum of $m_o(t)$ is an inverted version of $m(t)$, one more inversion of $m_o(t)$ should yield $m(t)$.

4.10. Example 4.3 analyzes the switching demodulator. Do the same for the nonlinear demodulator shown in Fig. 4.4*a*, with the output bandpass filter replaced by a lowpass filter.

4.11. Show that the AM wave $[A + m(t)] \cos \omega_c t$ can be synchronously demodulated regardless of the value of A.

4.12. Sketch the AM wave for the modulating signal $m(t)$ in Fig. P4.12 corresponding to the

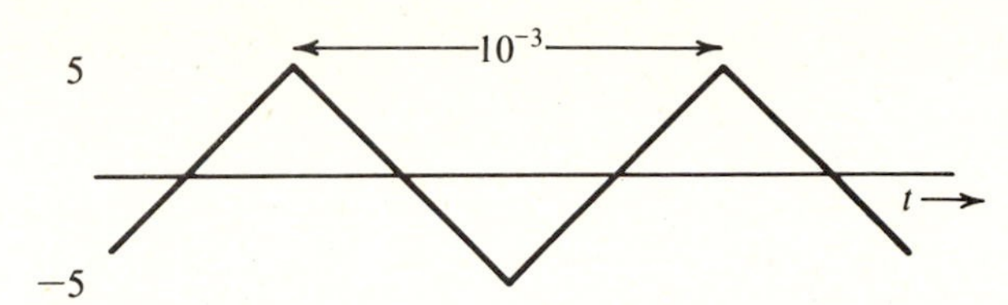

Figure P4.12

modulation index (a) $\mu = 0.4$, (b) $\mu = 0.8$, (c) $\mu = 1$, and (d) $\mu = 1.2$. Use $f_c = 100$ kHz.

4.13. **(a)** Sketch the DSB-SC wave for the modulating signal $m(t)$ shown in Fig. P4.13.

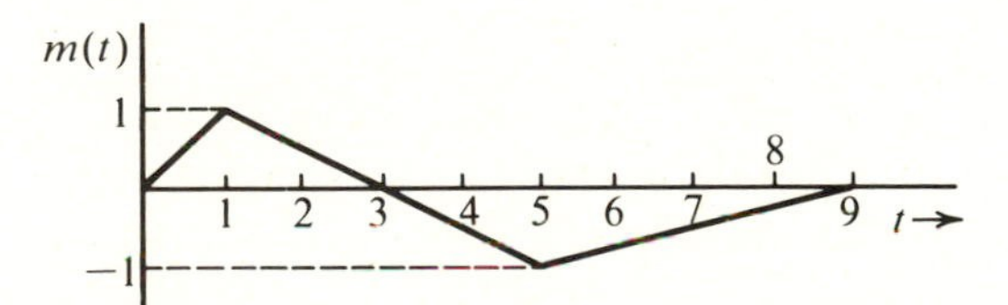

Figure P4.13

(b) This DSB wave is applied to the input of an envelope detector. Sketch the output waveform. Verify that this output is $|m(t)|$.

(c) Show that to be able to envelope detect an AM wave,

$$A \geq |m(t)_{\min}|$$

Hint: The envelope detector output is $|A + m(t)|$ [see part (b)]. Now show that $|A + m(t)| = A + m(t)$ if $A \geq |m(t)_{\min}|$.

4.14. The envelope detector in Fig. 4.14 appears equivalent to a half-wave rectifier followed by a lowpass RC filter. Hence, it may seem that the two methods of AM detection discussed in the text (rectifier detection and envelope detection) are not two different methods but two different points of view looking at the same thing. Show the fallacy in this statement.
(*Hint:* Show that the time constants RC required in each case are different. Moreover, from Example 4.6, show that the RC requirements in the envelope detector depend als· on the value of μ.)

4.15. Show that a squaring circuit followed by a lowpass filter of bandwidth $2B$ followed by a square rooter (Fig. P4.15) acts as an envelope detector for an AM wave. Show that if a DSB-SC signal $m(t) \cos \omega_c t$ is demodulated by this scheme, the output will be $|m(t)|$.

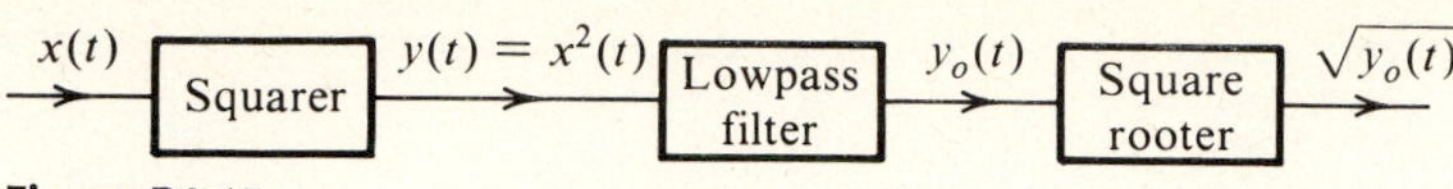

Figure P4.15

4.16. Show that the circuit in Fig. P4.16 acts as an envelope detector for a bandpass signal.

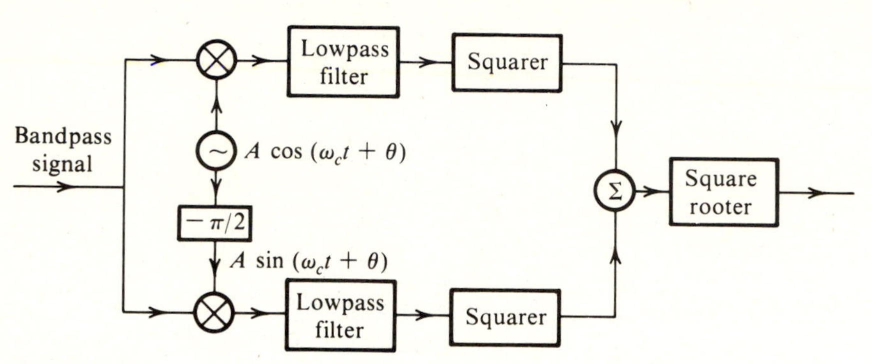

Figure P4.16

Verify that this circuit can indeed demodulate an AM wave. *Hint:* Consider a general bandpass signal $m_c(t) \cos \omega_c t + m_s(t) \sin \omega_c t$. Show that the output is the envelope $E(t) = \sqrt{m_c^2(t) + m_s^2(t)}$. Assume a narrowband signal.

4.17. For the AM wave corresponding to $m(t)$ in Fig. P4.12 (with $\mu = 0.8$, $f_c = 100$ Hz), determine the largest value of the time constant RC of the RC elements in the envelope detector that will enable the detector to follow the envelope of the modulating signal.

4.18. A received signal is given by $(1 + 2 \cos \omega_m t) \cos \omega_c t$ ($\omega_m = 5000$, $\omega_c = 10^5$).
(a) If this is considered as an AM waveform, what is the modulation index μ?
(b) If this signal is envelope detected, sketch the output.

4.19. A modulating signal $m(t)$ is given by
(a) $\cos 2000t$
(b) $\cos 2000t + 2 \cos 3000t$
(c) $2 \cos 2000t \cos 3000t$
If $m(t)$ SSB modulates the carrier $A \cos 10{,}000t$, in each case, find
(i) the USB signal $\varphi_{\text{USB}}(t)$ and sketch its spectrum.
(ii) the LSB signal $\varphi_{\text{LSB}}(t)$ and sketch its spectrum.
Find the answer by first determining the DSB-SC signal and then eliminating the unwanted sideband.

4.20. Solve Prob. 4.19 by using Eq. (4.19). *Hint:* See Example 4.7a to determine $m_h(t)$ of a sinusoid (also see the table in Prob. 4.25).

4.21. Figure P4.21 shows the spectrum $M(\omega)$.
(a) Sketch the spectrum of $m_h(t)$ using Eq. (4.17c).
(b) Sketch the spectra of $m(t) \cos \omega_c t$ and $m_h(t) \sin \omega_c t$.
(c) Sketch the spectrum of $m(t) \cos \omega_c t \pm m_h(t) \sin \omega_c t$. Show that it is indeed the SSB spectrum.

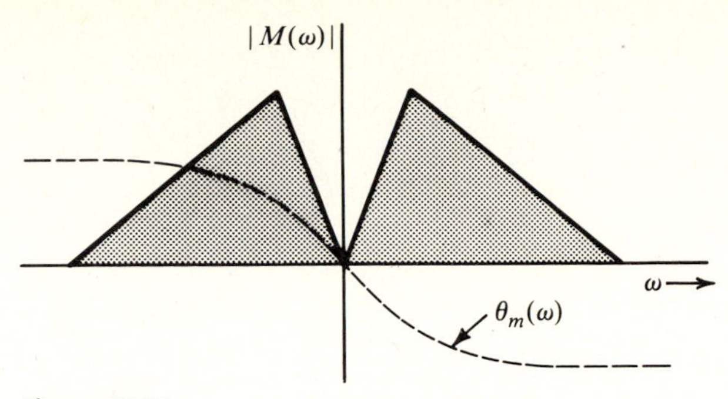

Figure P4.21

4.22. For a lowpass signal $m(t)$ bandlimited to B Hz, show that the Hilbert transform of $m(t) \cos(\omega_c t + \theta)$ is $m(t) \sin(\omega_c t + \theta)$ [assume $\omega_c \geq 2\pi B$].

4.23. A bandpass signal $m(t)$ is expressed in terms of quadrature components as

$$m(t) = m_c(t) \cos \omega_c t + m_s(t) \sin \omega_c t$$

Determine $m_+(t)$ and show that the envelope of $m(t)$ is given by $2|m_+(t)|$. Assume $\omega_c \gg 2\pi B$, where B is the bandwidth of $m(t)$. *Hint:* Use the results in Prob. 4.22 to determine $m_h(t)$. Find $m_c(t)$ and $m_s(t)$ in terms of $m(t)$ and $m_h(t)$.

4.24. Show that if $m_h(t)$ is the Hilbert transform of $m(t)$, then $-m(t)$ is the Hilbert transform of $m_h(t)$.

4.25. Derive the following Hilbert transform pairs:

$m(t)$	$m_h(t)$	
$e^{\pm j\omega_o(t)}$	$\mp je^{\pm j\omega_o t}$	
$\cos \omega_1 t \cos \omega_2 t$	$\cos \omega_1 t \sin \omega_2 t$	$\omega_2 > \omega_1$
$\delta(t)$	$1/\pi t$	
$1/t$	$-\pi\delta(t)$	
$\text{sinc}\ (at)$	$-\dfrac{\pi at}{2} \text{sinc}^2\ (at)$	
$\Pi\left(\dfrac{t}{T}\right)$	$\dfrac{1}{\pi} \ln \left\lvert\dfrac{t + (T/2)}{t - (T/2)}\right\rvert$	

4.26. The generation of an SSB signal by the phase-shifting method (Fig. 4.21) requires a circuit to shift the phases of all components of $m(t)$ by 90°. A slight deviation can be serious.

(a) Consider tone modulation [$m(t) = \cos \omega_m(t)$]. Show that for a phase error of 1° in the phase-shifting network, the undesired sideband (instead of being 0) will be nonzero and only 41 dB below the desired sideband.

(b) Show that even if the phase shifting of $m(t)$ is correct, an error of 1° in the carrier phase shifting will leave the undesired sideband only 41 dB below the desired sideband.

4.27. The third method of SSB generation (known as Weaver's method) is shown in Fig. P4.27, with ω_c the final carrier frequency ($\omega_c \gg \omega_o$).

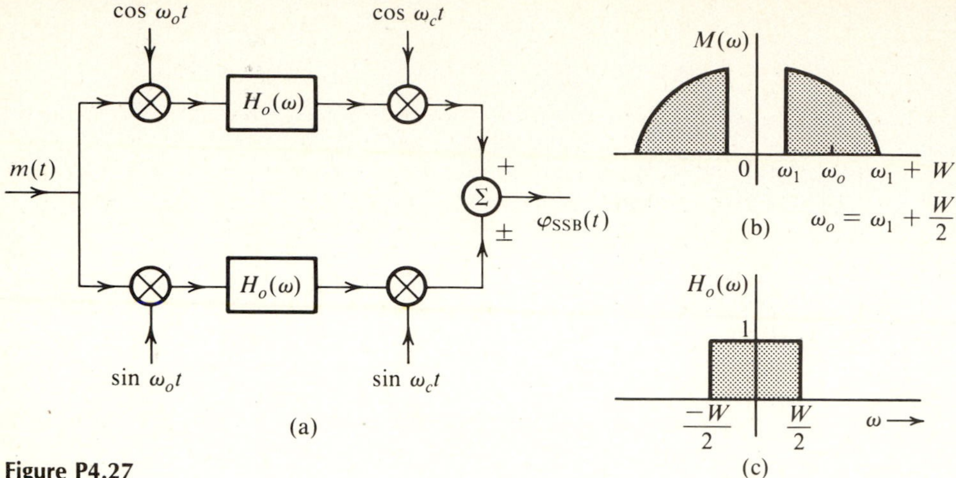

Figure P4.27

(a) Analyze this system and show that the output is an SSB signal. *Hint:* Use $m(t) = m_+(t) + m_-(t)$.

(b) Figure P4.27 shows the spectrum $M(\omega)$ with a notch at the origin. Show that the system works even without such a notch if ideal filters are used for $H_o(\omega)$.

(c) Show that if the notch at the origin is present, the filters $H_o(\omega)$ do not have to be ideal.

(d) What is the advantage of this scheme over the selective-filtering method?

4.28. Determine the percentage of second harmonic distortion in the envelope detection of an SSB + C signal. Consider tone modulation; that is, the input to the envelope detector is

$$\varphi_i(t) = A \cos \omega_c t + m(t) \cos \omega_c t + m_h(t) \sin \omega_c t$$

with $m(t) = \alpha \cos \omega_m t$.

To maintain the second-harmonic-distortion amplitude below 10 percent of the desired signal amplitude, determine what the ratio α/A must be.

4.29. In a QAM system (Fig. 4.22), the locally generated carrier has a frequency error $\Delta\omega$ and a phase error δ; for example, the carrier is $\cos [(\omega_c + \Delta\omega)t + \delta]$ or $\sin [(\omega_c + \Delta\omega)t + \delta]$.

Show that the output of channel 1 is

$$m_1(t) \cos [(\Delta\omega)t + \delta] - m_2(t) \sin [(\Delta\omega)t + \delta]$$

instead of $m_1(t)$ and the output of channel 2 is

$$m_1(t) \sin [(\Delta\omega)t + \delta] + m_2(t) \cos [(\Delta\omega)t + \delta]$$

instead of $m_2(t)$.

4.30. A DSB-SC signal can be demodulated by an envelope detector if a sufficient amount of carrier is reinserted at the receiver. Show that the distortion in the envelope-detected output caused by errors in the reinserted carrier is similar to that found in synchronous demodulation with similar errors [Eq. (4.23)] in the local carrier. Specifically, show that if the carrier $A \cos [(\omega_c + \Delta\omega)t + \delta]$ is reinserted in the received DSB-SC signal $m(t)$

$\cos \omega_c t$, and the resulting signal is envelope detected, then the output is

$$e_o(t) \simeq m(t) \cos [(\Delta\omega)t + \delta] \qquad A \gg |m(t)|$$

4.31. A DSB-SC signal is demodulated by carrier reinsertion and envelope detection. For tone modulation $[m(t) = \alpha \cos \omega_m t]$, the received signal is $\alpha \cos \omega_m t \cos \omega_c t$. The reinserted carrier is $A \cos (\omega_c t + \delta)$. Determine the ratio α/A in order to maintain the second-harmonic-distortion amplitude below 5 percent of the desired signal amplitude, assuming $\alpha/A \ll 1$ and $\delta \ll \pi/2$.

4.32. An SSB-SC signal can be demodulated by an envelope detector if a sufficient amount of carrier is reinserted at the receiver. Show that the distortion in the envelope-detected output caused by errors in the reinserted carrier is similar to that found in synchronous demodulation [Eq. (4.26)] with similar errors in the local carrier. Specifically, show that if the carrier $A \cos [(\omega_c + \Delta\omega)t + \delta]$ is reinserted in the received SSB signal $m(t) \cos \omega_c t + m_h(t) \sin \omega_c t$, and the resulting signal is envelope detected, then the output is

$$e_o(t) \simeq m(t) \cos [(\Delta\omega)t + \delta] - m_h(t) \sin [(\Delta\omega)t + \delta]$$

provided $A \gg |m(t)|$ and $|m_h(t)|$.

4.33. VSB modulation is a special case of the *asymmetrical sideband* (*ASB*) scheme. In ASB, one of the sidebands of the DSB spectrum is cut off gradually by a filter $H(\omega)$, as shown in Fig. 4.27*a*. For the special case of VSB, the shaping filter $H(\omega)$ must have vestigial symmetry [Eq. (4.31)]. For a general ASB case, $H(\omega)$ need not satisfy this condition. Instead, the output lowpass filter in the receiver (Fig. 4.27*a*) is appropriately shaped to yield correct $m(t)$. Show that the lowpass filter in the receiver in Fig. 4.27*a* must have a transfer function $1/[H(\omega + \omega_c) + H(\omega - \omega_c)]$ for $|\omega| \leq 2\pi B$.

4.34. Discuss the effects of frequency and phase errors in the locally generated carrier used in the synchronous demodulation of a VSB-SC signal. Assume the received signal to be $m(t) \cos \omega_c t + m_s(t) \sin \omega_c t$ and the local carrier to be $A \cos [(\omega_c + \Delta\omega)t + \delta]$.

4.35. If a VSB-SC signal is envelope detected after the reinsertion of a large carrier, discuss the effects on the output caused by frequency and phase errors in the reinserted carrier. Assume the received signal to be $m(t) \cos \omega_c t + m_s(t) \sin \omega_c t$ and the reinserted carrier to be $A \cos [(\omega_c + \Delta\omega)t + \delta]$, with $A \gg |m(t)|$ and $|m_s(t)|$.

4.36. Consider the VSB system in Example 4.10 (Fig. 4.29) with $\omega_c = 10^6$. Find the VSB-modulated signal if
(a) $m(t) = a \cos 4000t$
(b) $m(t) = b \sin 2000t$
(c) $m(t) = a \cos 3000t \cos 4000t$

4.37. For the baseband communication system shown in Fig. P4.37, $m(t)$ is a baseband signal of bandwidth B. It is transmitted over a channel with white noise of PSD $\mathcal{N}/2$. The

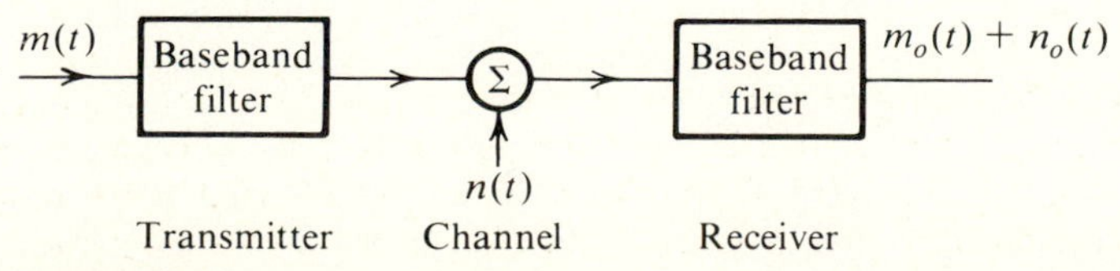

Figure P4.37

receiver consists of an ideal baseband filter of bandwidth B. Show that the output SNR is γ, that is, it is identical to that of DSB-SC and SSB-SC systems.

4.38. For a DSB-SC system with a white channel noise of PSD 10^{-8}, and the baseband signal $m(t)$ with a bandwidth of 4 kHz, the output SNR is required to be at least 30 dB.
(a) What must be the received message-signal power S_i?
(b) What must be the transmitted power if the channel transfer function $H_c(\omega) = 10^{-3}$?
(c) What is the output noise power N_o?

4.39. Repeat Prob. 4.38 for an SSB-SC case.

4.40. Assume that $[m(t)]_{\max} = -[m(t)]_{\min} = m_p$.
(a) Show that in AM, $m_p = \mu A$.
(b) Show that the output SNR in Eq. (4.42d) can be expressed as

$$\frac{S_o}{N_o} = \frac{\mu^2}{K_m^2 + \mu^2}\gamma$$

where K_m (the *peak factor*, or *crest factor*) is defined as

$$K_m^2 = \frac{m_p^2}{\widetilde{m^2(t)}}$$

(c) Using the result in part (b), show that for tone modulation with $\mu = 1$,

$$\frac{S_o}{N_o} = \frac{\gamma}{3}$$

4.41. For a gaussian $m(t)$ of zero mean and mean square value σ^2, if the peak amplitude m_p is taken as 3σ (3σ loading), show that for AM with $\mu = 1$, the output SNR $S_o/N_o = 0.1\gamma$.

4.42. Sketch the FM and PM waves for the modulating signal $m(t)$ shown in Fig. P4.42, given $\omega_c = 10^8$, $k_f = 10^5$, and $k_p = 25$.

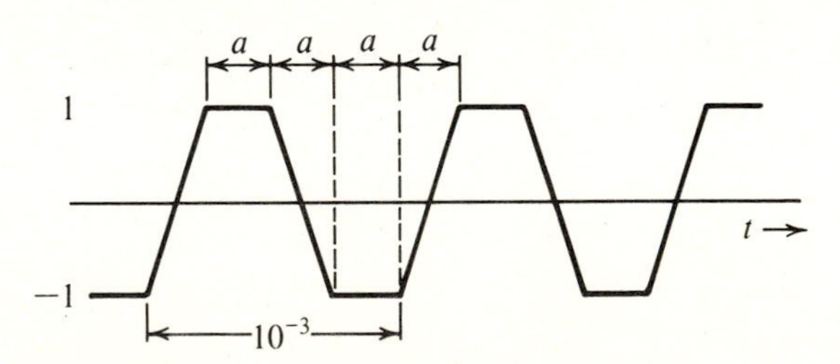

Figure P4.42

4.43. Sketch the FM and PM waves for the modulating signal $m(t)$ shown in Fig. P4.43, given $\omega_c = 10^6$, $k_f = 1000$, and $k_p = \pi/2$. Explain why it is necessary to use $k_p < \pi$ in this case.

4.44. Repeat Prob. 4.43 for the $m(t)$ shown in Fig. P4.44.

4.45. For the modulating signal

$$m(t) = \cos 1000t + 2 \sin 2000t$$

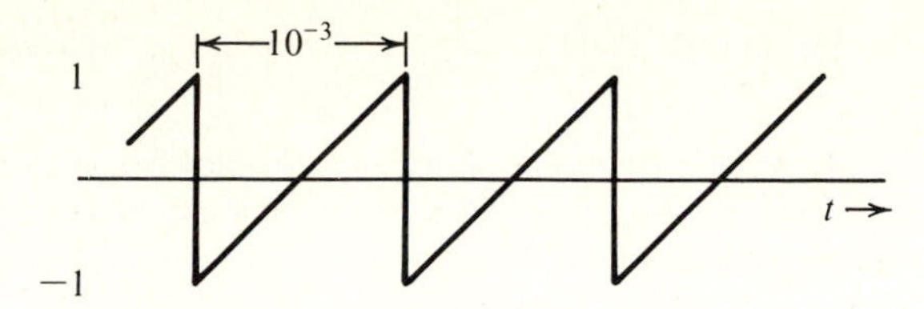

Figure P4.43

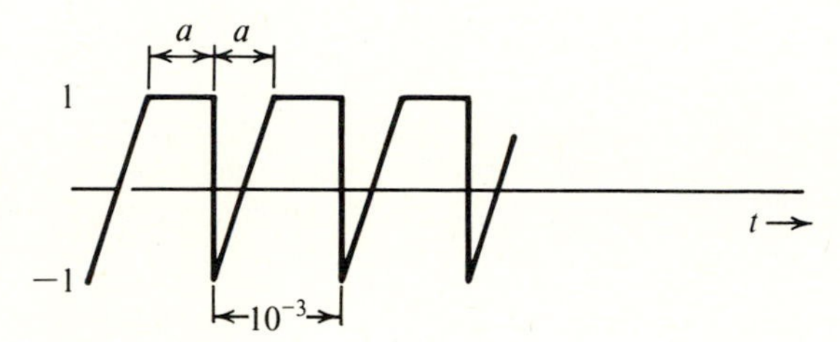

Figure P4.44

determine $\varphi_{FM}(t)$ and $\varphi_{PM}(t)$ when $\omega_c = 10^8$, $k_f = 100$, and $k_p = 10$. Estimate the bandwidths of these waves.

4.46. An angle-modulated wave is described by the equation

$$\varphi_{EM}(t) = 10 \cos (2\pi \times 10^6 t + 0.1 \sin 2000\pi t)$$

(a) Find the power of the modulated signal.
(b) Find the maximum frequency deviation.
(c) Find the maximum phase deviation.
(d) Find the estimate of the signal bandwidth.

4.47. Repeat Prob. 4.46 if

$$\varphi_{EM}(t) = 5 \cos (2\pi \times 10^6 t + 20 \sin 1000\pi t + 10 \sin 2000\pi t)$$

4.48. Estimate the bandwidth of the angle-modulated wave $\varphi_{EM}(t)$ in Prob. 4.46 in two ways:
(a) Consider $\varphi_{EM}(t)$ as a PM wave with $k_p = 10$. Find $m(t)$. Knowing k_p and $m(t)$, estimate the bandwidth of $\varphi_{EM}(t)$.
(b) Consider $\varphi_{EM}(t)$ as an FM wave with $k_f = 20\pi$. Find $m(t)$. Knowing k_f and $m(t)$, estimate the bandwidth of $\varphi_{EM}(t)$.

4.49. Estimate the bandwidths of the FM and PM waves in Prob. 4.42. *Hint:* The Fourier series for $m(t)$ shows that the amplitudes of the harmonics of $m(t)$ beyond the third harmonic are negligible. Hence, the bandwidth of $m(t)$ may be taken as its third harmonic frequency.

4.50. Estimate the bandwidths of the FM and PM waves in Prob. 4.43. *Hint:* The amplitudes of the harmonics of $m(t)$ beyond the seventh harmonic may be considered negligible.

4.51. Estimate the bandwidths of the FM and PM waves in Prob. 4.44. *Hint:* Same as Prob. 4.50.

4.52. A 10-MHz carrier is frequency modulated by a sinusoidal signal of 5 kHz so that the maximum frequency deviation is 1 MHz. Estimate the bandwidth of the FM carrier. Now estimate the bandwidth of the FM carrier if the modulating-signal amplitude is doubled. Estimate the bandwidth of the FM carrier if the frequency of the modulating signal is also doubled.

4.53. A 10-MHz carrier is phase modulated by a sinusoidal signal of frequency 5 kHz and of unit amplitude. The maximum phase deviation is 10 radians for a unit amplitude of the modulating signal. Estimate the bandwidth of the PM carrier. If the frequency of the modulating signal is now doubled to 10 kHz, estimate the new bandwidth of the PM carrier. If the frequency is the same as before (5 kHz) but the amplitude is doubled, estimate the bandwidth of the PM carrier.

4.54. In the direct generation of FM, the output contains distortion because of the approximation in Eq. (4.75b). Show that when this FM wave is demodulated, the ratio of the second-harmonic-distortion amplitude to the desired signal amplitude is $0.75\,\Delta\omega/\omega_c$ for tone modulation; that is, $m(t) = \alpha \cos \omega_m t$ in Eq. (4.75). Hence, show that for the case $\Delta f = 75$ kHz and $f_c = 100$ MHz, the second-harmonic-distortion amplitude is 0.056 percent of the desired signal amplitude. *Hint:* $\Delta\omega/\omega_c = k\alpha/2C_0$.

4.55. Show that when $m(t)$ has no jump discontinuities, an FM demodulator followed by an integrator (Fig. P4.55*a*) acts as a PM demodulator, and a PM demodulator followed by a differentiator (Fig. P4.56*b*) serves as an FM demodulator even if $m(t)$ has jump discontinuities.

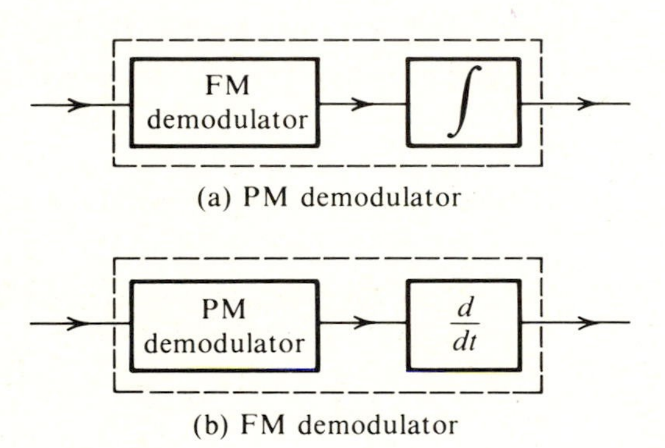

Figure P4.55

4.56. Consider the two modulating waveforms $m_1(t)$ and $m_2(t)$ shown in Fig. P4.56. A carrier of 100 MHz is phase modulated using these waveforms, with $k_p = \pi/2$.
(a) Sketch the PM waveforms corresponding to $m_1(t)$ and $m_2(t)$.
(b) Show that there is no way to demodulate, without ambiguity, the PM wave corresponding to $m_1(t)$.
(c) Show that the PM wave corresponding to $m_2(t)$ can be demodulated without ambiguity by using an FM demodulator followed by an integrator.

4.57. For an FM communication system with a white channel noise of PSD $S_n(\omega) = 10^{-10}$, and the baseband signal $m(t)$ with a bandwidth of 15 kHz, the output SNR is found to be 28 dB. It is given that $\widetilde{m^2(t)}/m_p^2 = 1/9$ and $\beta = 2$. The demodulator constant $\alpha = 10^{-4}$ (see footnote on page 321).
(a) Determine the output noise power N_o.
(b) Determine the output signal power S_o.

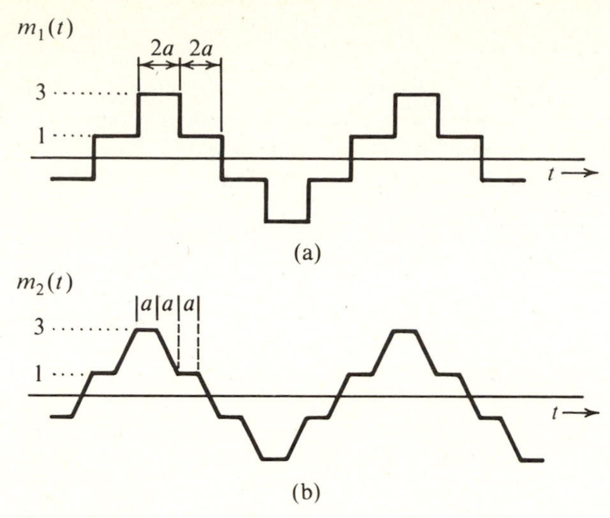

Figure P4.56

(c) Determine the received signal power S_i.
(d) Determine the transmitted power if the channel transfer function $|H_c(\omega)| = 10^{-3}$.

4.58. For the modulating signal $m(t)$ shown in Fig. 4.49*a*, show that PM is superior to FM by a factor of $3\pi^2/4$ from the output SNR point of view. *Hint:* See Example 4.13 to determine the bandwidth of $m(t)$.

4.59. For a modulating signal $m(t) = \cos^3 \omega_o t$, show that PM is superior to FM by a factor of 2.25 from the SNR point of view. *Hint:* Determine m'_p from $d/dt[\dot{m}(t)] = 0$.

4.60. In the text, it was shown that for tone modulation, FM is superior to PM by a factor of 3 from the output SNR point of view. For a multitone modulation, PM can be superior to FM. Show that for a two-tone case where $m(t) = \alpha_1 \cos \omega_1 t + \alpha_2 \cos \omega_2 t$, PM is superior to FM when $(1 + xy)^2 < (1 + x)^2/3$, where $x = \alpha_1/\alpha_2$ and $y = \omega_1/\omega_2$. Assume $\omega_2 > \omega_1$, and, consequently, the bandwidth of $m(t)$ is $B = \omega_2/2\pi$. For the purpose of computing m_p and m'_p, assume that two sinusoid peaks add in phase at some instant.

4.61. For a certain FM system with $\beta = 4$ and $\widetilde{m^2(t)}/m_p^2 = 1/9$, the output SNR is found to be 18 dB.
(a) Determine if the system is in threshold.
(b) If β is increased to 6 (keeping all other parameters unchanged), would the output SNR increase or decrease? Explain.
(c) If β is reduced to 2 (keeping all other parameters unchanged), would the output SNR increase or decrease? Explain.

4.62. In a certain FM system used in space communication, the output SNR is found to be 21.5 dB with $\beta = 2$. The bandwidth of the modulating signal $m(t)$ is 10 kHz, and $\widetilde{m^2(t)}/m_p^2 = 1/9$. The system with $\beta = 2$ is in the linear-operation region (small-noise region).

The output SNR is required to be at least 39.5 dB. Because power is at a premium in space communication, it is decided to increase the output SNR by increasing the transmission bandwidth only.

(a) What is the maximum value of β and the corresponding transmission bandwidth that can be used without running into the threshold? Determine the maximum output SNR that can be obtained by using this value of β.

(b) Determine the minimum increase in the transmitted power required to attain the SNR of 39.5 dB. What is the corresponding value of β and the transmission bandwidth?

Hint: From Eq. (4.114), $\beta = (\gamma - 40)/20$. Substitute this in Eq. (4.109c) to determine the minimum usable value of γ.

5

Probability Theory and Random Processes

PART I Probability Theory

Thus far, we have studied signals whose values at any instant t were known from their analytical or graphical description. Such signals are called *deterministic* signals, implying complete certainty about their values at any instant t. Such signals, which can be specified with certainty, can never convey information. It will be seen in Chapter 8 that information is related to uncertainty. The higher the uncertainty about a signal (or message) to be received, the higher its information content. If a message to be received is specified (i.e., if it is known beforehand), it cannot convey any information. Hence, signals that convey information must be unpredictable. Noise signals that perturb information signals are also unpredictable. These unpredictable signal and noise waveforms are examples of *random processes*.

Random phenomena arise either because of our partial ignorance of the generating mechanism (as in message or noise signals) or because the laws governing the phenomena may be fundamentally random (as in quantum mechanics). Yet in another situation, such as the outcome of rolling a die, it is possible to predict the outcome provided we know exactly all the conditions, such as the angle of the throw, the nature of the surface on which it is thrown, the force imparted by the player, and so on. The exact analysis, however, is so complex that it is impractical to carry it out, and we are content to accept the outcome prediction on an average basis. Here the random phenomenon arises from our unwillingness to carry out exact analysis because it is not worth the trouble.

We shall begin with a review of the basic concepts of the theory of probability, which forms the basis for describing random processes.

5.1 INTRODUCTION TO THE THEORY OF PROBABILITY

We begin by defining some important terms. An experiment is called a *random experiment* if its outcome cannot be predicted precisely because the conditions under which it is performed cannot be predetermined with sufficient accuracy and completeness. Tossing a coin, rolling a die, and drawing a card from a deck are some examples of random experiments. A random experiment may have several separately identifiable *outcomes*. For example, rolling a die has six possible identifiable outcomes (1, 2, 3, 4, 5, and 6). *Events* are sets of outcomes meeting some specifications. In the experiment of rolling a die, for example, the event "odd number on a throw" can result from any one of three outcomes (viz., 1, 3, and 5). Hence, this event is a set of three outcomes (1, 3, and 5). Thus, events are groupings of outcomes into classes among which we choose to distinguish. These ideas can be better understood by using the concepts of set theory.

We define the *sample space* $\mathcal{S}$ as a collection of all possible separately identifiable outcomes of a random experiment. Each outcome is an *element,* or *sample point,* of this space and can be conveniently represented by a point in the sample space. In the random experiment of rolling a die, for example, the sample space consists of six elements represented by six sample points ζ_1, ζ_2, ζ_3, ζ_4, ζ_5, and ζ_6, where ζ_i represents the outout "a number i is thrown" (Fig. 5.1). The event, on the other hand, is a subset of $\mathcal{S}$. The event "an odd number is thrown," denoted by A_o, is a subset of $\mathcal{S}$ (or a set of sample points ζ_1, ζ_3, and ζ_5). Similarly, the event A_e, "an even number is thrown," is another subset of $\mathcal{S}$ (or a set of sample points ζ_2, ζ_4, and ζ_6).

$$A_o = (\zeta_1, \zeta_3, \zeta_5) \qquad A_e = (\zeta_2, \zeta_4, \zeta_6)$$

Let us denote the event "a number equal to or less than 4 is thrown" as B. Thus $B = (\zeta_1, \zeta_2, \zeta_3, \zeta_4)$. These events are clearly marked in Fig. 5.1. Note that an outcome can also be an event, because an outcome is a subset of $\mathcal{S}$ with only one element.

The *complement* of any event A, denoted by A^c, is the event containing all points not in A. Thus, for the event B in Fig. 5.1, $B^c = (\zeta_5, \zeta_6)$, $A_o^c = A_e$, and $A_e^c = A_o$. An event that has no sample points is a *null event,* which is denoted by $\emptyset$ and is equal to $\mathcal{S}^c$.

The *union* of events A and B, denoted by $A \cup B$, is that event which contains all points in A and B. This is the event "A or B." For the events in Fig. 5.1,

$$A_o \cup B = (\zeta_1, \zeta_3, \zeta_5, \zeta_2, \zeta_4)$$

$$A_e \cup B = (\zeta_2, \zeta_4, \zeta_6, \zeta_1, \zeta_3)$$

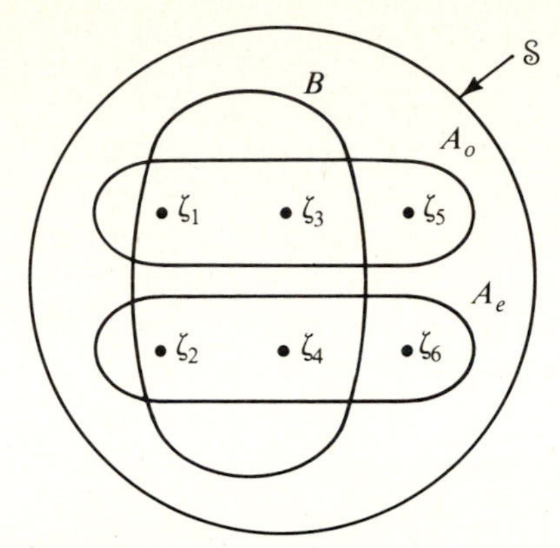

Figure 5.1 Sample space for a throw of a die.

Observe that

$$A \cup B = B \cup A \tag{5.1}$$

The *intersection* of events A and B, denoted as $A \cap B$ or simply AB, is the event that contains points common to A and B. This is the event "both A and B," also known as the *joint event AB*. Thus, the event A_eB, "a number that is even and equal to or less than 4 is thrown," is a set (ζ_2, ζ_4).

$$A_eB = (\zeta_2, \zeta_4) \qquad A_oB = (\zeta_1, \zeta_3)$$

Observe that

$$AB = BA \tag{5.2}$$

All these concepts can be demonstrated on a Venn diagram (Fig. 5.2). If the

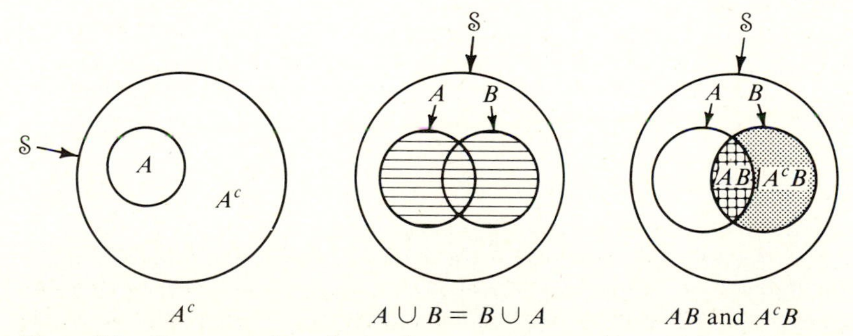

Figure 5.2 Representation of complement, union, and intersection of events.

events A and B are such that

$$AB = \emptyset \tag{5.3}$$

then A and B are said to be *disjoint*, or *mutually exclusive*, events. This means events A and B cannot occur simultaneously. In Fig. 5.1 events A_e and A_o are mutually exclusive, meaning that in any trial of the experiment if A_e occurs, A_o cannot occur at the same time and vice versa.

Relative Frequency and Probability

Although the outcome of a random experiment is unpredictable, there is a *statistical regularity* about the outcomes. For example, if a coin is tossed a large number of times, about half the times the outcome will be "heads," and the remaining half of the times it will be "tails." We may say that the relative frequency of the outcome heads or tails is one-half.

Let A be one of the events of a random experiment. If we conduct a sequence of N independent trials* of this experiment, and if the event A occurs in $N(A)$ out of these N trials, then the fraction

$$f(A) = \lim_{N\to\infty} \frac{N(A)}{N} \tag{5.4}$$

is called the *relative frequency* of the event A. Observe that for small N, the fraction $N(A)/N$ may vary widely with N. As N increases, however, the fraction will approach a limit because of statistical regularity.

The probability of an event has the same connotations as the relative frequency of that event. Hence, to each event we assign a probability that is equal to the relative frequency of that event.† Therefore, to an event A, we assign the probability $P(A)$ as

$$P(A) = \lim_{N\to\infty} \frac{N(A)}{N} \tag{5.5}$$

From Eq. (5.5), it follows that

$$0 \le P(A) \le 1 \tag{5.6}$$

■ EXAMPLE 5.1

Assign probabilities to each of the six outcomes in Fig. 5.1.

□ **Solution:** Because each of the six outcomes is equally likely in a large number of independent trials, each outcome will appear in one-sixth of the trials. Hence,

$$P(\zeta_i) = \frac{1}{6} \qquad i = 1, 2, 3, 4, 5, 6 \tag{5.7}$$

■

Consider now the two events A and B of a random experiment. Suppose we conduct N independent trials of this experiment and events A and B occur in $N(A)$ and $N(B)$ trials, respectively. If A and B are mutually exclusive (or disjoint), then if A

---

* Trials conducted under similar discernible conditions.

† Observe that we are not *defining* the probability by the relative frequency. To a given event, we *assign* a probability that is equal to the relative frequency of the event. Modern theory of probability, being a branch of mathematics, starts with certain axioms about probability [Eqs. (5.6), (5.8), and (5.11)]. It does not concern itself with how the probability is assigned to an event. It assumes that somehow these probabilities are assigned. We use relative frequency to assign probability because it is reasonable in the sense that it closely approximates the real-world behavior.

occurs, B cannot occur, and vice versa. Hence, the event $A \cup B$ occurs in $N(A) + N(B)$ trials and

$$\begin{aligned} P(A \cup B) &= \lim_{N\to\infty} \frac{N(A) + N(B)}{N} \\ &= P(A) + P(B) \qquad \text{if } AB = \emptyset \end{aligned} \tag{5.8}$$

This result can be extended to more than two mutually exclusive events.

■ EXAMPLE 5.2

Assign probabilities to the events A_e, A_o, B, A_eB, and A_oB in Fig. 5.1.

□ **Solution:** Because $A_e = (\zeta_2 \cup \zeta_4 \cup \zeta_6)$ where ζ_2, ζ_4, and ζ_6 are mutually exclusive,

$$P(A_e) = P(\zeta_2) + P(\zeta_4) + P(\zeta_6)$$

From Eq. (5.7) it follows that

$$P(A_e) = 1/2 \tag{5.9a}$$

Similarly,

$$P(A_o) = 1/2 \tag{5.9b}$$

$$P(B) = 2/3 \tag{5.9c}$$

From Fig. 5.1 we also observe that

$$A_eB = \zeta_2 \cup \zeta_4$$

and

$$P(A_eB) = P(\zeta_2) + P(\zeta_4) = 1/3 \tag{5.10a}$$

Similarly,

$$P(A_oB) = 1/3 \tag{5.10b}$$

In addition, we can show that

$$P(\mathcal{S}) = 1 \tag{5.11}$$

This result can be proved using the relative frequency. Let a random experiment be repeated N times (N, large). Because $\mathcal{S}$ is the union of all possible outcomes, $\mathcal{S}$ occurs in every trial. Hence, N out of N trials are favorable to $\mathcal{S}$ and the result follows. ■

■ EXAMPLE 5.3

Two dice are thrown. Determine the probability that a seven is thrown (i.e., the probability of the event that the sum of the numbers of dots showing on two dice is seven).

□ **Solution:** For this experiment, the sample space contains 36 sample points because 36 possible outcomes exist. All the outcomes are equally likely. Hence, the probability of each outcome is $1/36$.

A sum of seven can be obtained by the six combinations: (1, 6), (2, 5), (3, 4), (4, 3), (5, 2), and (6, 1). Hence, the event "a seven is thrown" is the union of six outcomes, each with probability 1/36. Therefore,

$$P\text{ ("a seven is thrown")} = \frac{1}{36} + \frac{1}{36} + \frac{1}{36} + \frac{1}{36} + \frac{1}{36} + \frac{1}{36} = \frac{1}{6}$$ ■

■ EXAMPLE 5.4

A coin is tossed four times in succession. Determine the probability of obtaining exactly two heads.

□ **Solution:** A total of $2^4 = 16$ distinct outcomes are possible, all of which are equally likely because of the symmetry of the situation. Hence, the sample space consists of 16 points, each with probability 1/16. The 16 outcomes are listed below:

	H H H H		T T T T
	H H H T		T T T H
	H H T H		T T H T
⟶	H H T T	⟶	T T H H
	H T H H		T H T T
⟶	H T H T	⟶	T H T H
⟶	H T T H	⟶	T H H T
	H T T T		T H H H

Six out of these 16 outcomes are favorable to the event "obtaining two heads" (shown by arrows). Because all of the six outcomes are disjoint (mutually exclusive),

$$P\text{ ("obtaining two heads")} = \frac{6}{16} = \frac{3}{8}$$ ■

This method of listing all possible outcomes quickly becomes unwieldy as the number of tosses increases. For example, if a coin is tossed just 10 times, the total number of outcomes is 1024. A mcre convenient approach would be to use the results of combinatorial analysis. If a coin is tossed k times, the number of ways in which j heads can occur is the same as the number of combinations of k things taken j at a time. This is given by $\binom{k}{j}$, where

$$\binom{k}{j} = \frac{k!}{j!\,(k-j)!} \tag{5.12}$$

This can be proved as follows. Consider an urn containing k distinguishable balls marked $1, 2, \ldots, k$. Suppose we draw j balls from this urn without replacement. The first ball could be any one of the k balls, the second ball could be any one of the remaining $(k - 1)$ balls, and so on. Hence, the total number of ways in which j balls can be drawn is

$$k(k-1)(k-2)\ldots(k-j+1) = k!/(k-j)!$$

Next, consider any one set of the j balls drawn. These balls can be ordered in different ways. We could choose any one of the j balls for number 1, and any one of the remaining $(j - 1)$ balls for number 2, and so on. This will give a total of $j(j - 1)(j - 2) \dots 1 = j!$ distinguishable patterns formed from the j balls. The total number of ways in which j things can be taken from k things is $k!/(k - j)!$. But many of these ways will use the same j things, but arranged in different order. The ways in which j things can be taken from k things without regard to order (unordered subset j taken from k things) is $k!/(k - j)!$ divided by $j!$. This is precisely $\binom{k}{j}$ defined by Eq. (5.12). Thus, the number of ways in which two heads can occur in four tosses is

$$\binom{4}{2} = \frac{4!}{2!\,2!} = 6$$

Conditional Probability and Independent Events

One often comes across a situation where the probability of one event is influenced by that of another event. As an example, consider drawing two cards in succession from a deck. Let A denote the event that the first card drawn is an ace. We do not replace the card drawn in the first trial. Let B denote the event that the second card drawn is an ace. It is evident that the probability of drawing an ace in the second trial will be influenced by the outcome of the first draw. If the first draw does not result in an ace, then the probability of obtaining an ace in the second trial is $4/51$. The probability of event B thus depends on whether or not event A occurs. We now introduce the *conditional probability* $P(B|A)$ to denote the probability of the event B when it is known that the event A has occurred. $P(B|A)$ is read as "probability of B given A."

Let an experiment be performed N times, in which the event A occurs n_1 times. Of *these* n_1 trials, the event B occurs n_2 times. It is clear that n_2 is the number of times that the joint event AB (Fig. 5.2*c*) occurs. That is

$$P(AB) = \lim_{N\to\infty} \left(\frac{n_2}{N}\right) = \lim_{N\to\infty} \left(\frac{n_1}{N}\right)\left(\frac{n_2}{n_1}\right)$$

Note that $\lim_{N\to\infty} (n_1/N) = P(A)$. Also, $\lim_{N\to\infty} (n_2/n_1) = P(B|A)$*, because B occurs n_2 of the n_1 times that A occurred. This represents the conditional probability of B given A. Therefore,

$$P(AB) = P(A)P(B|A) \tag{5.13}$$

and

$$P(B|A) = \frac{P(AB)}{P(A)} \qquad \text{provided } P(A) \neq 0 \tag{5.14a}$$

* Here we are implicitly using the fact that $n_1 \to \infty$ as $N \to \infty$. This is true provided the ratio $\lim_{N\to\infty} (n_1/N) \neq 0$ (that is, if $P(A) \neq 0$).

Using a similar argument, we obtain

$$P(A|B) = \frac{P(AB)}{P(B)} \qquad \text{provided } P(B) \neq 0 \tag{5.14b}$$

It follows from Eq. (5.14a and b) that

$$P(A|B) = \frac{P(A)P(B|A)}{P(B)} \tag{5.15a}$$

$$P(B|A) = \frac{P(B)P(A|B)}{P(A)} \tag{5.15b}$$

Equation (5.15a and b) are called *Bayes' rule*. In Bayes' rule, one conditional probability is expressed in terms of the reversed conditional probability.

■ EXAMPLE 5.5

A random experiment consists of drawing two cards from a deck in succession (without replacing the first card drawn). Assign a value to the probability of obtaining two red aces in two draws.

□ **Solution:** Let A and B be the events "red ace in the first draw" and "red ace in the second draw," respectively. We wish to determine $P(AB)$.

$$P(AB) = P(A)P(B|A)$$

and the relative frequency of A is $2/52 = 1/26$. Hence,

$$P(A) = 1/26$$

Also, $P(B|A)$ is the probability of drawing a red ace in the second draw given that the first draw was a red ace. The relative frequency of this event is $1/51$, so

$$P(B|A) = 1/51$$

Hence,

$$P(AB) = \left(\frac{1}{26}\right)\left(\frac{1}{51}\right) = \frac{1}{1326}$$ ■

Independent Events. Under conditional probability, we presented an example where the occurrence of one event was influenced by the occurrence of another. There are, of course, many examples where the two or more events are entirely independent; that is, the occurrence of one event in no way influences the occurrence of the other event. As an example, we again consider the drawing of two cards in succession, but in this case we replace the card obtained in the first draw and shuffle the deck before the second draw. In this case, the outcome of the second draw is in no way influenced by the outcome of the first draw. Thus $P(B)$, the probability of drawing an ace in the second draw, is independent of whether or not the event A (drawing an ace in the first trial) occurs. The events A and B are therefore independent. The conditional probability $P(B|A)$ is given by $P(B)$.

The event B is said to be *independent* of the event A if

$$P(B|A) = P(B) \tag{5.16a}$$

It can be seen from Eq. (5.15a and b) that if event B is independent of event A, then event A is also independent of B; that is,

$$P(A|B) = P(A) \tag{5.16b}$$

Note that if the events A and B are independent, it follows from Eqs. (5.14a) and (5.16a) that

$$P(AB) = P(A)P(B) \tag{5.16c}$$

■ EXAMPLE 5.6

A binary-symmetric-channel (BSC) has an error probability P_e (i.e., the probability of receiving **0** when **1** is transmitted, or vice versa, is P_e). Note that the channel behavior is symmetrical with respect to **0** and **1**. Thus,

$$P(\mathbf{0}|\mathbf{1}) = P(\mathbf{1}|\mathbf{0}) = P_e$$

and

$$P(\mathbf{0}|\mathbf{0}) = P(\mathbf{1}|\mathbf{1}) = 1 - P_e$$

where $P(y|x)$ denotes the probability of receiving y when x is transmitted.

A sequence of k binary digits is transmitted over this channel. Determine the probability of receiving exactly j digits in error.

□ **Solution:** If $P(y_i|x_i)$ denotes the probability of receiving y_i when x_i is transmitted, and $P(y_1y_2 \ldots y_k|x_1x_2 \ldots x_k)$ denotes the probability that the received digit sequence is $y_1y_2 \ldots y_k$ when the transmitted digit sequence is $x_1x_2 \ldots x_k$, then

$$P(y_1y_2 \ldots y_k|x_1x_2 \ldots x_k) = P(y_1|x_1)P(y_2|x_2) \ldots P(y_k|x_k)$$

This is because the reception of each digit is independent of all the remaining digits.

If in a certain received sequence j digits are in error, then $k - j$ digits are correct. Then the probability of receiving this sequence is $P_e^j(1 - P_e)^{k-j}$.

There are $\binom{k}{j}$ different ways in which j errors can occur in k digits [see Eq. (5.12)]. Hence,

$$P(\text{receiving } j \text{ out of } k \text{ digits in error}) = \binom{k}{j} P_e^j(1 - P_e)^{k-j} \tag{5.17}$$

For example, if $P_e = 10^{-5}$, the probability of receiving two digits wrong in a sequence of eight digits is

$$\binom{8}{2}(10^{-5})^2(1 - 10^{-5})^6 \simeq \frac{8!}{2!\,6!}\,10^{-10} = (2.8)\,10^{-9}$$ ■

EXAMPLE 5.7 PCM Repeater Error Probability

In PCM, regenerative repeaters are used to detect pulses (before they are lost in noise) and retransmit new, clean pulses. This combats the accumulation of noise and pulse distortion.

A certain PCM channel consists of k identical links in tandem (Fig. 5.3). The

Figure 5.3 A PCM repeater.

pulses are detected at the end of each link and clean new pulses are transmitted over the next link. If P_e is the probability of error in detecting a pulse over any one link, show that P_E, the probability of error in detecting a pulse over the entire channel (over the k links in tandem), is

$$P_E \simeq kP_e \qquad kP_e \ll 1$$

☐ **Solution:** The probabilities of detecting a pulse correctly over one link and over the entire channel (k links in tandem) are $1 - P_e$ and $1 - P_E$, respectively. A pulse can be detected correctly over the entire channel if either the pulse is detected correctly over every link or if errors are made over an even number of links only.

$$1 - P_E = P(\text{correct detection over all links}) + P(\text{error over two links only}) + P(\text{error over four links only}) + \cdots + P(\text{error over } \alpha \text{ links only})$$

where α is k or $k - 1$, depending on whether k is even or odd. Because pulse detection over each link is independent of the other links (see Example 5.6),

$$P(\text{correct detection over all } k \text{ links}) = (1 - P_e)^k$$

and

$$P(\text{error over } j \text{ links only}) = \frac{k!}{j!\,(k-j)!} P_e^j (1 - P_e)^{k-j}$$

Hence,

$$1 - P_E = (1 - P_e)^k + \sum_{j=2,4,6,\ldots}^{\alpha} \frac{k!}{j!\,(k-j)!} P_e^j (1 - P_e)^{k-j}$$

In practice, $P_e \ll 1$, so only the first two terms on the right side of the above equation are of significance. Also, $(1 - P_e)^{k-j} \simeq 1$, and

$$1 - P_E \simeq (1 - P_e)^k + \frac{k!}{2!\,(k-2)!} P_e^2 = (1 - P_e)^k + \frac{k(k-1)}{2} P_e^2$$

If $kP_e \ll 1$, then the second term can also be neglected, and

$$1 - P_E \simeq (1 - P_e)^k$$

and

$$\simeq 1 - kP_e \qquad kP_e \ll 1$$

$$P_E \simeq kP_e$$

■

■ EXAMPLE 5.8

In binary communication, one of the techniques used to increase the reliability of a channel is to repeat the message several times. For example, we can send each message (**0** or **1**) three times. Hence, the transmitted digits are **000** (for message **0**) or **111** (for message **1**). Because of channel noise, we may receive any one of the eight possible combinations of three binary digits. The decision as to which message is transmitted is made by the majority rule; that is, if at least two of the three detected digits are **0**, the decision is **0**, and so on. This scheme permits correct reception of data even if one out of three digits is in error. Detection error occurs only if at least two out of three digits are received in error. Thus $P(\epsilon)$, the detection error probability, is

$$P(\epsilon) = \sum_{j=2}^{3} \binom{3}{j} P_e^j (1 - P_e)^{3-j}$$

$$= 3P_e^2(1 - P_e) + P_e^3$$

In practice, $P_e \ll 1$, and

$$P(\epsilon) \simeq 3P_e^2$$

For instance, if $P_e = 10^{-4}$, $P(\epsilon) \simeq 3 \times 10^{-8}$. Thus, the error probability is reduced from 10^{-4} to 3×10^{-8}. We can use any odd number of repetitions for this scheme to function.

In this example, higher reliability is achieved at the cost of a reduction in the rate of information transmission by a factor of 3. We shall see in Chapter 9 that more efficient ways exist to trade off between the reliability and the rate of transmission.

■

5.2 RANDOM VARIABLES

The outcome of a random experiment may be a real number (as in the case of rolling a die), or it may be nonnumerical and describable by a phrase (such as "heads" or "tails" in tossing a coin). From a mathematical point of view, it is desirable to have numerical values for all outcomes. For this reason, we assign a real number to each sample point according to some rule. If there are m sample points $\zeta_1, \zeta_2, \ldots, \zeta_m$, then using some convenient rule, we assign a real number $\mathrm{x}(\zeta_i)$ to sample point ζ_i $(i = 1, 2, \ldots, m)$. In the case of tossing a coin, for example, we may assign the number 1 for the outcome heads and the number -1 for the outcome tails (Fig. 5.4).

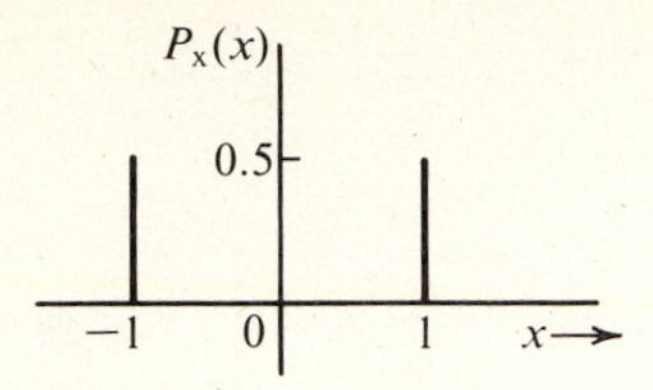

Figure 5.4 Probabilities in a coin tossing experiment.

Thus x (.) is a function that maps sample points $\zeta_1, \zeta_2, \ldots, \zeta_m$ into real numbers $x_1, x_2, \ldots, x_n$.* We now have a *random variable* x that takes on values $x_1, x_2, \ldots, x_n$. We shall use roman type (x) to denote a random variable (r.v.) and italic type (for example, $x_1, x_2, \ldots, x_n$, etc.) to denote the value it takes. The probability of an r.v. x taking a value x_i is $P_x(x_i)$.

Discrete Random Variables

A random variable is a discrete r.v. if there exists a denumerable sequence of distinct numbers x_i such that

$$\sum_i P_x(x_i) = 1 \tag{5.18}$$

Thus, a discrete r.v. can assume only certain discrete values. A random variable that can assume any value from a continuous interval is called a continuous random variable.

■ EXAMPLE 5.9

Two dice are thrown. The sum of the points appearing on the two dice is a random variable x. Find the values taken by x, and the corresponding probabilities.

□ **Solution:** x can take on all integral values from 2 through 12. Various probabilities can be determined by the method outlined in Example 5.3.

There are 36 sample points in all, each with probability 1/36. Dice outcomes for various values of x are shown in Table 5.1. Note that although there are 36 sample points, they all map into 11 values of x. This is because more than one sample point maps into the same value of x. For example, six sample points map into x = 7.

The reader can verify that $\sum_{i=2}^{12} P_x(x_i) = 1$. ■

The above discussion can be extended to two r.v.'s x and y. The joint probability $P_{xy}(x_i, y_j)$ is the probability that $x = x_i$ and $y = y_j$. Consider, for example, the case of a coin tossed twice in succession. If the outcomes of the first and second toss are mapped into r.v.'s x and y, then x and y each take values 1 and −1. Because the

* m is not necessarily equal to n. More than one sample point can map into one value of x.

Table 5.1

Value of x x_i	Dice Outcomes	$P_x(x_i)$
2	(1, 1)	1/36
3	(1, 2), (2, 1)	2/36 = 1/18
4	(1, 3), (2, 2), (3, 1)	3/36 = 1/12
5	(1, 4), (2, 3), (3, 2), (4, 1)	4/36 = 1/9
6	(1, 5), (2, 4), (3, 3), (4, 2), (5, 1)	5/36
7	(1, 6), (2, 5), (3, 4), (4, 3), (5, 2), (6, 1)	6/36 = 1/6
8	(2, 6), (3, 5), (4, 4), (5, 3), (6, 2)	5/36
9	(3, 6), (4, 5), (5, 4), (6, 3)	4/36 = 1/9
10	(4, 6), (5, 5), (6, 4)	3/36 = 1/12
11	(5, 6), (6, 5)	2/36 = 1/18
12	(6, 6)	1/36

outcomes of the two tosses are independent, x and y are independent, and

$$P_{xy}(x_i, y_j) = P_x(x_i)\, P_y(y_j)$$

and

$$P_{xy}(1, 1) = P_{xy}(1, -1) = P_{xy}(-1, 1) = P_{xy}(-1, -1) = \tfrac{1}{4}$$

These probabilities are plotted in Fig. 5.5.

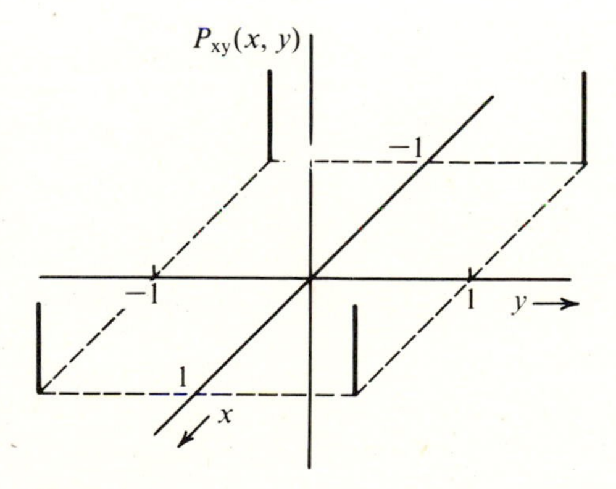

Figure 5.5

For a general case where the variable x can take values $x_1, x_2, \ldots, x_n$ and the variable y can take values $y_1, y_2, \ldots, y_m$, we have

$$\sum_i \sum_j P_{xy}(x_i, y_j) = 1 \qquad \textbf{(5.19)}$$

This follows from the fact that the summation on the left is the probability of the union of all possible outcomes and must be unity (a certain event).

Conditional Probabilities. If x and y are two r.v.'s, then the conditional probability of $\mathrm{x} = x_i$ given $\mathrm{y} = y_j$ is denoted by $P_{\mathrm{x|y}}(x_i|y_j)$. We must have

$$\sum_i P_{\mathrm{x|y}}(x_i|y_j) = \sum_j P_{\mathrm{y|x}}(y_j|x_i) = 1 \tag{5.20}$$

This can be proved by observing that probabilities $P_{\mathrm{x|y}}(\cdot|y_j)$ are specified over the sample space corresponding to the condition $\mathrm{y} = y_j$. Hence, $\Sigma_i P_{\mathrm{x|y}}(x_i|y_j)$ is the probability of the union of all possible outcomes of x (under the condition $\mathrm{y} = y_j$) and must be unity (a certain event). A similar argument applies to $\Sigma_j P_{\mathrm{y|x}}(y_j|x_i)$.

Corresponding to Eq. (5.13), we also have

$$P_{\mathrm{xy}}(x_i, y_j) = P_{\mathrm{x|y}}(x_i|y_j)P_{\mathrm{y}}(y_j) = P_{\mathrm{y|x}}(y_j|x_i)P_{\mathrm{x}}(x_i) \tag{5.21}$$

Bayes' rule follows from Eq. (5.21). Also from Eq. (5.21), we have

$$\begin{aligned}\sum_i P_{\mathrm{xy}}(x_i, y_j) &= \sum_i P_{\mathrm{x|y}}(x_i|y_j)P_{\mathrm{y}}(y_j) \\ &= P_{\mathrm{y}}(y_j)\sum_i P_{\mathrm{x|y}}(x_i|y_j) \\ &= P_{\mathrm{y}}(y_j)\end{aligned} \tag{5.22a}$$

Similarly,

$$P_{\mathrm{x}}(x_i) = \sum_j P_{\mathrm{xy}}(x_i, y_j) \tag{5.22b}$$

The probabilities $P_{\mathrm{x}}(x_i)$ and $P_{\mathrm{y}}(y_j)$ are called *marginal probabilities*. Equation (5.22) shows how to determine marginal probabilities from joint probabilities. Results of Eqs. (5.19) through (5.22) can be extended to more than two random variables.

■ EXAMPLE 5.10

A binary-symmetric-channel (BSC) error probability is P_e. The probability of transmitting **1** is Q, and that of transmitting **0** is $1 - Q$ (Fig. 5.6). Determine the probabilities of receiving **1** and **0** at the receiver.

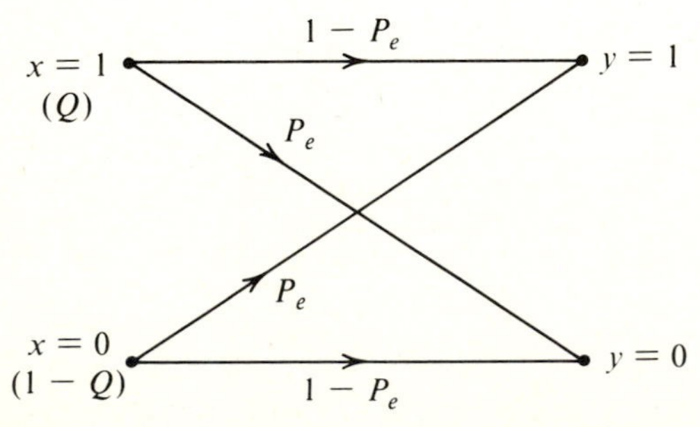

Figure 5.6 A binary symmetric channel (BSC).

☐ **Solution:** If x and y are the transmitted digit and the received digit, respectively, then for a BSC

$$P_{y|x}(\mathbf{0}|\mathbf{1}) = P_{y|x}(\mathbf{1}|\mathbf{0}) = P_e \qquad \text{and} \qquad P_{x|y}(\mathbf{0}|\mathbf{0}) = P_{y|x}(\mathbf{1}|\mathbf{1}) = 1 - P_e$$

Also,

$$P_x(\mathbf{1}) = Q \qquad \text{and} \qquad P_x(\mathbf{0}) = 1 - Q$$

We need to find $P_y(\mathbf{1})$ and $P_y(\mathbf{0})$.

Because

$$P_y(y_j) = \sum_i P_{xy}(x_i, y_j)$$

$$= \sum_i P_x(x_i)P_{y|x}(y_j|x_i)$$

$$P_y(\mathbf{1}) = P_x(\mathbf{0})P_{y|x}(\mathbf{1}|\mathbf{0}) + P_x(\mathbf{1})P_{y|x}(\mathbf{1}|\mathbf{1})$$

$$= (1 - Q)P_e + Q(1 - P_e)$$

Similarly, we find

$$P_y(\mathbf{0}) = (1 - Q)(1 - P_e) + QP_e$$

These answers seem almost obvious from Fig. 5.6.

Note that because of channel errors, the probability of receiving a digit **1** is not the same as that of transmitting **1**. The same is true of **0**. ■

■ EXAMPLE 5.11

Over a certain binary communication channel, the symbol **0** is transmitted with probability 0.4 and **1** is transmitted with probability 0.6. It is given that $P(\epsilon|\mathbf{0}) = 10^{-6}$ and $P(\epsilon|\mathbf{1}) = 10^{-4}$, where $P(\epsilon|x_i)$ is the probability of detection error given that x_i is transmitted. Determine $P(\epsilon)$, the error probability of the channel.

☐ **Solution:** If $P(\epsilon, x_i)$ is the joint probability that x_i is transmitted and it is detected wrongly, then [Eq. (5.22b)]

$$P(\epsilon) = \sum_i P(\epsilon, x_i)$$

$$= P(\epsilon, \mathbf{0}) + P(\epsilon, \mathbf{1})$$

$$= P_x(\mathbf{0})P(\epsilon|\mathbf{0}) + P_x(\mathbf{1})P(\epsilon|\mathbf{1})$$

$$= 0.4(10^{-6}) + 0.6(10^{-4})$$

$$= 0.604(10^{-4})$$

Note that $P(\epsilon|\mathbf{0}) = 10^{-6}$ means that on the average, one out of 1 million received **0**'s will be detected erroneously. Similarly, $P(\epsilon|\mathbf{1}) = 10^{-4}$ means that on the average, one out of 10,000 received **1**'s will be in error. But $P(\epsilon) = 0.604(10^{-4})$ indicates that on the average, one out of $1/0.604(10^{-4}) \simeq 16{,}556$ digits (regardless of whether they are **1**'s or **0**'s) will be received in error. ■

The Cumulative Distribution Function

The *Cumulative Distribution Function (CDF)* $F_x(x)$ of a random variable x is the probability that an r.v. x takes a value less than or equal to x; that is,

$$F_x(x) = P(\mathrm{x} \leq x) \tag{5.23}$$

We can show that a CDF $F_x(x)$ has the following four properties:

1. $F_x(x) \geq 0$ **(5.24a)**
2. $F_x(\infty) = 1$ **(5.24b)**
3. $F_x(-\infty) = 0$ **(5.24c)**
4. $F_x(x)$ is a nondecreasing function, that is,

$$F_x(x_1) \leq F_x(x_2) \text{ for } x_1 \leq x_2 \tag{5.24d}$$

The first property is obvious. The second and third properties are proved by observing that $F_x(\infty) = P(\mathrm{x} \leq \infty)$ and $F_x(-\infty) = P(\mathrm{x} \leq -\infty)$. To prove the fourth property, we have from Eq. (5.23)

$$\begin{aligned} F_x(x_2) &= P(\mathrm{x} \leq x_2) \\ &= P[(\mathrm{x} \leq x_1) \cup (x_1 < \mathrm{x} \leq x_2)] \end{aligned}$$

Because $\mathrm{x} \leq x_1$ and $x_1 < \mathrm{x} \leq x_2$ are disjoint, we have

$$\begin{aligned} F_x(x_2) &= P(\mathrm{x} \leq x_1) + P(x_1 < \mathrm{x} \leq x_2) \\ &= F_x(x_1) + P(x_1 < \mathrm{x} \leq x_2) \end{aligned} \tag{5.25}$$

Because $P(x_1 < \mathrm{x} \leq x_2)$ is nonnegative, the result follows.

■ EXAMPLE 5.12

In a random experiment, a trial consists of four successive tosses of a coin. If we define an r.v. x as the number of heads appearing in a trial, determine $P_x(x)$ and $F_x(x)$.

☐ **Solution:** A total of 16 distinct equiprobable outcomes are listed in Example 5.4. Various probabilities can be readily determined by counting the outcomes pertaining to a given value of x. For example, only one outcome maps into $\mathrm{x} = 0$, whereas six outcomes map into $\mathrm{x} = 2$. Hence $P_x(0) = 1/16$ and $P_x(2) = 6/16$. In the same way, we find

$$\begin{aligned} P_x(0) &= P_x(4) = 1/16 \\ P_x(1) &= P_x(3) = 4/16 = 1/4 \\ P_x(2) &= 6/16 = 3/8 \end{aligned}$$

The probabilities $P_x(x_i)$ and the corresponding CDF $F_x(x_i)$ are shown in Fig. 5.7. ■

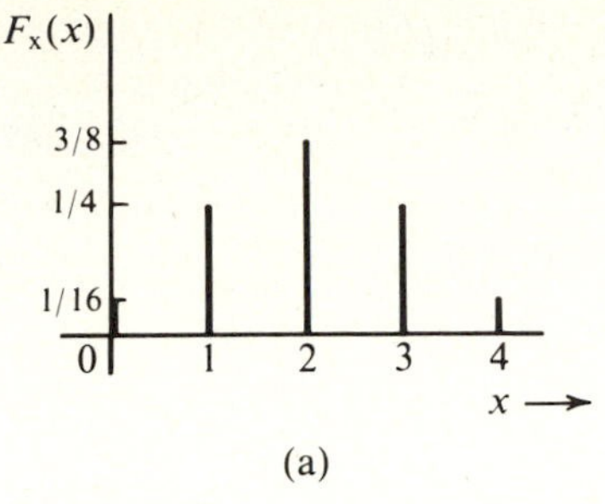

(a)

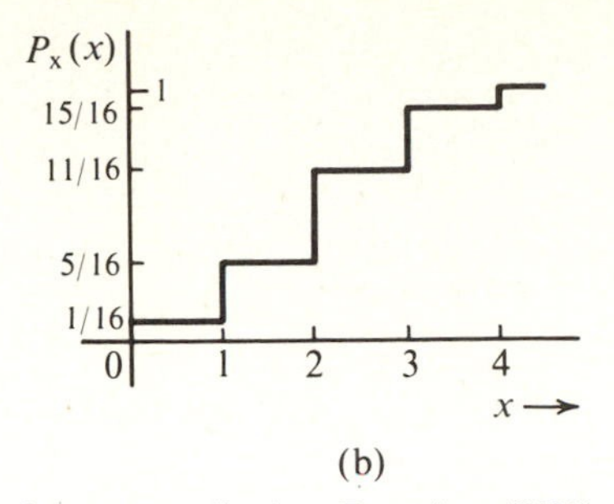

(b)

Figure 5.7 Probabilities $P_x(x_i)$ and the Cumulative Distribution Function (CDF).

Continuous Random Variables

A continuous random variable x can assume any value in a certain interval. In a continuum of any range, an uncountably infinite number of possible values exist, and $P_x(x_i)$, the probability that $x = x_i$, is one of the uncountably infinite values and is generally zero. Consider the case of a temperature T at a certain point. We may suppose that this temperature can assume any of a range of values. Thus, an uncountably infinite number of possible temperature values may prevail, and the probability that the r.v. T assumes a certain value T_i is zero. The situation is somewhat similar to that described on page 33 in connection with a continuously loaded beam (Fig. 2.9*b*). There is a loading along the beam at every point, but at any one point the load is zero. The meaningful measure in that case was the loading (or weight) not at a point but over a finite interval. Similarly, for a continuous r.v., the meaningful quantity is not the probability that $x = x_i$ but the probability that $x < \mathrm{x} \leq x + \Delta x$. For such a measure, the CDF is eminently suited because the latter probability is simply $F_x(x + \Delta x) - F_x(x)$ [see Eq. (5.25)]. Hence, we begin our study of continuous r.v.'s with the CDF.

Properties of the CDF [Eqs. (5.24) and (5.25)] derived earlier are general and are valid for continuous as well as discrete random variables.

The Probability Density Function. From Eq. (5.25), we have

$$F_x(x + \Delta x) = F_x(x) + P(x < \mathrm{x} \leq x + \Delta x) \tag{5.26a}$$

If $\Delta x \rightarrow 0$, then we can also express $F_x(x + \Delta x)$ via Taylor expansion as

$$F_x(x + \Delta x) \simeq F_x(x) + \frac{dF_x(x)}{dx} \Delta x \tag{5.26b}$$

From Eq. (5.26a and b), it follows that

$$\lim_{\Delta x \rightarrow 0} \frac{dF_x(x)}{dx} \Delta x = P(x < \mathrm{x} \leq x + \Delta x) \tag{5.27}$$

We designate the derivative of $F_x(x)$ with respect to x by $p_x(x)$ (Fig. 5.8).

$$\frac{dF_x(x)}{dx} = p_x(x) \tag{5.28}$$

The function $p_x(x)$ is called the *probability density function (PDF)* of the random variable x. It follows from Eq. (5.27) that the probability of observing the random variable x in the interval $(x, x + \Delta x)$ is $p_x(x)\,\Delta x(\Delta x \to 0)$. This is the area under the PDF $p_x(x)$ over the interval Δx, as shown in Fig. 5.8*b*.

From Eq. (5.28) it follows that

$$F_x(x) = \int_{-\infty}^{x} p_x(x)\,dx \tag{5.29}$$

Here we use the fact that $F_x(-\infty) = 0$. We also have from Eq. (5.25)

$$\begin{aligned} P(x_1 < x \le x_2) &= F_x(x_2) - F_x(x_1) \\ &= \int_{-\infty}^{x_2} p_x(x)\,dx - \int_{-\infty}^{x_1} p_x(x)\,dx \\ &= \int_{x_1}^{x_2} p_x(x)\,dx \end{aligned} \tag{5.30}$$

Thus, the probability of observing x in any interval (x_1, x_2) is given by the area under the PDF $p_x(x)$ over the interval (x_1, x_2), as shown in Fig. 5.8. Compare this with a

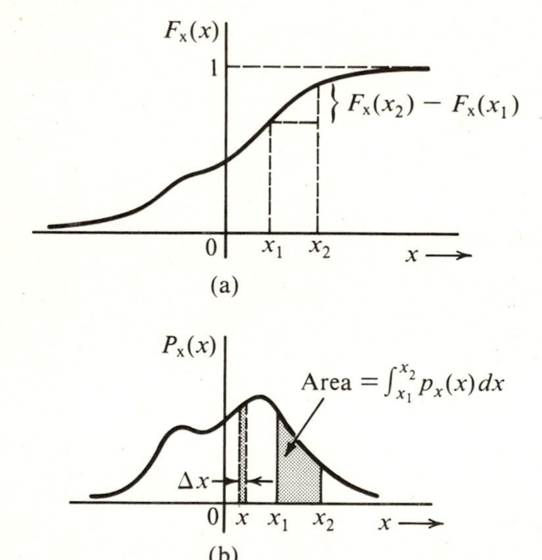

Figure 5.8 Cumulative Distribution Function (CDF) and the probability density function (PDF).

continuously loaded beam (Fig. 2.9), where the weight over any interval was given by an integral of the loading density over the interval.

Because $F_x(\infty) = 1$, we have

$$\int_{-\infty}^{\infty} p_x(x)\,dx = 1 \tag{5.31}$$

This also follows from the fact that the integral in Eq. (5.31) represents the probability of observing x in the interval $(-\infty, \infty)$. Every PDF must satisfy the condition in Eq. (5.31). It is also evident that the PDF must be nonnegative, that is,

$$p_x(x) \geq 0$$

Although it is true that the probability of an impossible event is zero and that of a certain event is one, the converse is not true. An event whose probability is zero is not necessarily an impossible event, and an event with a probability of one is not necessarily a certain event. This may be illustrated by the following example. The temperature T of a certain city on a summer day is a random variable taking on any value in the range 5 to 50°C. Because the PDF $p_T(T)$ is continuous, the probability that T = 34.56, for example, is zero. But this is not an impossible event. Similarly, the probability that T takes on any value but 34.56 is one, although this is not a certain event. In fact, a continuous r.v. x takes every value in a certain range. Yet $p_x(x)$, the probability that x = x, is zero for every x in that range.

We can also determine the PDF $p_x(x)$ for a discrete random variable. Because the CDF $F_x(x)$ for the discrete case is always a sequence of step functions (Fig. 5.7), the PDF (the derivative of the CDF) will consist of a train of impulses. If an r.v. x takes values $x_1, x_2, \ldots, x_n$ with probabilities $a_1, a_2, \ldots, a_n$, respectively, then

$$F_x(x) = a_1 u(x - x_1) + a_2 u(x - x_2) + \cdots + a_n u(x - x_n) \tag{5.32a}$$

This can be easily verified from Example 5.12 (Fig. 5.7). Hence,

$$\begin{aligned} p_x(x) &= a_1\,\delta(x - x_1) + a_2\,\delta(x - x_2) + \cdots + a_n\,\delta(x - x_n) \\ &= \sum_{r=1}^{n} a_r\,\delta(x - x_r) \end{aligned} \tag{5.32b}$$

It is, of course, possible to have a mixed case where a PDF may have a continous part and an impulsive part (see Prob. 5.21).

The Gaussian PDF. Consider a PDF (Fig. 5.9*a*)

$$p_x(x) = \frac{1}{\sqrt{2\pi}} e^{-x^2/2} \tag{5.33}$$

This is a case of the well-known *gaussian,* or *normal,* probability density.

The CDF $F_x(x)$ in this case is

$$F_x(x) = \frac{1}{\sqrt{2\pi}} \int_{-\infty}^{x} e^{-x^2/2}\, dx$$

This integral cannot be evaluated in a closed form and must be computed numerically. It is convenient to use the function $Q(.)$, defined as[1]

$$Q(y) = \frac{1}{\sqrt{2\pi}} \int_{y}^{\infty} e^{-x^2/2}\, dx \tag{5.34}$$

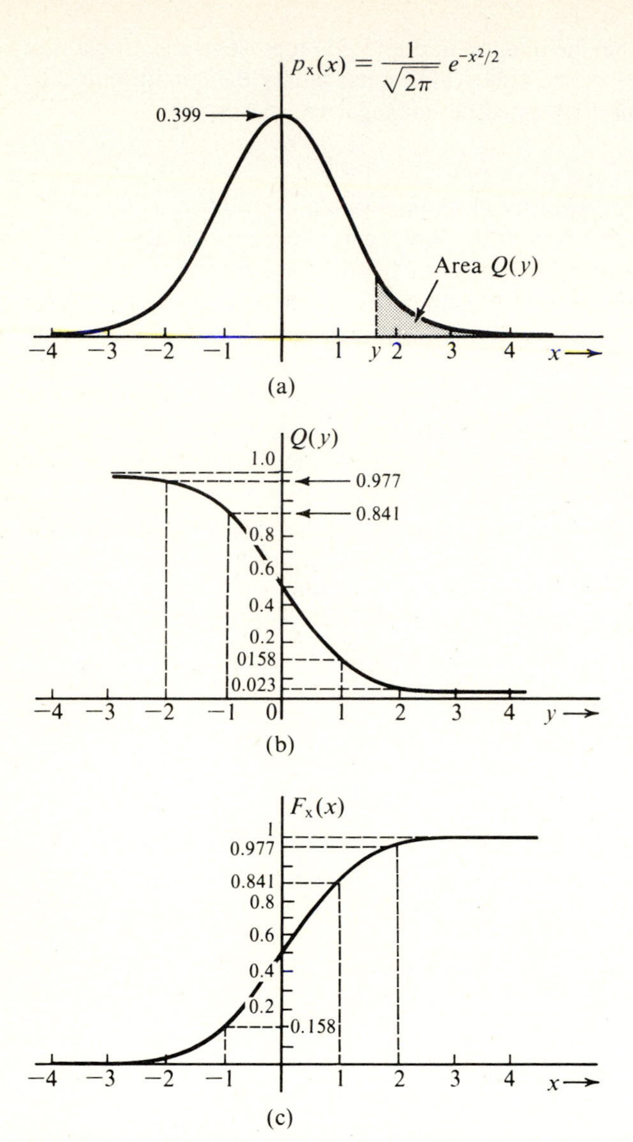

Figure 5.9 (a) Gaussian PDF. (b) Function $Q(y)$. (c) CDF of the gaussian PDF.

The area under $p_x(x)$ from y to ∞ (shown shaded in Fig. 5.9*a*) is* $Q(y)$. From the symmetry of $p_x(x)$ about the origin, and the fact that the total area under $p_x(x) = 1$,

---

* The function $Q(x)$ is closely related to functions erf (x) and erfc (x).

$$\text{erfc}\,(x) = \frac{2}{\sqrt{\pi}} \int_x^\infty e^{-y^2}\, dy = 2Q(x\sqrt{2})$$

Therefore

$$Q(x) = \frac{1}{2}\,\text{erfc}\left(\frac{x}{\sqrt{2}}\right) = \frac{1}{2}\left[1 - \text{erf}\left(\frac{x}{\sqrt{2}}\right)\right]$$

it follows that

$$Q(-y) = 1 - Q(y) \tag{5.35}$$

The function $Q(x)$ is tabulated in Table 5.2 and plotted in Fig. 5.11*d*. This function is widely tabulated and can be found in most of the standard mathematical tables.[1,2] It can be shown that[3] (see Appendix A, Sec. A.2)

$$Q(x) \simeq \frac{1}{x\sqrt{2\pi}} e^{-x^2/2} \qquad \text{for } x \gg 1 \tag{5.36a}$$

For example, when $x = 2$, the error in this approximation is 18.7 percent. But for $x = 4$ it is 5.4 percent and for $x = 6$ it is 2.3 percent.

A much better approximation to $Q(x)$ is

$$Q(x) \simeq \frac{1}{x\sqrt{2\pi}} \left(1 - \frac{0.7}{x^2}\right) e^{-x^2/2} \qquad x > 2 \tag{5.36b}$$

The error in this approximation is just within 1 percent for $x > 2.15$. For larger values of x the error approaches 0.

Observe that for the PDF in Fig. 5.9*a*, the CDF is given by (Fig. 5.9*c*)

$$F_x(x) = 1 - Q(x) \tag{5.37}$$

A more general gaussian density function is (Fig. 5.10)

$$p_x(x) = \frac{1}{\sigma\sqrt{2\pi}} e^{-(x-m)^2/2\sigma^2} \tag{5.38}$$

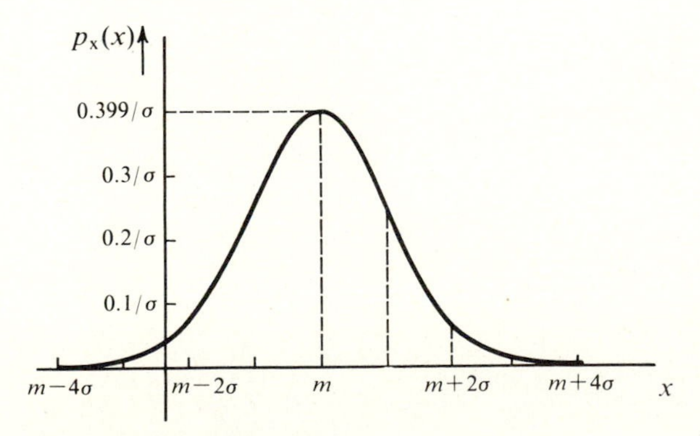

Figure 5.10 Gaussian PDF.

For this case,

$$F_x(x) = \frac{1}{\sigma\sqrt{2\pi}} \int_{-\infty}^{x} e^{-(x-m)^2/2\sigma^2}\, dx$$

x	0.00	0.01	0.02	0.03	0.04	0.05	0.06	0.07	0.08	0.09
0.0000	.5000	.4960	.4920	.4880	.4840	.4801	.4761	.4721	.4681	.4641
.1000	.4602	.4562	.4522	.4483	.4443	.4404	.4364	.4325	.4286	.4247
.2000	.4207	.4168	.4129	.4090	.4052	.4013	.3974	.3936	.3897	.3859
.3000	.3821	.3783	.3745	.3707	.3669	.3632	.3594	.3557	.3520	.3483
.4000	.3446	.3409	.3372	.3336	.3300	.3264	.3228	.3192	.3156	.3121
.5000	.3085	.3050	.3015	.2981	.2946	.2912	.2877	.2843	.2810	.2776
.6000	.2743	.2709	.2676	.2643	.2611	.2578	.2546	.2514	.2483	.2451
.7000	.2420	.2389	.2358	.2327	.2296	.2266	.2236	.2206	.2177	.2148
.8000	.2119	.2090	.2061	.2033	.2005	.1977	.1949	.1922	.1894	.1867
.9000	.1841	.1814	.1788	.1762	.1736	.1711	.1685	.1660	.1635	.1611
1.000	.1587	.1562	.1539	.1515	.1492	.1469	.1446	.1423	.1401	.1379
1.100	.1357	.1335	.1314	.1292	.1271	.1251	.1230	.1210	.1190	.1170
1.200	.1151	.1131	.1112	.1093	.1075	.1056	.1038	.1020	.1003	.9853E-01
1.300	.9680E-01	.9510E-01	.9342E-01	.9176E-01	.9012E-01	.8851E-01	.8691E-01	.8534E-01	.8379E-01	.8226E-01
1.400	.8076E-01	.7927E-01	.7780E-01	.7636E-01	.7493E-01	.7353E-01	.7215E-01	.7078E-01	.6944E-01	.6811E-01
1.500	.6681E-01	.6552E-01	.6426E-01	.6301E-01	.6178E-01	.6057E-01	.5938E-01	.5821E-01	.5705E-01	.5592E-01
1.600	.5480E-01	.5370E-01	.5262E-01	.5155E-01	.5050E-01	.4947E-01	.4846E-01	.4746E-01	.4648E-01	.4551E-01
1.700	.4457E-01	.4363E-01	.4272E-01	.4182E-01	.4093E-01	.4006E-01	.3920E-01	.3836E-01	.3754E-01	.3673E-01
1.800	.3593E-01	.3515E-01	.3438E-01	.3362E-01	.3288E-01	.3216E-01	.3144E-01	.3074E-01	.3005E-01	.2938E-01
1.900	.2872E-01	.2807E-01	.2743E-01	.2680E-01	.2619E-01	.2559E-01	.2500E-01	.2442E-01	.2385E-01	.2330E-01
2.000	.2275E-01	.2222E-01	.2169E-01	.2118E-01	.2068E-01	.2018E-01	.1970E-01	.1923E-01	.1876E-01	.1831E-01
2.100	.1786E-01	.1743E-01	.1700E-01	.1659E-01	.1618E-01	.1578E-01	.1539E-01	.1500E-01	.1463E-01	.1426E-01
2.200	.1390E-01	.1355E-01	.1321E-01	.1287E-01	.1255E-01	.1222E-01	.1191E-01	.1160E-01	.1130E-01	.1101E-01
2.300	.1072E-01	.1044E-01	.1017E-01	.9903E-02	.9642E-02	.9387E-02	.9137E-02	.8894E-02	.8656E-02	.8424E-02
2.400	.8198E-02	.7976E-02	.7760E-02	.7549E-02	.7344E-02	.7143E-02	.6947E-02	.6756E-02	.6569E-02	.6387E-02
2.500	.6210E-02	.6037E-02	.5868E-02	.5703E-02	.5543E-02	.5386E-02	.5234E-02	.5085E-02	.4940E-02	.4799E-02
2.600	.4661E-02	.4527E-02	.4396E-02	.4269E-02	.4145E-02	.4025E-02	.3907E-02	.3793E-02	.3681E-02	.3573E-02
2.700	.3467E-02	.3364E-02	.3264E-02	.3167E-02	.3072E-02	.2980E-02	.2890E-02	.2803E-02	.2718E-02	.2635E-02
2.800	.2555E-02	.2477E-02	.2401E-02	.2327E-02	.2256E-02	.2186E-02	.2118E-02	.2052E-02	.1988E-02	.1926E-02
2.900	.1866E-02	.1807E-02	.1750E-02	.1695E-02	.1641E-02	.1589E-02	.1538E-02	.1489E-02	.1441E-02	.1395E-02
3.000	.1350E-02	.1306E-02	.1264E-02	.1223E-02	.1183E-02	.1144E-02	.1107E-02	.1070E-02	.1035E-02	.1001E-02
3.100	.9676E-03	.9354E-03	.9043E-03	.8740E-03	.8447E-03	.8164E-03	.7888E-03	.7622E-03	.7364E-03	.7114E-03
3.200	.6871E-03	.6637E-03	.6410E-03	.6190E-03	.5976E-03	.5770E-03	.5571E-03	.5377E-03	.5190E-03	.5009E-03
3.300	.4834E-03	.4665E-03	.4501E-03	.4342E-03	.4189E-03	.4041E-03	.3897E-03	.3758E-03	.3624E-03	.3495E-03
3.400	.3369E-03	.3248E-03	.3131E-03	.3018E-03	.2909E-03	.2803E-03	.2701E-03	.2602E-03	.2507E-03	.2415E-03
3.500	.2326E-03	.2241E-03	.2158E-03	.2078E-03	.2001E-03	.1926E-03	.1854E-03	.1785E-03	.1718E-03	.1653E-03
3.600	.1591E-03	.1531E-03	.1473E-03	.1417E-03	.1363E-03	.1311E-03	.1261E-03	.1213E-03	.1166E-03	.1121E-03
3.700	.1078E-03	.1036E-03	.9961E-04	.9574E-04	.9201E-04	.8842E-04	.8496E-04	.8162E-04	.7841E-04	.7532E-04
3.800	.7235E-04	.6948E-04	.6673E-04	.6407E-04	.6152E-04	.5906E-04	.5669E-04	.5442E-04	.5223E-04	.5012E-04
3.900	.4810E-04	.4615E-04	.4427E-04	.4247E-04	.4074E-04	.3908E-04	.3747E-04	.3594E-04	.3446E-04	.3304E-04
4.000	.3167E-04	.3036E-04	.2910E-04	.2789E-04	.2673E-04	.2561E-04	.2454E-04	.2351E-04	.2252E-04	.2157E-04
4.100	.2066E-04	.1978E-04	.1894E-04	.1814E-04	.1737E-04	.1662E-04	.1591E-04	.1523E-04	.1458E-04	.1395E-04
4.200	.1335E-04	.1277E-04	.1222E-04	.1168E-04	.1118E-04	.1069E-04	.1022E-04	.9774E-05	.9345E-05	.8934E-05
4.300	.8540E-05	.8163E-05	.7801E-05	.7455E-05	.7124E-05	.6807E-05	.6503E-05	.6212E-05	.5934E-05	.5668E-05
4.400	.5413E-05	.5169E-05	.4935E-05	.4712E-05	.4498E-05	.4294E-05	.4098E-05	.3911E-05	.3732E-05	.3561E-05
4.500	.3398E-05	.3241E-05	.3092E-05	.2949E-05	.2813E-05	.2682E-05	.2558E-05	.2439E-05	.2325E-05	.2216E-05
4.600	.2112E-05	.2013E-05	.1919E-05	.1828E-05	.1742E-05	.1660E-05	.1581E-05	.1506E-05	.1434E-05	.1366E-05
4.700	.1301E-05	.1239E-05	.1179E-05	.1123E-05	.1069E-05	.1017E-05	.9680E-06	.9211E-06	.8765E-06	.8339E-06
4.800	.7933E-06	.7547E-06	.7178E-06	.6827E-06	.6492E-06	.6173E-06	.5869E-06	.5580E-06	.5304E-06	.5042E-06
4.900	.4792E-06	.4554E-06	.4327E-06	.4111E-06	.3906E-06	.3711E-06	.3525E-06	.3348E-06	.3179E-06	.3019E-06
5.000	.2867E-06	.2722E-06	.2584E-06	.2452E-06	.2328E-06	.2209E-06	.2096E-06	.1989E-06	.1887E-06	.1790E-06
5.100	.1698E-06	.1611E-06	.1528E-06	.1449E-06	.1374E-06	.1302E-06	.1235E-06	.1170E-06	.1109E-06	.1051E-06
5.200	.9964E-07	.9442E-07	.8946E-07	.8476E-07	.8029E-07	.7605E-07	.7203E-07	.6821E-07	.6459E-07	.6116E-07
5.300	.5790E-07	.5481E-07	.5188E-07	.4911E-07	.4647E-07	.4398E-07	.4161E-07	.3937E-07	.3724E-07	.3523E-07
5.400	.3332E-07	.3151E-07	.2980E-07	.2818E-07	.2664E-07	.2518E-07	.2381E-07	.2250E-07	.2127E-07	.2010E-07
5.500	.1899E-07	.1794E-07	.1695E-07	.1601E-07	.1512E-07	.1428E-07	.1349E-07	.1274E-07	.1203E-07	.1135E-07
5.600	.1072E-07	.1012E-07	.9548E-08	.9010E-08	.8503E-08	.8022E-08	.7569E-08	.7140E-08	.6735E-08	.6352E-08
5.700	.5990E-08	.5649E-08	.5326E-08	.5022E-08	.4734E-08	.4462E-08	.4206E-08	.3964E-08	.3735E-08	.3519E-08
5.800	.3316E-08	.3124E-08	.2942E-08	.2771E-08	.2610E-08	.2458E-08	.2314E-08	.2179E-08	.2051E-08	.1931E-08
5.900	.1818E-08	.1711E-08	.1610E-08	.1515E-08	.1425E-08	.1341E-08	.1261E-08	.1186E-08	.1116E-08	.1049E-08
6.000	.9866E-09	.9276E-09	.8721E-09	.8198E-09	.7706E-09	.7242E-09	.6806E-09	.6396E-09	.6009E-09	.5646E-09
6.100	.5303E-09	.4982E-09	.4679E-09	.4394E-09	.4126E-09	.3874E-09	.3637E-09	.3414E-09	.3205E-09	.3008E-09

6.200	.2823E-09	.2649E-09	.2486E-09	.2332E-09	.2188E-09	.2052E-09	.1925E-09	.1805E-09	.1693E-09	.1587E-09
6.300	.1488E-09	.1395E-09	.1308E-09	.1226E-09	.1149E-09	.1077E-09	.1009E-09	.9451E-10	.8854E-10	.8294E-10
6.400	.7769E-10	.7276E-10	.6814E-10	.6380E-10	.5974E-10	.5593E-10	.5235E-10	.4900E-10	.4586E-10	.4292E-10
6.500	.4016E-10	.3758E-10	.3515E-10	.3288E-10	.3076E-10	.2877E-10	.2690E-10	.2516E-10	.2352E-10	.2199E-10
6.600	.2056E-10	.1922E-10	.1796E-10	.1678E-10	.1568E-10	.1465E-10	.1369E-10	.1279E-10	.1195E-10	.1116E-10
6.700	.1042E-10	.9731E-11	.9086E-11	.8483E-11	.7919E-11	.7392E-11	.6900E-11	.6439E-11	.6009E-11	.5607E-11
6.800	.5231E-11	.4880E-11	.4552E-11	.4246E-11	.3960E-11	.3692E-11	.3443E-11	.3210E-11	.2993E-11	.2790E-11
6.900	.2600E-11	.2423E-11	.2258E-11	.2104E-11	.1960E-11	.1826E-11	.1701E-11	.1585E-11	.1476E-11	.1374E-11
7.000	.1280E-11	.1192E-11	.1109E-11	.1033E-11	.9612E-12	.8946E-12	.8325E-12	.7747E-12	.7208E-12	.6706E-12
7.100	.6238E-12	.5802E-12	.5396E-12	.5018E-12	.4667E-12	.4339E-12	.4034E-12	.3750E-12	.3486E-12	.3240E-12
7.200	.3011E-12	.2798E-12	.2599E-12	.2415E-12	.2243E-12	.2084E-12	.1935E-12	.1797E-12	.1669E-12	.1550E-12
7.300	.1439E-12	.1336E-12	.1240E-12	.1151E-12	.1068E-12	.9910E-13	.9196E-13	.8531E-13	.7914E-13	.7341E-13
7.400	.6809E-13	.6315E-13	.5856E-13	.5430E-13	.5034E-13	.4667E-13	.4326E-13	.4010E-13	.3716E-13	.3444E-13
7.500	.3191E-13	.2956E-13	.2739E-13	.2537E-13	.2350E-13	.2176E-13	.2015E-13	.1866E-13	.1728E-13	.1600E-13
7.600	.1481E-13	.1370E-13	.1268E-13	.1174E-13	.1086E-13	.1005E-13	.9297E-14	.8600E-14	.7954E-14	.7357E-14
7.700	.6803E-14	.6291E-14	.5816E-14	.5377E-14	.4971E-14	.4595E-14	.4246E-14	.3924E-14	.3626E-14	.3350E-14
7.800	.3095E-14	.2859E-14	.2641E-14	.2439E-14	.2253E-14	.2080E-14	.1921E-14	.1773E-14	.1637E-14	.1511E-14
7.900	.1395E-14	.1287E-14	.1188E-14	.1096E-14	.1011E-14	.9326E-15	.8602E-15	.7934E-15	.7317E-15	.6747E-15
8.000	.6221E-15	.5735E-15	.5287E-15	.4874E-15	.4492E-15	.4140E-15	.3815E-15	.3515E-15	.3238E-15	.2983E-15
8.100	.2748E-15	.2531E-15	.2331E-15	.2146E-15	.1976E-15	.1820E-15	.1675E-15	.1542E-15	.1419E-15	.1306E-15
8.200	.1202E-15	.1106E-15	.1018E-15	.9361E-16	.8611E-16	.7920E-16	.7284E-16	.6698E-16	.6159E-16	.5662E-16
8.300	.5206E-16	.4785E-16	.4398E-16	.4042E-16	.3715E-16	.3413E-16	.3136E-16	.2881E-16	.2646E-16	.2431E-16
8.400	.2232E-16	.2050E-16	.1882E-16	.1728E-16	.1587E-16	.1457E-16	.1337E-16	.1227E-16	.1126E-16	.1033E-16
8.500	.9480E-17	.8697E-17	.7978E-17	.7317E-17	.6711E-17	.6154E-17	.5643E-17	.5174E-17	.4744E-17	.4348E-17
8.600	.3986E-17	.3653E-17	.3348E-17	.3068E-17	.2811E-17	.2575E-17	.2359E-17	.2161E-17	.1979E-17	.1812E-17
8.700	.1659E-17	.1519E-17	.1391E-17	.1273E-17	.1166E-17	.1067E-17	.9763E-18	.8933E-18	.8174E-18	.7478E-18
8.800	.6841E-18	.6257E-18	.5723E-18	.5234E-18	.4786E-18	.4376E-18	.4001E-18	.3657E-18	.3343E-18	.3055E-18
8.900	.2792E-18	.2552E-18	.2331E-18	.2130E-18	.1946E-18	.1777E-18	.1623E-18	.1483E-18	.1354E-18	.1236E-18
9.000	.1129E-18	.1030E-18	.9404E-19	.8584E-19	.7834E-19	.7148E-19	.6523E-19	.5951E-19	.5429E-19	.4952E-19
9.100	.4517E-19	.4119E-19	.3756E-19	.3425E-19	.3123E-19	.2847E-19	.2595E-19	.2365E-19	.2155E-19	.1964E-19
9.200	.1790E-19	.1631E-19	.1486E-19	.1353E-19	.1232E-19	.1122E-19	.1022E-19	.9307E-20	.8474E-20	.7714E-20
9.300	.7022E-20	.6392E-20	.5817E-20	.5294E-20	.4817E-20	.4382E-20	.3987E-20	.3627E-20	.3299E-20	.3000E-20
9.400	.2728E-20	.2481E-20	.2255E-20	.2050E-20	.1864E-20	.1694E-20	.1540E-20	.1399E-20	.1271E-20	.1155E-20
9.500	.1049E-20	.9533E-21	.8659E-21	.7864E-21	.7142E-21	.6485E-21	.5888E-21	.5345E-21	.4852E-21	.4404E-21
9.600	.3997E-21	.3627E-21	.3292E-21	.2986E-21	.2709E-21	.2458E-21	.2229E-21	.2022E-21	.1834E-21	.1663E-21
9.700	.1507E-21	.1367E-21	.1239E-21	.1123E-21	.1018E-21	.9223E-22	.8358E-22	.7573E-22	.6861E-22	.6215E-22
9.800	.5629E-22	.5098E-22	.4617E-22	.4181E-22	.3786E-22	.3427E-22	.3102E-22	.2808E-22	.2542E-22	.2300E-22
9.900	.2081E-22	.1883E-22	.1704E-22	.1541E-22	.1394E-22	.1261E-22	.1140E-22	.1031E-22	.9323E-23	.8429E-23
10.00	.7620E-23	.6888E-23	.6225E-23	.5626E-23	.5084E-23	.4593E-23	.4150E-23	.3749E-23	.3386E-23	.3058E-23

Notes: E-01 should be read as $\times 10^{-1}$
E-02 should be read as $\times 10^{-2}$ and so on.

This table lists $Q(x)$ for x in the range of 0 to 10 in the increments of 0.01. To find $Q(5.36)$, for example, look up the row starting with $x = 5.3$. The sixth entry in this row (under 0.06) is the desired value 0.4161×10^{-7}.

Letting $(x - m)/\sigma = z$

$$F_{\mathrm{x}}(x) = \frac{1}{\sqrt{2\pi}} \int_{-\infty}^{(x-m)/\sigma} e^{-z^2/2}\, dz$$

$$= 1 - Q\left(\frac{x - m}{\sigma}\right) \tag{5.39a}$$

Therefore,

$$p(\mathrm{x} \le x) = 1 - Q\left(\frac{x - m}{\sigma}\right) \tag{5.39b}$$

and

$$P(\mathrm{x} > x) = Q\left(\frac{x - m}{\sigma}\right) \tag{5.39c}$$

The gaussian PDF is perhaps the most important PDF in the area of communication. The majority of the noise processes observed in practice are gaussian. The amplitude n of a gaussian noise signal is a random variable with a gaussian PDF. This means the probability of observing n in an interval $(n, n + \Delta n)$ is $p_{\mathrm{n}}(n)\Delta n$, where $p_{\mathrm{n}}(n)$ is of the form in Eq. (5.38) [with $m = 0$].

EXAMPLE 5.13 Threshold Detection

Over a certain binary channel, messages m = **0** and **1** are transmitted with equal probability using a positive and a negative pulse, respectively. The received pulse corresponding to **1** is $p(t)$, shown in Fig. 5.11*a*, and the received pulse corresponding to **0** will be $-p(t)$. Let the peak amplitude of $p(t)$ be A_p at $t = T_p$. Because of the channel noise n(t), the received pulses will be $\pm\, p(t) + \mathrm{n}(t)$ (Fig. 5.11*b*). To detect the pulses at the receiver, each pulse is sampled at its peak amplitude. In the absence of noise, the sampler output is either A_p (for m = **1**) or $-A_p$ (for m = **0**). Because of the channel noise, the sampler output is $\pm A_p + \mathrm{n}$, where n, the noise amplitude at the sampling instant (Fig. 5.11*b*), is a random variable. For gaussian noise, the PDF of n is (Fig. 5.11*c*)

$$p_{\mathrm{n}}(n) = \frac{1}{\sigma_n \sqrt{2\pi}} e^{-n^2/2\sigma_n^2} \tag{5.40}$$

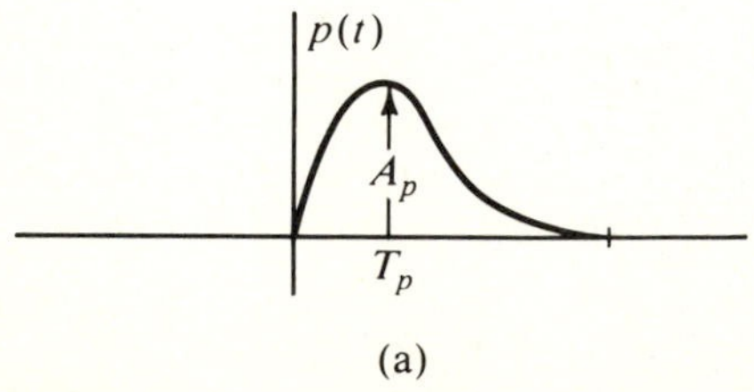

Figure 5.11 Error probability in threshold detection.

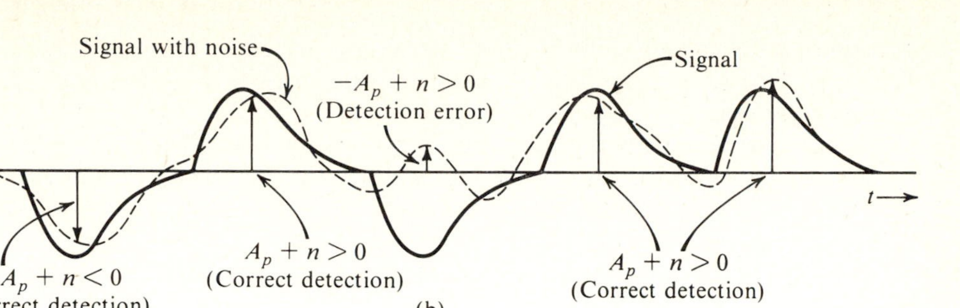

(b)

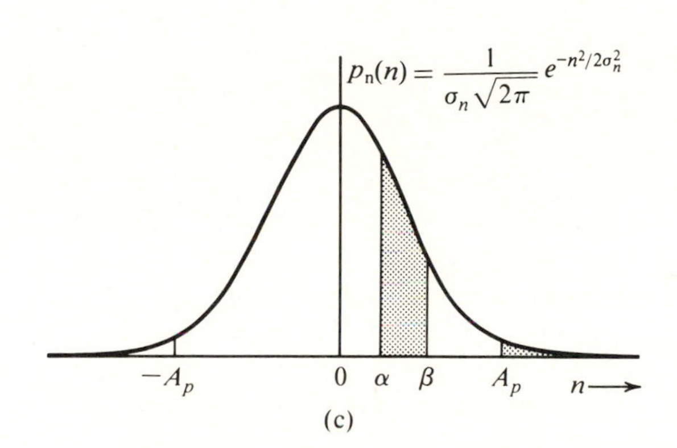

(c)

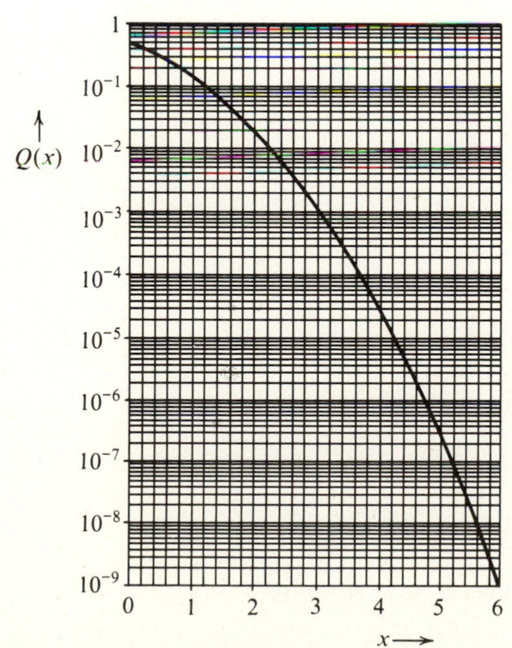

Figure 5.11 *(Continued)*

Because of the symmetry of the situation, the optimum detection threshold is zero; that is, the received pulse is detected as a **1** or **0** depending on whether the sample value is positive or negative.

Because noise amplitudes range from $-\infty$ to ∞, the sample value $-A_p + \mathrm{n}$ can occasionally be positive, causing the received **0** to be read as **1** (see Fig. 5.11*b*). Similarly, $A_p + \mathrm{n}$ can occasionally be negative, causing the received **1** to be read as **0**. If **0** is transmitted, it will be detected as **1** if $-A_p + \mathrm{n} > 0$, that is, if $\mathrm{n} > A_p$.

If $P(\epsilon|\mathbf{0})$ is the error probability given that **0** is transmitted, then

$$P(\epsilon|\mathbf{0}) = P(\mathrm{n} > A_p)$$

Because $P(\mathrm{n} > A_p)$ is the shaded area in Fig. 5.11*c*, from Eq. (5.39c) [with $m = \mathbf{0}$] it follows that

$$P(\epsilon|\mathbf{0}) = Q\left(\frac{A_p}{\sigma_n}\right) \tag{5.41a}$$

Similarly,

$$\begin{aligned} P(\epsilon|\mathbf{1}) &= P(\mathrm{n} < -A_p) \\ &= Q\left(\frac{A_p}{\sigma_n}\right) = P(\epsilon|\mathbf{0}) \end{aligned} \tag{5.41b}$$

and

$$\begin{aligned} P_e &= \sum_i P(\epsilon, m_i) \\ &= \sum_i P(m_i)P(\epsilon|m_i) \\ &= Q\left(\frac{A_p}{\sigma_n}\right) \sum_i P(m_i) \\ &= Q\left(\frac{A_p}{\sigma_n}\right) \end{aligned} \tag{5.41c}$$

The error probability P_e is plotted as a function of A_p/σ_n in Fig. 5.11*d*. ■

Joint Distribution. For two r.v.'s x and y, we define a CDF $F_{\mathrm{xy}}(x, y)$ as follows:

$$P(\mathrm{x} \le x \text{ and } \mathrm{y} \le y) = F_{\mathrm{xy}}(x, y) \tag{5.42}$$

and the joint PDF $p_{\mathrm{xy}}(x, y)$ as

$$p_{\mathrm{xy}}(x, y) = \frac{\partial^2}{\partial x\, \partial y} F_{\mathrm{xy}}(x, y) \tag{5.43}$$

Arguing along lines similar to those used for a single variable, we can show that

$$\lim_{\substack{\Delta x \to 0 \\ \Delta y \to 0}} p_{\mathrm{xy}}(x, y)\Delta x\, \Delta y = P(x < \mathrm{x} \le x + \Delta x, y < \mathrm{y} \le y + \Delta y) \tag{5.44}$$

Hence, the probability of observing the variable x in the interval $(x, x + \Delta x)$ and y in the interval $(y, y + \Delta y)$ jointly is given by the volume under the joint PDF $p_{xy}(x, y)$ over the region bounded by $(x, x + \Delta x)$ and $(y, y + \Delta y)$, as shown in Fig. 5.12*a*.

From Eq. (5.44), it follows that

$$P(x_1 < \mathrm{x} \leq x_2, y_1 < \mathrm{y} \leq y_2) = \int_{x_1}^{x_2} \int_{y_1}^{y_2} p_{xy}(x, y)\, dxdy \tag{5.45}$$

Thus, the probability of jointly observing x in the interval (x_1, x_2) and y in the interval (y_1, y_2) is the volume under the PDF over the region bounded by (x_1, x_2) and (y_1, y_2).

The event of observing x in the interval $(-\infty, \infty)$ and observing y in the interval $(-\infty, \infty)$ is a certainty. Hence,

$$\int_{-\infty}^{\infty} \int_{-\infty}^{\infty} p_{xy}(x, y)\, dxdy = 1 \tag{5.46}$$

Thus, the total volume under the joint PDF must be unity.

When we are dealing with two random variables x and y, the individual probability densities $p_x(x)$ and $p_y(y)$ can be obtained from the joint density $P_{xy}(x, y)$. These individual densities are also called *marginal densities*. To obtain these densities, we note that $p_x(x)\, \Delta x$ is the probability of observing x in the interval $(x, \mathrm{x} + \Delta x)$. The value of y may lie anywhere in the interval $(-\infty, \infty)$. Hence,

$$\begin{aligned}
\lim_{\Delta x \to 0} p_x(x)\, \Delta x &= \text{probability } (x < \mathrm{x} \leq x + \Delta x, -\infty < \mathrm{y} \leq \infty) \\
&= \lim_{\Delta x \to 0} \int_{x}^{x+\Delta x} \int_{-\infty}^{\infty} p_{xy}(x, y)\, dxdy \\
&= \lim_{\Delta x \to 0} \Delta x \int_{-\infty}^{\infty} p_{xy}(x, y)\, dy
\end{aligned}$$

The last step follows from the fact that $p_{xy}(x, y)$ is constant over $(x, x + \Delta x)$ because $\Delta x \to 0$. Therefore,

$$p_x(x) = \int_{-\infty}^{\infty} p_{xy}(x, y)\, dy \tag{5.47a}$$

Similarly,

$$p_y(y) = \int_{-\infty}^{\infty} p_{xy}(x, y)\, dx \tag{5.47b}$$

These results may be generalized for n random variables $\mathrm{x}_1, \mathrm{x}_2, \ldots, \mathrm{x}_n$.

Conditional Densities. The concept of conditional probabilities can be extended to the case of continuous random variables. We define the conditional PDF $p_{x|y}(x|y_j)$ as the PDF of x given that y has a value y_j. This is equivalent to saying that $p_{x|y}(x|y_j)\, \Delta x$ is the probability of observing x in the range $(x, x + \Delta x)$, given that $\mathrm{y} = y_j$. The probability density $p_{x|y}(x|y_j)$ is the intersection of the plane $\mathrm{y} = y_j$ with the joint PDF $p_{xy}(x, y)$ (Fig. 5.12*b*). Because every PDF must have unit area, however, we must

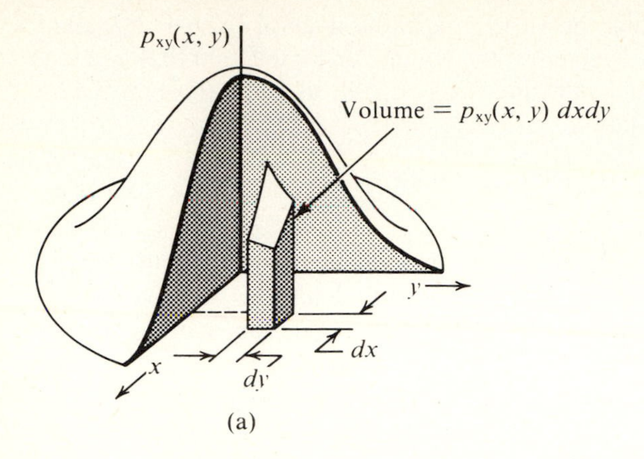

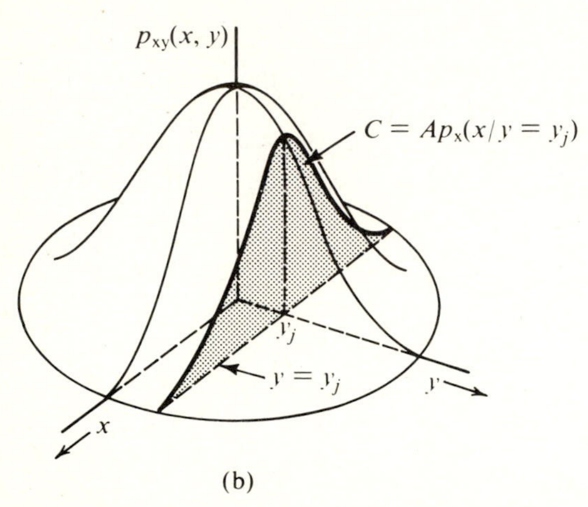

Figure 5.12 (a) Joint PDF. (b) Conditional PDF.

normalize the area under the intersection curve C to unity to get the desired PDF. Hence, C is $Ap_{x|y}(x \mid y)$, where A is the area under C. An extension of the results derived for the discrete case yields

$$p_{x|y}(x \mid y)p_y(y) = p_{xy}(x, y) \tag{5.48a}$$

$$p_{y|x}(y \mid x)p_x(x) = p_{xy}(x, y) \tag{5.48b}$$

and

$$p_{x|y}(x \mid y) = \frac{p_{y|x}(y \mid x)p_x(x)}{p_y(y)} \tag{5.49a}$$

Equation (5.49a) is Bayes' rule for continuous random variables. When we have mixed variables (i.e., discrete and continuous), the mixed form of Bayes' rule is

$$P_{x|y}(x \mid y)p_y(y) = P_x(x)p_{y|x}(y \mid x) \tag{5.49b}$$

where x is a discrete r.v. and y is a continuous r.v.

Continuous random variables x and y are said to be independent if

$$p_{x|y}(x|y) = p_x(x) \tag{5.50a}$$

From Eqs. (5.50a) and (5.49), it follows that

$$p_{y|x}(y|x) = p_y(y) \tag{5.50b}$$

This implies that for independent r.v.'s x and y,

$$p_{xy}(x, y) = p_x(x)p_y(y) \tag{5.50c}$$

■ EXAMPLE 5.14 Rayleigh Density

The Rayleigh density is characterized by the PDF (Fig. 5.13*b*)

$$p_r(r) = \begin{cases} \dfrac{r}{\sigma^2}\, e^{-r^2/2\sigma^2} & r \geq 0 \\ 0 & r < 0 \end{cases} \tag{5.51}$$

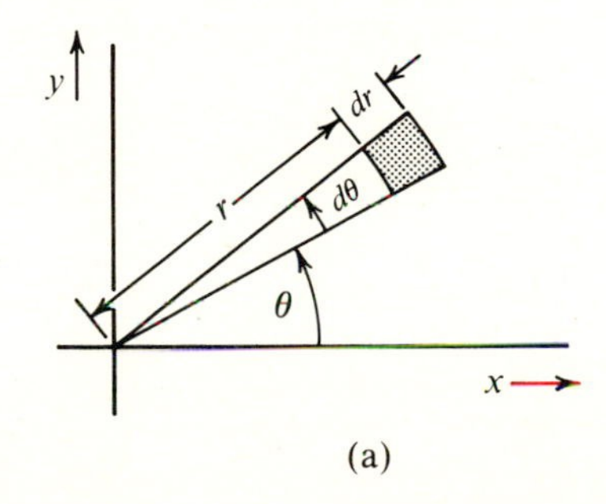

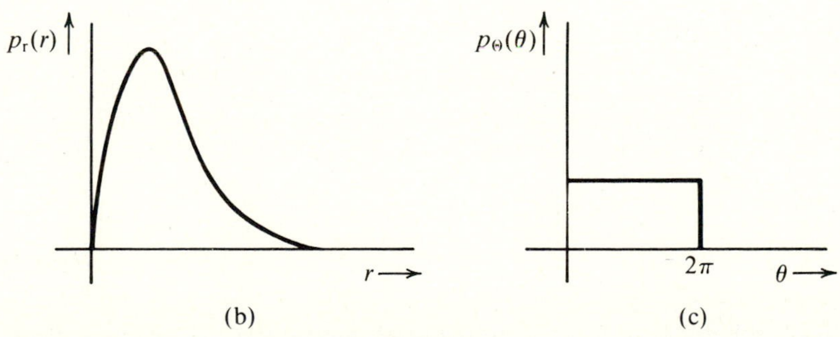

Figure 5.13 Derivation of Rayleigh density.

A Rayleigh r.v. can be derived from two independent gaussian r.v.'s as follows. Let x and y be independent gaussian variables with identical PDFs:

$$p_x(x) = \frac{1}{\sigma\sqrt{2\pi}}\, e^{-x^2/2\sigma^2}$$

$$p_y(y) = \frac{1}{\sigma\sqrt{2\pi}}\, e^{-y^2/2\sigma^2}$$

Then

$$p_{xy}(x, y) = p_x(x)p_y(y) = \frac{1}{2\pi\sigma^2} e^{-(x^2+y^2)/2\sigma^2} \tag{5.52}$$

The joint density appears somewhat like the bell-shaped surface shown in Fig. 5.12. The points in the x, y plane can also be described in polar coordinates as (r, θ), where (Fig. 5.13*a*)

$$\mathrm{r} = \sqrt{\mathrm{x}^2 + \mathrm{y}^2} \qquad \Theta = -\tan^{-1}\frac{\mathrm{y}}{\mathrm{x}}$$

In Fig. 5.13*a*, the shaded region represents $r < \mathrm{r} \leq r + dr$ and $\theta < \Theta \leq \theta + d\theta$ (where dr and $d\theta$ both $\to 0$). Hence, if $p_{r\Theta}(r, \theta)$ is the joint PDF of r and Θ, then by definition [Eq. (5.44)], the probability of observing r and Θ in this region is $p_{r\Theta}(r, \theta)\, drd\theta$. But we also know that this probability is $p_{xy}(x, y)$ times the area $rdrd\theta$ of the shaded region. Hence, [Eq. (5.52)]

$$\frac{1}{2\pi\sigma^2} e^{-(x^2+y^2)/2\sigma^2}\, rdrd\theta = p_{r\Theta}(r, \theta)\, drd\theta$$

and

$$p_{r\Theta}(r, \theta) = \frac{r}{2\pi\sigma^2} e^{-(x^2+y^2)/2\sigma^2}$$

$$= \frac{r}{2\pi\sigma^2} e^{-r^2/2\sigma^2} \tag{5.53}$$

and [Eq. (5.47a)]

$$p_r(r) = \int_{-\infty}^{\infty} p_{r\Theta}(r, \theta)\, d\theta$$

Because Θ exists only in the region $(0, 2\pi)$,

$$p_r(r) = \int_0^{2\pi} \frac{r}{2\pi\sigma^2} e^{-r^2/2\sigma^2}\, d\theta$$

$$= \frac{r}{\sigma^2} e^{-r^2/2\sigma^2} \qquad \mathrm{r} \geq 0 \tag{5.54a}$$

Note that r is always greater than 0. In a similar way, we find

$$p_\Theta(\theta) = \begin{cases} \dfrac{1}{2\pi} & 0 < \Theta < 2\pi \\ 0 & \text{otherwise} \end{cases} \tag{5.54b}$$

Random variables r and Θ are independent because $p_{r\Theta}(r, \theta) = p_r(r)p_\Theta(\theta)$. The PDF $p_r(r)$ is the *Rayleigh density function*. We shall later show that the envelope of narrowband gaussian noise has a Rayleigh density. Both $p_r(r)$ and $p_\Theta(\theta)$ are shown in Fig. 5.13*b* and *c*. ■

5.3 STATISTICAL AVERAGES (MEANS)

Averages are extremely important in the study of random variables. In order to find a proper definition for the average of a random variable x, consider the problem of determining the average height of the entire population of a country. Let us assume that we have enough resources to gather data about the height of every person. If the data is recorded within the accuracy of a centimeter, then the height x of every person will be approximated to one of the n numbers $x_1, x_2, \ldots, x_n$. If there are N_i persons of height x_i, then the average height $\overline{\mathrm{x}}$ is given by

$$\overline{\mathrm{x}} = \frac{N_1 x_1 + N_2 x_2 + \cdots + N_n x_n}{N}$$

where the total number of persons $N = \sum_i N_i$. Hence,

$$\overline{\mathrm{x}} = \frac{N_1}{N} x_1 + \frac{N_2}{N} x_2 + \cdots + \frac{N_n}{N} x_n$$

In the limit as $N \to \infty$, the ratio N_i/N approaches $P_{\mathrm{x}}(x_i)$ according to the relative-frequency definition of the probability. Hence,

$$\overline{\mathrm{x}} = \sum_{i=1}^{n} x_i P_{\mathrm{x}}(x_i)$$

The mean value is also called the *average value,* or the *expected value,* of the random variable x and is denoted by $E[\mathrm{x}]$. Thus,

$$\overline{\mathrm{x}} = E[\mathrm{x}] = \sum_i x_i P_{\mathrm{x}}(x_i) \tag{5.55a}$$

We shall use both these notations, depending upon the circumstances.

If the r.v. x is continuous, an argument similar to that used in arriving at Eq. (5.55a) yields

$$\overline{\mathrm{x}} = E[\mathrm{x}] = \int_{-\infty}^{\infty} x p_{\mathrm{x}}(x)\, dx \tag{5.55b}$$

This result can be derived by approximating the continuous variable x with a discrete variable by quantizing it in steps of Δx and then letting $\Delta x \to 0$.

Equation (5.55b) is more general and includes Eq. (5.55a), because the discrete r.v. can be considered as a continuous r.v. with an impulsive density. In such a case, Eq. (5.55b) reduces to Eq. (5.55a).

As an example, consider the general gaussian PDF given by (Fig. 5.10)

$$p_{\mathrm{x}}(x) = \frac{1}{\sigma\sqrt{2\pi}} e^{-(x-m)^2/2\sigma^2} \tag{5.56a}$$

From Eq. (5.55b) we have

$$\overline{\mathrm{x}} = \frac{1}{\sigma\sqrt{2\pi}} \int_{-\infty}^{\infty} x e^{-(x-m)^2/2\sigma^2}\, dx$$

Changing the variable to $x = y + m$ yields

$$\overline{\mathrm{x}} = \frac{1}{\sigma\sqrt{2\pi}} \int_{-\infty}^{\infty} (y + m)e^{-y^2/2\sigma^2}\, dy$$

$$= \frac{1}{\sigma\sqrt{2\pi}} \left[\int_{-\infty}^{\infty} ye^{-y^2/2\sigma^2}\, dy + m \int_{-\infty}^{\infty} e^{-y^2/2\sigma^2}\, dy\right]$$

The first integral inside the bracket is zero, because the integrand is an odd function of y. The second integral is found from standard tables[2] to be $\sigma\sqrt{2\pi}$. Hence,

$$\overline{\mathrm{x}} = m \tag{5.56b}$$

The Mean of a Function of a Random Variable

It is often necessary to find the mean value of a function of a random variable. For instance, in practice we are often interested in the mean square amplitude of a signal. The mean square amplitude is the mean of the square of the amplitude x, that is, $\overline{\mathrm{x}^2}$.

In general, we may seek the mean value of a random variable y that is a function of the random variable x; that is, we wish to find $\overline{\mathrm{y}}$ where y $= g(\mathrm{x})$. Let x be a discrete r.v. that takes values $x_1, x_2, \ldots, x_n$ with probabilities $P_{\mathrm{x}}(x_1), P_{\mathrm{x}}(x_2), \ldots, P_{\mathrm{x}}(x_n)$, respectively. But because y $= g(\mathrm{x})$, y takes values $g(x_1), g(x_2), \ldots, g(x_n)$ with probabilities $P_{\mathrm{x}}(x_1), P_{\mathrm{x}}(x_2), \ldots, P_{\mathrm{x}}(x_n)$, respectively. Hence, from Eq. (5.55a) we have

$$\overline{\mathrm{y}} = \overline{g(\mathrm{x})} = \sum_{i=1}^{n} g(x_i)P_{\mathrm{x}}(x_i) \tag{5.57a}$$

If x is a continuous random variable, a similar line of reasoning leads to

$$\overline{g(\mathrm{x})} = \int_{-\infty}^{\infty} g(x)p_{\mathrm{x}}(x)\, dx \tag{5.57b}$$

■ EXAMPLE 5.15

A sinusoid generator output voltage is $A \cos \omega t$. This output is sampled randomly (Fig. 5.14*a*). The sampled output is a random variable x, which can take on any value in the range $(-A, A)$. Determine the mean value ($\overline{\mathrm{x}}$) and the mean square value ($\overline{\mathrm{x}^2}$) of the sampled output x.

☐ **Solution:** If the output is sampled at a random instant t, the output x is a function of the random variable t:

$$\mathrm{x}(\mathrm{t}) = A \cos \omega \mathrm{t}$$

If we let $\omega \mathrm{t} = \Theta$, Θ is also a random variable, and if we consider only modulo $= 2\pi$ values of Θ, then the r.v. Θ lies in the range $(0, 2\pi)$. Because t is randomly chosen, Θ can take any value in the range $(0, 2\pi)$ with uniform probability. Because the area under the PDF must be unity, $p_\Theta(\theta)$ is as shown in Fig. 5.14*b*.

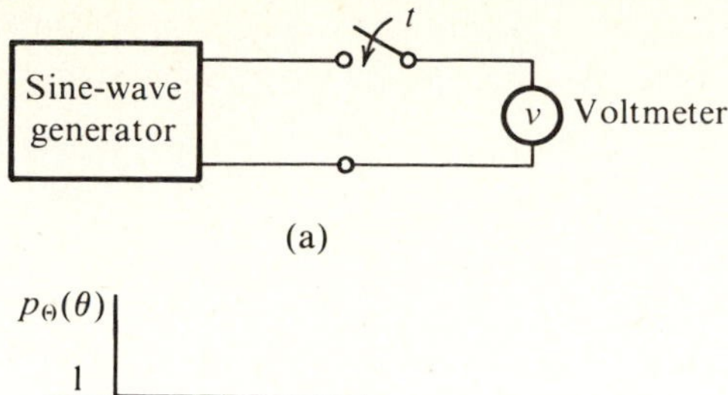

(a)

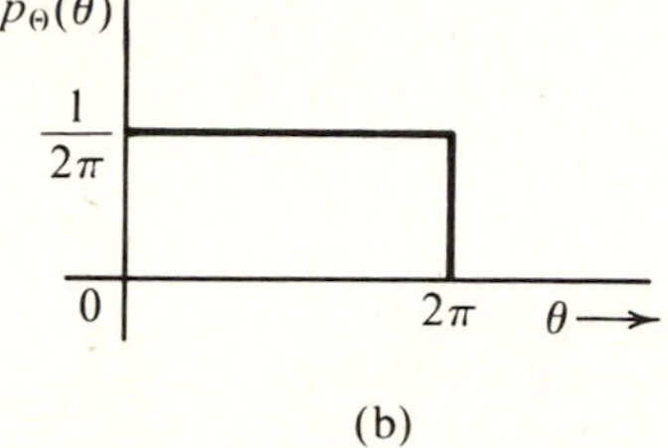

(b)

Figure 5.14 Random sampling of a sine-wave generator.

The r.v. x is thus a function of the r.v. Θ.

$$\mathrm{x} = A \cos \Theta$$

Hence, from Eq. (5.57b)

$$\overline{\mathrm{x}} = \int_0^{2\pi} x \, p_\Theta(\theta) \, d\theta = \frac{1}{2\pi} \int_0^{2\pi} A \cos \theta \, d\theta = 0$$

and

$$\overline{\mathrm{x}^2} = \int_0^{2\pi} x^2 p_\Theta(\theta) \, d\theta$$

$$= \frac{A^2}{2\pi} \int_0^{2\pi} \cos^2\theta \, d\theta = \frac{A^2}{2}$$ ■

Similarly, for the case of two variables x and y, we have

$$\overline{g(\mathrm{x}, \mathrm{y})} = \int_{-\infty}^{\infty} \int_{-\infty}^{\infty} g(x, y) p_{\mathrm{xy}}(x, y) \, dxdy \tag{5.58}$$

The Mean of the Sum. If $g_1(\mathrm{x}, \mathrm{y}), g_2(\mathrm{x}, \mathrm{y}), \ldots, g_n(\mathrm{x}, \mathrm{y})$ are functions of the r.v.'s x and y, then

$$\overline{g_1(\mathrm{x}, \mathrm{y}) + g_2(\mathrm{x}, \mathrm{y}) + \cdots + g_n(\mathrm{x}, \mathrm{y})} = \overline{g_1(\mathrm{x}, \mathrm{y})} + \overline{g_2(\mathrm{x}, \mathrm{y})} + \cdots + \overline{g_n(\mathrm{x}, \mathrm{y})} \tag{5.59a}$$

The proof is trivial and follows directly from Eq. (5.58).

Thus, the mean (expected value) of the sum is equal to the sum of the means. An important special case is

$$\overline{\mathrm{x} + \mathrm{y}} = \overline{\mathrm{x}} + \overline{\mathrm{y}} \tag{5.59b}$$

Equation (5.59a) can be extended to functions of any number of random variables.

The Mean of the Product of Two Functions. Unfortunately, there is no simple result [as Eq. (5.59)] for the product of two functions. For the special case where

$$g(\mathrm{x}, \mathrm{y}) = g_1(\mathrm{x})g_2(\mathrm{y}) \tag{5.60a}$$

$$\overline{g_1(\mathrm{x})g_2(\mathrm{y})} = \int_{-\infty}^{\infty} g_1(x)g_2(y)p_{\mathrm{xy}}(x, y)\, dxdy$$

If x and y are independent, then [Eq. (5.50c)]

$$p_{\mathrm{xy}}(x, y) = p_{\mathrm{x}}(x)p_{\mathrm{y}}(y)$$

and

$$\begin{aligned}\overline{g_1(\mathrm{x})g_2(\mathrm{y})} &= \int_{-\infty}^{\infty} g_1(x)p_{\mathrm{x}}(x)\, dx \int_{-\infty}^{\infty} g_2(y)p_{\mathrm{y}}(y)\, dy \\ &= \overline{g_1(\mathrm{x})}\;\overline{g_2(\mathrm{y})} \qquad \text{if x and y independent}\end{aligned} \tag{5.60b}$$

A special case of this is

$$\overline{\mathrm{xy}} = \overline{\mathrm{x}}\,\overline{\mathrm{y}} \qquad \text{if x and y are independent} \tag{5.60c}$$

Moments

The nth *moment* of a random variable x is defined as the mean value of x^n. Thus, the nth moment of x is

$$\overline{\mathrm{x}^n} = \int_{-\infty}^{\infty} x^n p_{\mathrm{x}}(x)\, dx \tag{5.61a}$$

The nth *central moment* of an r.v. x is defined as

$$\overline{(\mathrm{x} - \overline{\mathrm{x}})^n} = \int_{-\infty}^{\infty} (x - \overline{\mathrm{x}})^n p_{\mathrm{x}}(x)\, dx \tag{5.61b}$$

The second central moment of an r.v. x is of special importance. It is called the *variance* of x and is denoted by σ_{x}^2, where σ_{x} is known as the *standard deviation* (*S.D.*) of the r.v. x. By definition

$$\begin{aligned}\sigma_{\mathrm{x}}^2 &= \overline{(\mathrm{x} - \overline{\mathrm{x}})^2} \\ &= \overline{\mathrm{x}^2} - 2\overline{\mathrm{x}\overline{\mathrm{x}}} + \overline{\mathrm{x}}^2 = \overline{\mathrm{x}^2} - 2\overline{\mathrm{x}}^2 + \overline{\mathrm{x}}^2 \\ &= \overline{\mathrm{x}^2} - \overline{\mathrm{x}}^2\end{aligned} \tag{5.62}$$

Thus, the variance of x is equal to the mean square value minus the square of the mean. When the mean is zero, the variance is the mean square; that is $\overline{\mathrm{x}^2} = \sigma_{\mathrm{x}}^2$.

■ EXAMPLE 5.16

Find the mean, the mean square, and the variance of the gaussian r.v. with the PDF in Eq. (5.38) [see Fig. 5.10].

☐ **Solution:**

$$\overline{x^2} = \frac{1}{\sigma\sqrt{2\pi}} \int_{-\infty}^{\infty} x^2 \, e^{-(x-m)^2/2\sigma^2} \, dx$$

Changing the variable to $y = (x - m)/\sigma$ and integrating, we get

$$\overline{x^2} = \sigma^2 + m^2 \tag{5.63a}$$

Also, from Eqs. (5.62) and (5.56b)

$$\begin{aligned} \sigma_x^2 &= \overline{x^2} - \bar{x}^2 \\ &= (\sigma^2 + m^2) - (m)^2 \\ &= \sigma^2 \end{aligned} \tag{5.63b}$$

Hence, a gaussian r.v. described by the density in Eq. (5.56a) has mean m and variance σ^2. Observe that the gaussian density function is completely specified by the first moment ($\bar{x}$) and the second moment ($\overline{x^2}$). ■

■ EXAMPLE 5.17 The Mean Square of the Quantization Error in PCM

In the PCM scheme discussed in Chapter 3, a signal bandlimited to B Hz is sampled at a rate of $2B$ samples per second. The entire range $(-m_p, m_p)$ of the signal amplitudes is partitioned into L uniform intervals, each of magnitude $2m_p/L$ (Fig. 5.15*a*). Each sample is approximated to the midpoint of the interval in which it falls. Thus,

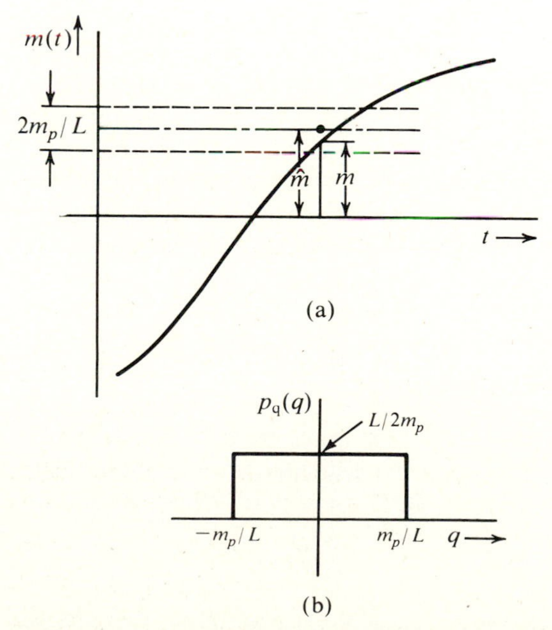

Figure 5.15 Quantization error in PCM and its PDF.

sample m in Fig. 5.15a is approximated by a value $\hat{m}$, the midpoint of the interval in which m falls. Each sample is thus approximated (quantized) to one of the L numbers.

The difference $\mathrm{q} = \mathrm{m} - \hat{\mathrm{m}}$ is the quantization error and is a random variable. We shall determine $\overline{\mathrm{q}^2}$, the mean square value of the quantization error. From Fig. 5.15a it can be seen that q is a continuous r.v. existing over the range $(-m_p/L, m_p/L)$ and is zero outside this range. If we assume that it is equally likely for the sample to lie anywhere in the quantizing interval,* then the PDF of q is uniform ($p_{\mathrm{q}}(q) = L/2m_p$) over the interval $(-m_p/L, m_p/L)$, as shown in Fig. 5.15b, and

$$\overline{\mathrm{q}^2} = \int_{-m_p/L}^{m_p/L} q^2 p_{\mathrm{q}}(q)\, dq$$

$$= \frac{L}{2m_p} \frac{q^3}{3} \Bigg|_{-m_p/L}^{m_p/L}$$

$$= \frac{1}{3}\left(\frac{m_p}{L}\right)^2 \tag{5.64a}$$

From Fig. 5.15b it can be seen that $\overline{\mathrm{q}} = 0$. Hence,

$$\sigma_{\mathrm{q}}^2 = \overline{\mathrm{q}^2} = \frac{1}{3}\left(\frac{m_p}{L}\right)^2 \tag{5.64b}$$

■

■ EXAMPLE 5.18 The Mean Square Error Caused by Channel Noise in PCM

The quantization noise is one of the sources of error in PCM. The other source of error is the channel noise. Each quantized sample is coded by a group of n binary pulses. Because of channel noise, some of these pulses are incorrectly detected at the receiver. Hence, the decoded sample value $\tilde{m}$ at the receiver will differ from the quantized sample value $\hat{m}$ that is transmitted. The error $\varepsilon = \hat{\mathrm{m}} - \tilde{\mathrm{m}}$ is a random variable. Let us calculate $\overline{\varepsilon^2}$, the mean square error in the sample value caused by the channel noise.

To begin with, let us determine the values that ε can take and the corresponding probabilities. Each sample is transmitted by n binary pulses. The value of ε depends on the position of the incorrectly detected pulse. Consider, for example, the case of $L = 16$ transmitted by four binary pulses ($n = 4$), as shown in Fig. 1.6. Here the transmitted code **1101** represents a value of 13. A detection error in the first digit changes the received code to **0101**, which is a value of 5. This causes an error $\varepsilon = 8$. Similarly, an error in the second digit gives $\varepsilon = 4$. Errors in the third and the fourth digit will give $\varepsilon = 2$ and $\varepsilon = 1$, respectively. In general, the error in the ith digit causes an error $\varepsilon_i = (2^{-i})16$. For a general case, the error $\varepsilon_i = (2^{-i})F$, where F is the full scale, that is, $2m_p$ in PCM. Thus,

$$\varepsilon_i = (2^{-i})(2m_p) \qquad i = 1, 2, \ldots, n$$

* Because the quantizing interval is generally very small, variations in the PDF of signal amplitudes over the interval are small and this assumption is reasonable.

Note that the error ε is a discrete random variable. Hence,*

$$\overline{\varepsilon^2} = \sum_{i=1}^{n} \varepsilon_i^2 P_\varepsilon(\varepsilon_i) \tag{5.65}$$

Because $P_\varepsilon(\varepsilon_i)$ is the probability that $\varepsilon = \varepsilon_i$, $P_\varepsilon(\varepsilon_i)$ is the probability of error in the detection of the ith digit. Because the error probability of detecting any one digit is the same as that of any other, that is, P_e,

$$\begin{aligned}\overline{\varepsilon^2} &= P_e \sum_{i=1}^{n} \varepsilon_i^2 \\ &= P_e \sum_{i=1}^{n} 4m_p^2(2^{-2i}) \\ &= 4m_p^2 P_e \sum_{i=1}^{n} 2^{-2i}\end{aligned}$$

This summation is a geometric progression with a common ratio $r = 2^{-2}$, with the first term $a_n = 2^{-2}$ and the last term $a_n = 2^{-2n}$. Hence, (see Appendix A, Sec. A.3)

$$\begin{aligned}\overline{\varepsilon^2} &= 4m_p^2 P_e \left[\frac{(2^{-2})2^{-2n} - 2^{-2}}{2^{-2} - 1}\right] \\ &= \frac{4m_p^2 P_e (2^{2n} - 1)}{3(2^{2n})}\end{aligned} \tag{5.66a}$$

Note that the magnitude of the error ε varies from $2^{-1}(2m_p)$ to $2^{-n}(2m_p)$. The error ε can be positive as well as negative. For example, $\varepsilon = 8$ because of a first-digit error in **1101**. But the corresponding error ε will be -8 if the transmitted code is **0101**. Of course the sign of ε does not matter in Eq. (5.65). It must be remembered, however, that ε varies from $-2^{-n}(2m_p)$ to $2^{-n}(2m_p)$ and its probabilities are symmetrical about $\varepsilon = 0$. Hence, $\overline{\varepsilon} = 0$ and

$$\sigma_\varepsilon^2 = \overline{\varepsilon^2} = \frac{4m_p^2 P_e (2^{2n} - 1)}{3(2^{2n})} \tag{5.66b}$$

■

The Variance of a Sum of Independent Random Variables. The variance of the sum of independent r.v.'s is the sum of the variances of those variables. Thus, if x and y are independent random variables and

$$\mathrm{z} = \mathrm{x} + \mathrm{y}$$

---

*Here we are assuming that the error can occur only in one of the n digits. But more than one digit may be in error. Because the digit error probability $P_e \ll 1$ (of the order 10^{-5} or less), however, the probability of more than one wrong digit is extremely small (see Example 5.6) and its contribution $\varepsilon_i^2 P_\varepsilon(\varepsilon_i)$ is negligible.

then

$$\sigma_z^2 = \sigma_x^2 + \sigma_y^2 \tag{5.67}$$

This can be shown as follows:

$$\begin{aligned}\sigma_z^2 &= \overline{(z - \bar{z})^2} = \overline{[x + y - (\bar{x} + \bar{y})]^2} \\ &= \overline{[(x - \bar{x}) + (y - \bar{y})]^2} \\ &= \overline{(x - \bar{x})^2} + \overline{(y - \bar{y})^2} + \overline{2(x - \bar{x})(y - \bar{y})} \\ &= \sigma_x^2 + \sigma_y^2 + \overline{2(x - \bar{x})(y - \bar{y})}\end{aligned}$$

Because x and y are independent r.v.'s, $(x - \bar{x})$ and $(y - \bar{y})$ are also independent r.v.'s. Hence, from Eq. (5.60b) we have

$$\overline{(x - \bar{x})(y - \bar{y})} = \overline{(x - \bar{x})}\,\overline{(y - \bar{y})}$$

But

$$\overline{(x - \bar{x})} = \bar{x} - \bar{\bar{x}} = \bar{x} - \bar{x} = 0$$

Similarly,

$$\overline{(y - \bar{y})} = 0$$

and

$$\sigma_z^2 = \sigma_x^2 + \sigma_y^2$$

This result can be extended to any number of variables. If r.v.'s x and y both have zero means (i.e., $\bar{x} = \bar{y} = 0$), then $\bar{z} = \bar{x} + \bar{y} = 0$. Also, because the variance equals the mean square value when the mean is zero, it follows that

$$\overline{z^2} = \overline{(x + y)^2} = \overline{x^2} + \overline{y^2} \tag{5.68}$$

provided $\bar{x} = \bar{y} = 0$, and provided x and y are independent r.v.'s.

EXAMPLE 5.19 Total Mean Square Error in PCM

In PCM, as seen in Examples 5.17 and 5.18, a signal sample m is transmitted as a quantized sample $\hat{m}$, causing a quantization error $q = m - \hat{m}$. Because of channel noise, the transmitted sample $\hat{m}$ is read as $\tilde{m}$, causing a detection error $\epsilon = \hat{m} - \tilde{m}$. Hence, the actual signal sample m is received as $\tilde{m}$ with a total error

$$m - \tilde{m} = (m - \hat{m}) + (\hat{m} - \tilde{m}) = q + \epsilon$$

where both q and ε are zero-mean r.v.'s. Because the quantization error q and the channel-noise error ε are independent, the mean square of the sum is [see Eq. (5.68)]

$$\begin{aligned}\overline{(m - \tilde{m})^2} &= \overline{(q + \varepsilon)^2} = \overline{q^2} + \overline{\varepsilon^2} \\ &= \frac{1}{3}\left(\frac{m_p}{L}\right)^2 + \frac{4m_p^2 P_e(2^{2n} - 1)}{3(2^{2n})}\end{aligned}$$

Also, because $L = 2^n$,

$$\overline{(\mathrm{m} - \tilde{\mathrm{m}})^2} = \overline{\mathrm{q}^2} + \overline{\varepsilon^2} = \frac{m_p^2}{3(2^{2n})}[1 + 4P_e(2^{2n} - 1)] \tag{5.69}$$

■

Chebyshev's Inequality

The standard deviation σ_{x} of an r.v. x is a measure of the width of its PDF. The larger the σ_{x}, the wider the PDF. Figure 5.16 illustrates this effect for a gaussian PDF.

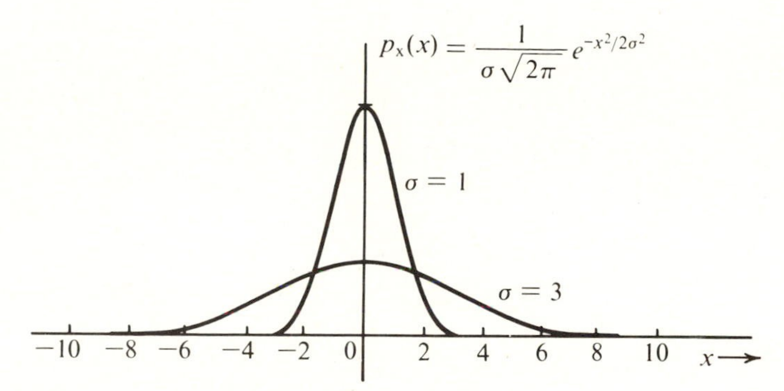

Figure 5.16 Gaussian PDF with standard deviation $\sigma = 1$ and $\sigma = 3$.

Chebyshev's inequality is a statement of this fact. It states that for a zero-mean r.v. x

$$P(|\mathrm{x}| \leq k\sigma_{\mathrm{x}}) \geq 1 - \frac{1}{k^2} \tag{5.70}$$

This means the probability of observing x within a few standard deviations is very high. For example, the probability of finding $|\mathrm{x}|$ within $3\sigma_{\mathrm{x}}$ is equal to or greater than 0.88. Thus, for a PDF with $\sigma_{\mathrm{x}} = 1$, $P(|\mathrm{x}| \leq 3) \geq 0.88$, whereas for a PDF with $\sigma_{\mathrm{x}} = 3$, $P(|\mathrm{x}| \leq 9) \geq 0.88$. It is clear that the PDF with $\sigma_{\mathrm{x}} = 3$ is spread out much more than the PDF with $\sigma_{\mathrm{x}} = 1$. Hence, σ_{x} or σ_{x}^2 is often used as a measure of the width of a PDF. In Chapter 6, we shall use this measure to estimate the bandwidth of a signal spectrum. The proof of Eq. (5.70) is as follows:

$$\sigma_{\mathrm{x}}^2 = \int_{-\infty}^{\infty} x^2 p_{\mathrm{x}}(x)\, dx$$

Because the integrand is positive,

$$\sigma_{\mathrm{x}}^2 \geq \int_{|x| \geq k\sigma_{\mathrm{x}}} x^2 p_{\mathrm{x}}(x)\, dx$$

If we replace x by its smallest value $k\sigma_{\mathrm{x}}$, the inequality still holds.

$$\sigma_{\mathrm{x}}^2 \geq k^2\sigma_{\mathrm{x}}^2 \int_{|x| \geq k\sigma_{\mathrm{x}}} p_{\mathrm{x}}(x)\, dx = k^2\sigma_{\mathrm{x}}^2\, P(|\mathrm{x}| \geq k\sigma_{\mathrm{x}})$$

or

$$P(|\mathrm{x}| \geq k\sigma_{\mathrm{x}}) \leq \frac{1}{k^2}$$

Hence,

$$P(|\mathrm{x}| \leq k\sigma_{\mathrm{x}}) \geq 1 - \frac{1}{k^2}$$

This inequality can be generalized for a nonzero-mean r.v. as:

$$P(|\mathrm{x} - \overline{\mathrm{x}}| \leq k\sigma_{\mathrm{x}}) \geq 1 - \frac{1}{k^2} \tag{5.71}$$

■ EXAMPLE 5.20

Estimate the width, or the spread, of a gaussian PDF [Eq. (5.56a)].

☐ **Solution:** For a gaussian r.v. [see Eqs. (5.34) and (5.39b)]

$$P(|\mathrm{x} - \overline{\mathrm{x}}| < \sigma) = 1 - 2Q(1) = 0.6826$$

$$P(|\mathrm{x} - \overline{\mathrm{x}}| < 2\sigma) = 1 - 2Q(2) = 0.9546$$

$$P(|\mathrm{x} - \overline{\mathrm{x}}| < 3\sigma) = 1 - 2Q(3) = 0.9974$$

This means that the area under the PDF over the interval ($\overline{\mathrm{x}} - 3\sigma, \overline{\mathrm{x}} + 3\sigma$) is 99.74 percent of the total area. A negligible fraction (0.26 percent) of the area lies outside this interval. Hence, the width, or the spread, of the gaussian PDF may be considered roughly $\pm 3\sigma$ about its mean, giving a total width of roughly 6σ. ■

5.4 THE CENTRAL-LIMIT THEOREM

Under centain conditions, the sum of a large number of independent r.v.'s tends to be a gaussian random variable, independent of the probability densities of the variables added.* The rigorous statement of this tendency is what is known as the *central-limit theorem.*† Proof of this theorem can be found in Refs. 3 and 4. We shall give here only a simple plausibility argument. Consider a sum of two r.v.'s x and y:

$$\mathrm{z} = \mathrm{x} + \mathrm{y}$$

Because $\mathrm{z} = \mathrm{x} + \mathrm{y}$, $\mathrm{y} = \mathrm{z} - \mathrm{x}$ regardless of the value of x. Hence, the event $\mathrm{z} \leq z$ is the joint event [$\mathrm{y} \leq z - x$ and x to have any value in the range $(-\infty, \infty)$]. Hence,

$$\begin{aligned} F_{\mathrm{z}}(z) &= P(\mathrm{z} \leq z) = P(\mathrm{x} \leq \infty, \mathrm{y} \leq z - x) \\ &= \int_{-\infty}^{\infty}\int_{-\infty}^{z-x} p_{\mathrm{xy}}(x, y)\, dxdy \\ &= \int_{-\infty}^{\infty} dx \int_{-\infty}^{z-x} p_{\mathrm{xy}}(x, y)\, dy \end{aligned}$$

* If the variables are gaussian, this is true even if the variables are not independent.

† Actually, a group of theorems collectively called the central-limit theorem.

and

$$p_z(z) = \frac{dF_z(z)}{dz} = \int_{-\infty}^{\infty} p_{xy}(x, z - x)\, dx$$

If x and y are independent r.v.'s, then

$$p_{xy}(x, z - x) = p_x(x)p_y(z - x)$$

and

$$p_z(z) = \int_{-\infty}^{\infty} p_x(x)p_y(z - x)\, dx \tag{5.72}$$

The PDF $p_z(z)$ is the convolution of PDFs $p_x(x)$ and $p_y(y)$. We can extend this result to a sum of n independent r.v.'s $x_1, x_2, \ldots, x_n$. If

$$z = x_1 + x_2 + \cdots + x_n$$

then the PDF $p_z(z)$ will be the convolution of PDFs $p_{x_1}(x_1), p_{x_2}(x_2), \ldots, p_{x_n}(x_n)$.

The tendency toward a gaussian distribution when a large number of functions are convolved is shown in Fig. 5.17. For simplicity, we assume all PDFs to be identical,

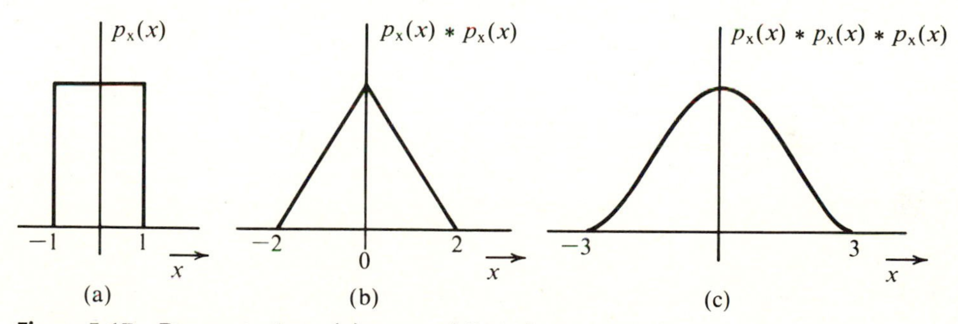

Figure 5.17 Demonstration of the central-limit theorem.

that is, a gate function $0.5\, \Pi(x/2)$. Figure 5.17 shows the successive convolutions of gate functions. The tendency toward a bell-shaped density is evident.

5.5 CORRELATION

Consider a random experiment with two outcomes described by r.v.'s x and y. We conduct several trials of this experiment and record values of x and y for each trial. From this data, it may be possible to determine the nature of the relationship between x and y. The covariance of r.v.'s x and y is one measure that is simple to compute and can yield useful information about the relationship between x and y.

The covariance σ_{xy} of two random variables is defined as

$$\sigma_{xy} = \overline{(x - \bar{x})(y - \bar{y})} \tag{5.73}$$

Note that the concept of covariance is a natural extension of the concept of variance, which is defined as

$$\sigma_{\mathrm{x}}^2 = \overline{(\mathrm{x} - \bar{\mathrm{x}})(\mathrm{x} - \bar{\mathrm{x}})}$$

Let us consider a case where the variables x and y are dependent such that they tend to vary in harmony; that is, if x increases y increases, and if x decreases y also decreases. For instance, x may be the income of a father and y the income of his son. It is reasonable to expect the two incomes to vary in harmony for a majority of the cases. Suppose we consider the following random experiment: pick randomly a family and record the income of the father as the value of x and the income of his working son as the value of y. We examine several families (several trials of the random experiment) and record the data x and y for each trial. We now plot points (x, y) for all the trials. This plot, known as the *scatter diagram,* may appear as shown in Fig. 5.18*a*. The plot shows that when x is large, y is likely to be large. Note the use of the word *likely*. It is not *always* true that y will be large if x is large, but it is true most of the time. In other words, a few cases will occur where a father with a low income has a son earning well, and vice versa. This is quite obvious from the scatter diagram in Fig. 5.18*a*.

To continue this example, the variable $\mathrm{x} - \bar{\mathrm{x}}$ represents the difference between the actual and the average value of x, and $\mathrm{y} - \bar{\mathrm{y}}$ represents the difference between the actual and the average value of y. It is more instructive to plot $(\mathrm{y} - \bar{\mathrm{y}})$ vs. $(\mathrm{x} - \bar{\mathrm{x}})$. This is the same as the scatter diagram in Fig. 5.18*a* with the origin shifted to $(\bar{\mathrm{x}}, \bar{\mathrm{y}})$ (see Fig. 5.18*b*). From this figure, we see that a father with an above-average income is likely to have his son's income above average, and a father with a below-average income is likely to have his son's income below average. That is, if $x - \bar{\mathrm{x}}$ is positive, $y - \bar{\mathrm{y}}$ is likely to be positive, and if $x - \bar{\mathrm{x}}$ is negative, $y - \bar{\mathrm{y}}$ is more likely to

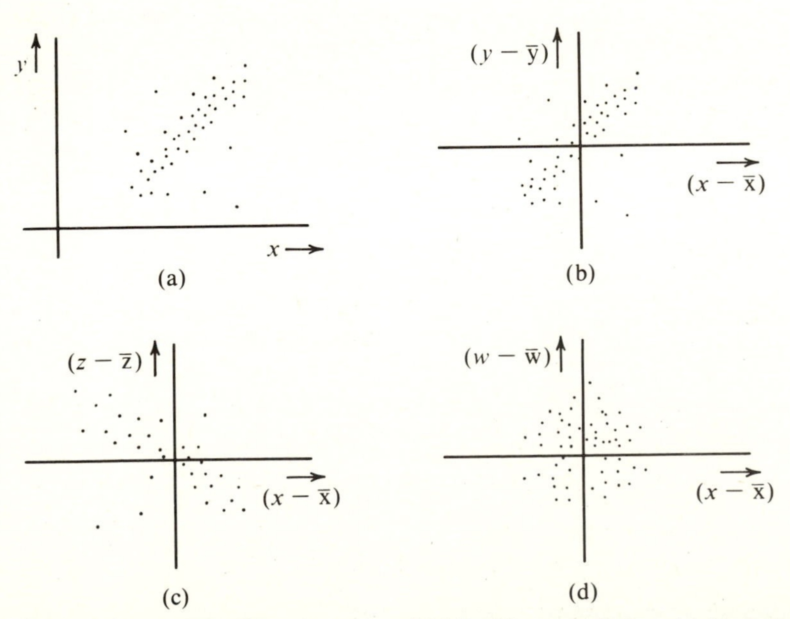

Figure 5.18 Scatter diagrams. (a) and (b) Positive correlation. (c) Negative correlation. (d) Zero correlation.

be negative. Thus, the quantity $(x - \bar{x})(y - \bar{y})$ will be positive for most trials. We compute this product for every pair, add these products, and then divide by the number of pairs. The result is the mean value of $(x - \bar{x})(y - \bar{y})$, that is, the covariance $\sigma_{xy} = \overline{(x - \bar{x})(y - \bar{y})}$. The covariance will be positive in the example under consideration. In such cases, we say that a positive correlation exists between variables x and y. We may conclude that a positive correlation implies variation of two variables in harmony (in the same direction, up or down).

Next, we consider the case where the two variables are x, the father's income, and z, the number of his children. If the results of several recent studies are to be believed, we shall find that as x (the father's income) increases, z (the number of children) tends to decrease. A hypothetical scatter diagram for this experiment is shown in Fig. 5.18*c*. Thus, if $x - \bar{x}$ is positive (above-average income), $z - \bar{z}$ is likely to be negative (below-average number of children). Similarly, when $x - \bar{x}$ is negative, $z - \bar{z}$ is likely to be positive. The product $(x - \bar{x})(z - \bar{z})$ will be negative for most of the trials, and the mean $\overline{(x - \bar{x})(z - \bar{z})} = \sigma_{xz}$ will be negative. In such a case, we say that negative correlation exists between x and y. It should be stressed here that negative correlation does not mean that x and y are unrelated. It means that they are dependent, but when one increases, the other decreases, and vice versa.

Lastly, consider the variables x (the father's income) and w (the father's height). Various studies in this area conclude that a person's income has little to do with his height. A hypothetical scatter diagram for this case will appear as shown in Fig. 5.18*d*. If $x - \bar{x}$ is positive, $w - \bar{w}$ is equally likely to be positive or negative. The product $(x - \bar{x})(w - \bar{w})$ is therefore equally likely to be positive or negative, and the mean $\overline{(x - \bar{x})(w - \bar{w})} = \sigma_{xw}$ will be zero. In such a case, we say that random variables x and w are *uncorrelated*.

To reiterate, if σ_{xy} is positive (or negative), then x and y are said to have a positive (or negative) correlation, and if $\sigma_{xy} = 0$, then the variables x and y are said to be uncorrelated.

From the above discussion, it appears that under suitable conditions, covariance can serve as a measure of the dependence of two variables. It often provides *some* information about the interdependence of the two r.v.'s and proves useful in a number of applications.

The covariance σ_{xy} may be expressed in another way, as follows. By definition

$$\begin{aligned} \sigma_{xy} &= \overline{(x - \bar{x})(y - \bar{y})} \\ &= \overline{xy} - \overline{\bar{x}y} - \overline{x\bar{y}} + \overline{\bar{x}\bar{y}} \\ &= \overline{xy} - \bar{x}\bar{y} - \bar{x}\bar{y} + \bar{x}\bar{y} \\ &= \overline{xy} - \bar{x}\bar{y} \end{aligned} \tag{5.74}$$

From the above equation, it follows that the variables x and y are uncorrelated ($\sigma_{xy} = 0$) if

$$\overline{xy} = \bar{x}\bar{y} \tag{5.75}$$

Note that for independent r.v.'s [Eq. (5.60c)]

$$\overline{xy} = \bar{x}\bar{y} \qquad \text{and} \qquad \sigma_{xy} = 0$$

Hence, independent random variables are uncorrelated. This supports the heuristic argument presented earlier. It should be noted that whereas independent variables are uncorrelated, the converse is not necessarily true—uncorrelated variables are not necessarily independent (see Prob. 5.38). Independence is, in general, a stronger and more restrictive condition than uncorrelatedness. For independent variables, we have shown [Eq. (5.60b)] that

$$\overline{g_1(\mathrm{x})g_2(\mathrm{y})} = \overline{g_1(\mathrm{x})}\,\overline{g_2(\mathrm{y})}$$

for any functions g_1 and g_2, whereas for uncorrelatedness, the only requirement is that

$$\overline{\mathrm{xy}} = \overline{\mathrm{x}}\,\overline{\mathrm{y}}$$

The *coefficient of correlation* ρ_{xy} is σ_{xy} normalized by $\sigma_x\sigma_y$

$$\rho_{xy} = \frac{\sigma_{xy}}{\sigma_x\sigma_y} \tag{5.76}$$

It can be shown that (see Prob. 5.36)

$$-1 \le \rho_{xy} \le 1 \tag{5.77}$$

The Mean Square of the Sum of Uncorrelated Variables

If x and y are uncorrelated, then for $z = x + y$,

$$\sigma_z^2 = \sigma_x^2 + \sigma_y^2 \tag{5.78}$$

That is, the variance of the sum is the sum of variances for uncorrelated random variables.

We have proved this result earlier for independent variables x and y. Following the development after Eq. (5.67), we have

$$\begin{aligned}\sigma_z^2 &= \overline{[(\mathrm{x} - \overline{\mathrm{x}}) + (\mathrm{y} - \overline{\mathrm{y}})]^2} \\ &= \overline{(\mathrm{x} - \overline{\mathrm{x}})^2} + \overline{(\mathrm{y} - \overline{\mathrm{y}})^2} + 2\overline{(\mathrm{x} - \overline{\mathrm{x}})(\mathrm{y} - \overline{\mathrm{y}})} \\ &= \sigma_x^2 + \sigma_y^2 + 2\sigma_{xy}\end{aligned}$$

Because x and y are uncorrelated, $\sigma_{xy} = 0$, and Eq. (5.78) follows. If x and y have zero means, then z also has a zero mean, and the mean square values of these variables are equal to their variances. Hence,

$$\overline{(\mathrm{x} + \mathrm{y})^2} = \overline{\mathrm{x}^2} + \overline{\mathrm{y}^2} \tag{5.79}$$

if x and y are uncorrelated and have zero means.

PART II Random Processes

5.6 FROM RANDOM VARIABLE TO RANDOM PROCESS

The notion of a random process is an extension of the random variable. Consider, for example, the temperature x of a certain city at 12 A.M. The temperature x is a random variable and takes on different values every day. To get the complete statistics of x, we need to record values of x over many days (a large number of trials). From this data, we can determine $p_x(x)$, the PDF of the r.v. x.

But the temperature is also a function of time. At 1 P.M., for example, the temperature may have entirely different distribution from that of the temperature at 12 A.M. Thus, the r.v. x is a function of time and can be expressed as $x(t)$. A random variable that is a function of time* is called a *random process* (or *stochastic process*).

To specify a random variable x, we repeat the experiment a large number of times and from the outcomes determine $p_x(x)$. Similarly, to specify the random process $x(t)$, we do the same thing for each value of t. To continue with our example of the random process $x(t)$, the temperature of the city, we need to record daily temperatures for each value of t (for each time of the day). This can be done by recording temperatures at every instant of the day. This gives one waveform $x(t, \zeta_i)$ where ζ_i indicates the day for which the record was taken. We need to repeat this procedure every day for a large number of days. The collection of all possible waveforms is known as the *ensemble* of the random process $x(t)$, and a waveform in this collection is a *sample function* (rather than a sample point) of the random process (Fig. 5.19). Sample-function amplitudes at some instant $t = t_1$ are the values taken by the random variable $x(t_1)$ in various trials.

We can view a random process in another way. In the case of a random variable, the outcome of each trial of the random experiment is a number. We can view a random process, also, as an outcome of a random experiment, where the outcome of each trial is a waveform (a sample function) that is a function of t. The number of waveforms in an ensemble may be finite or infinite. In the case of the random process $x(t)$ (the temperature of a city), the ensemble has infinite waveforms. On the other hand, if we consider the output of a binary signal generator (over the period 0 to 10 T), there are at most 2^{10} waveforms in this ensemble (Fig. 5.20).

One fine point that needs clarification is that the waveforms (sample functions) in the ensemble are not random. They are deterministic. Randomness in this situation is associated not with the waveform but with the uncertainty as to which waveform will occur in a given trial. This is completely analogous to the situation of a random variable. For example, in the experiment of tossing a coin four times in succession (Example 5.4), 16 possible outcomes exist, all of which are known. The randomness in this situation is associated not with the outcomes but with the uncertainty as to which of the 16 outcomes will occur in a given trial.

The next important question is how to specify a random process. If we continue with the idea that a random process $x(t)$ is a random variable x that is a function of

---

* Actually, to qualify as a random process, x could be a function of any other variable, such as distance. A random process may also be a function of more than one variable.

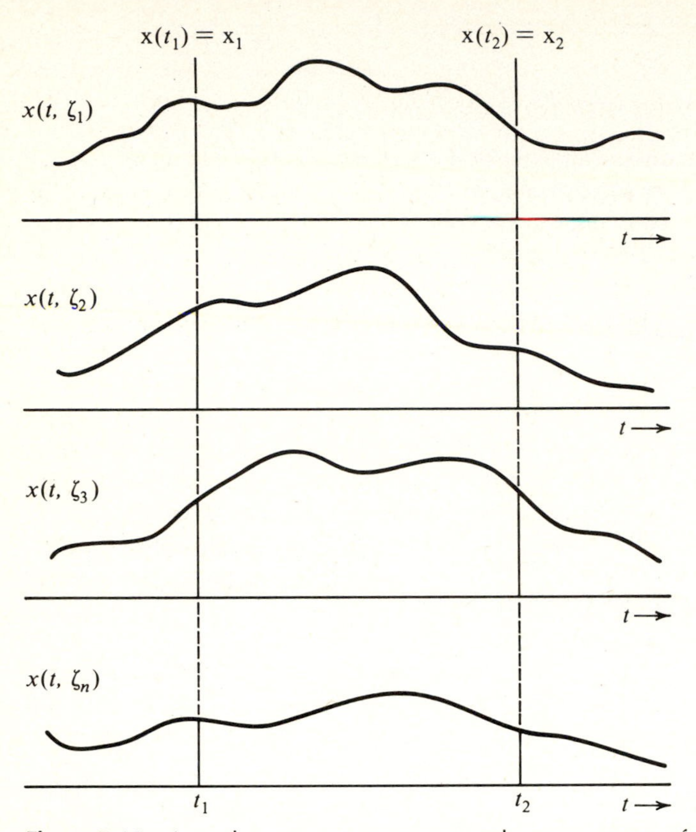

Figure 5.19 A random process to represent the temperature of a city.

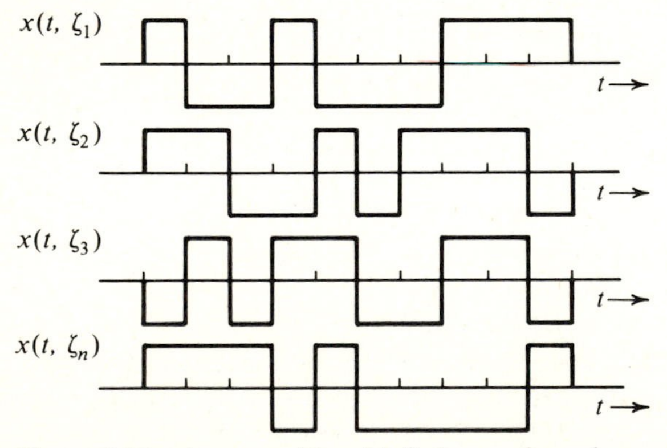

Figure 5.20 An ensemble with finite number of sample functions.

time, we shall come to a conclusion that $x(t)$ is completely specified if the PDF of x is specified for each value of t. We shall soon see that things are not quite this simple. But let us begin with the idea of specifying an r.v. x for each value of t. For the random process $x(t)$, representing the temperature of a city, this will imply considering sample function amplitudes at some instant $t = t_1$. The value $x(t_1, \zeta_i)$ represents the temperature at instant t_1 on the ζ_ith day and is the outcome of the ζ_ith trial. Thus, all the sample-function amplitudes at $t = t_1$ represent values taken by the random variable x at $t = t_1$; that is, $\mathrm{x}(t_1)$. From this data, using the relative-frequency measure, we can determine the PDF of $\mathrm{x}(t_1)$. We can do this for each value of t. The PDF may be different for different values of t in general. To indicate this fact, the PDF of x at instant t is expressed as $p_{\mathrm{x}}(x; t)$. Thus, to specify a random process, we need ensemble statistics.

To underscore the importance of ensemble statistics, consider the problem of threshold detection in Example 5.13. A **1** is transmitted by $p(t)$ and **0** is transmitted by $-p(t)$ [polar signaling]. The peak pulse amplitude is A_p. Received pulses are sampled at their peaks to obtain $\pm A_p + \mathrm{n}$, where n is the noise. It was shown that the optimum detection threshold is 0; that is, if the pulse sampled at its peak has a value > 0, the decision is **1**, and if it is < 0, the decision is **0**. Let us try to interpret $P(\epsilon|\mathbf{1})$, the error probability given that **1** is transmitted. If **1** is transmitted, the sampler output at the receiver is $A_p + \mathrm{n}$. If $A_p + \mathrm{n} > 0$, we make a correct decision, and if $A_p + \mathrm{n} < 0$ or $\mathrm{n} < -A_p$, we make a wrong decision. Interpreting probability in terms of relative frequency, if we repeat the experiment (of transmitting and receiving **1**) N times ($N \to \infty$), and if N_ϵ times the noise sample goes negative enough for $A_p + \mathrm{n}$ to be negative (that is, $\mathrm{n} < -A_p$), then

$$P(\epsilon|\mathbf{1}) = \frac{N_\epsilon}{N}$$

Let us examine the noise signal at the sampling instant t_s. In each trial, we have a new noise signal (sample function) from the noise ensemble and a different value of n at the sampling instant t_s (Fig. 5.21). Every time, we record the value of the noise signal at the sampling instant t_s, and if $\mathrm{n} < -A_p$ say 100 times out of 100 million trials, then the error probability $P(\epsilon|1) = 100/(100)10^6 = 10^{-6}$. But 10^{-6} is precisely the probability that $\mathrm{n} < -A_p$ where n is an r.v. formed by amplitudes at $t = t_s$ of the sample

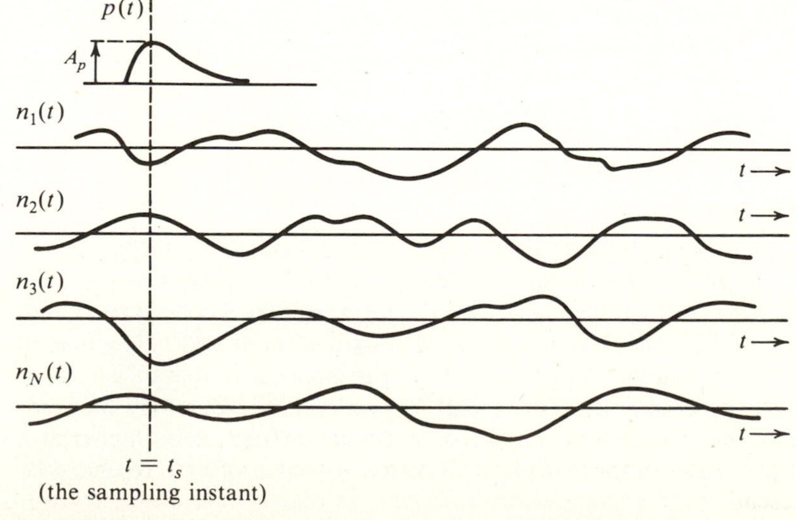

Figure 5.21 A random process to represent a channel noise.

functions in the ensemble of the random process n(t). This is the random variable n(t_s) whose PDF is $p_{\mathrm{n}}(n;\, t_s)$.

The importance of ensemble statistics is clear from this example. The PDF $p_{\mathrm{x}}(x;\, t)$ is known as the first-order PDF. Unfortunately, knowledge of the first-order PDF is insufficient to specify a random process. To show this, consider a random process x(t) whose ensemble is shown in Fig. 5.22. For the random process x(t), suppose that the

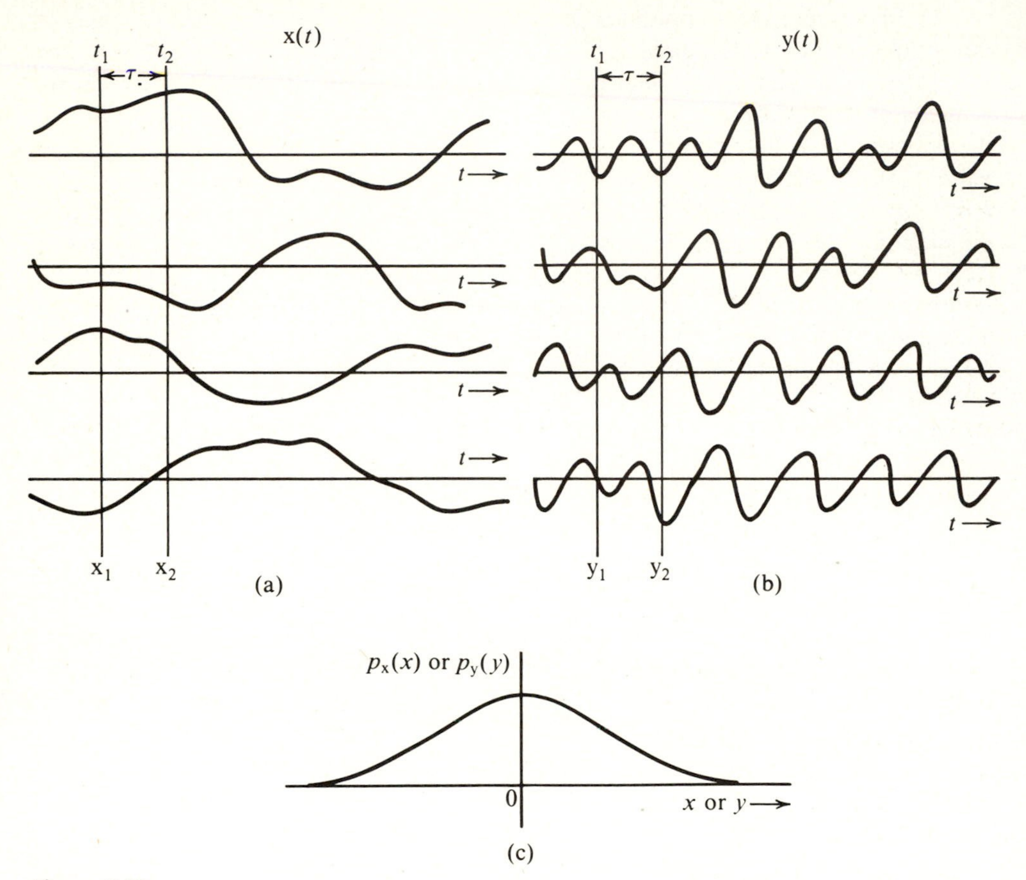

Figure 5.22

amplitude distribution at any instant t is the same; that is, $p_{\mathrm{x}}(x;\, t)$ is independent of t, and $p_{\mathrm{x}}(x;\, t) = p_{\mathrm{x}}(x)$, as shown in Fig. 5.22$c$. Further, suppose that we time compress the process x(t) by a factor $k(k > 1)$ to form another process y(t) in Fig. 5.22b. A little reflection shows that the amplitude distribution of y(t) is identical to that of x(t), and, hence, the first-order PDF of y(t) is identical to that of x(t) [Fig. 5.22]. Hence, both processes have the same first-order PDF. But they are very different in nature. The process y(t) contains components of frequencies higher than those of x(t). In fact, the spectrum of y(t) will be the spectrum of x(t) expanded by a factor k.

It is clear that the first-order PDF is not sufficient to specify completely a random process. The frequency content of a process depends upon the rapidity of the amplitude change with time. This can be measured by correlating amplitudes at t_1 and $t_1 + \tau$. If the process is slowly varying, amplitudes at t_1 and $t_1 + \tau$ may be similar (Fig. 5.22*a*). On the other hand, if the process varies rapidly, amplitudes at t_1 and $t_1 + \tau$ may have no resemblance (Fig. 5.22*b*). We can use correlation to measure the similarity of amplitudes at t_1 and $t_2 = t_1 + \tau$. If the random variables $\mathrm{x}(t_1)$ and $\mathrm{x}(t_2)$ are denoted by x_1 and x_2, respectively, then for a real random process,* the *autocorrelation function* $R_\mathrm{x}(t_1, t_2)$ is defined as

$$R_\mathrm{x}(t_1, t_2) = \overline{\mathrm{x}(t_1)\mathrm{x}(t_2)} = \overline{\mathrm{x}_1\mathrm{x}_2} \tag{5.80a}$$

This is the correlation of r.v.'s $\mathrm{x}(t_1)$ and $\mathrm{x}(t_2)$ and is computed by multiplying amplitudes at t_1 and t_2 of a sample function and then averaging this product over the ensemble. It can be seen that for a small τ, the product $\mathrm{x}_1\mathrm{x}_2$ will be positive for most sample functions of $\mathrm{x}(t)$, but the product $\mathrm{y}_1\mathrm{y}_2$ will be equally likely to be positive or negative. Hence, $\overline{\mathrm{x}_1\mathrm{x}_2}$ will be larger than $\overline{\mathrm{y}_1\mathrm{y}_2}$. Moreover, x_1 and x_2 will show correlation for considerably larger values of τ, whereas y_1 and y_2 will lose correlation quickly even for small τ, as shown in Fig. 5.23. Thus, $R_\mathrm{x}(t_1, t_2)$, the autocorrelation

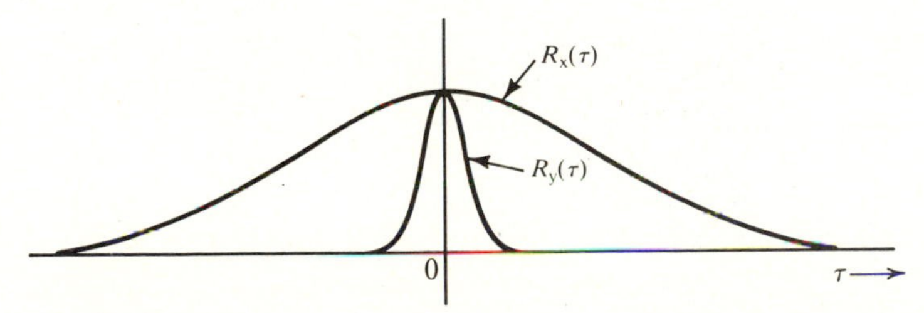

Figure 5.23 Autocorrelation functions for a slowly varying and a rapidly varying random process.

function of $\mathrm{x}(t)$, provides valuable information about the frequency content of the process. In fact, we shall show that the PSD of $\mathrm{x}(t)$ is the Fourier transform of its autocorrelation function.

$$R_\mathrm{x}(t_1, t_2) = \overline{\mathrm{x}_1\mathrm{x}_2}$$

$$= \int_{-\infty}^{\infty}\int_{-\infty}^{\infty} x_1x_2p_{\mathrm{x}_1\mathrm{x}_2}(x_1, x_2)\,dx_1dx_2 \tag{5.80b}$$

Hence, $R_\mathrm{x}(t_1, t_2)$ can be derived from the joint PDF of x_1 and x_2. This is the second-order PDF. In short, to specify a random process, we not only need the first-order PDF $p_\mathrm{x}(x; t)$ but also the second-order PDF $p_{\mathrm{x}_1\mathrm{x}_2}(x_1, x_2; t_1, t_2)$. In general, we need the measure of interdependence of n variables $\mathrm{x}_1, \mathrm{x}_2, \ldots, \mathrm{x}_n$ at instants t_1,

---

* For a complex random process $\mathrm{x}(t)$, the autocorrelation function is defined as

$$R_\mathrm{x}(t_1, t_2) = \overline{\mathrm{x}^*(t_1)\mathrm{x}(t_2)}$$

$t_2, \ldots, t_n$. This is specified by the nth-order PDF $p_{x_1x_2\ldots x_n}(x_1, x_2, \ldots, x_n; t_1, t_2, \ldots, t_n)$ for all n and any $t_1, t_2, \ldots, t_n$. Determining this PDF is a formidable task. Fortunately, when we are dealing with random processes in conjunction with linear systems, we can get by with only the first- and second-order statistics.

We can always derive a lower-order PDF from a higher-order PDF by simple integration. For instance,

$$p_{x_1}(x_1) = \int_{-\infty}^{\infty} p_{x_1x_2}(x_1, x_2)\, dx_2$$

Hence, when the nth-order PDF is available, there is no need to specify PDFs of order lower than n. The mean $\overline{x(t)}$ of a random process $x(t)$ can be determined from the first-order PDF as

$$\overline{x(t)} = \int_{-\infty}^{\infty} x p_x(x; t)\, dx \tag{5.81}$$

Classification of Random Processes

Stationary and Nonstationary Random Processes. A random process whose statistical characteristics do not change with time is classified as a *stationary random process*. For a stationary process, we can say that a shift of time origin will be impossible to detect; the process will appear to be the same. Suppose we determine $p_{x_1}(x_1; t_1)$, shift the origin by t_0, and again determine $p_{x_1}(x_1; t_1)$. The instant t_1 in the new frame of reference is $t_2 = t_1 + t_0$ in the old frame of reference. Hence, the PDFs of x at t_1 and $t_2 = t_1 + t_0$ must be the same. Therefore, for a stationary process, $p_{x_1}(x_1; t_1)$ and $p_{x_2}(x_2; t_2)$ must be identical. This is possible only if $p_{x_i}(x_i; t_i)$ is independent of t. Thus, the first-order density of a stationary random process can be expressed as

$$p_x(x; t) = p_x(x) \tag{5.82a}$$

In a similar way, we can see that the autocorrelation function $R_x(t_1, t_2)$ must be a function of only $t_2 - t_1$. If not, we could determine a unique time origin. Hence, for a real stationary process

$$R_x(t_1, t_2) = R_x(t_2 - t_1) \tag{5.82b}$$

$$= R_x(\tau) \qquad \tau = t_2 - t_1 \tag{5.82c}$$

and

$$R_x(\tau) = \overline{x(t)x(t + \tau)} \tag{5.83}$$

For a stationary process, the joint probability density function for x_1 and x_2 must also depend only on $t_2 - t_1$. Similarly, higher-order probability density functions, such as $p_{x_1x_2\ldots x_n}(x_1, x_2, \ldots, x_n)$ where $x_i = x(t_i)$, are all independent of the choice of origin.

The random process $x(t)$ representing the temperature of a city is an example of a nonstationary random process, because the temperature statistics (mean value, for example) depend on the time of the day. On the other hand, the noise process in Fig. 5.21 is stationary, because its statistics (the mean and the mean square values, for example) do not change with time. In general, it is not easy to determine if a process is stationary, because it involves investigation of the nth-order ($n = \infty$) statistics. In practice, we can ascertain stationarity if there is no change in the signal-generating mechanism. Such is the case for the noise process in Fig. 5.21.

Wide-Sense (or Weakly) Stationary Processes. A process may not be stationary in the *strict* sense, as discussed above, yet it may have a mean value and autocorrelation function that are independent of the shift of time origin. This means

$$\overline{\mathrm{x}(t)} = \text{constant}$$

$$R_{\mathrm{x}}(t_1, t_2) = R_{\mathrm{x}}(\tau) \qquad \tau = t_2 - t_1 \tag{5.84}$$

Such a process is known as a *wide-sense stationary* (or *weakly stationary*) *process*. Note that stationarity is a much stronger condition than wide-sense stationarity. All stationary processes are wide-sense stationary, but the converse is not necessarily true.

Just as no sinusoidal signals exist in actual practice, no truly stationary process can occur in real life. All processes in practice are nonstationary because they must begin at some finite time and must terminate at some finite time. A truly stationary process must start at $t = -\infty$ and go on forever. Many processes appear to be stationary for the time interval of interest, however, and the stationarity assumption allows a manageable mathematical model. The use of a stationary model is analogous to the use of a sinusoidal model in deterministic analysis.

■ EXAMPLE 5.21

Show that the random process

$$\mathrm{x}(t) = A \cos(\omega_c t + \theta)$$

where Θ is an r.v. uniformly distributed in the range $(0, 2\pi)$, is a wide-sense stationary process.

□ **Solution:** The ensemble (Fig. 5.24) consists of sinusoids of constant amplitude (A) and constant frequency (ω_c), but the phase Θ is random. For any sample function, the phase is equally likely to have any value in the range $(0, 2\pi)$. Because Θ is an r.v. uniformly distributed over the range $(0, 2\pi)$, one can determine[5] $p_{\mathrm{x}}(x, t)$ and, hence, $\overline{\mathrm{x}(t)}$ as in Eq. (5.81). For this particular case, however, $\overline{\mathrm{x}(t)}$ can be determined directly as follows:

$$\overline{\mathrm{x}(t)} = \overline{A \cos(\omega_c t + \Theta)} = A\, \overline{\cos(\omega_c t + \Theta)}$$

Because $\cos(\omega_c t + \Theta)$ is a function of an r.v. Θ, we have [Eq. (5.57b)]

$$\overline{\cos(\omega_c t + \Theta)} = \int_0^{2\pi} \cos(\omega_c t + \theta) p_\Theta(\theta)\, d\theta$$

Because $p_\Theta(\theta) = 1/2\pi$ over $(0, 2\pi)$ and is 0 outside this range,

$$\overline{\cos(\omega_c t + \Theta)} = \frac{1}{2\pi} \int_0^{2\pi} \cos(\omega_c t + \theta)\, d\theta = 0$$

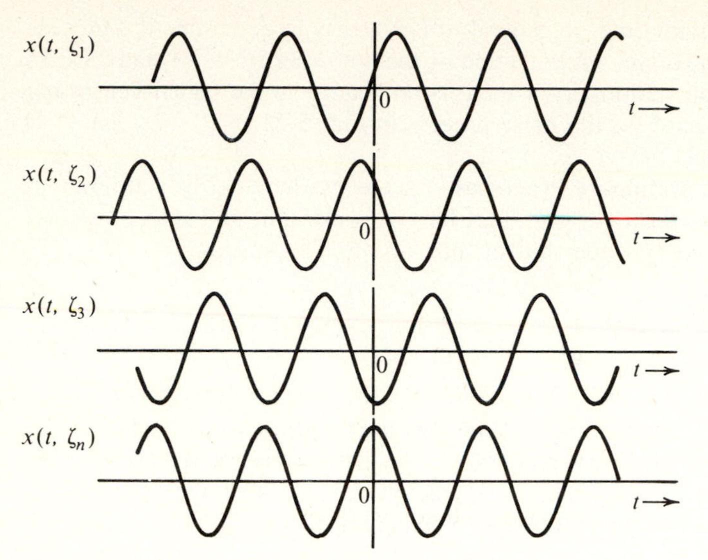

Figure 5.24 Random process $A \cos(\omega_c t + \Theta)$.

Hence,

$$\overline{\mathrm{x}(t)} = 0 \tag{5.85a}$$

Thus, the ensemble mean of sample-function amplitudes at any instant t is zero.

The autocorrelation function $R_{\mathrm{x}}(t_1, t_2)$ for this process also can be determined directly from Eq. (5.80a).

$$\begin{aligned} R_{\mathrm{x}}(t_1, t_2) &= \overline{A^2 \cos(\omega_c t_1 + \Theta) \cos(\omega_c t_2 + \Theta)} \\ &= A^2\, \overline{\cos(\omega_c t_1 + \Theta) \cos(\omega_c t_2 + \Theta)} \\ &= \frac{A^2}{2}\left[\overline{\cos[\omega_c(t_1 - t_2)]} + \overline{\cos[\omega_c(t_1 + t_2) + 2\Theta]}\right] \end{aligned}$$

The first term on the right-hand side contains no random variable. Hence, the mean of $\cos \omega_c(t_1 - t_2)$ is $\cos \omega_c(t_1 - t_2)$ itself. The second term is a function of the r.v. Θ, and its mean is

$$\overline{\cos[\omega_c(t_1 + t_2) + 2\Theta]} = \frac{1}{2\pi}\int_0^{2\pi} \cos[\omega_c(t_1 + t_2) + 2\theta]\, d\theta = 0$$

Hence,

$$R_{\mathrm{x}}(t_1, t_2) = \frac{A^2}{2} \cos[\omega_c(t_2 - t_1)] \tag{5.85b}$$

or

$$R_x(\tau) = \frac{A^2}{2} \cos \omega_c \tau \qquad \tau = t_2 - t_1 \tag{5.85c}$$

From Eq. (5.85a and b) it is clear that $x(t)$ is a wide-sense stationary process. ■

Ergodic Processes. We have studied the mean and the autocorrelation function of a random process. These are ensemble averages of some kind. For example, $\overline{x(t)}$ is the ensemble average of sample-function amplitudes at t, and $R_x(t_1, t_2) = \overline{x_1 x_2}$ is the ensemble average of the product of sample-function amplitudes $x(t_1)$ and $x(t_2)$.

We can also define time averages for each sample function. For example a time mean, $\widetilde{x(t)}$, of a sample function $x(t)$ is*

$$\widetilde{x(t)} = \lim_{T\to\infty} \frac{1}{T} \int_{T/2}^{T/2} x(t)\, dt \tag{5.86a}$$

Similarly, the time-autocorrelation function $\mathcal{R}_x(\tau)$ defined in Eq. (2.113b) is

$$\mathcal{R}_x(\tau) = \widetilde{x(t)x(t + \tau)} = \lim_{T\to\infty} \frac{1}{T} \int_{T/2}^{T/2} x(t)x(t + \tau)\, dt \tag{5.86b}$$

For *ergodic processes*, ensemble averages are equal to time averages of any sample function. Thus, for an ergodic process $x(t)$

$$\overline{x(t)} = \widetilde{x(t)} \tag{5.87a}$$

$$R_x(\tau) = \mathcal{R}_x(\tau) \tag{5.87b}$$

These are just two of the many possible averages. For an ergodic process, all possible ensemble averages are equal to the corresponding time averages of one of its sample functions. Because a time average cannot be a function of time, it is evident that an ergodic process is necessarily a stationary process; but the converse is not true (see Fig. 5.25).

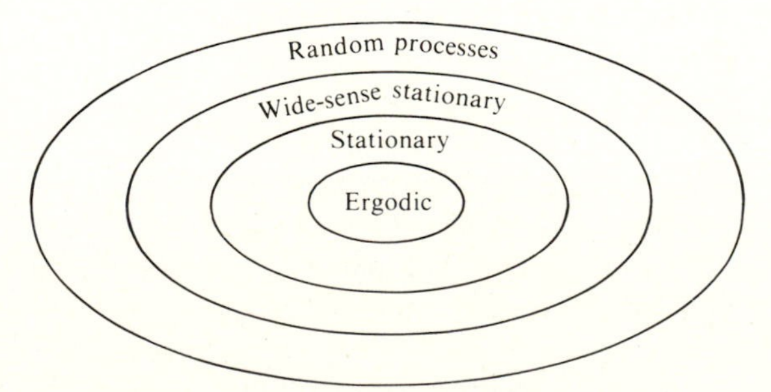

Figure 5.25 Classification of random processes.

---

* Here a sample function $x(t, \zeta_i)$ is represented by $x(t)$ for convenience.

It is difficult to test whether a process is ergodic or not, because we must test all possible orders of time and ensemble averages. Nevertheless, in practice many of the stationary processes are usually ergodic with respect to at least second-order averages, such as the mean and the autocorrelation. For the process in Example 5.21 (Fig. 5.24), we can show that $\widetilde{x(t)} = 0$ and $\mathcal{R}_x(\tau) = (A^2/2) \cos \omega_c \tau$ (see Example 2.35). It is actually ergodic, but we have shown ergodicity with respect to first- and second-order averages only.

The ergodicity concept can be explained by a simple example of traffic lights in a city. Suppose the city is well planned, with all its streets in E–W and N–S directions only and with traffic lights at each intersection. Assume that each light stays green for 0.75 second in the E–W direction and 0.25 second in the N–S direction and that switching of any light is independent of the other lights.

If we consider a certain person driving a car arriving at any traffic light randomly in the E–W direction, the probability that he will have a green light is 0.75; that is, on the average, 75 percent of the time he will observe a green light. On the other hand, if we consider a large number of drivers arriving at any traffic light in the E–W direction simultaneously at some instant t, then 75 percent of the drivers will have a green light, and the remaining 25 percent will have a red light. Thus, the experience of a single driver arriving randomly many times at a traffic light will contain the same statistical information (sample-function statistics) as that of a large number of drivers arriving simultaneously at various traffic lights (ensemble statistics at one instant).

The ergodicity notion is extremely important, because we do not have a large number of sample functions available in practice from which to compute ensemble averages. If the process is known to be ergodic, then we need only one sample function to compute ensemble averages. As mentioned earlier, many of the stationary processes encountered in practice are ergodic with respect to at least second-order averages. As we shall see in dealing with stationary processes in conjunction with linear systems, we need only first- and second-order averages. This means that in most cases we can get by with a single sample function.

5.7 THE POWER SPECTRAL DENSITY OF A RANDOM PROCESS

An electrical engineer commonly thinks of signals in terms of their frequency content. Linear systems are characterized by their frequency response (the transfer function), and signals are expressed in terms of the relative amplitudes and phases of their frequency components (the Fourier transform). From a knowledge of the input spectrum and transfer function, the response of a linear system to a given signal can be obtained in terms of the frequency content of that signal. This is an important procedure for deterministic signals. We may wonder if similar methods may be found for random processes. Ideally, all the sample functions of a random process are assumed to exist over the entire time interval $(-\infty, \infty)$ and, thus, are power signals.* We therefore inquire about the existence of a power spectral density. Superficially, the concept of a PSD of a random process may appear ridiculous for the following

* As we shall soon see, for the PSD to exist, the process must be stationary (at least in the wide sense). Stationary processes, because their statistics do not change with time, are power signals.

reasons. In the first place, we may not be able to describe a sample function analytically. Secondly, for a given process, every sample function may be different from another. Hence, even if a PSD does exist for each sample function, it may be different for different sample functions. Fortunately, both problems can be neatly resolved, and it is possible to define a meaningful PSD for a stationary (at least in the wide sense) random process. For nonstationary processes, the PSD does not exist.

Whenever randomness is involved, our inquiries can at best provide answers in terms of means. When tossing a coin, for instance, the most we can say about the outcome is that on the average we will obtain heads in about half the trials and tails in the remaining half of the trials. For random signals or random variables, we do not have enough information to predict the outcome with certainty, and we must accept answers in terms of averages. It is not possible to transcend this limit of knowledge, because of the fundamental ignorance of the process. It seems reasonable to define the PSD of a random process as a weighed mean of the PSDs of the sample functions. This is the only sensible solution, because we do not know exactly which of the sample functions may occur. We must be prepared for any sample function. Consider, for example, the problem of filtering a certain random process. We would not want to design a filter with respect to any one particular sample function because any of the sample functions in the ensemble may be present at the input. A sensible approach is to design the filter with respect to the mean parameters of the input process. In designing a system to perform certain operations, one must design it with respect to the whole ensemble. We are therefore justified in defining the PSD $S_x(\omega)$ of a random process $x(t)$ as the ensemble average of the PSDs of all sample functions. Thus [see Eq. (2.111)],

$$S_x(\omega) = \lim_{T\to\infty} \frac{\overline{|X_T(\omega)|^2}}{T} \tag{5.88a}$$

where $X_T(\omega)$ is the Fourier transform of the truncated random process $x(t)\Pi(t/T)$ and the bar represents ensemble average. Note that the averaging must be done before the limiting operation in order to ensure convergence of $S_x(\omega)$. We shall now show that the PSD as defined in Eq. (5.88a) is the Fourier transform of the autocorrelation function $R_x(\tau)$ of the process $x(t)$; that is,

$$R_x(\tau) \leftrightarrow S_x(\omega) \tag{5.88b}$$

This can be proved as follows:

$$X_T(\omega) = \int_{-\infty}^{\infty} x_T(t)e^{-j\omega t}\,dt = \int_{-T/2}^{T/2} x(t)e^{-j\omega t}\,dt \tag{5.89}$$

Thus, for real $x(t)$,

$$|X_T(\omega)|^2 = X_T(-\omega)X_T(\omega) = \int_{-T/2}^{T/2} x(t_1)e^{j\omega t_1}\,dt_1 \int_{-T/2}^{T/2} x(t_2)e^{-j\omega t_2}\,dt_2$$

$$= \int_{-T/2}^{T/2}\int_{-T/2}^{T/2} x(t_1)x(t_2)e^{-j\omega(t_2 - t_1)}\,dt_1 dt_2$$

and

$$S_x(\omega) = \lim_{T\to\infty} \frac{\overline{|X_T(\omega)|^2}}{T}$$

$$= \lim_{T\to\infty} \frac{1}{T} \overline{\int_{-T/2}^{T/2}\int_{-T/2}^{T/2} x(t_1)x(t_2)e^{-j\omega(t_2 - t_1)}\, dt_1 dt_2} \tag{5.90}$$

Interchanging the operation of integration and ensemble averaging,* we get

$$S_x(\omega) = \lim_{T\to\infty} \frac{1}{T} \int_{-T/2}^{T/2}\int_{-T/2}^{T/2} \overline{x(t_1)x(t_2)}e^{-j\omega(t_2-t_1)}\, dt_1 dt_2$$

$$= \lim_{T\to\infty} \frac{1}{T} \int_{-T/2}^{T/2}\int_{-T/2}^{T/2} R_x(t_2 - t_1)e^{-j\omega(t_2-t_1)}\, dt_1 dt_2$$

Here we are assuming that the process $x(t)$ is at least wide-sense stationary, so that $\overline{x(t_1)x(t_2)} = R_x(t_2 - t_1)$. For convenience, let

$$R_x(t_2 - t_1)e^{-j\omega(t_2-t_1)} = \varphi(t_2 - t_1) \tag{5.91}$$

then

$$S_x(\omega) = \lim_{T\to\infty} \frac{1}{T} \int_{-T/2}^{T/2}\int_{-T/2}^{T/2} \varphi(t_2 - t_1)\, dt_1 dt_2 \tag{5.92}$$

The integral on the right-hand side is a double integral over the range $(-T/2, T/2)$ for each of the variables t_1 and t_2. The square region of integration in the $t_1 - t_2$ plane is shown in Fig. 5.26. The integral in Eq. (5.92) is a volume under the surface

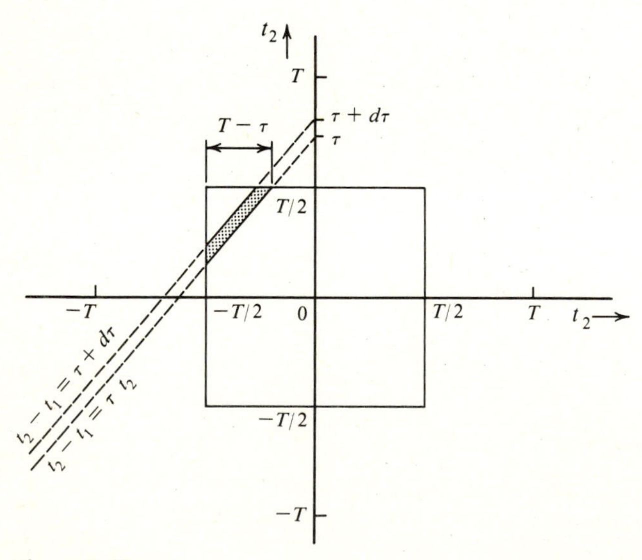

Figure 5.26

---

*The operation of ensemble averaging is also an operation of integration. Hence, interchanging integration with ensemble averaging is equivalent to interchanging the order of integration.

$\varphi(t_2 - t_1)$ over the square region in Fig. 5.26. The double integral in Eq. (5.92) can be converted to a single integral by observing that $\varphi(t_2 - t_1)$ is constant along any line $t_2 - t_1 =$ constant in the $t_1 - t_2$ plane (Fig. 5.26).

Let us consider two such lines, $t_2 - t_1 = \tau$ and $t_2 - t_1 = \tau + \Delta\tau$. If $\Delta\tau \to 0$, $\varphi(t_2 - t_1) \simeq \varphi(\tau)$ over the shaded region whose area is $(T - \tau)\,\Delta\tau$. Hence, the volume under the surface $\varphi(t_2 - t_1)$ over the shaded region is $\varphi(\tau)(T - \tau)\,\Delta\tau$. If τ were negative, the volume would be $\varphi(\tau)(T + \tau)\,\Delta\tau$. Hence, in general, the volume over the shaded region is $\varphi(\tau)(T - |\tau|)\,\Delta\tau$. The desired volume over the square region in Fig. 5.26 is the sum of the volumes over the shaded strips and is obtained by integrating $\varphi(\tau)(T - |\tau|)$ over the range of τ, which is $(-T, T)$ (see Fig. 5.26). Hence,

$$\begin{aligned} S_x(\omega) &= \lim_{T\to\infty} \frac{1}{T} \int_{-T}^{T} \varphi(\tau)(T - |\tau|)\, d\tau \\ &= \lim_{T\to\infty} \int_{-T}^{T} \varphi(\tau)\left(1 - \frac{|\tau|}{T}\right) d\tau \\ &= \int_{-\infty}^{\infty} \varphi(\tau)\, d\tau \end{aligned}$$

provided $\int_{-\infty}^{\infty} |\tau| \varphi(\tau)\, d\tau$ is bounded. Substituting Eq. (5.91) in the above, we have

$$S_x(\omega) = \int_{-\infty}^{\infty} R_x(\tau) e^{-j\omega\tau}\, d\tau \qquad \textbf{(5.93)}$$

provided $\int_{-\infty}^{\infty} |\tau| R_x(\tau) e^{-j\omega\tau}\, d\tau$ is bounded. Thus, the PSD of a wide-sense stationary random process is the Fourier transform of its autocorrelation function.*

$$R_x(\tau) \leftrightarrow S_x(\omega) \qquad \textbf{(5.94)}$$

This is the well-known Wiener-Khinchine relation.

From the discussion thus far, the autocorrelation function emerges as one of the most significant quantities in the spectral analysis of a random process. Earlier, we showed heuristically how the autocorrelation function is connected with the frequency content of a random process.

The autocorrelation function $R_x(\tau)$ for real processes is an even function τ. This can be proved in two ways. First, because $|X_T(\omega)|^2 = |X_T(\omega)X_T^*(\omega)| = |X_T(\omega)X_T(-\omega)|$ is an even function of ω, $S_x(\omega)$ is an even function of ω, and $R_x(\tau)$, its inverse transform, is also an even function of τ (see Prob. 2.6). Alternately, we may argue that

$$R_x(\tau) = \overline{x(t)x(t + \tau)} \qquad \text{and} \qquad R_x(-\tau) = \overline{x(t)x(t - \tau)}$$

Letting $t - \tau = \sigma$, we have

$$R_x(-\tau) = \overline{x(\sigma)x(\sigma + \tau)} = R_x(\tau) \qquad \textbf{(5.95)}$$

The PSD $S_x(\omega)$ is also a real and even function of ω.

---

* It can be shown that Eq. (5.93) holds also for complex random processes, for which we define $R_x(\tau) = \overline{x^*(t)x(t + \tau)}$.

The mean square value $\overline{\mathrm{x}^2(t)}$ of the random process $\mathrm{x}(t)$ is $R_{\mathrm{x}}(0)$.

$$R_{\mathrm{x}}(0) = \overline{\mathrm{x}(t)\mathrm{x}(t)} = \overline{\mathrm{x}^2(t)} = \overline{\mathrm{x}^2} \tag{5.96}$$

The mean square value $\overline{\mathrm{x}^2}$ is the ensemble average of the amplitude squares of sample functions at any instant t.

The Power of a Random Process

The mean square value $\overline{\mathrm{x}^2}$ of a wide-sense stationary random process $\mathrm{x}(t)$ is the power P_{x} (average power) of $\mathrm{x}(t)$. It is important to clarify the meaning of the power, or the mean square value, of a random process.

From Eq. (5.94),

$$R_{\mathrm{x}}(\tau) = \frac{1}{2\pi}\int_{-\infty}^{\infty} S_{\mathrm{x}}(\omega)e^{j\omega\tau}\,d\omega$$

Hence,

$$P_{\mathrm{x}} = \overline{\mathrm{x}^2} = R_{\mathrm{x}}(0) = \frac{1}{2\pi}\int_{-\infty}^{\infty} S_{\mathrm{x}}(\omega)\,d\omega \tag{5.97a}$$

Because $S_{\mathrm{x}}(\omega)$ is an even function of ω, we have

$$P_{\mathrm{x}} = \overline{\mathrm{x}^2} = 2\int_{0}^{\infty} S_{\mathrm{x}}(\omega)\,df \tag{5.97b}$$

This is the same relationship as that derived for deterministic signals in Chapter 2 [Eq. (2.112)]. The power P_{x} is the area under the PSD. Also, $P_{\mathrm{x}} = \overline{\mathrm{x}^2}$ is the ensemble mean of the square amplitudes of the sample functions at any instant.

It is helpful to repeat here, once again, that the PSD does not exist for processes that are not wide-sense stationary. All our future discussion will be restricted to processes that are at least wide-sense stationary. Hence, in future discussion random processes will be assumed to be wide-sense stationary unless specifically stated otherwise.

■ EXAMPLE 5.22

Determine the autocorrelation function $R_{\mathrm{x}}(\tau)$ and the power P_{x} of a lowpass random process with a white PSD $S_{\mathrm{x}}(\omega) = \mathcal{N}/2$ (Fig. 5.27*a*).

□ Solution:

$$S_{\mathrm{x}}(\omega) = \frac{\mathcal{N}}{2}\Pi\left(\frac{\omega}{4\pi B}\right) \tag{5.98a}$$

Hence, from Table 2.1 (pair 12)

$$R_{\mathrm{x}}(\tau) = \mathcal{N}B\ \mathrm{sinc}\,(2Bt) \tag{5.98b}$$

This is shown in Fig. 5.27*b*:

$$P_{\mathrm{x}} = \overline{\mathrm{x}^2} = R_{\mathrm{x}}(0) = \mathcal{N}B \tag{5.98c}$$

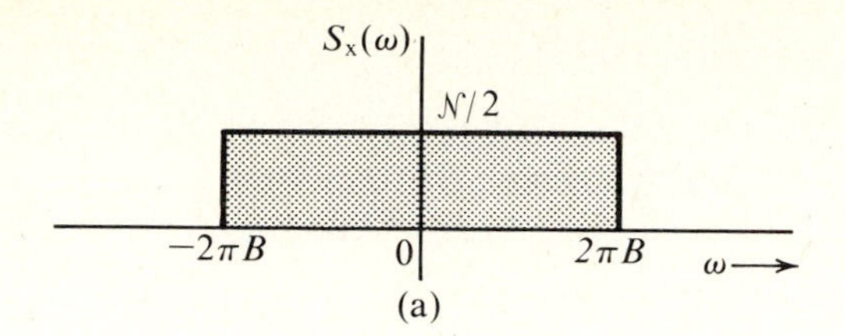

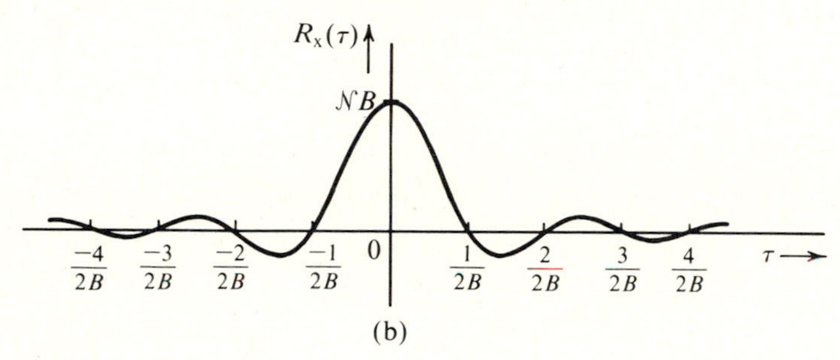

Figure 5.27 Bandpass white noise PSD and its autocorrelation function.

Alternately,

$$\begin{aligned} P_x = \overline{x^2} &= 2\int_0^\infty S_x(\omega)\, df \\ &= 2\int_0^B \frac{\mathcal{N}}{2}\, df \\ &= \mathcal{N}B \end{aligned} \tag{5.98d}$$

■

■ EXAMPLE 5.23

Determine the PSD and the mean square value of a random process

$$x(t) = A\cos(\omega_c t + \Theta) \tag{5.99a}$$

where Θ is an r.v. uniformly distributed over $(0, 2\pi)$. For this case $R_x(\tau)$ is already determined [Eq. (5.85c)].

$$R_x(\tau) = \frac{A^2}{2}\cos \omega_c \tau \tag{5.99b}$$

Hence,

$$S_x(\omega) = \pi A^2[\delta(\omega + \omega_c) + \delta(\omega - \omega_c)] \tag{5.99c}$$

$$P_x = \overline{x^2} = R_x(0) = \frac{A^2}{2} \tag{5.99d}$$

Thus, the power, or the mean square value, of the process $x(t) = A\cos(\omega_c t + \Theta)$ is $A^2/2$. The power P_x can also be obtained by integrating $S_x(\omega)$ with respect to f. ■

■ EXAMPLE 5.24 Amplitude Modulation

Determine the autocorrelation function and the PSD of the DSB-SC-modulated process $\mathrm{m}(t) \cos(\omega_c t + \Theta)$, where $\mathrm{m}(t)$ is a wide-sense stationary random process and Θ is a random variable uniformly distributed over $(0, 2\pi)$ and is independent of $\mathrm{m}(t)$.

Let

$$\varphi(t) = \mathrm{m}(t) \cos(\omega_c t + \Theta)$$

$$R_\varphi(\tau) = \overline{[\mathrm{m}(t) \cos(\omega_c t + \Theta)][\mathrm{m}(t+\tau) \cos(\omega_c (t+\tau) + \Theta)]}$$

$$= \overline{\mathrm{m}(t)\mathrm{m}(t+\tau) \cos(\omega_c t + \Theta) \cos[\omega_c (t+\tau) + \Theta]}$$

Because $\mathrm{m}(t)$ and Θ are independent, we can write [see Eq. (5.60b)]

$$R_\varphi(\tau) = \overline{\mathrm{m}(t)\mathrm{m}(t+\tau)}\ \overline{\cos(\omega_c t + \Theta) \cos[\omega_c (t+\tau) + \Theta]}$$

$$= \frac{1}{2} R_{\mathrm{m}}(\tau) \cos \omega_c \tau \qquad \textbf{(5.100a)}$$

and*

$$S_\varphi(\omega) = \frac{1}{4}[S_{\mathrm{m}}(\omega + \omega_c) + S_{\mathrm{m}}(\omega - \omega_c)] \qquad \textbf{(5.100b)}$$

From Eq. (5.100a) it follows that

$$\overline{\varphi^2(t)} = R_\varphi(0) = \frac{1}{2} R_{\mathrm{m}}(0) = \frac{1}{2}\overline{\mathrm{m}^2(t)} \qquad \textbf{(5.100c)}$$

Hence, the power of the DSB-SC-modulated signal is half the power of the modulating signal. We derived the same result earlier [Eq. (2.118)] for deterministic signals. ■

■ EXAMPLE 5.25 Random Binary Process

In this example we shall consider a random binary process for which a typical sample function is shown in Fig. 5.28*a*. The signal can assume only two states (values), 1 or -1 with equal probability. The transition from one state to another can take place only at node points, which occur every T_o seconds. The probability of a transition from one state to the other is 0.5. The first node is equally likely to be situated at any instant within the interval 0 to T_o from the origin.† The amplitudes at t represent r.v. x_1, and those at $t + \tau$ represent r.v. x_2. Note that x_1 and x_2 are discrete and each can assume

---

*We obtain the same result even if $\varphi(t) = \mathrm{m}(t) \sin(\omega_c t + \Theta)$.

†Analytically, we can represent $\mathrm{x}(t)$ as

$$\mathrm{x}(t) = \sum_n \mathrm{a}_n p(t - nT_o - \alpha)$$

where α is an r.v. uniformly distributed over the range $(0, T_o)$ and $p(t)$ is the basic pulse (in this case $\Pi[(t - T_o/2)/T_o]$. Note that α is the distance of the first node from the origin, and it varies randomly from sample function to sample function. In addition, a_n is random, taking values 1 or -1 with equal probability.

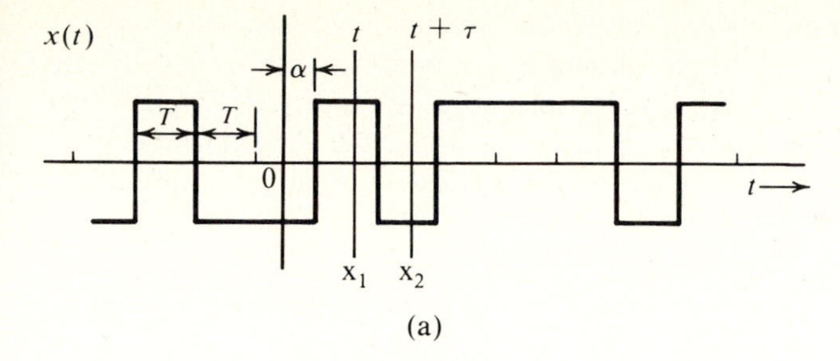

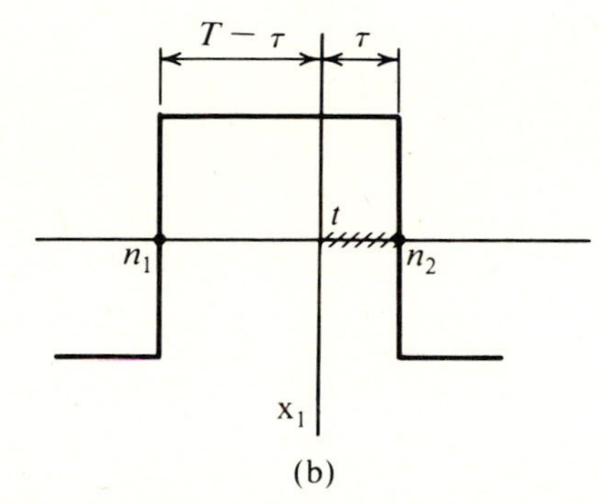

Figure 5.28 Derivation of autocorrelation function of a random binary process.

only two values: -1 and 1. Hence,

$$R_{\mathrm{x}}(\tau) = \overline{\mathrm{x}_1 \mathrm{x}_2} = \sum_{\mathrm{x}_1} \sum_{\mathrm{x}_2} x_1 x_2 P_{\mathrm{x}_1 \mathrm{x}_2}(x_1, x_2)$$

$$= P_{\mathrm{x}_1\mathrm{x}_2}(1, 1) + P_{\mathrm{x}_1\mathrm{x}_2}(-1, -1) - P_{\mathrm{x}_1\mathrm{x}_2}(-1, 1) - P_{\mathrm{x}_1\mathrm{x}_2}(1, -1) \quad \textbf{(5.101a)}$$

By symmetry, the first two terms and the last two terms on the right-hand side are equal. Therefore,

$$R_{\mathrm{x}}(\tau) = 2[P_{\mathrm{x}_1\mathrm{x}_2}(1, 1) - P_{\mathrm{x}_1\mathrm{x}_2}(1, -1)] \quad \textbf{(5.101b)}$$

Using Bayes' rule, we have

$$R_{\mathrm{x}}(\tau) = 2P_{\mathrm{x}_1}(1)[P_{\mathrm{x}_2|\mathrm{x}_1}(1 \mid 1) - P_{\mathrm{x}_2|\mathrm{x}_1}(-1 \mid 1)]$$

$$= P_{\mathrm{x}_2|\mathrm{x}_1}(1 \mid 1) - P_{\mathrm{x}_2|\mathrm{x}_1}(-1 \mid 1) \quad \textbf{(5.101c)}$$

We shall first consider the case $\tau < T_o$, where, at most, one node is in the interval t to $t + \tau$. In this case, the event $\mathrm{x}_2 = -1$ given $\mathrm{x}_1 = 1$ is a joint event AB, where the event A is "a node in the interval $(t, t + \tau)$" and B is "the state change at this node." Because A and B are independent events,

$$P_{\mathrm{x}_2|\mathrm{x}_1}(-1 \mid \mathrm{x}_1 = 1) = P \text{ (a node lies in } t \text{ to } t + \tau)P \text{ (state change)}$$

$$= \tfrac{1}{2}P \text{ (a node lies in } t \text{ to } t + \tau)$$

Figure 5.28*b* shows adjacent nodes n_1 and n_2, between which t lies. We mark off the interval τ from the node n_2. If t lies anywhere in this interval (shown shaded), the node n_2 lies within t and $t + \tau$. But because the instant t is arbitrarily chosen between nodes n_1 and n_2, it is equally likely to be at any instant over the T_o seconds between n_1 and n_2, and the probability that t lies in the shaded interval is simply τ/T_o. Therefore,

$$P_{\mathrm{x}_2|\mathrm{x}_1}(-1 \mid \mathrm{x}_1 = 1) = \frac{1}{2}\left(\frac{\tau}{T_o}\right) \tag{5.102a}$$

and

$$P_{\mathrm{x}_2|\mathrm{x}_1}(1 \mid \mathrm{x}_1 = 1) = 1 - \frac{\tau}{2T_o} \tag{5.102b}$$

Therefore,

$$R_{\mathrm{x}}(\tau) = 1 - \frac{\tau}{T_o} \qquad \tau < T_o \tag{5.103}$$

Because $R_{\mathrm{x}}(\tau)$ is an even function of τ, we have

$$R_{\mathrm{x}}(\tau) = 1 - \frac{|\tau|}{T_o} \qquad |\tau| < T_o \tag{5.104}$$

Next, consider the range $\tau > T_o$. In this case at least one node lies in the interval t to $t + \tau$. Hence, x_1 and x_2 become independent, and

$$R_{\mathrm{x}}(\tau) = \overline{\mathrm{x}_1\mathrm{x}_2} = \overline{\mathrm{x}}_1\,\overline{\mathrm{x}}_2 = 0 \qquad \tau > T_o$$

where, by inspection, we observe that $\overline{\mathrm{x}}_1 = \overline{\mathrm{x}}_2 = 0$ (Fig. 5.28*a*). This result can also be obtained by observing that for $|\tau| > T_o$, x_1 and x_2 are independent, and it is equally likely that $\mathrm{x}_2 = 1$ or -1 given that $\mathrm{x}_1 = 1$ (or -1). Hence, all four probabilities in Eq. (5.101a) are equal to 1/4, and

$$R_{\mathrm{x}}(\tau) = 0 \qquad \tau > T_o$$

Therefore,

$$R_{\mathrm{x}}(\tau) = \begin{cases} 1 - \dfrac{|\tau|}{T_o} & |\tau| < T_o \\ 0 & |\tau| > T_o \end{cases} \tag{5.105a}$$

and

$$S_{\mathrm{x}}(\omega) = T_o \operatorname{sinc}^2\left(\frac{\omega T_o}{2\pi}\right) \tag{5.105b}$$

The autocorrelation function and the PSD are shown in Fig. 5.29. Observe that $\overline{\mathrm{x}^2} = R_{\mathrm{x}}(0) = 1$ as expected.

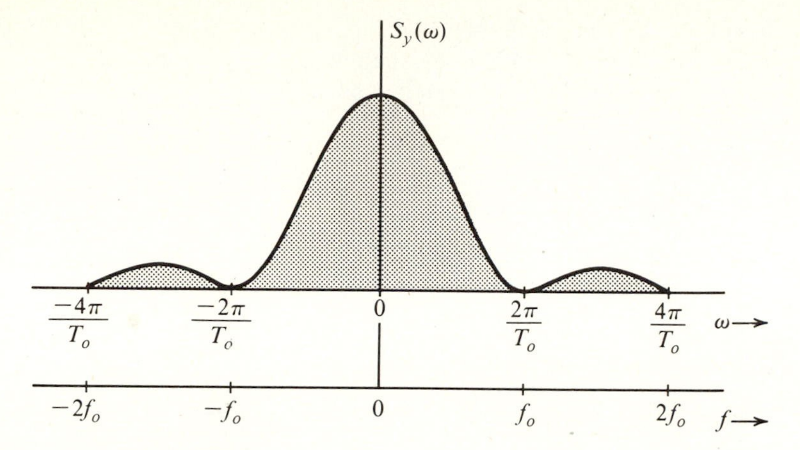

Figure 5.29 Autocorrelation function and PSD of a random binary process.

Let us now consider a more general case of the impulse train x(t), discussed in Sec. 3.2 (Fig. 3.8b). From the knowledge of the PSD of this train, we can derive the PSD of on-off, polar, bipolar, duobinary, split-phase, and many more important digital signals. ■

■ EXAMPLE 5.26 Random PAM Impulse Train

Find the autocorrelation function and PSD of a random impulse train whose sample function is shown in Fig. 5.30. Impulses are spaced T_o seconds apart, and their

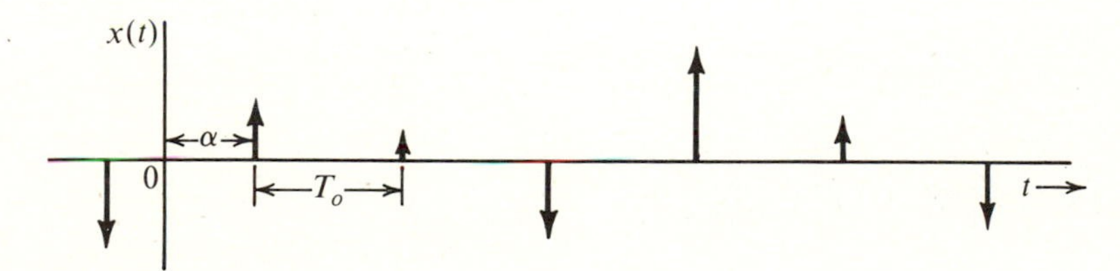

Figure 5.30

strengths a_k are random. The distance α of the first impulse from the origin is equally likely to be any value in the range (0, T_o). The random process x(t) can be described as

$$x(t) = \sum_k a_k \, \delta(t - kT_o - \alpha)$$

where α, the distance of the first impulse from the origin, is an r.v. uniformly distributed in the interval (0, T_o). Thus, α is different for each sample function.

☐ **Solution:** In solving this problem, we shall approximate a unit impulse by a narrow rectangular pulse $b(t)$ of width ϵ and height $h = 1/\epsilon$, with $\epsilon \to 0$ (Fig. 5.31*a*). The impulse train x(t) in Fig. 5.30 can be approximated by a train of narrow

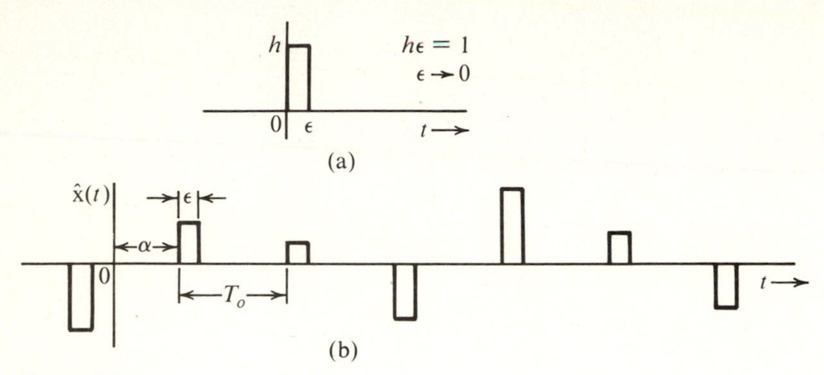

Figure 5.31

rectangular pulses, as shown in Fig. 5.31*b*. The random process $\hat{x}(t)$ can be described as

$$\hat{x}(t) = \sum_k a_k\, b(t - kT_o - \alpha)$$

and

$$R_{\hat{x}}(\tau) = \overline{\hat{x}(t)\hat{x}(t+\tau)}$$

$$= \overline{\sum_k a_k b(t - kT_o - \alpha) \sum_m a_m b(t + \tau - mT_o - \alpha)}$$

$$= \sum_k \sum_m \overline{a_k a_m b(t - kT_o - \alpha) b(t + \tau - mT_o - \alpha)}$$

Because a_k and a_m are independent of α,

$$R_{\hat{x}}(\tau) = \sum_k \sum_m \overline{a_k a_m}\; \overline{b(t - kT_o - \alpha) b(t + \tau - mT_o - \alpha)} \tag{5.106}$$

Let us consider the range of $|\tau| < \epsilon$. Random variables $\hat{x}(t)$ and $\hat{x}(t+\tau)$ are amplitude values τ seconds apart (Fig. 5.31*a*). Because we are considering the range of small $\tau(|\tau| < \epsilon$ and $\epsilon \to 0)$, the product $\hat{x}(t)\hat{x}(t+\tau)$ will be nonzero only if t and $t + \tau$ fall on the same pulse. Hence for this range of τ, the product $b(t - kT_o - \alpha)b(t + \tau - mT_o - \alpha)$ will certainly be zero when $m \neq k$. As a result, Eq. (5.106) reduces to a simple sum:

$$R_{\hat{x}}(\tau) = \sum_k \overline{a_k^2}\; \overline{b(t - kT_o - \alpha) b(t + \tau - kT_o - \alpha)}$$

Also, $b(t - kT_o - \alpha)$ and $b(t + \tau - kT_o - \alpha)$ are both r.v.'s that can take on values 0 or h. Let us denote these by variables b_1 and b_2, respectively. Then

$$R_{\hat{x}}(\tau) = \sum_k \overline{a_k^2}\,\overline{\mathrm{b}_1\mathrm{b}_2}$$

where

$$\overline{\mathrm{b}_1\mathrm{b}_2} = \sum_{\mathrm{b}_1}\sum_{\mathrm{b}_2} b_1 b_2 P_{\mathrm{b}_1\mathrm{b}_2}(b_1, b_2)$$

Because b_1 and b_2 each can take on values 0 and h, the summation on the right-hand side will have four terms, out of which only one is nonzero.

$$\begin{aligned}\overline{\mathrm{b}_1\mathrm{b}_2} &= h^2 P_{\mathrm{b}_1\mathrm{b}_2}(h, h)\\ &= h^2 P_{\mathrm{b}_1}(h) P_{\mathrm{b}_2|\mathrm{b}_1}(h|h)\end{aligned}$$

The instant t is chosen arbitrarily, and $P_{\mathrm{b}_1}(h)$ is the probability that the instant t will fall on the kth pulse $b(t - kT_o - \alpha)$. The sample function exists over the interval $(-T/2, T/2)$, with $T \to \infty$. Let $T = NT_o$, that is, there are N pulses $(N \to \infty)$ in the sample function. The probability that t will fall on a pulse of width ϵ over the entire interval T is ϵ/T. Hence,

$$P_{\mathrm{b}_1}(h) = \frac{\epsilon}{T} = \frac{\epsilon}{NT_o}$$

To calculate $P_{\mathrm{b}_2|\mathrm{b}_1}(h|h)$, consider Fig. 5.32*a*. The desired probability $P_{\mathrm{b}_2|\mathrm{b}_1}(h|h)$ is the probability that the instant $t + \tau$ lies on the kth pulse given that t lies somewhere on that pulse. To find this, we mark off an interval τ from the lagging edge of the kth pulse (Fig. 5.32*a*). If t lies anywhere in the hatched interval, $t + \tau$ will be out of the pulse. Hence, $P_{\mathrm{b}_2|\mathrm{b}_1}(h|h)$ is the probability that t lies in the pulse interval not hatched. Because t is arbitrary and is likely to be anywhere on the pulse, this probability is $(\epsilon - \tau)/\epsilon$. Hence,

$$\begin{aligned}P_{\mathrm{b}_1\mathrm{b}_2}(h, h) &= P_{\mathrm{b}_1}(h) P_{\mathrm{b}_2|\mathrm{b}_1}(h|h)\\ &= \frac{\epsilon}{NT_o}\left(\frac{\epsilon - \tau}{\epsilon}\right) = \frac{\epsilon - \tau}{NT_o}\end{aligned}$$

and

$$R_{\hat{x}}(\tau) = \sum_k \overline{\mathrm{a}_k^2}\, h^2 \frac{\epsilon - \tau}{NT_o}$$

Also, because $R_{\hat{x}}(\tau)$ is an even function of τ, and $h = 1/\epsilon$,

$$R_{\hat{x}}(\tau) = \frac{\epsilon - |\tau|}{\epsilon^2 NT_o} \sum_k \overline{\mathrm{a}_k^2}$$

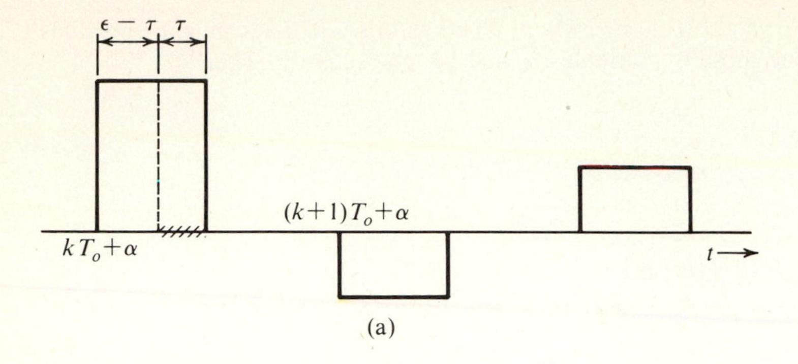

(a)

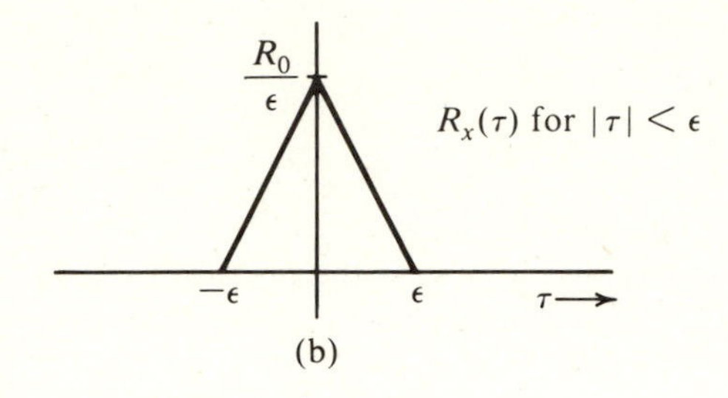

(b)

Figure 5.32

The summation is over N values of k. Hence,

$$R_{\hat{x}}(\tau) = \frac{\epsilon - |\tau|}{\epsilon^2 N T_o} N(\overline{a_k^2})$$

$$= \frac{R_0}{\epsilon T_o}\left[1 - \frac{|\tau|}{\epsilon}\right]$$

where R_0 is the ensemble average of a_k^2, that is,

$$R_0 = \overline{a_k^2}$$

Figure 5.32*b* shows $R_x(\tau)$ for $|\tau| < \epsilon$. Observe that $R_x(\tau) \to 0$ as $\tau \to \epsilon$. This is expected because as τ increases, the chances of t and $t + \tau$ falling on the same pulse diminish, until at $\tau = \epsilon$, the chance is zero. But as τ is increased further in the vicinity of T_o, t and $t + \tau$ can lie on neighboring pulses; that is, t lies on the kth pulse and $t + \tau$ lies on the $(k + 1)$st pulse, and $P_{b_1 b_2}(h, h)$ is again nonzero. Thus, $R_{\hat{x}}(\tau)$ is again nonzero in the vicinity of $|\tau| = T_o$. The same thing repeats for $|\tau| = 2T_o, 3T_o, \ldots,$ and so on. To compute $R_{\hat{x}}(\tau)$ for a general case, when $|\tau|$ is in the vicinity of nT_o, we follow similar arguments to obtain

$$R_{\hat{x}}(\tau) = \sum_k \overline{a_k a_{k+n}}\ \overline{b_1 b_2}$$

$$= NR_n \overline{b_1 b_2} \qquad nT_o - \epsilon < |\tau| < nT_o + \epsilon$$

where $R_n = \overline{a_k a_{k+n}}$.

Using similar arguments, we find that $N R_n \overline{b_1 b_2}$ yields a triangular pulse of height $R_n/\epsilon T_o$, width 2ϵ, and centered at $|\tau| = nT_o$, as shown in Fig. 5.33. Note that the area

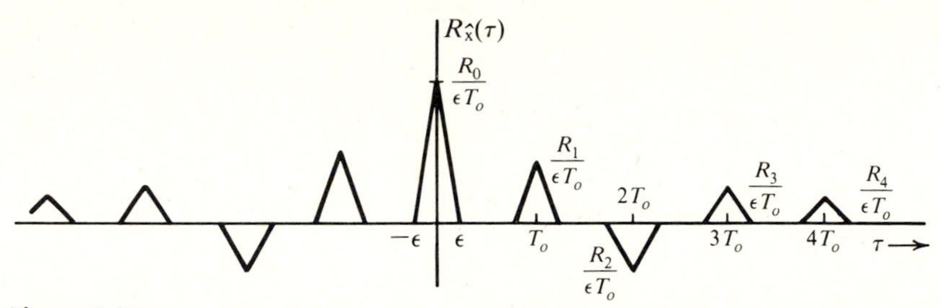

Figure 5.33

under the pulse centered at nT_o is $\frac{1}{2}(2\epsilon)(R_n)/(\epsilon T_o) = R_n/T_o$. Hence, in the limit as $\epsilon \to 0$, the pulse centered at $|\tau| = nT_o$ becomes an impulse of strength R_n/T_o and $R_x(\tau)$ becomes a sequence of impulses, as shown in Fig. 5.34.

$$R_x(\tau) = \frac{1}{T_o} \sum_{n=-\infty}^{\infty} R_n\, \delta(\tau - nT_o) \tag{5.107}$$

where

$$R_n = \overline{a_k a_{k+n}} \tag{5.108}$$

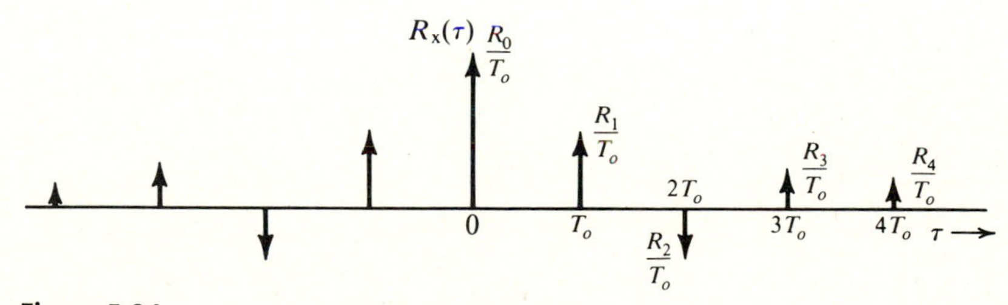

Figure 5.34

Compare this result with that in Eq. (3.4). They are identical except that in Eq. (3.4), $\mathcal{R}_x(\tau)$ is the time-autocorrelation function of a sample function $x(t)$, and R_n is the time average $\widetilde{a_k a_{k+n}}$. This shows that the random process under consideration can be considered ergodic at least with respect to the first- and second-order averages. ■

■ EXAMPLE 5.27

Determine $R_x(\tau)$ and $S_x(\omega)$ for a polar binary signal where **1** is transmitted by an impulse $\delta(t)$ and **0** is transmitted by an impulse $-\delta(t)$ (Fig. 5.35). The digits **1** and **0** are equally likely, and digits are transmitted every T_o seconds. Each digit is independent of the remaining digits.

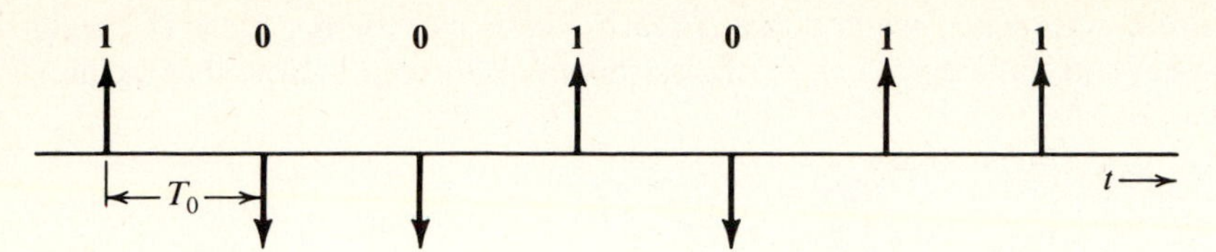

Figure 5.35

□ **Solution:** In this case, a_n can take on values 1 and -1 with probability 0.5 each. Hence,

$$\overline{a_k} = 0$$

$$R_0 = \overline{a_k^2} = 1$$

and because each digit is independent of the remaining digits,

$$R_n = \overline{a_k a_{k+n}} = \overline{a_k}\,\overline{a_{k+n}} = 0 \qquad n \geq 1$$

Hence [Eq. (5.107)],

$$R_x(\tau) = \frac{1}{T_o}\,\delta(\tau) \qquad S_x(\omega) = \frac{1}{T_o}$$ ■

5.8 MULTIPLE RANDOM PROCESSES

For two real random processes x(t) and y(t), we define the *cross-correlation function** $R_{xy}(t_1, t_2)$ as

$$R_{xy}(t_1, t_2) = \overline{x(t_1)y(t_2)} \tag{5.109a}$$

The two processes are said to be *jointly stationary* (at least in the wide sense) if

$$R_{xy}(t_1, t_2) = R_{xy}(t_2 - t_1)$$
$$= R_{xy}(\tau) \tag{5.109b}$$

Uncorrelated, Orthogonal (Incoherent), and Independent Processes

Two processes x(t) and y(t) are said to be *uncorrelated* if their cross-correlation function is equal to the product of their means; that is,

$$R_{xy}(\tau) = \overline{x(t)y(t+\tau)} = \overline{x}\,\overline{y} \tag{5.110a}$$

This implies that r.v.'s x(t) and y$(t + \tau)$ are uncorrelated for all t and τ.

Processes x(t) and y(t) are said to be *incoherent*, or *orthogonal*, if

$$R_{xy}(\tau) = 0 \tag{5.110b}$$

Incoherent, or orthogonal, processes are uncorrelated processes with $\overline{x}$ and/or $\overline{y} = 0$.

---

* For complex random processes, the cross-correlation function is defined as

$$R_{xy}(t_1, t_2) = \overline{x^*(t_1)y(t_2)}$$

Processes $x(t)$ and $y(t)$ are *independent* random processes if the group of r.v.'s $x(t_1), x(t_2), \ldots, x(t_n)$ is independent of the group of r.v.'s $y(t'_1), y(t'_2), \ldots, y(t'_m)$ for any n, m and $t_1, t_2, \ldots, t_n, t'_1, t'_2, \ldots, t'_m$. We define group $x_1, x_2, \ldots, x_n$ as independent of the group $y_1, y_2, \ldots, y_m$ if

$$p_{x_1x_2\ldots x_ny_1y_2\cdots y_m} = p_{x_1x_2\ldots x_n}p_{y_1y_2\cdots y_n} \tag{5.111}$$

Cross-Power Spectral Density

We define the *cross-power spectral density* $S_{xy}(\omega)$ for two random processes $x(t)$ and $y(t)$ as

$$S_{xy}(\omega) = \lim_{T\to\infty} \frac{\overline{X_T^*(\omega)Y_T(\omega)}}{T} \tag{5.112}$$

where $X_T(\omega)$ and $Y_T(\omega)$ are the Fourier transforms of the truncated processes $x(t)\Pi(t/T)$ and $y(t)\Pi(t/T)$, respectively. Proceeding along the lines of the derivation of Eq. (5.94), it can be shown that*

$$R_{xy}(\tau) \leftrightarrow S_{xy}(\omega) \tag{5.113a}$$

It can be seen from Eq. (5.109a) that for real random processes $x(t)$ and $y(t)$,

$$R_{xy}(\tau) = R_{yx}(-\tau)$$

Therefore,

$$S_{xy}(\omega) = S_{yx}(-\omega) \tag{5.113b}$$

5.9 TRANSMISSION OF RANDOM PROCESSES THROUGH LINEAR SYSTEMS

If a random process $x(t)$ is applied at the input of a linear time-invariant system (Fig. 5.36) with transfer function $H(\omega)$, we can determine the autocorrelation function and

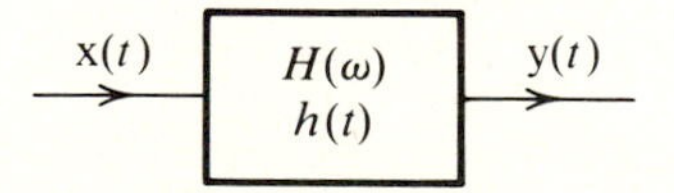

Figure 5.36

the PSD of the output process $y(t)$. It can be shown that

$$R_y(\tau) = h(\tau) * h(-\tau) * R_x(\tau) \tag{5.114}$$

and

$$S_y(\omega) = |H(\omega)|^2 S_x(\omega) \tag{5.115}$$

---

*Equation (5.113a) is valid for complex processes as well.

To prove this, we observe that

$$\mathrm{y}(t) = \int_{-\infty}^{\infty} h(\alpha)\mathrm{x}(t - \alpha)\, d\alpha$$

and

$$\mathrm{y}(t + \tau) = \int_{-\infty}^{\infty} h(\alpha)\mathrm{x}(t + \tau - \alpha)\, d\alpha$$

Hence,*

$$\begin{aligned} R_{\mathrm{y}}(\tau) = \overline{\mathrm{y}(t)\mathrm{y}(t + \tau)} &= \overline{\int_{-\infty}^{\infty} h(\alpha)\mathrm{x}(t - \alpha)\, d\alpha \int_{-\infty}^{\infty} h(\beta)\mathrm{x}(t + \tau - \beta)\, d\beta} \\ &= \int_{-\infty}^{\infty}\int_{-\infty}^{\infty} h(\alpha)h(\beta)\overline{\mathrm{x}(t - \alpha)\mathrm{x}(t + \tau - \beta)}\, d\alpha\, d\beta \\ &= \int_{-\infty}^{\infty}\int_{-\infty}^{\infty} h(\alpha)h(\beta)R_{\mathrm{x}}(\tau + \alpha - \beta)\, d\alpha\, d\beta \end{aligned}$$

This double integral is precisely the double convolution $h(\tau) * h(-\tau) * R_{\mathrm{x}}(\tau)$. Hence Eqs. (5.114) and (5.115) follow.

■ EXAMPLE 5.28

Find the autocorrelation function and PSD of a random pulse train y(t), where a basic pulse $p(t)$ repeats every T_o seconds. The nth pulse is $\mathrm{a}_n p(t)$, where a_n is random (Fig. 5.37).

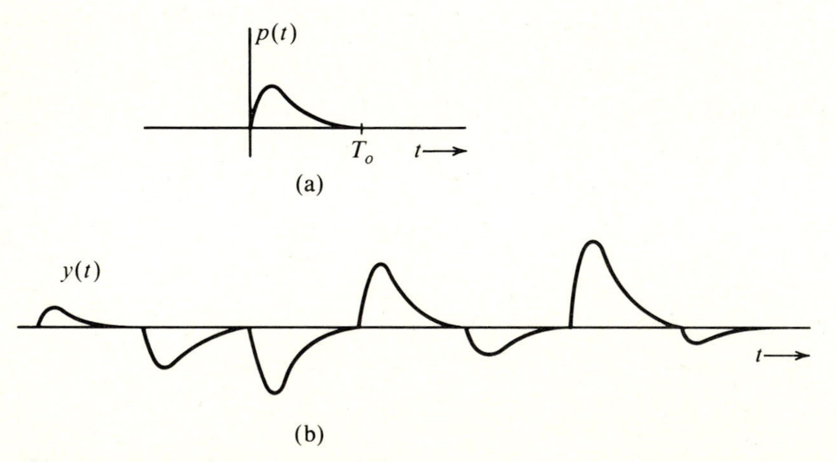

Figure 5.37 Random PAM process.

* In this development, we interchange the operations of averaging and integrating. Because averaging is really an operation of integration, we are really changing the order of integration, and we assume that such a change is permissible. See footnote on page 54.

☐ **Solution:** The pulse train y(t) can be obtained by passing the impulse train x(t) in Fig. 5.30 through a filter with unit impulse response $p(t)$. Hence,

$$S_y(\omega) = |P(\omega)|^2 S_x(\omega)$$

where $S_x(\omega)$ is the PSD of x(t) in Fig. 5.30. From Eq. (5.107),

$$S_x(\omega) = \frac{1}{T_o} \sum_n R_n e^{-jn\omega T_o} \tag{5.116a}$$

From Eq. (3.7) it follows that $S_x(\omega)$ can also be expressed as

$$S_x(\omega) = \frac{1}{T_o}\left[R_0 + 2\sum_{n=1}^{\infty} R_n \cos n\omega T_o\right] \tag{5.116b}$$

Hence*

$$S_y(\omega) = \frac{|P(\omega)|^2}{T_o}\left[R_0 + 2\sum_{n=1}^{\infty} R_n \cos n\omega_o T_o\right] \tag{5.117a}$$

where $R_n = \overline{a_k a_{k+n}}$. **(5.117b)**

We can use these results to derive the PSDs of a variety of digital signals (e.g., on-off, polar, bipolar, duobinary, split-phase, HDBn, etc.). This topic is thoroughly discussed in Sec. 3.2. Here we shall repeat the case of polar signaling. Consider the case of binary random transmission in Example 5.25 where the basic pulse $p(t)$ is a rectangular pulse $\Pi(t/T_o)$. The PSD for this process was already determined in Example 5.25. It will be instructive to rederive it using the results in Eq. (5.117). In Example 5.27, we showed that for this case $R_0 = 1$ and $R_n = 0 (n \geq 1)$. Hence [Eq. 5.117a)],

$$S_y(\omega) = \frac{|P(\omega)|^2}{T_o} = T_o \operatorname{sinc}^2\left(\frac{\omega T_o}{2\pi}\right)$$

This is exactly the result derived earlier in Example 5.25. ■

---

*This result may seem to be at variance with that derived by W. R. Bennett in "Statistics of Regenerative Digital Transmission," *Bell Syst. Tech. J.*, vol. 37, pp. 1501–1543, Nov. 1958. The reason is that Bennett defines y(t) in a way slightly different from ours. We define y(t) directly as

$$y(t) = \sum_k a_k p(t - kT_o - \alpha)$$

where α is an r.v. uniformly distributed in the range $(0, T_o)$, whereas Bennett defines y(t) as

$$y(t) = \sum_k (a_k - m_1) p(t - kT_o)$$

where m_1 is the ensemble average of a_k, that is, $m_1 = \overline{a_k}$.

■ EXAMPLE 5.29 Thermal Noise

Random thermal motion of electrons in a resistor R causes a random voltage across its terminals. This voltage $n(t)$ is known as the *thermal noise*. Its PSD $S_n(\omega)$ is practically flat over a very large band (up to 1000 GHz at room temperature) and is given by[5]

$$S_n(\omega) = 2kTR \tag{5.118}$$

where k is the Boltzmann constant (1.38×10^{-23}) and T is the ambient temperature in Kelvin units. A resistor R at a temperature T°K can be represented by a noiseless resistor R in series with a random white-noise voltage source (thermal noise) of PSD $2kTR$ (Fig. 5.38*a*). Observe that the thermal noise power over a band Δf is $2\Delta f\,(2kTR) = 4kTR\,\Delta f$.

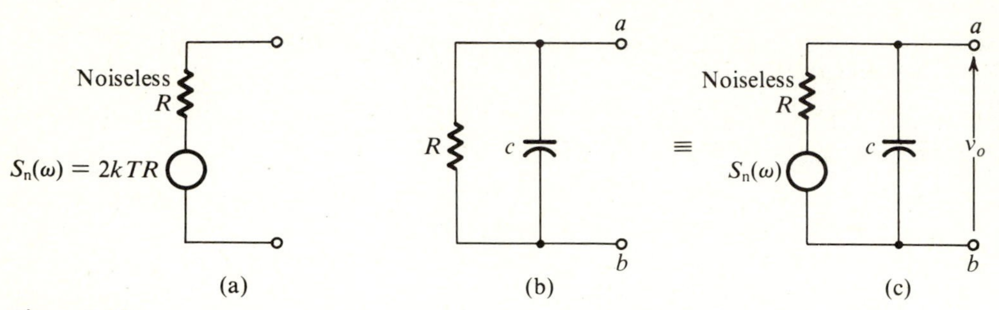

Figure 5.38

Let us calculate the thermal noise voltage (rms value) across the simple RC circuit in Fig. 5.38*b*. The resistor R is replaced by an equivalent noiseless resistor in series with the thermal-noise voltage source. The transfer function $H(\omega)$ relating the voltage v_o at terminals ab to the thermal noise voltage is given by

$$H(\omega) = \frac{1/j\omega C}{R + \dfrac{1}{j\omega C}} = \frac{1}{1 + j\omega RC}$$

If $S_o(\omega)$ is the power spectral density of the voltage v_o, then from Eq. (5.115) we have

$$S_o(\omega) = \left|\frac{1}{1 + j\omega RC}\right|^2 2kTR$$

$$= \frac{2kTR}{1 + \omega^2 R^2 C^2}$$

The mean square value $\overline{v_o^2}$ is given by

$$\overline{v_o^2} = \frac{1}{2\pi}\int_{-\infty}^{\infty} \frac{2kTR}{1 + \omega^2 R^2 C^2}\, d\omega$$
$$= \frac{kT}{C} \tag{5.119}$$

The rms thermal noise voltage across the capacitor is $\sqrt{kT/C}$. ■

The Sum of Random Processes

If two stationary processes (at least in the wide sense) $x(t)$ and $y(t)$ are added to form the process $z(t)$, the statistics of $z(t)$ can be determined in terms of those of $x(t)$ and $y(t)$. Thus, if

$$z(t) = x(t) + y(t) \tag{5.120a}$$

then

$$R_z(\tau) = \overline{z(t)z(t+\tau)} = \overline{[x(t) + y(t)][x(t+\tau) + y(t+\tau)]}$$

$$= R_x(\tau) + R_y(\tau) + R_{xy}(\tau) + R_{yx}(\tau) \tag{5.120b}$$

If $x(t)$ and $y(t)$ are uncorrelated, then from Eq. (5.110a),

$$R_{xy}(\tau) = R_{yx}(\tau) = \overline{x}\,\overline{y}$$

and

$$R_z(\tau) = R_x(\tau) + R_y(\tau) + 2\overline{x}\,\overline{y} \tag{5.121}$$

Most processes of interest in communications problems have zero means. If processes $x(t)$ and $y(t)$ are uncorrelated with either $\overline{x}$ or $\overline{y} = 0$ (that is, if $x(t)$ and $y(t)$ are incoherent), then

$$R_z(\tau) = R_x(\tau) + R_y(\tau) \tag{5.122a}$$

and

$$S_z(\omega) = S_x(\omega) + S_y(\omega) \tag{5.122b}$$

It also follows from Eqs. (5.122a) and (5.97) that

$$\overline{z^2} = \overline{x^2} + \overline{y^2} \tag{5.122c}$$

Hence, the mean square of a sum of incoherent processes is equal to the sum of the mean squares of these processes.

EXAMPLE 5.30

Two independent random voltage processes $x_1(t)$ and $x_2(t)$ are applied to an *RC* network, as shown in Fig. 5.39. It is given that

$$S_{x_1}(\omega) = K \qquad S_{x_2}(\omega) = \frac{2\alpha}{\alpha^2 + \omega^2}$$

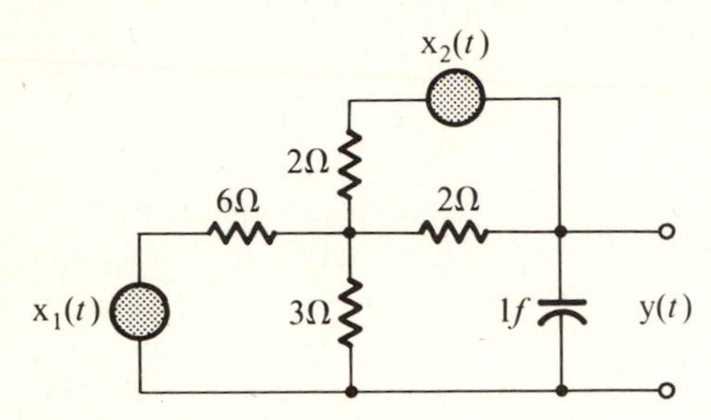

Figure 5.39 Thermal noise in resistors.

Determine the PSD and the power P_y of the output random process $y(t)$. Assume that the resistors in the circuit contribute negligible thermal noise (i.e., assume that they are noiseless).

☐ **Solution:** Because the network is linear, the output voltage $y(t)$ can be expressed as

$$y(t) = y_1(t) + y_2(t)$$

where $y_1(t)$ is the output from input $x_1(t)$ [assuming $x_2(t) = 0$] and $y_2(t)$ is the output from input $x_2(t)$ [assuming $x_1(t) = 0$]. The transfer functions relating $y(t)$ to $x_1(t)$ and $x_2(t)$ are $H_1(\omega)$ and $H_2(\omega)$, respectively, given by

$$H_1(\omega) = \frac{1}{3(3j\omega + 1)} \qquad H_2(\omega) = \frac{1}{2(3j\omega + 1)}$$

Hence,

$$S_{y_1}(\omega) = |H_1(\omega)|^2 S_{x_1}(\omega) = \frac{K}{9(1 + 9\omega^2)}$$

and

$$S_{y_2}(\omega) = |H_2(\omega)|^2 S_{x_2}(\omega) = \frac{\alpha}{2(1 + 9\omega^2)(\alpha^2 + \omega^2)}$$

Because the input processes $x_1(t)$ and $x_2(t)$ are independent, the outputs $y_1(t)$ and $y_2(t)$ generated by them will also be independent. Also, the PSDs of $y_1(t)$ and $y_2(t)$ have no impulses at $\omega = 0$, implying that they have no *dc* components (i.e., $\overline{y_1(t)} =$

$\overline{y_2(t)} = 0$). Hence, $y_1(t)$ and $y_2(t)$ are incoherent, and

$$S_y(\omega) = S_{y_1}(\omega) + S_{y_2}(\omega)$$
$$= \frac{2K(\alpha^2 + \omega^2) + 9\alpha}{18(1 + 9\omega^2)(\alpha^2 + \omega^2)}$$

The power P_y (or the mean square value $\overline{y^2}$) can be determined in two ways. We can find $R_y(\tau)$ by taking the inverse transforms of $S_{y_1}(\omega)$ and $S_{y_2}(\omega)$ as

$$R_y(\tau) = \underbrace{\frac{K}{54}e^{-|\tau|/3}}_{R_{y_1}(\tau)} + \underbrace{\frac{3\alpha - e^{-\alpha|\tau|}}{4(9\alpha^2 - 1)}}_{R_{y_2}(\tau)}$$

and

$$P_y = \overline{y^2} = R_y(0) = \frac{K}{54} + \frac{3\alpha - 1}{4(9\alpha^2 - 1)}$$

Alternatively, we can determine $\overline{y^2}$ by integrating $S_y(\omega)$ with respect to f [see Eq. (5.97)]. ■

5.10 BANDPASS RANDOM PROCESSES

If the PSD of a random process is confined to a certain passband (Fig. 5.40), the

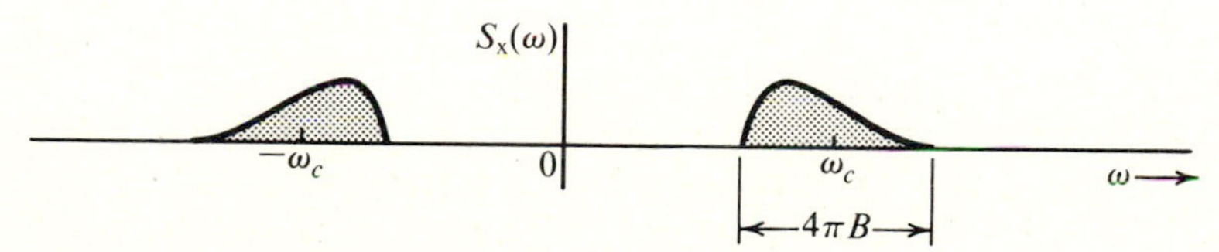

Figure 5.40 PSD of a bandpass random process.

process is a *bandpass* random process. Just as a bandpass signal can be represented in terms of quadrature components [see Eq. (2.61a)], we can express a bandpass random process $x(t)$ in terms of quadrature components as follows:

$$x(t) = x_c(t) \cos \omega_c t + x_s(t) \sin \omega_c t \qquad \textbf{(5.123)}$$

This can be proved by considering the system in Fig. 5.41*a*, where $H_o(\omega)$ is an ideal lowpass filter (Fig. 5.41*b*) with unit impulse response $h_o(t)$. First we show that the system in Fig. 5.41*a* is an ideal bandpass filter with the transfer function $H(\omega)$ shown in Fig. 5.41*c*. This can be conveniently done by computing the response $h(t)$ to the unit impulse input $\delta(t)$. Because the system contains time varying multipliers, however, we must also test whether it is a time varying or a time-invariant system. It is therefore appropriate to consider the system response to an input $\delta(t - \alpha)$. This is an impulse at $t = \alpha$. Signals at various points are as follows.

Signal at a_1	$2\cos(\omega_c\alpha + \theta)\,\delta(t - \alpha)$
a_2	$2\sin(\omega_c\alpha + \theta)\,\delta(t - \alpha)$
b_1	$2\cos(\omega_c\alpha + \theta)h_o(t - \alpha)$
b_2	$2\sin(\omega_c\alpha + \theta)h_o(t - \alpha)$
c_1	$2\cos(\omega_c\alpha + \theta)\cos(\omega_c t + \theta)h_o(t - \alpha)$
c_2	$2\sin(\omega_c\alpha + \theta)\sin(\omega_c t + \theta)h_o(t - \alpha)$
d	$2h_o(t - \alpha)[\cos(\omega_c\alpha + \theta)\cos(\omega_c t + \theta) + \sin(\omega_c\alpha + \theta)\sin(\omega_c t + \theta)]$
	$= 2h_o(t - \alpha)\cos[\omega_c(t - \alpha)]$

Thus, the system response to the input $\delta(t - \alpha)$ is $2h_o(t - \alpha)\cos\omega_c(t - \alpha)$. Clearly,

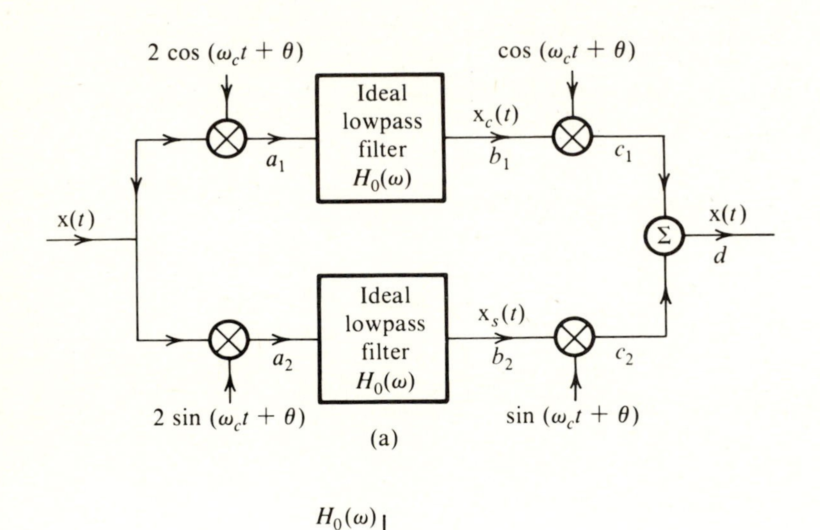

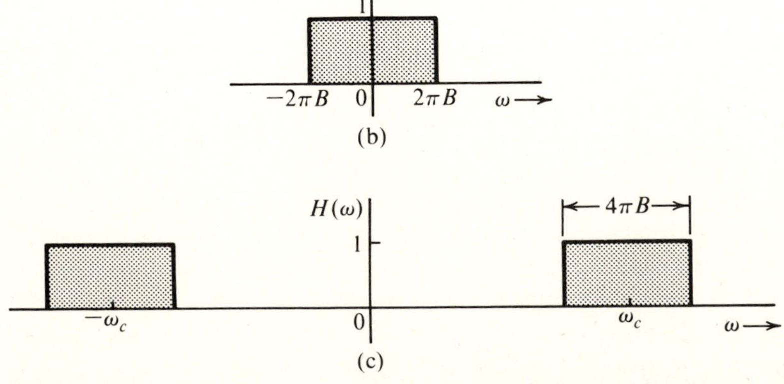

Figure 5.41 (a) An equivalent circuit of an ideal bandpass filter. (b) Ideal lowpass filter frequency response. (c) Ideal bandpass filter frequency response.

the system is linear time-invariant, with impulse response

$$h(t) = 2h_o(t) \cos \omega_c t$$

and

$$H(\omega) = H_o(\omega + \omega_c) + H_o(\omega - \omega_c)$$

The transfer function $H(\omega)$ [Fig. 5.41*c*] represents an ideal bandpass filter.

If we apply the bandpass process $\mathrm{x}(t)$ (Fig. 5.40) to the input of this system, the output will be $\mathrm{x}(t)$. If the processes at points b_1 and b_2 (lowpass filter outputs) are denoted by $\mathrm{x}_c(t)$ and $\mathrm{x}_s(t)$, respectively, then the output $\mathrm{x}(t)$ can be written as

$$\mathrm{x}(t) = \mathrm{x}_c(t) \cos (\omega_c t + \theta) + \mathrm{x}_s(t) \sin (\omega_c t + \theta) \tag{5.124}$$

where $\mathrm{x}_c(t)$ and $\mathrm{x}_s(t)$ are lowpass random processes bandlimited to B (because they are the outputs of lowpass filters of bandwidth B). Because Eq. (5.124) is valid for any value of θ, by substituting $\theta = 0$, we get the desired representation in Eq. (5.123).

In order to characterize $\mathrm{x}_c(t)$ and $\mathrm{x}_s(t)$ in Eq. (5.123), consider once again Fig. 5.41*a* with the input $\mathrm{x}(t)$. Let θ be an r.v. uniformly distributed over the range $(0, 2\pi)$, that is, where each sample function θ will be different. In this case, $\mathrm{x}(t)$ is represented as in Eq. (5.124). We observe that $\mathrm{x}_c(t)$ is obtained by multiplying $\mathrm{x}(t)$ by $2 \cos (\omega_c t + \theta)$, and then passing the result through a lowpass filter. The PSD of $2\,\mathrm{x}(t) \cos (\omega_c t + \theta)$ is [see Eq. (5.100b)]

$$4(\tfrac{1}{4})[S_{\mathrm{x}}(\omega + \omega_c) + S_{\mathrm{x}}(\omega - \omega_c)]$$

This PSD is $S_{\mathrm{x}}(\omega)$ shifted up and down by ω_c, as shown in Fig. 5.42*a*. When this is passed through a lowpass filter, the resulting PSD of $\mathrm{x}_c(t)$ is as shown in Fig. 5.42*b*. It is clear that

$$S_{\mathrm{x}_c}(\omega) = \begin{cases} S_{\mathrm{x}}(\omega + \omega_c) + S_{\mathrm{x}}(\omega - \omega_c) & |\omega| < 2\pi B \\ 0 & |\omega| > 2\pi B \end{cases} \tag{5.125a}$$

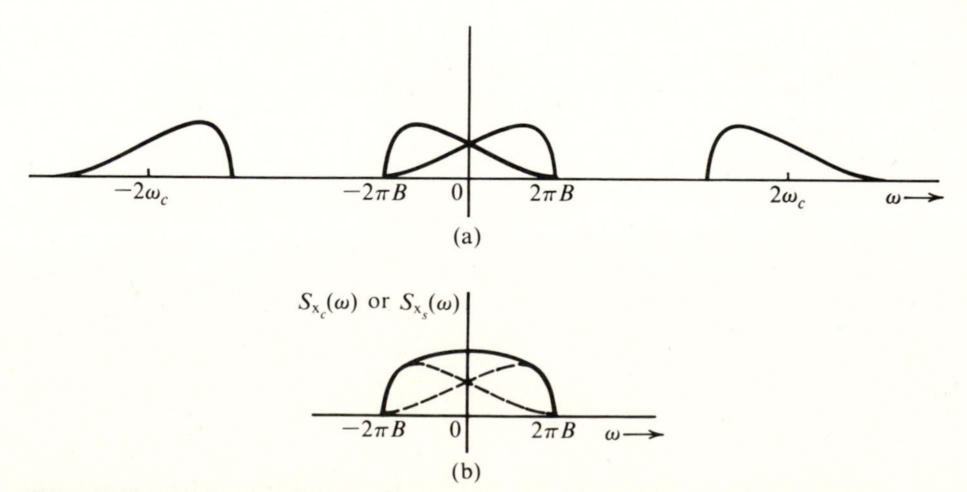

Figure 5.42 Derivation of PSDs of quadrature components of a bandpass random process.

We can obtain $S_{x_s}(\omega)$ in the same way. As far as the PSD is concerned, multiplication by $\cos(\omega_c t + \theta)$ or $\sin(\omega_c t + \theta)$ makes no difference (see footnote on page 416), and we get

$$S_{x_c}(\omega) = S_{x_s}(\omega) = \begin{cases} S_x(\omega + \omega_c) + S_x(\omega - \omega_c) & |\omega| < 2\pi B \\ 0 & |\omega| > 2\pi B \end{cases} \tag{5.125b}$$

From Figs. 5.40 and 5.42*b*, we make the interesting observation that the areas under the PSDs $S_x(\omega)$, $S_{x_c}(\omega)$, and $S_{x_s}(\omega)$ are equal. Hence, it follows that

$$\overline{x_c^2(t)} = \overline{x_s^2(t)} = \overline{x^2(t)} \tag{5.125c}$$

Thus, the mean square values (or powers) of $x_c(t)$ and $x_s(t)$ are identical to that of $x(t)$.

These results are derived by assuming Θ to be an r.v. For the representation in Eq. (5.123), $\Theta = 0$, and Eq. (5.125b and c) may not be true. Fortunately, Eq. (5.125b and c) hold even for the case of $\Theta = 0$. The proof is rather long and cumbersome and will not be given here.[3,5,6]

It can also be shown that[3,5,6]

$$\overline{x_c(t)x_s(t)} = R_{x_c x_s}(0) = 0 \tag{5.126}$$

That is, the amplitudes x_c and x_s at any given instant are uncorrelated. Moreover, if $S_x(\omega)$ is symmetrical about ω_c (as well as $-\omega_c$), then

$$R_{x_c x_s}(\tau) = 0 \tag{5.127}$$

■ EXAMPLE 5.31

The PSD of a bandpass white noise $n(t)$ is $\mathcal{N}/2$ (Fig. 5.43*a*). Represent this process

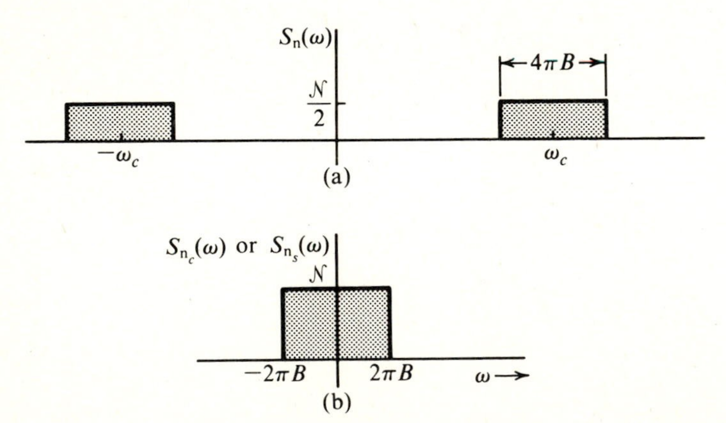

Figure 5.43 (a) PSD of a bandpass white noise process. (b) PSD of its quadrature components.

in terms of quadrature components. Derive $S_{n_c}(\omega)$ and $S_{n_s}(\omega)$, and verify that $\overline{n_c^2} = \overline{n_s^2} = \overline{n^2}$.

□ **Solution:**

$$\mathrm{n}(t) = \mathrm{n}_c(t)\cos\omega_c t + \mathrm{n}_s(t)\sin\omega_c t \tag{5.128}$$

where

$$S_{\mathrm{n}_c}(\omega) = S_{\mathrm{n}_s}(\omega) = \begin{cases} S_{\mathrm{n}}(\omega+\omega_c) + S_{\mathrm{n}}(\omega-\omega_c) & |\omega| < 2\pi B \\ 0 & |\omega| > 2\pi B \end{cases}$$

It follows from this equation and from Fig. 5.43 that

$$S_{\mathrm{n}_c}(\omega) = S_{\mathrm{n}_s}(\omega) = \begin{cases} \mathcal{N} & |\omega| < 2\pi B \\ 0 & |\omega| > 2\pi B \end{cases} \tag{5.129}$$

Also,

$$\overline{\mathrm{n}^2} = 2\int_{f_c-B}^{f_c+B} \frac{\mathcal{N}}{2}\,df = 2\mathcal{N}B \tag{5.130a}$$

From Fig. 5.43*b* it follows that

$$\overline{\mathrm{n}_c^2} = \overline{\mathrm{n}_s^2} = 2\int_0^B \mathcal{N}\,df = 2\mathcal{N}B \tag{5.130b}$$

Hence,

$$\overline{\mathrm{n}_c^2} = \overline{\mathrm{n}_s^2} = \overline{\mathrm{n}^2} = 2\mathcal{N}B \tag{5.130c}$$

■

Nonuniqueness of the Quadrature Representation

No unique center frequency exists for a bandpass signal. For the spectrum in Fig. 5.44*a*, for example, we may consider the spectrum to have a bandwidth $2B$ centered

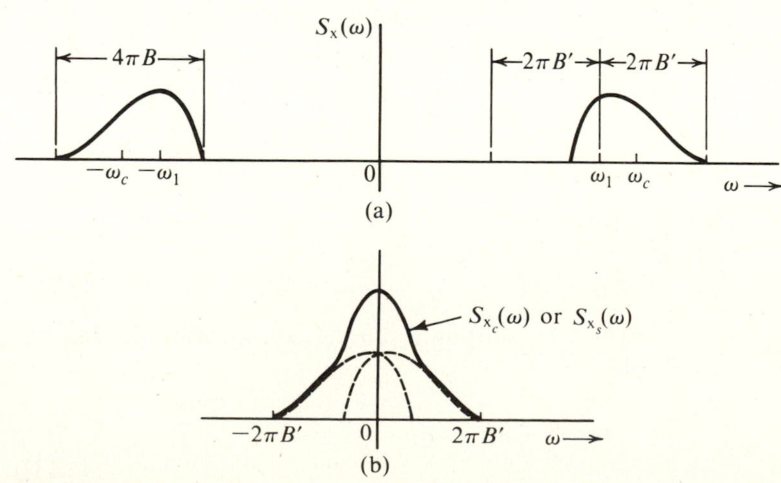

Figure 5.44

at ω_c. The same spectrum can be considered to have a bandwidth $2B'$ centered at ω_1, as also shown in Fig. 5.44*a*. The quadrature representation [Eq. (5.123)] is also possible for center frequency ω_1:

$$\mathrm{x}(t) = \mathrm{x}_{c_1}(t) \cos \omega_1 t + \mathrm{x}_{s_1}(t) \sin \omega_1 t$$

where

$$S_{\mathrm{x}_{c_1}}(\omega) = S_{\mathrm{x}_{s_1}}(\omega) = \begin{cases} S_{\mathrm{x}}(\omega + \omega_1) + S_{\mathrm{x}}(\omega - \omega_1) & |\omega| < 2\pi B' \\ 0 & |\omega| > 2\pi B' \end{cases} \tag{5.131}$$

This is shown in Fig. 5.44*b*. Thus, the quadrature representation of a bandpass process is not unique. An infinite number of possible choices exist for the center frequency, and corresponding to each center frequency is a distinct quadrature representation.

■ EXAMPLE 5.32

A bandpass white noise PSD of an SSB channel (lower side band) is shown in Fig. 5.45*a*. Represent this signal in terms of quadrature components with the carrier frequency ω_c.

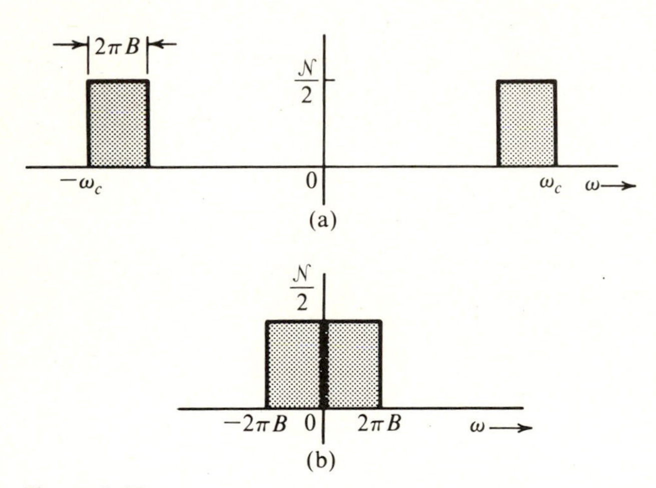

Figure 5.45

□ **Solution:** The true center frequency of this PSD is not ω_c, but we can still use ω_c as a center frequency, as discussed earlier.

$$\mathrm{n}(t) = \mathrm{n}_c(t) \cos \omega_c t + \mathrm{n}_s(t) \sin \omega_c t \tag{5.132}$$

The PSD $S_{\mathrm{n}_c}(\omega)$ or $S_{\mathrm{n}_s}(\omega)$ obtained by shifting $S_{\mathrm{n}}(\omega)$ up and down by ω_c [see Eq. 5.131)] is shown in Fig. 5.45*b*.

$$S_{\mathrm{n}_c}(\omega) = S_{\mathrm{n}_s}(\omega) = \begin{cases} \dfrac{\mathcal{N}}{2} & |\omega| < 2\pi B \\ 0 & |\omega| > 2\pi B \end{cases} \tag{5.133}$$

From Fig. 5.45*a* it follows that

$$\overline{n^2} = \mathcal{N}B \tag{5.134a}$$

Similarly, from Fig. 5.45*b* we have

$$\overline{n_c^2} = \overline{n_s^2} = \mathcal{N}B \tag{5.134b}$$

Hence,

$$\overline{n_c^2} = \overline{n_s^2} = \overline{n^2} = \mathcal{N}B \tag{5.134c}$$

■

Bandpass White Gaussian Random Process

Thus far we have avoided defining a gaussian random process. The gaussian random process is perhaps the single most important random process in the area of communication. It requires a rather careful and unhurried discussion. Fortunately, we do not need to know much about the gaussian process at this point; therefore, its detailed discussion is postponed until Chapter 7 in order to avoid unnecessary digression. All we need to know here is that a random variable $x(t)$ formed by sample function amplitudes at instant t of a gaussian process is gaussian, with a PDF of the form in Eq. (5.38).

A gaussian random process with a uniform PSD is called a white gaussian random process. A bandpass white gaussian process $n(t)$ with PSD $\mathcal{N}/2$ centered at ω_c and with a bandwidth $2B$ (Fig. 5.43*a*) can be expressed in terms of quadrature components as

$$n(t) = n_c(t) \cos \omega_c t + n_s(t) \sin \omega_c t \tag{5.135}$$

where, from Eq. (5.129), we have

$$S_{n_c}(\omega) = S_{n_s}(\omega) = \begin{cases} \mathcal{N} & |\omega| < 2\pi B \\ 0 & |\omega| > 2\pi B \end{cases}$$

Also, from Eq. (5.130),

$$\overline{n_c^2} = \overline{n_s^2} = \overline{n^2} = 2\mathcal{N}B \tag{5.136}$$

The bandpass signal can also be expressed in polar form [see Eq. (2.61b)]:

$$n(t) = E(t) \cos (\omega_c t + \Theta) \tag{5.137a}$$

where

$$E(t) = \sqrt{n_c^2(t) + n_s^2(t)} \tag{5.137b}$$

$$\Theta(t) = -\tan^{-1} \frac{n_s(t)}{n_c(t)} \tag{5.137c}$$

The r.v.'s $n_c(t)$ and $n_s(t)$ are uncorrelated [see Eq. (5.126)] gaussian r.v.'s with zero means and variance $2\mathcal{N}B$ [Eq. (5.136)]. Hence, their PDFs are identical:

$$p_{n_c}(\alpha) = p_{n_s}(\alpha) = \frac{1}{\sigma\sqrt{2\pi}} e^{-\alpha^2/2\sigma^2} \tag{5.138a}$$

where

$$\sigma^2 = 2\mathcal{N}B \tag{5.138b}$$

It will also be shown in Chapter 7 that if two gaussian r.v.'s are uncorrelated, they are independent. In such a case, as shown in Example 5.14, E(t) has a Rayleigh density

$$p_{\mathrm{E}}(E) = \frac{E}{\sigma^2} e^{-E^2/2\sigma^2} \qquad \sigma^2 = 2\mathcal{N}B \tag{5.139}$$

and Θ in Eq. (5.137a) is uniformly distributed over $(0, 2\pi)$.

The Sinusoidal Signal in Noise

Another case of interest is a sinusoid plus a narrowband gaussian noise. If $A \cos(\omega_c t + \varphi)$ is a sinusoid mixed with n(t), a gaussian bandpass noise centered at ω_c, then the sum y(t) is given by

$$\mathrm{y}(t) = A\cos(\omega_c t + \varphi) + \mathrm{n}(t)$$

Using Eq. (5.124) to represent the bandpass noise, we have

$$\mathrm{y}(t) = [A + \mathrm{n}_c(t)]\cos(\omega_c t + \varphi) + \mathrm{n}_s(t)\sin(\omega_c t + \varphi) \tag{5.140a}$$

$$= \mathrm{E}(t)\cos[\omega_c t + \Theta(t) + \varphi] \tag{5.140b}$$

where E(t) is the envelope (E(t) > 0) and $\Theta(t)$ is the phase of y(t).

$$\mathrm{E}(t) = \sqrt{[A + \mathrm{n}_c(t)]^2 + \mathrm{n}_s^2(t)} \tag{5.141a}$$

$$\Theta(t) = -\tan^{-1}\frac{\mathrm{n}_s(t)}{A + \mathrm{n}_c(t)} \tag{5.141b}$$

Both $\mathrm{n}_c(t)$ and $\mathrm{n}_s(t)$ are gaussian, with variance σ^2. For white gaussian noise, $\sigma^2 = 2\mathcal{N}B$ [Eq. (5.138b)]. Arguing in a manner analogous to that used in deriving Eq. (5.53), and observing that

$$\begin{aligned}\mathrm{n}_c^2 + \mathrm{n}_s^2 &= \mathrm{E}^2 - A^2 - 2A\mathrm{n}_c \\ &= \mathrm{E}^2 - 2A(A + \mathrm{n}_c) + A^2 \\ &= \mathrm{E}^2 - 2A\mathrm{E}\cos\Theta(t) + A^2\end{aligned}$$

we have

$$p_{\mathrm{E}\Theta}(E, \theta) = \frac{E}{2\pi\sigma^2} e^{-(E^2 - 2AE\cos\theta + A^2)/2\sigma^2} \tag{5.142}$$

where σ^2 is the variance of n_c (or n_s) and is equal to $2\mathcal{N}B$ for white noise. From Eq. (5.142) we have

$$\begin{aligned}p_{\mathrm{E}}(E) &= \int_{-\pi}^{\pi} p_{\mathrm{E}\Theta}(E, \theta)\, d\theta \\ &= \frac{E}{\sigma^2} e^{-(E^2 + A^2)/2\sigma^2}\left[\frac{1}{2\pi}\int_{-\pi}^{\pi} e^{(AE/\sigma^2)\cos\theta}\, d\theta\right]\end{aligned} \tag{5.143}$$

The bracketed term on the right-hand side of Eq. (5.143) is similar in form to that encountered in Eq. (4.65) and is given by $I_o(AE/\sigma^2)$, where I_o is the *modified zero-order Bessel function* of the first kind. Thus,

$$p_E(E) = \frac{E}{\sigma^2} e^{-(E^2+A^2)/2\sigma^2} I_o\left(\frac{AE}{\sigma^2}\right) \tag{5.144a}$$

This is known as the *Rice density* (or *Rician density*). For a large sinusoidal signal ($A \gg \sigma$), it can be shown that[7]

$$I_o\left(\frac{AE}{\sigma^2}\right) \simeq \sqrt{\frac{\sigma^2}{2\pi AE}} e^{AE/\sigma^2}$$

and

$$p_E(E) \simeq \sqrt{\frac{E}{2\pi A\sigma^2}} e^{-(E-A)^2/2\sigma^2} \tag{5.144b}$$

Because $A \gg \sigma$, $E \simeq A$, and $p_E(E)$ in Eq. (5.144b) is very nearly a gaussian density with mean A and variance σ.

$$p_E(E) \simeq \frac{1}{\sigma\sqrt{2\pi}} e^{-(E-A)^2/2\sigma^2} \tag{5.144c}$$

Figure 5.46 shows the PDF of the normalized r.v. E/σ. Note that for $A/\sigma = 0$, we obtain the Rayleigh density.

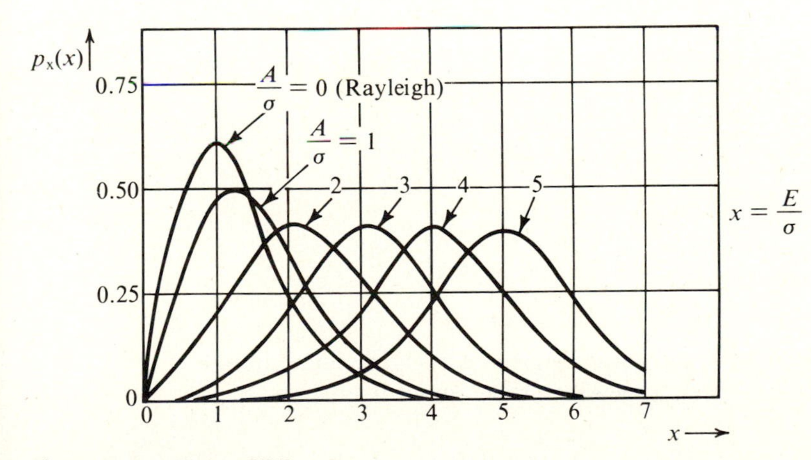

Figure 5.46 Rician PDF.

From the joint PDF $p_{E\Theta}(E, \theta)$, we can also obtain $p_\Theta(\theta)$, the PDF of the phase Θ, by integrating the joint PDF with respect to θ.

$$p_\Theta(\theta) = \int_0^\infty p_{E\Theta}(E, \theta)\, dE$$

Although the integration is straightforward, there are a number of involved steps, and for this reason it will not be repeated here. The final result is

$$p_\Theta(\theta) = \frac{1}{2\pi} e^{-A^2/2\sigma^2}\left[1 + \frac{A}{\sigma}\sqrt{2\pi}\cos\theta\, e^{A^2\cos^2\theta/2\sigma^2}\left[1 - Q\left(\frac{A\cos\theta}{\sigma}\right)\right]\right] \tag{5.144d}$$

5.11 OPTIMUM FILTERING: THE WIENER-HOPF FILTER

When a desired signal is mixed with noise, the SNR can be improved by passing it through a filter that suppresses frequency components where the signal is weak but the noise is strong. The SNR improvement in this case can be explained qualitatively by considering a case of white noise mixed with a signal m(t) whose PSD decreases at high frequencies. If the filter attenuates higher frequencies more, the signal will be reduced—in fact, distorted. The distortion component $m_\varepsilon(t)$ may be considered as an added noise. Thus, attenuation of higher frequencies will cause additional noise (from signal distortion), but, in compensation, it will reduce the channel noise, which is strong at high frequencies. Because at higher frequencies the signal has a small power content, the distortion component will be small compared to the reduction in channel noise, and the total noise may be smaller than before.

Let $H_{op}(\omega)$ be the optimum filter (Fig. 5.47a). This filter, not being ideal, will

m(t) + n(t) → $H_{op}(\omega)$ → m(t) + $m_\epsilon(t)$ + $n_o(t)$

(a)

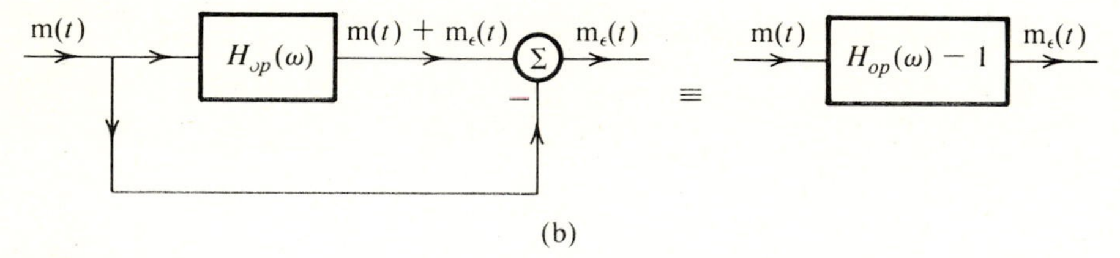

(b)

Figure 5.47

cause signal distortion. The distortion signal $m_\varepsilon(t)$ can be found from Fig. 5.47b. The distortion signal power, N_D, appearing at the output is given by

$$N_D = \frac{1}{2\pi}\int_{-\infty}^{\infty} S_m(\omega)\,|H_{op}(\omega) - 1|^2\, d\omega$$

where $S_m(\omega)$ is the signal PSD at the input of the receiving filter. The channel noise power, N_c, appearing at the filter output is given by

$$N_c = \frac{1}{2\pi}\int_{-\infty}^{\infty} S_n(\omega)\,|H_{op}(\omega)|^2\, d\omega$$

where $S_n(\omega)$ is the noise PSD appearing at the input of the receiving filter. The distortion component acts as a noise. Because the signal and the channel noise are incoherent, the total noise, N_o, at the receiving filter output is the sum of the channel noise N_c and the distortion noise N_D.

$$N_o = N_c + N_D$$

$$= \frac{1}{2\pi}\int_{-\infty}^{\infty} [|H_{op}(\omega)|^2 S_n(\omega) + |H_{op}(\omega) - 1|^2 S_m(\omega)]\, d\omega \tag{5.145a}$$

Using the fact that $|A + B|^2 = (A + B)(A^* + B^*)$, and noting that both $S_m(\omega)$ and $S_n(\omega)$ are real, Eq. (5.145a) can be rearranged as

$$N_o = \frac{1}{2\pi}\int_{-\infty}^{\infty} \left[\left| H_{op}(\omega) - \frac{S_m(\omega)}{S_r(\omega)} \right|^2 S_r(\omega) + \frac{S_m(\omega)S_n(\omega)}{S_r(\omega)} \right] d\omega \tag{5.145b}$$

where $S_r(\omega) = S_m(\omega) + S_n(\omega)$. The integrand on the right-hand side of Eq. (5.145b) is nonnegative. Moreover, it is a sum of two nonnegative terms. Hence, to minimize N_o, we must minimize each term. Because the second term $S_m(\omega)S_n(\omega)/S_r(\omega)$ is independent of $H_{op}(\omega)$, only the first term can be minimized. From Eq. (5.145b) it is obvious that this term is minimum when

$$H_{op}(\omega) = \frac{S_m(\omega)}{S_r(\omega)}$$

$$= \frac{S_m(\omega)}{S_m(\omega) + S_n(\omega)} \tag{5.146a}$$

For the optimum choice, the output noise power N_o is given by

$$N_o = \frac{1}{2\pi}\int_{-\infty}^{\infty} \frac{S_m(\omega)S_n(\omega)}{S_m(\omega) + S_n(\omega)}\, d\omega \tag{5.146b}$$

The optimum filter is known as the *Wiener-Hopf filter* in the literature.

Comments on the Optimum Filter

If the SNR at the filter input is reasonably large—e.g., $S_m(\omega) > 100 S_n(\omega)$ (SNR of 20 dB)—the optimum filter [Eq. (5.146a)] in this case is practically an ideal filter, and N_o [Eq. (5.146b)] is given by

$$N_o \simeq \frac{1}{2\pi}\int_{-\infty}^{\infty} S_n(\omega)\, d\omega$$

Hence for a large input SNR, optimization yields insignificant improvement. The Wiener-Hopf filter is therefore practical only when the input SNR is small (large-noise case).

Another issue is the realizability of the optimum filter in Eq. (5.146a). Because $S_m(\omega)$ and $S_n(\omega)$ are both even functions of ω, the optimum filter $H_{op}(\omega)$ is an even function of ω. Hence, the unit impulse response $h_{op}(t)$ is an even function of t (see Prob. 2.6). This makes $h_{op}(t)$ noncausal and the filter unrealizable. As noted earlier, such a filter can be realized approximately if we are willing to tolerate some delay in the output. If delay cannot be tolerated, the derivation of $H_{op}(\omega)$ must be repeated with a realizability constraint. Note that the realizable optimum filter can never be superior to the unrealizable optimum filter [Eq. (5.146a)]. Thus, the filter in Eq. (5.146a) gives the upper bound on performance (output SNR). Discussion of realizable optimum filters can be readily found in the literature.[5,6]

EXAMPLE 5.33

A random process m(t) [the signal] is mixed with a white channel noise n(t). Given

$$S_m(\omega) = \frac{2\alpha}{\alpha^2 + \omega^2} \quad \text{and} \quad S_n(\omega) = \frac{\mathcal{N}}{2}$$

find the Wiener-Hopf filter to maximize the SNR. Find the resulting output noise power N_o.

☐ **Solution:** From Eq. (5.146a),

$$H_{op}(\omega) = \frac{4\alpha}{4\alpha + \mathcal{N}(\alpha^2 + \omega^2)}$$

$$= \frac{4\alpha}{\mathcal{N}(\beta^2 + \omega^2)} \qquad \beta^2 = \frac{4\alpha}{\mathcal{N}} + \alpha^2 \tag{5.147a}$$

Hence,

$$h_{op}(t) = \frac{2\alpha}{\mathcal{N}\beta} e^{-\beta|t|} \tag{5.147b}$$

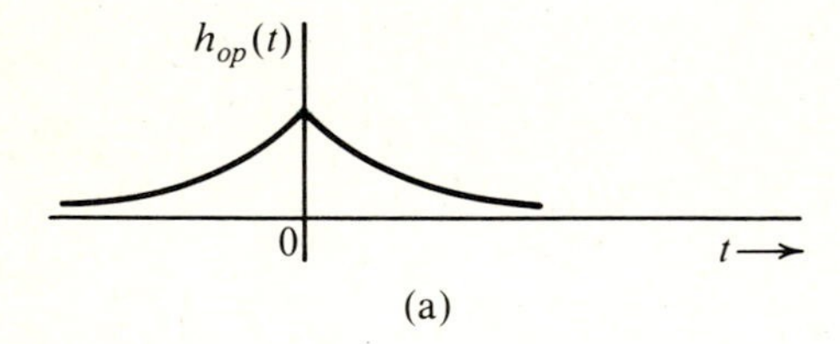

(a)

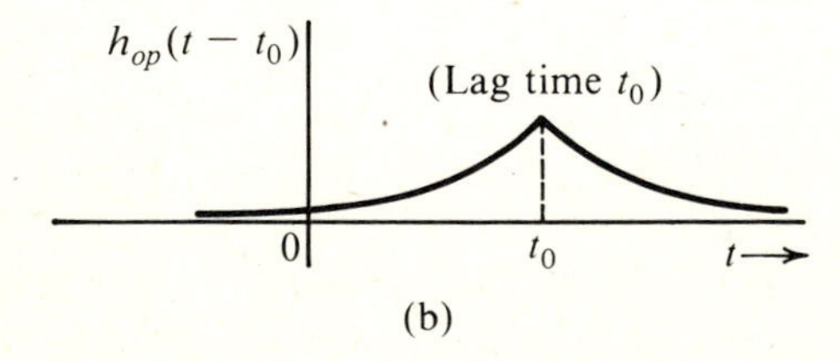

(b)

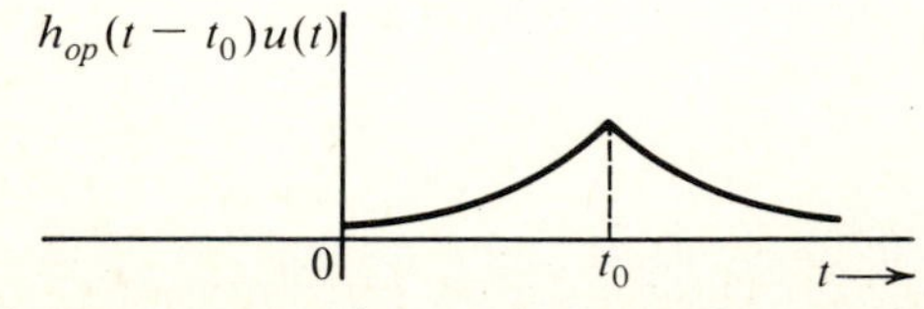

Figure 5.48 Close realization of a nonrealizable filter using delay.

Figure 5.48*a* shows $h_{op}(t)$. It is evident that this is an unrealizable filter. However, a delayed version (Fig. 5.48*b*) of this filter—that is, $h_{op}(t - t_o)$, is closely realizable if we make $t_o \geq 3/\beta$ and eliminate the tail for $t < 0$ (Fig. 5.48*c*).

The output noise power N_o is [Eq. (5.146b)]

$$N_o = 2\int_0^\infty \frac{2\alpha}{\beta^2 + \omega^2}\, df$$

$$= 2\int_0^\infty \frac{2\alpha}{\beta^2 + 4\pi^2 f^2}\, df = \frac{\alpha}{\beta}$$

$$= \frac{\alpha}{\sqrt{\alpha^2 + \dfrac{4\alpha}{\mathcal{N}}}} \tag{5.148}$$

■

REFERENCES

1. M. Abromowitz and I. A. Stegum, eds., *Handbook of Mathematical Functions,* National Bureau of Standards, Washington, D.C., 1964, Sec. 26.
2. The Chemical Rubber Co., *CRC Standard Mathematical Tables,* 26th ed., 1980.
3. J. M. Wozencraft and I. M. Jacobs, *Principles of Communication Engineering,* Wiley, New York, 1965, p. 83.
4. J. V. Uspenski, *Introduction to Mathematical Probability,* McGraw-Hill, New York, 1937.
5. B. P. Lathi, *An Introduction to Random Signals and Communication Theory,* International Textbook Co., Scranton, Pa., 1968.
6. A. Papoulis, *Probability, Random Variables, and Stochastic Processes,* McGraw-Hill, New York, 1965, Chap. 10.
7. S. O. Rice, "Mathematical Analysis of Random Noise," *Bell Syst. Tech. J.,* vol. 23, pp. 282–332, July 1944, and vol. 24, pp. 46–156, Jan. 1945.
8. J. J. Freeman, *Principles of Noise,* Wiley, New York, 1958.

PROBLEMS

5.1. A card is drawn randomly from a regular deck of cards. Assign a probability to the event that the card drawn is (a) a red card, (b) a black queen, (c) a picture card (count an ace as a picture card), or (d) a number card with the number 7.

5.2. Three regular dice are thrown. Assign probabilities to the following events: the sum of the points appearing on the three dice is (a) 3, and (b) 9.

5.3. The probability that the number i appears on a throw of a certain loaded die is ki ($i = 1, 2, \ldots, 6$). Assign probabilities to all six outcomes.

5.4. An urn contains three red balls, marked r_1, r_2, and r_3, and two white balls, marked w_1 and w_2. Two balls are drawn without replacement.

(a) How many outcomes are possible, that is, how many points are in the sample space? List all the outcomes and assign probabilities to each of them.

(b) Express the following events as unions of the outcomes in part (a): (i) one ball drawn is red and the other is white, (ii) both balls are white, (iii) both balls are red, and (iv) both balls are of the same color. Assign probabilities to each of these events.

5.5. Find the probabilities in Prob. 5.4(b) using Eq. (5.13).

5.6. In Prob. 5.4, determine the probability that
(a) the second draw is a red ball given that the first draw is a white ball.
(b) the second draw is a red ball given that the first draw is a red ball.

5.7. In Example 5.5, page 364, determine:
(a) $P(B)$, the probability of drawing a red ace in the second draw, and
(b) $P(A|B)$, the probability that the first draw was a red ace given that the second draw is a red ace.
Hint: Event B can occur in two ways: the first draw is a red ace and the second draw is a red ace, or the first draw is not a red ace and the second draw is a red ace. This is $AB \cup A^C B$ (see Fig. 5.2).

5.8. Two dice are tossed. One die is regular and the other is biased with probabilities:

$$P(1) = P(6) = \tfrac{1}{3},\ P(2) = P(4) = \tfrac{1}{6},\ \text{and}\ P(3) = P(5) = 0$$

Determine the probabilities of obtaining a sum (a) 4, and (b) 5.

5.9. A regular coin is tossed 10 times in succession. Assign probabilities to the following events: the outcome (a) has exactly two tails and eight heads, and (b) has at least four heads.

5.10. What is the probability that a bridge hand will contain
(a) all 13 cards in one suit?
(b) all card values (for example, $A, K, Q, \ldots, 3, 2$) regardless of the suit?

5.11. An urn contains one red ball and nine white balls. A person draws a ball randomly, and if he draws the red ball he wins.
(a) What is the probability of winning in one trial (i.e., in one draw)?
(b) What is the probability of winning in five trials?

5.12. A system consists of 10 subsystems $s_1, s_2, \ldots, s_{10}$ in series (Fig. P5.12). If any one of the subsystems fails, the entire system fails. The probability of failure of any one of the subsystems is 0.01.
(a) What is the probability of the failure of the system?
Hint: Consider the probability of the event that none of the subsystems fails.
(b) The reliability of a system is the probability of not failing. If the system reliability is required to be 0.99, what must be the failure probability of each subsystem?

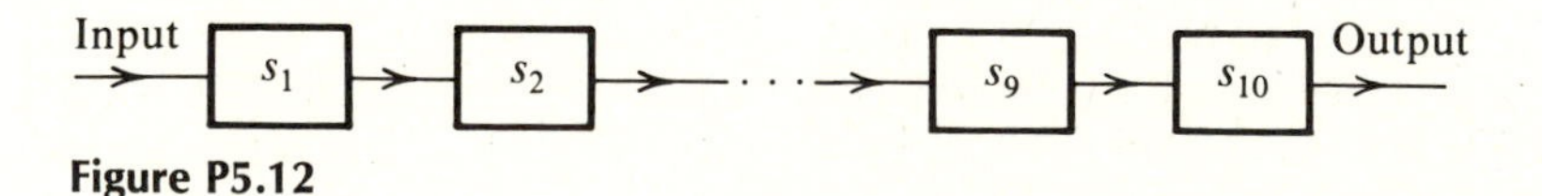

Figure P5.12

5.13. System reliability is improved by using redundant systems. Consider the system in Prob. 5.12 (Fig. P5.12). The reliability of this system can be improved if we use two such systems in parallel (Fig. P5.13). Thus if one fails, the other will still function.

(a) Using the data in Prob. 5.12, determine the reliability of the system in Fig. P5.13.
(b) If the reliability of the system in Fig. P5.13 is required to be 0.999, what must be the failure probability of each of the subsystems s_i?

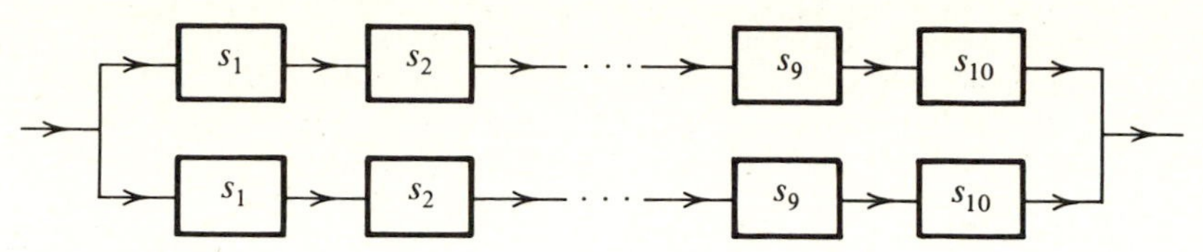

Figure P5.13

5.14. Compare the reliability of the two systems shown in Fig. P5.14 given that the probability of failure of subsystems s_1 and s_2 is p each.

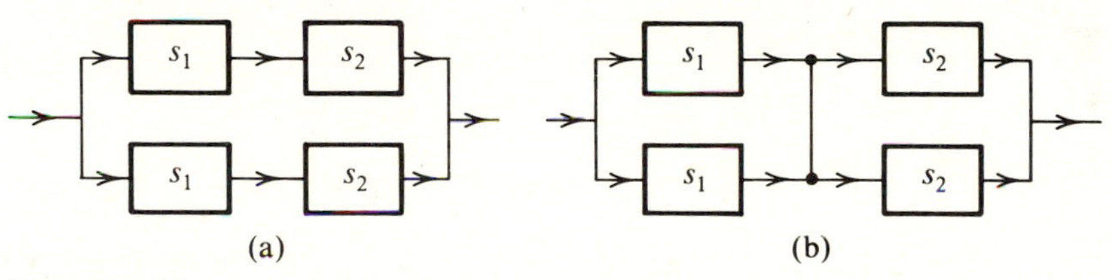

Figure P5.14

5.15. A binary source generates digits **1** and **0** randomly with probabilities $P(\mathbf{1}) = 0.8$ and $P(\mathbf{0}) = 0.2$.
(a) What is the probability that two **1**'s and three **0**'s will occur in a five-digit sequence?
(b) What is the probability that at least three **1**'s will occur in a five-digit sequence?

5.16. In a binary communication channel, the receiver detects binary pulses with an error probability P_e. What is the probability that out of 100 received digits, no more than three digits are in error?

5.17. Example 5.8 in the text considers the possibility of improving reliability by three repetitions of a digit. Repeat this analysis for five repetitions.

5.18. A PCM channel is made up of 10 links, with a regenerative repeater at the beginning of each link. If the detection error probabilities of the 10 detectors are $p_1, p_2, \ldots, p_{10}$, respectively, determine the detection error probability of the entire channel assuming the probability of errors over more than one link is negligible.

5.19. A binary symmetric channel (see Example 5.6) has an error probability P_e. The probability of transmitting **1** is Q, and the probability of transmitting **0** is $1 - Q$. If the receiver detects an incoming digit as **1**, what is the probability that the corresponding transmitted digit was (a) **1**, and (b) **0**?
Hint: If x is the transmitted digit and y is the received digit, you are given $P_{y|x}(\mathbf{0}|\mathbf{1}) = P_{y|x}(\mathbf{1}|\mathbf{0}) = P_e$. Now using Bayes' rule, find $P_{x|y}(\mathbf{1}|\mathbf{1})$ and $P_{x|y}(\mathbf{0}|\mathbf{1})$.

5.20. The PDF of amplitude x of a certain signal x(t) is given by $p_x(x) = xe^{-x}u(x)$.
(a) Find the probability that $x \geq 1$.
(b) Find the probability that $1 < x \leq 2$.

5.21. The PDF of an amplitude x of a gaussian signal x(*t*) is given by

$$p_x(x) = \frac{1}{\sigma\sqrt{2\pi}} e^{-x^2/2\sigma^2}$$

This signal is applied to the input of a half-wave rectifier circuit (Fig. P5.21). Assuming an ideal diode, determine $F_y(y)$ and $p_y(y)$ of the output signal amplitude y.

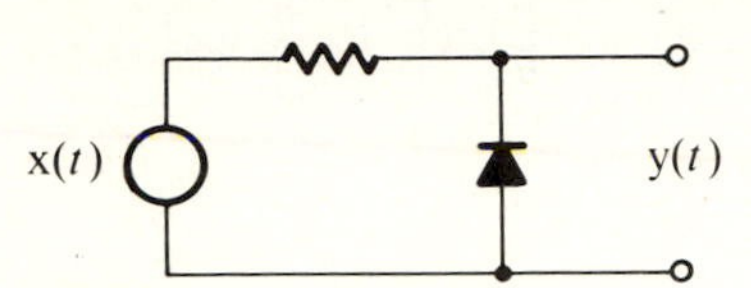

Figure P5.21

5.22. The PDF of a gaussian variable x is given by

$$p_x(x) = \frac{1}{4\sqrt{2\pi}} e^{-(x-3)^2/32}$$

Determine (a) $P(x \geq 7)$, (b) $P(x \geq 0)$, and (c) $P(x \leq -5)$.

5.23. In the example on threshold detection (Example 5.13), it was assumed that the digits **1** and **0** were transmitted with equal probability. If $P_x(\mathbf{1})$ and $P_x(\mathbf{0})$, the probabilities of transmitting **1** and **0**, respectively, are not equal, show that the optimum threshold is not 0 but is a, where

$$a = \frac{\sigma_n^2}{2A_p} \ln \frac{P_x(\mathbf{0})}{P_x(\mathbf{1})}$$

Hint: Assume that the optimum threshold is a, and write P_e in terms of the Q functions. For the optimum case, $dP_e/da = 0$. Use the fact that

$$Q(x) = 1 - \frac{1}{2\pi}\int_{-\infty}^{x} e^{-y^2/2}\, dy \qquad \text{and} \qquad \frac{dQ(x)}{dx} = -\frac{1}{2\pi} e^{-x^2/2}$$

5.24. The joint PDF $p_{xy}(x, y)$ of two continuous random variables is given by

$$p_{xy}(x, y) = xy\, e^{-(x^2+y^2)/2} u(x)u(y)$$

(a) Find

$$p_x(x),\ p_y(y),\ p_{x|y}(x|y),\ \text{and}\ p_{y|x}(y|x)$$

(b) Are x and y independent?

5.25. Random variables x and y are said to be jointly gaussian if their joint PDF is given by

$$p_{xy}(x, y) = \frac{1}{2\pi\sqrt{M}} e^{-(ax^2+by^2-2cxy)/2M}$$

where $M = ab - c^2$. Show that $p_x(x)$, $p_y(y)$, $p_{x|y}(x|y)$, and $p_{y|x}(y|x)$ are all gaussian and that $\overline{x_1^2} = b$, $\overline{x_2^2} = a$, and $\overline{x_1 x_2} = c$. *Hint:* Use

$$\int_{-\infty}^{\infty} e^{-px^2+qx}\, dx = \sqrt{\frac{\pi}{p}}\, e^{q^2/4p}$$

5.26. The joint PDF of random variables x and y is given by

$$p_{xy}(x, y) = ke^{-(x^2+xy+y^2)}$$

Determine (a) the constant k and (b) $p_x(x)$, (c) $p_y(y)$, (d) $p_{x|y}(x, y)$, and (e) $p_{y|x}(y|x)$. Are x and y independent?

5.27. If an amplitude x of a gaussian signal x(t) has a mean value of 1 and an rms value of 2, determine its PDF.

5.28. Determine the mean, the mean square, and the variance of the random variable x in Prob. 5.20.

5.29. For a gaussian PDF $p_x(x) = (1/\sigma_x\sqrt{2\pi})e^{-x^2/2\sigma_x^2}$, show that

$$\overline{x^n} = \begin{cases} (1)(3)(5)\ldots(n-1)\sigma_x^n & n \text{ even} \\ 0 & n \text{ odd} \end{cases}$$

Hint: See appropriate definite integrals in any standard mathematical table.

5.30. Determine the mean, the mean square, and the variance of the random variable x whose PDF is shown in Fig. P5.30.

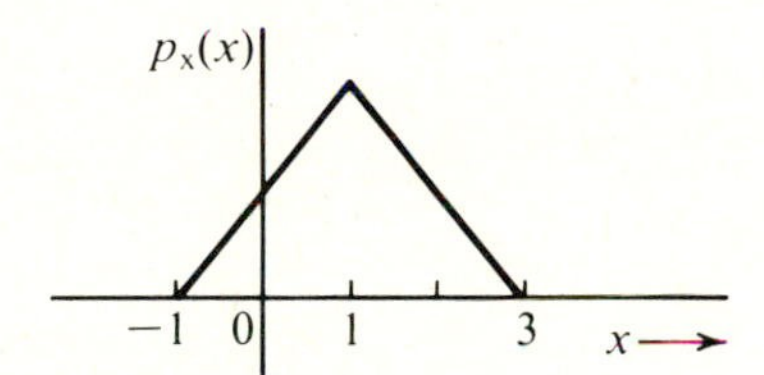

Figure P5.30

5.31. The sum of the points on two tossed dice is a discrete random variable x, as analyzed in Example 5.9. Determine the mean, the mean square, and the variance of x.

5.32. The random binary signal x(t), shown in Fig. P5.32, can take on only two values, 1 and −1, with equal probability. A gaussian channel noise n(t) is added to this signal, giving the received signal y(t). The PDF of the noise amplitude n is gaussian with a zero mean and an rms value of 2. Determine and sketch the PDF of the amplitude y.

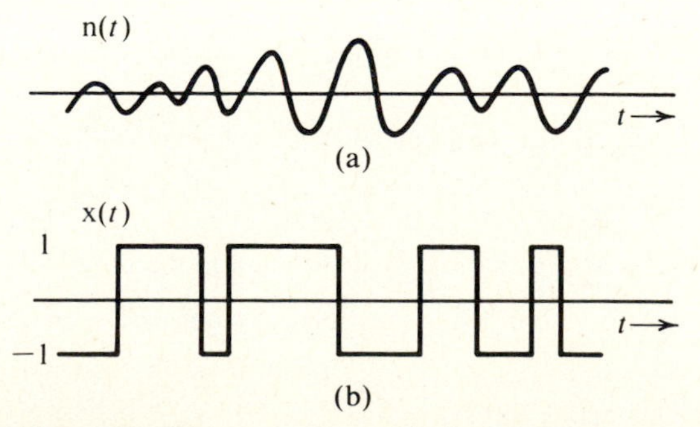

Figure P5.32

5.33. Repeat Prob. 5.32 if the amplitudes 1 and -1 of $\mathrm{x}(t)$ are not equiprobable but $P_{\mathrm{x}}(1) = Q$ and $P_{\mathrm{x}}(-1) = 1 - Q$.

5.34. Repeat Prob. 5.33 if $\mathrm{x}(t)$ and $\mathrm{n}(t)$ are both independent binary signals with

$$P_{\mathrm{x}}(1) = Q = 1 - P_{\mathrm{x}}(-1)$$

$$P_{\mathrm{n}}(1) = P = 1 - P_{\mathrm{n}}(-1)$$

5.35. If $\mathrm{z} = \mathrm{x} + \mathrm{y}$ where x and y are independent gaussian random variables with

$$p_{\mathrm{x}}(x) = \frac{1}{\sigma_{\mathrm{x}}\sqrt{2\pi}} e^{-(x-\overline{\mathrm{x}})^2/2\sigma_{\mathrm{x}}^2} \quad \text{and} \quad p_{\mathrm{y}}(y) = \frac{1}{\sigma_{\mathrm{x}}\sqrt{2\pi}} e^{-(y-\overline{\mathrm{y}})^2/2\sigma_{\mathrm{y}}^2}$$

Then show that z is also gaussian with

$$\overline{\mathrm{z}} = \overline{\mathrm{x}} + \overline{\mathrm{y}} \quad \text{and} \quad \sigma_{\mathrm{z}}^2 = \sigma_{\mathrm{x}}^2 + \sigma_{\mathrm{y}}^2$$

Hint: Use frequency-domain analysis to convolve $p_{\mathrm{x}}(x)$ and $p_{\mathrm{y}}(y)$. See pair 17 in Table 2.1.

5.36. Show that $|\rho_{\mathrm{xy}}| \le 1$, where ρ_{xy} is the correlation coefficient [Eq. (5.76)] of random variables x and y. *Hint:* For any real number a,

$$\overline{[a(\mathrm{x} - \overline{\mathrm{x}}) - (\mathrm{y} - \overline{\mathrm{y}})]^2} \ge 0$$

The discriminant of this quadratic in a is nonpositive.

5.37. Show that if two random variables x and y are related by

$$\mathrm{y} = k_1\mathrm{x} + k_2$$

where k_1 and k_2 are arbitrary constants, the correlation coefficient $\rho_{\mathrm{xy}} = 1$ if k_1 is positive, and $\rho_{\mathrm{xy}} = -1$ if k_1 is negative.

5.38. Given $\mathrm{x} = \cos\Theta$ and $\mathrm{y} = \sin\Theta$ where Θ is an r.v. uniformly distributed in the range $(0, 2\pi)$, show that x and y are uncorrelated but are not independent.

5.39. **(a)** Sketch the ensemble of a random process

$$\mathrm{x}(t) = \mathrm{a}\cos(\omega t + \Theta)$$

where ω and Θ are constants and a is an r.v. uniformly distributed in the range $(-A, A)$.

(b) Just by observing the ensemble, determine if this is a stationary or nonstationary process. Give your reasons.

5.40. Repeat Prob. 5.39(a) if a and Θ are constants but ω is an r.v. uniformly distributed in the range (0, 10).

5.41. **(a)** Sketch the ensemble of a random process

$$\mathrm{x}(t) = \mathrm{a}t + b$$

where b is a constant and a is an r.v. uniformly distributed in the range $(-1, 1)$.

(b) Just by observing the ensemble, state if this is a stationary or nonstationary process.

5.42. Determine $\overline{\mathrm{x}(t)}$ and $R_{\mathrm{x}}(t_1, t_2)$ for the random process in Prob. 5.39, and determine whether this is a wide-sense stationary process.

5.43. Repeat Prob. 5.42 for the process $x(t)$ in Prob. 5.40.

5.44. Repeat Prob. 5.42 for the process $x(t)$ in Prob. 5.41.

5.45. Given a random process $x(t) = k$ where k is an r.v. uniformly distributed in the range $(-1, 1)$.

(a) Sketch the ensemble of this process.
(b) Determine $\overline{x(t)}$.
(c) Determine $R_x(t_1, t_2)$.
(d) Is the process wide-sense stationary?
(e) Is the process ergodic?
(f) If the process is wide-sense stationary, what is its power P_x [or its mean square value $\overline{x^2(t)}$]?

5.46. Repeat Prob. 5.45 for the random process

$$x(t) = a \cos(\omega_c t + \Theta)$$

where ω_c is a constant and a and Θ are independent r.v.'s uniformly distributed in the ranges $(-1, 1)$ and $(0, 2\pi)$, respectively.

5.47. For each of the following functions, state whether it can be a valid PSD of a real random process.

(a) $\dfrac{\omega^2}{\omega^2 + 4}$

(b) $\dfrac{1}{\omega^2 - 4}$

(c) $\dfrac{\omega}{\omega^2 + 4}$

(d) $\delta(\omega) + \dfrac{1}{\omega^2 + 4}$

(e) $\delta(\omega + \omega_o) - \delta(\omega - \omega_o)$

(f) $j[\delta(\omega + \omega_o) + \delta(\omega - \omega_o)]$

(g) $\dfrac{j\omega^2}{\omega^2 + 4}$

5.48. Find the mean square value (or power) for the processes whose PSDs are given below:

(a) $\dfrac{1}{\omega^2 + 4}$

(b) $\dfrac{1}{\omega^4 + 10\omega^2 + 9}$

(c) $\dfrac{1}{\omega^4 + 9\omega^2 + 10}$

(d) $\dfrac{\omega^2 + 2}{\omega^4 + 13\omega^2 + 36}$

5.49. Show that for a wide-sense stationary process $x(t)$,

(a) $R_x(0) \geq |R_x(\tau)| \qquad \tau \neq 0$

Hint:

$$\overline{(x_1 + x_2)^2} = \overline{x_1^2} + \overline{x_2^2} + \overline{2x_1x_2} \geq 0$$

Let $x_1 = x(t_1)$ and $x_2 = x(t_2)$.

(b) $\lim_{\tau \to \infty} R_x(\tau) = \bar{x}^2$

Hint: As $\tau \to \infty$, x_1 and x_2 tend to become independent.

5.50. Show that if the PSD of a random process $\mathrm{x}(t)$ is bandlimited to B Hz, and if

$$R_{\mathrm{x}}\left(\frac{n}{2B}\right) = 0 \qquad n = \pm 1, \pm 2, \pm 3, \ldots$$

then $\mathrm{x}(t)$ is a white bandlimited process; that is, $S_{\mathrm{x}}(\omega) = k\Pi(\omega/4\pi B)$.

5.51. For the random binary process in Example 5.25 (Fig. 5.28*a*), determine $R_{\mathrm{x}}(\tau)$ and $S_{\mathrm{x}}(\omega)$ if the probability of transition (from 1 to -1 or vice versa) at each node is 0.4 instead of 0.5.

5.52. A random process $\mathrm{x}(t)$ consists of unit impulses located at random instants (Fig. P5.52). There are an average of α impulses /second, and the location of any impulse is independent of the locations of other impulses. Show that $R_{\mathrm{x}}(\tau) = \alpha\,\delta(\tau) + \alpha^2$.

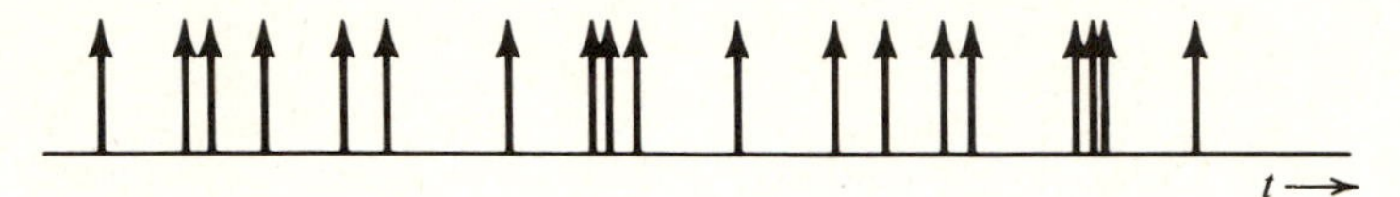

Figure P5.52

5.53. A wide-sense stationary white process $\mathrm{m}(t)$ bandlimited to B Hz is sampled at the Nyquist rate. Each sample is transmitted by a basic pulse $p(t)$ multiplied by the sample value. This is a PAM signal. Show that the PSD of the PAM signal is $2BR_{\mathrm{m}}(0)\,|P(\omega)|^2$. *Hint:* Use Eq. (5.117a). Show that Nyquist samples a_k and a_{k+n} $(n \geq 1)$ are independent.

5.54. Random binary digits are transmitted every T_o seconds, using on-off signaling. The resulting signal constitutes a random process $\mathrm{y}(t)$. Determine $S_{\mathrm{y}}(\omega)$ if the basic pulse used is a rectangular pulse $\Pi(t/T_o)$, and $P(\mathbf{0}) = P(\mathbf{1}) = 0.5$. *Hint:* Use Eq. (5.117a). See Sec. 3.2 for the description of on-off signaling.

5.55. Repeat Prob. 5.54 if $P(\mathbf{1}) = Q$ and $P(\mathbf{0}) = 1 - Q$.

5.56. Repeat Prob. 5.54 for bipolar signaling with $P(\mathbf{0}) = P(\mathbf{1}) = 0.5$. (See Sec. 3.2 for the description of bipolar signaling.)

5.57. Repeat Prob. 5.54 for duobinary signaling with $P(\mathbf{0}) = P(\mathbf{1}) = 0.5$. (See Sec. 3.2 for the description of duobinary signaling.)

5.58. A sample function of a random process $\mathrm{x}(t)$ is shown in Fig. P5.58. The signal $\mathrm{x}(t)$ changes abruptly in amplitude at random instants. There are an average of β amplitude changes (or shifts) per second. The probability that there will be no amplitude shift in τ seconds is given by $P_0(\tau) = e^{-\beta\tau}$. The amplitude after a shift is independent of the amplitude before the shift. The amplitudes are randomly distributed, with a PDF $p_{\mathrm{x}}(x)$.

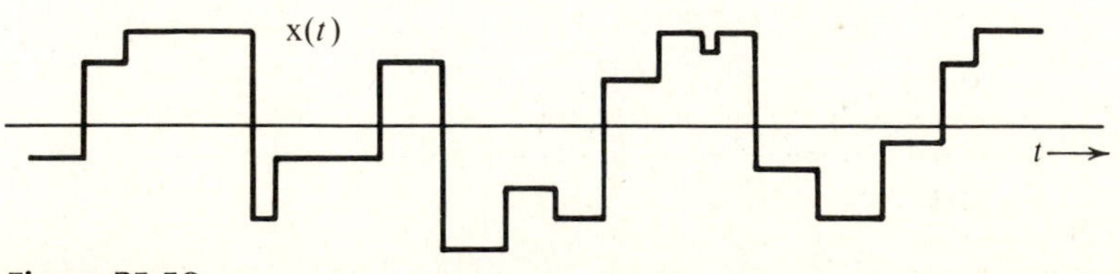

Figure P5.58

Show that

$$R_x(\tau) = \overline{x^2} e^{-\beta|\tau|} \qquad \text{and} \qquad S_x(\omega) = \frac{2\beta\overline{x^2}}{\beta^2 + \omega^2}$$

This process represents a model for thermal noise.[5,8]

5.59. Show that for jointly wide-sense stationary, real, random processes $x(t)$ and $y(t)$,

$$|R_{xy}(\tau)| \leq [R_x(0)R_y(0)]^{1/2}$$

Hint: For any real number a, $\overline{(ax - y)^2} \geq 0$.

5.60. If $x(t)$ and $y(t)$ are two incoherent random processes, and two new processes $u(t)$ and $v(t)$ are formed as follows:

$$u(t) = x(t) + y(t) \qquad v(t) = 2x(t) + 3y(t)$$

find $R_u(\tau)$, $R_v(\tau)$, $R_{uv}(\tau)$, and $R_{vu}(\tau)$ in terms of $R_x(\tau)$ and $R_y(\tau)$.

5.61. Two random processes $x(t)$ and $y(t)$ are

$$x(t) = A\cos(\omega_o t + \varphi) \qquad \text{and} \qquad y(t) = B\cos(n\,\omega_o t + n\varphi)$$

where A, B, and ω_o are constants and φ is an r.v. uniformly distributed in the range $(0, 2\pi)$. Show that the two processes are incoherent.

5.62. A sample function of a periodic random process $x(t)$ is shown in Fig. P5.62. The interval b where the first pulse begins is an r.v. uniformly distributed in the range $(0, T_o)$. If a sample function with $b = 0$ is expressed by a compact trigonometric Fourier series [Eq. (2.5)]

$$x(t) = C_0 + \sum_{n=1}^{\infty} C_n \cos(n\omega_o t + \theta_n)$$

then show that

$$R_x(\tau) = C_0^2 + \frac{1}{2}\sum_{n=1}^{\infty} C_n^2 \cos n\omega_o\tau \qquad \omega_o = \frac{2\pi}{T_o}$$

Find the PSD $S_x(\omega)$.

Hint:

$$x(t) = C_0 + \sum_{n=1}^{\infty} C_n \cos[n\omega_o(t - b) + \theta_n]$$

where b is an r.v. uniformly distributed in the range $(0, T_o)$. Now use the results in Prob. 5.61.

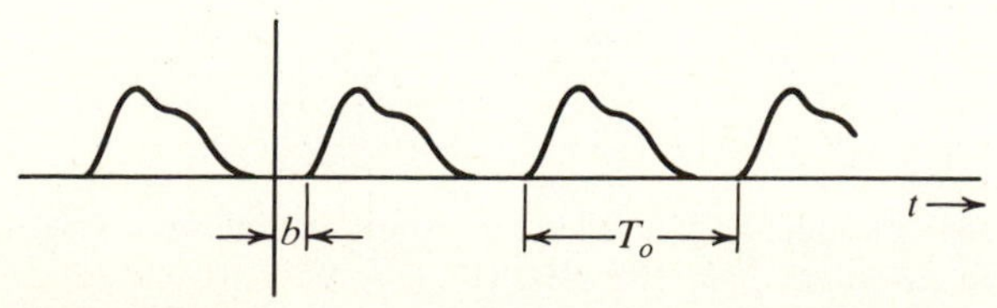

Figure P5.62

5.63. A simple RC circuit has two resistors R_1 and R_2 in parallel (Fig. P5.63a). Calculate the rms value of the thermal noise voltage v_o across the capacitor in two ways:

(a) Consider resistors R_1 and R_2 as two separate resistors, with respective thermal noise voltages of PSD $2kTR_1$ and $2kTR_2$ (Fig. P5.63b). Note that the two sources are incoherent.

(b) Consider the parallel combination of R_1 and R_2 as a single resistor of value $R_1R_2/R_1 + R_2$, with its thermal-noise voltage source of PSD $2kTR_1R_2/(R_1 + R_2)$ (Fig. P5.63c). Comment.

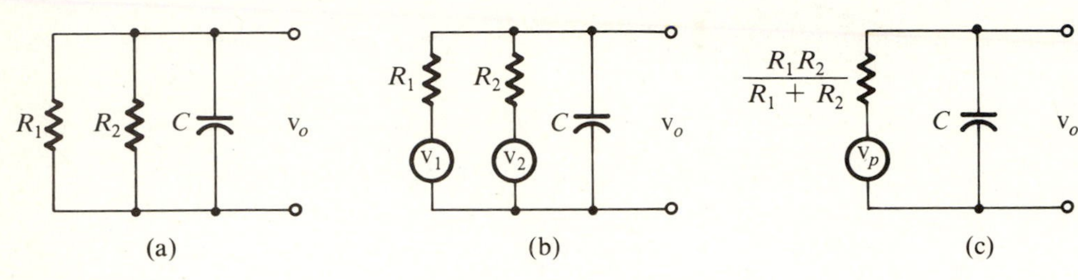

Figure P5.63

5.64. Show that $R_{xy}(\tau)$, the cross-correlation function of the input process $x(t)$ and the output process $y(t)$ in Fig. 5.36, is

$$R_{xy}(\tau) = h(\tau) * R_x(\tau) \qquad \text{and} \qquad S_{xy}(\omega) = H(\omega)S_x(\omega)$$

Hence, show that for the thermal noise $n(t)$ and the output $v_o(t)$ in Fig. 5.38 (Example 5.29),

$$S_{nv_o}(\omega) = \frac{2kTR}{1 + j\omega RC} \qquad \text{and} \qquad R_{nv_o}(\tau) = \frac{2kT}{C}e^{-\tau/RC}u(\tau)$$

5.65. A white process $n(t)$ of PSD $\mathcal{N}/2$ is transmitted through a bandpass filter $H(\omega)$ (Fig. P5.65). Represent the filter output in terms of quadrature components, and determine $S_{n_c}(\omega)$, $S_{n_s}(\omega)$, $\overline{n_c^2}$, $\overline{n_s^2}$, and $\overline{n^2}$ when the center frequency used in this representation is 100 kHz (that is, $\omega_c = 200\pi \times 10^3$).

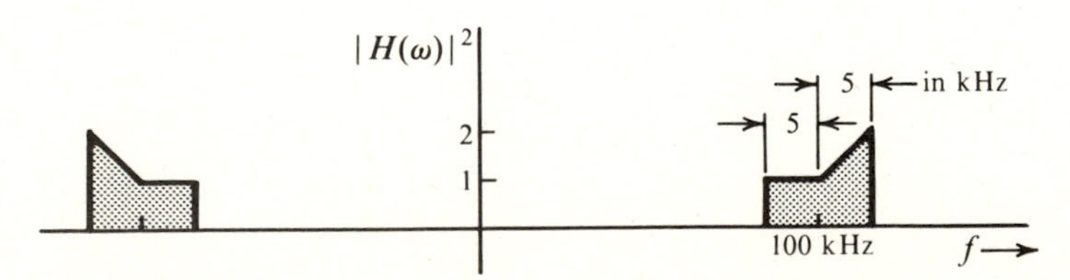

Figure P5.65

5.66. Repeat Prob. 5.65 if the center frequency ω_c used in the representation is not a true center frequency. Consider three cases: (a) $f_c = 105$ kHz, (b) $f_c = 95$ kHz, and (c) $f_c = 90$ kHz.

5.67. A random process $x(t)$ with the PSD shown in Fig. P5.67*a* is passed through a bandpass filter (Fig. 5.67*b*). Determine the PSDs and mean square values of the quadrature components of the output process. Assume the center frequency in the representation to be 0.5 MHz.

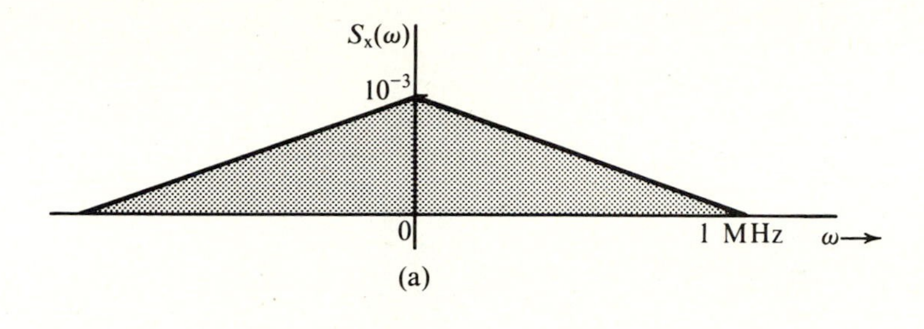

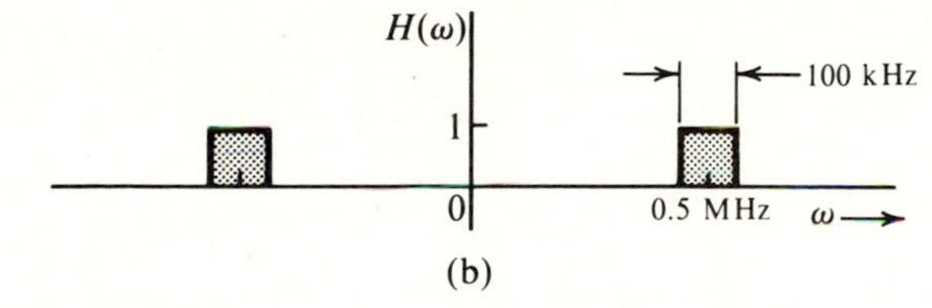

Figure P5.67

5.68. A signal process $m(t)$ is mixed with a channel noise $n(t)$. The respective PSDs are

$$S_m(\omega) = \frac{1}{1+\omega^2} \quad \text{and} \quad S_n(\omega) = \frac{8}{16+\omega^2}$$

(a) Find the optimum Wiener-Hopf filter.
(b) Sketch its unit impulse response.
(c) Estimate the amount of delay necessary to make this filter closely realizable.
(d) Compute the noise power at the input and the output of the filter.
(e) What is the improvement in the SNR realized by the use of the filter?
Hint: Note that because of the special way in which the problem of optimum filtering is formulated, the signal power at the input and that at the output of the filter are identical (see Fig. 5.47).

6

Behavior of Communication Systems in the Presence of Noise

PART I Analog Systems

In this chapter we shall analyze the behavior of analog and digital communication systems in the presence of noise. The behavior of analog communication systems in the presence of noise has already been studied in Chapter 4. We shall derive similar results in this chapter. The only difference is that in Chapter 4, we analyzed the behavior using time averages. The power of a signal, for example, was defined as the time average of the signal squared (time mean square value). This means in Chapter 4, we examined the SNR behavior for individual sample functions. In this chapter we shall derive the results using ensemble means; that is, the signal power is considered to be the ensemble average of the squared signal values. Not surprisingly, the results derived here are the same as those derived in Chapter 4 with the time means replaced by the ensemble means.

Figure 6.1 shows a schematic of a communication system. A certain signal power S_T is transmitted over a channel.* The transmitted signal is corrupted by channel noise during transmission. We shall assume channel noise to be additive (Fig. 6.1). The

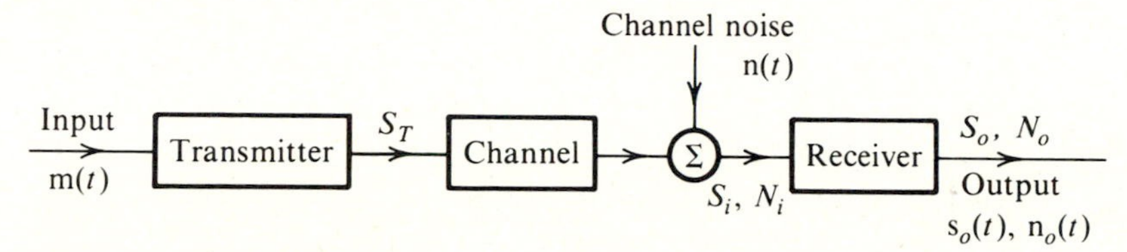

Figure 6.1 A communication system model.

channel will attenuate (and may also distort) the signal. At the receiver input, we have a signal mixed with noise. The signal and the noise powers at the receiver input are S_i and N_i, respectively. The receiver processes (filters, demodulates, etc.) the signal to yield the desired signal plus noise. The signal and the noise powers at the receiver output are S_o and N_o, respectively. In analog systems, the quality of the received signal is determined by the output SNR S_o/N_o. Hence, we shall focus our attention on this parameter. But S_o/N_o can be increased as much as desired simply by increasing the transmitted power S_T. In practice, however, the maximum value of S_T is limited by other considerations, such as the transmitter cost, channel capability, interference with other channels, and so on. Hence, the value of S_o/N_o for a given transmitted power is an appropriate figure of merit in an analog communication system. In practice, it is more convenient to deal with the received power S_i rather than the transmitted power S_T. From Fig. 6.1, it is apparent that S_i is proportional to S_T. Hence the value of S_o/N_o for a given S_i will serve equally well as a figure of merit.

6.1 BASEBAND SYSTEMS

In baseband systems, the signal is transmitted directly without any modulation. This mode of communication is suitable over a pair of wires or over coaxial cables. It is mainly used in short-haul links. Although baseband sytems are not widely used, their study is important because many of the basic concepts and parameters encountered in baseband systems are carried over directly to modulation systems. Secondly, baseband systems serve as a basis against which other systems may be compared.

For a baseband system, the transmitter and the receiver are simple baseband filters (Fig. 6.2). The lowpass filter $H_p(\omega)$ at the transmitter limits the input signal spectrum to a given bandwidth. The lowpass filter $H_d(\omega)$ at the receiver eliminates the out-of-band noise and other channel interference. These filters can also serve an additional

* Here the channel is used in the sense of a transmission medium.

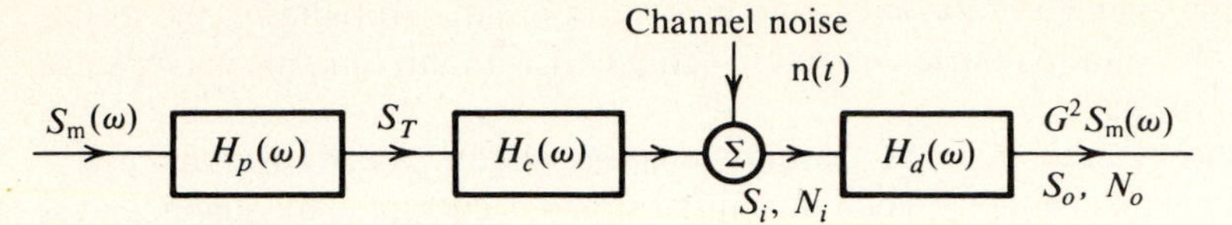

Figure 6.2 A baseband system.

purpose, that of preemphasis and deemphasis, which optimizes the signal-to-noise ratio at the receiver (or minimizes the channel noise interference).

The baseband signal m(t) is assumed to be a zero-mean, wide-sense stationary random process bandlimited to B Hz. To begin with, we shall consider the case of ideal lowpass (or baseband) filters with bandwidth B at the transmitter and the receiver (Fig. 6.2). The channel is assumed to be distortionless. For this case,

$$S_o = S_i \tag{6.1a}$$

and

$$N_o = 2 \int_0^B S_n(\omega)\, df \tag{6.1b}$$

where $S_n(\omega)$ is the PSD of the channel noise. For the case of a white noise, $S_n(\omega) = \mathcal{N}/2$, and

$$N_o = 2 \int_0^B \frac{\mathcal{N}}{2}\, df$$

$$= \mathcal{N}B \tag{6.1c}$$

and

$$\frac{S_o}{N_o} = \frac{S_i}{\mathcal{N}B} \tag{6.1d}$$

As in Chapter 4, we defined the parameter γ as

$$\gamma = \frac{S_i}{\mathcal{N}B} \tag{6.2}$$

From Eqs. (6.1d) and (6.2) we have

$$\frac{S_o}{N_o} = \gamma \tag{6.3}$$

The parameter γ is directly proportional to S_i and, therefore, directly proportional to S_T. Hence, a given S_T (or S_i) implies a given γ. Equation (6.3) is precisely the result we are looking for. It gives the receiver output SNR for a given S_T (or S_i).

The value of SNR in Eq. (6.3) will serve as a standard against which the output SNR of other systems will be measured.

The power, or the mean square value, of m(t) is $\overline{\text{m}^2}$, given by

$$\overline{\text{m}^2} = 2 \int_0^B S_{\text{m}}(\omega)\, df \tag{6.4}$$

In analog signals, the SNR is basic in specifying the signal quality. For voice signals, an SNR of 5 to 10 dB at the receiver implies a barely intelligible signal. Telephone quality signals have an SNR of 25 to 35 dB, whereas for television, an SNR of 45 to 55 dB is required.

6.2 AMPLITUDE-MODULATED SYSTEMS

We shall analyze DSB-SC, SSB-SC, and AM systems separately.

DSB-SC

A basic DSB-SC system is shown in Fig. 6.3.* The modulated signal is a bandpass signal centered at ω_c with a bandwidth $2B$. The channel noise is assumed to be additive (Fig. 6.3). The channel and the filters in Fig. 6.3 are assumed to be ideal.

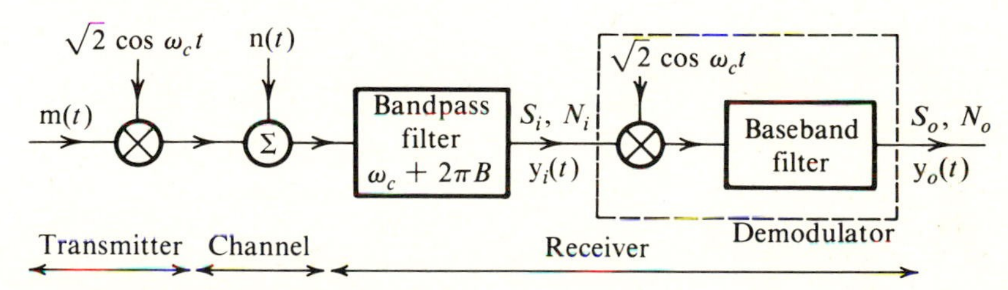

Figure 6.3 DSB-SC system.

Let S_i and S_o represent the useful signal powers at the input and the output of the demodulator, and let N_o represent the noise power at the demodulator output. The signal at the demodulator input is $\sqrt{2}\,\text{m}(t) \cos \omega_c t + \text{n}_i(t)$, where $\text{n}_i(t)$ is the bandpass channel noise. Its spectrum is centered at ω_c, and has a bandwidth $2B$. Hence, the input signal power S_i is the power of the modulated signal† $\sqrt{2}\,\text{m}(t) \cos \omega_c t$. From Eq. (5.100c),

$$S_i = \overline{[\sqrt{2}\,\text{m}(t) \cos \omega_c t]^2} = \overline{\text{m}^2(t)} = \overline{\text{m}^2} \tag{6.5}$$

The reader may now appreciate our use of $\sqrt{2} \cos \omega_c t$ (rather than $\cos \omega_c t$) in the modulator (Fig. 6.3). This was done to make the received power equal to that in the

---

* The use of an input bandpass filter in the receiver may appear redundant because the out-of-band noise components will be suppressed by the final baseband filter. In practice, an input filter is useful because by removing the out-of-band noise, it reduces the probability of nonlinear distortion from overload effects.

† The modulated signal also has a random phase Θ, which is uniformly distributed in the range $(0, 2\pi)$. This random phase [which is independent of m(t)] does not affect the final results and, hence, is ignored in this discussion.

baseband system in order to facilitate comparison. We shall use a similar artifice in our analysis of the SSB system.

To determine the output powers S_o and N_o, we note that the signal at the demodulator input is

$$y_i(t) = \sqrt{2}\, m(t) \cos \omega_c t + n_i(t)$$

Because $n_i(t)$ is a bandpass signal centered at ω_c, we can express it in terms of quadrature components, as in Eq. (5.128). This gives

$$y_i(t) = [\sqrt{2}\, m(t) + n_c(t)] \cos \omega_c t + n_s(t) \sin \omega_c t$$

When this signal is multiplied by $\sqrt{2} \cos \omega_c t$ (synchronous demodulation) and then lowpass filtered, the demodulator output $y_o(t)$ is

$$y_o(t) = m(t) + \frac{1}{\sqrt{2}} n_c(t)$$

Hence,

$$S_o = \overline{m^2} = S_i \tag{6.6a}$$

$$N_o = \tfrac{1}{2} \overline{n_c^2(t)} \tag{6.6b}$$

For white noise with power density $\mathcal{N}/2$, we have [Eq. (5.130b)]

$$\overline{n_c^2(t)} = \overline{n_i^2(t)} = 2\mathcal{N}B$$

and

$$N_o = \mathcal{N}B \tag{6.7}$$

Hence, from Eqs. (6.6a and 6.7) we have

$$\frac{S_o}{N_o} = \frac{S_i}{\mathcal{N}B} = \gamma \tag{6.8}$$

Comparison of Eq. (6.8) with Eq. (6.1d) shows that for a fixed transmitted power (which also implies a fixed signal power at the demodulator input), the SNR at the demodulator output is the same for the baseband and DSB-SC systems. Moreover, quadrature multiplexing in DSB-SC can render its bandwidth requirement identical to that of baseband. Thus, theoretically, baseband and DSB-SC systems have identical capabilities.

SSB-SC

An SSB-SC system is shown in Fig. 6.4. The SSB signal* $\varphi_{\text{SSB}}(t)$ can be expressed as [see Eq. (4.19)]

$$\varphi_{\text{SSB}}(t) = m(t) \cos \omega_c t + m_h(t) \sin \omega_c t \tag{6.9}$$

The spectrum of $\varphi_{\text{SSB}}(t)$ is shown in Fig. 4.16. This signal can be obtained (Fig. 6.4) by multiplying $m(t)$ by $2 \cos \omega_c t$ and then suppressing the unwanted sideband. The

* This is LSB. The discussion is valid for USB as well.

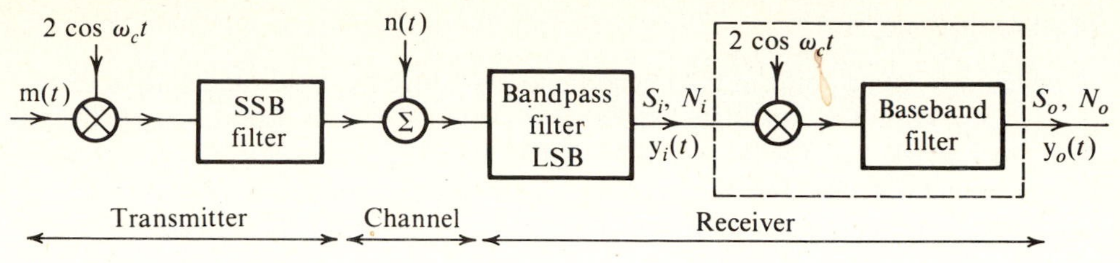

Figure 6.4 SSB-SC system.

power of the modulated signal 2 m(t) cos $\omega_c t$ is 2 $\overline{m^2}$ (four times the power of m(t) cos $\omega_c t$). Suppression of one sideband halves the power. Hence S_i, the power of $\varphi_{SSB}(t)$, is

$$S_i = \overline{m^2} \tag{6.10}$$

Expressing the channel bandpass noise in terms of quadrature components as in Eq. (5.132) [Example 5.32, Fig. 5.45], $y_i(t)$, the signal at the detector input, is

$$y_i(t) = [m(t) + n_c(t)] \cos \omega_c t + [m_h(t) + n_s(t)] \sin \omega_c t$$

This signal is multiplied by 2 cos $\omega_c t$ (synchronous demodulation) and then lowpass filtered to yield the demodulator output

$$y_o(t) = [m(t) + n_c(t)]$$

Hence,

$$S_o = \overline{m^2} = S_i$$

$$N_o = \overline{n_c^2} \tag{6.11}$$

We have already found $\overline{n_c^2}$ for the SSB channel noise (lower sideband) in Eq. (5.134b) [Example 5.32, Fig. 5.45]

$$N_o = \overline{n_c^2} = \mathcal{N}B$$

Thus,

$$\frac{S_o}{N_o} = \frac{S_i}{\mathcal{N}B} = \gamma \tag{6.12}$$

This shows that baseband, DSB-SC, and SSB-SC systems perform identically.

VSB-SC

The performance of VSB-SC is almost the same as that of SSB-SC. The signal power S_i in VSB is somewhat smaller than that in SSB. The output power S_o in VSB is identical to that in SSB. The noise N_o in VSB, however, is slightly larger than that in SSB, because the transmission bandwidth of VSB is slightly larger than that of SSB. The reduced value of S_i compensates partly for the increased value of N_o, and the SNR in VSB is close to that of SSB. Therefore,

$$\frac{S_o}{N_o} \simeq \gamma \tag{6.13}$$

■ EXAMPLE 6.1

A signal m(t) with a uniform PSD bandlimited to 4 kHz DSB-SC modulates a carrier of f_c = 500 kHz. The modulated signal is transmitted over a distortionless channel with a noise PSD $S_n(\omega) = 1/(\omega^2 + a^2)$, $(a = 10^6\pi)$. The useful signal power at the receiver input is 1μW. The received signal is bandpass filtered, multiplied by 2 cos $\omega_c t$, and then lowpass filtered to obtain the output $s_o(t) + n_o(t)$. Determine the output SNR.

□ **Solution:** If the received signal is $k\text{m}(t) \cos \omega_c t$, the demodulator input is $[k\text{m}(t) + n_c(t)] \cos \omega_c t + n_s(t) \sin \omega_c t$. When this is multiplied by $2 \cos \omega_c t$ and lowpass filtered, the output is

$$s_o(t) + n_o(t) = k\text{m}(t) + n_c(t)$$

Hence,

$$S_o = k^2\overline{\text{m}^2} \qquad \text{and} \qquad N_o = \overline{n_c^2}$$

But the power of the received signal k m(t) cos $\omega_c t$ is 1 μW. Hence,

$$\frac{k^2\overline{\text{m}^2}}{2} = 10^{-6}$$

and

$$S_o = k^2\overline{\text{m}^2} = 2 \times 10^{-6}$$

To compute $\overline{n_c^2}$, we use Eq. (5.125c):

$$\overline{n_c^2} = \overline{n^2}$$

where $\overline{n^2}$ is the power of the incoming bandpass noise of bandwidth 8 kHz centered at 500 kHz; that is,

$$\overline{n^2} = 2\int_{496{,}000}^{504{,}000} \frac{1}{\omega^2 + a^2}\, df$$

$$= 2\int_{496{,}000}^{504{,}000} \frac{1}{4\pi^2 f^2 + 10^{12}\pi^2}\, df$$

$$= 8.25 \times 10^{-10} = N_o$$

Therefore,

$$\frac{S_o}{N_o} = \frac{2 \times 10^{-6}}{8.25 \times 10^{-10}} = 2.42 \times 10^3$$

$$= 33.83 \text{ dB}$$

■

AM

AM signals can be demodulated synchronously or by envelope detection. The former is of theoretical interest only. It is useful, however, for comparing the noise performance of the envelope detector. For this reason, we shall consider both of these methods.

The Coherent, or Synchronous, Demodulation of AM. Coherent AM is identical to DSB-SC in every respect except for the additional carrier.

If the received signal $\sqrt{2}[A + \mathrm{m}(t)] \cos \omega_c t$ is multiplied by $\sqrt{2} \cos \omega_c t$, the demodulator output is m(t). Hence,

$$S_o = \overline{\mathrm{m}^2}$$

The output noise will be exactly the same as that in DSB-SC [Eq. (6.6b)]:

$$N_o = \overline{\mathrm{n}_o^2} = \mathcal{N}B$$

The received signal is $\sqrt{2}[A + \mathrm{m}(t)] \cos \omega_c t$. Hence,

$$\begin{aligned} S_i &= (\sqrt{2})^2 \frac{\overline{[A + \mathrm{m}(t)]^2}}{2} \\ &= \overline{[A + \mathrm{m}(t)]^2} \\ &= \overline{A^2} + \overline{\mathrm{m}^2(t)} + 2A\,\overline{\mathrm{m}(t)} \end{aligned}$$

Because m(t) is assumed to have a zero mean,

$$S_i = \overline{A^2} + \overline{\mathrm{m}^2(t)}$$

and

$$\begin{aligned} \frac{S_o}{N_o} &= \frac{\overline{\mathrm{m}^2}}{\mathcal{N}B} \\ &= \frac{\overline{\mathrm{m}^2}}{A^2 + \overline{\mathrm{m}^2}} \frac{S_i}{\mathcal{N}B} \\ &= \frac{\overline{\mathrm{m}^2}}{A^2 + \overline{\mathrm{m}^2}} \gamma \end{aligned} \tag{6.14}$$

If $\mathrm{m}(t)_{\max} = m_p$, then $A \geq m_p$. For the maximum SNR, $A = m_p$, and

$$\left(\frac{S_o}{N_o}\right)_{\max} = \frac{\overline{\mathrm{m}^2}}{m_p^2 + \overline{\mathrm{m}^2}} \gamma$$

$$= \frac{1}{(m_p^2/\overline{\mathrm{m}^2}) + 1} \gamma \tag{6.15a}$$

Because $(m_p^2/\overline{\mathrm{m}^2}) \geq 1$,

$$\frac{S_o}{N_o} \leq \frac{\gamma}{2} \tag{6.15b}$$

It can be seen that the SNR in AM is at least 3 dB (and usually about 6 dB in practice) worse than that in DSB-SC and SSB-SC (depending on the modulation index and the signal waveform). For example, when m(t) is sinusoidal, $m_p^2/\overline{\mathrm{m}^2} = 2$, and AM requires three times as much power (4.77 dB) as that needed for DSB-SC or SSB-SC.

In many communication systems the transmitter is limited by peak power rather than average power transmitted. In such a case, AM fares even worse. It can be shown (see Prob. 6.8) that in tone modulation, for a fixed peak power transmitted, the output SNR of AM is 6 dB below that of DSB-SC and 9 dB below that of SSB-SC.* For this reason, volume compression and peak limiting are generally used in AM transmission in order to have full modulation most of the time.

AM Envelope Detection. Assuming the received signal to be $[A + \mathrm{m}(t)] \cos \omega_c t$, the demodulator input is

$$\mathrm{y}_i(t) = [A + \mathrm{m}(t)] \cos \omega_c t + \mathrm{n}_i(t)$$

Using the quadrature component representation for $\mathrm{n}_i(t)$, we have

$$\mathrm{y}_i(t) = [A + \mathrm{m}(t) + \mathrm{n}_c(t)] \cos \omega_c t + \mathrm{n}_s(t) \sin \omega_c t \tag{6.16a}$$

The desired signal at the demodulator input is $[A + \mathrm{m}(t)] \cos \omega_c t$. Hence, the signal power S_i is [see Eq. (5.100c)]

$$S_i = \frac{\overline{[A + \mathrm{m}(t)]^2}}{2} = \frac{A^2 + \overline{\mathrm{m}^2}}{2}$$

To compute S_o and N_o, we need the envelope of $\mathrm{y}_i(t)$.

$$\mathrm{y}_i(t) = \mathrm{E}_i(t) \cos [\omega_c t + \Theta_i(t)] \tag{6.16b}$$

where the envelope $\mathrm{E}_i(t)$ is

$$\mathrm{E}_i(t) = \sqrt{[A + \mathrm{m}(t) + \mathrm{n}_c(t)]^2 + \mathrm{n}_s^2(t)} \tag{6.16c}$$

The envelope detector output is $\mathrm{E}_i(t)$ [Eq. (6.16c)]. We shall consider two extreme cases: small noise and large noise.

The Small-Noise Case. If $[A + \mathrm{m}(t)] \gg \mathrm{n}_i(t)$ for almost all t, then $[A + \mathrm{m}(t)] \gg \mathrm{n}_c(t)$ and $\mathrm{n}_s(t)$ for almost all t.† In this case $\mathrm{E}_i(t)$ in Eq. (6.16c) can be approximated by $[A + \mathrm{m}(t) + \mathrm{n}_c(t)]$.

$$\mathrm{E}_i(t) \simeq A + \mathrm{m}(t) + \mathrm{n}_c(t)$$

The dc component A of the envelope detector output E_i is blocked by a capacitor, yielding $\mathrm{m}(t)$ as the useful signal and $\mathrm{n}_c(t)$ as the noise. Hence,

$$S_o = \overline{\mathrm{m}^2}$$

* These results are valid under conditions most favorable to AM, that is, with modulation index $\mu = 1$. For $\mu < 1$, AM would be worse than this.

† Here we use the term "almost all t" because $\mathrm{n}_c(t)$ and $\mathrm{n}_s(t)$ are both gaussian (amplitude range $-\infty$ to ∞), and in some instances $\mathrm{n}_c(t)$ or $\mathrm{n}_s(t)$ or both will exceed $A + \mathrm{m}(t)$, no matter how large $A + \mathrm{m}(t)$ is. For large signals, however, this occurs only over relatively short time intervals.

and from Eq. (5.130b)

$$N_o = \overline{n_c^2(t)} = 2\mathcal{N}B$$

and

$$\begin{aligned}\frac{S_o}{N_o} &= \frac{\overline{m^2}}{2\mathcal{N}B} \\ &= \frac{\overline{m^2}}{A^2 + \overline{m^2}} \frac{S_i}{\mathcal{N}B} \\ &= \frac{\overline{m^2}}{A^2 + \overline{m^2}} \gamma \qquad \textbf{(6.17)}\end{aligned}$$

which is identical to the result for AM with synchronous demodulation [Eq. (6.14)]. Therefore for AM, when the noise is small compared to the signal, the performance of the envelope detector is identical to that of the synchronous detector.

The Large-Noise Case. In this case $n_i(t) \gg [A + m(t)]$. Hence, $n_c(t)$ and $n_s(t) \gg [A + m(t)]$ for almost all t. Under this condition Eq. (6.16c) becomes

$$\begin{aligned}E_i(t) &\simeq \sqrt{n_c^2(t) + n_s^2(t) + 2n_c(t)[A + m(t)]} \\ &= E_n(t)\sqrt{1 + \frac{2[A + m(t)]}{E_n(t)} \cos \Theta_n(t)}\end{aligned}$$

where $E_n(t)$ and $\Theta_n(t)$ represent the envelope and the phase of the noise $n_i(t)$:

$$E_n(t) = \sqrt{n_c^2(t) + n_s^2(t)}$$

$$\Theta_n(t) = -\tan^{-1}\left[\frac{n_s(t)}{n_c(t)}\right]$$

Because $E_n(t) \gg A + m(t)$, $E_i(t)$ may be further approximated as

$$\begin{aligned}E_i(t) &\simeq E_n(t)\left[1 + \frac{A + m(t)}{E_n(t)} \cos \Theta_n(t)\right] \\ &= E_n(t) + [A + m(t)] \cos \Theta_n(t) \qquad \textbf{(6.18)}\end{aligned}$$

A glance at Eq. (6.18) shows that the output contains no term proportional to $m(t)$. The signal $m(t) \cos \Theta_n(t)$ represents $m(t)$ multiplied by a time-varying function (actually a noise signal) $\cos \Theta_n(t)$ and, hence, is of no use in recovering $m(t)$. In all previous cases, the output signal contained a term of the form $am(t)$, where a was constant. Furthermore, the output noise was additive (even for envelope detection with small noise). In Eq. (6.18), the noise is multiplicative. In this situation the useful signal is badly mutilated. This is the threshold phenomenon, where the signal quality at the output undergoes disproportionately rapid deterioration when the input noise increases beyond a certain level (i.e., when γ drops below a certain value).

Calculation of the SNR for the intermediate case (transition region) is quite complex.[1] Here we shall state the final results only:

$$\frac{S_o}{N_o} \simeq 0.916 A^2 \overline{m^2} \gamma^2 \tag{6.19}$$

Figure 6.5 shows the plot of S_o/N_o as a function of γ for AM with synchronous

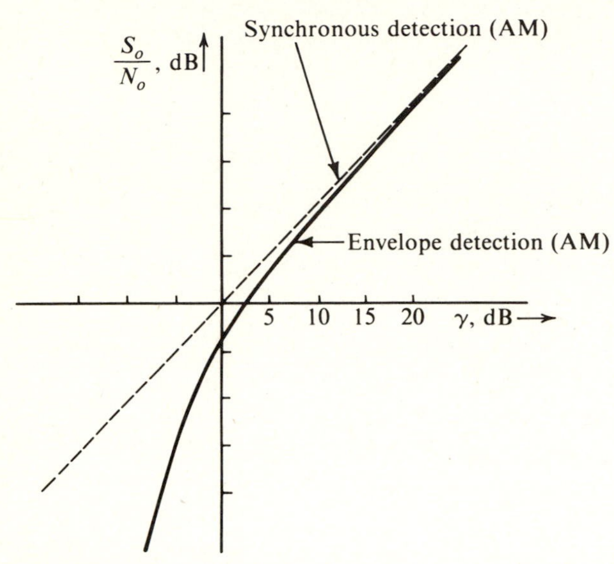

Figure 6.5 Performance of AM (synchronous detection and envelope detection).

detection and AM with envelope detection. The threshold effect is clearly seen from this figure. The threshold occurs when γ is of the order of 10 or less. For a reasonable quality AM signal, γ should be of the order of 1000 (30 dB), and the threshold is rarely a limiting condition.

■ EXAMPLE 6.2

Find γ_{thresh}, the value of γ at the threshold, in tone-modulated AM with $\mu = 1$ if the onset of the threshold is when $E_n > A$ with probability 0.01, E_n being the noise envelope.

☐ **Solution:** Because E_n has a Rayleigh PDF with variance σ_n^2

$$P(E_n \geq A) = \int_A^\infty \frac{E_n}{\sigma_n^2} e^{-E_n^2/2\sigma_n^2}\, dE_n$$

$$= e^{-A^2/2\sigma_n^2} = 0.01$$

Hence, at the onset of the threshold,

$$\frac{A^2}{2\sigma_n^2} = 4.605$$

The variance σ_n^2 of the bandpass noise of PSD $\mathcal{N}/2$ and bandwidth $2B$ centered at ω_c is $2(2B)(\mathcal{N}/2) = 2\mathcal{N}B$. Hence, at the onset of the threshold,

$$\frac{A^2}{4\mathcal{N}B} = 4.605$$

For tone modulation

$$\begin{aligned} m(t) &= \mu A \cos(\omega_m t + \Theta) \\ &= A \cos(\omega_m t + \Theta) \qquad (\mu = 1) \end{aligned}$$

and

$$S_i = \frac{A^2 + \overline{m^2}}{2} = \frac{A^2 + 0.5A^2}{2} = \frac{3A^2}{4}$$

Hence,

$$\gamma_{\text{thresh}} = \frac{S_i}{\mathcal{N}B} = \frac{3A^2}{4\mathcal{N}B} = 3(4.605) = 13.8 \text{ (or } 11.4 \text{ dB)}$$ ■

Square-Law Detection. The AM signal with noise is squared and then passed through a lowpass filter. The incoming signal (AM plus bandpass noise) is $y_i(t)$ in Eq. (6.16b). The output of the squarer is

$$\begin{aligned} y_i^2(t) &= E_i^2(t) \cos^2[\omega_c t + \Theta_i(t)] \\ &= \tfrac{1}{2}E_i^2(t)[1 + \cos(2\omega_c t + 2\Theta_i(t))] \end{aligned}$$

The lowpass filter suppresses the high-frequency term, yielding the output $y_o(t)$ as

$$\begin{aligned} y_o(t) &= \tfrac{1}{2}E_i^2(t) \\ &= \tfrac{1}{2}\{[A + m(t) + n_c(t)]^2 + n_s^2(t)\} \\ &= \tfrac{1}{2}[A^2 + m^2(t) + 2Am(t) + 2n_c(t)[A + m(t)] + n_c^2(t) + n_s^2(t)] \end{aligned}$$

The term $m^2(t)$ represents the signal distortion and is assumed to be small. Similarly, A^2 is the dc term. The useful signal is $Am(t)$, and the rest is the noise signal $n_o(t)$. Thus,

$$\begin{aligned} s_o(t) &= Am(t) \\ n_o(t) &= n_c(t)[A + m(t)] + \tfrac{1}{2}n_c^2(t) + \tfrac{1}{2}n_s^2(t) \end{aligned}$$

Here $n_c(t)$ and $n_s(t)$ are both gaussian, with identical PSDs. Moreover, $n_c(t)$ and $n_s(t)$ are incoherent. Hence, $n_c^2(t)$ and $n_s^2(t)$ are incoherent. Also, because n_c is gaussian, $\overline{n_c^3(t)} = 0$ (see Prob. 5.29). Hence, all the three terms in $n_o(t)$ are incoherent, and the power of $n_o(t)$ is the sum of the powers of each of the three terms. Moreover, the powers of $n_c^2(t)$ and $n_s^2(t)$ are identical. In addition, we note that $n_c^2(t)$ has a dc component $\overline{n_c^2}$. The dc noise component is eliminated by a blocking capacitor. Therefore, the effective power of $n_c^2(t)$ is $\overline{[n_c^2(t) - \overline{n_c^2}]^2}$. Hence,

$$\begin{aligned} N_o &= \overline{n_c^2(t)[A + m(t)]^2} + 2\tfrac{1}{4}\overline{[n_c^2(t) - \overline{n_c^2}]^2} \\ &= \overline{n_c^2}\,\overline{[A + m(t)]^2} + \tfrac{1}{2}[\overline{n_c^4} - (\overline{n_c^2})^2] \end{aligned}$$

Because n_c is gaussian (Prob. 5.29),

$$\overline{n_c^4} = 3(\overline{n_c^2})^2$$

and

$$N_o = \overline{n_c^2}[A^2 + \overline{m^2}] + \frac{1}{2}[3(\overline{n_c^2})^2 - (\overline{n_c^2})^2]$$
$$= \overline{n_c^2}[A^2 + \overline{m^2}] + (\overline{n_c^2})^2$$

Because

$$\overline{n_c^2} = 2\mathcal{N}B$$
$$N_o = 2\mathcal{N}B[A^2 + \overline{m^2}] + 4\mathcal{N}^2B^2$$

Also,

$$S_o = A^2\overline{m^2}$$

and

$$\frac{S_o}{N_o} = \frac{A^2\overline{m^2}}{2\mathcal{N}B(A^2 + \overline{m^2}) + 4\mathcal{N}^2B^2}$$

$$= \left(\frac{A^2}{A^2 + \overline{m^2}}\right)\left(\frac{\overline{m^2}}{A^2 + \overline{m^2}}\right)\frac{\dfrac{A^2 + \overline{m^2}}{2\mathcal{N}B}}{1 + \dfrac{2\mathcal{N}B}{A^2 + \overline{m^2}}}$$

$$= \left(\frac{A^2}{A^2 + \overline{m^2}}\right)\left(\frac{\overline{m^2}}{A^2 + \overline{m^2}}\right)\frac{\gamma}{1 + 1/\gamma}$$

$$= \left(\frac{A^2}{A^2 + \overline{m^2}}\right)\left(\frac{\overline{m^2}}{A^2 + \overline{m^2}}\right)\gamma \qquad \gamma \gg 1$$

$$= \left(\frac{A^2}{A^2 + \overline{m^2}}\right)\left(\frac{\overline{m^2}}{A^2 + \overline{m^2}}\right)\gamma^2 \qquad \gamma \ll 1$$

Note that for $\gamma \gg 1$, the behavior of the square-law detector is similar to that of the envelope detector except for the factor $(A^2/A^2 + \overline{m^2})$. Thus for tone modulation, the square-law detector is inferior to the envelope detector by a factor of 1.5 (or 1.8 dB). For small γ, the output SNR decreases rapidly with γ as γ^2. This is the threshold phenomenon.

The square-law detector not only causes distortion but its output SNR is inferior by 1.8 dB (for tone modulation) as compared to the envelope detector.

6.3 ANGLE-MODULATED SYSTEMS

A block diagram of an angle-modulated system is shown in Fig. 6.6. The angle-modulated (or exponentially modulated) carrier, $\varphi_{EM}(t)$, can be written as

$$\varphi_{EM}(t) = A\cos[\omega_c t + \psi(t)] \qquad \textbf{(6.20a)}$$

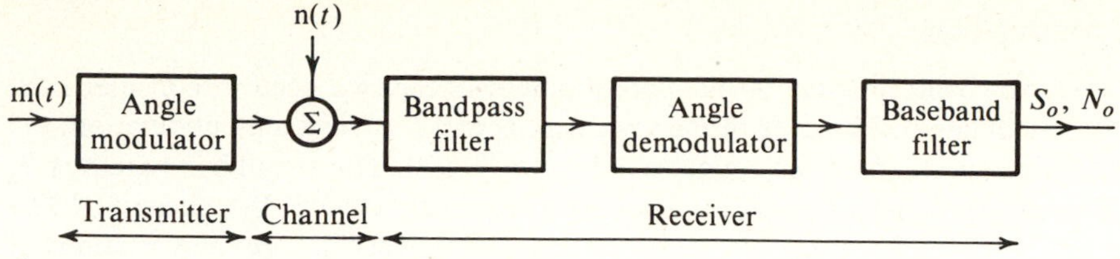

Figure 6.6 An angle-modulated system.

where

$$\psi(t) = k_p \mathrm{m}(t) \qquad \text{for PM} \tag{6.20b}$$

$$= k_f \int_{-\infty}^{t} \mathrm{m}(\alpha)\, d\alpha \qquad \text{for FM} \tag{6.20c}$$

and m(t) is the message signal. The channel noise, $\mathrm{n}_i(t)$, at the demodulator input is a bandpass noise with PSD $S_\mathrm{n}(\omega)$ and bandwidth $2(\Delta f + kB)$, where k is between 1 and 2, depending upon whether we have narrowband or wideband modulation. The noise $\mathrm{n}_i(t)$ can be expressed in terms of quadrature components as

$$\mathrm{n}_i(t) = \mathrm{n}_c(t) \cos \omega_c t + \mathrm{n}_s(t) \sin \omega_c t \tag{6.21a}$$

where $\mathrm{n}_c(t)$ and $\mathrm{n}_s(t)$ are lowpass signals of bandwidth $(\Delta f + kB)$. The noise may also be expressed in terms of the envelope $\mathrm{E}_\mathrm{n}(t)$ and phase $\Theta_\mathrm{n}(t)$ as

$$\mathrm{n}(t) = \mathrm{E}_\mathrm{n}(t) \cos [\omega_c t + \Theta_\mathrm{n}(t)] \tag{6.21b}$$

where

$$\mathrm{E}_\mathrm{n}(t) = \sqrt{\mathrm{n}_c^2(t) + \mathrm{n}_s^2(t)} \tag{6.21c}$$

$$\Theta_\mathrm{n}(t) = -\tan^{-1} \frac{\mathrm{n}_s(t)}{\mathrm{n}_c(t)} \tag{6.21d}$$

and

$$\mathrm{n}_c(t) = \mathrm{E}_\mathrm{n}(t) \cos \Theta_\mathrm{n}(t) \tag{6.21e}$$

$$\mathrm{n}_s(t) = \mathrm{E}_\mathrm{n}(t) \sin \Theta_\mathrm{n}(t) \tag{6.21f}$$

Angle modulation (and particularly wideband angle modulation) is a nonlinear type of modulation. Hence, superposition does not apply. In AM, the signal output can be calculated by assuming the channel noise to be zero, and the noise output can be calculated by assuming the modulating signal to be zero. This is a consequence of linearity. The signal and noise do not form intermodulation components. Unfortunately, exponential modulation is nonlinear, and we cannot use superposition to calculate the output as can be done in AM. We shall show that because of special circumstances, however, even in angle modulation the noise output can be calculated by assuming the modulating signal to be zero. To prove this we shall first consider the case of PM and then extend those results to FM.

Phase Modulation

Because narrowband modulation is approximately linear, we need to consider only wideband modulation. The crux of the argument is that for wideband modulation, the signal $\mathrm{m}(t)$ changes very slowly relative to the noise $\mathrm{n}_i(t)$. The signal bandwidth is B, and the noise bandwidth is $2(\Delta f + 2B)$, with $\Delta f \gg B$. Hence, the phase and frequency variations of the modulated carrier are much slower than are the variations of $\mathrm{n}_i(t)$. The modulated carrier appears to have constant frequency and phase over several cycles, and, hence, the carrier appears to be unmodulated. We may therefore calculate the output noise by assuming $\mathrm{m}(t)$ to be zero. This is a qualitative justification for the linearity argument. A quantitative justification is given in the following development.

To calculate the signal and noise powers at the output, we shall first construct a phasor diagram of the signal $\mathrm{y}_i(t)$ at the demodulator input.

$$\begin{aligned} \mathrm{y}_i(t) &= A \cos [\omega_c t + \psi(t)] + \mathrm{n}_i(t) \\ &= A \cos [\omega_c t + \psi(t)] + \mathrm{E}_\mathrm{n}(t) \cos [\omega_c t + \Theta_\mathrm{n}(t)] \\ &= \mathrm{R}(t) \cos [\omega_c t + \psi(t) + \Delta\psi(t)] \end{aligned} \tag{6.22}$$

where

$$\psi(t) = k_p \mathrm{m}(t) \qquad \text{for PM} \tag{6.23}$$

Figure 6.7 shows the phasor diagram depicting Eq. (6.22). For the small-noise case,

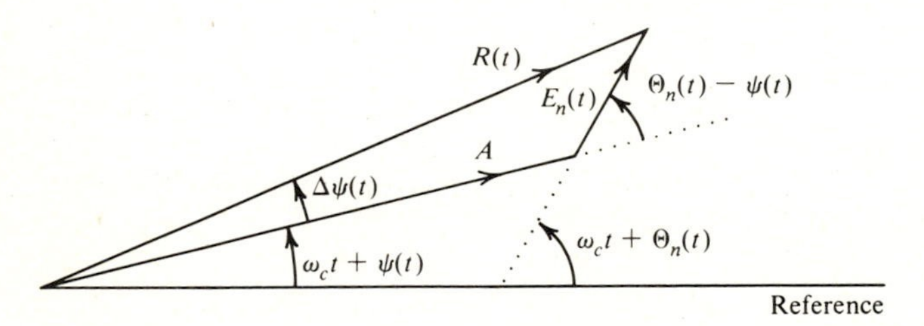

Figure 6.7 Phasor representation of signals in angle-modulated system.

where $\mathrm{E}_\mathrm{n}(t) \ll A$ for "almost all t," $\Delta\psi(t) \ll \pi/2$ for "almost all t," and

$$\Delta\psi(t) \simeq \frac{\mathrm{E}_\mathrm{n}(t)}{A} \sin [\Theta_\mathrm{n}(t) - \psi(t)] \tag{6.24}$$

The demodulator detects the phase of the input $\mathrm{y}_i(t)$. Hence, the demodulator output is

$$\mathrm{y}_o(t) = \psi(t) + \Delta\psi(t) \tag{6.25a}$$

$$= k_p \mathrm{m}(t) + \frac{\mathrm{E}_\mathrm{n}(t)}{A} \sin [\Theta_\mathrm{n}(t) - \psi(t)] \tag{6.25b}$$

Note that the noise term $\Delta\psi(t)$ involves the signal $\psi(t)$ because of the nonlinear nature of angle modulation.

$$\Delta\psi(t) = \frac{\mathrm{E_n}(t)}{A} \sin\left[\Theta_\mathrm{n}(t) - \psi(t)\right] \tag{6.26}$$

Because $\psi(t)$ [the baseband signal] varies much more slowly than $\Theta_\mathrm{n}(t)$ [the wideband noise], we can approximate $\psi(t)$ by a constant ψ.

$$\begin{aligned}\Delta\psi(t) &\simeq \frac{\mathrm{E_n}(t)}{A} \sin\left[\Theta_\mathrm{n}(t) - \psi\right] \\ &= \frac{\mathrm{E_n}(t)}{A} \sin\Theta_\mathrm{n}(t)\cos\psi + \frac{\mathrm{E_n}(t)}{A}\cos\Theta_\mathrm{n}(t)\sin\psi \\ &= \frac{\mathrm{n}_s(t)}{A}\cos\psi + \frac{\mathrm{n}_c(t)}{A}\sin\psi\end{aligned}$$

Also, because $\mathrm{n}_c(t)$ and $\mathrm{n}_s(t)$ are incoherent for white noise,

$$\begin{aligned}S_{\Delta\psi}(\omega) &= \frac{\cos^2\psi}{A^2} S_{\mathrm{n}_s}(\omega) + \frac{\sin^2\psi}{A^2} S_{\mathrm{n}_c}(\omega) \\ &= \frac{S_{\mathrm{n}_s}(\omega)}{A^2}\end{aligned}$$

For a white channel noise with PSD $\mathcal{N}/2$ [Eq. (5.129)],

$$S_{\Delta\psi}(\omega) = \begin{cases} \dfrac{\mathcal{N}}{A^2} & |f| < \Delta f + kB \\ 0 & \text{otherwise} \end{cases} \tag{6.27a}$$

Note that if we had assumed $\psi(t) = 0$ (zero message signal) in Eq. (6.26), we would have obtained exactly the same result.*

The demodulated noise bandwidth is $\Delta f + kB$. But because the useful signal bandwidth is only B, the demodulator output is passed through a lowpass filter of bandwidth B to remove the out-of-band noise. Hence, the PSD of the lowpass filter output noise is

$$S_{\mathrm{n}_o}(\omega) = \begin{cases} \dfrac{\mathcal{N}}{A^2} & |\omega| < 2\pi B \\ 0 & |\omega| > 2\pi B \end{cases} \tag{6.27b}$$

and

$$N_o = 2B\left(\frac{\mathcal{N}}{A^2}\right) = \frac{2\mathcal{N}B}{A^2} \tag{6.28a}$$

From Eq. (6.25b) we have

$$S_o = k_p^2 \overline{\mathrm{m}^2} \tag{6.28b}$$

---

* This follows from the fact that $\mathrm{E_n}(t)\sin\Theta_\mathrm{n}(t) = \mathrm{n}_s(t)$. Hence, $\Delta\psi(t) = \mathrm{n}_s(t)/A$, and $S_{\Delta\psi}(\omega) = S_{\mathrm{n}_s}(\omega)/A^2$. A somewhat more rigorous derivation of this result can be found in Sakrison.[2]

Thus,

$$\frac{S_o}{N_o} = (Ak_p)^2 \frac{\overline{\mathrm{m}^2}}{2\mathcal{N}B} \tag{6.28c}$$

These results are valid for small noise, and they apply to both WBPM and NBPM. We also have

$$\gamma = \frac{S_i}{\mathcal{N}B} = \frac{A^2/2}{\mathcal{N}B} = \frac{A^2}{2\mathcal{N}B}$$

and

$$\frac{S_o}{N_o} = k_p^2 \overline{\mathrm{m}^2}\, \gamma \tag{6.29}$$

Also, for PM (Eq. 4.69),

$$\Delta\omega = k_p m_p'$$

Hence,

$$\frac{S_o}{N_o} = (\Delta\omega)^2 \left(\frac{\overline{\mathrm{m}^2}}{m_p'^2}\right) \gamma \tag{6.30}$$

Note that the bandwidth of the angle-modulated waveform is about $2\Delta f$ (for the wideband case). Thus, the output SNR increases with the square of the transmission bandwidth; that is, the output SNR increases by 6 dB for each doubling of the transmission bandwidth. Remember, however, that this result is valid only when the noise power is much smaller than the carrier power. Because the noise power is proportional to the transmission bandwidth, the output SNR obtainable by increasing the transmission bandwidth is limited. When the noise power becomes comparable to the carrier power, the threshold appears, and a further increase in bandwidth would actually reduce the output SNR instead of increasing it.

Let us apply Eq. (6.30) to tone modulation, where $\mathrm{m}(t) = \alpha \cos \omega_m t$. For this case $\overline{\mathrm{m}^2} = \alpha^2/2$, and $m_p' = |\dot{\mathrm{m}}(t)|_{\max} = \alpha\omega_m$. Hence,

$$\frac{S_o}{N_o} = \frac{1}{2}\left(\frac{\Delta\omega}{\omega_m}\right)^2 \gamma \tag{6.31}$$

Note that Eqs. (6.28c), (6.29), (6.30), and (6.31) are valid for both NBPM and WBPM.

Frequency Modulation

Frequency modulation may be considered as a special case of phase modulation, where the modulating signal is $k_f \int_{-\infty}^{t} \mathrm{m}(\alpha)\, d\alpha$ (Fig. 6.8). At the receiver, we can demodulate FM with a PM demodulator followed by a differentiator to yield the output signal $k_f \mathrm{m}(t)$. Thus,

$$S_o = k_f^2 \overline{\mathrm{m}^2} \tag{6.32}$$

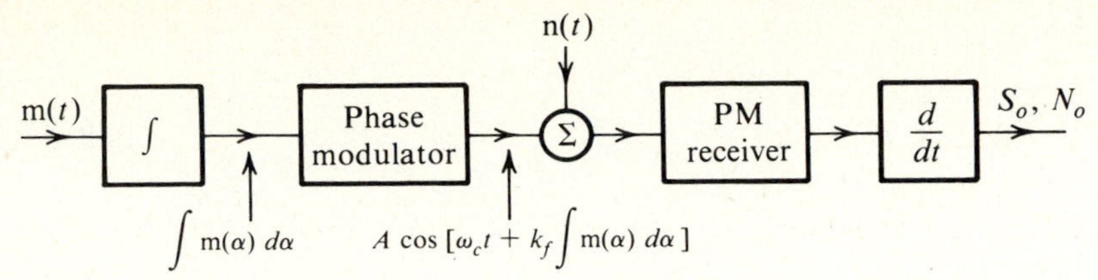

Figure 6.8 FM system as a special case of PM system.

The phase demodulator output noise will be identical to that calculated earlier, with PSD $\mathcal{N}/A^2$ for white channel noise. This noise is passed through an ideal differentiator whose transfer function is $j\omega$. Hence, the PSD $S_{n_o}(\omega)$ of the output noise is $|j\omega|^2$ times the PSD in Eq. (6.27b).

$$S_{n_o}(\omega) = \begin{cases} \dfrac{\mathcal{N}}{A^2}\,\omega^2 & |\omega| \le 2\pi B \\ 0 & |\omega| > 2\pi B \end{cases} \tag{6.33}$$

The PSD of the output noise is parabolic (Fig. 6.9), and the output noise power is

$$\begin{aligned} N_o &= 2\int_0^B \frac{\mathcal{N}}{A^2}(2\pi f)^2\, df \\ &= \frac{8\pi^2 \mathcal{N} B^3}{3A^2} \end{aligned} \tag{6.34}$$

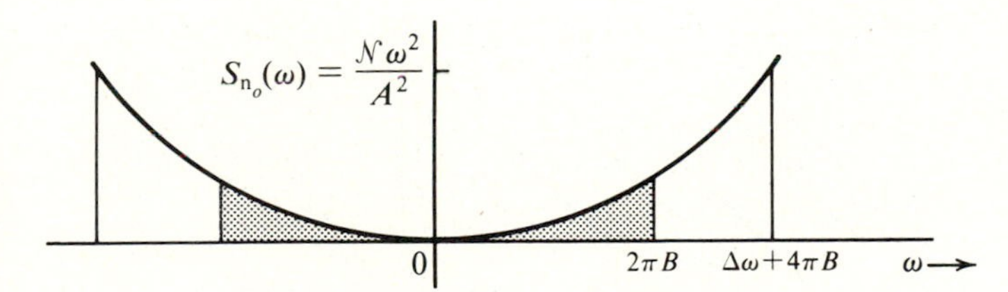

Figure 6.9 PSD of output noise in FM receiver.

Hence, the output SNR is

$$\begin{aligned} \frac{S_o}{N_o} &= 3\left(\frac{k_f^2\overline{\mathrm{m}^2}}{(2\pi B)^2}\right)\left(\frac{A^2/2}{\mathcal{N}B}\right) \\ &= 3\left[\frac{k_f^2\overline{\mathrm{m}^2}}{(2\pi B)^2}\right]\gamma \end{aligned} \tag{6.35}$$

Because $\Delta\omega = k_f m_p$,

$$\frac{S_o}{N_o} = 3\left(\frac{\Delta f}{B}\right)^2 \left(\frac{\overline{m^2}}{m_p^2}\right) \gamma \tag{6.36}$$

$$= 3\beta^2\gamma \left(\frac{\overline{m^2}}{m_p^2}\right) \tag{6.37}$$

Observe that the transmission bandwidth is about $2\Delta f$. Hence, for each doubling of the bandwidth, the output SNR increases by 6 dB. Just as in the case of PM, the output SNR does not increase indefinitely because threshold appears as the increased bandwidth makes the channel noise power comparable to the carrier power.

For tone modulation, $\overline{m^2}/m_p^2 = 0.5$ and

$$\frac{S_o}{N_o} = \frac{3}{2}\beta^2\gamma \tag{6.38}$$

The output SNR S_o/N_o (in dB) in Eq. (6.38) is plotted as a function of γ (in dB) for various values of β in Fig. 6.10. The dotted portion of the curves indicates the threshold region (discussed later in this section). Although the curves in Fig. 6.10 are valid for tone modulation only ($\overline{m^2}/m_p^2 = 0.5$), they can be used for any other

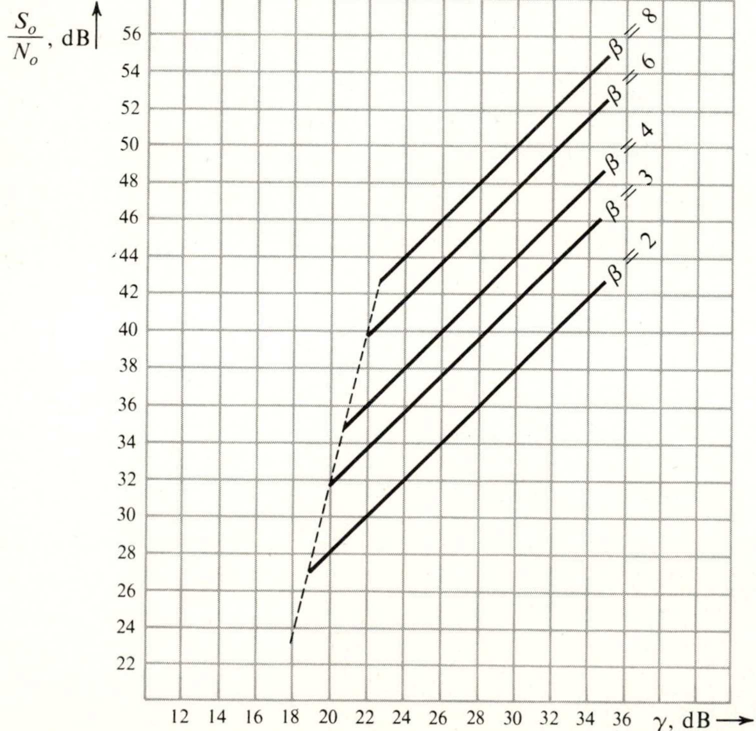

Figure 6.10 Performance of FM system.

modulating signal $m(t)$ simply by shifting them vertically by a factor $(\overline{m^2}/m_p^2)/0.5 = 2\overline{m^2}/m_p^2$.

From Eqs. (6.31) and (6.38), we observe that for tone modulation FM is superior to PM by a factor of 3. This does not mean that FM is superior to PM for other modulating signals as well. In fact, PM proves to be superior to FM for most of the practical signals (this was discussed in great detail in Sec. 4.15). From Eqs. (6.30) and (6.36), it can be seen that

$$\frac{(S_o/N_o)_{\text{PM}}}{(S_o/N_o)_{\text{FM}}} = \frac{(2\pi B)^2 m_p^2}{3m_p'^2} \tag{6.39}$$

Hence, if $(2\pi B)^2 m_p^2 > 3m_p'^2$, PM is superior to FM. If the PSD of $m(t)$ is concentrated at lower frequencies, low-frequency components predominate in $m(t)$, and m_p' is small. This favors PM. Therefore, in general, when $S_m(\omega)$ is concentrated at lower frequencies, PM is superior to FM, and when $S_m(\omega)$ is concentrated at higher frequencies, FM is superior to PM. This explains why for tone modulation, FM is superior to PM (see Fig. 4.67). The reader should reread Example 4.17, which shows a case of $m(t)$ for which PM is superior to FM by a factor of 4.

■ EXAMPLE 6.3

For a zero-mean gaussian random process $m(t)$ as the baseband signal, determine the output SNR for FM, assuming white gaussian channel noise.

☐ **Solution:** For a gaussian $m(t)$, $m_p = \infty$. But because the probability that amplitude m lies beyond $3\sigma_m$ ($|m| \geq 3\sigma_m$) is about 0.0027, one may consider* m_p to be $3\sigma_m$. Hence,

$$\overline{m^2} = \sigma_m^2 \qquad \text{and} \qquad m_p^2 = (3\sigma_m)^2$$

From Eq. (6.37) it follows that

$$\frac{S_o}{N_o} = \frac{1}{3}\beta^2\gamma$$

■

Narrowband Modulation. The equations derived thus far are valid for both narrowband and wideband modulation. A word about narrowband exponential modulation (NBEM) is in order here, however. We observed in Chapter 4 that this modulation is approximately linear and is very similar to AM. In fact, the output SNRs for NBEM and AM are similar. To see this, consider the cases of NBPM [Eq. (4.53)] and AM [Eq. (4.9a)].

$$\begin{aligned}\varphi_{\text{AM}}(t) &= A\cos\omega_c t + m(t)\cos\omega_c t \\ \varphi_{\text{NBPM}}(t) &= A\cos\omega_c t - Ak_p m(t)\sin\omega_c t \\ &= A\cos\omega_c t - m_1(t)\sin\omega_c t\end{aligned}$$

* This is known as "3σ loading." In the literature, "4σ loading" ($m_p = 4\sigma_m$) is also used.

where $m_1(t) = Ak_p m(t)$. Both φ_{AM} and φ_{NBPM} contain a carrier and a DSB term. In φ_{NBPM} the carrier and the DSB component are out of phase by $\pi/2$ radians, whereas in φ_{AM} they are in phase. But the $\pi/2$-radian phase difference has no effect on the power. Thus, $m(t)$ in φ_{AM} is analogous to $m_1(t)$ in φ_{NBPM}.

Now let us compare the output SNRs for AM and NBPM. For AM [Eq. (6.17)]

$$\left(\frac{S_o}{N_o}\right)_{AM} = \frac{\overline{m^2}}{A^2 + \overline{m^2}}\gamma$$

whereas for NBPM [Eq. (6.29)]

$$\left(\frac{S_o}{N_o}\right)_{PM} = k_p^2\overline{m^2}\,\gamma$$

$$= \frac{\overline{m_1^2}}{A^2}\gamma$$

Note that for NBPM, we require that $k_p m(t) \ll 1$, that is, $m_1(t)/A \ll 1$. Hence,

$$A^2 \simeq A^2 + \overline{m_1^2}$$

and

$$\left(\frac{S_o}{N_o}\right)_{PM} \simeq \frac{\overline{m_1^2}}{A^2 + \overline{m_1^2}}\gamma$$

which is of the same form as $(S_o/N_o)_{AM}$. Hence, NBPM is very similar to AM. Under the best possible conditions, however, AM outperforms NBPM because for AM, we need only to satisfy the conditions $[A + m(t)] > 0$ [Eq. (4.10a)], which implies $[m(t)]_{max} \leq A$. Thus for tone modulation, the modulation index for AM can be nearly equal to unity. For NBPM, however, the narrowband condition would be equivalent to requiring $\mu \ll 1$. Hence, although AM and NBPM have identical performance for a given value of μ, AM has the edge over NBPM.

It is interesting to look for the line (in terms of Δf) that separates narrowband and wideband FM. We may consider the dividing line to be that value of Δf for which the output SNR for FM given in Eq. (6.37) is equal to the maximum output SNR for AM. The maximum SNR for AM occurs when $\mu = 1$, or when $A = m_p$. Hence, equating Eq. (6.37) with Eq. (6.17) [with $A = m_p$],

$$3\beta^2\gamma\left(\frac{\overline{m^2}}{m_p^2}\right) = \frac{\overline{m^2}}{m_p^2 + \overline{m^2}}\gamma$$

or

$$\beta^2 = \frac{1}{3}\left[\frac{1}{1 + (\overline{m^2}/m_p^2)}\right] \tag{6.40}$$

Usually, $\overline{m^2}/m_p^2 \ll 1$ for practical signals, and

$$B^2 \simeq \frac{1}{3}$$

or

$$B \simeq 0.6 \tag{6.41a}$$

This gives

$$\Delta f = 0.6B \tag{6.41b}$$

Mean Square Bandwidth and Estimation of Angle-Modulation Bandwidth

The bandwidth of a signal is a measure of the width of the signal spectrum. Several definitions of bandwidth have appeared in the literature. All are meaningful and useful in different situations. A 3-dB bandwidth is commonly used in electronic circuits.

We will now introduce another useful measure of bandwidth. In Chapter 5, we noted that the standard deviation σ is a good measure of the width of a probability density function. We can extend this idea to the PSD by normalizing the spectrum to a unit area. The variance of such a normalized spectrum is known as the *mean square bandwidth* of the spectrum. The mean square bandwidth is meaningful, extremely useful, and mathematically tractable for angle-modulated signals.

For a baseband signal m(t) with PSD $S_{\text{m}}(\omega)$, the normalized PSD is $S_{\text{m}}(\omega)/\int_{-\infty}^{\infty} S_{\text{m}}(\omega)\, df$. The normalized PSD has unit area (when integrated with respect to f). Because the normalized PSD is symmetrical about the vertical axis ($f = 0$), it has a zero mean (in the sense of the PDF), and its variance $\overline{B_{\text{m}}^2}$ is*

$$\overline{B_{\text{m}}^2} = \frac{\int_{-\infty}^{\infty} f^2 S_{\text{m}}(2\pi f)\, df}{\int_{-\infty}^{\infty} S_{\text{m}}(2\pi f)\, df} \tag{6.42a}$$

$$= \frac{1}{\overline{\text{m}^2}} \int_{-\infty}^{\infty} f^2 S_{\text{m}}(2\pi f)\, df \tag{6.42b}$$

■ EXAMPLE 6.4

For a lowpass signal with PSD $S_{\text{m}}(\omega) = \Pi(\omega/4\pi B)$, show that $\overline{B_{\text{m}}^2} = B^2/3$.

☐ **Solution:** Because $S_{\text{m}}(\omega) = 1$ for $|f| < B$ and 0 for $|f| > B$, we have [Eq. (6.42)]

$$\overline{B_{\text{m}}^2} = \frac{\int_{-B}^{B} f^2\, df}{\int_{-B}^{B} df} = \frac{B^2}{3}$$

■

■ EXAMPLE 6.5

For a gaussian PSD $S_{\text{m}}(\omega) = ke^{-\omega^2/2\sigma^2}$ show that

$$\overline{B_{\text{m}}^2} = (\sigma/2\pi)^2$$

---

* For $\overline{B_{\text{m}}^2}$ to exist, $S_{\text{m}}(\omega)$ must approach zero at a rate faster than $1/\omega^2$ for large values of ω.

☐ **Solution:**

$$\overline{B_{\mathrm{m}}^2} = \frac{k \int_{-\infty}^{\infty} f^2 e^{-4\pi^2 f^2/2\sigma^2}\, df}{k \int_{-\infty}^{\infty} e^{-4\pi^2 f^2/2\sigma^2}\, df} = \frac{\sigma^2}{4\pi^2}$$ ■

We shall now investigate the dependence of the FM wave PSD on instantaneous frequency. Instantaneous frequency f_i in the range $(f, f + df)$ gives rise to power spectral components (components of the PSD) in the range $(f, f + df)$. The power contribution of these components is proportional to the relative time that f_i remains in this range. If $p_{\mathrm{f}_i}(f)$ is the PDF of instantaneous frequency f_i, then $p_{\mathrm{f}_i}(f)\, df$ is the probability of observing f_i in the range $(f, f + df)$. This probability is proportional to the time that f_i remains in the range $(f, f + df)$ and, hence, is proportional to the power contributed by spectral components in the range $(f, f + df)$. If $S_{\mathrm{FM}}(\omega)$ is the PSD of an FM wave, then the above argument implies*

$$S_{\mathrm{FM}}(2\pi f) = k p_{\mathrm{f}_i}(f) \qquad f > 0 \tag{6.43}$$

where k is a constant of proportionality.†

To find the mean square bandwidth of $S_{\mathrm{FM}}(2\pi f)$, we observe from Eq. (6.43) that $p_{\mathrm{f}_i}(f)$ is precisely $S_{\mathrm{FM}}(2\pi f)$ normalized to unit area. Hence, the mean square bandwidth of $S_{\mathrm{FM}}(2\pi f)$ is the variance $\sigma_{\mathrm{f}_i}^2$ of f_i. Because

$$\mathrm{f}_i = \frac{\omega_i}{2\pi} = \frac{1}{2\pi}[\omega_c + k_f \mathrm{m}(t)]$$

$$\overline{\mathrm{f}_i} = \frac{1}{2\pi}[\omega_c + k_f \overline{\mathrm{m}(t)}]$$

$$= f_c \qquad (\text{because } \overline{\mathrm{m}(t)} = 0)$$

and

$$\sigma_{\mathrm{f}_i}^2 = \overline{(\mathrm{f}_i - f_c)^2}$$

$$= \frac{1}{4\pi^2}\overline{[k_f \mathrm{m}(t)]^2}$$

$$= \frac{1}{4\pi^2} k_f^2 \overline{\mathrm{m}^2}$$

In other words,

$$\overline{B_{\mathrm{FM}}^2} = \frac{1}{4\pi^2} k_f^2 \overline{\mathrm{m}^2} \tag{6.44}$$

For phase modulation,

$$\mathrm{f}_i = \frac{1}{2\pi}\omega_i = \frac{1}{2\pi}[\omega_c + k_p \dot{\mathrm{m}}(t)]$$

---

* Because $p_{\mathrm{f}_i}(f) = 0$ for $f < 0$, $S_{\mathrm{FM}}(\omega)$ in Eq. (6.43) is the unilateral PSD. This means $S_{\mathrm{FM}}(\omega) = 0$ for $\omega < 0$ and is twice the bilateral PSD for $\omega > 0$.

† By integrating both sides of Eq. (6.43) over $(0, \infty)$, it can be shown that $k = A^2/2$.

and

$$\sigma_{f_i}^2 = \frac{1}{(2\pi)^2}\overline{(f_i - f_c)^2}$$

$$= \frac{1}{4\pi^2}k_p^2\overline{[\dot{m}(t)]^2}$$

Because the PSD of $\dot{m}(t)$ is $4\pi^2 f^2 S_m(2\pi f)$,

$$\sigma_{f_i}^2 = \frac{1}{4\pi^2}k_p^2 \int_{-\infty}^{\infty} 4\pi^2 f^2 S_m(2\pi f)\, df$$

Using Eq. (6.42) in the above,

$$\sigma_{f_i}^2 = k_p^2\overline{m^2}\,\overline{B_m^2}$$

that is,

$$\overline{B_{PM}^2} = k_p^2\overline{m^2}\,\overline{B_m^2} \tag{6.45}$$

From Eqs. (6.44) and (6.45) we observe that the bandwidth of the FM wave is independent of the modulating signal spectrum, whereas the bandwidth of the PM wave is strongly influenced by the modulating signal spectrum.

The output SNRs in Eqs. (6.29) and (6.35) can now be expressed in terms of mean square bandwidths.

For PM:

$$\frac{S_o}{N_o} = \frac{\overline{B_{PM}^2}}{\overline{B_m^2}}\gamma \tag{6.46}$$

For FM:

$$\frac{S_o}{N_o} = 3\frac{\overline{B_{FM}^2}}{B^2}\gamma \tag{6.47}$$

Quantities $\overline{B_{PM}^2}$ and $\overline{B_{FM}^2}$ are the variances of the normalized PSDs of the PM and FM waves, respectively. As seen earlier, the actual transmission bandwidth will be several times the standard deviation $\sqrt{\overline{B_{PM}^2}}$ or $\sqrt{\overline{B_{FM}^2}}$. For example, when the modulated signal PSD has a gaussian form,* 99.74 percent of the total power resides within 3σ of the carrier frequency. Hence, the bandwidth in this case may be taken as 6σ, or $6\sqrt{\overline{B_{PM}^2}}$ for PM and $6\sqrt{\overline{B_{FM}^2}}$ for FM. From Eqs. (6.46) and (6.47), it follows that in PM as well as in FM the output SNR improves by 6 dB for each doubling of the transmission bandwidth. It should be stressed that these results are valid only for small noise.

* It can be shown[3] that for wideband angle modulation, the PSD of the modulated carrier is gaussian when the modulating random process is gaussian.

Which Is Superior: PM or FM? From Eqs. (6.46) and (6.47), we have

$$\frac{(S_o/N_o)_{\text{PM}}}{(S_o/N_o)_{\text{FM}}} = \left(\frac{B^2}{3\overline{B_{\text{m}}^2}}\right)\left(\frac{\overline{B_{\text{PM}}^2}}{\overline{B_{\text{FM}}^2}}\right) \tag{6.48}$$

It will be instructive to compare the performance of PM and FM for the same transmission bandwidth (the same mean square bandwidth), that is, for $\overline{B_{\text{PM}}^2} = \overline{B_{\text{FM}}^2}$. This gives

$$\frac{(S_o/N_o)_{\text{PM}}}{(S_o/N_o)_{\text{FM}}} = \frac{B^2}{3\overline{B_{\text{m}}^2}} \tag{6.49a}$$

Thus, if

$$B^2 > 3\overline{B_{\text{m}}^2} \tag{6.49b}$$

PM is superior to FM. Otherwise, FM is superior to PM. We showed in Example 6.4 that when the PSD of m(t) is uniform, $B^2 = 3\overline{B_{\text{m}}^2}$. Hence, PM and FM perform equally well in that case. If the spectrum falls off with frequency, as it does for all real signals, $\overline{B_{\text{m}}^2}$ is less than $B^2/3$, and PM is superior to FM. If, on the other hand, the spectrum is weighted heavily at higher frequencies, $\overline{B_{\text{m}}^2}$ is greater than $B^2/3$, and FM is superior to PM. This is exactly what happens for tone modulation, where the spectrum is concentrated at one frequency B, with no power at lower frequencies, making $\overline{B_{\text{m}}^2} = B^2$. This is why for tone modulation FM is superior to PM by a factor of 3. Tone modulation proves to be grossly misleading in the SNR analysis of angle modulation. For practical signals including audio, the spectrum falls off with frequency, and PM is superior to FM. Actually, so-called FM broadcast is not pure FM but is FM modified by preemphasis-deemphasis, as discussed in Sec. 4.16.

We have derived two different criteria [Eqs. (6.39) and (6.49)] for comparing PM and FM performance. In Eq. (6.39), the output SNRs are compared for a given transmission bandwidth, whereas in Eq. (6.49), the output SNRs are compared for a given mean square transmission bandwidth. Consequently, we may get slightly different answers by using these two criteria (see Probs. 6.14 and 6.18). The criterion in Eq. (6.39) is preferred over that in Eq. (6.49). In most practical cases, however, it is impossible to determine the parameters required in Eq. (6.39), and the only way to make the comparison is through Eq. (6.49).

■ EXAMPLE 6.6

If a baseband signal m(t) has a gaussian-shaped PSD, show that PM is superior to FM by a factor of 3 (4.77 dB) when the bandwidth B is taken as 3σ, where σ is the standard deviation of the normalized PSD* of m(t).

□ **Solution:**

$$\overline{B_{\text{m}}^2} = \sigma^2 \qquad \text{and} \qquad B = 3\sigma$$

Hence, from Eq. (6.49), it follows that PM is superior to FM by a factor of 3. ■

* $S_{\text{m}}(f) = ke^{-f^2/2\sigma^2}$. This gives $\overline{B_{\text{m}}^2} = \sigma^2$ and $B = 3\sigma$.

The Threshold in Angle Modulation

When the noise power at the demodulator input is comparable to the carrier power, the threshold phenomenon (discussed earlier for AM systems) appears. In FM this effect is much more pronounced than in AM. Let us discuss qualitatively how the threshold effect appears. We refer back to Fig. 6.7 and observe that the phasor E_n rotates from the terminal of the phasor A. When $E_n \ll A$, the angle $\Delta\psi(t)$ is quite small, and because $\Theta_n(t)$ is random with uniform distribution in the range $(0, 2\pi)$, $\Delta\psi(t)$ assumes positive as well as negative values (Fig. 6.11*a*), which are usually much smaller than 2π. When E_n is large (of the order of A or greater), however, the resultant phasor is much more likely to rotate around the origin, and $\Delta\psi$ is more likely to go through changes of 2π (Fig. 6.11*c*) in a relatively short time, because the noise varies much faster than the modulating signal. The noise at the FM demodulator is given by $\Delta\dot{\psi}(t)$. This is shown in Fig. 6.11 for large and small noise. For large noise,

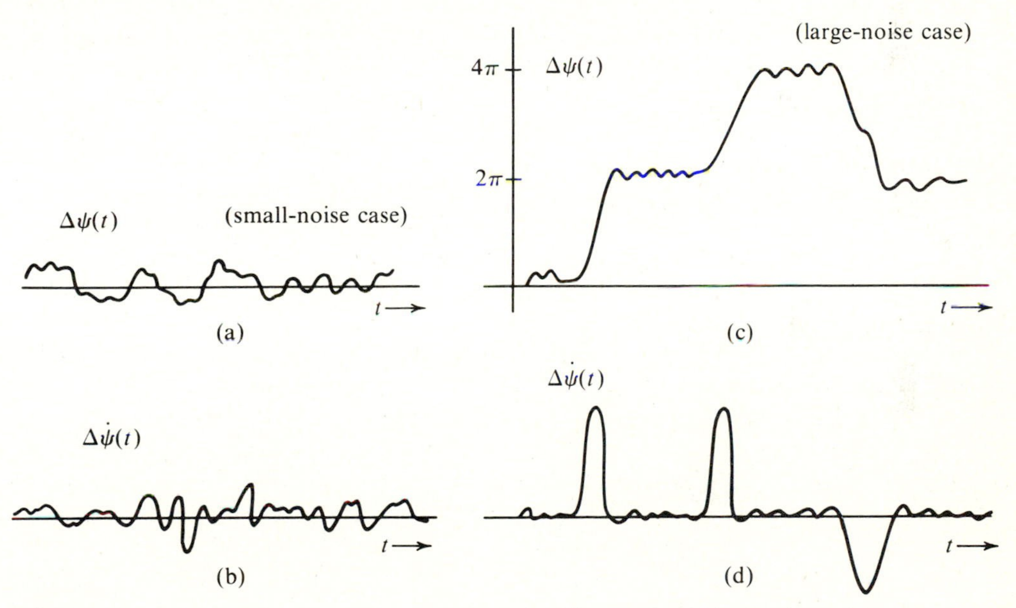

Figure 6.11 Nature of output noise in FM receiver for small and large channel noise.

we observe the appearance of spikes (of area 2π), which give rise to a crackling sound. When the noise is small ($E_n \ll A$), the power spectral density of the output noise $\Delta\dot{\psi}(t)$ is parabolic, and most of its power is in the frequencies greater than B and is therefore filtered out by the baseband filter at the output. For the large-noise case, on the other hand, we have the presence of spikes, which are like impulses. Consequently, they have considerable power at lower frequencies. Hence, a spike will contribute much more noise at the output. For this reason, when E_n approaches the order of A, the output noise starts increasing disproportionately (Fig. 6.12). This is precisely the phenomenon of threshold.

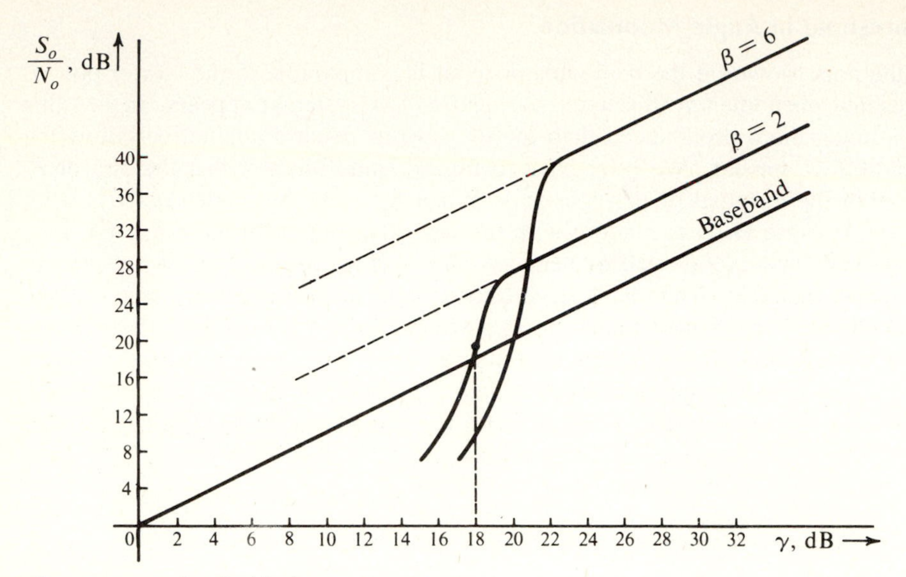

Figure 6.12 Threshold phenomenon in FM.

It has been shown that[4] N_s, the noise power caused by the spikes, is

$$N_s = \frac{8\pi^2 B_{\text{FM}} B}{\sqrt{3}} Q\left(\sqrt{\frac{2B}{B_{\text{FM}}}\gamma}\right) \tag{6.50a}$$

and N_T, the total noise power, is the sum of N_o in Eq. (6.34) and N_s in Eq. (6.50a). The output SNR is

$$\frac{S_o}{N_o + N_s} = \frac{3\beta^2\gamma(\overline{m^2}/m_p^2)}{1 + \dfrac{2\sqrt{3}\,B_{\text{FM}}\gamma}{B} Q\left(\sqrt{\dfrac{2B}{B_{\text{FM}}}\gamma}\right)}$$

$$= \frac{3\beta^2\gamma(\overline{m^2}/m_p^2)}{1 + 4\sqrt{3}(\beta + k)\gamma Q[\sqrt{\gamma/(\beta + k)}]} \tag{6.50b}$$

The behavior of an FM system in threshold is explained in Sec. 4.16. The onset of the threshold is when the carrier power is 10 times the channel noise power. We have shown that [Eq. (4.114)] γ_{thresh}, the value of γ at the onset of threshold, is

$$\gamma_{\text{thresh}} = 20(\beta + 2) \tag{6.51}$$

■ EXAMPLE 6.7

A gaussian m(t) with 4σ loading (that is, $m_p = 4\sigma_{\text{m}}$) frequency modulates a carrier using $\beta = 4$. The output SNR is found to be 20.5 dB.

Determine if the system is in threshold.

☐ **Solution:** For $\beta = 4$, γ_{thresh} is

$$\gamma_{\text{thresh}} = 20(\beta + 2) = (20)(6) = 120$$

For this value of γ [Eq. (6.37)],

$$\left(\frac{S_o}{N_o}\right) = 3(16)(120)\left(\frac{1}{16}\right) = 360$$

Because 20.5 dB is a ratio of 160, the SNR is below 360, and the system is in threshold. ■

Threshold Extension in Angle Modulation. The problem of threshold is rather serious in angle modulation. The ability to communicate even at low power levels simply by increasing the transmission bandwidth is the very raison d' être of angle-modulated systems, and threshold deprives angle modulation of its very essence. For this reason, much attention has been paid to the problem of pushing the threshold level further down and extending the useful operating range. The two principal methods used for this purpose utilize frequency compression feedback (FCF) and phase-lock loop (PLL). We shall look briefly into these techniques.

Frequency Compression Feedback (FCF). The basic circuit is shown in Fig. 6.13. Let us consider a frequency-modulated wave $A \cos [\omega_c t + \psi(t)]$ at the input.

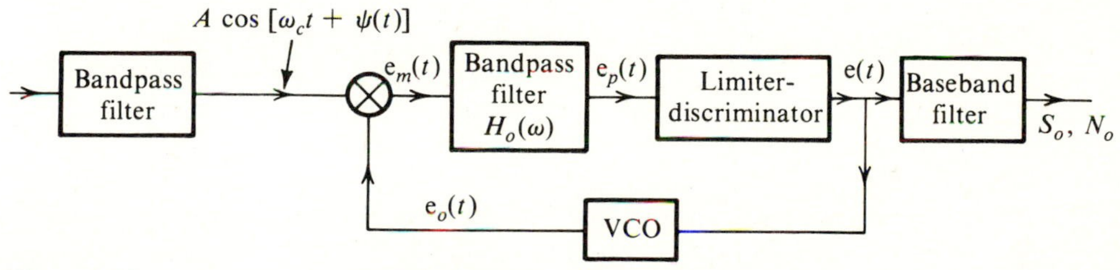

Figure 6.13 Frequency compression feedback (FCF).

The quiescent frequency of the VCO is $\omega_c - \omega_o$. The instantaneous frequency, ω_i, of the VCO is given by

$$\omega_i = (\omega_c - \omega_o) + \alpha \mathrm{e}(t)$$

and the VCO output, $\mathrm{e}_o(t)$, is given by* (assuming initial instant $t = 0$)

$$\mathrm{e}_o(t) = 2 \cos [(\omega_c - \omega_o)t + \alpha \int_0^t \mathrm{e}(\tau)\, d\tau] \tag{6.52}$$

The bandpass filter $H_o(\omega)$ is centered at ω_o. The multiplier output, $\mathrm{e}_m(t)$, has a spectrum centered at the sum and difference frequencies. The sum frequencies are centered at $2\omega_c - \omega_o$ and are suppressed by the bandpass filter $H_o(\omega)$, which allows only the difference frequencies centered at ω_o to pass. The output, $\mathrm{e}_p(t)$, of the bandpass filter is

$$\mathrm{e}_p(t) = A \cos [\omega_o t + \psi(t) - \alpha \int_0^t \mathrm{e}(\tau)\, d\tau] \tag{6.53}$$

*The amplitude of $\mathrm{e}_o(t)$ is immaterial in our discussion. For convenience, it is considered to be 2.

The limiter-discriminator frequency demodulates the signal $e_p(t)$. Therefore, the output $e(t)$ is given by

$$e(t) = \frac{d}{dt}[\psi(t) - \alpha \int_0^t e(\tau)\, d\tau]$$
$$= \dot{\psi}(t) - \alpha e(t)$$

Hence,

$$e(t) = \frac{\dot{\psi}(t)}{1 + \alpha}$$

Substitution of this value of $e(t)$ in Eq. (6.53) yields

$$e_p(t) = A \cos \left[\omega_o t + \frac{\psi(t)}{1 + \alpha} \right] \tag{6.54}$$

This is a rather interesting result. The signal $e_p(t)$ is another frequency-modulated signal, with carrier frequency ω_o. This is similar to the incoming carrier $A \cos [\omega_c t + \psi(t)]$. The difference is that the angle is $\psi(t)/(1 + \alpha)$ instead of $\psi(t)$. This implies a reduction in the frequency deviation and, consequently, a reduction in the bandwidth of the modulated signal by a factor* of $(1 + \alpha)$. Hence, the bandwidth of $H_o(\omega)$ need only be $2(\Delta f + 2B)/(1 + \alpha)$. The second conclusion is that when $e_p(t)$ is applied to a limiter-discriminator, the output $e(t)$ is $\dot{\psi}(t)/(1 + \alpha)$. Hence, the FCF demodulator indeed frequency demodulates the incoming frequency-modulated carrier.

Let us now see what happens when the input signal is a frequency-modulated carrier plus bandpass channel noise. The signal plus noise can be expressed as $R(t) \cos [\omega_c t + \psi(t) + \Delta\psi(t)]$ [Eq. (6.22)]. The amplitude variations $R(t)$ are eventually eliminated in the limiter-discriminator (see Sec. 4.15), and

$$e_p(t) = A \cos \left[\omega_o t + \frac{\psi(t) + \Delta\psi(t)}{1 + \alpha} \right] \tag{6.55}$$

Similarly,

$$e(t) = \frac{1}{1 + \alpha}[\dot{\psi}(t) + \Delta\dot{\psi}(t)]$$

where $\dot{\psi}$ is the useful signal and $\Delta\dot{\psi}$ is the noise. The output of the FCF demodulator is identical to that of the conventional demodulator except for the multiplicative factor $1/(1 + \alpha)$. Hence, the output SNR of the FCF demodulator is identical to that of the conventional demodulator. The advantage is gained, however, in extending the threshold. This can be seen from the fact that the FM wave passes unmolested through the filter $H_o(\omega)$ [Eq. (6.55)]. The channel noise, however, which had hitherto a bandwidth of $2(\Delta f + 2B)$, has to pass through $H_o(\omega)$ having $1/(1 + \alpha)$ times its former band-

* In order to avoid distortion of the modulating signal, the bandwidth $(\Delta f + kB)/1 + \alpha$ must be greater than B.

width. Hence, the carrier-power-to-noise ratio at the input of the limiter-discriminator is enhanced by a factor of $1 + \alpha$. As seen earlier, this does not affect the SNR, but it does extend the threshold because of the relative enhancement of the carrier power in relation to noise power. It can be seen that the carrier power can now afford to be reduced by a factor of $1 + \alpha$ before the onset of the threshold. In other words, the threshold is extended roughly by $10 \log \alpha$ dB. Because of practical problems, however, this benefit is not completely realized. In practical FCF demodulators, threshold extension of about 5 to 7 dB can be realized.

Phase-Lock Loop (PLL). The functioning of the PLL was discussed in Chapter 4. For the small-noise case, the PLL performance is identical to that of a conventional demodulator (just as in FCF). For large noise, however, the PLL extends the threshold just as the FCF does, by reducing the filter bandwidth. The detailed analysis is beyond our scope.[5] A practical PLL extends the threshold region by about 3 to 6 dB.

6.4 PULSE-MODULATED SYSTEMS

Among pulse-modulated systems (PAM, PWM, PPM, and PCM), only PCM is of practical importance. The other systems are rarely used in practice. For this reason we shall discuss in detail only PCM. It can be shown that PAM performance is similar to that of AM-SC systems (i.e., output SNR is equal to γ). The PWM and PPM systems are capable of exchanging the transmission bandwidth with output SNR, as in angle-modulated systems. In PWM, the output SNR is proportional to the transmission bandwidth B_T. This performance is clearly inferior to that of angle-modulated systems, where the output SNR increases as ${B_T}^2$. In PPM systems under optimum conditions, the output SNR increases as ${B_T}^2$ but is still inferior to FM by a factor of 6. For in-depth treatment of PAM, PWM, and PPM, the reader is referred to Rowe[6] or Panter.[1]

Another pulse-modulation system that deserves mention is the delta-modulation (DM) system discussed in Chapter 3. For speech signals, this system's performance is comparable to that of PCM for the bandwidth expansion ratio B_T/B of 7 to 8. For $B_T/B > 8$, PCM is superior to DM, and for $B_T/B < 8$, DM is superior to PCM.

Pulse-Code Modulation

In PCM, a baseband signal $m(t)$ bandlimited to B Hz and with amplitudes in the range $-m_p$ to m_p is sampled at a rate of $2B$ samples/second. The sample amplitudes are quantized into L levels, each separated by $2m_p/L$. Each quantized sample is encoded into n binary digits ($2^n = L$). The binary signal is transmitted over a channel. The receiver detects the binary signal and reconstructs quantized samples (decoding). The quantized samples are then passed through a lowpass filter to obtain the desired signal $m(t)$.

There are two sources of error in PCM: (1) quantization or "rounding off" error, and (2) detection error. The latter is caused by error in detection of the binary signal at the receiver.

As usual, $m(t)$ is assumed to be a wide-sense stationary random process. The random variable $m(kT_s)$, formed by sample-function amplitudes at $t = kT_s$, will be denoted by m_k. The kth sample m_k is rounded off, or quantized, to a value $\hat{\mathrm{m}}_k$, which is encoded and transmitted as binary digits. Because of the channel noise, some of the digits may be detected erroneously at the receiver, and the reconstructed sample will be $\widetilde{\mathrm{m}}_k$ instead of $\hat{\mathrm{m}}_k$. If q_k and ϵ_k are the quantization and detection errors, respectively, then

$$\mathrm{q}_k = \mathrm{m}_k - \hat{\mathrm{m}}_k \tag{6.56}$$

$$\epsilon_k = \hat{\mathrm{m}}_k - \widetilde{\mathrm{m}}_k$$

and

$$\mathrm{m}_k - \widetilde{\mathrm{m}}_k = \mathrm{q}_k + \epsilon_k \tag{6.57}$$

The total error, $\mathrm{m}_k - \widetilde{\mathrm{m}}_k$, at the receiver is $\mathrm{q}_k + \epsilon_k$. The receiver reconstructs the signal $\widetilde{\mathrm{m}}(t)$ from samples $\widetilde{\mathrm{m}}_k$ according to Eq. (2.84).

$$\begin{aligned}
\widetilde{\mathrm{m}}(t) &= \sum_k \widetilde{\mathrm{m}}_k \operatorname{sinc}(2Bt - k) \\
&= \sum_k [\mathrm{m}_k - (\mathrm{q}_k + \epsilon_k)] \operatorname{sinc}(2Bt - k) \\
&= \sum_k \mathrm{m}_k \operatorname{sinc}(2Bt - k) - \sum_k (\mathrm{q}_k + \epsilon_k) \operatorname{sinc}(2Bt - k) \\
&= \mathrm{m}(t) - \mathrm{e}(t)
\end{aligned} \tag{6.58a}$$

where

$$\mathrm{e}(t) = \sum_k (\mathrm{q}_k + \epsilon_k) \operatorname{sinc}(2Bt - k) \tag{6.58b}$$

The receiver therefore receives the signal $\mathrm{m}(t) - \mathrm{e}(t)$ instead of $\mathrm{m}(t)$. The error signal $\mathrm{e}(t)$ is a random process with kth sample $\mathrm{q}_k + \epsilon_k$. Because the process is wide-sense stationary, the mean square value of the process is the same as the mean square value at any instant. Because $\mathrm{q}_k + \epsilon_k$ is the value of $\mathrm{e}(t)$ at $t = kT_s$,

$$\overline{\mathrm{e}^2(t)} = \overline{(\mathrm{q}_k + \epsilon_k)^2}$$

Because q_k and ϵ_k are independent with zero-mean r.v.'s (see Examples 5.17, 5.18, and 5.19),

$$\overline{\mathrm{e}^2(t)} = \overline{\mathrm{q}_k^2} + \overline{\epsilon_k^2}$$

We have already derived $\overline{\mathrm{q}_k^2}$ and $\overline{\epsilon_k^2}$ in Examples 5.17 and 5.18 [Eqs. (5.64b) and (5.66b)]. Hence,

$$\overline{\mathrm{e}^2(t)} = \frac{1}{3}\left(\frac{m_p}{L}\right)^2 + \frac{4m_p^2 P_e (L^2 - 1)}{3L^2} \tag{6.59a}$$

where P_e is the detection error probability. For binary coding, each sample is encoded into n binary digits. Hence $2^n = L$, and

$$\overline{e^2(t)} = \frac{m_p^2}{3(2^{2n})}[1 + 4P_e(2^{2n} - 1)] \tag{6.59b}$$

As seen from Eq. (6.58a), the output $m(t) - e(t)$ contains the signal $m(t)$ and noise $e(t)$. Hence,

$$S_o = \overline{m^2} \qquad N_o = \overline{e^2(t)}$$

and

$$\frac{S_o}{N_o} = \frac{3(2^{2n})}{1 + 4P_e(2^{2n} - 1)}\left(\frac{\overline{m^2}}{m_p^2}\right) \tag{6.60}$$

The error probability P_e depends on A_p, the peak pulse amplitude, and the channel noise power, σ_n^2 [Eq. (5.41)].*

$$P_e = Q\left(\frac{A_p}{\sigma_n}\right)$$

It will be shown in Sec. 6.6 that A_p/σ_n can be maximized (that is, P_e can be minimized) by passing the incoming digital signal through an optimum filter (known as the *matched filter*). It will be shown that for polar signaling [Eq. (6.97)]

$$\left(\frac{A_p}{\sigma_n}\right)_{\max} = \sqrt{\frac{2E_p}{\mathcal{N}}}$$

and

$$(P_e)_{\min} = Q\left(\sqrt{\frac{2E_p}{\mathcal{N}}}\right) \tag{6.61}$$

where E_p is the energy of the received binary pulse and the channel noise is assumed to be white with PSD $\mathcal{N}/2$. Because there are n binary pulses/sample and there are $2B$ samples/second, there are a total of $2Bn$ pulses/second. Hence, the received signal power $S_i = 2BnE_p$, and

$$P_e = Q\left(\sqrt{\frac{S_i}{n\mathcal{N}B}}\right)$$

$$= Q\left(\sqrt{\frac{\gamma}{n}}\right) \tag{6.62}$$

---

* This assumes polar signaling. Bipolar (or pseudoternary) signaling requires 3 dB more power than polar to achieve the same P_e. In practice, bipolar rather than polar signaling is used for PCM to realize a dc null in the signal PSD (see Sec. 3.2).

and

$$\frac{S_o}{N_o} = \frac{3(2^{2n})}{1 + 4(2^{2n} - 1)Q\left(\sqrt{\frac{\gamma}{n}}\right)} \left(\frac{\overline{m^2}}{m_p^2}\right) \tag{6.63}$$

Figure 6.14 shows a plot of the output SNR as a function of γ for tone modulation

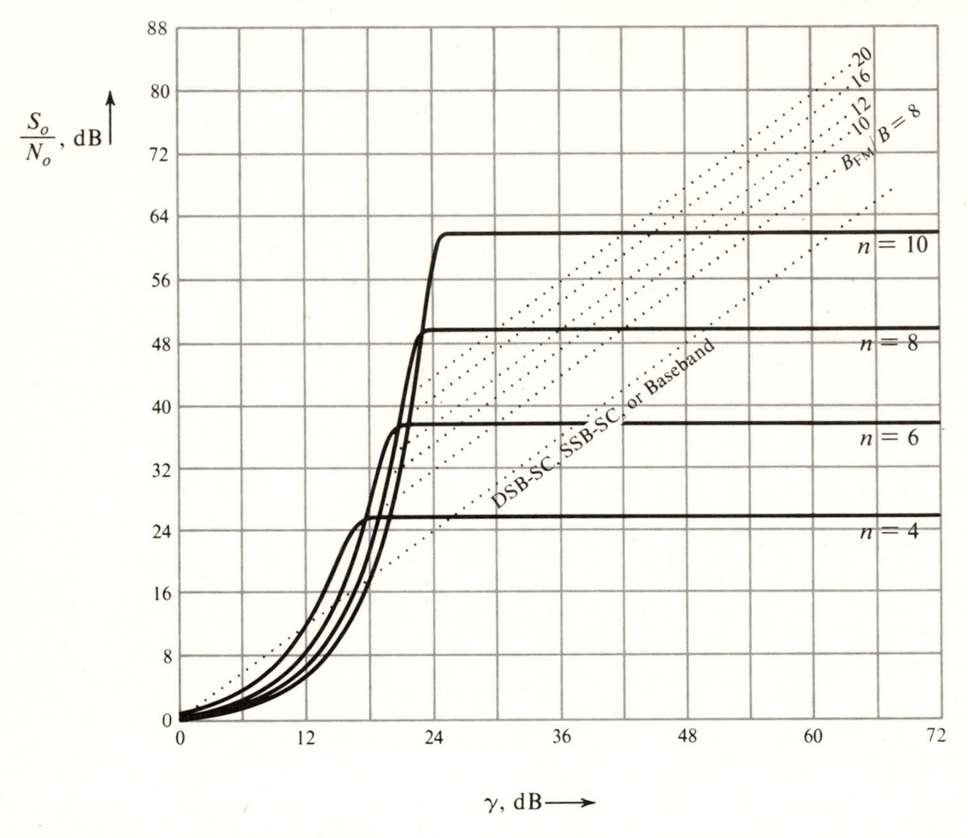

Figure 6.14 Performance of PCM.

($\overline{m^2}/m_p^2 = 0.5$). But it can be used for a general case, with $\overline{m^2}/m_p^2 = \lambda$. For this case the SNR is $\lambda/0.5 = 2\lambda$ times that in Fig. 6.14. Hence, we simply shift the curves in Fig. 6.14 upwards by 10 log (2λ) dB.

Figure 6.14 shows two features that need further comment: the threshold and the saturation. When γ is too small, a large pulse-detection error results, and the decoded pulse sequence yields a sample value that has no relation to the actual sample transmitted. The received signal is thus meaningless, and we have the phenomenon of threshold. To explain saturation, we observe that when γ is sufficiently large (implying sufficiently large pulse amplitude), the detection error $P_e \rightarrow 0$, and Eq. (6.60) becomes

$$\frac{S_o}{N_o} = 3L^2\left(\frac{\overline{m^2}}{m_p^2}\right) \tag{6.64a}$$

$$= 3(2^{2n})\left(\frac{\overline{m^2}}{m_p^2}\right) \tag{6.64b}$$

The SNR in this case is practically independent of γ. Because the detection error approaches zero, the output noise now consists entirely of the quantization noise, which depends only on L. Because the pulse amplitude is large enough so that there is very little probability of making a detection error, a further increase in γ by increasing the pulse amplitude buys no advantage, and we have the saturation effect.

In the saturation region

$$\left(\frac{S_o}{N_o}\right)_{\text{dB}} = 10\,[\log 3 + 2n \log 2 + \log (\overline{m^2}/m_p^2)]$$

$$= \alpha + 6n \tag{6.64c}$$

where $\alpha = 4.77 + 10 \log_{10} (\overline{m^2}/m_p^2)$.

■ EXAMPLE 6.8

For PCM with $n = 8$, determine the output SNR for a gaussian m(t). Assume the saturation region of operation.

□ **Solution:** For a gaussian signal, $m_p = \infty$. In practice, however, we may clip amplitudes $> 3\sigma_m$ or $4\sigma_m$, depending upon the accuracy desired. For example, in the case of 3σ loading,

$$p(|m| > 3\sigma_m) = 2Q(3) = 0.0026$$

and for 4σ loading

$$p(|m| > 4\sigma_m) = 2Q(4) = 6 \times 10^{-5}$$

If we take the case of 3σ loading,

$$\overline{m^2}/m_p^2 = \sigma_m^2/(3\sigma_m) = \tfrac{1}{9}$$

and

$$\frac{S_o}{N_o} = 3(2)^{16}\left(\frac{1}{9}\right) = 21{,}845 = 43.4 \text{ dB}$$

For 4σ loading,

$$\frac{S_o}{N_o} = 3(2)^{16}\left(\frac{1}{16}\right) = 12{,}288 = 40.9 \text{ dB}$$ ■

To facilitate the comparison of PCM with other types of modulation, the SNRs of FM and DSB-SC are superimposed on the SNR of PCM in Fig. 6.14. The theoretical bandwidth expansion ratio for PCM is $B_{PCM}/B = n$. In practice, this can be achieved by using duobinary signaling. Today's PCM systems use bipolar signaling, however, requiring $B_{PCM}/B = 2n$. Moreover, P_e in Eq. (6.62) is valid only for polar signaling. Bipolar signaling requires twice as much power. Hence, the plot in Fig. 6.14 is valid for bipolar signaling if 3 dB is added to each value of γ.

From Eq. (6.64b) we have

$$\frac{S_o}{N_o} = 3\left(\frac{\overline{m^2}}{m_p^2}\right)2^{2B_{PCM}/kB} \tag{6.65}$$

where $1 \le k \le 2$. For duobinary $k = 1$, and for bipolar $k = 2$.

It is clear from Eq. (6.65) that in PCM, the output SNR increases exponentially with the transmission bandwidth. Compare this with angle modulation, where the SNR increases as a square of transmission bandwidth. In angle modulation, doubling the transmission bandwidth quadruples the output SNR. From Eq. (6.64c) we see that in PCM, increasing n by 1 quadruples the SNR. But increasing n by 1 increases the bandwidth only by the fraction $1/n$. For $n = 8$, a mere 12.5 percent increase in the transmission bandwidth quadruples the SNR. Therefore in PCM, the exchange of SNR for bandwidth is much more efficient than that in angle modulation. This is particularly evident for large values of n, as can be seen from Fig. 6.14. For smaller values of n, the difference between FM and PCM is not as impressive as that for large values of n.*

Actually, to do full justice to PCM, we must consider the use of regenerative repeaters, which cannot be used for angle modulation. The results for PCM derived thus far apply for a single link. If k regenerative repeaters are used along the path, the noise power over each link is reduced by the factor k, and hence P_e, the error probability over each link, is

$$P_e = Q\left(\frac{k\gamma}{n}\right)$$

But because k links are now in tandem, the overall error probability is kP_e (see Example 5.7), and

$$\frac{S_o}{N_o} = \frac{3(2^{2n})}{1 + 4(2^{2n} - 1)kQ\left(\sqrt{\dfrac{k\gamma}{n}}\right)}\left(\frac{\overline{m^2}}{m_p^2}\right) \tag{6.66}$$

To maintain a given S_o/N_o, the value of γ decreases as k, the number of repeaters, increases. Figure 6.15 shows γ vs. k needed to maintain $S_o/N_o = 49.5$ dB for $n = 8$.

Companded PCM. The output SNR of PCM is proportional to $\overline{m^2}/m_p^2$, where $\overline{m^2}$ is the power of the baseband signal $m(t)$ and m_p is the peak value of $m(t)$. It may appear that $\overline{m^2}/m_p^2$ will remain more or less constant regardless of the speech level, because $\overline{m^2}$ is proportional to m_p^2. Unfortunately, m_p^2 is a constant of the quantizer with a

* The FM plots in Fig. 6.14 are without preemphasis and deemphasis.

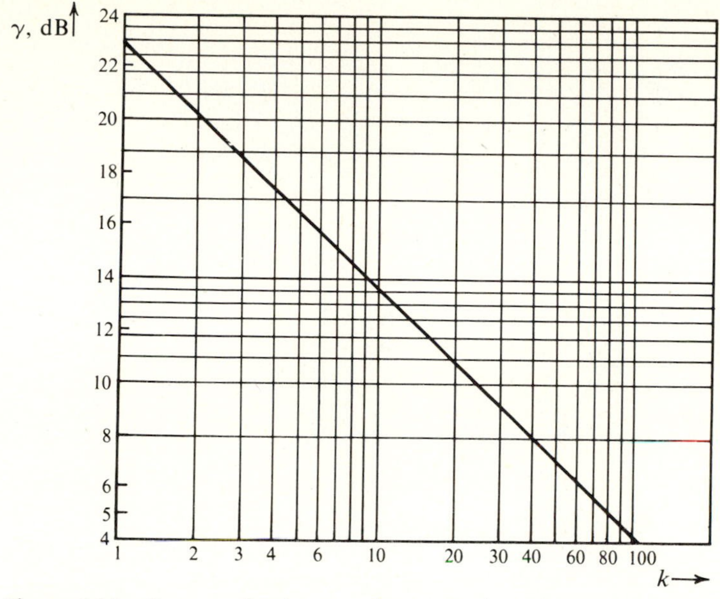

Figure 6.15 Power reduction as a function of number of repeaters in PCM.

quantization range $(-m_p, m_p)$. Once a quantizer is designed, m_p is fixed, and $\overline{\mathrm{m}^2}/m_p^2$ is proportional to the speech signal power $\overline{\mathrm{m}^2}$ only. This can vary from talker to talker (or even for the same talker) by as much as 40 dB, causing the output SNR to vary widely. This problem can be mitigated, and a relatively constant SNR over a large dynamic range of $\overline{\mathrm{m}^2}$ can be obtained, either by nonuniform quantization or by signal companding. Both methods are equivalent, but the latter is simpler to implement. In this method, the signal amplitudes are nonlinearly compressed.

Figures 3.41 and 3.42 show the input-output characteristics of the two most commonly used compressors (the μ-law and the A-law). For convenience, let us denote

$$\mathrm{x} = \frac{\mathrm{m}}{m_p}$$

Thus, x and y are the normalized input and output of the compressor, each with unit peak value (Fig. 6.16). The input-output characteristics have an odd symmetry about $x = 0$. For convenience, we have only shown the region $\mathrm{x} \geq 0$. The output signal samples in the range $(-1, 1)$ are uniformly quantized into L levels, with a quantization interval of $2/L$. Figure 6.16 shows the jth quantization interval for the output y as well as the input x. All input sample amplitudes that lie in the range Δ_j are mapped into y_j. For the input sample value x in the range Δ_j, the quantization error is $q = (x - x_j)$, and

$$2 \int_{x_j-(\Delta_j/2)}^{x_j+(\Delta_j/2)} (x - x_j)^2 p_{\mathrm{x}}(x)\, dx$$

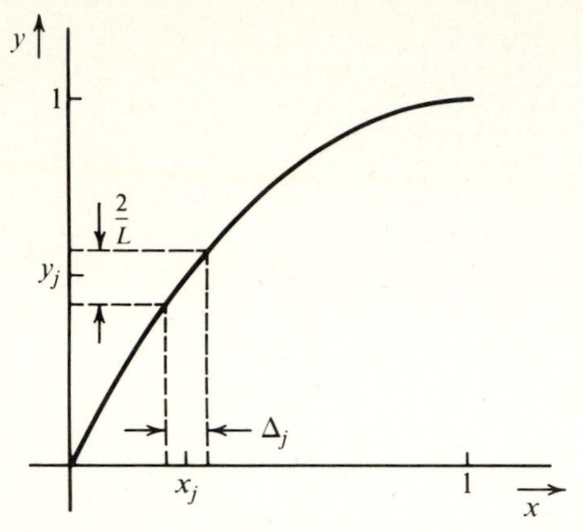

Figure 6.16 Input-output characteristic of a PCM compressor.

is the part of $\overline{q^2}$ (the mean square quantizing error) contributed by x in the region Δ_j. The factor 2 appears because there is an equal contribution from negative amplitudes of x centered at $-x_j$. Thus,

$$\overline{q^2} = 2 \sum_j \int_{x_j-(\Delta_j/2)}^{x_j+(\Delta_j/2)} (x - x_j)^2 p_x(x) dx$$

Because $L \gg 1$, the quantizing interval $(2/L)$ and Δ_j are very small, and $p_x(x)$ can be assumed to be constant over each interval. Hence,

$$\overline{q^2} = 2 \sum_j p_x(x_j) \int_{x_j-(\Delta_j/2)}^{x_j+(\Delta_j/2)} (x - x_j)^2 \, dx$$

$$= 2 \sum_j \frac{p_x(x_j)\, \Delta_j^3}{12} \tag{6.67}$$

Because $2/L$ and Δ_j are very small, the compression characteristics can be assumed to be linear over each Δ_j, and

$$\dot{y}(x_j) \simeq \frac{2/L}{\Delta_j}$$

Substituting this in Eq. (6.67), we have

$$\overline{q^2} \simeq \frac{2}{3L^2} \sum_j \frac{p_x(x_j)}{[\dot{y}(x_j)]^2} \Delta_j$$

For L large enough, the above sum can be approximated by an integral

$$\overline{q^2} \simeq \frac{2}{3L^2} \int_0^1 \frac{p_x(x)}{[\dot{y}(x)]^2} dx \tag{6.68}$$

For the μ-law [Eq. (3.54a)],

$$y = \frac{\ln(1 + \mu x)}{\ln(1 + \mu)} \qquad 0 \leq x \leq 1$$

and

$$\dot{y}(x) = \frac{\mu}{\ln(1+\mu)}\left(\frac{1}{1+\mu x}\right)$$

and

$$\overline{q^2} = \left(\frac{2}{3L^2}\right)\left(\frac{\ln(1+\mu)}{\mu}\right)^2 \int_0^1 (1+\mu x)^2 p_x(x)\,dx \qquad \textbf{(6.69)}$$

If $p_x(x)$ is symmetrical about $x = 0$,

$$\sigma_x^2 = 2\int_0^1 x^2 p_x(x)\,dx \qquad \textbf{(6.70a)}$$

and $\overline{|x|}$, the mean of the rectified x, is

$$\overline{|x|} = 2\int_0^1 x p_x(x)\,dx \qquad \textbf{(6.70b)}$$

We can express $\overline{q^2}$ as

$$\overline{q^2} = \left[\frac{\ln(1+\mu)}{\mu}\right]^2\left[\frac{1+\mu^2\sigma_x^2 + 2\mu\overline{|x|}}{3L^2}\right] \qquad \textbf{(6.71a)}$$

$$= \frac{[\ln(1+\mu)]^2}{3L^2}\left(\sigma_x^2 + \frac{2\overline{|x|}}{\mu} + \frac{1}{\mu^2}\right) \qquad \textbf{(6.71b)}$$

It should be noted that $\overline{q^2}$ in Eq. (6.71) is the normalized quantization error. The actual error is $m_p^2\overline{q^2}$. The normalized output signal is $x(t)$, and, hence, the normalized output power $S_o = \sigma_x^2 = \overline{m^2}/m_p^2$. The actual S_o will be $m_p^2\sigma_x^2$. Hence,

$$\frac{S_o}{N_o} = \frac{\sigma_x^2}{\overline{q^2}} = \frac{3L^2}{[\ln(1+\mu)]^2}\,\frac{\sigma_x^2}{\left(\sigma_x^2 + \dfrac{2\overline{|x|}}{\mu} + \dfrac{1}{\mu^2}\right)} \qquad \textbf{(6.72a)}$$

$$= \frac{3L^2}{[\ln(1+\mu)]^2}\,\frac{1}{\left(1 + \dfrac{2\overline{|x|}}{\mu\sigma_x^2} + \dfrac{1}{\mu^2\sigma_x^2}\right)} \qquad \textbf{(6.72b)}$$

To get an idea of the relative importance of various terms in the parentheses in Eq. (6.72b), we note that x is an r.v. distributed in the range $(-1, 1)$. Hence, σ_x^2 and $\overline{|x|}$ are both less than 1, and $\overline{|x|}/\sigma_x$ is typically in the range of 0.7 to 0.9. The values of μ used in practice are greater than 100. For example, the D2 channel bank used in conjunction with the T-1 carrier system has $\mu = 255$. It is therefore evident that the second and third terms in the parentheses in Eq. (6.72b) are small compared to 1 if σ_x^2 is not too small, and so

$$\frac{S_o}{N_o} \simeq \frac{3L^2}{[\ln(1+\mu)]^2} \qquad \textbf{(6.72c)}$$

which is independent of σ_x^2. The exact expression in Eq. (6.72b) has a weak dependence on σ_x^2 over a broad range of σ_x. Note that the SNR in Eq. (6.72b) also depends on the signal statistics $\overline{|x|}$ and σ_x^2. But for most of the practical PDFs, $\overline{|x|}/\sigma_x$ is practically the same (in the range of 0.7 to 0.9). Hence, S_o/N_o depends only on σ_x^2. This means the plot of S_o/N_o vs. σ_x^2 will be practically independent of the PDF of x. Figure 6.17 shows the plot of S_o/N_o vs. σ_x^2 for two different PDFs, namely the

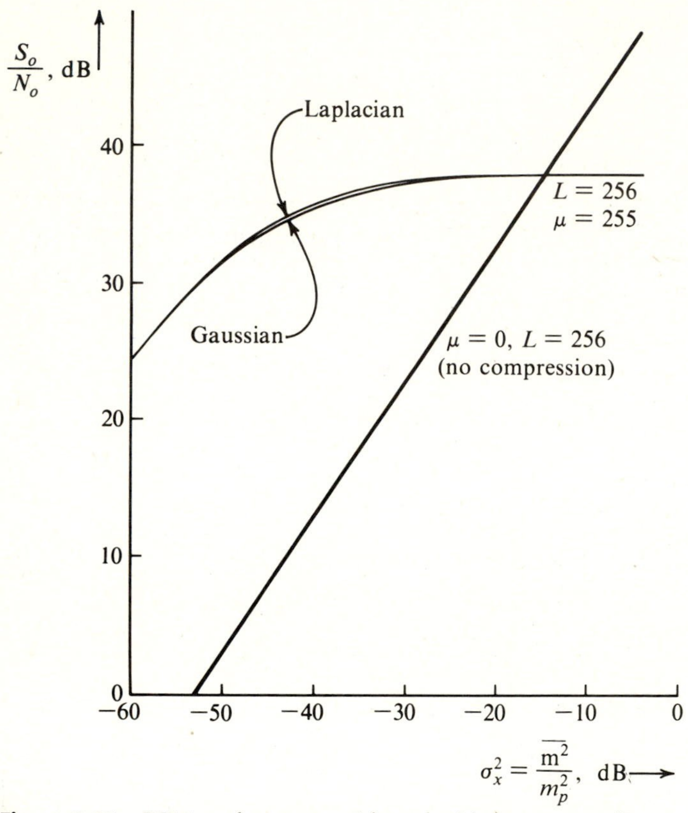

Figure 6.17 PCM performance with and without companding.

laplacian and gaussian (see the next example). It can be seen that there is hardly any difference between the two curves.

Because $x = m/m_p$, $\sigma_x^2 = \sigma_m^2/m_p^2$ and $\overline{|x|} = \overline{|m|}/m_p$, Eq. (6.72a) becomes

$$\frac{S_o}{N_o} = \frac{3L^2}{[\ln(1+\mu)]^2}\left[\frac{\sigma_m^2/m_p^2}{\dfrac{\sigma_m^2}{m_p^2} + \dfrac{2\overline{|m|}}{\mu m_p} + \dfrac{1}{\mu^2}}\right] \tag{6.73a}$$

One should be careful in interpreting m_p in the above equation. Once the system is designed for some $\mathrm{m}(t)$, m_p is fixed. Hence, m_p is a constant of the system, not of the signal $\mathrm{m}(t)$ that may be subsequently transmitted.

■ EXAMPLE 6.9

A voice-signal-amplitude PDF can be closely modeled by Laplace density*:

$$p_{\mathrm{m}}(m) = \frac{1}{\sigma_{\mathrm{m}}\sqrt{2}} e^{-\sqrt{2}|m|/\sigma_{\mathrm{m}}}$$

For a voice PCM system with $n = 8$ and $\mu = 255$, find and sketch the output SNR as a function of the normalized voice power $\sigma_{\mathrm{m}}^2/m_p^2$.

It is straightforward to show that the variance of the above Laplace PDF is σ_{m}^2. In practice, the speech amplitude will be limited either by 3σ loading or 4σ loading. In either case, the probability of observing m beyond this limit will be negligible, and in computing $\overline{|\mathrm{m}|}$ (etc.), we may use the limits 0 to ∞.

$$\overline{|\mathrm{m}|} = 2\int_0^{\infty} \frac{m}{\sigma_{\mathrm{m}}\sqrt{2}} e^{-\sqrt{2}m/\sigma_x}\, dm = 0.707\sigma_{\mathrm{m}}$$

Hence, from Eq. (6.73a),

$$\frac{S_o}{N_o} = \frac{(6383)(\sigma_{\mathrm{m}}^2/m_p^2)}{(\sigma_{\mathrm{m}}^2/m_p^2) + 0.00555(\sigma_{\mathrm{m}}/m_p) + 1.53 \times 10^{-5}} \tag{6.74}$$

This is plotted as a function of $(\sigma_{\mathrm{m}}^2/m_p^2)$ in Fig. 6.17. ■

■ EXAMPLE 6.10

Repeat Example 6.9 for the gaussian $\mathrm{m}(t)$.

☐ **Solution:** In this case,

$$\overline{|\mathrm{m}|} = 2\int_0^{\infty} \frac{m}{\sigma_{\mathrm{m}}\sqrt{2\pi}} e^{-m^2/2\sigma_{\mathrm{m}}^2}\, dm = 0.798\sigma_{\mathrm{m}}$$

and

$$\frac{S_o}{N_o} = = \frac{6383(\sigma_{\mathrm{m}}^2/m_p^2)}{(\sigma_{\mathrm{m}}^2/m_p^2) + 0.0063(\sigma_{\mathrm{m}}/m_p) + 1.53 \times 10^{-5}} \tag{6.75}$$

---

* A better but more complex model for speech signal amplitude m is the gamma density[7]

$$p_{\mathrm{m}}(m) = \sqrt{\frac{k}{4\pi|\mathrm{m}|}}\, e^{-k|m|}$$

The SNR here is practically the same as that in Eq. (6.74). The plot of the SNR vs. σ_m^2/m_p^2 (Fig. 6.17) is practically indistinguishable from that in Example 6.9. ■

6.5 OPTIMUM PREEMPHASIS-DEEMPHASIS SYSTEMS

It is possible to increase the output SNR by deliberate distortion of the transmitted signal (preemphasis) and corresponding compensation (deemphasis) at the receiver. This can be shown qualitatively in the case of white channel noise and a signal m(t) whose PSD decreases with frequency. In this case, we can boost the high-frequency components of m(t) at the transmitter (preemphasis). Because the signal has relatively less power at high frequencies, this preemphasis will require only a small increase in transmitted power.* At the receiver, the high-frequency components are attenuated (or deemphasized) in order to undo the preemphasis at the transmitter. This will restore the useful signal to its original form. It is an entirely different story with the channel noise. Because the noise is added after the transmitter, it does not undergo preemphasis. At the receiver, however, it does undergo deemphasis (i.e., attenuation of high-frequency components). Thus, at the receiver output, the signal power is restored but the noise power is reduced. The output SNR is therefore increased.

We shall first consider a baseband system and then extend the discussion to modulated systems. A baseband system with a preemphasis filter $H_p(\omega)$ at the transmitter and the corresponding complementary deemphasis filter $H_d(\omega)$ at the receiver is shown in Fig. 6.18. The channel transfer function is $H_c(\omega)$, and the PSD of the input

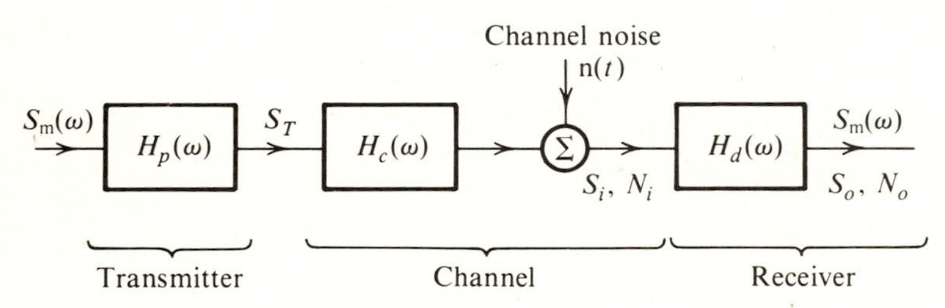

Figure 6.18 Optimum PDE in baseband system.

signal m(t) is $S_m(\omega)$. We shall determine the optimum preemphasis-deemphasis (PDE) filters $H_p(\omega)$ and $H_d(\omega)$ required for distortionless transmission of the signal m(t).

For distortionless transmission,

$$|H_p(\omega)H_c(\omega)H_d(\omega)| = G \text{ (a constant)} \tag{6.76a}$$

and

$$\theta_p(\omega) + \theta_c(\omega) + \theta_d(\omega) = -\omega t_d \tag{6.76b}$$

We want to maximize the output SNR, S_o/N_o, for a given transmitted power S_T.

* Actually, the transmitted power is maintained constant by attenuating the preemphasized signal slightly.

Referring to Fig. 6.18, we have

$$S_T = \int_{-\infty}^{\infty} S_m(\omega)|H_p(\omega)|^2 \, df \tag{6.77a}$$

Because $|H_p(\omega)H_c(\omega)H_d(\omega)| = G$, the signal power S_o at the receiver output is

$$S_o = G^2 \int_{-\infty}^{\infty} S_m(\omega) \, df \tag{6.77b}$$

The noise power N_o at the receiver output is

$$N_o = \int_{-\infty}^{\infty} S_n(\omega)|H_d(\omega)|^2 \, df \tag{6.77c}$$

Thus,

$$\frac{S_o}{N_o} = \frac{G^2 \int_{-\infty}^{\infty} S_m(\omega) \, df}{\int_{-\infty}^{\infty} S_n(\omega)|H_d(\omega)|^2 \, df} \tag{6.78}$$

We wish to maximize this ratio subject to the condition in Eq. (6.77a) with S_T as a given constant. We can include this constraint by multiplying the numerator and the denominator of the right-hand side of Eq. (6.78) by the left-hand side and the right-hand side, respectively, of Eq. (6.77a). This gives

$$\frac{S_o}{N_o} = \frac{G^2 S_T \int_{-\infty}^{\infty} S_m(\omega) \, df}{\int_{-\infty}^{\infty} S_n(\omega)|H_d(\omega)|^2 \, df \int_{-\infty}^{\infty} S_m(\omega)|H_p(\omega)|^2 \, df} \tag{6.79}$$

The numerator of the right side of Eq. (6.79) is fixed. Hence, to maximize S_o/N_o, we need only minimize the denominator of the right side of Eq. (6.79). To do this, we use the Schwarz inequality [Eq. (2.78b)], as follows: We have

$$\int_{-\infty}^{\infty} S_m(\omega)|H_p(\omega)|^2 \, df \int_{-\infty}^{\infty} S_n(\omega)|H_d(\omega)|^2 \, df \geq \left| \int_{-\infty}^{\infty} [S_m(\omega)S_n(\omega)]^{1/2} |H_p(\omega)H_d(\omega)| \, df \right|^2 \tag{6.80}$$

The equality holds only if

$$S_m(\omega)|H_p(\omega)|^2 = K^2 S_n(\omega)|H_d(\omega)|^2 \tag{6.81}$$

where K is an arbitrary constant. Thus to maximize S_o/N_o, Eq. (6.81) must be satisfied. Substitution of Eq. (6.76a) in Eq. (6.81) yields

$$|H_p(\omega)|^2_{\text{opt}} = GK \frac{\sqrt{S_n(\omega)/S_m(\omega)}}{|H_c(\omega)|} \tag{6.82a}$$

$$|H_d(\omega)|^2_{\text{opt}} = \frac{G}{K} \frac{\sqrt{S_m(\omega)/S_n(\omega)}}{|H_c(\omega)|} \tag{6.82b}$$

for the preemphasis and deemphasis filters. The constant K in Eq. (6.82) can be determined from Eqs. (6.77a) and (6.82a). The output SNR under optimum preemphasis and deemphasis can be calculated by substituting Eq. (6.82a and b) in Eq. (6.79). The result is

$$\left(\frac{S_o}{N_o}\right)_{\text{opt}} = \frac{S_T \int_{-\infty}^{\infty} S_{\text{m}}(\omega)\, df}{\left(\int_{-\infty}^{\infty} \frac{\sqrt{S_{\text{m}}(\omega) S_{\text{n}}(\omega)}}{|H_c(\omega)|}\, df\right)^2} \tag{6.82c}$$

Equation (6.82a and b) gives the magnitudes of the optimum filters $H_p(\omega)$ and $H_d(\omega)$. The phase functions must be chosen to satisfy the condition of distortionless transmission [Eq. (6.76b)].

Observe that the preemphasis filter in Eq. (6.82a) boosts frequency components where the signal is weak and suppresses frequency components where the signal is strong. The deemphasis filter in Eq. (6.82b) does exactly the opposite. Thus, the signal is unchanged but the noise is reduced.

EXAMPLE 6.11

Consider the case

$$S_{\text{m}}(\omega) = \begin{cases} \dfrac{C}{\omega^2 + (1400\pi)^2} & |\omega| \leq 8000\pi \\ 0 & |\omega| \geq 8000\pi \end{cases} \tag{6.83a}$$

The channel noise is white with PSD

$$S_{\text{n}}(\omega) = \frac{\mathcal{N}}{2} \tag{6.83b}$$

The channel is assumed to be ideal ($H_c(\omega) = 1$ and $G = 1$) over the band of interest (0 to 4000 Hz).

Without preemphasis-deemphasis, we have

$$S_o = \int_{-4000}^{4000} S_{\text{m}}(\omega)\, df$$

$$= 2 \int_0^{4000} \frac{C}{4\pi^2 f^2 + (1400\pi)^2}\, df$$

$$= 10^{-4} C \tag{6.84a}$$

Also, because $G = 1$, the transmitted power $S_T = S_o$

$$S_o = S_T = 10^{-4} C \tag{6.84b}$$

and

$$N_o = \mathcal{N} B = 4000\mathcal{N} \tag{6.84c}$$

Therefore,

$$\frac{S_o}{N_o} = \frac{(2.5C)10^{-8}}{\mathcal{N}} \tag{6.85}$$

The optimum transmitting and receiving filters are given by [Eq. (6.82a and b)]

$$|H_p(\omega)|^2 = K\sqrt{\frac{\mathcal{N}}{2C}[\omega^2 + (1400\pi)^2]} \tag{6.86a}$$

$$|H_d(\omega)|^2 = \frac{1}{K}\sqrt{\frac{2C}{\mathcal{N}[\omega^2 + (1400\pi)^2]}} \tag{6.86b}$$

The arbitrary constant K may be determined by substituting Eq. (6.86a) in Eq. (6.77a). The output SNR using optimum preemphasis and deemphasis is given by [Eq. (6.82c)]

$$\left(\frac{S_o}{N_o}\right)_{\text{opt}} = \frac{(10^{-4}C)^2}{\dfrac{\mathcal{N}C}{2}\left[\displaystyle\int_{-4000}^{4000} \frac{df}{\sqrt{4\pi^2 f^2 + (1400\pi)^2}}\right]^2}$$

$$= \frac{3.3(10^{-8}C)}{\mathcal{N}} \tag{6.87}$$

Comparison of Eq. (6.85) with the above result shows that preemphasis-deemphasis has increased the output SNR by a factor of 1.32. ■

Optimum Preemphasis-Deemphasis in AM Systems

Because the channel noise in a DSB system is in the band $\omega_c \pm 2\pi B$, the PDE should be carried out in this band. This means PDE filters $H_p(\omega)$ and $H_d(\omega)$ are bandpass filters located as shown in Fig. 6.19. The optimization is localized to the subsystem

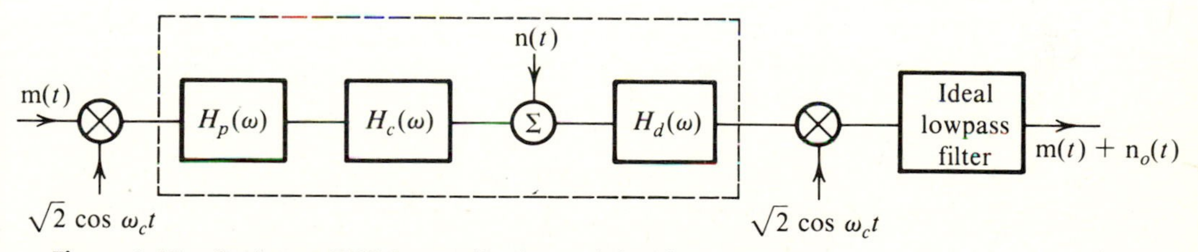

Figure 6.19 Optimum PDE in amplitude-modulated systems.

shown in the dashed box. This subsystem is identical to that in Fig. 6.18. Hence, $H_p(\omega)$ and $H_d(\omega)$ can be obtained from Eq. (6.82a and b). However, because the signal PSD at the input of the subsystem in Fig. 6.19 is $\frac{1}{2}[S_m(\omega + \omega_c) + S_m(\omega - \omega_c)]$, we should use this PSD in place of $S_m(\omega)$ in Eq. (6.82a and b). The same argument applies to SSB and VSB systems. In these cases, $H_p(\omega)$ and $H_d(\omega)$ can be computed from Eq. (6.82a and b) by replacing $S_m(\omega)$ with the appropriate SSB or VSB power spectral density.

Optimum Preemphasis-Deemphasis in Angle Modulation

We shall consider the case of PDE in the baseband, that is, preemphasis before modulation and deemphasis after demodulation (Fig. 6.20a). The channel is assumed to be ideal, that is, $H_c(\omega) = 1$.

The problem here is very different from the baseband communication case. In the baseband case, the preemphasis filter changes the transmitted power but not its bandwidth, and optimization was performed under the constraint of a fixed transmitted power (because bandwidth remained fixed, it was of no concern). In FM, on the other hand, the transmitted power is always fixed ($A^2/2$), but the transmission bandwidth is modified by the preemphasis filter. But optimization must be performed under the constraint of a fixed transmission bandwidth. Which bandwidth should be used? The coventional ($2\Delta f$) or the mean square? A look at Eq. (6.36) shows that the SNR in terms of conventional bandwidth involves the constant m_p. When $\mathrm{m}(t)$ is passed through $H_p(\omega)$, m_p will change, and no simple relationship exists between $\mathrm{m}(t)$, $H_p(\omega)$, and the new m_p. The problem is mathematically intractable. No such difficulty appears when we use Eq. (6.47), which uses the mean square bandwidth. Hence, we shall optimize the ratio S_o/N_o with the constraint of a fixed mean square bandwidth of transmission. Fortunately, this constraint turns out to be similar to that used in the baseband system. This can be seen from Eq. (6.44):

$$\overline{B_{\mathrm{FM}}^2} = \frac{1}{4\pi^2} k_f^2 \overline{\mathrm{m}^2}$$

Thus to maintain $\overline{B_{\mathrm{FM}}^2}$ fixed, $\overline{\mathrm{m}^2}$, the power of the modulating signal, must be the same with or without preemphasis. This is exactly the constraint of the baseband system [Eq. (6.77a)]. If the input to the system in the dotted box (Fig. 6.20a) is $\mathrm{x}(t)$, its output

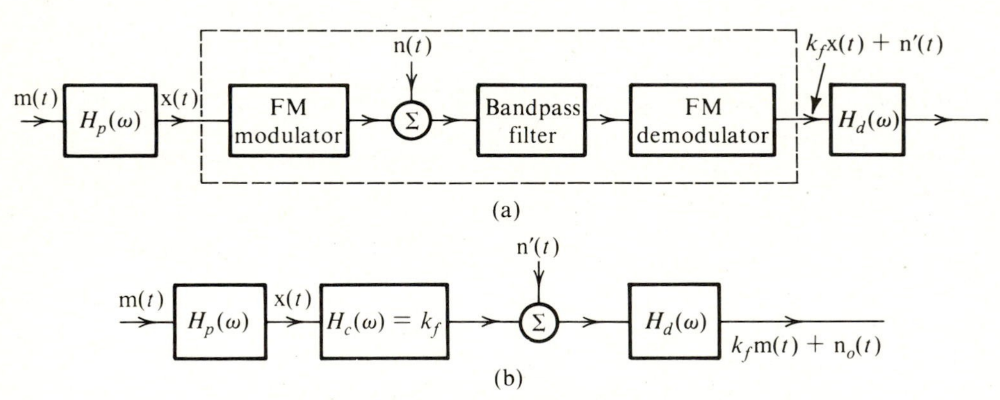

Figure 6.20 (a) Optimum PDE in FM system. (b) Equivalent of system in part (a).

is $k_f\mathrm{x}(t)$ plus the parabolic noise $\mathrm{n}'(t)$ with the PSD in Eq. (6.33). Hence, the system in Fig. 6.20b is the equivalent of the system in Fig. 6.20a. Our problem is the optimization of the output SNR with the constraint that the output of $H_p(\omega)$ has a given mean square value. This problem is identical to the optimization of the baseband system with $G = H_c(\omega) = k_f$. Hence, from Eq. (6.82a, b, and c) with $S_{\mathrm{n}'}(\omega) = \mathcal{N}\omega^2/A^2$ and $S_T = \overline{\mathrm{m}^2}$, we have

$$|H_p(\omega)|_{\mathrm{opt}}^2 = \frac{K\omega}{A}\sqrt{\mathcal{N}/S_{\mathrm{m}}(\omega)} \tag{6.88a}$$

$$|H_d(\omega)|^2_{\text{opt}} = \frac{A}{K\omega}\sqrt{S_m(\omega)/\mathcal{N}} \tag{6.88b}$$

and

$$\left(\frac{S_o}{N_o}\right)_{\text{opt}} = \frac{\left(\frac{A^2k_f^2}{4\pi^2\mathcal{N}}\right)\overline{m^2}\int_0^B S_m(2\pi f)\,df}{2\left[\int_0^B f\sqrt{S_m(2\pi f)}\,df\right]^2} \tag{6.88c}$$

Because $\gamma = A^2/2\mathcal{N}B$,

$$\left(\frac{S_o}{N_o}\right)_{\text{opt}} = \left[\frac{B^3\overline{m^2}}{6\left[\int_0^B f\sqrt{S_m(2\pi f)}\,df\right]^2}\right]\left(\frac{3\overline{B^2_{\text{FM}}}}{B^2}\gamma\right) \tag{6.88d}$$

Comparison of Eq. (6.88d) with Eq. (6.47) shows that PDE in FM improves the SNR by the factor inside the brackets on the right-hand side of Eq. (6.88d). The constant K in Eq. (6.88a and b) can be obtained from the constraint

$$\int_{-B}^{B} S_m(\omega)\,df = \int_{-B}^{B} S_m(\omega)|H_p(\omega)|^2\,df$$

Substituting Eq. (6.88a) in the right-hand side and solving it gives the value of K.

Optimum preemphasis and deemphasis is not used in commercial FM broadcasting for historical and practical reasons. The relatively simple suboptimum scheme, discussed in Sec. 4.16, is used instead.

PART II Digital Systems

In analog systems, the chief objective is the fidelity of reproduction of waveforms, and, hence, the suitable performance criterion is the output signal-to-noise ratio. The choice of this criterion stems from the fact that the signal-to-noise ratio is related to the ability of the listener to interpret a message. In digital communication systems, the transmitter input is chosen from a finite set of possible symbols, or messages. The objective at the receiver is not to reproduce the waveform with fidelity, because the possible waveforms are already known exactly, and the details of the waveform really are not important. Our goal is to determine, from the received signal masked by channel noise, which of the finite number of waveforms has been transmitted. Logically, the appropriate figure of merit in a digital communication system is not the signal-to-noise ratio but the probability of error in making the decision at the receiver.

6.6 OPTIMUM THRESHOLD DETECTION: THE BINARY CASE

In the threshold detection method discussed in Example 5.13, the received pulse is sampled at its peak amplitude A_p. Because of channel noise, the sampled value is not A_p but is $A_p + \mathrm{n}$. The decision is made from the value $A_p + \mathrm{n}$. It was shown in Example 5.13 that for the polar binary case the error probability P_e is

$$P_e = Q(\rho) \tag{6.89}$$

where

$$\rho = \frac{A_p}{\sigma_{\mathrm{n}}} \tag{6.90a}$$

and σ_{n}^2 is the variance of the received noise.

Because $Q(\rho)$ decreases monotonically with ρ, to minimize P_e, we need to maximize ρ.

Let the received pulse $p(t)$ be time limited to T_o (Fig. 6.21), and let the pulses be transmitted every T_o interval. There is a possibility of increasing $\rho = A_p/\sigma_{\mathrm{n}}$ by passing

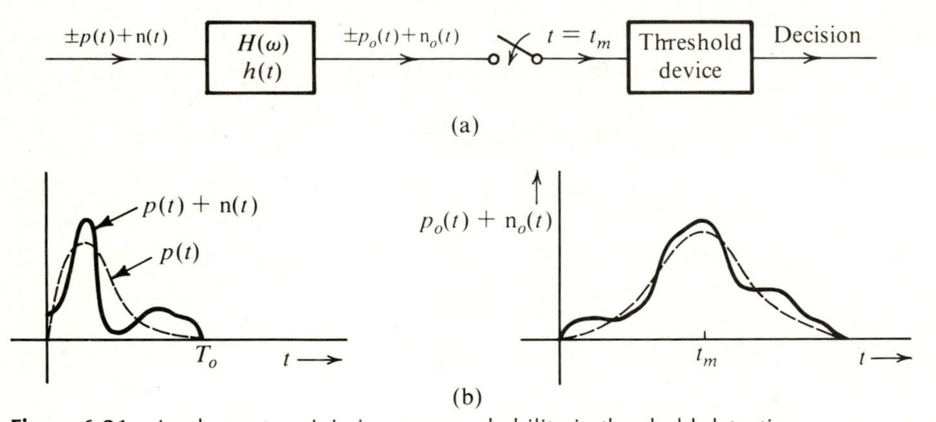

Figure 6.21 A scheme to minimize error probability in threshold detection.

the received pulse plus noise through a filter that enhances the pulse amplitude at some instant t_m and simultaneously reduces the noise power σ_{n}^2 (Fig. 6.21). We thus seek a filter with a transfer function $H(\omega)$ that maximizes ρ where

$$\rho^2 = \frac{p_o^2(t_m)}{\sigma_{\mathrm{n}}^2} \tag{6.90b}$$

Because

$$p_o(t) = \mathcal{F}^{-1}[P(\omega)H(\omega)]$$

$$= \frac{1}{2\pi}\int_{-\infty}^{\infty} P(\omega)H(\omega)e^{j\omega t}\, d\omega$$

we have

$$p_o(t_m) = \frac{1}{2\pi}\int_{-\infty}^{\infty} P(\omega)H(\omega)e^{j\omega t_m}\,d\omega \tag{6.91}$$

Also,

$$\sigma_n^2 = \overline{n_o^2(t)} = \frac{1}{2\pi}\int_{-\infty}^{\infty} S_n(\omega)|H(\omega)|^2\,d\omega \tag{6.92}$$

Hence,

$$\rho^2 = \frac{1/2\pi\left[\int_{-\infty}^{\infty} H(\omega)P(\omega)e^{j\omega t_m}\,d\omega\right]^2}{\int_{-\infty}^{\infty} S_n(\omega)|H(\omega)|^2\,d\omega} \tag{6.93}$$

In the Schwarz inequality [Eq. (2.78b)], if we identify $G_1(\omega) = H(\omega)\sqrt{S_n(\omega)}$ and $G_2(\omega) = P(\omega)e^{j\omega t_m}/\sqrt{S_n(\omega)}$, then it follows from Eq. (6.93) that

$$\rho^2 \le \frac{1}{2\pi}\int_{-\infty}^{\infty} \frac{|P(\omega)|^2}{S_n(\omega)}\,d\omega \tag{6.94a}$$

with equality only if

$$H(\omega)\sqrt{S_n(\omega)} = k\left[\frac{P(\omega)e^{j\omega t_m}}{\sqrt{S_n(\omega)}}\right]^* = \frac{kP(-\omega)e^{-j\omega t_m}}{\sqrt{S_n(\omega)}}$$

or

$$H(\omega) = k\,\frac{P(-\omega)e^{-j\omega t_m}}{S_n(\omega)} \tag{6.94b}$$

where k is an arbitrary constant.

For white channel noise $S_n(\omega) = \mathcal{N}/2$, and Eq. (6.94a and b) becomes

$$\rho_{\max}^2 = \frac{1}{\pi\mathcal{N}}\int_{-\infty}^{\infty} |P(\omega)|^2\,d\omega = \frac{2E_p}{\mathcal{N}} \tag{6.95a}$$

where E_p is the energy of $p(t)$, and

$$H(\omega) = k'P(-\omega)e^{-j\omega t_m} \tag{6.95b}$$

where $k' = 2k/\mathcal{N}$ is an arbitrary constant.

The unit impulse response, $h(t)$, of the optimum filter is given by

$$h(t) = \mathcal{F}^{-1}[k'P(-\omega)e^{-j\omega t_m}]$$

Note that $p(-t) \leftrightarrow P(-\omega)$ and $e^{-j\omega t_m}$ represents the time delay of t_m seconds. Hence,

$$h(t) = k'p(t_m - t) \tag{6.95c}$$

The signal $p(t_m - t)$ is the signal $p(-t)$ delayed by t_m. Three cases, $t_m < T_o$, $t_m = T_o$, and $t_m > T_o$, are shown in Fig. 6.22. The first case, $t_m < T_o$, yields noncausal impulse

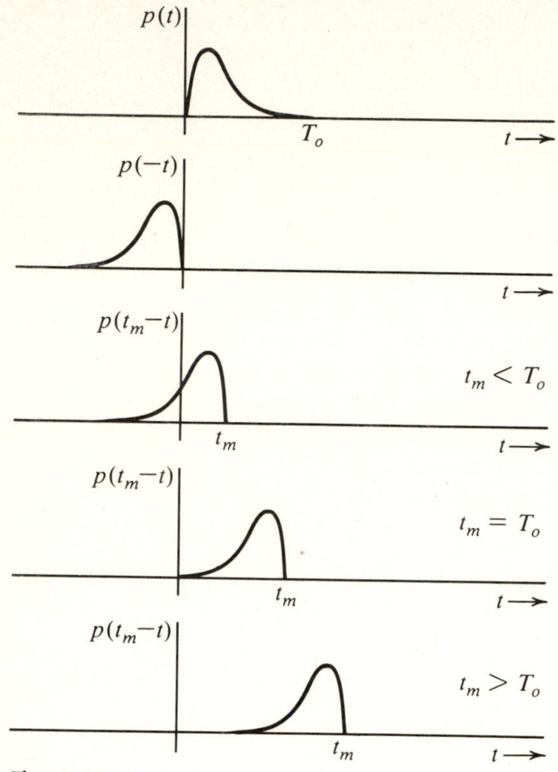

Figure 6.22

response and is therefore physically unrealizable.* Although the last two cases yield physically realizable filters, the last case, $t_m > T_o$, delays the decision-making instant t_m an unnecessary length of time. The case $t_m = T_o$ gives the minimum delay for decision making using a relizable filter.

Observe that both $p(t)$ and $h(t)$ have a width of T_o seconds. Hence $p_o(t)$, which is a convolution of $p(t)$ and $h(t)$, has a width of $2T_o$ seconds, with its peak occurring at $t = T_o$. Also, because $P_o(\omega) = P(\omega)H(\omega) = k' \, |P(\omega)|^2 \, e^{-j\omega T_o}$, $p_o(t)$ is symmetrical about† $t = T_o$ (Fig. 6.21). The output from the previous input pulse terminates and has a zero value at $t = T_o$. Similarly, the output from the following pulse starts and has a zero value at $t = T_o$. Hence, at the decision-making instant T_o, no intersymbol interference occurs.

The arbitrary constant k' in Eq. (6.95) multiplies both the signal and the noise by the same factor and does not affect the ratio ρ. Hence, the error probability, or the

* It is easy to understand why the optimum filter is unrealizable when the decision-making instant is $t_m < T_o$. In this case we are forced to make a decision even before the complete pulse is fed to the filter ($t_m < T_o$). This calls for a prophetic filter, which can respond to inputs even before they are applied. As we know, only unrealizable filters can do this job.

† This follows from the fact that because $|P(\omega)|^2$ is an even function of ω, its inverse transform is symmetrical about $t = 0$ (see Prob. 2.6).

system performance, is independent of the value of k'. For convenience we choose $k' = 1$. This gives

$$h(t) = p(T_o - t) \tag{6.96a}$$

and

$$H(\omega) = P(-\omega)e^{-j\omega T_o} \tag{6.96b}$$

The optimum filter in Eq. (6.96) is known as the *matched filter*.* At the output of this filter, the signal-to-rms-noise-amplitude ratio is a maximum at the decision-making instant $t = T_o$.

The matched filter is optimum in the sense that it maximizes the signal-to-rms-noise-amplitude ratio at the decision-making instant. Although it is reasonable to assume that the maximization of the signal-to-rms-noise amplitude will minimize the detection error probability, we have not proved that threshold detection (sample and decide) is the optimum method from the detection error point of view. It will be shown in Chapter 7 that when the channel noise is white gaussian, the matched-filter receiver is indeed the optimum receiver that minimizes the detection error probability. The maximum value of the signal-to-rms-noise ratio attained by the matched filter is given in Eq. (6.95a). The output A_p is obtained by substituting Eq. (6.96b) [with $k' = 1$] in Eq. (6.91):

$$A_p = \frac{1}{2\pi}\int_{-\infty}^{\infty} |P(\omega)|^2 \, d\omega = E_p \tag{6.97a}$$

The noise power σ_n^2 is obtained by substituting Eq. (6.96b) [with $k' = 1$] in Eq. (6.92):

$$\sigma_n^2 = \frac{\mathcal{N}}{4\pi}\int_{-\infty}^{\infty} |P(\omega)|^2 \, d\omega = \frac{\mathcal{N}E_p}{2} \tag{6.97b}$$

Hence,

$$\rho_{\max}^2 = \frac{A_p^2}{\sigma_n^2} = \frac{2E_p}{\mathcal{N}} \tag{6.97c}$$

and

$$P_e = Q(\rho_{\max}) = Q\left(\sqrt{\frac{2E_p}{\mathcal{N}}}\right) \tag{6.97d}$$

Equations (6.97a, b, c, and d) are truly remarkable. They show that at the decision-making instant, the signal amplitude and the rms noise amplitude depend on the waveform $p(t)$ only through its energy E_p. As far as the system performance is concerned, when the matched-filter receiver is used, all the waveforms used for $p(t)$ are equivalent as long as they have the same energy.

* It is important to remember that the optimum filter is the matched filter only when the channel noise is white. For a general case, the optimum filter is given in Eq. (6.94b).

If the transmission rate is f_o pulses/second, the received power S_i is

$$S_i = E_p f_o$$

and

$$\frac{E_p}{\mathcal{N}} = \frac{S_i}{\mathcal{N} f_o} \tag{6.98}$$

Observe that $S_i/\mathcal{N}f_o$ is similar to the parameter γ (signal-to-noise ratio $S_i/\mathcal{N}B$) used earlier. In fact, for a scheme that uses Nyquist's first criterion, $f_o = B$ (the rate of transmission is equal to the bandwidth). Let us define λ as

$$\lambda = \frac{S_i}{\mathcal{N} f_o} \tag{6.99}$$

Equation (6.97d) can now be expressed as

$$P_e = Q(\sqrt{2\lambda}) \tag{6.100a}$$

Figure 6.23 shows the plot of P_e as a function of λ (in dB).

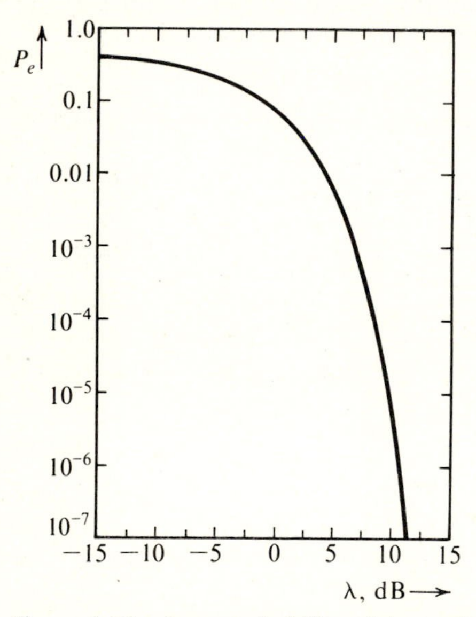

Figure 6.23 Error probability of the optimum receiver for polar signaling.

Using an asymptotic approximation [Eq. (5.36)] for $Q(\sqrt{2\lambda})$, we have

$$P_e \simeq \frac{1}{2\sqrt{\pi\lambda}} e^{-\lambda} \qquad \lambda \gg 1 \tag{6.100b}$$

The parameter λ is directly proportional to the received power S_i [Eq. (6.99)]. Hence, comparing P_e for a given S_i is the same as comparing P_e for a given λ.

The matched filter may also be realized by the alternative arrangement shown in

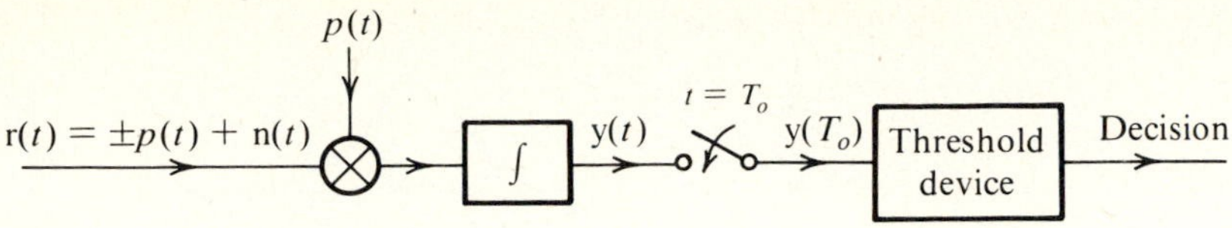

Figure 6.24 Correlator detector.

Fig. 6.24. If the input to the matched filter is $r(t)$, then the output $y(t)$ is given by

$$y(t) = \int_{-\infty}^{\infty} r(x)h(t - x)\,dx$$

where $h(t) = p(T_o - t)$ and

$$h(t - x) = p[T_o - (t - x)] = p(x + T_o - t)$$

Hence,

$$y(t) = \int_{-\infty}^{\infty} r(x)p(x + T_o - t)\,dx$$

At the decision-making instant $t = T_o$, we have

$$y(T_o) = \int_{-\infty}^{\infty} r(x)p(x)\,dx \tag{6.101a}$$

Because the input $r(x)$ is assumed to start at $x = 0$ and $p(x) = 0$ for $x > T_o$,

$$y(T_o) = \int_{0}^{T_o} r(x)p(x)\,dx \tag{6.101b}$$

We can implement Eq. (6.101) as shown in Fig. 6.24. This type of arrangement is known as the *correlation receiver* and is equivalent to the matched-filter receiver.

Suboptimum Filters

Another approach to designing a receiving filter is to assume a particular filter form and adjust its parameters to maximize ρ. Such filters are inferior to the optimum filter but may be simpler to design. The loss in ρ_{max} caused by the use of a suboptimum filter is investigated in the example below.

■ EXAMPLE 6.12

If $p(t)$ is a rectangular pulse of height A and width T_o (Fig. 6.25*a*), determine ρ_{max} if, instead of the matched filter, a one-stage RC filter with $H(\omega) = 1/(1 + j\omega RC)$ (Fig. 6.25*b*) is used.

□ **Solution:** The output $p_o(t)$ of such a filter is given by (Fig. 6.25*c*)

$$\begin{aligned} p_o(t) &= A(1 - e^{-t/RC}) \qquad 0 < t < T_o \\ &= A(1 - e^{-T_o/RC})\,e^{-(t-T_o)/RC} \qquad t > T_o \end{aligned}$$

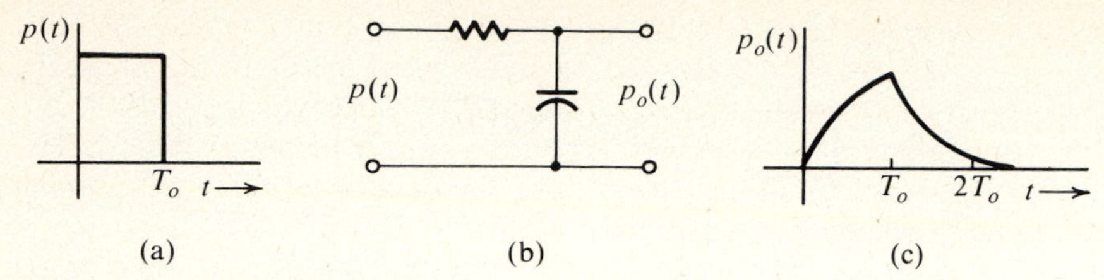

Figure 6.25 Suboptimum filter.

The maximum value of $p_o(t)$ is A_p, which occurs at T_o:

$$A_p = p_o(T_o) = A(1 - e^{-T_o/RC})$$

$$\sigma_n^2 = \frac{1}{2\pi} \cdot \frac{\mathcal{N}}{2} \int_{-\infty}^{\infty} \frac{d\omega}{1 + \omega^2 R^2 C^2} = \frac{\mathcal{N}}{4RC}$$

and

$$\rho^2 = \frac{A_p^2}{\sigma_n^2} = \frac{4A^2 RC(1 - e^{-T_o/RC})^2}{\mathcal{N}}$$

$$= \frac{4A^2 T_o}{\mathcal{N}} \cdot \frac{(1 - e^{-T_o/RC})^2}{T_o/RC}$$

We now maximize ρ^2 with respect to RC. Letting $x = T_o/RC$, we have

$$\rho^2 = \frac{4A^2 T_o}{\mathcal{N}} \cdot \frac{(1 - e^{-x})^2}{x}$$

and

$$\frac{d\rho^2}{dx} = \frac{2xe^{-x}(1 - e^{-x}) - (1 - e^{-x})^2}{x^2} = 0$$

This gives

$$2xe^{-x} = 1 - e^{-x} \qquad \text{or} \qquad 1 + 2x = e^x$$

and

$$x \simeq 1.26 \qquad \text{or} \qquad \frac{1}{RC} = \frac{1.26}{T_o}$$

Hence,

$$\rho_{\max}^2 = (0.816) \frac{2A^2 T_o}{\mathcal{N}}$$

Observe that for the matched filter,

$$\rho_{\max}^2 = \frac{2E_p}{\mathcal{N}} = \frac{2A^2 T_o}{\mathcal{N}}$$

Hence, using a simple *RC* filter, ρ^2 is reduced by a factor of 0.816, or about 0.9 dB. This is really insignificant. If we use more complex filters, such as multistage *RC* filters, we find that ρ^2 approximates that of the matched filter within 0.5 dB.

The suboptimal filter considered here, however, does have the problem of ISI. The decaying exponential tail of the pulse (Fig. 6.25*c*) interferes with the following pulse at its sampling instant. For an optimum *RC* ($RC = T_o/1.26$), the interference at the next sampling instant is about -11 dB. The choice of a nonoptimum value of *RC* can reduce the interference, but it also reduces ρ. Thus, for a choice of $RC = T_o/2.3$, the interference is about -20 dB, but ρ^2 is now 0.7 times the ρ^2 for the matched filter (a loss of 1.55 dB). The choice of $RC = T_o/3.43$ reduces intersymbol interference to -30 dB but also reduces ρ^2 by 2.63 dB from that of the matched filter (see Prob. 6.33).

One way to avoid intersymbol interference in suboptimal filters is to use an auxiliary circuit to discharge the filter at the end of each pulse (immediately after the decision-making instant). One outstanding practical application of this scheme is the so-called *integrate-and-dump* filter, which turns out to be the matched filter for rectangular pulses (see Prob. 6.35). ■

Pulse Shaping and Optimization with the Constraints of a Given Transmitted Power and Transmission Bandwidth

In the case of the matched filter, optimization was carried out without any constraint on the transmitted power or transmission bandwidth. The pulses were assumed to be time limited, which requires a large transmission bandwidth (theoretically infinite). It would be more realistic to maximize the signal-to-rms-noise-amplitude ratio under the condition of a given transmitted signal power and a given finite transmission bandwidth. A finite transmission bandwidth, however, causes dispersion, or spreading out, of the transmitted pulses and the consequent ISI. But by using proper pulse shaping, it is possible to transmit kB pulses/second ($1 \le k \le 2$) with zero ISI over a bandwidth B (see Sec. 3.3). This means the transmitted pulses should be such that after passing through the channel $H_c(\omega)$ and the receiving filter, they will emerge as pulses that satisfy Nyquist criterion for zero ISI.

Actually, for optimum performance we should consider joint optimization of the transmitting (or preemphasis) filter and the receiving (or deemphasis) filter. The problem is that of joint optimization of $H_p(\omega)$ and $H_d(\omega)$ given that the input to $H_p(\omega)$ is a pulse of a given form $p(t)$ and the pulse at the output of $H_d(\omega)$ is another pulse $p_o(t)$ (such as a raised-cosine pulse for zero ISI), and the signal-amplitude-to-rms-noise ratio at the output of $H_d(\omega)$ is maximum at the sampling instant for a given transmitted power. An analysis of this problem can be found in the literature.[8,9] The results are similar to those derived here for the case of time-limited pulses. Indeed, for the special case of white gaussian channel noise and $H_c(\omega) = 1$, the optimum receiving filter $H_d(\omega)$ is a matched filter with

$$\rho^2_{\max} = \frac{2E_p}{\mathcal{N}}$$

The raised-cosine pulse at the detector input is jointly shaped by the transmitting filter $H_p(\omega)$ and the receiving filter $H_d(\omega)$.

Thus far we have discussed signaling where only one basic pulse $p(t)$ is used. In general, we may use two distinct pulses $p(t)$ and $q(t)$ to represent the two symbols. The optimum receiver for such a case will now be discussed.

The Optimum Binary Receiver

Let $q(t)$ and $p(t)$ be the two pulses used to transmit the two binary symbols **0** and **1**. The optimum receiver structure considered here is shown in Fig. 6.26a. The incoming

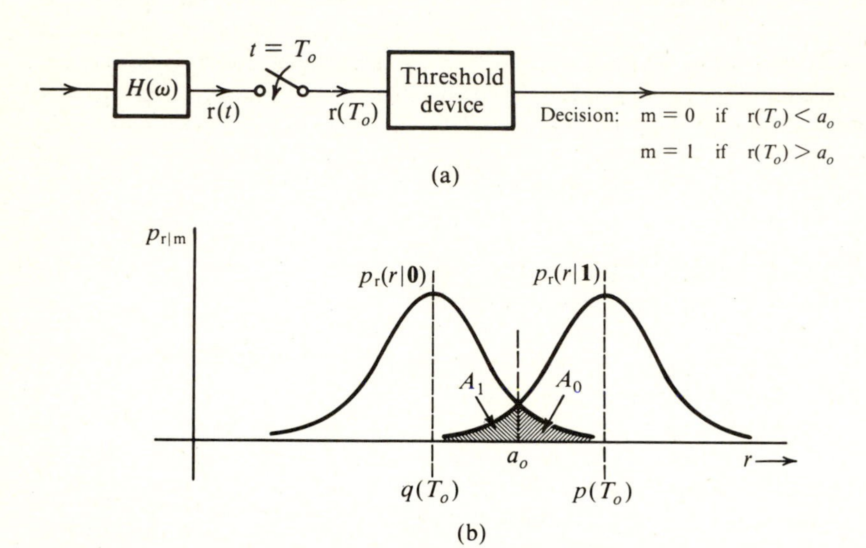

Figure 6.26 Optimum binary threshold detection scheme using two types of pulses.

pulse is transmitted through a filter $H(\omega)$, and the output r(t) is sampled at T_o. The decision as to whether **0** or **1** was present at the input depends on whether r(T_o) < or > a_o, where a_o is the optimum threshold.

Let $q_o(t)$ and $p_o(t)$ be the response of $H(\omega)$ to inputs $q(t)$ and $p(t)$, respectively. From Eq. (6.91) it follows that

$$q_o(T_o) = \frac{1}{2\pi}\int_{-\infty}^{\infty} Q(\omega)H(\omega)e^{j\omega T_o}\, d\omega \tag{6.102a}$$

$$p_o(T_o) = \frac{1}{2\pi}\int_{-\infty}^{\infty} P(\omega)H(\omega)e^{j\omega T_o}\, d\omega \tag{6.102b}$$

and σ_{n}^2, the variance (or the power) of the noise at the filter output, is

$$\sigma_{\mathrm{n}}^2 = \frac{1}{2\pi}\int_{-\infty}^{\infty} S_{\mathrm{n}}(\omega)|H(\omega)|^2\, d\omega \tag{6.102c}$$

If n is the noise output at T_o, then the sampler output $\mathrm{r}(T_o) = q_o(T_o) + \mathrm{n}$ or $p_o(T_o) + \mathrm{n}$, depending on whether $\mathrm{m} = \mathbf{0}$ or $\mathrm{m} = \mathbf{1}$ is received. Hence, r is a gaussian r.v. of variance σ_{n}^2 and mean $q_o(T_o)$ or $p_o(T_o)$, depending on whether $\mathrm{m} = \mathbf{0}$ or $\mathbf{1}$. Thus, the conditional PDFs of the sampled output $\mathrm{r}(T_o)$ are

$$p_{\mathrm{r|m}}(r|\mathbf{0}) = \frac{1}{\sigma_{\mathrm{n}}\sqrt{2\pi}} e^{-[r-q_o(T_o)]^2/2\sigma_{\mathrm{n}}^2}$$

$$p_{\mathrm{r|m}}(r|\mathbf{1}) = \frac{1}{\sigma_{\mathrm{n}}\sqrt{2\pi}} e^{-[r-p_o(T_o)]^2/2\sigma_{\mathrm{n}}^2}$$

The two PDFs are shown in Fig. 6.26*b*. If a_o is the optimum threshold of detection, then the decision is $\mathrm{m} = \mathbf{0}$ if $\mathrm{r} < a_o$ and $\mathrm{m} = \mathbf{1}$ if $\mathrm{r} > a_o$. The conditional probability $P(\epsilon|\mathrm{m} = \mathbf{0})$ is the probability of making a wrong decision when $\mathrm{m} = \mathbf{0}$ is transmitted. This is simply the area A_0 under $p_{\mathrm{r|m}}(r|\mathbf{0})$ from a_o to ∞. Similarly, $P(\epsilon|\mathrm{m} = \mathbf{1})$ is the area A_1 under $p_{\mathrm{r|m}}(r|\mathbf{1})$ from $-\infty$ to a_o (Fig. 6.26*b*), and

$$P_e = \sum_i P(\epsilon, m_i) = \sum_i P(\epsilon|m_i)p(m_i)$$

Assuming both symbols to be equiprobable,

$$P_e = \tfrac{1}{2}\sum_i P(\epsilon|m_i) = \tfrac{1}{2}(A_0 + A_1)$$

From Fig. 6.26*b* it can be seen that the sum $A_0 + A_1$ of the shaded areas is minimized by choosing a_o at the intersection of the two PDFs. Thus,

$$a_o = \frac{p_o(T_o) + q_o(T_o)}{2} \tag{6.103a}$$

and the corresponding P_e is

$$\begin{aligned} P_e &= P(\epsilon|\mathbf{0}) = P(\epsilon|\mathbf{1}) \\ &= \frac{1}{\sigma_n\sqrt{2\pi}}\int_{a_o}^{\infty} e^{-[r-q_o(T_o)]^2/2\sigma_n^2}\,dr \\ &= Q\left[\frac{a_o - q_o(T_o)}{\sigma_{\mathrm{n}}}\right] \\ &= Q\left[\frac{p_o(T_o) - q_o(T_o)}{2\sigma_{\mathrm{n}}}\right] \qquad (6.103\text{b}) \\ &= Q\left(\frac{\beta}{2}\right) \qquad (6.103\text{c}) \end{aligned}$$

where we define

$$\beta = \frac{p_o(T_o) - q_o(T_o)}{\sigma_{\mathrm{n}}} \tag{6.104}$$

Substituting Eq. (6.102a, b, and c) in Eq. (6.104), we get

$$\beta^2 = \frac{\frac{1}{2\pi}\left[\int_{-\infty}^{\infty} [P(\omega) - Q(\omega)]H(\omega)\, e^{j\omega T_o}\, d\omega\right]^2}{\int_{-\infty}^{\infty} S_n(\omega)|H(\omega)|^2\, d\omega}$$

This equation is of the same form as Eq. (6.93). Hence,

$$\beta^2_{\max} = \frac{1}{2\pi}\int_{-\infty}^{\infty} \frac{|P(\omega) - Q(\omega)|^2}{S_n(\omega)}\, d\omega \tag{6.105a}$$

and the optimum filter $H(\omega)$ is given by

$$H(\omega) = k\frac{[P(-\omega) - Q(-\omega)]e^{-j\omega T_o}}{S_n(\omega)} \tag{6.105b}$$

where k is an arbitrary constant.

For white noise $S_n(\omega) = \mathcal{N}/2$, and the optimum filter $H(\omega)$ is given by*

$$H(\omega) = [P(-\omega) - Q(-\omega)]e^{-j\omega T_o} \tag{6.106a}$$

and

$$h(t) = p(T_o - t) - q(T_o - t) \tag{6.106b}$$

This is a filter matched to the pulse $p(t) - q(t)$. The corresponding β is [Eq. (6.105a)]

$$\beta^2_{\max} = \frac{1}{\pi\mathcal{N}}\int_{-\infty}^{\infty} |P(\omega) - Q(\omega)|^2\, d\omega \tag{6.107a}$$

$$= \frac{2}{\mathcal{N}}\int_0^{T_o} [p(t) - q(t)]^2\, dt \tag{6.107b}$$

$$= \frac{E_p + E_q - 2E_{pq}}{\mathcal{N}/2} \tag{6.107c}$$

where E_p and E_q are the energies of $p(t)$ and $q(t)$, respectively, and

$$E_{pq} = \int_0^{T_o} p(t)q(t)\, dt \tag{6.108}$$

and P_e is [Eq. (6.103c)]

$$P_e = Q\left(\frac{\beta_{\max}}{2}\right) \tag{6.109a}$$

$$= Q\left(\sqrt{\frac{E_p + E_q - 2E_{pq}}{2\mathcal{N}}}\right) \tag{6.109b}$$

---

* Because k in Eq. (6.105b) is arbitrary, we choose $k = \mathcal{N}/2$ for convenience.

The optimum threshold a_o is obtained by substituting Eqs. (6.102a and b) and (6.106a) in Eq. (6.103) and recognizing that [see Eq. (2.77a and b)]

$$\frac{1}{2\pi}\int_{-\infty}^{\infty} P(\omega)Q(-\omega)\, d\omega = \frac{1}{2\pi}\int_{-\infty}^{\infty} P(-\omega)Q(\omega)\, d\omega = E_{pq}$$

This gives

$$a_o = \tfrac{1}{2}[E_p - E_q] \tag{6.110}$$

In deriving the optimum binary receiver, we assumed a certain receiver structure (the threshold-detection receiver in Fig. 6.26). It is not clear yet whether there exists another structure that may have better performance than that in Fig. 6.26. It will be shown in the next chapter that for a gaussian noise, the receiver derived here is the absolute optimum. Equation (6.109b) gives P_e for the optimum receiver when the channel noise is white. For the case of nonwhite noise, P_e is obtained by substituting $\beta_{\max}$ in Eq. (6.105a) in Eq. (6.109a).

Equivalent Optimum Binary Receivers. For the optimum receiver in Fig. 6.26*a*,

$$H(\omega) = P(-\omega)\, e^{-j\omega T_o} - Q(-\omega)\, e^{-j\omega T_o}$$

This filter can be realized as a parallel combination of two filters matched to $p(t)$ and $q(t)$, respectively, as shown in Fig. 6.27*a*. Yet another equivalent form is shown in Fig. 6.27*b*. Because the threshold is $(E_p - E_q)/2$, we subtract $E_p/2$ and $E_q/2$ from the two matched filter outputs, respectively. This is equivalent to shifting the threshold to 0. In the case where $E_p = E_q$, we need not subtract $E_p/2$ and $E_q/2$ from the two outputs, and the receiver simplifies to that shown in Fig. 6.27*c*.

On-Off Signaling. In this case, $q(t) = 0$ and $H(\omega) = P(-\omega)\, e^{-j\omega T_o}$. This filter is identical to that used for the polar case. The optimum threshold a_o is

$$a_o = \frac{E_p}{2} \qquad \text{and} \qquad P_e = Q\left(\sqrt{\frac{E_p}{2\mathcal{N}}}\right)$$

In on-off signaling, if the rate is f_o pulses, half of the pulses are zero, and the remaining half have energy E_p each. Hence,

$$S_i = \frac{E_p f_o}{2} \qquad \text{and} \qquad \lambda = \frac{S_i}{\mathcal{N} f_o} = \frac{E_p}{2\mathcal{N}}$$

and

$$P_e = Q(\sqrt{\lambda}) \tag{6.111a}$$

$$\simeq \frac{1}{\sqrt{2\pi\lambda}}\, e^{-\lambda/2} \qquad \lambda \gg 1 \tag{6.111b}$$

A comparison of this with Eq. (6.100) shows that on-off signaling requires 3 dB more power to achieve the same performance (i.e., the same P_e) as that of polar signaling.

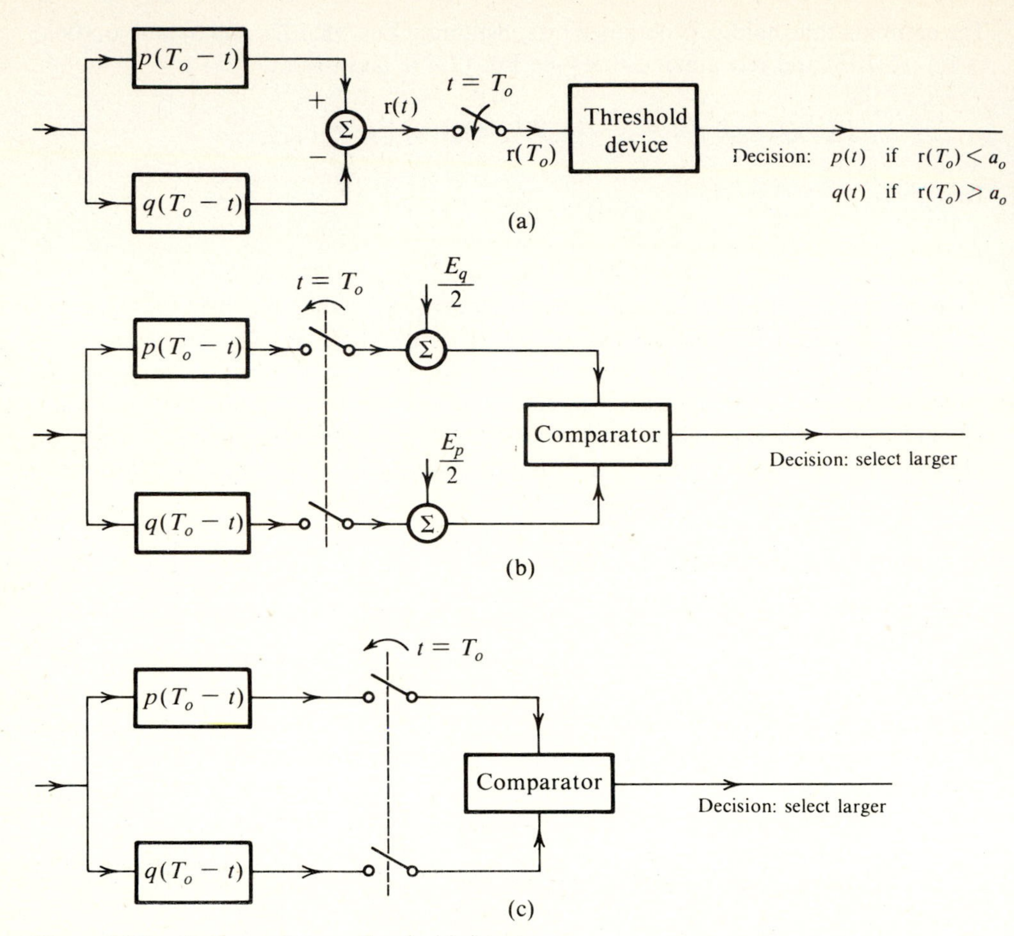

Figure 6.27 Optimum binary threshold detector.

Orthogonal Signaling. In orthogonal signaling, $p(t)$ and $q(t)$ are selected to be orthogonal over the interval $(0, T_o)$. This gives

$$E_{pq} = \int_0^{T_o} p(t)q(t)\, dt = 0$$

An example of two orthogonal pulses is shown in Fig. 6.28. From Eq. (6.109),

$$P_e = Q\left(\sqrt{\frac{E_p + E_q}{2\mathcal{N}}}\right) \qquad \textbf{(6.112)}$$

Because

$$S_i = \frac{E_p f_o}{2} + \frac{E_q f_o}{2} = \frac{f_o}{2}(E_p + E_q)$$

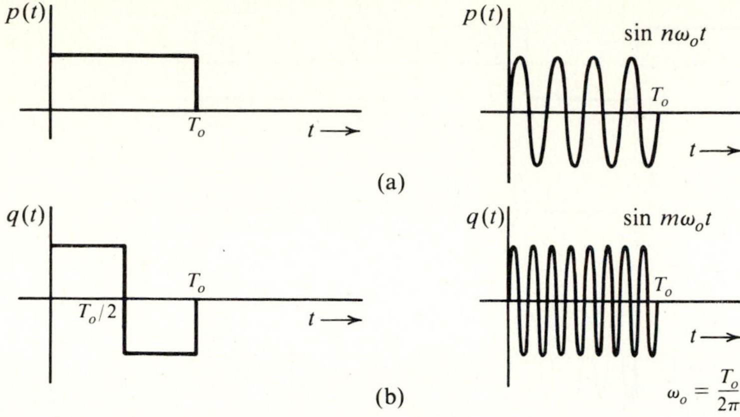

Figure 6.28 Examples of orthogonal signals.

and

$$\lambda = \frac{S_i}{\mathcal{N} f_o} = \frac{E_p + E_q}{2\mathcal{N}}$$

we have

$$P_e = Q(\sqrt{\lambda}) \tag{6.113a}$$

$$\simeq \frac{1}{\sqrt{2\pi\lambda}} e^{-\lambda/2} \tag{6.113b}$$

This shows that the performance of orthogonal signaling is inferior to that of polar signaling by 3 dB, but it is identical to that of on-off signaling. This is not surprising—on-off signaling *is* orthogonal signaling because $E_{pq} = 0$.

Carrier Systems: ASK, FSK, PSK, and DPSK

In digital carrier systems, baseband pulses modulate a high-frequency carrier. We have already briefly discussed amplitude-shift keying (ASK), frequency-shift keying (FSK), and phase-shift keying (PSK).

Figure 6.29 shows the three schemes, using a rectangular baseband pulse. The baseband pulse may be specially shaped (e.g., a raised cosine) to eliminate intersymbol interference and to have a finite bandwidth.

The error probability of the optimum detector depends only on the pulse energy, not on the pulse shape. Hence, as far as the error probability is concerned, the performance of a modulated scheme will be identical to that of the baseband scheme of the same energy.

The incoming modulated pulses can be demodulated either coherently (synchronously) or noncoherently (by envelope detection). The former method is the optimum and requires much more sophisticated equipment. Naturally, it has a superior performance in comparison to the latter method.

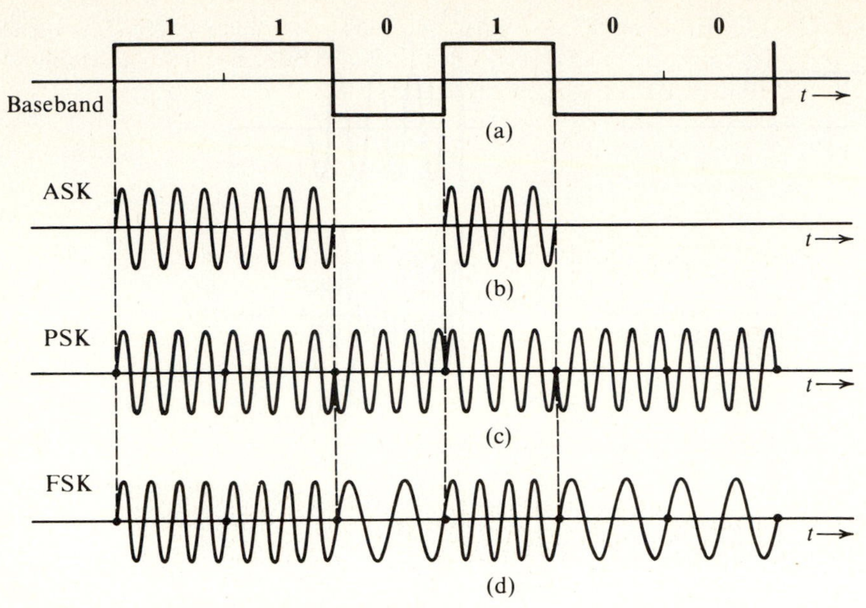

Figure 6.29 Digital modulated waveforms.

Coherent Detection. Let the RF pulse $p(t) = \sqrt{2}\, p'(t) \cos \omega_c t$, where $p'(t)$ is a baseband pulse. The RF pulse can be detected by a filter matched to the RF pulse $p(t)$ followed by a sampler (Fig. 6.30*a*). In this case [Eq. (6.95a)]

$$\rho^2 = \frac{2E_p}{\mathcal{N}}$$

where E_p is the energy of $p(t)$.

We may also detect the RF pulse by first demodulating it coherently by multiplying it by $\sqrt{2} \cos \omega_c t$. The product is the baseband pulse* $p'(t)$ plus a baseband

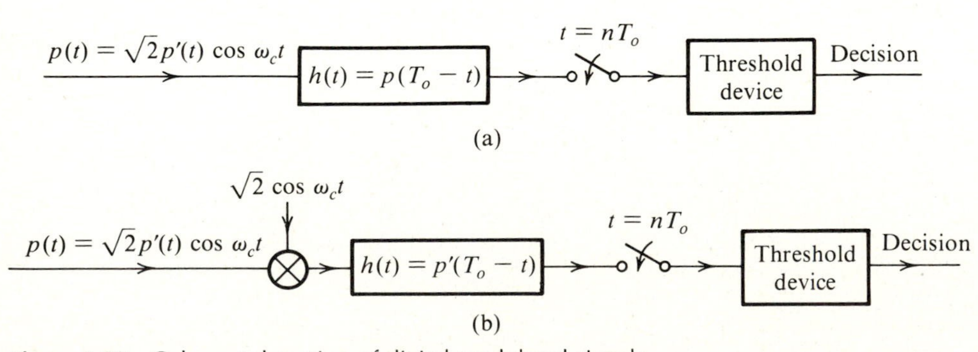

Figure 6.30 Coherent detection of digital modulated signals.

* There is also a spectrum of $p'(t)$ centered at $2\omega_c$, which is eventually eliminated by the filter matched to $p'(t)$.

noise with PSD $\mathcal{N}/2$ (see the discussion on page 458), and this is applied to a filter matched to the baseband pulse $p'(t)$ as shown in Fig. 6.30*b*. Because the energy of $p'(t)$ is E_p, the same as that of $p(t)$, in this case also

$$\rho^2 = \frac{2E_p}{\mathcal{N}}$$

Hence, the two schemes are equivalent.

Let us consider the cases of PSK, ASK, and FSK, individually.

Phase-Shift Keying. This is a case of polar signaling, and the results derived earlier [Eq. (6.100)] apply. The optimum detector is shown in Fig. 6.30. From Eq. (6.100a and b),

$$P_e = Q(\sqrt{2\lambda}) \tag{6.114a}$$

$$\simeq \frac{1}{2\sqrt{\pi\lambda}} e^{-\lambda} \qquad \lambda \gg 1 \tag{6.114b}$$

Amplitude-Shift Keying. This is a case of on-off signaling, and Eq. (6.111) applies. The optimum detector is the same as that for PSK (Fig. 6.30). From Eq. (6.111a and b) we have

$$P_e = Q(\sqrt{\lambda}) \tag{6.115a}$$

$$\simeq \frac{1}{2\sqrt{\pi\lambda}} e^{-\lambda/2} \qquad \lambda \gg 1 \tag{6.115b}$$

Comparison of Eq. (6.115) with Eq. (6.114) shows that ASK requires 3 dB more power than PSK for the same performance. Hence, when using coherent detection, PSK is always preferable to ASK. For this reason, ASK is of no practical importance in coherent detection. But ASK can be useful in noncoherent (envelope) detection. In PSK, the information lies in the phase, and, hence, it cannot be detected noncoherently.

The baseband pulses used in carrier systems should be shaped to minimize the ISI. The bandwidth of the PSK or ASK signal is twice that of the corresponding baseband signal because of modulation.*

Frequency-Shift Keying. In FSK, binary **0** and **1** are transmitted by RF pulses $\sqrt{2}\,p'(t)\cos[\omega_c - (\Delta\omega/2)]t$ and $\sqrt{2}\,p'(t)\cos[\omega_c + (\Delta\omega/2)]t$, respectively. Such a waveform may be considered to be two interleaved ASK waves. Hence, the PSD will consist of two PSDs, centered at $[f_c - (\Delta f/2)]$ and $[f_c + (\Delta f/2)]$. For a large $\Delta f/f_c$, the PSD will consist of two nonoverlapping PSDs. For a small $\Delta f/f_c$, the two spectra merge, and the bandwidth decreases. But in no case is the bandwidth less that of ASK or PSK.

---

* We can also use QAM (quadrature multiplexing) to double the data rate.

The receiver in Fig. 6.27*a* or *b* can serve as the optimum receiver. But because the pulses have equal energy, the simplified form of the optimum receiver in Fig. 6.27*c* is the most convenient. The filters $p(T_o - t)$ and $q(T_o - t)$ are matched to the two RF pulses and can be replaced by respective synchronous demodulators followed by filters matched to the baseband pulse $p'(t)$.

Consider the case:

$$q(t) = \sqrt{2}\, A \cos\left(\omega_c - \frac{\Delta\omega}{2}\right)t$$

$$p(t) = \sqrt{2}\, A \cos\left(\omega_c - \frac{\Delta\omega}{2}\right)t$$

To compute P_e from Eq. (6.109b), we need E_{pq}.

$$\begin{aligned}
E_{pq} &= \int_0^{T_o} p(t)q(t)\, dt \\
&= 2A^2 \int_0^{T_o} \cos\left(\omega_c - \frac{\Delta\omega}{2}\right)t \cos\left(\omega_c + \frac{\Delta\omega}{2}\right)t\, dt \\
&= A^2\left[\int_0^{T_o} \cos(\Delta\omega)t\, dt + \int_0^{T_o} \cos 2\omega_c t\, dt\right] \\
&= A^2 T_o\left[\frac{\sin(\Delta\omega)T_o}{(\Delta\omega)T_o} + \frac{\sin 2\omega_c T_o}{2\omega_c T_o}\right]
\end{aligned}$$

In practice $\omega_c T_o \gg 1$, and the second term on the right-hand side can be ignored. Therefore,

$$E_{pq} = A^2 T_o \operatorname{sinc}(2\Delta f)T_o$$

Figure 6.31*a* shows E_{pq} as a function of $(\Delta f)T_o$.

To minimize P_e [Eq. (6.109b)], E_{pq} must be minimized. From Fig. 6.31*a*, the minimum value of E_{pq} is $-0.217\, A^2 T_o$ and occurs at $(\Delta f)T_o = 0.715$ or when

$$\Delta f = \frac{0.715}{T_o} = 0.715\, f_o$$

Since $E_p = E_q = A^2 T_o$ and $E_{pq} = -0.217\, A^2 T_o$,

$$\begin{aligned}
P_e &= Q\left(\sqrt{\frac{1.217\, A^2 T_o}{\mathcal{N}}}\right) \\
&= Q(\sqrt{1.217\, \lambda}) \qquad \textbf{(6.116a)}
\end{aligned}$$

When $E_{pq} = 0$, we have the case of orthogonal signaling. From Fig. 6.31*a*, it is clear that $E_{pq} = 0$ for $\Delta f = n/2T_o$ where n is any integer. Larger Δf means wider separation between signaling frequencies $\omega_c - (\Delta\omega/2)$ and $\omega_c + (\Delta\omega/2)$, and consequently larger transmission bandwidth. To minimize the bandwidth, Δf should be as small as possible. The minimum value of Δf that can be used for orthogonal signaling is $1/2T_o$. FSK using this value of Δf is known as *minimum-shift keying* (*MSK*) also known as *fast-frequency-shift keying*.

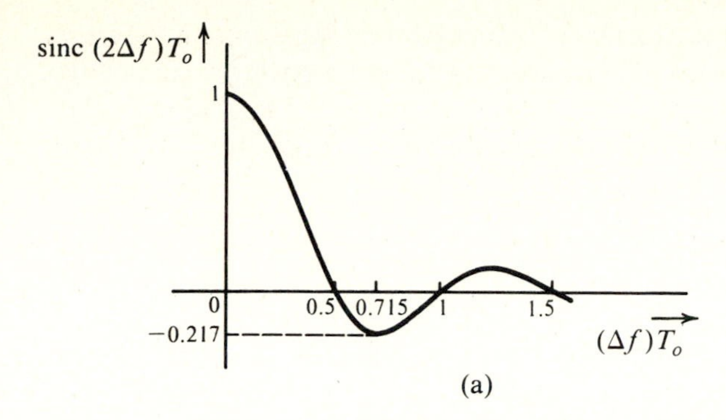

(a)

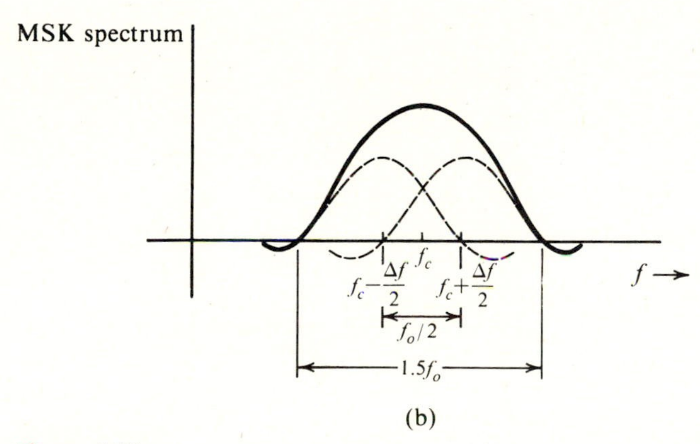

(b)

Figure 6.31

In MSK, abrupt phase changes at the bit transition instants characteristic of other FSK implementations are avoided. FSK schemes where phase continuity is maintained are known as *continuous phase FSK* (*CP-FSK*) of which MSK is one example. These schemes have rapid spectral roll-off and improved efficiency.

To maintain phase continuity in CP-FSK (or MSK), the phase at every bit transition is made dependent on the past data sequence. Consider, for example, the data sequence **1001**. . . starting at $t = 0$. The first pulse corresponding to first bit **1** is $\cos [\omega_c + (\Delta\omega/2)]t$ over the interval 0 to T_o seconds. At $t = T_o$, this pulse ends with a phase $[\omega_c + (\Delta\omega/2)]T_o$. The next pulse corresponding to the second data bit **0** is $\cos [\omega_c - (\Delta\omega/2)]t$. This pulse is given additional phase $\theta = [(\omega_c + (\Delta\omega/2)]T_o$ in order to maintain phase continuity at the transition instant. We continue this way at each transition.

The MSK, being an orthogonal scheme, the error probability is given by

$$P_e = Q(\sqrt{\lambda}) \tag{6.116b}$$

This performance appears inferior to that of the optimum case in Eq. (6.116a). Closer examination shows that MSK is actually superior to the so-called optimum case. If MSK is coherently detected as ordinary FSK using an observation interval of T_o, MSK would have $P_e = Q(\sqrt{\lambda})$. But recall that MSK is CP-FSK, where the phase of each pulse is dependent on the past data sequence. Hence, better performance may be obtained by observing the received waveform over a period longer than T_o. It can be shown that if MSK is detected using observation interval $2T_o$, the performance of MSK is identical to that of PSK, that is,

$$P_e = Q(\sqrt{2\lambda}) \tag{6.116c}$$

MSK also has other useful properties. It has self-synchronization capabilities and its bandwidth is only $1.5f_o$ as shown in Fig. 6.31*b*. This is only 50 percent higher than the duobinary signaling. Moreover, the MSK spectrum decays much more rapidly as $1/f^4$ in contrast to duobinary (or PSK) spectrum which decays only as $1/f^2$ [see Eq. (3.20)]. Because of these properties, MSK has received a great deal of attention recently. More discussion of MSK can be found in references 10 and 11.

Noncoherent Detection. If the phase Θ in the received RF pulse $\sqrt{2}\,p'(t)\cos(\omega_c t + \Theta)$ is unknown, we cannot use coherent detection techniques but must rely on noncoherent techniques, such as envelope detection. It can be shown [5,12] that when the phase Θ of the received pulse is random and uniformly distributed over $(0, 2\pi)$, the optimum detector is a filter matched to the RF pulse $\sqrt{2}\,p'(t)\cos\omega_c t$ followed by an envelope detector, a sampler (to sample at $t = T_o$), and a comparator to make the decision (Fig. 6.32).

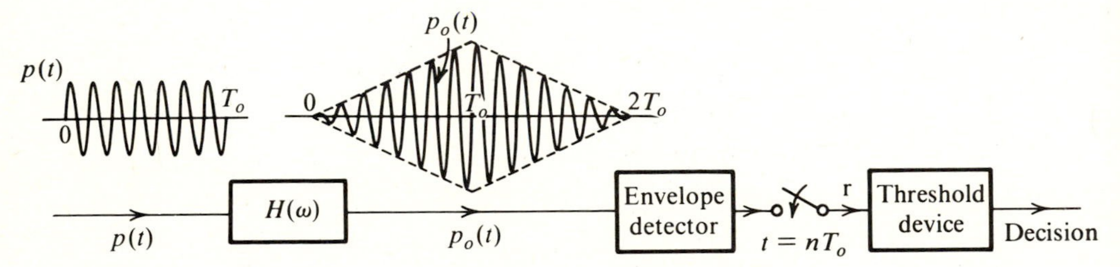

Figure 6.32 Noncoherent detection of digital modulated signals.

Amplitude-Shift Keying. The incoherent detector for ASK is shown in Fig. 6.32. The filter $H(\omega)$ is a filter matched to the RF pulse, ignoring the phase. This means the filter output amplitude A_p will not necessarily be maximum at the sampling instant. But the envelope will be close to maximum at the sampling instant (Fig. 6.32). The matched filter output is now detected by an envelope detector. The envelope is sampled at $t = T_o$ for making the decision.

When a **1** is transmitted, the output of the envelope detector at $t = T_o$ is an envelope of a sine wave of amplitude A_p in a gaussian noise of variance σ_n^2. In this case, the envelope, r, has a rician density, given by [Eq. (5.144a)]

$$p_r(r|m = 1) = \frac{r}{\sigma_n^2} e^{-(r^2 + A_p^2)/2\sigma_n^2} I_0\left(\frac{rA_p}{\sigma_n^2}\right) \tag{6.117a}$$

Also, when $A_p \gg \sigma_n$ (small-noise case) from Eq. (5.144b), we have

$$p_r(r|\mathrm{m} = \mathbf{1}) \simeq \sqrt{\frac{r}{2\pi A_p \sigma_n^2}}\, e^{-(r-A_p)^2/2\sigma_n^2} \tag{6.117b}$$

$$\simeq \frac{1}{\sigma_n\sqrt{2\pi}} e^{-(r-A_p)^2/2\sigma_n^2} \tag{6.117c}$$

Observe that for small noise, the PDF of r is practically gaussian, with mean A_p and variance σ_n^2. When **0** is transmitted, the output of the envelope detector is an envelope of a gaussian noise of variance σ_n^2. The envelope in this case has a Rayleigh density, given by [Eq. (5.139)]

$$p_r(r|\mathrm{m} = \mathbf{0}) = \frac{r}{\sigma_n^2} e^{-r^2/2\sigma_n^2}$$

Both $p_r(r|\mathrm{m} = \mathbf{1})$ and $p_r(r|\mathrm{m} = \mathbf{0})$ are shown in Fig. 6.33. Using the argument used

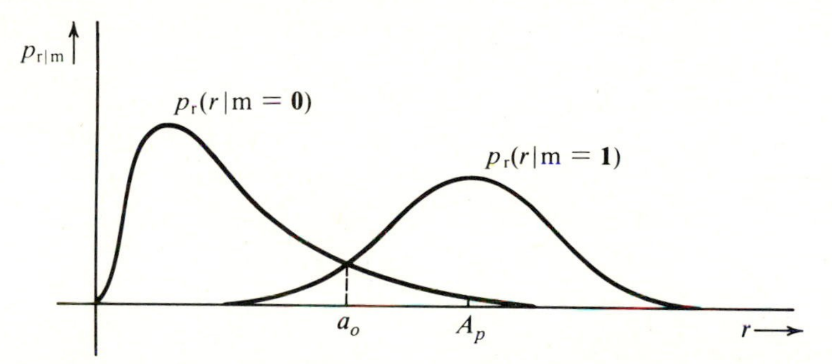

Figure 6.33 Conditional PDFs in noncoherent detection of ASK signals.

earlier (see Fig. 6.26), the optimum threshold is found to be the point where the two densities intersect. Hence, the optimum threshold is a_o, given by

$$\frac{a_o}{\sigma_n^2} e^{-(a_o^2+A_p^2)/2\sigma_n^2} I_0\left(\frac{A_p a_o}{\sigma_n^2}\right) = \frac{a_o}{\sigma_n^2} e^{-a_o^2/2\sigma_n^2}$$

or

$$e^{-A_p^2/2\sigma_n^2} I_0\left(\frac{A_p a_o}{\sigma_n^2}\right) = 1$$

This equation is satisfied to a close approximation for

$$a_o = \frac{A_p}{2}\sqrt{1 + \frac{8\sigma_n^2}{A_p^2}}$$

Because the matched filter is used, $A_p = E_p$ and $\sigma_n^2 = \mathcal{N}E_p/2$ [see Eq. (6.97a and b)]. Also, for ASK there are, on the average, only $f_o/2$ pulses/second. Hence,

$$\left(\frac{A_p}{\sigma_n}\right)^2 = \frac{2E_p}{\mathcal{N}} = 4\frac{E_p(f_o/2)}{\mathcal{N}f_o} = 4\frac{S_i}{\mathcal{N}f_o} = 4\lambda$$

and

$$a_o = \frac{A_p}{2}\sqrt{1 + \frac{2}{\lambda}} \tag{6.118a}$$

Observe that the optimum threshold is not constant but depends on λ. This is a serious drawback in a fading channel. For a strong signal, $\lambda \gg 1$, and

$$a_o \simeq \frac{A_p}{2} \tag{6.118b}$$

and

$$\begin{aligned} P(\epsilon|\mathrm{m} = \mathbf{0}) &= \int_{A_p/2}^{\infty} p_{\mathrm{r}}(r|\mathrm{m} = \mathbf{0})\, dr \\ &= \int_{A_p/2}^{\infty} \frac{r}{\sigma_{\mathrm{n}}^2} e^{-r^2/2\sigma_{\mathrm{n}}^2}\, dr \\ &= e^{-A_p^2/8\sigma_{\mathrm{n}}^2} \\ &= e^{-\lambda/2} \end{aligned} \tag{6.119}$$

Also,

$$P(\epsilon|\mathrm{m} = \mathbf{1}) = \int_{-\infty}^{A_p/2} p_{\mathrm{r}}(r|\mathrm{m} = \mathbf{1})\, dr$$

Evaluation of this integral is somewhat cumbersome.[13] For a strong signal (that is, for $\lambda \gg 1$), the rician PDF can be approximated by the gaussian PDF [Eq. (5.144c)], and

$$\begin{aligned} P(\epsilon|\mathrm{m} = \mathbf{1}) &= \frac{1}{\sigma_{\mathrm{n}}\sqrt{2\pi}} \int_{-\infty}^{A_p/2} e^{-(r-A_p)^2/2\sigma_{\mathrm{n}}^2}\, dr \\ &= Q\left(\frac{A_p}{2\sigma_{\mathrm{n}}}\right) \\ &= Q(\sqrt{\lambda}) \end{aligned} \tag{6.120}$$

Hence,

$$P_e = P_{\mathrm{m}}(\mathbf{1})P(\epsilon|\mathrm{m} = \mathbf{1}) + P_{\mathrm{m}}(\mathbf{0})P(\epsilon|\mathrm{m} = \mathbf{0})$$

Assuming $P_{\mathrm{m}}(\mathbf{1}) = P_{\mathrm{m}}(\mathbf{0}) = 0.5$,

$$P_e = \tfrac{1}{2}[e^{-\lambda/2} + Q(\sqrt{\lambda})] \tag{6.121a}$$

Using the approximation in Eq. (5.36),

$$P_e \simeq \frac{1}{2}\left(1 + \frac{1}{\sqrt{2\pi\lambda}}\right) e^{-\lambda/2} \qquad \lambda \gg 1 \tag{6.121b}$$

$$\simeq \frac{1}{2} e^{-\lambda/2} \tag{6.121c}$$

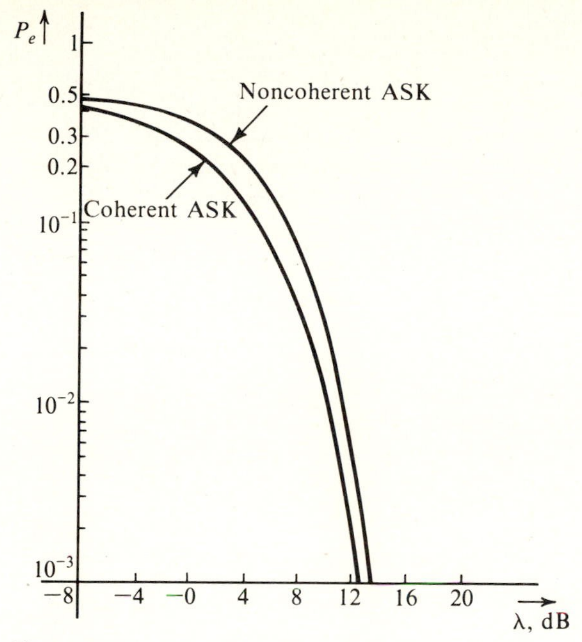

Figure 6.34 Error probability of ASK.

Note that in an optimum receiver, for $\lambda \gg 1$, $P(\epsilon|\mathrm{m} = \mathbf{1})$ is much smaller than $P(\epsilon|\mathrm{m} = \mathbf{0})$. For example, at $\lambda = 10$, $P(\epsilon|\mathrm{m} = \mathbf{0}) \simeq 8.7\, P(\epsilon|\mathrm{m} = \mathbf{1})$. Hence, mistaking **0** for **1** is the type of error that predominates.

The timing information in noncoherent detection is extracted from the envelope of the received signal by methods discussed in Sec. 3.5.

For a coherent detector,

$$P_e = Q(\sqrt{\lambda})$$

$$\simeq \frac{1}{\sqrt{2\pi\lambda}}\, e^{-\lambda/2} \qquad \text{for } \lambda \gg 1 \qquad \mathbf{(6.122)}$$

This appears similar to Eq. (6.121c) [the noncoherent case]. Thus for a large λ, the performances of the coherent detector and the envelope detector are similar (Fig. 6.34). This is similar to the behavior observed in the case of analog signals.

Frequency-Shift Keying. The noncoherent receiver for FSK is shown in Fig. 6.35. The filters $H_0(\omega)$ and $H_1(\omega)$ are matched to the two RF pulses corresponding to **0** and **1**, respectively. The outputs of the envelope detectors at $t = T_o$ are r_0 and r_1, respectively. The noise components of outputs of filters $H_0(\omega)$ and $H_1(\omega)$ are the gaussian r.v.'s n_0 and n_1, respectively, with $\sigma_{\mathrm{n}_0} = \sigma_{\mathrm{n}_1} = \sigma_{\mathrm{n}}$.

If **1** is transmitted (m = **1**), then at the sampling instant, the envelope r_1 has the rician PDF*

$$p_{\mathrm{r}_1}(r_1) = \frac{r_1}{\sigma_n^2}\, e^{-(r_1^2+A_p^2)/2\sigma_n^2}\, I_0\left(\frac{r_1 A_p}{\sigma_n^2}\right)$$

* An orthogonal FSK is assumed. This ensures that r_0 and r_1 have Rayleigh and Rice densities respectively when **1** is transmitted.

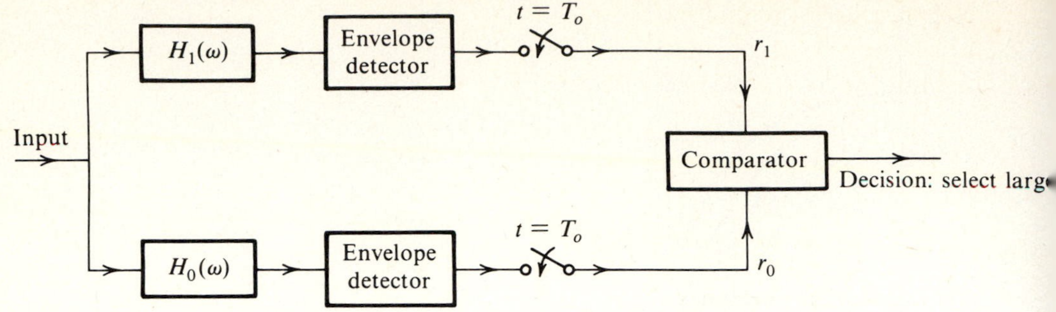

Figure 6.35 Noncoherent detection of binary FSK.

and r_0 is the noise envelope with Rayleigh density

$$P_{r_0}(r_0) = \frac{r_0}{\sigma_n^2} e^{-r_0^2/2\sigma_n^2}$$

The decision is m = **1** if $r_1 > r_0$ and m = **0** if $r_1 < r_0$. Hence, when binary **1** is transmitted, an error is made if $r_0 > r_1$.

$$P(\epsilon | m = \mathbf{1}) = P(r_0 > r_1)$$

The event $r_0 > r_1$ is the same as the joint event "r_1 has any positive value* and r_0 has a value greater than r_1." This is simply the joint event ($0 < r_1 < \infty$, $r_0 > r_1$). Hence,

$$P(\epsilon | m = \mathbf{1}) = P(0 < r_1 < \infty, r_0 > r_1)$$

$$= \int_0^\infty \int_{r_1}^\infty p_{r_1 r_0}(r_1, r_0)\, dr_1\, dr_0$$

Because r_1 and r_0 are independent, $p_{r_1 r_0} = p_{r_1} p_{r_0}$. Hence,

$$P(\epsilon | m = \mathbf{1}) = \int_0^\infty \frac{r_1}{\sigma_n^2} e^{-(r_1^2 + A_p^2)/2\sigma_n^2} I_0\left(\frac{r_1 A_p}{\sigma_n^2}\right) \int_{r_1}^\infty \frac{r_0}{\sigma_n^2} e^{-r_0^2/2\sigma_n^2}\, dr_1\, dr_0$$

$$= \int_0^\infty \frac{r_1}{\sigma_n^2} e^{-(2r_1^2 + A_p^2)/2\sigma_n^2} I_0\left(\frac{r_1 A_p}{\sigma_n^2}\right) dr_1$$

Letting $x = \sqrt{2}\, r_1$ and $\alpha = A_p/\sqrt{2}$, we have

$$P(\epsilon | m = \mathbf{1}) = \frac{1}{2} e^{-A_p^2/4\sigma_n^2} \int_0^\infty \frac{x}{\sigma_n^2} e^{-(x^2+\alpha^2)/2\sigma_n^2} I_0\left(\frac{x\alpha}{\sigma_n^2}\right) dx$$

Observe that the integrand is a rician density, and, hence, its integral is unity. Therefore,

$$P(\epsilon | m = \mathbf{1}) = \tfrac{1}{2} e^{-A_p^2/4\sigma_n^2} \qquad \textbf{(6.123a)}$$

* r_1 is the envelope detector output and can take only positive values.

Note that for a matched filter

$$\rho^2_{\max} = \frac{A_p^2}{\sigma_n^2} = \frac{2E_p}{\mathcal{N}}$$

and

$$\lambda = \frac{S_i}{\mathcal{N} f_o} = \frac{f_o E_p}{\mathcal{N} f_o} = \frac{A_p^2}{2\sigma_n^2}$$

and Eq. (6.123a) becomes

$$P(\epsilon|\mathrm{m} = \mathbf{1}) = \tfrac{1}{2}\, e^{-\lambda/2} \tag{6.123b}$$

Similarly,

$$P(\epsilon|\mathrm{m} = \mathbf{0}) = \tfrac{1}{2}\, e^{-\lambda/2} \tag{6.123c}$$

and

$$P_e = \tfrac{1}{2}\, e^{-\lambda/2} \tag{6.124}$$

This behavior is similar to that of noncoherent ASK [Eq. (6.121c)].

Again we observe that for $\lambda \gg 1$, the performance of coherent and noncoherent FSK are essentially similar.

From the practical point of view, FSK is to be preferred over ASK because FSK has a fixed optimum threshold, whereas the optimum threshold of ASK depends on λ (the signal level). Hence, ASK is particularly susceptible to signal fading. Because, the decision requires comparison between r_0 and r_1, this problem does not arise in FSK. This is the outstanding advantage of noncoherent FSK over noncoherent ASK. In addition, unlike noncoherent ASK, probabilities $P(\epsilon|\mathrm{m} = \mathbf{1})$ and $P(\epsilon|\mathrm{m} = \mathbf{0})$ are equal in noncoherent FSK. The disadvantage of FSK is that it requires a larger bandwidth than that of ASK.

Differentially Coherent PSK. Just as it is impossible to demodulate a DSB-SC signal with an envelope detector, it is also impossible to demodulate PSK (which is really DSB-SC) incoherently. We can, however, demodulate PSK without the synchronous, or coherent, local carrier by using what is known as *differentially coherent PSK (DPSK)*.

The optimum receiver is shown in Fig. 6.36. This receiver is very much like a correlation detector (Fig. 6.24), which is equivalent to a matched-filter detector. In a correlation detector, we multiply pulse $p(t)$ by a locally generated pulse $p(t)$. In the

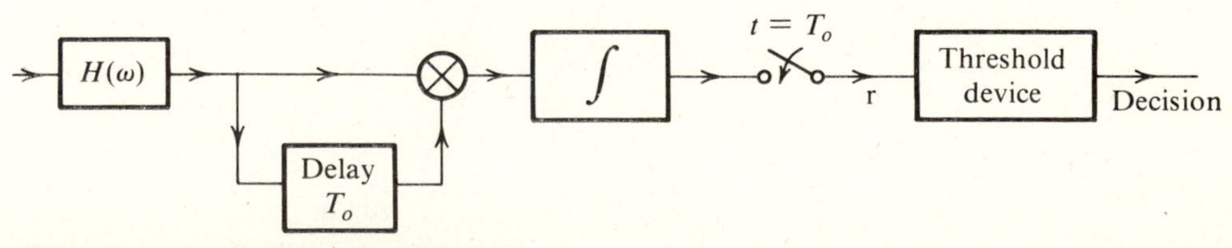

Figure 6.36 Differential PSK detection.

case of DPSK, we take advantage of the fact that the two RF pulses used in transmission are identical except for the sign. In the detector in Fig. 6.36, we multiply the incoming pulse by the preceding pulse. Hence, the preceding pulse serves as a substitute for the locally generated pulse. The only difference is that the preceding pulse is noisy because of channel noise, and this tends to degrade the performance in comparison to coherent PSK. When the output r is positive, the present pulse is identical to the previous one, and when r is negative, the present pulse is the negative of the previous pulse. Hence, from the knowledge of the first reference digit, it is possible to detect all the received digits. Detection is facilitated by using so-called *differential coding* as discussed in Sec. 4.7.

In order to derive the DPSK error probability, we observe that DPSK using differential coding is essentially an orthogonal signaling scheme. A binary **1** is transmitted by a sequence of two pulses (p, p) or $(-p, -p)$ over $2T_o$ seconds (no transition). Similarly, a binary **0** is transmitted by a sequence of two pulses $(p, -p)$ or $(-p, p)$ over $2T_o$ seconds (transition). Either of the pulse sequences used for binary **1** is orthogonal to either of the pulse sequences used for binary **0**. Because no local carrier is generated for demodulation, the detection is noncoherent, with an effective pulse energy equal to $2E_p$ (twice the energy of pulse p). The actual energy transmitted per digit is only E_p, however, the same as in noncoherent FSK. Consequently, the performance of DPSK is 3 dB superior to that of noncoherent FSK. Hence from Eq. (6.124), we can write P_e for DPSK as

$$P_e = \frac{1}{2} e^{-\lambda} \tag{6.125}$$

This error probability (Fig. 6.37) is superior to that of noncoherent FSK by 3 dB and

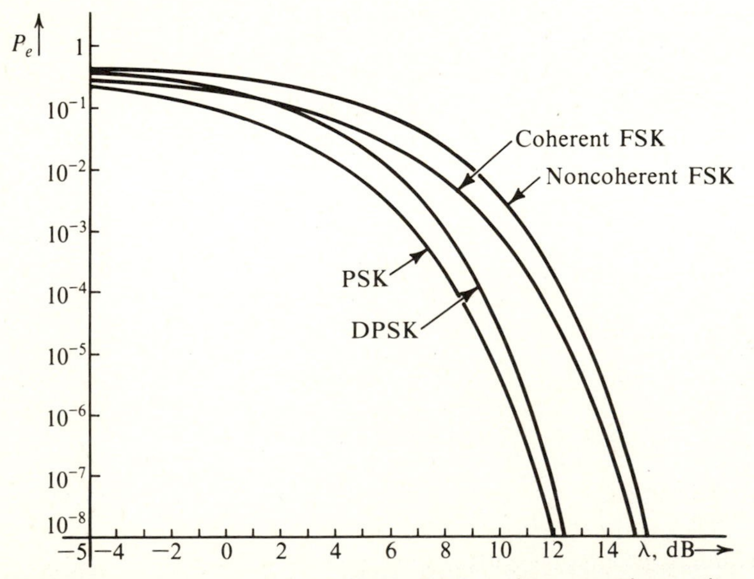

Figure 6.37 Error probability of PSK, DPSK, coherent, and noncoherent FSK.

is essentially similar to coherent PSK for $\lambda \gg 1$ [Eq. (6.114b)]. This is as expected, because we saw earlier that DPSK appears similar to PSK. Rigorous derivation of Eq. (6.125) can be found in the literature.[14,15]

In deriving Eq. (6.125) it was assumed that the received pulse and the preceding pulse differ at most in sign. For an RF pulse, this implies that the phase drift over the channel is slow enough so that it is essentially unchanged over the interval $2T_o$.

A drawback in DPSK is that errors tend to occur in pairs.

6.7 *M*-ARY COMMUNICATION

Thus far we have stressed binary communication, which happens to be perhaps the single most important mode of communication in practice today. In the binary case, only two symbols are used, whereas in the *M*-ary case, the total number of symbols used is M. Each *M*-ary symbol carries as much information as $\log_2 M$ binary digits (see Sec. 3.7). In compensation, we may have to increase the transmitted power (as in the multiamplitude or multiphase case) or transmission bandwidth (as in the multitone case) in order to maintain a given performance level.

Multiamplitude Signaling

In the binary case, we transmit two symbols, consisting of the pulses $p(t)$ and $-p(t)$, where $p(t)$ may be either a baseband pulse or a carrier modulated by a baseband pulse. In the multiamplitude (MASK) case, the M symbols are transmitted by M pulses $\pm p(t)$, $\pm 3p(t)$, $\pm 5p(t)$, . . . , $\pm(M-1)p(t)$. Thus, to transmit f_o *M*-ary digits/second, we are required to transmit f_o pulses/second of the form $kp(t)$. If E_p is the energy of pulse $p(t)$, then assuming that pulses $\pm p(t)$, $\pm 3p(t)$, $\pm 5p(t)$, . . . , $\pm(M-1)p(t)$ are equally likely, the average pulse energy E_{pM} is given by

$$E_{pM} = \frac{2}{M}[E_p + 9E_p + 25E_p + \cdots + (M-1)^2 E_p]$$

$$= \frac{2E_p}{M} \sum_{k=0}^{\frac{M-2}{2}} (2k+1)^2$$

$$= \frac{M^2-1}{3} E_p \tag{6.126a}$$

$$\simeq \frac{M^2}{3} E_p \qquad M \gg 1 \tag{6.126b}$$

The power S_i is given by

$$S_i = f_o E_{pM} = \frac{M^2-1}{3} E_p f_o \tag{6.126c}$$

Because the transmission bandwidth depends only on the shape of the pulses and not their amplitudes, the *M*-ary bandwidth is the same as in the binary case for the given rate of pulses, yet it carries more information. This means for a given information rate, the MASK bandwidth is less than that of the binary case by a factor $\log_2 M$.

To calculate the error probability, we observe that because we are dealing with the same basic pulse $p(t)$, the optimum M-ary receiver is a filter matched to $p(t)$. When the input pulse is $kp(t)$, the output $\mathrm{r}(T_o)$ at the sampling instant will be $kA_p + \mathrm{n}_o(T_o)$. Note that $A_p = E_p$, the energy of $p(t)$, and that σ_{n}^2, the variance of $\mathrm{n}_o(t)$, is $\mathcal{N}E_p/2$. Thus, the optimum receiver for the multiamplitude M-ary signaling case is identical to that of the polar binary case (Fig. 6.21). The sampler has M possible outputs $\pm kA_p + \mathrm{n}_o(T_o)$ $[k = 1, 3, 5, \ldots, M-1]$ that we wish to detect. The conditional PDFs $p(r|m)$ are gaussian with mean kA_p and variance σ_{n}^2, as shown in Fig. 6.38a. Let P_{eM} be the error probability of detection and $P(\epsilon|m)$ be the error probability, given that the symbol m is transmitted.

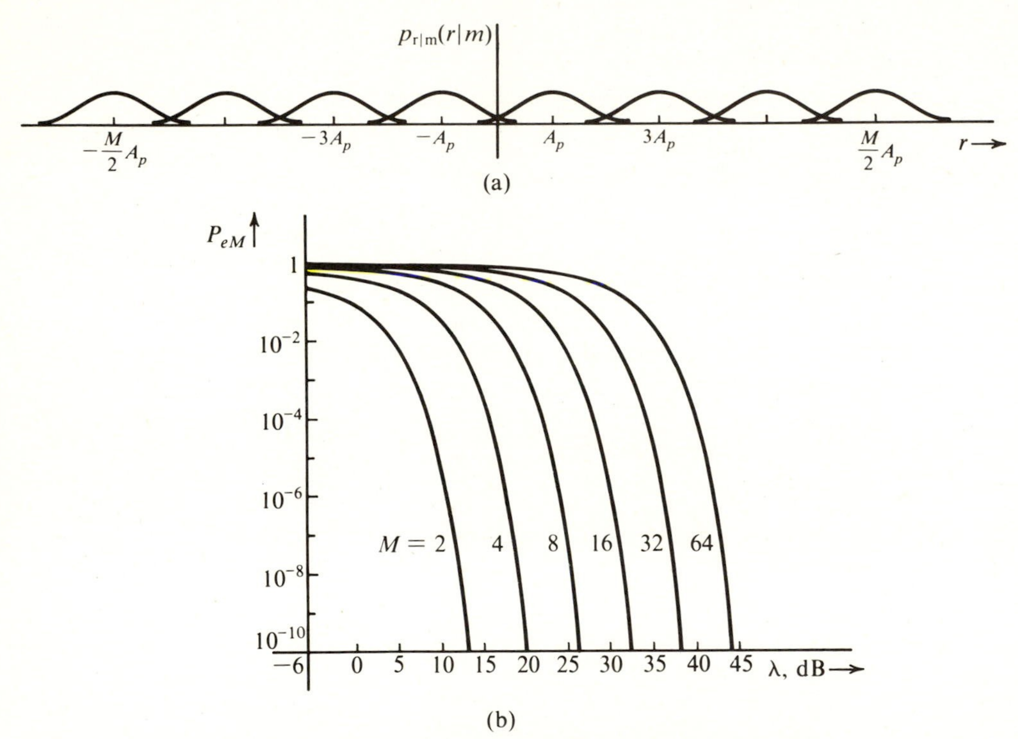

Figure 6.38 (a) Conditional PDFs in MASK. (b) Error probability in MASK.

To calculate P_{eM}, we observe that the case of the two extreme symbols [represented by $\pm(M-1)p(t)$] is similar to the binary case because they have to guard against only one neighbor. As for the remaining symbols, they must guard against neighbors on both sides, and, hence, $P(\epsilon|m)$ in this case is twice that of the extreme symbol. From Fig. 6.38a it is evident that $P(\epsilon|m_i)$ is $Q(A_p/\sigma_{\mathrm{n}})$ for the two extreme signals and is

$2Q\,(A_p/\sigma_{\rm n})$ for the remaining $(M-2)$ symbols. Hence,

$$P_{eM} = \sum_{i=1}^{M} P(m_i)P(\epsilon|m_i)$$

$$= \frac{1}{M}\sum_{i=1}^{M} P(\epsilon|m_i)$$

$$= \frac{1}{M}\left[Q\left(\frac{A_p}{\sigma_{\rm n}}\right) + Q\left(\frac{A_p}{\sigma_{\rm n}}\right) + (M-2)2Q\left(\frac{A_p}{\sigma_{\rm n}}\right)\right]$$

$$= \frac{2(M-1)}{M}\,Q\left(\frac{A_p}{\sigma_{\rm n}}\right) \tag{6.127a}$$

For a matched-filter receiver, $A_p/\sigma_{\rm n} = 2E_p/\mathcal{N}$, and

$$P_{eM} = 2\,\frac{M-1}{M}\,Q\left(\sqrt{\frac{2E_p}{\mathcal{N}}}\right) \tag{6.127b}$$

$$= 2\,\frac{M-1}{M}\,Q\left(\sqrt{\frac{6E_{pM}}{(M^2-1)\mathcal{N}}}\right) \tag{6.127c}$$

For this case, $\lambda = S_i/\mathcal{N}f_o = E_{pM}f_o/\mathcal{N}f_o = E_{pM}/\mathcal{N}$. Hence,

$$P_{eM} = 2\left(\frac{M-1}{M}\right)Q\left(\sqrt{\frac{6\lambda}{(M^2-1)}}\right) \tag{6.127d}$$

$$\simeq 2Q\left(\sqrt{\frac{6\lambda}{M^2}}\right) \qquad M \gg 1 \tag{6.127e}$$

Figure 6.38*b* shows P_{eM} as a function of λ for several values of M. It can be seen from Eq. (6.127) that for a given P_{eM}, the power transmitted increases as M^2.

It is somewhat unfair to compare M-ary signaling on the basis of P_{eM}, the error probability of an M-ary symbol, which conveys the information of $k = \log_2 M$ binary digits. This weighs unfairly against large values of M. A more logical basis of comparison would be the error probability per binary digit of information. Let us consider the problem of transmitting f_o binary digits/second by using two different schemes: the first uses binary signaling at a rate of f_o digits/second, and the second uses M-ary signaling that transmits a group of k binary source digits ($k = \log_2 M$) by one M-ary symbol. Because the type of errors that predominate are those where a symbol is mistaken for its immediate neighbors (see Fig. 6.38*a*), it would be logical to assign neighboring M-ary symbols k binary digit words that differ in the least possible digits.

The Gray code* is suitable for this because adjacent binary combinations in this code differ only by one digit. Hence, an error in one M-ary symbol detection will cause only one error in k binary digits transmitted by the M-ary symbol. As an example, let us consider the case of $M = 4$ ($k = 2$). In this case two binary digits are transmitted by one 4-ary symbol. If the 4-ary symbols are transmitted by pulses of amplitudes $3A/2$, $A/2$, $-A/2$, and $-3A/2$, and these are assigned binary digits **10**, **11**, **01**, and **00**, respectively, then we observe that the groups of binary digits assigned to any two neighboring levels differ only by one digit. Thus, if an error is made in detecting an M-ary symbol, the error will be in only one of the k binary digits transmitted by the symbol. Hence, the binary digit error probability is only $P_{eM}/\log_2 M$. Therefore, for a fair comparison, we should compare $P_{eM}/\log_2 M$ (rather than P_{eM}) as a function of λ for various values of M. In general, however, the factor $\log_2 M$ is negligible compared to other factors that vary exponentially with M in P_{eM} [see Eq. (6.127e)] and we can ignore it. But one must always remember that in comparing M-ary signaling on the basis of P_{eM} we are underestimating its performance by a factor $\log_2 M$ in the error probability.

To compare the binary case with the M-ary case, we note that because each M-ary pulse transmits $\log_2 M$ binary digits, in order to maintain a given information rate, the pulse transmission rate in the M-ary case is reduced by the factor $\log_2 M$. This means the bandwidth of the M-ary case is reduced by the factor $\log_2 M$, and the transmitted power is increased by the factor $M^2/\log_2 M$. On the other hand, if we maintain a given transmission bandwidth, the information rate in the M-ary case is increased by the factor $k = \log_2 M$. But the transmitted power increases by the factor $M^2 = 2^{2k}$. Thus, the power increases exponentially with the information-rate increase factor. In high-powered radio systems, such an increase may not be tolerable. Multiamplitude systems are attractive where bandwidth is at a premium. Because the voice channels of a telephone network have a fixed bandwidth, multiamplitude (or multiphase or a combination of both) signaling appears to be an attractive method of increasing the information rate, particularly because telephone lines can be made to have a high degree of channel stability and low additive noise.

All the results derived above apply to baseband as well as modulated digital systems with coherent detection. For noncoherent detection, similar relationships exist between the binary and M-ary systems.†

---

* The Gray code can be constructed as follows: Construct an n-digit binary code corresponding to 2^n decimal numbers. If $b_1 b_2 \ldots b_n$ is a code word in this code, then the corresponding Gray code word $g_1 g_2 \ldots g_n$ is obtained by the rule

$$g_1 = b_1$$
$$g_k = b_k \oplus b_{k-1} \qquad k \geq 2$$

Thus for $n = 3$, the binary code **000**, **001**, **010**, **011**, **100**, **101**, **110**, **111** is transformed into the Gray code **000**, **001**, **011**, **010**, **110**, **111**, **101**, **100**.

† For the noncoherent case, the baseband pulses must be of the same polarity, for example, 0, $p(t)$, $2p(t)$, . . . , $(M - 1)p(t)$.

Multiphase Signaling

In binary PSK (BPSK), the two basic pulses are $p'(t)\cos\omega_c t$ and $p'(t)\cos(\omega_c t + \pi)$. We generalize the idea for the M-ary case (MPSK) using M pulses with the kth pulse $p'(t)\cos[\omega_c t + (2\pi/M)k]$. Thus, the phases of successive pulses are $2\pi/M$ radians apart (Fig. 6.39). From the symmetry of this scheme it is evident that if, because of

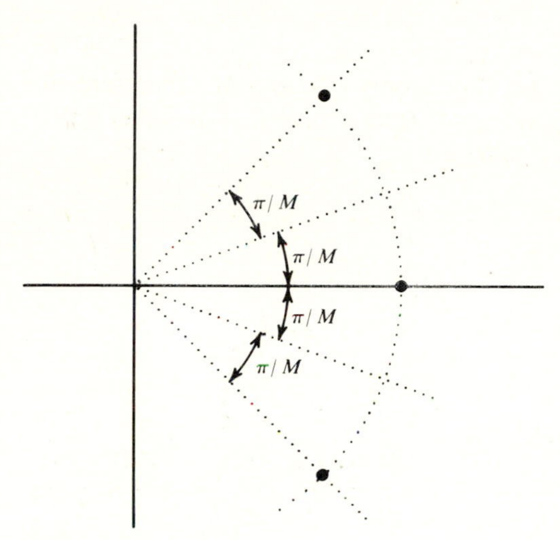

Figure 6.39 MPSK signals.

channel noise, the phase of any pulse deviates by more than π/M radians, an error is made. Hence, the error probability P_{eM} is the probability that the phase of any pulse deviates by more than π/M radians.

At the receiver, coherent phase reference is available. The basic function of the receiver is to detect the phase of the received pulse. This can be done by two phase detectors with $\cos\omega_c t$ and $\sin\omega_c t$ as respective references, along with a logic circuit to determine the ratio of the two detected components.*

To compute P_{eM}, the error probability, we note that detection error results if the phase of any pulse deviates by more than π/M. The PDF $p_\Theta(\theta)$ of the phase Θ of a sinusoid plus a bandpass gaussian noise is found in Eq. (5.144d). Hence,

$$P_{eM} = 1 - \int_{-\pi/M}^{\pi/M} p_\Theta(\theta)$$

The PDF $p_\Theta(\theta)$ in Eq. (5.144d) involves A (the sinusoid amplitude) and σ^2 (the noise variance). Assuming matched filtering and white noise, $A^2/\sigma^2 = 2E_p/\mathcal{N}$ [Eq.

*This can be done by two matched filters matched to $p'(t)\cos\omega_c t$ and $p'(t)\sin\omega_c t$, respectively.

(6.97c)], and

$$\lambda = \frac{S_i}{\mathcal{N} f_o} = \frac{E_p f_o}{\mathcal{N} f_o} = \frac{E_p}{\mathcal{N}} = \frac{A^2}{2\sigma^2}$$

Hence,

$$P_{eM} = 1 - \frac{1}{2\pi} \int_{-\pi/M}^{\pi/M} e^{-\lambda}[1 + \sqrt{4\pi\lambda} \cos\theta\, e^{\lambda \cos^2\theta}[1 - Q(\sqrt{2\lambda}\cos\theta)]]\, d\theta \tag{6.128}$$

The error probability is numerically computed and plotted in Fig. 6.40.

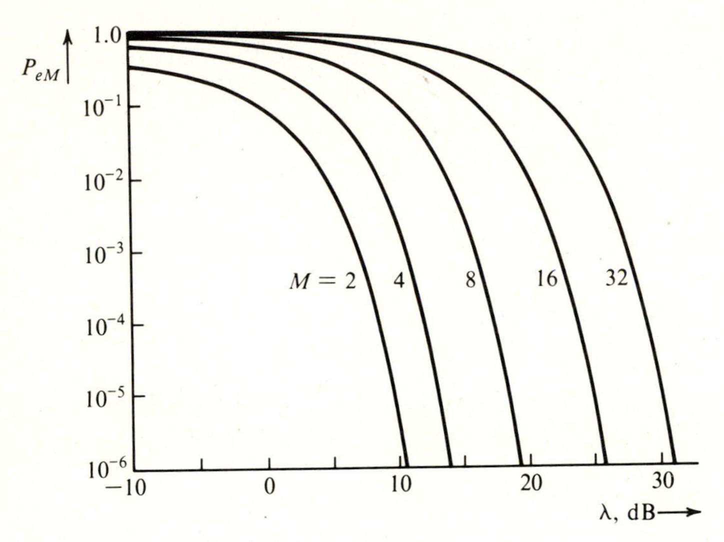

Figure 6.40 Error probability of MPSK.

For $\lambda \gg 1$ (weak noise) and $M \gg 2$, Eq. (6.128) can be approximated as[14]

$$P_{eM} \simeq 2Q\left(\sqrt{2\lambda \sin^2 \frac{\pi}{M}}\right) \tag{6.129a}$$

$$\simeq 2Q\left(\sqrt{\frac{2\pi^2\lambda}{M^2}}\right) \tag{6.129b}$$

An alternate expression for P_{eM} is derived in Chapter 7 [Eq. (7.59a)]. Comparison of Eq. (6.129b) with Eq. (6.127e) shows that the behavior of MPSK is very similar to that of MASK, at least for large λ and $M \gg 2$. Hence, all the comments made in reference to MASK apply to MPSK also.

One of the schemes frequently used in practice is the case of $M = 4$. This is four-phase, or quadriphase, PSK (QPSK). This case can be thought of as two binary PSK systems in parallel in which the carriers are in phase quadrature. Thus, one system uses pulses $\pm p(t) \cos \omega_c t$, and the other uses pulses $\pm p(t) \sin \omega_c t$. Because the carriers are in quadrature, both of these signals can be transmitted over the same channel without interference. Addition of these two streams yields four possible pulses, $\pm\sqrt{2}p(t) \cos(\omega_c t \pm \pi/4)$. This is precisely QPSK. Thus, QPSK doubles the transmission rate without increasing the bandwidth or the transmitted power per binary digit.

An example of QPSK is found in the SPADE multiple-access communications system used for PCM via the Intelsat satellite communication global network, which attains a rate of 64 kbits/second over a transmission bandwidth of 38 kHz. The scheme uses Nyquist's first criterion with a roll-off factor of 19 percent (see Example 3.2). Regular telephone lines with a bandwidth of the order of 2400 Hz (from 600 to 3000 Hz) are used to transmit binary data at a rate of 1200 digits/second. This rate can be increased by using MPSK. For example, to transmit data at respective rates of 2400 and 4800 digits/second, QPSK and 8-ary PSK are used. Higher rates (9600 digits/second) are obtained by using 16-ary QAM (discussed next).

Quadrature Amplitude Modulation

A QAM signal can be written as

$$\begin{aligned} p_i(t) &= p'(t)[a_i \cos \omega_c t + b_i \sin \omega_c t] \\ &= p'(t)[r_i \cos(\omega_c t + \theta_i)] \qquad i = 1, 2, \ldots, M \end{aligned} \tag{6.130}$$

where $r_i = \sqrt{a_i^2 + b_i^2}$ and $\theta_i = -\tan^{-1} b_i/a_i$. Graphically, signal $p_i(t)$ can be mapped into a point (a_i, b_i) or (r_i, θ_i). One such signal set is shown in Fig. 6.41. Because each M-ary pulse consists of a sum of pulses modulated by carriers in quadrature (quadrature multiplexing), this is the case of quadrature amplitude modulation.

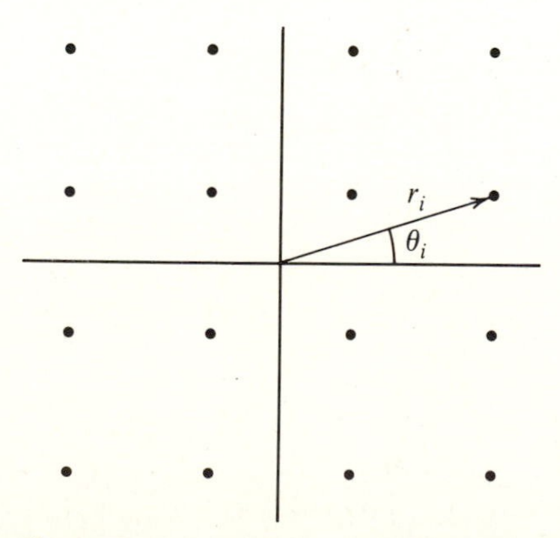

Figure 6.41 16-ary (or 16 point) QAM.

Techniques for analyzing such general signals are discussed in Chapter 7. For the 16-ary case in Fig. 6.41, it can be shown that (see Example 7.3)

$$P_{eM} \simeq 3Q\left(\sqrt{\frac{\lambda}{5}}\right) \tag{6.131}$$

The 16-point QAM in Fig. 6.41 is used in telephone lines.

Multitone Signaling (MFSK)

In this case, M symbols are transmitted by M orthogonal pulses of frequencies ω_1, ω_2, . . . , ω_M, respectively, each of duration T_o. Thus, the M transmitted pulses are of the form* $\sqrt{2}p'(t) \cos \omega_k t$, where $\omega_k = 2\pi(N + k)/T_o$. The receiver (Fig. 6.42) is

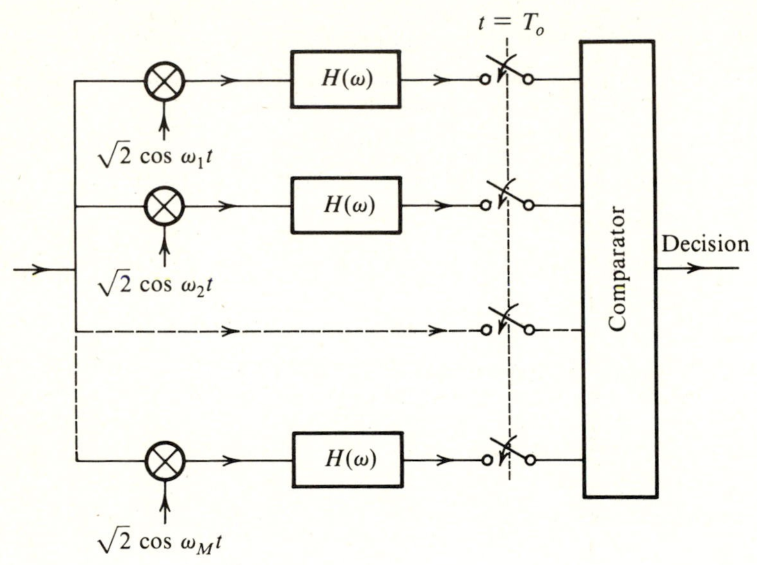

Figure 6.42 Coherent MFSK receiver.

a simple extension of the binary receiver. The incoming pulse is multiplied by corresponding references $\sqrt{2} \cos \omega_i t$ ($i = 1, 2, \ldots, M$). The filter $H(\omega)$ is matched to the baseband pulse $p'(t)$. The same result is obtained if in the ith bank, instead of using a multiplier and $H(\omega)$, we use a filter matched to the RF pulse $p'(t) \cos \omega_i t$. The M bank outputs sampled at $t = T_o$ are $\mathrm{r}_1, \mathrm{r}_2, \ldots, \mathrm{r}_M$, respectively. They are compared, and the decision is $\mathrm{m} = j$ if $\mathrm{r}_j > \mathrm{r}_i$ for all $i \neq j$. If a pulse $p'(t) \cos \omega_i t$ is received, it will cause an output only in the ith bank and will be completely suppressed in the

* A better scheme is to use pulses $\sqrt{2}p'(t) \sin \omega_k t$ and $\sqrt{2}p'(t) \cos \omega_k t$, $k = 1, 2, \ldots, M/2$. Because these are all orthogonal pulses, we required only $M/2$ distinct frequencies, and the bandwidth is reduced by a factor of 2 (see Prob. 6.42). This scheme cannot be used for noncoherent detection for obvious reasons.

outputs of all the remaining banks.* This is true if all the M RF pulses are orthogonal.

The bandwidth of MFSK increases as M. When $\mathrm{m} = \mathbf{1}$ is transmitted, the corresponding sampler output will be $A_p + \mathrm{n}_1$, and the other sampler outputs are n_2, $\mathrm{n}_3, \ldots, \mathrm{n}_M$, respectively, where $\mathrm{n}_1, \mathrm{n}_2, \ldots, \mathrm{n}_M$ are all mutually independent and all have the same variance σ_{n}^2. An error is made if $\mathrm{n}_j > A_p + \mathrm{n}_1$ ($j \neq 1$). Suppose the value of $\mathrm{r}_1 = r_1$. Now the correct decision will be made if $\mathrm{n}_2, \mathrm{n}_3, \ldots, \mathrm{n}_M$ are all less than r_1. Because all these variables have identical PDFs (gaussian with variance σ_n), the probability of $\mathrm{n}_j < r_1$ is

$$\int_{-\infty}^{r_1} \frac{1}{\sigma_n\sqrt{2\pi}} e^{-n_j^2/2\sigma_n^2}\, dn_j$$

If $P(C|\mathrm{m} = \mathbf{1})$ is the probability of making a correct decision given that $\mathrm{m} = \mathbf{1}$ is transmitted, then

$$\begin{aligned}
P(C|\mathrm{m} = \mathbf{1}) &= \mathrm{P}(\mathrm{r}_1 < \infty, \mathrm{n}_2 < \mathrm{r}_1, \mathrm{n}_3 < \mathrm{r}_1, \ldots, \mathrm{n}_M < \mathrm{r}_1) \\
&= \int_{-\infty}^{\infty} p_{\mathrm{r}_1}(r_1)\, dr_1 \left[\prod_{j=2}^{M} \int_{-\infty}^{r_1} \frac{1}{\sigma_n\sqrt{2\pi}} e^{-n_j^2/2\sigma_n^2}\, dn_j\right] \\
&= \frac{1}{(2\pi\sigma_n^2)^{M/2}} \int_{-\infty}^{\infty} e^{-(r_1-A_p)^2/2\sigma_n^2}\, dr_1 \left[\int_{-\infty}^{r_1} e^{-x^2/2\sigma_n^2}\, dx\right]^{M-1} \\
&= \frac{1}{\sigma_n\sqrt{2\pi}} \int_{-\infty}^{\infty} e^{-(r_1-A_p)^2/2\sigma_n^2}\left[1 - Q\left(\frac{r_1}{\sigma_n}\right)\right]^{M-1} dr_1
\end{aligned}$$

Because of symmetry, $P(C|\mathrm{m} = \mathbf{1}) = P(C|\mathrm{m} = \mathbf{2}) = \cdots = P(C|\mathrm{m} = \mathbf{M})$. Hence, P_{CM}, the probability of correct decision, is $P(C|\mathrm{m} = \mathbf{1})$ provided all symbols are equiprobable. Because the filters are matched [Eq. (6.97)],

$$\frac{A_p^2}{\sigma_{\mathrm{n}}^2} = \frac{2E_p}{\mathcal{N}} = \frac{2S_i/f_o}{\mathcal{N}} = 2\lambda$$

and we can express

$$\begin{aligned}
P_{eM} &= 1 - P_{CM} = 1 - P(C|\mathrm{m} = 1) \\
&= 1 - \frac{1}{\sqrt{2\pi}} \int_{-\infty}^{\infty} e^{-(y-\sqrt{2\lambda})^2/2}[1 - Q(y)]^{M-1}\, dy
\end{aligned} \tag{6.132}$$

---

*This can be shown as follows: The filter matched to $p(t)$ has impulse response $h(t) = p(T_o - t)$. If $q(t)$ is applied at the input of this filter, the response $r(t)$ is

$$r(t) = h(t) * q(t) = \int_0^{\infty} p(T_o - x)q(t - x)\, dx$$

and

$$r(T_o) = \int_0^{\infty} p(T_o - x)q(T_o - x)\, dx = \int_0^{\infty} p(t)q(t)\, dt$$

If $p(t)$ and $q(t)$ are orthogonal, $r(T_o) = 0$.

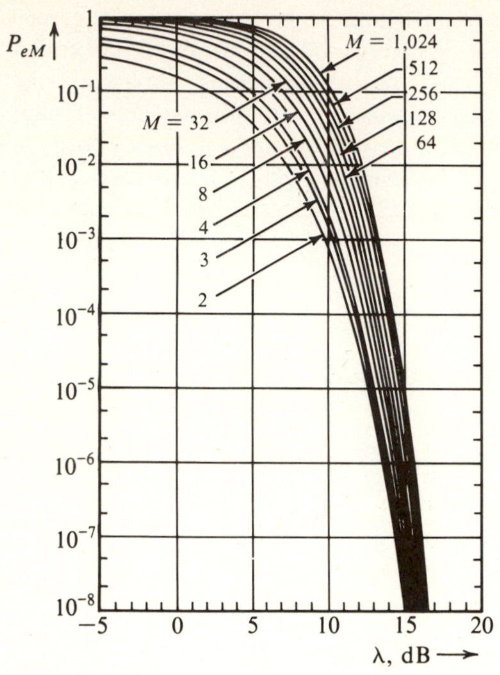

Figure 6.43 Error probability of coherent MFSK.

The integral appearing on the right-hand side of Eq. (6.132) is computed and plotted in Fig. 6.43 (P_{eM} versus λ).

From the practical point of view, the phase coherence of M frequencies is difficult to maintain. Hence in practice, coherent MFSK is rarely used. Noncoherent MFSK is more common. The receiver for noncoherent MFSK is similar to that for binary noncoherent FSK (Fig. 6.35), but with M banks corresponding to M frequencies. The filter $H_i(\omega)$ is matched to the RF pulse $p(t)\cos\omega_i t$. The analysis is straightforward. If $\mathrm{m} = \mathbf{1}$ is transmitted, then r_1 is the envelope of a sinusoid of amplitude A_p plus bandpass gaussian noise, and $\mathrm{r}_j\,(j = 2, 3, \ldots, M)$ is the envelope of the bandpass gaussian noise. Hence, r_1 has rician density, and $\mathrm{r}_2, \mathrm{r}_3, \ldots, \mathrm{r}_M$ have Rayleigh density. Using the same arguments as in the coherent case, we have

$$P_{CM} = P(C|\mathrm{m} = \mathbf{1}) = P(\mathrm{r}_1 < \infty, \mathrm{n}_2 < \mathrm{r}_1, \mathrm{n}_3 < \mathrm{r}_1, \ldots, \mathrm{n}_M < \mathrm{r}_1)$$

$$= \int_{-\infty}^{\infty} \frac{r_1}{\sigma_n^2} I_0\left(\frac{r_1 A_p}{\sigma_n^2}\right) e^{-(r_1^2+A_p^2)/2\sigma_n^2}\, dr_1 \left[\int_{-\infty}^{r_1} \frac{x}{\sigma_n^2} e^{-x^2/2\sigma_n^2}\, dx\right]^{M-1} dx$$

$$= \int_{-\infty}^{\infty} \frac{r_1}{\sigma_n^2} I_0\left(\frac{r_1 A_p}{\sigma_n^2}\right) e^{-(r_1^2+A_p^2)/2\sigma_n^2} [e^{-r_1^2/2\sigma_n^2}]^{M-1}\, dr_1$$

$$= \int_{-\infty}^{\infty} y I_0(y\sqrt{2\lambda})\, e^{-(y^2+2\lambda)/2} [e^{-y^2/2}]^{M-1}\, dy \tag{6.133}$$

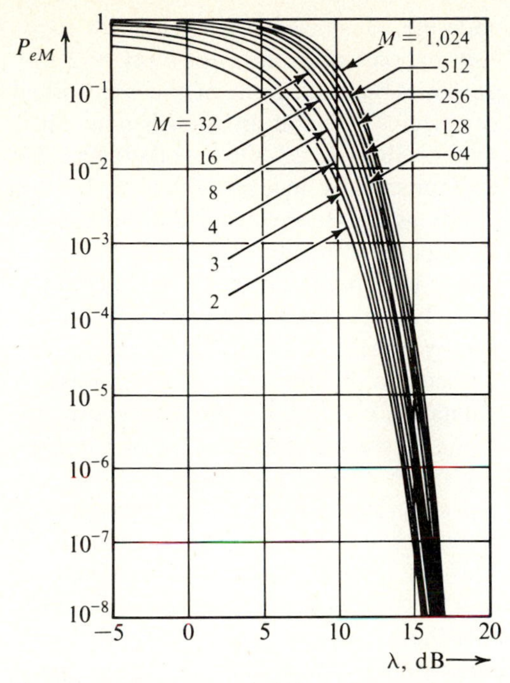

Figure 6.44 Error probability of noncoherent MFSK.

and

$$P_{eM} = 1 - P_{CM}$$

The integral appearing in Eq. (6.133) is computed numerically. The error probability P_{eM} as a function of λ is shown in Fig. 6.44. It can be seen that the performance of noncoherent MFSK is only slightly inferior to coherent MFSK, particularly for large M.

Observe that in multitone signaling the transmitted power depends very little on M. Thus, at $P_{eM} \simeq 10^{-6}$, the power for $M = 1024$ is just about 3 dB more than that for $M = 2$. But the transmission bandwidth increases as M. Thus, multitone signaling is radically different from multiamplitude or multiphase signaling. In the latter, the bandwidth is independent of M, but the transmitted power increases as M^2; that is, the power increases exponentially with the information-rate increase factor k ($M^2 = 2^{2k}$). In the former, on the other hand, the transmitted power is practically independent of M, but the transmission bandwidth increases as M; that is, the bandwidth increases exponentially with the information-rate increase factor k ($M = 2^k$).

Comments on *M*-ary Signaling

M-ary signaling provides us with additional means of exchanging, or trading, the transmission rate, transmission bandwidth, and transmitted power. It therefore gives us flexibility in matching a given source to a channel of given characteristics. Thus, for a given rate of transmission, we can trade the transmission bandwidth for transmitted power. We can also increase the information rate by a factor k ($M = 2^k$) by paying a suitable price in terms of the transmission bandwidth or the transmitted power. For multiamplitude and multiphase systems, the transmission bandwidth is fixed, but the transmitted power increases exponentially as 2^{2k}. On the other hand, for multitone signaling, the transmitted power is practically independent of k, but the transmission bandwidth increases exponentially as 2^k. Hence, we should use multiamplitude or multiphase signaling if the bandwidth is at premium (as in telephone lines) and use multitone signaling when power is at premium (as in space communication). A compromise exists between these two extremes. Let us investigate the possibility of increasing the information rate by a factor k simply by increasing the number of binary pulses transmitted by a factor k. In this case, the transmission bandwidth increases linearly with k. To maintain a given P_e in a binary system, we must maintain a given λ. Because $\lambda = S_i/\mathcal{N}f_o$, increasing f_o by a factor of k implies an increase in the transmitted power S_i by a factor of k. Thus in this case, we can increase the information rate by a factor of k by increasing both the transmission bandwidth and the transmitted power linearly with k, thus avoiding the phantom of the exponential increase that was required in the M-ary system. But here we must increase both the bandwidth and the power, whereas in the M-ary case the increase in information rate can be achieved by increasing either the bandwidth or the power. We have thus a great flexibility in trading various parameters and thus in our ability to match a given source to a given channel.

■ EXAMPLE 6.13

It is required to transmit 2.08×10^6 binary digits/second with a binary digit error probability of $P_e \leq 10^{-6}$. Five possible schemes are considered:

1. Binary
2. 16-ary ASK
3. 16-ary PSK
4. 16-ary QAM
5. 16-ary noncoherent MFSK

The channel noise PSD is $S_n(\omega) = 10^{-8}$. Determine the transmission bandwidth and the signal power required at the receiver input in each case.

□ **Solution:**

1. Binary

We shall consider polar signaling (the most efficient scheme).

$$P_e = 10^{-6} = Q(\sqrt{2\lambda})$$

This yields $\lambda = 11.35$.

$$\lambda = 11.35 = \frac{S_i}{\mathcal{N}f_o}$$

Hence,

$$S_i = 11.35\,\mathcal{N}f_o = 11.35(2 \times 10^{-8})(2.08 \times 10^6) = 0.47 \text{ W}$$

Assuming raised-cosine baseband pulses, the bandwidth B_T is

$$B_T = f_o = 2.08 \text{ MHz}$$

2. 16-ary ASK

Because each 16-ary symbol carries the information equivalent of $\log_2 16 = 4$ binary digits, we need to transmit only $(2.08 \times 10^6)/4 = 0.52 \times 10^6$ 16-ary pulses/second. This requires a B_T of 520 kHz for baseband pulses and 1.04 MHz for modulated pulses (assuming raised-cosine pulses).

Also,

$$P_e = 10^{-6} = \frac{P_{eM}}{\log_2 16}$$

Therefore $P_{eM} = 4 \times 10^{-6}$.

$$P_{eM} = 4 \times 10^{-6} = 2Q\left(\sqrt{\frac{6\lambda}{256}}\right)$$

This yields $\lambda = 903$.

$$\lambda = 903 = \frac{S_i}{\mathcal{N} f_o}$$

$$S_i = (903)(2 \times 10^{-8})(0.52 \times 10^6) = 9.39 \text{ W}$$

3. 16-ary PSK

We need to transmit only 0.52×10^6 pulses/second. For baseband pulses, this will require a bandwidth of 520 kHz. But PSK is a modulated signal, and the required bandwidth is $2(0.52 \times 10^6) = 1.04$ MHz.

Also,

$$P_{eM} = 4P_e = 4 \times 10^{-6} = 2Q\left(\sqrt{\frac{2\pi\lambda}{256}}\right)$$

This yields $\lambda = 862$ and

$$S_i = \mathcal{N} f_o \lambda = (2 \times 10^{-8})(0.52 \times 10^6)(862) = 8.96 \text{ W}$$

4. QAM

For 16-point QAM, we need to transmit only 0.52×10^6 pulses/second. Hence, the transmission bandwidth will be $2 \times (0.52 \times 10^6) = 1.04$ MHz.

$$P_{eM} = 4 \times 10^{-6} = 3Q\left(\sqrt{\frac{\lambda}{5}}\right)$$

This yields $\lambda = 110.5$.

$$S_i = \mathcal{N} f_o \lambda = (2 \times 10^{-8})(0.52 \times 10^6)(110.5) = 1.15 \text{ W}$$

5. 16-ary Noncoherent FSK

The pulse rate is 0.52×10^6, and T_o, the pulse width, is 1.92×10^{-6}. We use RF pulses of the form $\cos [(2\pi k/T_o)t]$, $\sin [(2\pi k/T_o)t]$ and there are 8 distinct frequencies, separated by $1/T_o$. Hence, the total bandwidth (ignoring the spectrum spread at the extremes) is $8/T_o = 4.16$ MHz.

From Fig. 6.44, we find $\lambda \simeq 15$ dB for $P_{eM} = 4 \times 10^{-6}$. This gives $\lambda = 31.6$ and

$$S_i = \mathcal{N} f_o \lambda = (2 \times 10^{-8})(0.52 \times 10^6)(31.6) = 0.328 \text{ W}$$

6.8 SYNCHRONIZATION

In synchronous, or coherent, detection, we need to achieve synchronization at three different levels: (1) carrier synchronization, (2) bit synchronization, and (3) word synchronization. For noncoherent detection, we need only the second and the third level of synchronization—which were discussed in Chapter 3. Here we shall consider only carrier synchronization.

Carrier synchronization is similar to bit synchronization. The problem is much more difficult, however. In bit synchronization, the problem is to achieve synchronism from bit interval to bit interval—which is of the order T_o. In carrier synchronization, we must achieve synchronism within a fraction of a cycle, and because the duration of one carrier cycle is $1/f_c \ll T_o$, the problem is severe. It should be remembered that the phase error θ that can be tolerated is much less than $\pi/2$. For example, if we are transmitting data at a rate of 2 Mbits/second, the bit interval is 0.5 μs. If this data is transmitted by PSK with a carrier frequency of 100 MHz, a phase of $\pi/2$ corresponds to 2.5 ns; that is, the synchronization must be achieved within an interval of much less than 2.5 ns!

Carrier synchronization is achieved by three general methods that are similar to those used for bit synchronization (see timing extraction in Sec. 3.5):

1. Using a primary or a secondary standard (i.e., transmitter and receiver slaved to a master timing source).
2. Transmitting a separate synchronizing signal (a pilot).
3. Self-synchronization, where the timing information is extracted from the received signal itself.

The first method is expensive and is suitable only for large data systems, not for point-to-point systems.

The second method uses part of the channel capacity to transmit timing information and causes some degradation in performance (see Prob. 6.34). But this is a widely used method for point-to-point communication systems. A pilot may be transmitted by frequency-division multiplexing (by choosing a pilot of frequency at which the signal PSD has a null) or by time-division multiplexing (in which the modulated signal is interrupted for a short period of time, during which the synchronizing signal is transmitted).

A baseband signaling scheme with a dc null—such as bipolar, duobinary, or split-phase—is preferred, because such signals after modulation have a spectral null at the carrier frequency. This facilitates the separation of the pilot at the receiver.

The self-synchronization method extracts the carrier by squaring the incoming signal or by using a Costas loop, as discussed in Sec. 4.5. But because these methods yield sign ambiguities (see Sec. 4.5), they cannot be used for PSK unless differential coding is used.

REFERENCES

1. P. F. Panter, *Modulation, Noise, and Spectral Analysis,* McGraw-Hill, New York, 1965.
2. D. J. Sakrison, *Communication Theory: Transmission of Waveforms and Digital Information,* Wiley, New York, 1968.
3. N. Abramson, "Bandwidth and Spectra of Phase- and Frequency-Modulated Waves," *IEEE Trans. Commun. Syst.*, vol. CS-11, no. 4, pp. 407–414, Dec. 1963.
4. S. O. Rice, "Mathematical Analysis of Random Noise," *Bell Syst. Tech. J.*, vol. 23, pp. 282–332, July 1944, and vol. 24, pp. 46–156, Jan. 1945.
5. A. J. Viterbi, *Principles of Coherent Communication,* McGraw-Hill, New York, 1966.
6. H. E. Rowe, *Signals and Noise in Communication Systems,* Van Nostrand, Princeton, N.J., 1965.
7. M. D. Paez and T. H. Glissom, "Minimum Mean Square Error Quantization in Speech, PCM, and DPCM Systems," *IEEE Trans. Commun. Technol.*, vol. COM-20, pp. 225–230, April 1972.
8. W. R. Bennett and J. R. Davey, *Data Transmission,* McGraw-Hill, New York, 1965, Chap. 8.
9. E. D. Sunday, *Communication Systems Engineering Theory,* Wiley, New York, 1969, Chap. 4.
10. S. Pasupathy, "Minimum Shift Keying: A Spectrally Efficient Modulation," *IEEE Commun. Soc. Mag.*, vol. 17, no. 4, pp. 14–22, July 1979.
11. J. J. Spilker, *Digital Communications by Satellite,* Prentice-Hall, Englewood Cliffs, N. J., 1977.
12. J. M. Wozeneraft and I. M. Jacobs, *Principles of Communication Engineering,* Wiley, New York, 1965, Chap. 7.
13. J. I. Marcum, "Statistical Theory of Target Detection by Pulsed Radar," *IRE Trans. Inf. Theory,* vol. IT-6, no. 2, pp. 59–267, April 1960.
14. S. Stein and J. J. Jones, *Modern Communication Principles,* McGraw-Hill, New York, 1967.
15. M. Schwartz, W. R. Bennett, and S. Stein, *Communication Systems and Techniques,* McGraw-Hill, New York, 1966.

PROBLEMS

6.1. A certain telephone channel has $H_c(\omega) \simeq 10^{-3}$ over the signal band. The message signal PSD is $S_m(\omega) = \beta\Pi(\omega/2\alpha)$, with $\alpha = 8000\pi$. The channel noise PSD is $S_n(\omega) = 10^{-8}$. If the output SNR at the receiver is required to be at least 30 dB, what is the minimum transmitted power required? Calculate the value of β corresponding to this power.

6.2. A signal m(t) with PSD $S_m(\omega) = \beta\Pi(\omega/2\alpha)$ with $\alpha = 8000\pi$ is transmitted over a telephone channel with transfer function $H_c(\omega) = 10^{-3}\alpha/j\omega + \alpha$. The channel noise PSD is $S_n(\omega) = 10^{-10}$. To compensate for the channel distortion, the receiver filter transfer function is chosen to be

$$H_d(\omega) = \left(\frac{j\omega + \alpha}{\alpha}\right)\Pi\left(\frac{\omega}{2\alpha}\right)$$

The receiver output SNR is required to be at least 35 dB. Determine the minimum required value of β and the corresponding transmitted power S_T and the power S_i received at the receiver input.

6.3. For a DSB-SC system with a channel noise PSD of $S_n(\omega) = 10^{-10}$ and a baseband signal of bandwidth 4 kHz, the receiver output SNR is required to be at least 30 dB. The receiver is as shown in Fig. 6.3.

(a) What must be the signal power S_i received at the receiver input?

(b) What is the receiver output noise power N_o?

(c) What is the minimum transmitted power S_T if the channel transfer function is $H_c(\omega) = 10^{-4}$ over the transmission band?

6.4. Repeat Prob. 6.3 for SSB-SC.

6.5. Determine the output SNR of each of the two quadrature multiplexed channels and compare the results with those of DSB-SC and SSB-SC.

6.6. Assume $[\text{m}(t)]_{\text{max}} = -[\text{m}(t)]_{\text{min}} = m_p$.

(a) Show that for AM,

$$m_p = \mu A$$

(b) Show that the output SNR for AM [Eq. (6.14)] can be expressed as

$$\frac{S_o}{N_o} = \frac{\mu^2}{k^2 + \mu^2}\gamma$$

where $k^2 = m_p^2/\overline{\text{m}^2}$.

(c) Using the result in part (b), show that for tone modulation with $\mu = 1$, $S_o/N_o = \gamma/3$.

(d) Show that if S_T and S_T' are the AM and DSB-SC transmitted powers, respectively, required to attain a given output SNR, then

$$S_T \simeq k^2 S_T' \text{ for } k^2 \gg 1 \qquad \text{and} \qquad \mu = 1.$$

6.7. A gaussian baseband random process m(t) is transmitted by AM.

(a) Show that for 3σ loading (that is, $m_p = 3\sigma$), the output SNR is $\gamma/10$ when $\mu = 1$.

(b) Show that for 3σ loading and $\mu = 0.5$, the output SNR $\simeq \gamma/36$.

6.8. In many communication systems, the transmitted signal is limited by peak power rather than by average power. Under such limitation, AM fares even worse than DSB-SC or SSB-SC. Show that for tone modulation for a fixed peak power transmitted, the output SNR of AM is 6 dB below that of DSB-SC and 9 dB below that of SSB-SC.

6.9. Determine γ_{thresh} in AM with $\mu = 1$ if the onset of the threshold is when $E_n > A$ with probability 0.01, where E_n is the noise envelope. Assume the modulating signal m(t) to be gaussian and use 4σ loading.

6.10. An AM signal with a gaussian m(t) is demodulated by a square-law detector (Fig. P6.10). Determine the ratio of the distortion signal power to the useful signal power. Assume 3σ loading (that is, $m_p = 3\sigma_m$) and $\mu = 1$. *Hint:* The distortion signal is ½ $\text{m}^2(t)$, and the useful signal is $A\text{m}(t)$.

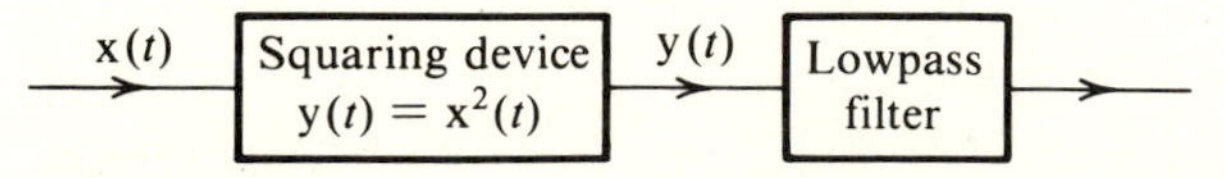

Figure P6.10

6.11. For an FM communication system with $\beta = 2$ and white channel noise with PSD $S_n(\omega) = 10^{-10}$, the output SNR is found to be 28 dB. The baseband signal m(t) is gaussian and bandlimited to 15 kHz, and 3σ loading is used. The demodulator constant $\alpha = 10^{-4}$ (see footnote on page 321), i.e., the signal at the demodulator output is αk_fm(t). The output noise is also multiplied by α.
(a) Determine the received signal power S_i.
(b) Determine the output signal power S_o.
(c) Determine the output noise power N_o.

6.12. For the modulating signal m(t) shown in Fig. P6.12, show that PM is superior to FM by a factor $3\pi^2/4$ from the SNR point of view. *Hint:* See Example 4.13 to determine the bandwidth of m(t).

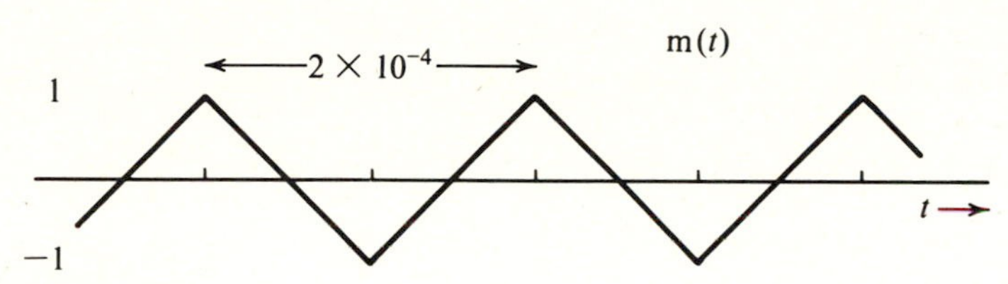

Figure P6.12

6.13. For a modulating signal m(t) = $\cos^3 \omega_o t$, show that PM is superior to FM by a factor 2.25 from the SNR point of view. *Hint:* Determine m_p' from $\ddot{m}(t) = 0$.

6.14. For m(t) $= a_1 \cos \omega_1 t + a_2 \cos \omega_2 t$, show that PM is superior to FM from the SNR point of view when $(1 + xy)^2 < (1 + x)^2/3$, where $x = a_1/a_2$ and $y = \omega_1/\omega_2$. *Hint:* Use Eq. (6.39).

6.15. Show that $\overline{B_m^2}$ in Eq. (6.42a) can also be expressed directly in terms of the time domain as

$$\overline{B_m^2} = \frac{\int_{-\infty}^{\infty} [\dot{m}(t)]^2 \, dt}{\int_{-\infty}^{\infty} m^2(t) \, dt}$$

for a waveform m(t).

6.16. Show that when the modulating signal m(t) has a Butterworth PSD, that is,

$$S_m(\omega) = \frac{1}{1 + \left(\dfrac{\omega}{2\pi f_o}\right)^{2k}}$$

then the mean square bandwidth $\overline{B_m^2}$ is given by

$$\overline{B_m^2} = f_o^2 \frac{\sin\left(\dfrac{\pi}{2k}\right)}{\sin\left(\dfrac{3\pi}{2k}\right)} \qquad k \geq 2$$

and as $k \to \infty$, $\overline{B_m^2} \to \frac{1}{3} f_o^2$.

6.17. A modulating signal PSD is given by

$$S_{\mathrm{m}}(\omega) = \frac{|\omega|}{\sigma^2} e^{-\omega^2/2\sigma^2}$$

The effective bandwidth B of m(t) is that bandwidth which contains χ percent of the total power of m(t). Determine which of the angle modulations is superior if (a) $\chi = 99$, (b) $\chi = 95$, and (c) $\chi = 90$.

6.18. In Prob. 6.14, FM and PM were compared for a given transmission bandwidth. If we compare them for a given mean square transmission bandwidth, we get a slightly different result. Show that PM is superior to FM if $(1 + x^2y^2) < (1 + x^2)/3$ when they are compared for a given mean square transmission bandwidth. This example shows that two signal spectra with the same mean square bandwidth can have different conventional bandwidths.

Hint: In this case $S_{\mathrm{m}}(\omega)$ is discrete. Normalize it, and then take the second moment about the origin. Show that

$$\overline{B_{\mathrm{m}}^2} = (a_1^2 f_1^2 + a_2^2 f_2^2)/a_1^2 + a_2^2$$

6.19. In a certain FM system used in space communication, the output SNR is found to be 21.5 dB with $\beta = 2$. The modulating signal m(t) is gaussian with a bandwidth of 10 kHz, and 3σ loading is used. The system with $\beta = 2$ is in the nonthreshold region of operation.

The output SNR is required to be at least 39.5 dB. Because power is at premium in space communication, it is decided to increase the output SNR by increasing β (i.e., increasing the transmission bandwidth) as much as is possible.

(a) What is the maximum value of β and the corresponding transmission bandwidth that can be used without running into the threshold? What is the corresponding output SNR?

(b) What must be the minimum increase in the transmitted power required to attain an output SNR of 39.5 dB? What is the corresponding value of β and the transmission bandwidth?

Hint: From Eq. (6.51), $\beta = (\gamma - 40)/20$. Substitute this in Eq. (6.37) to determine the minimum usable γ.

6.20. Show that (a) for tone modulation, the dividing line between narrowband and wideband modulation is $\beta = 0.47$; (b) for a gaussian modulating signal with 3σ loading, the dividing line is at $\beta = 0.55$; and (c) for 4σ loading it is at $\beta = 0.56$.

6.21. For FM stereophonic broadcasting (Fig. 4.73), show that the $(L - R)$ channel is about 22 dB noisier than the $L + R$ channel. *Hint:* $L + R$ is the baseband signal. It is preemphasized to obtain $(L + R)'$. The signal $(L - R)$ is preemphasized to obtain $(L - R)'$, which is used to obtain $(L - R)' \cos \omega_c t$. The sum $(L + R)' + (L - R)' \cos \omega_c t$ now frequency modulates a carrier. At the receiver, after frequency demodulation, $(L + R)'$ and $(L - R)' \cos \omega_c t$ are separated. $(L + R)'$ is deemphasized. $(L - R)' \cos \omega_c t$ is multiplied by $2 \cos \omega_c t$ to obtain $(L - R)'$, which is then deemphasized.

6.22. In M-ary PCM, pulses can take M distinct amplitudes (in contrast to two for binary PCM). Show that the signal-to-quantization-noise ratio for M-ary PCM is

$$\frac{S_o}{N_q} = 3M^{2n}\left(\frac{\overline{\mathrm{m}^2}}{m_p^2}\right)$$

6.23. A TV signal bandlimited to 4.5 MHz is to be transmitted by binary PCM. The receiver output signal-to-quantization-noise ratio is required to be at least 55 dB.

(a) If all brightness levels are assumed to be equally likely, that is, amplitudes of m(t) are uniformly distributed in the range $(-m_p, m_p)$, find the minimum number of quantization levels L required. Select the nearest value of L to satisfy $L = 2^n$.

(b) For this value of L, compute the receiver output SNR and the transmission bandwidth, assuming the nonthreshold region of operation.

(c) If the output SNR is required to be increased by 6 dB (four times), what is the new value of L and the corresponding transmission bandwidth?

6.24. A modulating signal m(t) bandlimited to 4 kHz is sampled at a rate of 8000 samples/second. The samples are quantized into 256 levels, binary coded, and transmitted over a channel with $S_n(\omega) = 10^{-7}$. Each received pulse has energy $E_p = 4 \times 10^{-6}$. Given that $m_p = 1$ and $\overline{m^2} = 1/9$,

(a) Find the transmission bandwidth.

(b) Find the output SNR.

(c) If the transmitted power is reduced by 10 dB, find the new SNR.

(d) At the reduced power level in part (c), is it possible to increase the output SNR by changing the value of L. Determine the maximum output SNR achievable and the corresponding value of L and the transmission bandwidth.

6.25. In a PCM channel using k identical regenerative links, we have shown that the error probability P_E of the overall channel is kP_e, where P_e is the error probability of an individual link (see Example 5.7). This shows that P_e is cumulative.

(a) Show that if $k - 1$ links are identical with error probability P_e and the remaining one link has an error probability P_e', then

$$P_E = (k - 1)P_e + P_e'$$

(b) For a certain chain of repeaters with $k = 100$, it is found that γ over 99 links is 25 dB, and over the remaining link γ is 23 dB. Calculate P_e and P_e' using Eq. (6.62) [with $n = 8$]. Now compute P_E and show that P_E is primarily determined by the single weakest link in the chain.

(c) Compare this behavior with an analog signal repeater chain if the noise over one link is 2 dB greater than that over one of the identical 99 links in the chain.

6.26. For companded PCM with $n = 8$, $\mu = 255$, and amplitude m uniformly distributed in the range $(-A, A)$ $[A \leq m_p]$, show that

$$\frac{S_o}{N_o} = \frac{6383\ (\sigma_{\mathrm{m}}^2/m_p^2)}{(\sigma_{\mathrm{m}}^2/m_p^2) + 0.0068\ (\sigma_{\mathrm{m}}/m_p) + 1.53 \times 10^{-5}}$$

Note that m_p is a constant of the system, not of the signal. The peak signal A can vary from talker to talker, whereas m_p is fixed for a given system.

6.27. For a certain telephone channel, $H_c(\omega) = 10^{-3}\ \alpha/(j\omega + \alpha)$ where $\alpha = 8000\pi$. Assume the message signal PSD to be an ideal lowpass spectrum of bandwidth 4 kHz, that is, $S_{\mathrm{m}}(\omega) = \beta\Pi(\omega/2\alpha)$. The channel noise is white with power spectral density $S_n(\omega) = 2 \times 10^{-10}$.

(a) If the overall loss is $G = 10^{-3}$, and the transmitted power is $S_T = 1000$ W, find the optimum transmitting and receiving filters and the output SNR.

(b) Instead of using optimum filters, if it is decided to use a receiving filter

$H_d(\omega) = (j\omega + \alpha)/\alpha$ (to compensate for the nonideal channel) and an ideal transmitting filter $H_p(\omega) = k$, where k is adjusted so that the transmitted power $S_T = 1000$ W, find the output SNR.

(c) If we use an ideal filter at the receiver and use $H_p(\omega) = c(j\omega + \alpha)$, where the constant c is adjusted to maintain $S_T = 1000$ W, find the output SNR.

6.28. A message signal m(t) with

$$S_{\mathrm{m}}(\omega) = \frac{\alpha^2}{\omega^2 + \alpha^2} \qquad (\alpha = 3000\pi)$$

DSB-SC modulates a carrier of 100 kHz. Assume an ideal channel with $H_c(\omega) = 10^{-3}$ and the channel noise PSD $S_{\mathrm{n}}(\omega) = 2 \times 10^{-9}$. The transmitted power is required to be 1 kW, and $G = 10^{-2}$.

(a) Determine transfer functions of optimum preemphasis and deemphasis filters.

(b) Determine the output signal power and noise power and the output SNR.

(c) Determine γ at the demodulator input.

6.29. Repeat Prob. 6.28 for the SSB (USB) case.

6.30. It was shown in the text that when the baseband m(t) is bandlimited with a uniform PSD, PM and FM have identical performance from the SNR point of view. For such m(t), show that optimum PDE filters in angle modulation can improve the output SNR by a factor of 4/3 (or 1.3 dB) only. Find the optimum PDE filter transfer functions. *Hint:* Use Eq. (6.88).

6.31. The so-called integrate-and-dump filter is shown in Fig. P6.31. The feedback amplifier is an ideal integrator. The switch s_1 closes momentarily and then opens at the instant $t = T_o$, thus dumping all the charge on C and causing the output to go to zero. The switch s_2 samples the output immediately before the dumping action.

(a) Sketch the output $p_o(t)$ when a square pulse $p(t)$ is applied to the input of this filter.

(b) Sketch the output $p_o(t)$ of the filter matched to the square pulse $p(t)$.

(c) Show that the performance of the integrate-and-dump filter is identical to that of the matched filter; that is, show that ρ in both cases is identical.

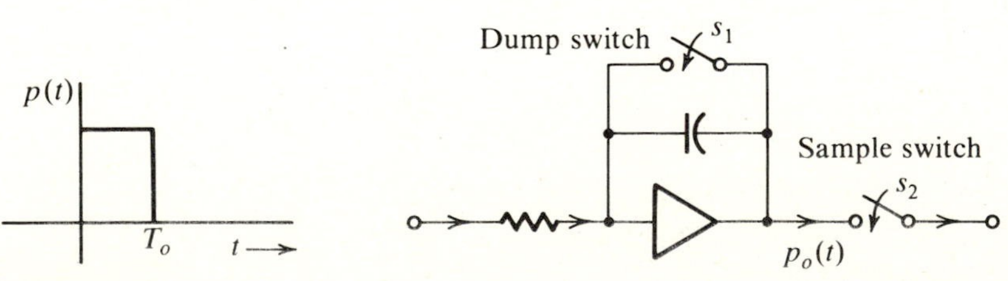

Figure P6.31

6.32. If the ideal integrator in Fig. P6.31 is replaced by a nonideal RC integrator (Fig. P6.32), we have a suboptimal filter that dumps the output at the termination of the input pulse $p(t)$, thus avoiding intersymbol interference with the subsequent pulses. Show that the performance of this filter is identical to that of the simple RC suboptimal filter discussed in the text.

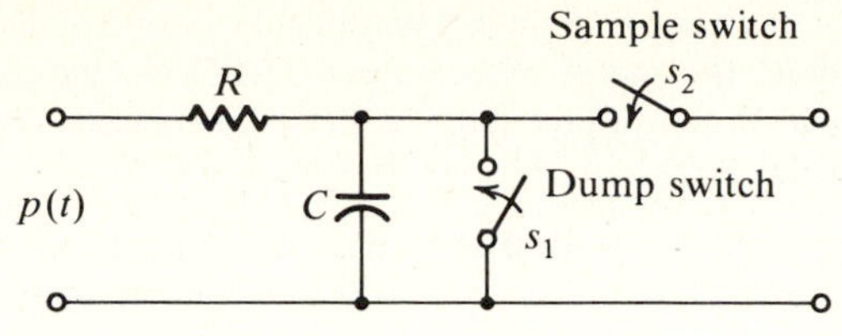

Figure P6.32

6.33. If the simple RC suboptimal filter in Fig. P6.32 is used without the dumping action, show that for optimum RC ($RC = T/1.26$), the interference at the next pulse-sampling instant is -11 dB, that is, at $t = 2T_o$ the ratio of the output amplitude from the second pulse alone (starting at $t = T_o$) to the output amplitude from the first input pulse alone (starting at $t = 0$) is 11 dB. Similarly, show that the ISI is -20 dB when $RC = T_o/2.31$ and is -30 dB when $RC = T_o/3.43$, and that the corresponding reductions in ρ are 1.55 dB and 2.63 dB, respectively.

6.34. As discussed in Sec. 6.8, in coherent schemes, a small pilot is added for synchronization. Because the pilot does not carry information, it causes degradation in P_e. Consider coherent PSK using the following two pulses of duration T_o each:

$$p(t) = A\sqrt{1 - m^2} \cos \omega_c t + Am \sin \omega_c t$$

$$q(t) = -A\sqrt{1 - m^2} \cos \omega_c t + Am \sin \omega_c t$$

where $Am \sin \omega_c t$ is the pilot. Show that when the channel noise is white gaussian,

$$P_e = Q(\sqrt{2\lambda(1 - m^2)})$$

Hint: Use Eqs. (6.103c) and (6.107).

6.35. For polar binary communication systems, each error in the decision has some cost. Suppose that when m $=$ **1** is transmitted and we read it as m $=$ **0** at the receiver, a quantitative penalty, or cost, C_{10} is assigned to such an error, and, similarly, a cost C_{01} is assigned when m $=$ **0** is transmitted and we read it as m $=$ **1**. For the polar case where $P_m(\mathbf{0}) = P_m(\mathbf{1}) = 0.5$, show that for white gaussian channel noise the optimum threshold that minimizes the overall cost is not 0 but is a_o, given by

$$a_o = \frac{\mathcal{N}}{4} \ln \frac{C_{01}}{C_{10}}$$

6.36. For a polar binary system with unequal message probabilities, show that the optimum decision threshold a_o is given by

$$a_o = \frac{\mathcal{N}}{4} \ln \frac{P_m(\mathbf{0})\, C_{01}}{P_m(\mathbf{1})\, C_{10}}$$

where C_{01} and C_{10} are the cost of the errors as explained in Prob. 6.35, and $P_m(\mathbf{0})$ and $P_m(\mathbf{1})$ are the probabilities of transmitting **0** and **1**, respectively.

6.37. For 3-ary communication, messages are chosen from any one of three symbols, m_{-1}, m_0, and m_1, which are transmitted by pulses $-p(t)$, 0, and $p(t)$, respectively. A filter matched to $p(t)$ is used at the receiver. If r is the matched filter output at T_o, plot $p_r(r \mid m_i)$

[$i = -1, 0$, and 1], and if $P(m_{-1}) = P(m_0) = P(m_1)$, determine the optimum decision thresholds and the error probability P_e. The energy of the pulse $p(t)$ is E_p and the channel noise PSD is $S_n(\omega) = \mathcal{N}/2$. Compare this error probability with the binary case. Assume all three symbols to be equiprobable.

6.38. Binary data is transmitted by using a pulse $p(t)$ for **0** and a pulse $2p(t)$ for **1**. Assuming a matched-filter receiver, determine P_e as a function of λ if **0** and **1** are equiprobable.

6.39. A binary source emits data at a rate of 128,000 bits/second. Multiamplitude signaling schemes with $M = 2, 4$, and 8 are considered. In each case, determine the signal power required at the receiver input and the transmission bandwidth if $S_n(\omega) = 10^{-8}$ and P_e, the binary digit error probability, is required to be less than 10^{-7}.

6.40. Repeat Prob. 6.39 if M-ary PSK is used instead.

6.41. Repeat Prob. 6.39 if M-ary coherent FSK is used instead. *Hint:* See Prob. 6.42 for the coherent FSK bandwidth.

6.42. Show that for M-ary coherent FSK using orthogonal pulses of width T_o, the transmission bandwidth $B_T = M/2T_o$. *Hint:* Consider M to be mutually orthogonal pulses $A \cos m\omega_o t$ and $A \sin m\omega_o t$, with $m = 0, 1, 2, 3, \ldots, (M/2) - 1$ and $\omega_o = 2\pi/T_o$.

7

Optimum Signal Detection

We discussed signal detection in the latter part of Chapter 6. Our scope there was rather limited, however. In the first place, we restricted ourselves primarily to binary systems with a receiver of assumed (linear) form. Although we optimized the linear type of receiver, we have no assurance that there does not exist another type of receiver that might be superior to the optimum linear receiver derived in Chapter 6. Secondly, we did not derive the optimum linear receiver for the M-ary case in general. In Chapter 6 we only analyzed a few M-ary schemes with an assumed receiver structure.

In this chapter, we shall analyze the problem of digital signal detection from a more fundamental point of view. We shall determine the optimum receiver (in the sense of minimizing the error probability) for general M-ary signaling in the presence of additive white gaussian noise (AWGN). We shall place no constraints on the receiver. Rather, we shall try to answer the question: What receiver will yield the minimum error probability?

Such an analysis is greatly facilitated by a geometrical representation of signals. We shall now show that a signal is in reality an n-dimensional vector and can be represented by a point in an n-dimensional hyperspace.

7.1 GEOMETRICAL REPRESENTATION OF SIGNALS: THE SIGNAL SPACE

We are used to 3-dimensional physical space and 3-dimensional vectors in this space. There is no reason, however, to restrict ourselves to three dimensions only. We can extend the concept to an n-dimensional space, although it may be hard to visualize the space for $n > 3$.

To begin with, we note that the familiar 3-dimensional vectors are nothing but entities specified by three numbers (x_1, x_2, x_3) in a certain order (ordered 3-tuple)—nothing more, nothing less. Extending this concept, we say that any entity specified by n numbers in a certain order (ordered n-tuple) is an n-dimensional vector. Thus, if an entity is specified by an ordered n-tuple $(x_1, x_2, \ldots, x_n)$, it is an n-dimensional vector $\boldsymbol{x}$. We define n unit vectors $\boldsymbol{\Phi}_1, \boldsymbol{\Phi}_2, \ldots, \boldsymbol{\Phi}_n$ as

$$\begin{aligned}\boldsymbol{\Phi}_1 &= (1, \ 0, \ 0, \ \ldots, \ 0)\\ \boldsymbol{\Phi}_2 &= (0, \ 1, \ 0, \ \ldots, \ 0)\\ &\cdots\cdots\cdots\cdots\\ \boldsymbol{\Phi}_n &= (0, \ 0, \ 0, \ \ldots, \ 1)\end{aligned} \tag{7.1}$$

Any vector $\boldsymbol{x} = (x_1, x_2, \ldots, x_n)$ can be expressed as a linear combination of n unit vectors.

$$\boldsymbol{x} = x_1\boldsymbol{\Phi}_1 + x_2\boldsymbol{\Phi}_2 + \cdots + x_n\boldsymbol{\Phi}_n \tag{7.2a}$$

$$= \sum_{k=1}^{n} x_k\boldsymbol{\Phi}_k \tag{7.2b}$$

We define the scalar product $\boldsymbol{x} \cdot \boldsymbol{y}$ as

$$\boldsymbol{x} \cdot \boldsymbol{y} = \sum_{k=1}^{n} x_k y_k \tag{7.3}$$

where $\boldsymbol{y} = (y_1, y_2, \ldots, y_n)$ is another n-dimensional vector in the same space. Vectors $\boldsymbol{x}$ and $\boldsymbol{y}$ are said to be *orthogonal* if

$$\boldsymbol{x} \cdot \boldsymbol{y} = 0 \tag{7.4}$$

The *length* of a vector $\boldsymbol{x}$ is $|\boldsymbol{x}|$, defined by

$$|\boldsymbol{x}|^2 = \boldsymbol{x} \cdot \boldsymbol{x} = \sum_{k=1}^{n} x_k^2 \tag{7.5}$$

A set of n-dimensional vectors is said to be independent if none of the vectors in this set can be represented as a linear combination of the remaining vectors in the set. Thus, if $\boldsymbol{x}_1, \boldsymbol{x}_2, \ldots, \boldsymbol{x}_m$ is an independent set, then it is impossible to find constants $a_1, a_2, \ldots, a_m$ (not all zero) such that

$$a_1\boldsymbol{x}_1 + a_2\boldsymbol{x}_2 + \cdots + a_m\boldsymbol{x}_m = 0 \tag{7.6}$$

An n-dimensional space can have at most n independent vectors. If a space has a maximum of n independent vectors, then every vector $\boldsymbol{x}$ in this space can be expressed as a linear combination of these n independent vectors. If not, then $\boldsymbol{x}$ becomes a member of the independent vector space. But this is not possible with the assumption that there exists a maximum of n independent vectors. Thus, any vector in this space can be specified by n numbers and, hence, is an n-dimensional space. It also follows from this discussion that the dimensionality of a space can be, at most, as large as the total number of vectors in the space.

A subset of vectors in a given n-dimensional space can have dimensionality less than n. Thus, in a 3-dimensional space, all vectors lying in one plane can be specified by two dimensions, and all vectors lying along a line can be specified by one dimension.

The n independent vectors in an n-dimensional space are called *basis vectors* because every vector in this space can be expressed as a linear combination of these n independent vectors. A set of basis vectors form coordinate axes and are not unique. The n unit vectors in Eq. (7.1) are independent and can serve as basis vectors. These vectors have an additional property in that they are all mutually *orthogonal,* that is

$$\boldsymbol{\Phi}_j \cdot \boldsymbol{\Phi}_k = \begin{cases} 0 & j \neq k \\ 1 & j = k \end{cases} \tag{7.7}$$

Such a set is an *orthonormal* set of vectors because in addition to being orthogonal, their lengths are unity.

A vector $\boldsymbol{x}(x_1, x_2, \ldots, x_n)$ can be represented as

$$\boldsymbol{x} = x_1\boldsymbol{\Phi}_1 + x_2\boldsymbol{\Phi}_2 + \cdots + x_n\boldsymbol{\Phi}_n$$

If we know $\boldsymbol{x}$, its components x_k can be found by taking the scalar product of both sides with $\boldsymbol{\Phi}_k$. Using Eq. (7.7) we get

$$\boldsymbol{x} \cdot \boldsymbol{\Phi}_k = x_k \qquad k = 1, 2, \ldots, n \tag{7.8}$$

One final observation: In order to represent an n-dimensional vector completely, we need n basis vectors (independent vectors). In general, an n-dimensional vector cannot be completely represented by less than n basis vectors of the space. A set of n basis vectors forms a *complete set.*

The Signal as a Vector

Any entity that can be represented by an n-tuple is an n-dimensional vector. If a signal $x(t)$ can be specified by an n-tuple, it, too, is a vector. As we know, a signal $x(t)$ bandlimited to B Hz can be specified by the values of its Nyquist samples. Consequently, such a signal, too, is a vector. We can express $x(t)$ as [see Eq. (2.84)]

$$x(t) = \sum_k x\left(\frac{k}{2B}\right) \operatorname{sinc}(2Bt - k) \tag{7.9}$$

Thus, $x(t)$ can be specified by an n-tuple (its Nyquist sample values). With this introduction, let us now turn to a systematic development of the signal-space concept.

We define n signals $\varphi_1(t), \varphi_2(t), \ldots, \varphi_n(t)$ as independent if none of these n signals can be represented by a linear combination of the remaining $(n - 1)$ signals.

This means it is impossible to find constants $a_1, a_2, \ldots, a_n$ (not all zero), such that

$$a_1\varphi_1(t) + a_2\varphi_2(t) + \cdots + a_n\varphi_n(t) = 0 \tag{7.10}$$

Suppose that a signal $x(t)$ can be represented by a linear combination of n independent signals $\{\varphi_k(t)\}$ as

$$x(t) = x_1\varphi_1(t) + x_2\varphi_2(t) + \cdots + x_n\varphi_n(t) \tag{7.11a}$$

$$= \sum_{k=1}^{n} x_k\varphi_k(t) \tag{7.11b}$$

If every signal in a certain signal space can be represented by a linear combination of n independent signals $\{\varphi_k(t)\}$, then we have an n-dimensional signal space.

Once the basis signals $\{\varphi_k(t)\}$ are specified, we can represent the signal $x(t)$ by an n-tuple $(x_1, x_2, \ldots, x_n)$. Alternately we may represent this signal geometrically by a point $(x_1, x_2, \ldots, x_n)$ in an n-dimensional space. We can now associate a vector $\boldsymbol{x}(x_1, x_2, \ldots, x_n)$ with the signal $x(t)$. Note that the basis signal $\varphi_1(t)$ is represented by the corresponding basis vector $\boldsymbol{\Phi}_1(1, 0, 0, \ldots, 0)$, and $\varphi_2(t)$ is represented by $\boldsymbol{\Phi}_2(0, 1, 0, \ldots, 0)$, and so on.

The signal set $\{\varphi_k(t)\}$ may or may not be orthogonal.* If

$$\int_{-\infty}^{\infty} \varphi_j(t)\varphi_k(t)\, dt = \begin{cases} 0 & j \neq k \\ k_j & j = k \end{cases} \tag{7.12}$$

then the set is an *orthogonal* signal set. If $k_j = 1$ for all j, then the set is *orthonormal*. For an orthonormal set, the coefficients x_k in Eq. (7.11) can be obtained by multiplying both sides of Eq. (7.11) by $\varphi_k(t)$, then integrating and using Eq. (7.12) [with $k_j = 1$]:

$$x_k = \int_{-\infty}^{\infty} x(t)\varphi_k(t)\, dt \tag{7.13}$$

Turning our attention to Eq. (7.9), we see that in this case the basis signals are $\{\text{sinc}\,(2Bt - k)\}$. This set is orthogonal because [see Example 2.25]

$$\int_{-\infty}^{\infty} \text{sinc}\,(2Bt - j)\,\text{sinc}\,(2Bt - k)\, dt = \begin{cases} 0 & j \neq k \\ 2B & j = k \end{cases} \tag{7.14}$$

We can normalize this set by multiplying each member by $\sqrt{2B}$. Thus, if we define

$$\varphi_k(t) = \sqrt{2B}\,\text{sinc}\,(2Bt - k) \tag{7.15}$$

---

* If $\{\varphi_k(t)\}$ is complex, orthogonality implies

$$\int_{-\infty}^{\infty} \varphi_j(t)\varphi_k^*(t)\, dt = 0$$

and Eq. (7.13) becomes

$$x_k = \int_{-\infty}^{\infty} x(t)\varphi_k^*(t)\, dt$$

then

$$x(t) = \sum_k x_k \varphi_k(t) \tag{7.16a}$$

where

$$x_k = \frac{1}{\sqrt{2B}} x\left(\frac{k}{2B}\right) \tag{7.16b}$$

and $\{\varphi_k(t)\}$ is an orthonormal set of signals. Thus, any bandlimited signal $x(t)$ can be represented by a point

$$(\ldots, x_{-k}, \ldots, x_{-2}, x_{-1}, x_0, x_1, x_2, \ldots, x_k, \ldots)$$

where x_k is the kth Nyquist sample of $x(t)$ divided by $1/\sqrt{2B}$. To determine the dimensionality of this signal space, recall that a bandlimited signal cannot be time limited (i.e., it exists over an infinite time interval). Hence, the total number of samples at a Nyquist rate of $2B$ samples/second will be infinite, and the dimensionality is infinite. Higher dimensions, however, can be ignored, because their contribution is negligible. For example, a bandlimited signal with a finite energy may exist over an infinite time interval. But its amplitude for large values of t must approach 0, or the energy

$$E_x = \int_{-\infty}^{\infty} x^2(t)\, dt$$

will not be finite. Hence, the amplitude of every bandlimited signal is negligible beyond some $|t| = T/2$. Such a signal is essentially time limited to T seconds. We can argue in the same way that a signal time limited to T seconds is essentially bandlimited to B Hz. Because the Nyquist rate is $2B$ samples/second, the total number of samples over T seconds is* $2BT + 1$, where T and B are interpreted as above. If $x(t)$ is bandlimited to B, then T is its essential time duration, and if $x(t)$ is time limited to T, B is its essential bandwidth. A rigorous development of this result, as well as an estimation of the error in ignoring higher dimensions (those beyond $2BT + 1$), can be found in the paper by Landau and Pollak.[1]

Just as there are infinite possible sets of basis vectors for a vector space, there are infinite possible sets of basis signals for a given signal space. For a bandlimited signal space, $\{\text{sinc}\,(2Bt - k)\}$ is one possible set of basis signals.

The Scalar Product. In a certain signal space, let $x(t)$ and $y(t)$ be two signals represented by vectors $\mathbf{x}(x_1, x_2, \ldots, x_n)$ and $\mathbf{y}(y_1, y_2, \ldots, y_n)$. If $\{\varphi_k(t)\}$ are the orthonormal basis signals, then

$$x(t) = \sum_i x_i \varphi_i(t)$$

$$y(t) = \sum_j y_j \varphi_j(t)$$

---

* Including end samples.

Hence,

$$\int_{-\infty}^{\infty} x(t)y(t)\,dt = \int_{-\infty}^{\infty} \left[\sum_i x_i\varphi_i(t)\right]\left[\sum_j y_j\varphi_j(t)\right] dt$$

The basis signals are orthonormal. Hence,

$$\int_{-\infty}^{\infty} \varphi_i(t)\varphi_j(t)\,dt = \begin{cases} 1 & \text{for } i = j \\ 0 & \text{for } i \neq j \end{cases}$$

Use of this result yields

$$\int_{-\infty}^{\infty} x(t)y(t)\,dt = \sum_k x_k y_k \tag{7.17a}$$

The right-hand side of Eq. (7.17a), however, is by definition the scalar product of vectors $\boldsymbol{x}$ and $\boldsymbol{y}$:

$$\boldsymbol{x}\cdot\boldsymbol{y} = \sum_k x_k y_k$$

Hence, we have

$$\boldsymbol{x}\cdot\boldsymbol{y} = \int_{-\infty}^{\infty} x(t)y(t)\,dt \tag{7.17b}$$

If $x(t)$ and $y(t)$ are mutually orthogonal, then it follows from Eq. (7.17b) that the corresponding vectors $\boldsymbol{x}$ and $\boldsymbol{y}$ are also orthogonal. We conclude that the integral of the product of two signals is equal to the scalar product of the corresponding vectors.

Energy of a Signal. For a signal $x(t)$, the energy E_x is given by

$$E_x = \int_{-\infty}^{\infty} x^2(t)\,dt$$

It follows from Eq. (7.17b) that

$$E_x = \boldsymbol{x}\cdot\boldsymbol{x} = |\boldsymbol{x}|^2 \tag{7.18}$$

where $|\boldsymbol{x}|^2$ is the square of the length of the vector $\boldsymbol{x}$. Hence, the signal energy is given by the square of the length of the corresponding vector.

■ EXAMPLE 7.1

A signal space consists of four signals $s_1(t)$, $s_2(t)$, $s_3(t)$, and $s_4(t)$, as shown in Fig. 7.1. Determine a suitable set of basis vectors and the dimensionality of the signals. Represent these signals geometrically in the vector space.

□ **Solution:** The two rectangular pulses $\varphi_1(t)$ and $\varphi_2(t)$ in Fig. 7.1*b* are suitable as a basis signal set. In terms of this set, the vectors $\boldsymbol{s}_1$, $\boldsymbol{s}_2$, $\boldsymbol{s}_3$, and $\boldsymbol{s}_4$ corresponding to signals $s_1(t)$, $s_2(t)$, $s_3(t)$, and $s_4(t)$ are $\boldsymbol{s}_1 = (1, -0.5)$, $\boldsymbol{s}_2 = (-0.5, 1)$, $\boldsymbol{s}_3 = (0, -1)$, and $\boldsymbol{s}_4 = (0.5, 1)$. These points are plotted in Fig. 7.1*c*. Observe that

$$\boldsymbol{s}_1\cdot\boldsymbol{s}_4 = 0.5 - 0.5 = 0$$

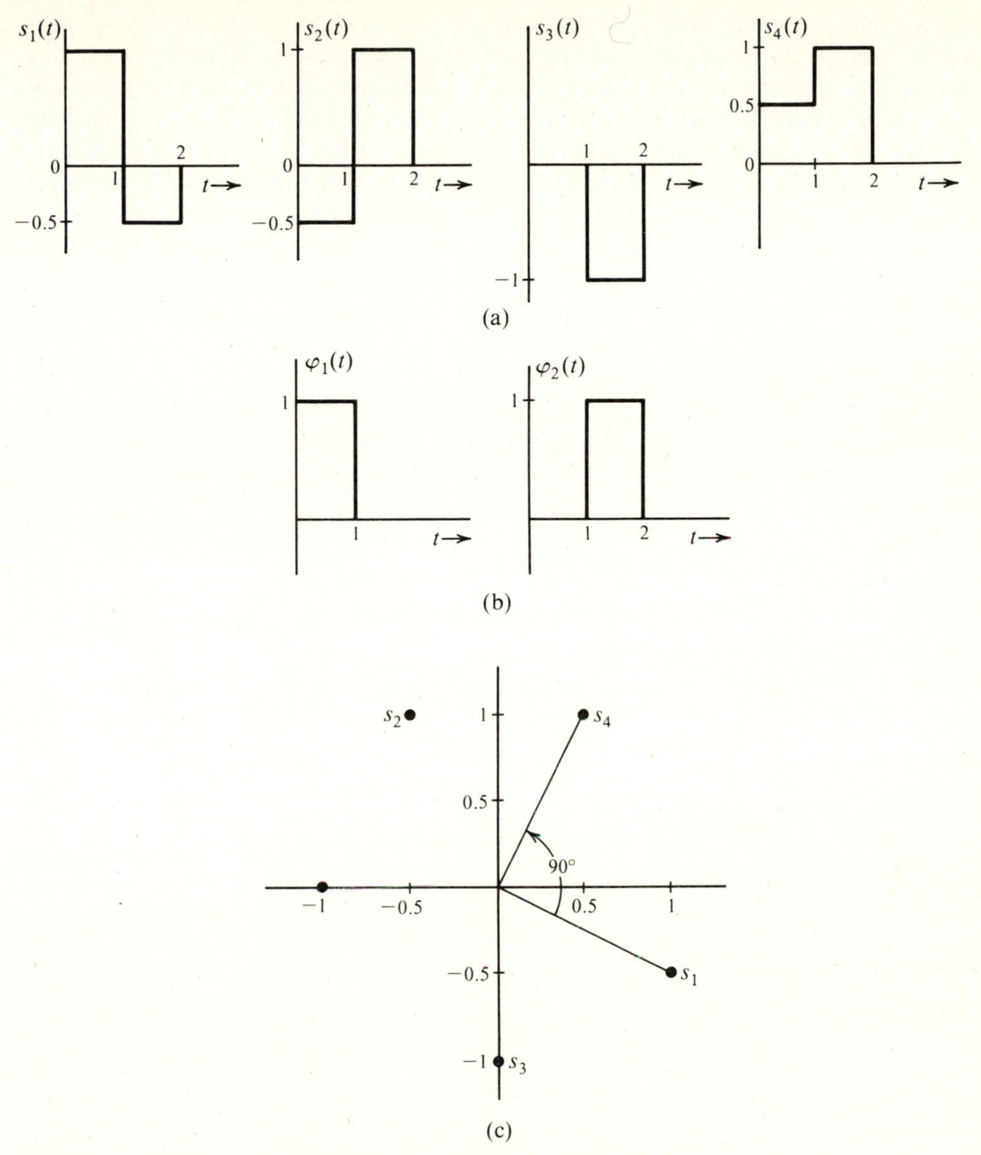

Figure 7.1 Signals and their representation in signal space.

Hence, s_1 and s_4 are orthogonal. This result may be verified from the fact that

$$\int_{-\infty}^{\infty} s_1(t)s_4(t)\, dt = 0$$

Note that each point in the signal space in Fig. 7.1c corresponds to some waveform. ■

Systematic Determination of an Orthogonal Basis Set. We have shown that the dimensionality of a vector space is equal to the maximum number of independent vectors in the space. Thus, in an n-dimensional space, there can be no more than n

vectors that are independent. Alternatively, it is always possible to find a set of n vectors that are independent. Once such a set (basis set) is chosen, any vector in this space can be expressed in terms of (as a linear combination of) the vectors in this set. This set of n independent vectors is by no means unique. The reader is familiar with this fact in the physical space of three dimensions, where one can find an infinite number of independent sets of three vectors. This is obvious from the fact that we have infinite possible coordinate systems. The orthogonal set, however, is of special interest, because it is easier to deal with as compared to nonorthogonal sets. If we are given a set of n independent vectors, it is possible to obtain from this set another set of n independent vectors that is orthogonal. This is done by the *Gram-Schmidt orthogonalization process* discussed in Appendix 7.1.

A deterministic signal can be represented by one point in a signal space. A random process, on the other hand, consists of an ensemble of waveforms, each of which maps into a point in a signal space. Hence, a random process appears as an ensemble of points in a signal space. Let us consider in detail the gaussian random process and its representation in a signal space.

7.2 THE GAUSSIAN RANDOM PROCESS

In order to specify a gaussian random process, we need to familiarize ourselves first with jointly gaussian random variables.

Jointly Gaussian Random Variables

Random variables $\mathrm{x}_1, \mathrm{x}_2, \ldots, \mathrm{x}_n$ are said to be jointly gaussian if their joint PDF is given by

$$p_{\mathrm{x}_1\mathrm{x}_2\ldots\mathrm{x}_n}(x_1, x_2, \ldots, x_n)$$
$$= \frac{1}{(2\pi)^{n/2}\sqrt{|K|}} \cdot \exp\left[\frac{1}{2|K|}\sum_i \sum_j \Delta_{ij}(x_i - \overline{\mathrm{x}}_i)(x_j - \overline{\mathrm{x}}_j)\right] \tag{7.19}$$

where K is the covariance matrix

$$K = \begin{bmatrix} \sigma_{11} & \sigma_{12} & \cdots & \sigma_{1n} \\ \sigma_{21} & \sigma_{22} & \cdots & \sigma_{2n} \\ \vdots & \vdots & \vdots & \vdots \\ \sigma_{n1} & \sigma_{n2} & \cdots & \sigma_{nn} \end{bmatrix} \tag{7.20a}$$

and

$$\sigma_{ij} = \overline{(\mathrm{x}_i - \overline{\mathrm{x}}_i)(\mathrm{x}_j - \overline{\mathrm{x}}_j)} \qquad \text{(the covariance of } \mathrm{x}_i \text{ and } \mathrm{x}_j\text{)} \tag{7.20b}$$

and $|K|$ is the determinant of matrix K and Δ_{ij} is the cofactor for the element σ_{ij}. Note that $\sigma_{ii} = \sigma_i^2$.

Gaussian variables are important not only because they are frequently observed, but also because they have certain properties that simplify many mathematical operations that are impossible or very difficult for other types of r.v.'s. These properties are as follows:

Property 1: The gaussian density is completely specified by only the first and second moments (means and covariances). This follows from Eq. (7.19). It can be shown that, in general, the probability density of an r.v. depends on all the moments $\overline{x^n}$ ($n = 1, 2, \ldots$) of the variable. A gaussian variable is a special case where only the first two moments (means) and the second moments (covariances) are necessary to specify the distribution.

Property 2: If n jointly gaussian variables $x_1, x_2, \ldots, x_n$ are uncorrelated, they are independent.

If the n variables are uncorrelated, $\sigma_{ij} = 0 (i \neq j)$, and Eq. (7.19) reduces to

$$p_{x_1 x_2 \cdots x_n}(x_1, x_2, \ldots, x_n) = \prod_{i=1}^{n} \frac{1}{\sqrt{2\pi\sigma_i^2}} \exp\left[(x_i - \bar{x}_i)^2 / 2\sigma_i^2\right] \tag{7.21a}$$

$$= p_{x_1}(x_1) p_{x_2}(x_2) \ldots p_{x_n}(x_n) \tag{7.21b}$$

As we observed earlier, independent variables are always uncorrelated, but uncorrelated variables are not necessarily independent. For the case of jointly gaussian random variables, however, uncorrelatedness implies independence.

Property 3: When $x_1, x_2, \ldots, x_n$ are jointly gaussian, all the marginal densities, such as $p_{x_i}(x_i)$, and all the conditional densities, such as $p_{x_i x_j}(x_i, x_j | x_k, x_l, \ldots, x_p)$ are gaussian. This property can be easily verified (see Prob. 5.25).

Property 4: Linear combinations of jointly gaussian variables are also jointly gaussian. Thus, if we form m variables $y_1, y_2, \ldots, y_m$ ($m \leq n$) such that

$$y_i = \sum_{k=1}^{n} a_{ik} x_k$$

then $y_1, y_2, \ldots, y_m$ are also jointly gaussian variables. By this property, one can show that if the input signal to a linear system is gaussian, the output signal will also be gaussian.

The Gaussian Random Process

A random process $x(t)$ is said to be gaussian if the r.v.'s $x(t_1), x(t_2), \ldots, x(t_n)$ are jointly gaussian [Eq. (7.19)] for every n and for every set $(t_1, t_2, \ldots, t_n)$.

For convenience, the r.v. $x(t_i)$ will be denoted by x_i. Hence, the joint PDF of r.v.'s $x_1, x_2, \ldots, x_n$ of a gaussian random process is given by Eq. (7.19). Observe that

$$\begin{aligned}
\sigma_{ij} &= \overline{(x_i - \bar{x}_i)(x_j - \bar{x}_j)} \\
&= \overline{x_i x_j} - \bar{x}_i \bar{x}_j - \bar{x}_i \bar{x}_j + \bar{x}_i \bar{x}_j \\
&= \overline{x_i x_j} - \bar{x}_i \bar{x}_j \\
&= \overline{x(t_i) x(t_j)} - \bar{x}_i \bar{x}_j \\
&= R_x(t_i, t_j) - \overline{x(t_i)}\,\overline{x(t_j)}
\end{aligned} \tag{7.22}$$

The Stationary Gaussian Random Process

Our discussion of the gaussian process thus far applies to stationary and nonstationary processes. We have shown that the gaussian process is completely specified by its autocorrelation function $R_x(t_i, t_j)$ and its mean value function.* An important corollary of this statement is that if the autocorrelation function $R_x(t_i, t_j)$ and the means are unaffected by a shift of the time origin, then all of the statistics of the process are unaffected by a shift of the time origin. In other words, the process is stationary. Thus, if

$$R_x(t_i, t_j) = R_x(t_i - t_j) \tag{7.23}$$

and

$$\overline{x(t)} = \text{constant for all } t$$

the process is stationary if it is gaussian. But the condition in Eq. (7.23) defines wide-sense stationarity. Hence, *for a gaussian process, wide-sense stationarity implies stationarity in the strict sense.*

For a stationary gaussian process, the covariance σ_{ij} becomes

$$\sigma_{ij} = R_x(t_j, t_i) - \bar{x}^2 \tag{7.24}$$

We shall once again stress the point that distinguishes the gaussian from the nongaussian process. The complete statistics of a gaussian process are determined from its autocorrelation function (which is a second-order parameter) and its mean value. This is the property that simplifies the study of gaussian processes. In general, for nongaussian processes higher-order statistics cannot be determined from lower-order statistics. For the gaussian process, however, the mean and second-order parameter σ_{ij} determine all the higher-order statistics. This very property also enables us to state that a gaussian process is strictly stationary if it is wide-sense stationary.

Transmission of a Gaussian Process Through a Linear System

Another significant property of the gaussian process is that the response of a linear system to a gaussian process is also a gaussian process. This can be shown as follows. Let $x(t)$ be a gaussian process applied to the input of a linear system whose unit impulses response is $h(t)$. If $y(t)$ is the output (response) process, then

$$y(t) = \int_{-\infty}^{\infty} x(t - \tau)h(\tau)\, d\tau$$

$$= \lim_{\Delta\tau \to 0} \sum_{k=-\infty}^{\infty} x(t - \tau_k)h(\tau_k)\, \Delta\tau \qquad \tau_k = k\Delta\tau$$

Because $x(t)$ is a gaussian process, all the variables $x(t - \tau_k)$ are jointly gaussian (by definition). Hence, the variables $y(t_1), y(t_2), \ldots, y(t_n)$ for all n and every set (t_1, t_2,

* For nonstationary processes (or processes that are not wide-sense stationary), the mean value is a function of t.

$\ldots, t_n$) are linear combinations of variables that are jointly gaussian. Therefore, the variables $y(t_1), y(t_2), \ldots, y(t_n)$ must be jointly gaussian, according to the earlier discussion. It follows that the process $y(t)$ is a gaussian process.

To summarize, the gaussian random process has the following properties:

1. A gaussian random process is completely specified by its autocorrelation function and mean value.
2. If a gaussian random process is wide-sense stationary, then it is stationary in the strict sense.
3. The response of a linear system to a gaussian random process is also a gaussian random process.

Geometrical Representation of a Random Process

Let $x(t)$ be a random process and $\varphi_1(t), \varphi_2(t), \ldots, \varphi_n(t)$ be a complete set of orthonormal basis signals for this space. We can express $x(t)$ as

$$x(t) = x_1\varphi_1(t) + x_2\varphi_2(t) + \cdots + x_n\varphi_n(t)$$

$$= \sum_{k=1}^{n} x_k\varphi_k(t) \tag{7.25}$$

where

$$x_k = \int_{-\infty}^{\infty} x(t)\varphi_k(t)\, dt \tag{7.26}$$

Note that $x(t)$ is a random process, consisting of an ensemble of sample functions. Coefficients x_k in Eq. (7.25) will be different for each sample function. Consequently, $x_1, x_2, \ldots, x_n$ appearing in Eq. (7.25) are random variables. Each sample function will have a specific set $(x_1, x_2, \ldots, x_n)$ and will map into one point in the signal space. Hence, the ensemble of sample functions will map into an ensemble of points in the signal space, as shown in Fig. 7.2. This figure shows only a 3-dimensional graph because it is not possible to show a higher-dimensional graph. But it is sufficient to indicate the idea.

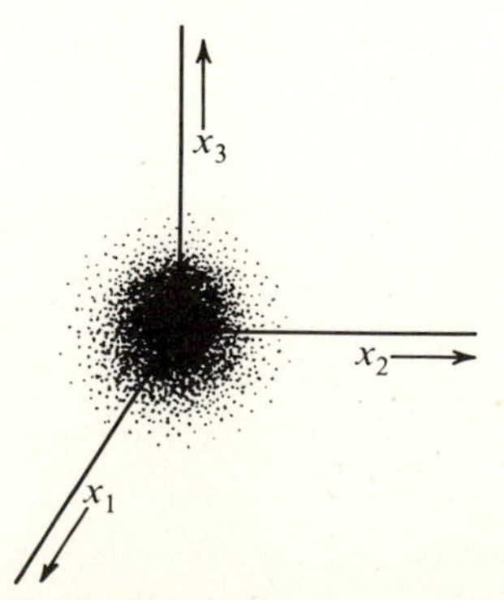

Figure 7.2 Geometrical representation of a gaussian random process.

Each time the random experiment is repeated, the outcome (the sample function) is a certain point $\boldsymbol{x}$. The ensemble of points in the signal space appears as a cloud, with the density of points directly proportional to the probability of observing $\mathbf{x}$ in that region. If we denote the joint PDF of $\mathrm{x}_1, \mathrm{x}_2, \ldots, \mathrm{x}_n$ by $p_{\mathbf{x}}(\boldsymbol{x})$, then

$$p_{\mathbf{x}}(\boldsymbol{x}) = p_{\mathrm{x}_1\mathrm{x}_2\cdots\mathrm{x}_n}(x_1, x_2, \ldots, x_n)$$

Thus, $p_{\mathbf{x}}(\boldsymbol{x})$ has a certain value at each point in the signal space, and $p_{\mathbf{x}}(\boldsymbol{x})$ represents the relative probability (cloud density) of observing $\mathbf{x}$ at that point.

A Word about Notation. We shall briefly discuss the notation used here to avoid confusion later. As before, we use roman type to denote a random variable or a random process. Thus, x or x(t) represents a random variable or a random process. A particular value assumed by the random variable in a certain trial is denoted by italic type. Thus, x represents the value assumed by x. Similarly, $x(t)$ represents a particular sample function of the random process x(t). In the case of random vectors, we follow the same convention; a random vector is denoted by roman boldface type, and a particular value assumed by the vector in a certain trial is represented by boldface italic type. Thus, **r** denotes a random vector representing a random process r(t), but $\boldsymbol{r}$ is a particular value of **r** and represents a particular received waveform (sample function) $r(t)$ in some trial. Note that the roman type represents random entities and italic type represents particular values (which are, of course, nonrandom).

White Gaussian Noise. Consider a white noise process $\mathrm{n}_w(t)$ with PSD $\mathcal{N}/2$. Let $\varphi_1(t), \varphi_2(t), \ldots$ be a complete set of orthonormal basis signals for this space. We can express $\mathrm{n}_w(t)$ as

$$\begin{aligned}\mathrm{n}_w(t) &= \mathrm{n}_1\varphi_1(t) + \mathrm{n}_2\varphi_2(t) + \cdots \\ &= \sum_k \mathrm{n}_k\varphi_k(t)\end{aligned}$$

White noise has infinite bandwidth. Consequently, the dimensionality of the signal space is infinity.

We shall now show that random variables $\mathrm{n}_1, \mathrm{n}_2, \ldots$ are independent, with variance $\mathcal{N}/2$. Because [see Eq. (7.13)]

$$\mathrm{n}_k = \int_{-\infty}^{\infty} \mathrm{n}_w(t)\varphi_k(t)\,dt$$

$$\begin{aligned}\overline{\mathrm{n}_j\mathrm{n}_k} &= \overline{\int_{-\infty}^{\infty} \mathrm{n}_w(\alpha)\varphi_j(\alpha)\,d\alpha \int_{-\infty}^{\infty} \mathrm{n}_w(\beta)\varphi_k(\beta)\,d\beta} \\ &= \int_{-\infty}^{\infty}\int_{-\infty}^{\infty} \overline{\mathrm{n}_w(\alpha)\mathrm{n}_w(\beta)}\varphi_j(\alpha)\varphi_k(\beta)\,d\alpha\,d\beta \\ &= \int_{-\infty}^{\infty}\int_{-\infty}^{\infty} R_{\mathrm{n}_w}(\beta - \alpha)\varphi_j(\alpha)\varphi_k(\beta)\,d\alpha\,d\beta\end{aligned}$$

Because $R_{n_w}(\tau) = \mathcal{F}^{-1}(\mathcal{N}/2) = (\mathcal{N}/2)\,\delta(\tau)$

$$\overline{n_j n_k} = \int_{-\infty}^{\infty}\int_{-\infty}^{\infty} \frac{\mathcal{N}}{2}\,\delta(\beta - \alpha)\varphi_j(\alpha)\varphi_k(\beta)\,d\alpha\,d\beta$$

$$= \frac{\mathcal{N}}{2}\int_{-\infty}^{\infty} \varphi_j(\alpha)\varphi_k(\alpha)\,d\alpha$$

$$= \begin{cases} 0 & j \neq k \\ \dfrac{\mathcal{N}}{2} & j = k \end{cases} \tag{7.27}$$

This proves the result.

For the time being, assume that we are considering an N-dimensional case. The joint PDF of independent joint gaussian r.v.'s $n_1, n_2, \ldots, n_N$, each with zero mean and variance $\mathcal{N}/2$, is [see Eq. (7.20)]

$$p_{\mathbf{n}}(\boldsymbol{n}) = \prod_j \frac{1}{\sqrt{\left(2\pi\frac{\mathcal{N}}{2}\right)^N}}\, e^{-n_j^2/2(\mathcal{N}/2)}$$

$$= \frac{1}{(\pi\mathcal{N})^{N/2}}\, e^{-(n_1^2 + n_2^2 + \cdots + n_N^2)/\mathcal{N}} \tag{7.28a}$$

$$= \frac{1}{(\pi\mathcal{N})^{N/2}}\, e^{-|\boldsymbol{n}|^2/\mathcal{N}} \tag{7.28b}$$

This shows that the PDF $p_{\mathbf{n}}(\boldsymbol{n})$ depends only on $|\boldsymbol{n}|$, the magnitude of the noise vector $\mathbf{n}$ in the hyperspace, and is therefore spherically symmetrical if plotted in the N-dimensional hyperspace.

7.3 THE OPTIMUM RECEIVER[2–5]

We shall now consider, from a more fundamental point of view, the problem of M-ary communication in the presence of additive white gaussian channel noise (AWGN). No constraint shall be placed on the optimum structure, and we shall try to answer the fundamental question: What receiver will yield the minimum error probability?

The comprehension of the signal-detection problem is greatly facilitated by geometrical representation of signals. In a signal space, we can represent a signal by a fixed point (or a vector). A random process can be represented by a random point (or a random vector). The region in which the random point may lie will be shown shaded, with the shading intensity proportional to the probability of observing the signal in that region. In the M-ary scheme, we use M symbols, or messages, $m_1, m_2, \ldots, m_M$. Each of these symbols is represented by a specified waveform. Let the corresponding waveforms be $s_1(t), s_2(t), \ldots, s_M(t)$. Thus, the symbol (or message) m_k is sent by transmitting the waveform $s_k(t)$. These waveforms are corrupted by AWGN $n_w(t)$

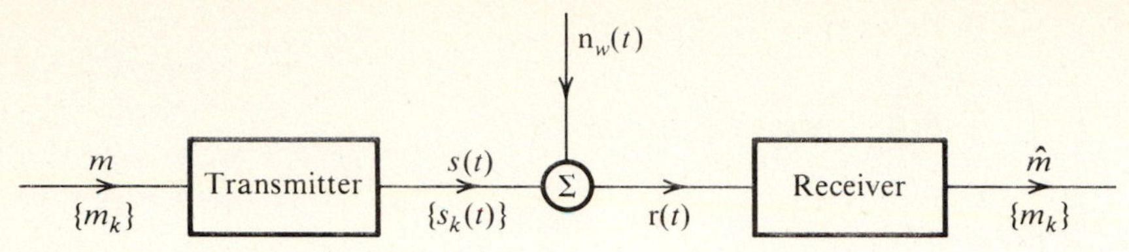

Figure 7.3 *M*-ary communication system.

(Fig. 7.3) with PSD

$$S_n(\omega) = \frac{\mathcal{N}}{2}$$

At the receiver, the received signal $r(t)$ consists of one of the M message waveforms $s_k(t)$ plus the channel noise. For the sake of generality, the message waveform will be denoted by $s(t)$, where it is understood that $s(t)$ is one of the m waveforms $s_k(t)$:

$$r(t) = s(t) + n_w(t) \tag{7.29a}$$

Let us consider the message m_k. Corresponding to this message, the received signal is

$$r(t) = s_k(t) + n_w(t)$$

We can represent $r(t)$ in a signal space by letting $\mathbf{r}$, s_k, and n_w be the points (or vectors) representing signals $r(t)$, $s_k(t)$, and $n_w(t)$, respectively. Then it is evident that

$$\mathbf{r} = s_k + n_w \tag{7.29b}$$

The vector s_k is a fixed vector, because the waveform $s_k(t)$ is nonrandom. The vector $\mathbf{n}_w$ (or point $\mathbf{n}_w$) is random. Hence, the vector $\mathbf{r}$ is also random. Because $n_w(t)$ is a gaussian white noise, the probability distribution of $\mathbf{n}_w$ has spherical symmetry in the signal space (see Sec. 7.2). Hence, the distribution of $\mathbf{r}$ is a spherical distribution centered at a fixed point s_k, as shown in Fig. 7.4. Whenever the message m_k is transmitted, the probability of observing the received signal $r(t)$ in a given region is indicated by the intensity of the shading in Fig. 7.4. Actually, because the noise is

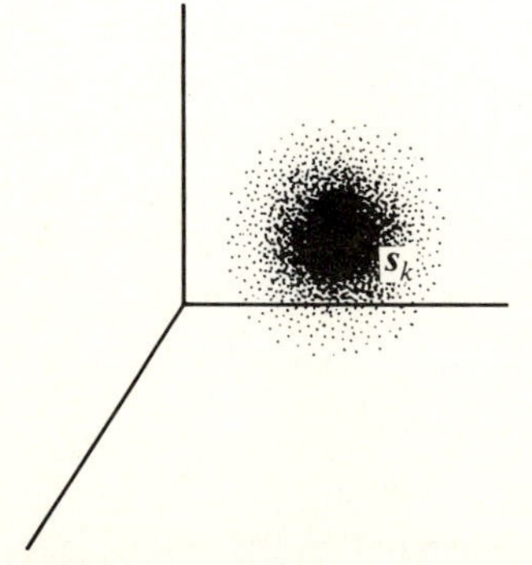

Figure 7.4 Effect of gaussian channel noise on the received signal.

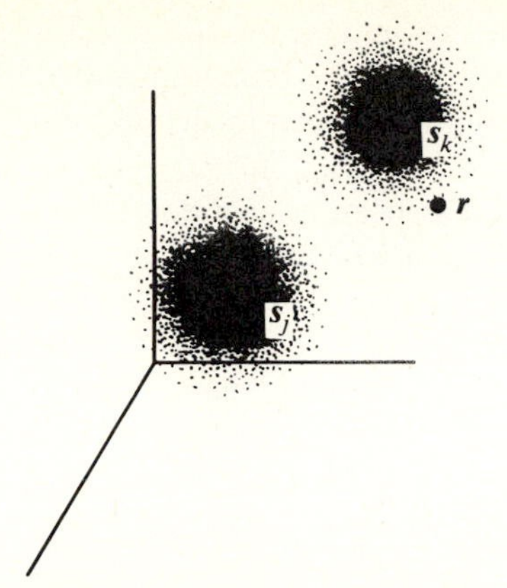

Figure 7.5 Binary communication in the presence of noise (high signal-to-noise ratio).

white, the space has an infinite number of dimensions. For simplicity, however, we have shown the space to be 3-dimensional. This will suffice to indicate our line of reasoning. We can draw similar regions for various points $s_1, s_2, \ldots, s_M$. Figure 7.5 shows the regions for two messages m_j and m_k when s_j and s_k are widely separated in signal space. In this case, there is virtually no overlap between the two regions. If either m_j or m_k is transmitted, the received signal will lie in one of the two regions. From the position of the received signal, one can decide with a very small probability of error whether m_j or m_k was transmitted. Note that theoretically each region extends to infinity, although the probability of observing the received signal diminishes rapidly as one moves away from the center. Hence, there will always be an overlap between the two regions, resulting in a nonzero error probability. In Fig. 7.5, the received signal $\boldsymbol{r}$ is much closer to s_k than s_j. It is therefore more likely that m_k was transmitted.

Figure 7.6 illustrates the case when the points s_j and s_k are spaced closely together. In this case, there is a considerable overlap between the two regions. Because the

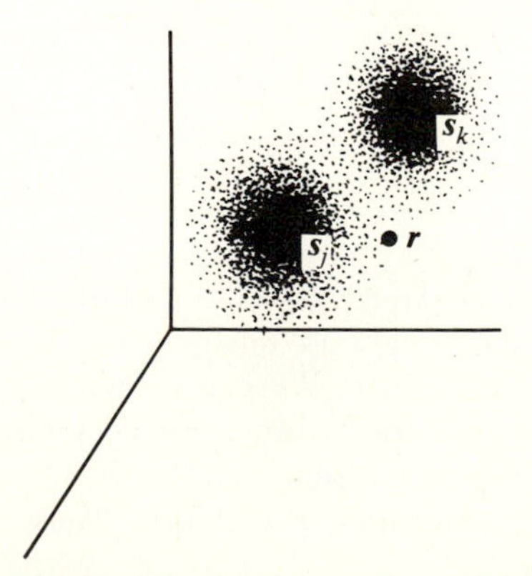

Figure 7.6 Binary communication in the presence of noise (small signal-to-noise ratio).

received signal $\boldsymbol{r}$ is closer to $\boldsymbol{s}_j$ than $\boldsymbol{s}_k$, it is more likely that m_j was transmitted. But in this case there is also a considerable probability that m_k may have been transmitted. Hence in this situation, there will be a much higher probability of error in any decision scheme.

The optimum receiver must decide which message has been transmitted from a knowledge of $\boldsymbol{r}$. The signal space must be divided into M nonoverlapping, or disjoint, regions $R_1, R_2, \ldots, R_M$, corresponding to the M messages $m_1, m_2, \ldots, m_M$. If $\boldsymbol{r}$ falls in the region R_k, the decision is m_k. The problem of designing the receiver then reduces to choosing the boundaries of these regions $R_1, R_2, \ldots, R_M$ such that the probability of error in decision making is minimum.

Before proceeding any further, it will be helpful to recapitulate the problem: A transmitter transmits a sequence of messages from a set of M messages $m_1, m_2, \ldots, m_M$. These messages are represented by finite energy waveforms $s_1(t), s_2(t), \ldots, s_M(t)$, respectively. One waveform is transmitted every T_o seconds. We assume that the receiver is time synchronized with the transmitter. The waveforms are corrupted during transmissions by an AWGN of PSD $\mathcal{N}/2$. Knowing the received waveform, the receiver must make a decision as to which waveform was transmitted. The merit criterion of the receiver is the minimum probability of error in making this decision.

Let us now discuss the dimensionality of the signal space in our problem. If there was no noise, we would be dealing with only M waveforms $s_1(t), s_2(t), \ldots, s_M(t)$. In this case a signal space of, at most, M dimensions would suffice. This is because the dimensionality of a signal space is always equal to or less than the number of independent signals in the space (see Sec. 7.1). For the sake of generality we shall assume the space to have N dimensions ($N \leq M$). Let $\varphi_1(t), \varphi_2(t), \ldots, \varphi_N(t)$ be the orthonormal basis set for this space. Such a set can be constructed by using the Gram-Schmidt procedure discussed in Appendix 7.1. We can then represent the signal waveform $s_k(t)$ as

$$\begin{aligned} s_k(t) &= s_{k_1}\varphi_1(t) + s_{k_2}\varphi_2(t) + \cdots + s_{kN}\varphi_N(t) \\ &= \sum_{j=1}^{N} s_{kj}\varphi_j(t) \end{aligned} \tag{7.30a}$$

where

$$s_{kj} = \int_{-\infty}^{\infty} s_k(t)\varphi_j(t)\,dt \tag{7.30b}$$

Now consider the white gaussian channel noise $\mathrm{n}_w(t)$. This signal has an infinite bandwidth ($B = \infty$). It has an infinite number of dimensions and obviously cannot be represented in the N-dimensional signal space discussed above. We can, however, split $\mathrm{n}_w(t)$ into two components: (1) the projection of $\mathrm{n}_w(t)$ on the N-dimensional signal space, and (2) the remaining component, which will be orthogonal to the N-dimensional signal space. Let us denote the two components by $\mathrm{n}(t)$ and $\mathrm{n}_0(t)$. Thus,

$$\mathrm{n}_w(t) = \mathrm{n}(t) + \mathrm{n}_0(t) \tag{7.31}$$

where

$$\mathrm{n}(t) = \sum_{k=1}^{N} \mathrm{n}_j \varphi_j(t) \tag{7.32a}$$

and

$$\mathrm{n}_j = \int_{-\infty}^{\infty} \mathrm{n}(t)\varphi_j(t)\, dt \tag{7.32b}$$

Because $\mathrm{n}_0(t)$ is orthogonal to the N-dimensional space, it is orthogonal to every signal in that space, Hence,

$$\int_{-\infty}^{\infty} \mathrm{n}_0(t)\varphi_j(t)\, dt = 0 \qquad \text{for } j = 1, 2, \ldots, N$$

Hence,

$$\begin{aligned} \mathrm{n}_j &= \int_{-\infty}^{\infty} [\mathrm{n}(t) + \mathrm{n}_0(t)]\varphi_j(t)\, dt \\ &= \int_{-\infty}^{\infty} \mathrm{n}_w(t)\varphi_j(t)\, dt \end{aligned} \tag{7.33}$$

From Eqs. (7.33) and (7.32a) it is evident that we can filter out the component $\mathrm{n}_0(t)$ from $\mathrm{n}_w(t)$. This can be seen from the fact that $\mathrm{r}(t)$, the received signal, can be expressed as

$$\begin{aligned} \mathrm{r}(t) &= s_k(t) + \mathrm{n}_w(t) \\ &= s_k(t) + \mathrm{n}(t) + \mathrm{n}_0(t) \\ &= \mathrm{q}(t) + \mathrm{n}_0(t) \end{aligned} \tag{7.34}$$

where $\mathrm{q}(t)$ is the projection of $\mathrm{r}(t)$ on the N-dimensional space. Thus

$$\mathrm{q}(t) = s_k(t) + \mathrm{n}(t) \tag{7.35}$$

We can obtain the projection $\mathrm{q}(t)$ from $\mathrm{r}(t)$ by observing that [see Eqs. (7.30) and (7.32a)]

$$\mathrm{q}(t) = \sum_{j=1}^{M} (s_{kj} + \mathrm{n}_j)\varphi_j(t) \tag{7.36}$$

From Eqs. (7.30b), (7.33), and (7.36) it follows that if we feed the received signal $\mathrm{r}(t)$ into the system shown in Fig. 7.7, the resultant outcome will be $\mathrm{q}(t)$. Thus, the orthogonal noise component can be filtered out without disturbing the message signal. The question here is: Would such filtering help in our decision making? We can easily show that it cannot hurt us. The noise $\mathrm{n}_w(t)$ is independent of the signal waveform $s_k(t)$. Therefore, its component $\mathrm{n}_0(t)$ is also independent of $s_k(t)$. Thus, $\mathrm{n}_0(t)$ contains no information about the transmitted signal, and discarding such a component from

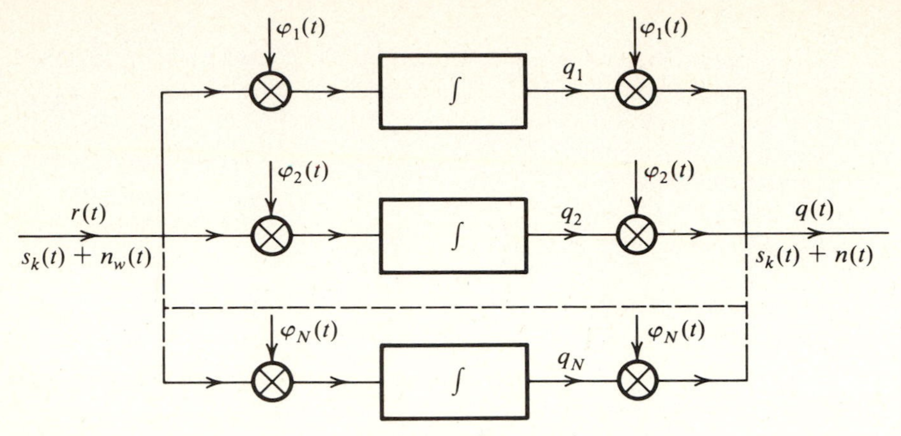

Figure 7.7 Eliminating the noise orthogonal to signal space.

the received signal $\mathrm{r}(t)$ will not cause any loss of information regarding the signal waveform $s_k(t)$. This, however, is not enough. We must also make sure that the noise being discarded $[\mathrm{n}_0(t)]$ is not in any way related to the remaining noise component $\mathrm{n}(t)$. If $\mathrm{n}_0(t)$ and $\mathrm{n}(t)$ are related in any way, it will be possible to obtain some information about $\mathrm{n}(t)$ from $\mathrm{n}_0(t)$, thus enabling us to detect that signal with less error probability. If the components $\mathrm{n}_0(t)$ and $\mathrm{n}(t)$ are independent random processes, the component $\mathrm{n}_0(t)$ does not carry any information about $\mathrm{n}(t)$ and can be discarded. Under these conditions, $\mathrm{n}_0(t)$ is irrelevant to the decision making at the receiver.

The process $\mathrm{n}(t)$ is represented by components $\mathrm{n}_1, \mathrm{n}_2, \ldots, \mathrm{n}_N$ along $\varphi_1(t), \varphi_2(t), \ldots, \varphi_N(t)$, and $\mathrm{n}_0(t)$ is represented by the remaining components (infinite number) along the remaining basis signals in the complete set, $\{\varphi_k(t)\}$. From Eq. (7.27) we observe that all the components are independent. Hence, the components representing $\mathrm{n}_0(t)$ are independent of components representing $\mathrm{n}(t)$. Consequently, $\mathrm{n}_0(t)$ is independent of $\mathrm{n}(t)$ and is irrelevant data.

The received signal $\mathrm{r}(t)$ is now reduced to signal $\mathrm{q}(t)$, which contains the desired signal waveform and the projection of the channel noise on the N-dimensional signal space. Thus, the signal $\mathrm{q}(t)$ can be completely represented in the signal space. Let the vectors representing $\mathrm{n}(t)$ and $\mathrm{q}(t)$ be denoted by $\mathbf{n}$ and $\mathbf{q}$. Thus,

$$\mathbf{q} = s + \mathbf{n}$$

where s may be any one of vectors $s_1, s_2, \ldots, s_M$.

The random vector $\mathbf{n}(\mathrm{n}_1, \mathrm{n}_2, \ldots, \mathrm{n}_N)$ is represented by N independent gaussian variables, each with zero mean and variance $\sigma^2 = \mathcal{N}/2$. The joint probability density function of vector $\mathbf{n}$ in such a case has a spherical symmetry, as shown in Eq. (7.28).

$$p_{\mathbf{n}}(\boldsymbol{n}) = \frac{1}{(\pi\mathcal{N})^{N/2}} e^{-|\boldsymbol{n}|^2/\mathcal{N}} \tag{7.37a}$$

Note that this actually represents

$$p_{\mathrm{n}_1,\mathrm{n}_2,\ldots,\mathrm{n}_N}(n_1, n_2, \ldots, n_N) = \frac{1}{(\pi\mathcal{N})^{N/2}} e^{-(n_1^2+n_2^2+\cdots+n_N^2)/\mathcal{N}} \tag{7.37b}$$

The Decision Procedure

Our problem is now considerably simplified. The irrelevant noise component has been filtered out. The residual signal $\mathrm{q}(t)$ can be represented in the N-dimensional signal space. We proceed to determine the M decision regions $R_1, R_2, \ldots, R_M$ in this space. The regions must be so chosen that the probability of error in making the decision is minimized.

Suppose the received vector $\mathbf{q} = \boldsymbol{q}$. Then if the receiver decides $\hat{m} = m_k$, the conditional probability of making the correct decision, given that $\mathbf{q} = \boldsymbol{q}$ is

$$P(C|\mathbf{q} = \boldsymbol{q}) = P(m_k|\mathbf{q} = \boldsymbol{q}) \tag{7.38}$$

where $P(C|\mathbf{q} = \boldsymbol{q})$ is the conditional probability of making the correct decision given $\mathbf{q} = \boldsymbol{q}$, and $P(m_k|\mathbf{q} = \boldsymbol{q})$ is the conditional probability that m_k was transmitted given $\mathbf{q} = \boldsymbol{q}$. The unconditional probability $P(C)$ is given by

$$P(C) = \int_{\boldsymbol{q}} P(C|\mathbf{q} = \boldsymbol{q})p_{\mathbf{q}}(\boldsymbol{q})\, d\boldsymbol{q} \tag{7.39}$$

where the integration is performed over the entire region occupied by $\mathbf{q}$. Note that this is an N-fold integration with respect to the variables $q_1, q_2, \ldots, q_N$ over the range $(-\infty, \infty)$. Also, because $\mathrm{p}_{\mathbf{q}}(\boldsymbol{q}) \geq 0$, this integral is maximum when $P(C|\mathbf{q} = \boldsymbol{q})$ is maximum. From Eq. (7.38) it now follows that if a decision $\hat{m} = m_k$ is made, the error probability is minimized if

$$P(m_k|\mathbf{q} = \boldsymbol{q})$$

is maximized. The probability $P(m_k|\mathbf{q} = \boldsymbol{q})$ is called the *a posteriori probability* of m_k. This is because it represents the probability that m_k was transmitted when $\boldsymbol{q}$ is received.

The decision procedure is now clear. Once we receive $\mathbf{q} = \boldsymbol{q}$, we evaluate all M a posteriori probabilities. Then we make the decision in favor of that message for which the a posteriori probability is highest—that is, the receiver decides that $\hat{m} = m_k$ if

$$P(m_k|\mathbf{q} = \boldsymbol{q}) > P(m_j|\mathbf{q} = \boldsymbol{q}) \qquad \text{for all } j \neq k \tag{7.40}$$

Thus, the detector that minimizes error probability is the *maximum a posteriori probability detector*, also known as the *MAP detector*.

We can use Bayes' mixed rule [Eq. (5.49b)] to determine the a posteriori probabilities. We have

$$P(m_k|\mathbf{q} = \boldsymbol{q}) = \frac{P(m_k)p_{\mathbf{q}}(\boldsymbol{q}|m_k)}{p_{\mathbf{q}}(\boldsymbol{q})} \tag{7.41}$$

Hence, the receiver decides $\hat{m} = m_k$ if the decision function

$$\frac{P(m_i)p_{\mathbf{q}}(\boldsymbol{q}|m_i)}{p_{\mathbf{q}}(\boldsymbol{q})} \qquad i = 1, 2, \ldots, M$$

is maximum for $i = k$.

Note that the denominator $\mathrm{p}_{\mathbf{q}}(\boldsymbol{q})$ is common to all decision functions and, hence, may be ignored. Thus, the receiver sets $\hat{m} = m_k$ if the decision function

$$P(m_i)p_{\mathbf{q}}(q \mid m_i) \qquad i = 1, 2, \ldots, M \tag{7.42}$$

is maximum for $i = k$.

Thus, once q is obtained, we compute the decision function [Eq. (7.42)] for all messages $m_1, m_2, \ldots, m_M$ and decide that the message for which the function is maximum is the one most likely to have been sent.

We now turn our attention to computing the decision functions. The a priori probability $P(m_i)$ represents the probability that the message m_i will be transmitted. These probabilities must be known if the above criterion is used.* The term $p_{\mathbf{q}}(q \mid m_i)$ represents the PDF of $\mathbf{q}$ when $s(t) = s_i(t)$. Under this condition,

$$\mathbf{q} = s_i + \mathbf{n}$$

and

$$\mathbf{n} = \mathbf{q} - s_i$$

The point s_i is constant, and $\mathbf{n}$ is a random point. Obviously, $\mathbf{q}$ is a random point with the same distribution as $\mathbf{n}$ but centered at the points s_i.

Alternately, the probability $\mathbf{q} = q$ (given m $= m_i$) is the same as the probability $\mathbf{n} = q - s_i$. Hence [Eq. (7.37a)],

$$p_{\mathbf{q}}(q \mid m_i) = p_{\mathbf{n}}(q - s_i) = \frac{1}{(\pi \mathcal{N})^{N/2}} e^{|q - s_i|^2/\mathcal{N}} \tag{7.43}$$

The decision function in Eq. (7.42) now becomes

$$\frac{P(m_i)}{(\pi \mathcal{N})^{N/2}} e^{-|q - s_i|^2/\mathcal{N}} \tag{7.44}$$

Note that the decision function is always nonnegative for all values of i. Hence, comparing these functions is equivalent to comparing their logarithms, because the logarithm is a monotonic function for the positive argument. Hence for convenience, the decision function will be chosen as the logarithm of Eq. (7.44). In addition, the factor $(\pi \mathcal{N})^{N/2}$ is common for all i and can be left out. Hence, the decision function is

$$\ln P(m_i) - \frac{1}{\mathcal{N}} \ln |q - s_i|^2 \tag{7.45}$$

Note that $|q - s_i|^2$ is the square of the length of the vector $q - s_i$. Hence,

$$\begin{aligned} |q - s_i|^2 &= (q - s_i) \cdot (q - s_i) \\ &= |q|^2 + |s_i|^2 - 2q \cdot s_i \end{aligned} \tag{7.46}$$

*In case these probabilities are unknown, one must use other merit criteria, such as maximum likelihood or minimax, as discussed in later sections.

Hence, the decision function in Eq. (7.45) becomes (after multiplying throughout by $\mathcal{N}/2$)

$$\frac{\mathcal{N}}{2} \ln P(m_i) - \frac{1}{2}[|\boldsymbol{q}|^2 + |\boldsymbol{s}_i|^2 - 2\boldsymbol{q}\cdot\boldsymbol{s}_i] \tag{7.47}$$

Note that the term $|\boldsymbol{s}_i|^2$ is the square of the length of $\boldsymbol{s}_i$ and represents E_i, the energy of signal $s_i(t)$. The terms $\mathcal{N} \ln P(m_i)$ and E_i are constants in the decision function. Let

$$a_i = \tfrac{1}{2}[\mathcal{N} \ln P(m_i) - E_i] \tag{7.48}$$

Now the decision function in Eq. (7.47) becomes

$$a_i + \boldsymbol{q}\cdot\boldsymbol{s}_i - \tfrac{1}{2}|\boldsymbol{q}|^2$$

The term $(1/2)|\boldsymbol{q}|^2$ is common to all M decision functions and can be omitted for the purpose of comparison. Thus, the new decision function b_i is

$$b_i = a_i + \boldsymbol{q}\cdot\boldsymbol{s}_i \tag{7.49}$$

We compute this function b_i for $i = 1, 2, \ldots, N$, and the receiver decides that $\hat{m} = m_k$ if this function is the largest for $i = k$. If the signal $q(t)$ is applied at the input terminals of a system whose impulse response is $h(t)$, the output at $t = T_o$ is given by

$$\int_{-\infty}^{\infty} q(\lambda)h(T_o - \lambda)\, d\lambda$$

If we choose a filter matched to $s_i(t)$, that is, $h(t) = s_i(T_o - t)$,

$$h(T_o - \lambda) = s_i(\lambda)$$

and the output is

$$\int_{-\infty}^{\infty} q(\lambda)s_i(\lambda)\, d\lambda = \boldsymbol{q}\cdot\boldsymbol{s}_i$$

Hence, $\boldsymbol{q}\cdot\boldsymbol{s}_i$ is the output at $t = T_o$ of a filter matched to $s_i(t)$ when $q(t)$ is applied to its input.

Actually, we do not have $q(t)$. The incoming signal $r(t)$ is given by

$$\begin{aligned} r(t) &= s_i(t) + n_w(t) \\ &= \underbrace{s_i(t) + n(t)}_{q(t)} + \underbrace{n_0(t)}_{\substack{\text{irrelevant}\\ \text{noise}}} \end{aligned}$$

where $n_0(t)$ is the (irrelevant) component of $n_w(t)$ orthogonal to the N-dimensional signal space. Because $n_0(t)$ is orthogonal to this space, it is orthogonal to every signal in this space. Hence, it is orthogonal to the signal $s_i(t)$, and

$$\int_{-\infty}^{\infty} n_0(t)s_i(t)\, dt = 0$$

and

$$\mathbf{q} \cdot \mathbf{s}_i = \int_{-\infty}^{\infty} q(t)s_i(t)\,dt + \int_{-\infty}^{\infty} n_0(t)s_i(t)\,dt$$

$$= \int_{-\infty}^{\infty} [q(t) + n_0(t)]s_i(t)\,dt$$

$$= \int_{-\infty}^{\infty} r(t)s_i(t)\,dt \tag{7.50}$$

Hence, it is immaterial whether we use $q(t)$ or $r(t)$ at the input. We thus apply the incoming signal $r(t)$ to a parallel bank of matched filters, and the output of the filters

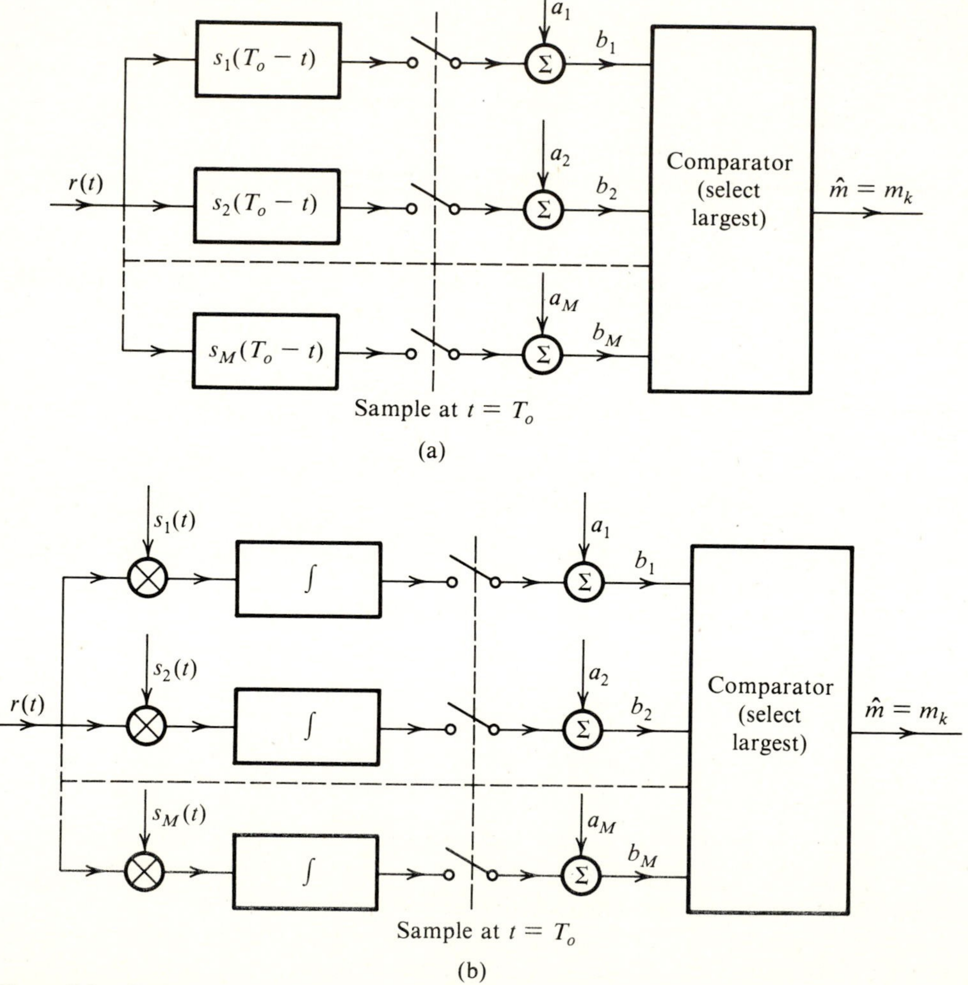

Figure 7.8 Optimum M-ary receiver. (a) Matched-filter detector. (b) Correlation detector.

is sampled at $t = T_o$. To this a constant a_i is added to the ith filter output sample, and the resulting outputs are compared. The decision is made in favor of the signal for which this output is the largest. The receiver implementation for this decision procedure is shown in Fig. 7.8*a*. As shown in Chapter 6, a matched filter is equivalent to a correlator. One may therefore use correlators instead of matched filters. Such an arrangement is shown in Fig. 7.8*b*.

We have shown that in the presence of AWGN, the matched-filter receiver is the optimum receiver when the merit criterion is the minimum error probability. Note that the system is linear, although it was not constrained to be so. It is therefore obvious that for white gaussian noise the optimum receiver happens to be linear.

The matched filter obtained in Chapter 6 and the decision procedure are identical to those derived here.

The optimum receiver can be implemented in another way. From Eq. (7.50), we have

$$\boldsymbol{q} \cdot \boldsymbol{s}_i = \boldsymbol{r} \cdot \boldsymbol{s}_i$$

From Eq. (7.3), we can rewrite this as

$$\boldsymbol{q} \cdot \boldsymbol{s}_i = \sum_{j=1}^{N} r_j s_{ij}$$

The term $\boldsymbol{q} \cdot \boldsymbol{s}_i$ is computed according to this equation by first generating r_j's and then computing the sum of $r_j s_{ij}$ (remember that the s_{ij}'s are known), as shown in Fig. 7.9*a*. The M correlator detectors in Fig. 7.8*b* can be replaced by N filters matched to $\varphi_1(t)$, $\varphi_2(t), \ldots, \varphi_N(t)$, as shown in Fig. 7.9*b*. Both types of optimum receivers (Figs. 7.8 and 7.9) perform identically. The choice will depend on the circumstances. For example, if $N < M$ and signals $\{\varphi_j(t)\}$ are easier to generate than $\{s_j(t)\}$, then the choice of Fig. 7.9 is obvious.

Decision Regions and Error Probability

In order to compute the error probability of the optimum receiver, we need to determine decision regions in the signal space first. As mentioned earlier, the signal space is divided into M nonoverlapping, or disjoint, decision regions $R_1, R_2, \ldots, R_M$, corresponding to M messages. If $\boldsymbol{q}$ falls in the region R_k, the decision is that m_k was transmitted. The decision regions are so chosen that the probability of error of the receiver is minimum. In the light of this geometrical representation, we shall now try to interpret how the optimum receiver sets these decision regions.

The decision function is given by Eq. (7.45). The optimum receiver sets $\hat{m} = m_k$ if the decision function

$$\mathcal{N} \ln P(m_i) - |\boldsymbol{q} - \boldsymbol{s}_i|^2$$

is maximum for $i = k$. This equation defines the decision regions.

For simplicity, let us first consider the case of equiprobable messages—that is, $P(m_i) = 1/M$ for all i. In this case, the first term in the decision function is the same for all i and, hence, can be dropped. Thus, the receiver decides that $\hat{m} = m_k$ if the term $-|\boldsymbol{q} - \boldsymbol{s}_i|^2$ has its largest (numerically the smallest) value for $i = k$. Alternatively,

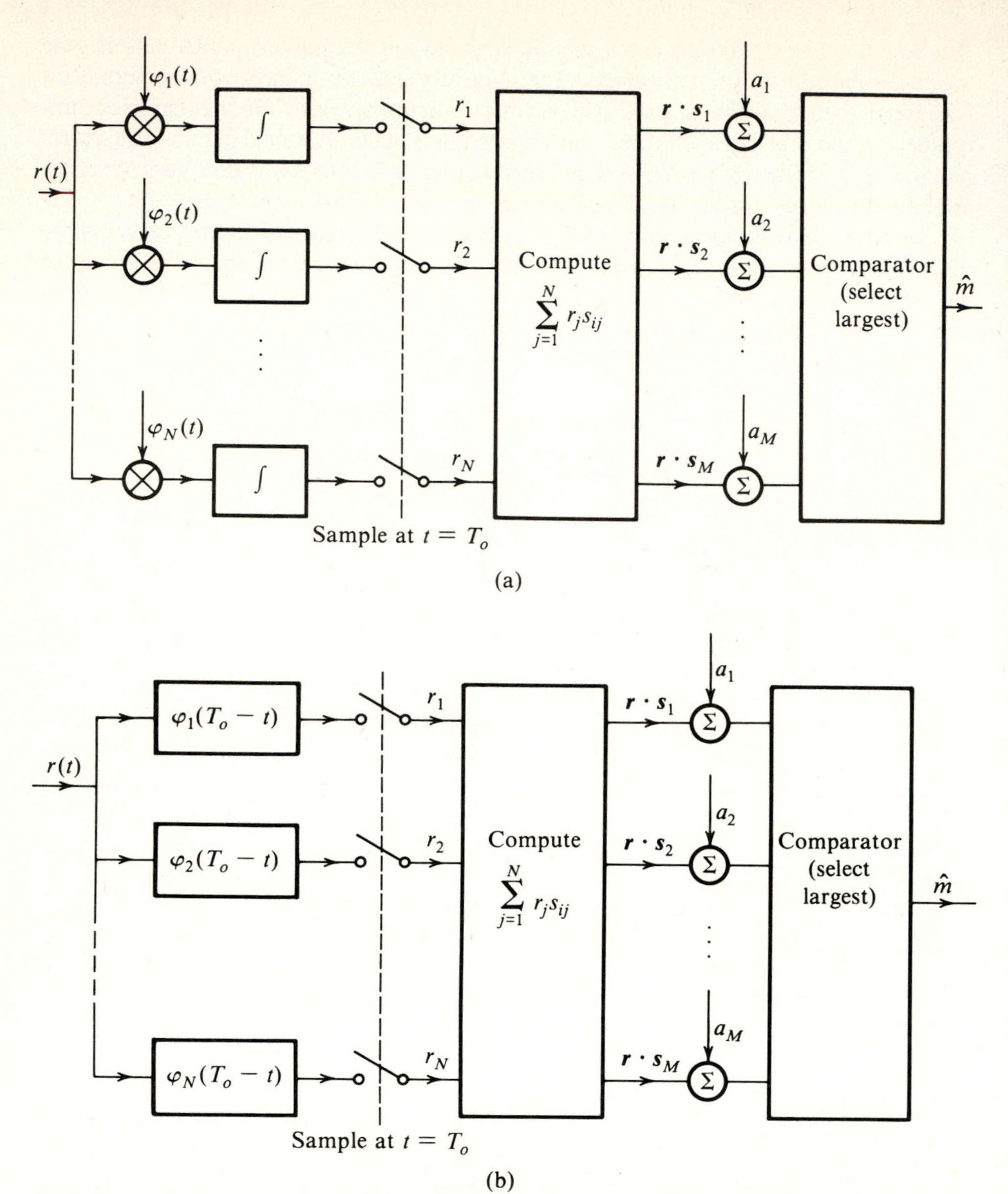

Figure 7.9 Another form of optimum M-ary receiver. (a) Correlator detector. (b) Matched-filter detector.

this may also be stated as follows: the receiver decides that $\hat{m} = m_k$ if the decision function $|\boldsymbol{q} - \boldsymbol{s}_i|^2$ is minimum for $i = k$. Note that $|\boldsymbol{q} - \boldsymbol{s}_i|$ is the distance of point $\boldsymbol{q}$ from point $\boldsymbol{s}_i$. Thus, the decision procedure in this case has a simple interpretation in geometrical space. We take the projection of the received signal $r(t)$ in the signal space. This is represented by point $\boldsymbol{q}$. Then the decision is made in favor of that signal which is closest to $\boldsymbol{q}$. This result is expected on qualitative grounds for gaussian noise, because the gaussian noise has a spherical symmetry. If, however, the messages are not equiprobable, we cannot go too far on purely qualitative grounds. Nevertheless, we can draw certain broad conclusions. If a particular message m_i is more likely than the others, one will be safer in deciding more often in favor of m_i than other messages. Hence, in such a case the decision regions will be biased, or weighted, in favor of m_i. This is shown by the appearance of the term $\ln P(m_i)$ in the decision function. To better understand this point, let us consider a 2-dimensional signal space and two signals $\boldsymbol{s}_1$ and $\boldsymbol{s}_2$, as shown in Fig. 7.10*a*. In this figure, the decision regions R_1 and R_2 are shown

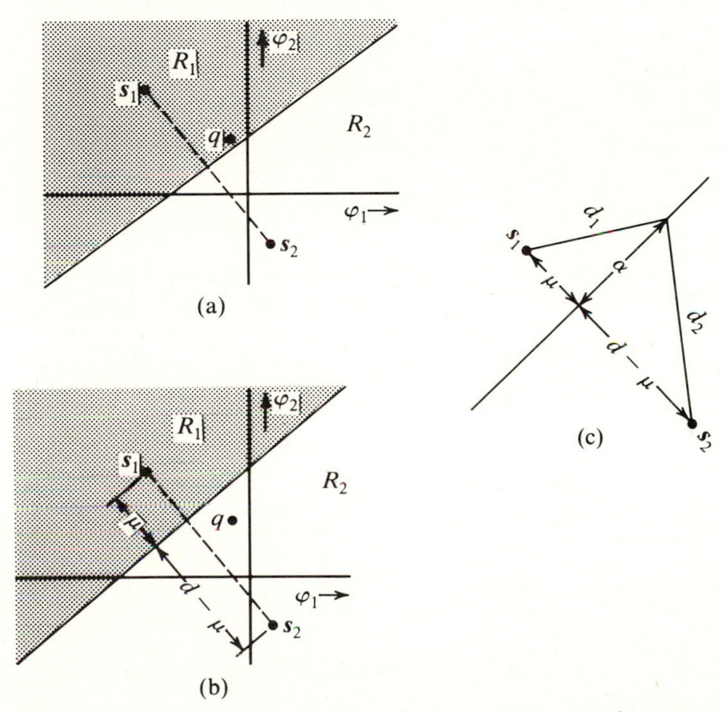

Figure 7.10 Determining optimum decision regions in a binary case.

for equiprobable messages; $P(m_1) = P(m_2) = 0.5$. It is obvious from Fig. 7.10*a* that the decision is made in favor of that message that is closest to $\boldsymbol{q}$. The boundary of the decision region is the perpendicular bisector of the line joining points $\boldsymbol{s}_1$ and $\boldsymbol{s}_2$. Note that any point on the boundary is equidistant from $\boldsymbol{s}_1$ and $\boldsymbol{s}_2$. If $\boldsymbol{q}$ happens to fall on the boundary, we just "flip a coin" and decide whether to select m_1 or m_2. Figure 7.10*b* shows the case when the two messages are not equiprobable. To delineate the boundary of the decision regions, we use Eq. (7.45). The decision is m_1 if

$$|\boldsymbol{q} - \boldsymbol{s}_1|^2 - \mathcal{N} \ln P(m_1) < |\boldsymbol{q} - \boldsymbol{s}_2|^2 - \mathcal{N} \ln P(m_2)$$

Otherwise, the decision is m_2.

Note that $|\boldsymbol{q} - \boldsymbol{s}_1|$ and $|\boldsymbol{q} - \boldsymbol{s}_2|$ represent d_1 and d_2, the distance of $\boldsymbol{q}$ from $\boldsymbol{s}_1$ and

s_2, respectively. Thus, the decision is m_1 if

$$d_1^2 - d_2^2 < \mathcal{N} \ln \frac{P(m_1)}{P(m_2)}$$

The right-hand side of the above inequality is a constant c:

$$c = \mathcal{N} \ln \frac{P(m_1)}{P(m_2)}$$

Thus, the decision is m_1 if

$$d_1^2 - d_2^2 < c$$

The decision is m_2 if

$$d_1^2 - d_2^2 > c$$

On the boundary of the decision regions,

$$d_1^2 - d_2^2 = c$$

We can easily show that such a boundary is given by a straight line, as shown in Fig. 7.10*b*. This line is perpendicular to line $\mathbf{s}_1\mathbf{s}_2$ and passes through $\mathbf{s}_1\mathbf{s}_2$ at a distance μ from $\mathbf{s}_1$, where

$$\mu = \frac{c + d^2}{2d} = \frac{\mathcal{N}}{2d} \ln \left[\frac{P(m_1)}{p(m_2)}\right] + \frac{d}{2} \tag{7.51}$$

where d is the distance between $\mathbf{s}_1$ and $\mathbf{s}_2$. To prove this, we redraw the pertinent part of Fig. 7.10*b* as Fig. 7.10*c*. It is evident from this figure that

$$d_1^2 = \alpha^2 + \mu^2$$

$$d_2^2 = \alpha^2 + (d - \mu)^2$$

Hence,

$$d_1^2 - d_2^2 = 2d\mu - d^2 = c$$

Therefore,

$$\mu = \frac{c + d^2}{2d}$$

This is the desired result. Thus, along the decision boundary $d_1^2 - d_2^2$ is constant and equal to c. The boundaries of the decision regions for $M > 2$ may be determined along similar lines. The decision regions for the case of three equiprobable 2-dimensional signals are shown in Fig. 7.11. The boundaries of the decision regions are perpendicular bisectors of the lines joining the original transmitted signals. If the signals are not equiprobable, then the boundaries will be shifted away from the signals with larger probabilities of occurrence.

For signals in N-dimensional space, the decision regions will be N-dimensional hypercones.

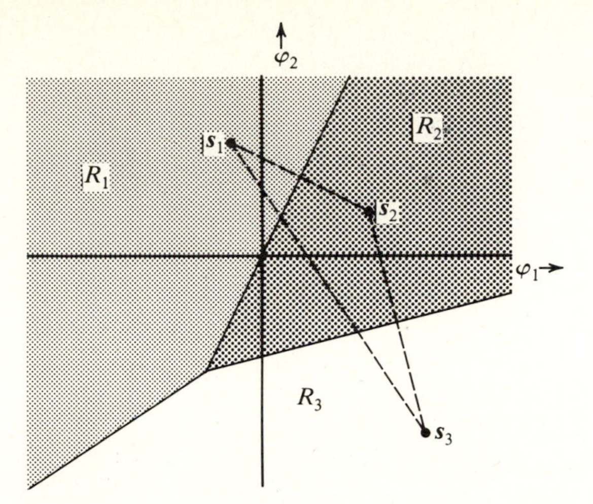

Figure 7.11 Determining optimum decision regions.

If there are M messages $m_1, m_2, \ldots, m_M$ with decision regions $R_1, R_2, \ldots, R_M$, respectively, then $P(C \mid m_i)$, the probability of correct decision when m_i is transmitted, is given by

$$P(C \mid m_i) = \text{probability that } \boldsymbol{q} \text{ lies in } R_1 \tag{7.52}$$

and $P(C)$, the probability of a correct decision, is given by

$$P(C) = \sum_{i=1}^{M} P(m_i)P(C \mid m_i) \tag{7.53a}$$

and P_{eM}, the probability of error, is given by

$$P_{eM} = 1 - P(C) \tag{7.53b}$$

■ EXAMPLE 7.2

Binary data is transmitted over an AWGN channel with noise PSD $\mathcal{N}/2$. The two signals used are

$$s_1(t) = -s_2(t) = \sqrt{E}\,\varphi(t)$$

This is polar signaling. The symbol probabilities $p(m_1)$ and $p(m_2)$ are unequal.

Design the optimum receiver and determine the corresponding error probability.

□ **Solution:** The two signals are represented graphically in Fig. 7.12a. The distance of each signal from the origin is $\sqrt{E}$. Hence, the energy of each signal is E. The distance d between the two signals is

$$d = 2\sqrt{E}$$

The decision regions R_1 and R_2 are shown in Fig. 7.12. The distance μ is given by Eq.

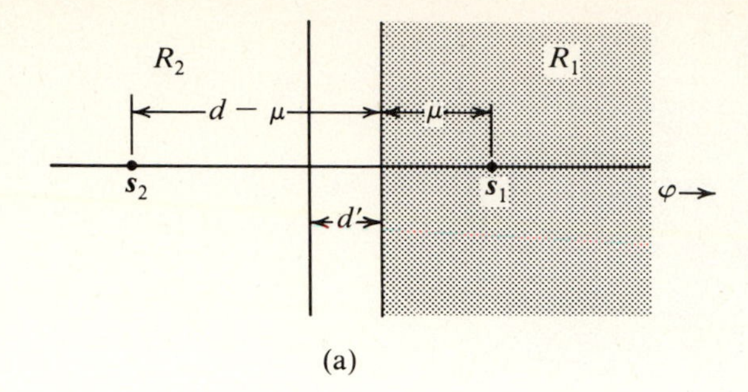

(a)

$r(t)$ → $\varphi(T_o - t)$ → switch → r → Threshold device → Decision:

m_1 if $r > \dfrac{\mathcal{N}}{4\sqrt{E}} \ln \dfrac{P(m_2)}{P(m_1)}$

m_2 if $r < \dfrac{\mathcal{N}}{4\sqrt{E}} \ln \dfrac{P(m_2)}{P(m_1)}$

(b)

Figure 7.12

(7.51). Also,

$$
\begin{aligned}
P(C|\mathrm{m} = m_1) &= P\text{ (noise vector originating at } s_1 \text{ remains in } R_1) \\
&= P(n > -\mu) \\
&= 1 - Q\left(\frac{\mu}{\sigma_{\mathrm{n}}}\right) \\
&= 1 - Q\left(\frac{\mu}{\sqrt{\mathcal{N}/2}}\right)
\end{aligned}
$$

Similarly,

$$
P(C|\mathrm{m} = m_2) = 1 - Q\left(\frac{(d - \mu)}{\sqrt{\mathcal{N}/2}}\right)
$$

and

$$
\begin{aligned}
P(C) &= P(m_1)\left[1 - Q\left(\frac{\mu}{\sqrt{\mathcal{N}/2}}\right)\right] + P(m_2)\left[1 - Q\left(\frac{d - \mu}{\sqrt{\mathcal{N}/2}}\right)\right] \\
&= 1 - P(m_1)Q\left(\frac{\mu}{\sqrt{\mathcal{N}/2}}\right) - P(m_2)Q\left(\frac{d - \mu}{\sqrt{\mathcal{N}/2}}\right)
\end{aligned}
$$

and

$$
P_e = 1 - P(C) = P(m_1)Q\left(\frac{\mu}{\sqrt{\mathcal{N}/2}}\right) + P(m_2)Q\left(\frac{d - \mu}{\sqrt{\mathcal{N}/2}}\right) \tag{7.54a}
$$

where

$$d = 2\sqrt{E} \tag{7.54b}$$

and

$$\mu = \frac{\mathcal{N}}{4\sqrt{E}} \ln \frac{P(m_1)}{P(m_2)} + \sqrt{E} \tag{7.54c}$$

When $P(m_1) = P(m_2) = 0.5$, $\mu = \sqrt{E} = d/2$, and Eq. (7.54a) reduces to

$$P_e = Q\left(\sqrt{\frac{2E}{\mathcal{N}}}\right) \tag{7.54d}$$

In this problem, $N = 1$ and $M = 2$. We shall consider the optimum receiver in Fig. 7.9*b*. We have

$$s_{11} = \sqrt{E}, \; s_{21} = -\sqrt{E}$$

and

$$\mathbf{r} \cdot \mathbf{s}_1 = r_1 s_{11} = \sqrt{E}\, r$$

$$\mathbf{r} \cdot \mathbf{s}_2 = r_1 s_{21} = -\sqrt{E}\, r$$

where r is the output of the filter $\varphi(T_o - t)$ at T_o. The decision criterion is [Eq. (7.49) or Fig. 7.9]

$$\sqrt{E}\, r + a_1 > -\sqrt{E}\, r + a_2 \qquad \text{select } m_1$$

or

$$r > \frac{a_2 - a_1}{2\sqrt{E}} \qquad \text{select } m_1$$

that is,

$$r > \frac{\mathcal{N}}{4\sqrt{E}} \ln \frac{P(m_2)}{P(m_1)} \qquad \text{select } m_1$$

This same result could have been obtained in a simpler way from Fig. 7.12*a*. The receiver output $r = q$. Hence if $r > d'$, q is in R_1 (Fig. 7.12*a*), and the decision is m_1. But

$$d' = \sqrt{E} - \mu = \frac{\mathcal{N}}{4\sqrt{E}} \ln \frac{P(m_2)}{P(m_1)}$$

Thus follows the result.

The final form of the receiver is shown in Fig. 7.12*b*. When $P(m_1) = P(m_2) = 0.5$, the decision threshold is zero. This is precisely the result derived in Chapter 6 for polar signaling. ■

EXAMPLE 7.3 QAM

Design the optimum receiver and compute the corresponding error probability for the 16-point QAM configuration shown in Fig. 7.13*a*, assuming all signals to be equiprobable and assuming an AWGN channel.

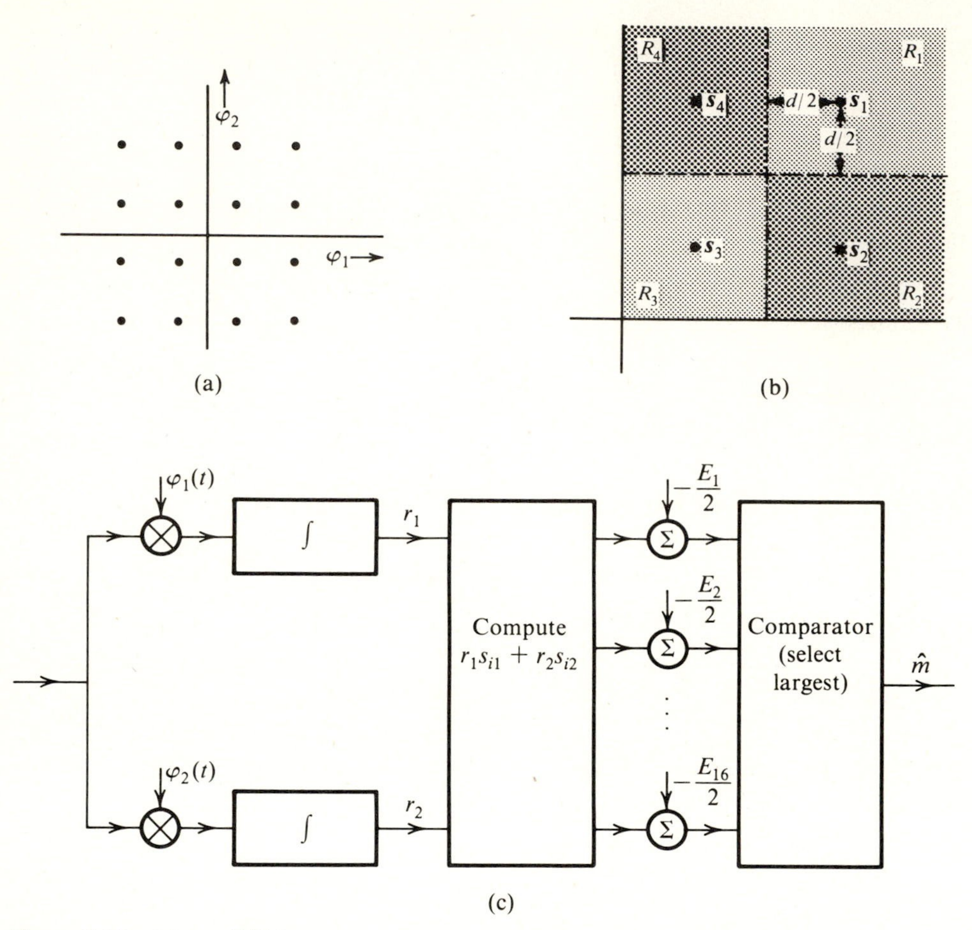

Figure 7.13 16-ary QAM.

☐ **Solution:** Let us first calculate the error probability. The first quadrant of the signal space is reproduced in Fig. 7.13*b*. Because all the signals are equiprobable, the decision region boundaries will be perpendicular bisectors joining various signals, as shown in Fig. 7.13*b*.

From Fig. 7.13*b* it follows that

$$\begin{aligned} P(C|m_1) &= P \text{ (noise vector originating at } \mathbf{s}_1 \text{ lies within } R_1) \\ &= P\left(\mathrm{n}_1 > -\frac{d}{2}, \mathrm{n}_2 > -\frac{d}{2}\right) \\ &= P\left(\mathrm{n}_1 > -\frac{d}{2}\right) P\left(\mathrm{n}_2 > -\frac{d}{2}\right) \\ &= \left[1 - Q\left(\frac{d/2}{\sigma_\mathrm{n}}\right)\right]^2 \\ &= \left[1 - Q\left(\frac{d}{\sqrt{2\mathcal{N}}}\right)\right]^2 \end{aligned}$$

For convenience, let us define

$$p = 1 - Q\left(\frac{d}{\sqrt{2\mathcal{N}}}\right) \tag{7.55}$$

Hence,

$$P(C|m_1) = p^2$$

Using similar arguments, we have

$$\begin{aligned} P(C|m_2) = P(C|m_4) &= \left[1 - Q\left(\frac{d}{\sqrt{2\mathcal{N}}}\right)\right]\left[1 - 2Q\left(\frac{d}{\sqrt{2\mathcal{N}}}\right)\right] \\ &= p(2p - 1) \end{aligned}$$

and

$$P(C|m_3) = (2p - 1)^2$$

Because of the symmetry of the signals in all four quadrants, we get similar probabilities for the four signals in each quadrant. Hence,

$$\begin{aligned} P(C) &= \sum_{i=1}^{16} P(C|m_i)P(m_i) \\ &= \frac{1}{16}\sum_{i=1}^{16} P(C|m_i) \\ &= \frac{1}{16}[4p^2 + 4p(2p - 1) + 4p(2p - 1) + 4(2p - 1)^2] \\ &= \frac{1}{4}[9p^2 - 6p + 1] \\ &= \left(\frac{3p - 1}{2}\right)^2 \end{aligned} \tag{7.56a}$$

and

$$P_{eM} = 1 - P(C) = \frac{9}{4}\left(p + \frac{1}{3}\right)(1 - p)$$

In practice $P_{eM} \to 0$, and, hence, $P(C) \to 1$. This means $p \simeq 1$ [see Eq. (7.56a)], and

$$P_{eM} \simeq 3(1 - p) = 3Q\left(\frac{d}{\sqrt{2\mathcal{N}}}\right) \tag{7.56b}$$

To express this in terms of the received power S_i, we determine $\bar{E}$, the average energy of the signal set in Fig. 7.13. Because E_i, the energy of S_i, is the square of the distance of S_i from the origin,

$$E_1 = \left(\frac{3d}{2}\right)^2 + \left(\frac{3d}{2}\right)^2 = \frac{9}{2}d^2$$

$$E_2 = \left(\frac{3}{2}d\right)^2 + \left(\frac{d}{2}\right)^2 = \frac{5}{2}d^2$$

Similarly,

$$E_3 = \frac{d^2}{2} \qquad \text{and} \qquad E_4 = \frac{5}{2}d^2$$

Hence,

$$\bar{E} = \frac{1}{4}\left[\frac{9}{2}d^2 + \frac{5}{2}d^2 + \frac{d^2}{2} + \frac{5}{2}d^2\right] = \frac{5}{2}d^2$$

and $d^2 = 0.4\bar{E}$. Also,

$$S_i = \bar{E}f_o$$

and

$$\lambda = \frac{S_i}{\mathcal{N}f_o} = \frac{\bar{E}}{\mathcal{N}} = \frac{5d^2}{2\mathcal{N}}$$

Hence,

$$P_{eM} = 3Q\left(\frac{d}{\sqrt{2\mathcal{N}}}\right)$$

$$= 3Q\left(\sqrt{\frac{\lambda}{5}}\right) \tag{7.57}$$

A comparison of this with binary PSK [Eq. (6.100a)] shows that 16-point QAM requires almost 10 times as much power as does binary PSK; but the rate of transmission is increased by a factor of $4 (M = 16)$.

In this case, $N = 2$ and $M = 16$. Hence, the receiver in Fig. 7.9 will be preferable. Such a receiver is shown in Fig. 7.13*c*. Note that because all signals are equiprobable,

$$a_i = -E_i/2$$

For QAM, $\varphi_1(t) = \sqrt{2/T_o} \cos \omega_o t$, and $\varphi_2(t) = \sqrt{2/T_o} \sin \omega_o t$. ■

■ EXAMPLE 7.4 MPSK

Determine the error probability of the optimum receiver for equiprobable MPSK signals each with energy E.

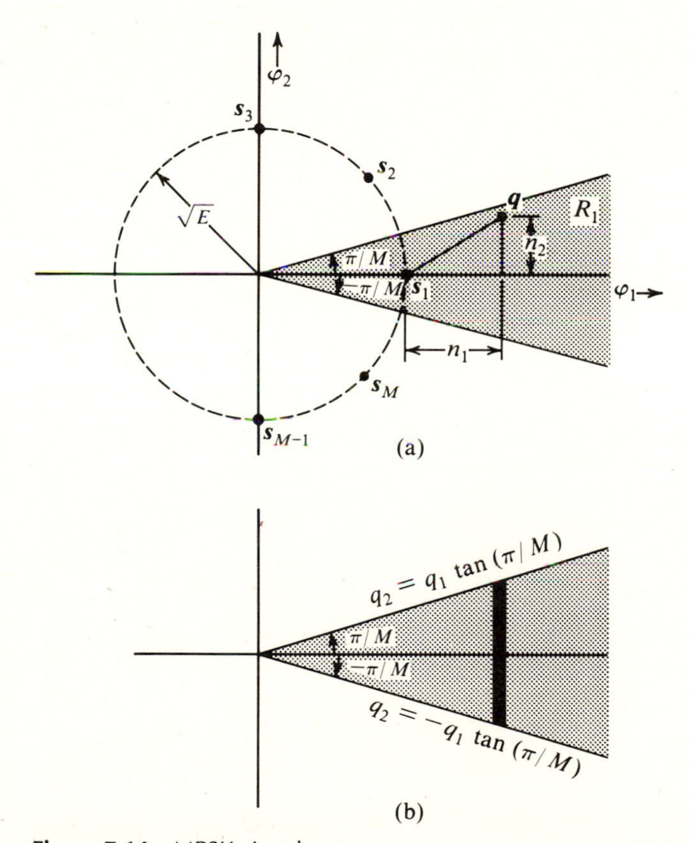

Figure 7.14 MPSK signals.

□ **Solution:** Figure 7.14*a* shows the MPSK signal configuration for $M = 8$. Because all the signals are equiprobable, the decision regions are conical, as shown in Fig. 7.14*a*. If m_1 is transmitted and the received signal is $\mathbf{q}$ (the projection of $\mathbf{r}$ in the $\varphi_1 - \varphi_2$ plane), then

$$\mathbf{q} = (s_1 + \mathrm{n}_1, \mathrm{n}_2) = (\sqrt{E} + \mathrm{n}_1, \mathrm{n}_2)$$

where n_1 and n_2 are noise components in the φ_1 and φ_2 dimensions. Also,

$$P(C|m_1) = P(\mathbf{q} \text{ lies in } R_1)$$

This is simply the volume under the conical region of the joint PDF of q_1 and q_2 (components of $\mathbf{q}$ in φ_1 and φ_2) where

$$q_1 = \sqrt{E} + n_1 \qquad q_2 = n_2$$

Because n_1 and n_2 are independent gaussian r.v.'s with variance $\mathcal{N}/2$, q_1 and q_2 are independent gaussian variables with means $\sqrt{E}$ and 0, respectively, and each with variance $\mathcal{N}/2$. Hence,

$$p_{q_1q_2}(q_1,q_2) = \left[\frac{1}{\sqrt{\pi\mathcal{N}}} e^{-(q_1-\sqrt{E})^2/\mathcal{N}}\right]\left[\frac{1}{\sqrt{\pi\mathcal{N}}} e^{-q_2^2/\mathcal{N}}\right]$$

and

$$P(C|m_1) = \frac{1}{\pi\mathcal{N}} \iint_{R_1} e^{-[(q_1-\sqrt{E})^2+q_2^2]/\mathcal{N}}\, dq_1\, dq_2 \tag{7.58}$$

Transferring to polar coordinates with $q_1 = \rho\sqrt{\mathcal{N}/2}\cos\theta$ and $q_2 = \rho\sqrt{\mathcal{N}/2}\sin\theta$, the limits on ρ are $(0, \infty)$ and those on θ are $-\pi/M$ to π/M. Hence,

$$P(C|m_1) = \frac{1}{2\pi}\int_{-\pi/M}^{\pi/M} d\theta \int_0^\infty \rho e^{-[\rho^2-2\rho\sqrt{2E/\mathcal{N}}\cos\theta+2E/\mathcal{N}]/2}\, d\rho \tag{7.59a}$$

Because of the symmetry of signal configuration, $P(C|m_i)$ is the same for all i. Hence,

$$P(C) = P(C|m_1)$$

and

$$P_{eM} = 1 - P(C|m_1) \tag{7.59b}$$

P_{eM} is numerically computed. The plot is shown in Fig. 6.40 (with* $\lambda = E/\mathcal{N}$).

The error probability can also be computed by using rectangular coordinates. The pertinent part of Fig. 7.14a is reproduced in Fig. 7.14b.

$$P(C|m_1) = \frac{1}{\pi\mathcal{N}}\int_{q_1}\left(\int_{q_2} e^{-q_2^2/\mathcal{N}}\, dq_2\right) e^{-(q_1-\sqrt{E})^2/\mathcal{N}}\, dq_1$$

To integrate over R_1, we first integrate over the shaded strip in Fig. 7.14b. Along the border of R_1

$$q_2 = \pm\left(\tan\frac{\pi}{M}\right)q_1$$

---

* Since the received power $S_i = Ef_o$,

$$\lambda = \frac{S_i}{\mathcal{N}f_o} = \frac{E}{\mathcal{N}}$$

Hence,

$$P(C|m_1) = \frac{1}{\pi\mathcal{N}} \int_0^\infty \left(\int_{-q_1 \tan \pi/M}^{q_1 \tan \pi/M} e^{-q_2^2/\mathcal{N}} \, dq_2 \right) e^{-(q_1 - \sqrt{E})^2/\mathcal{N}} \, dq_1$$

$$= \frac{1}{\sqrt{\pi\mathcal{N}}} \int_0^\infty \left[1 - 2Q\left(\frac{q_1 \tan \pi/M}{\sqrt{\mathcal{N}/2}} \right) \right] e^{-(q_1 - \sqrt{E})^2/\mathcal{N}} \, dq_1$$

Changing the variable to $x = \sqrt{2/\mathcal{N}}\, q_1$, we get

$$P(C|m_1) = \frac{1}{\sqrt{2\pi}} \int_0^\infty \left[1 - 2Q\left(x \tan \frac{\pi}{M} \right) \right] e^{-(x - \sqrt{2E/\mathcal{N}})^2/2} \, dx \tag{7.60}$$

and

$$P_{eM} = 1 - P(C|m_1)$$

Still another form of the integral for P_{eM} was found in Chapter 6 [Eq. (6.128)]. For MPSK, $\varphi_i(t) = \sqrt{2/T_o} \cos(\omega_o t + \theta_i)$, where $\theta_i = 2\pi i/M$, and the optimum receiver turns out to be just a phase detector similar to that shown in Fig. 7.13 (see Prob. 7.8). ■

The Significance of the Geometrical Configuration of Signals. From this discussion, one very interesting fact emerges. Whenever the optimum receiver is used, the error probability does not depend upon specific signal waveforms but depends only on their geometrical configuration in the signal space. A given configuration in a signal space can represent infinite possible signal sets, depending on the choice of the basis signals used. A given configuration *does* specify the average energy of the set. This means the error probability depends on signal waveforms only through the average energy of the set. Thus, the average signal energy (or the power) emerges as a fundamental parameter that determines the error probability.

This discussion also vividly demonstrates the insight provided by the use of signal space in the study of optimum receivers.

General Expression for Error Probability

Thus far we have considered rather simple schemes where the decision regions can be easily found. The method of computing error probabilities from the knowledge of decision regions has also been discussed. When the dimensions of the signal space increase, it becomes difficult to visualize the decision regions graphically, and as a result the method loses its power. We shall now develop the analytical expression for computing error probability for a general M-ary scheme.

From the structure of the optimum receiver in Fig. 7.8, we observe that if m_1 is transmitted, then the correct decision will be made only if

$$\mathrm{b}_1 > \mathrm{b}_2, \mathrm{b}_3, \ldots, \mathrm{b}_M$$

In other words

$$P(C|m_1) = \text{probability } (\mathrm{b}_1 > \mathrm{b}_2, \mathrm{b}_3, \ldots, \mathrm{b}_M | m_1) \tag{7.61}$$

If m_1 is transmitted, then (Fig. 7.8)

$$\mathrm{b}_k = \int_0^{T_o} [s_1(t) + \mathrm{n}(t)]s_k(t)\, dt + a_k \tag{7.62}$$

Let

$$\rho_{ij} = \int_0^{T_o} s_i(t)s_j(t)\, dt \qquad i, j = 1, 2, \ldots, M \tag{7.63}$$

ρ_{ij} are known as *cross-correlation coefficients*. Thus (if m_1 is transmitted),

$$\mathrm{b}_k = \rho_{1k} + \int_0^{T_o} \mathrm{n}(t)s_k(t)\, dt + a_k \tag{7.64a}$$

$$= \rho_{1k} + a_k + \sum_{j=1}^{N} s_{kj}\mathrm{n}_j \tag{7.64b}$$

where n_j is the component of $\mathrm{n}(t)$ along $\varphi_j(t)$. Note that $\rho_{1k} + a_k$ is a constant, and variables n_j $(j = 1, 2, \ldots, N)$ are independent jointly gaussian variables, each with zero mean and a variance of $\mathcal{N}/2$. Thus, variables b_k are a linear combination of jointly gaussian variables. It follows that the variables $\mathrm{b}_1, \mathrm{b}_2, \ldots, \mathrm{b}_M$ are also jointly gaussian. The probability of making a correct decision when m_1 is transmitted can be computed from Eq. (7.61). Note that b_1 can lie anywhere in the range $(-\infty, \infty)$. More precisely, if $p(b_1, b_2, \ldots, b_M | m_1)$ is the joint PDF of $\mathrm{b}_1, \mathrm{b}_2, \ldots, \mathrm{b}_M$, then Eq. (7.61) can be expressed as

$$P(C|m_1) = \int_{-\infty}^{\infty} \int_{-\infty}^{b_1} \cdots \int_{-\infty}^{b_1} p(b_1, b_2, \ldots, b_m | m_1)\, db_1, db_2, \ldots, db_M \tag{7.65a}$$

where the limits of integration of b_1 are $(-\infty, \infty)$, and for the remaining variables the limits are $(-\infty, b_1)$. Thus,

$$P(C|m_1) = \int_{-\infty}^{\infty} db_1 \int_{-\infty}^{b_1} db_2 \cdots \int_{-\infty}^{b_1} p(b_1, b_2, \ldots, b_M | m_1)\, db_M \tag{7.65b}$$

Similarly, $P(C|m_2), \ldots, P(C|m_M)$ can be computed, and

$$P(C) = \sum_{j=1}^{M} P(C|m_j)P(m_j)$$

and

$$P_{eM} = 1 - P(C)$$

■ EXAMPLE 7.5 Orthogonal Signal Set

In this set all M signals $s_1(t), s_2(t), \ldots, s_M(t)$ are mutually orthogonal. As an example, a signal set for $M = 3$ is shown in Fig. 7.15.

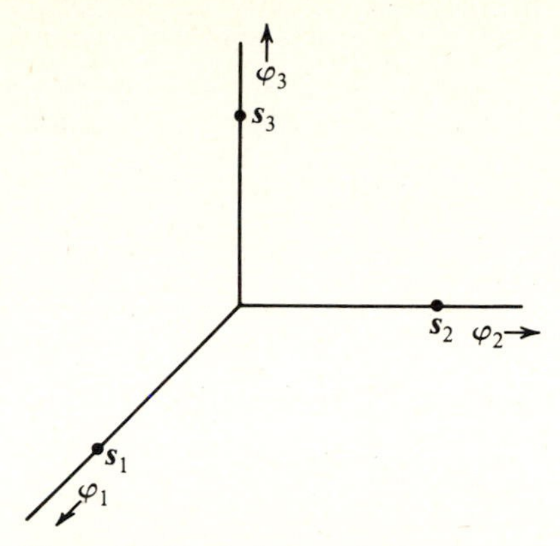

Figure 7.15 Orthogonal signals.

The orthogonal set $\{s_k(t)\}$ is characterized by the fact that

$$\mathbf{s}_j \cdot \mathbf{s}_k = \begin{cases} 0 & j \neq k \\ E & j = k \end{cases} \tag{7.66}$$

Hence

$$\rho_{ij} = \mathbf{s}_i \cdot \mathbf{s}_j = \begin{cases} 0 & i \neq j \\ E & i = j \end{cases} \tag{7.67}$$

Further, we shall assume all signals to be equiprobable. This yields

$$\begin{aligned} a_k &= \frac{1}{2}\left[\mathcal{N} \ln\left(\frac{1}{M}\right) - E_k\right] \\ &= -\frac{1}{2}[\mathcal{N} \ln M + E] \end{aligned}$$

where E is the energy of each signal. Note that a_k is the same for all values of k. Let this constant be a.

$$a_k = a \qquad k = 1, 2, \ldots, M \tag{7.68}$$

For an orthogonal set

$$s_k(t) = \sqrt{E}\, \varphi_k(t)$$

Therefore,

$$s_{kj} = \begin{cases} \sqrt{E} & k = j \\ 0 & k \neq j \end{cases} \tag{7.69}$$

Hence, from Eqs. (7.64b), (7.67), (7.68), and (7.69), we have (when m_1 is transmitted)

$$\mathrm{b}_k = \begin{cases} E + a + \sqrt{E}\,\mathrm{n}_1 & k = 1 \\ a + \sqrt{E}\,\mathrm{n}_k & k = 2, 3, \ldots, M \end{cases} \tag{7.70}$$

Note that $n_1, n_2, \ldots, n_M$ are independent gaussian variables, each with zero mean and variance $\mathcal{N}/2$. Obviously, b_k's that are of the form $(\alpha n_k + \beta)$ are also independent gaussian variables. From Eq. (7.70) it is evident that the variable b_1 has a mean $E + a$ and variance $(\sqrt{E})^2(\mathcal{N}/2) = \mathcal{N}E/2$. Hence,

$$p_{b_1}(b_1) = \frac{1}{\sqrt{\pi \mathcal{N} E}} e^{-(b_1 - E - a)^2/\mathcal{N}E}$$

$$p_{b_k}(b_k) = \frac{1}{\sqrt{\pi \mathcal{N} E}} e^{-(b_k - a)^2/\mathcal{N}E} \qquad k = 2, 3, \ldots, M$$

Because $b_1, b_2, \ldots, b_M$ are independent, the joint probability density is the product of the individual densities:

$$p(b_1, b_2, \ldots, b_M | m_1) = \frac{1}{\sqrt{\pi \mathcal{N} E}} e^{-(b_1 - E - a)^2/\mathcal{N}E} \prod_{k=2}^{M} \left[\frac{1}{\sqrt{\pi \mathcal{N} E}} e^{-(b_k - a)^2/\mathcal{N}E} \right]$$

and

$$P(C|m_1) = \frac{1}{\sqrt{\pi \mathcal{N} E}} \int_{-\infty}^{\infty} db_1 [e^{-(b_1 - E - a)^2/\mathcal{N}E}] \prod_{k=2}^{M} \left[\int_{-\infty}^{b_1} \frac{1}{\sqrt{\pi \mathcal{N} E}} e^{-(b_k - a)^2/\mathcal{N}E} \, db_k \right]$$

$$= \frac{1}{\sqrt{\pi \mathcal{N} E}} \int_{-\infty}^{\infty} db_1 [e^{-(b_1 - E - a)^2/\mathcal{N}E}] \left[\int_{-\infty}^{b_1} \frac{1}{\sqrt{\pi \mathcal{N} E}} e^{-(x - a)^2/\mathcal{N}E} \, dx \right]^{M-1}$$

$$= \frac{1}{\sqrt{\pi \mathcal{N} E}} \int_{-\infty}^{\infty} \left[1 - Q\left(\frac{b_1 - a}{\sqrt{\mathcal{N}E/2}} \right) \right]^{M-1} e^{-(b_1 - E - a)^2/\mathcal{N}E} \, db_1 \qquad \textbf{(7.71a)}$$

Changing the variable so that

$$\frac{b_1 - E - a}{\sqrt{\mathcal{N}E/2}} = y$$

the integral becomes

$$P(C|m_1) = \frac{1}{\sqrt{2\pi}} \int_{-\infty}^{\infty} \left[1 - Q\left(y + \sqrt{\frac{2E}{\mathcal{N}}} \right) \right]^{M-1} e^{-y^2/2} \, dy \qquad \textbf{(7.71b)}$$

Note that because this signal set is geometrically symmetrical,

$$P(C|m_1) = P(C|m_2) = \cdots = P(C|m_M)$$

Hence,

$$P(C) = P(C|m_1)$$

and

$$P_{eM} = 1 - P(C)$$

If we change the variable to $x = y + \sqrt{2E/\mathcal{N}}$ and recognize that for the orthogonal case

$$\lambda = \frac{S_i}{\mathcal{N} f_o} = \frac{E}{\mathcal{N}}$$

we get P_{eM} as in Eq. (6.132). P_{eM} as computed from Eq. (7.71) is plotted in Fig. 6.30. ■

Bandwidth of M-ary Signals

As discussed in Sec. 7.1, the dimensionality of a signal is $2BT_o + 1$, where T_o is the signal duration and B is its essential bandwidth. It follows that for an N-dimensional signal space ($N \leq M$), the bandwidth is $B = (N - 1)/2T_o$. Thus, reducing N reduces the bandwidth.

We can verify that N-dimensional signals can be transmitted over $(N - 1)/2T_o$ Hz by constructing a specific signal set. Let us choose signals

$$\begin{aligned}
\varphi_0(t) &= \frac{1}{\sqrt{T_o}} \\
\varphi_1(t) &= \sqrt{\frac{2}{T_o}} \sin \omega_o t \\
\varphi_2(t) &= \sqrt{\frac{2}{T_o}} \cos \omega_o t \qquad \omega_o = \frac{2\pi}{T_o} \\
\varphi_3(t) &= \sqrt{\frac{2}{T_o}} \sin 2\omega_o t \qquad 0 \leq t \leq T_o \\
\varphi_4(t) &= \sqrt{\frac{2}{T_o}} \cos 2\omega_o t \\
&\cdots\cdots\cdots \\
\varphi_{k-1}(t) &= \sqrt{\frac{2}{T_o}} \sin \left(\frac{k}{2}\, \omega_o t\right) \\
\varphi_k(t) &= \sqrt{\frac{2}{T_o}} \cos \left(\frac{k}{2}\, \omega_o t\right)
\end{aligned} \tag{7.72}$$

These are $k + 1$ orthogonal pulses with a total bandwidth of $(k/2)f_o = k/2T_o$ Hz. Hence, when $k + 1 = N$, the bandwidth* is $(N - 1)/2T_o$. Thus, $N = 2T_oB + 1$.

To attain a given error probability, there is a trade-off between the average energy of the signal set and its bandwidth. If we reduce the signal space dimensionality, the transmission bandwidth is reduced. But the signals are now closer together, because of the reduced dimensionality. This will increase P_{eM}. Hence, to maintain a given P_{eM}, we must move the signals farther apart—that is, increase the energy. Thus, the cost of reduced bandwidth is paid in terms of increased energy. In Chapter 8 we shall study

* Here we are ignoring the bandspreading at the edge. This spread is about $1/T_o$ Hz. The actual bandwidth is larger than $(N - 1)/2T_o$ by this amount.

the ideal trade-off between the bandwidth and signal energy and compare the performance of various signal schemes in light of this relationship.

7.4 EQUIVALENT SIGNAL SETS

The computation of error probabilities is greatly facilitated by translation and rotation of coordinate axes. We shall now show that these operations are permissible.

Consider the signal set with its corresponding decision regions shown in Fig. 7.16. The conditional probability $P(C|m_1)$ is the probability that the noise vector drawn from s_1 lies within R_1. Note that this probability does not depend upon the origin

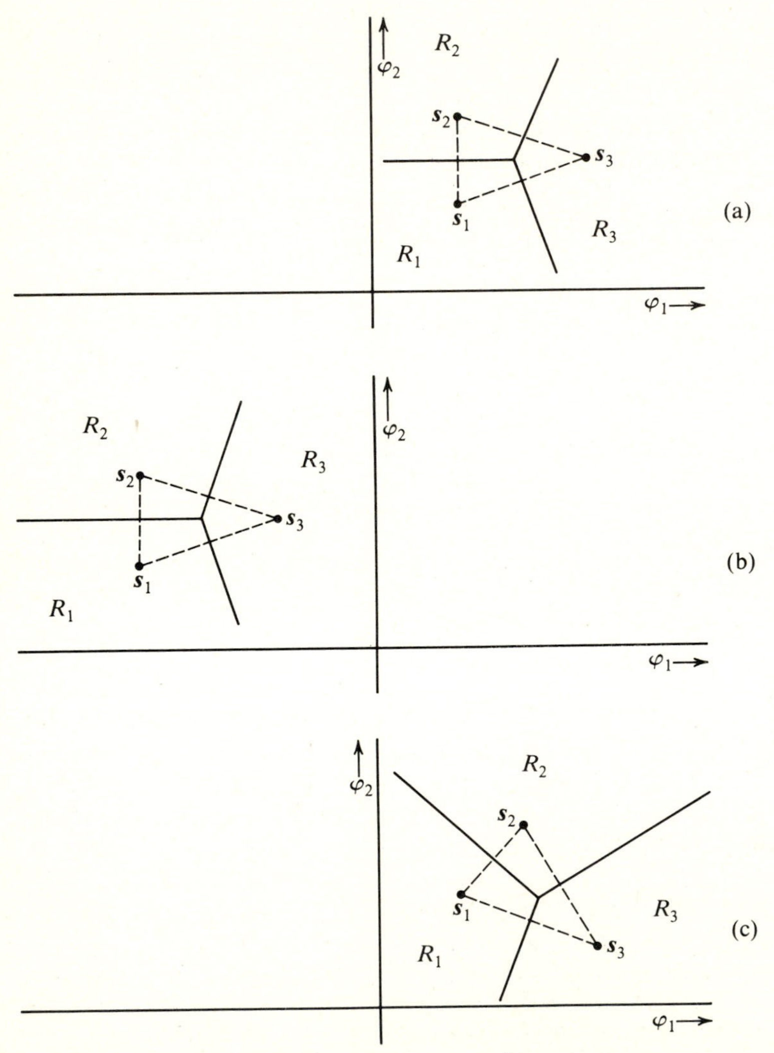

Figure 7.16 Translation and rotation of coordinate axes.

of the coordinate system. We may translate the coordinate system any way we wish. This is equivalent to translating the signal set and the corresponding decision regions. Thus, the $P(C|m_i)$ for the translated system shown in Fig. 7.16*b* is identical to that of the system in Fig. 7.16*a*.

In the case of gaussian noise, we make another important observation. The rotation of the coordinate system does not affect the error probability because the noise-vector probability density has spherical symmetry. To show this we shall consider Fig. 7.16*c*, where the signal set in Fig. 7.16*a* is shown translated and rotated. Note that a rotation of the coordinate system is equivalent to a rotation of the signal set in the opposite sense. Here for convenience we rotate the signal set instead of the coordinate system. It can be seen that the probability that the noise vector **n** drawn from s_1 lies in R_1 is the same in Fig. 7.16*a* and 7.16*c*, because this probability is given by the integral of the noise probability density $p_{\mathbf{n}}(\mathbf{n})$ over the region R_1. Because $p_{\mathbf{n}}(\mathbf{n})$ has a spherical symmetry for gaussian noise, the probability will remain unaffected by a rotation of the region R_1.

We therefore conclude that for additive gaussian channel noise, translation and rotation of the coordinate system (or translation and rotation of the signal set) does not affect the error probability. Note that when we rotate or translate a set of signals, the resulting set represents an entirely different set of signals. Yet the error probabilities of the two sets are identical. Such sets are called *equivalent sets*.

The following example demonstrates the utility of translation and rotation of a signal set in the computation of error probability.

EXAMPLE 7.6

A quaternary *PSK* (*QPSK*) signal set is shown in Fig. 7.17*a*.

$$s_1 = -s_2 = \sqrt{E}\,\varphi_1$$
$$s_3 = -s_4 = \sqrt{E}\,\varphi_2$$

Assuming all symbols to be equiprobable, determine P_{eM} for an AWGN channel with noise PSD $\mathcal{N}/2$.

□ **Solution:** This problem has already been solved in Example 7.4 for a general value of M. Here we shall solve it for $M = 4$ to demonstrate the power of rotation of axes.

Because all the symbols are equiprobable, the decision region boundaries will be perpendicular bisectors of lines joining various signal points (Fig. 7.17*a*). Now

$$P(C|m_1) = P \text{ (that noise vector originating at } s_1 \text{ remains in } R_1) \tag{7.73}$$

This can be found by integrating the joint PDF of components n_1 and n_2 (originating at s_1) over the region R_1. This double integral can be found by using suitable limits, as in Eq. (7.60). The problem is greatly simplified, however, if we rotate the signal set by 45°, as shown in Fig. 7.17*b*. The decision regions are rectangular, and if n_1 and

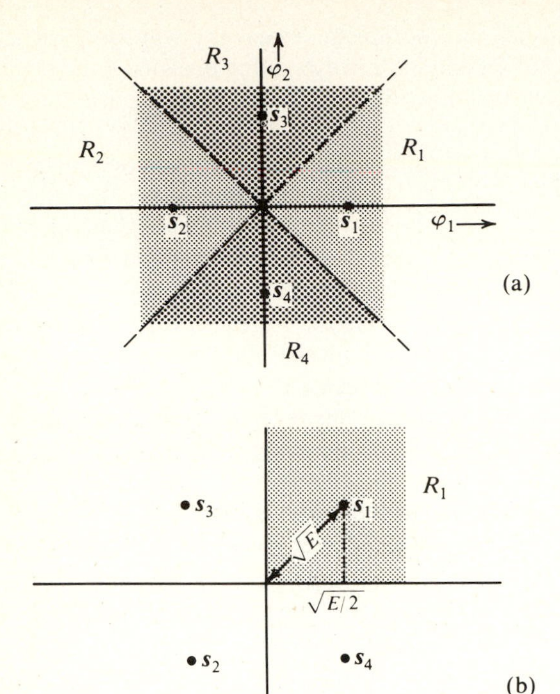

Figure 7.17 QPSK.

n_2 are noise components along φ_1 and φ_2, then Eq. (7.73) can be expressed as

$$
\begin{aligned}
P(C|m_1) &= P(n_1 > -\sqrt{E/2},\ n_2 > -\sqrt{E/2}) \\
&= P(n_1 > -\sqrt{E/2})\, P(n_2 > -\sqrt{E/2}) \\
&= \left[1 - Q\left(\sqrt{\frac{E}{2\sigma_n^2}}\right)\right]^2 \\
&= \left[1 - Q\left(\sqrt{\frac{E}{\mathcal{N}}}\right)\right]^2
\end{aligned} \tag{7.74}
$$

■

Minimum-Energy Signal Set

As noted earlier, an infinite number of possible equivalent signal sets exist. Because signal energy depends upon its distance from the origin, however, equivalent sets do not necessarily have the same average energy. Thus, among the infinite possible equivalent signal sets, the one in which the signals are closest to the origin has the minimum average signal energy (or transmitted power).

Let $m_1, m_2, \ldots, m_M$ be M messages with waveforms $s_1(t), s_2(t), \ldots, s_M(t)$ represented by points $\mathbf{s}_1, \mathbf{s}_2, \ldots, \mathbf{s}_M$ in the signal space. The mean energy of these signals is $\bar{E}$, given by

$$\bar{E} = \sum_{i=1}^{M} P(m_i)\,|\boldsymbol{s}_i|^2$$

Translation of this signal set is equivalent to subtracting some vector $\boldsymbol{a}$ from each signal. Let this operation yield the minimum-mean-energy set. We now wish to find the vector $\boldsymbol{a}$ such that the new mean energy

$$\bar{E} = \sum_{i=1}^{M} P(m_i)\,|\boldsymbol{s}_i - \boldsymbol{a}|^2 \tag{7.75}$$

is minimum. We can show that $\boldsymbol{a}$ must be the center of gravity of M points located at $\boldsymbol{s}_1, \boldsymbol{s}_2, \ldots, \boldsymbol{s}_M$ with masses $P(m_1), P(m_2), \ldots, P(m_M)$, respectively:

$$\boldsymbol{a} = \sum_{i=1}^{M} P(m_i)\,\boldsymbol{s}_i = \overline{\boldsymbol{s}_i} \tag{7.76}$$

To prove this, suppose the mean energy is minimum for some translation $\boldsymbol{b}$. Then

$$\begin{aligned}
\bar{E} &= \sum_{i=1}^{M} P(m_i)\,|\boldsymbol{s}_i - \boldsymbol{b}|^2 \\
&= \sum_{i=1}^{M} P(m_i)\,|(\boldsymbol{s}_i - \boldsymbol{a}) + (\boldsymbol{a} - \boldsymbol{b})|^2 \\
&= \sum_{i=1}^{M} P(m_i)\,|\boldsymbol{s}_i - \boldsymbol{a}|^2 + 2(\boldsymbol{a} - \boldsymbol{b}) \cdot \sum_{i=1}^{M} P(m_i)(\boldsymbol{s}_i - \boldsymbol{a}) + \sum_{i=1}^{M} P(m_i)\,|\boldsymbol{a} - \boldsymbol{b}|^2
\end{aligned}$$

Observe that the second term in the above expression vanishes because of the relationship in Eq. (7.76). Hence,

$$\bar{E} = \sum_{i=1}^{M} P(m_i)\,|\boldsymbol{s}_i - \boldsymbol{a}|^2 + \sum_{i=1}^{M} P(m_i)\,|\boldsymbol{a} - \boldsymbol{b}|^2$$

This is minimum when $\boldsymbol{b} = \boldsymbol{a}$. Note that the rotation of the coordinates does not change the energy, and, hence, there is no need to rotate the signal set to minimize the energy.

■ EXAMPLE 7.7

For the binary orthogonal signal set shown in Fig. 7.18*a*, determine the minimum-energy equivalent signal set.

As a concrete example, let us choose the basis signals as:

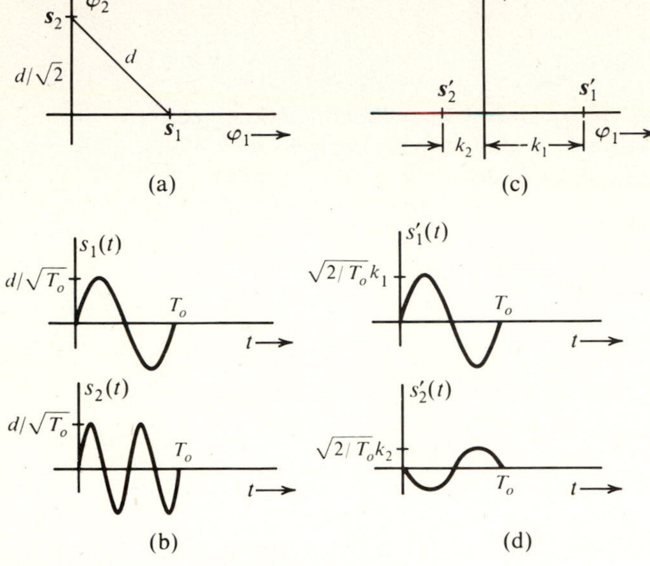

Figure 7.18 Equivalent signal sets.

$$
\left.
\begin{aligned}
\varphi_1(t) &= \sqrt{\frac{2}{T_o}} \sin \omega_o t \\
\varphi_2(t) &= \sqrt{\frac{2}{T_o}} \sin \omega_o t
\end{aligned}
\right\} \quad 0 \leq t < T_o
$$

Hence,

$$
\left.
\begin{aligned}
s_1(t) &= \frac{d}{\sqrt{2}} \varphi_1(t) = \frac{d}{\sqrt{T_o}} \sin \omega_o t \\
s_2(t) &= \frac{d}{\sqrt{2}} \varphi_2(t) = \frac{d}{\sqrt{T_o}} \sin \omega_o t
\end{aligned}
\right\} \quad 0 \leq t < T_o
$$

The signals $s_1(t)$ and $s_2(t)$ are shown in Fig. 7.18*b*, and the geometrical representation is shown in Fig. 7.18*a*. Both signals are located at a distance $d/\sqrt{2}$ from the origin, and the distance between the signals is d.

The minimum-energy set for this case is shown in Fig. 7.18*c*. The origin lies at the center of gravity of the signals. We have also rotated the signals for convenience. The distances k_1 and k_2 must be such that

$$k_1 + k_2 = d$$

and

$$k_1 P(m_1) = k_2 P(m_2)$$

that is,

$$\frac{k_1}{k_2} = \frac{P(m_2)}{P(m_1)}$$

Hence,

$$k_1 = P(m_2)d$$

and

$$k_2 = P(m_1)d$$

The new signals $s_1'(t)$ and $s_2'(t)$ for this set are given by

$$\left.\begin{aligned} s_1'(t) &= \sqrt{\frac{2}{T_o}}\,P(m_2)d \sin \omega_o t \\ s_2'(t) &= \sqrt{\frac{2}{T_o}}\,P(m_1)d \sin \omega_o t \end{aligned}\right\} \quad 0 \le t < T_o$$

These signals are sketched in Fig. 7.18*d*.

Both sets have the same error probability, but the latter has a smaller mean energy. If $\bar{E}$ and $\overline{E'}$ are the respective mean energies of the two sets, then

$$\bar{E} = P(m_1)\frac{d^2}{2} + P(m_2)\frac{d^2}{2} = \frac{d^2}{2}$$

and

$$\begin{aligned} \overline{E'} &= P(m_1)k_1^2 + P(m_2)k_2^2 \\ &= P(m_1)P^2(m_2)d^2 + P(m_2)P^2(m_1)d^2 \\ &= P(m_1)P(m_2)d^2 \end{aligned}$$

Note that the product $P(m_1)P(m_2)$ is maximum when $P(m_1) = P(m_2) = 1/2$, and in this case

$$P(m_1)P(m_2) = \frac{1}{4}$$

Hence,

$$\overline{E'} \le \frac{d^2}{4}$$

It is obvious that

$$\overline{E'} \le \frac{\bar{E}}{2}$$

and for the case of equiprobable signals,

$$\overline{E'} = \frac{\bar{E}}{2}$$

In this case

$$k_1 = k_2 = \frac{d}{2}$$

$$\bar{E} = \frac{d^2}{2} \quad \text{and} \quad \overline{E'} = \frac{d^2}{4}$$

The signals in Fig. 7.18*c* are called *antipodal signals* when $k_1 = k_2$. The error probability of the signal set in Fig. 7.18*a* (and Fig. 7.18*c*) is equal to that in Fig. 7.12*a* and can be found from Eq. (7.54a). ■

■ EXAMPLE 7.8 The Simplex Signal Set

A minimum-energy equivalent set of an equiprobable orthogonal set is called a *simplex* (or *transorthogonal*) *signal set*. Derive the simplex set corresponding to the orthogonal set in Eq. (7.66).

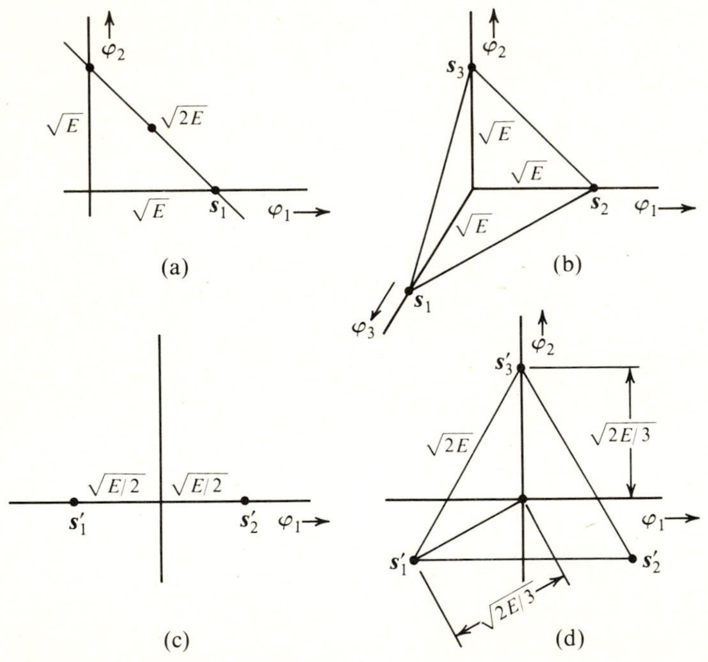

Figure 7.19 Simplex signals.

□ **Solution:** To obtain the minimum-energy set, the origin should be shifted to the center of gravity of the signal set. For the 2-dimensional case (Fig. 7.19*a*), the simplex set is shown in Fig. 7.19*c*, and for the 3-dimensional case (Fig. 7.19*b*), the simplex set is shown in Fig. 7.19*d*. Note that the dimensionality of the simplex signal set is less than that of the orthogonal set by one. This is true in general for any value of M.

It can be shown that the simplex signal set is the optimum (minimum error probability) for the case of equiprobable signals embedded in white gaussian noise when energy is constrained.[6,7]

We can calculate the mean energy of the simplex set by noting that it is obtained by translating the orthogonal set by a vector $\boldsymbol{a}$, given by Eq. (7.76):

$$\boldsymbol{a} = \frac{1}{M} \sum_{i=1}^{M} \mathbf{s}_i$$

For orthogonal signals,

$$\mathbf{s}_i = \sqrt{E}\,\boldsymbol{\Phi}_i$$

Therefore

$$\boldsymbol{a} = \frac{\sqrt{E}}{M} \sum_{i=1}^{M} \boldsymbol{\Phi}_i$$

where E is the energy of each signal in the orthogonal set and $\boldsymbol{\Phi}_i$ is the unit vector along the ith coordinate axis. The signals in the simplex set are given by

$$\begin{aligned} \mathbf{s}'_k &= \mathbf{s}_k - \boldsymbol{a} \\ &= \sqrt{E}\,\boldsymbol{\Phi}_k - \frac{\sqrt{E}}{M} \sum_{i=1}^{M} \boldsymbol{\Phi}_i \end{aligned} \tag{7.77}$$

The energy E' of signal $\mathbf{s}'_k$ is given by $|\mathbf{s}'_k|^2$.

$$E' = \mathbf{s}'_k \cdot \mathbf{s}'_k \tag{7.78}$$

Substituting Eq. (7.77) in Eq. (7.78) and observing that the set $\boldsymbol{\Phi}_i$ is orthonormal, we have

$$\begin{aligned} E' &= E - \frac{E}{M} \\ &= E\left(1 - \frac{1}{M}\right) \end{aligned} \tag{7.79}$$

Hence, for the same performance (error probability), the mean energy of the simplex signal set is $(1 - 1/M)$ times that of the orthogonal signal set. For $M \gg 1$, the difference is not significant. For this reason and because they are easier to generate, orthogonal rather than simplex signals are used in practice whenever M exceeds 4 or 5.

In the next chapter, we shall show that in the limit as $M \to \infty$, the orthogonal (as well as the simplex) signals attain the upper bound of performance predicted by Shannon's theorem. ■

7.5 NONWHITE (COLORED) CHANNEL NOISE

Thus far we have restricted our discussion exclusively to white gaussian channel noise. Our discussion can be extended with relative ease to nonwhite, or colored, gaussian channel noise.

If the noise PSD $S_n(\omega)$ is not white, we use a noise-whitening filter $H(\omega)$ at the input of the receiver, where

$$|H(\omega)| = \frac{1}{\sqrt{S_n(\omega)}}$$

Consider a signal set $\{s_i(t)\}$ and a channel noise n(t) that is not white [$S_n(\omega)$ is not constant]. At the input of the receiver, we use a noise-whitening filter $H(\omega)$, that renders the colored noise into white noise (Fig. 7.20). But it also alters the signal set $\{s_i(t)\}$ to $\{s_i'(t)\}$, where

$$s_i'(t) = s_i(t) * h(t)$$

or

$$S_i'(\omega) = S_i(\omega)H(\omega)$$

We now have a new signal set $\{s_i'(t)\}$ mixed with white gaussian noise, for which the optimum receiver and corresponding error probability can be determined by the method discussed earlier.

In general, a whitening filter may be unrealizable. In such a case we need to allow a sufficient delay in $H(\omega)$ to be able to closely realize the filter.

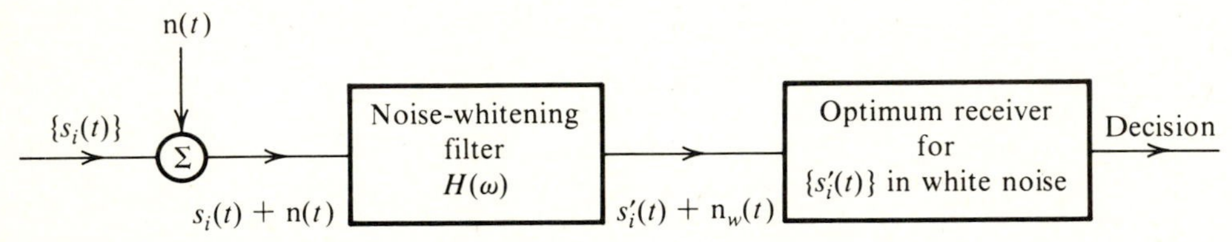

Figure 7.20 Optimum M-ary receiver for nonwhite channel noise.

7.6 OTHER USEFUL PERFORMANCE CRITERIA

The optimum receiver uses the decision strategy that makes the best possible use of the observed data and any a priori information available. The strategy will also depend upon the weights assigned to various types of errors. In this chapter we have thus far assumed that all errors have equal weight (or equal cost). This assumption may not be justified in some cases, and we may therefore have to alter the decision rule.

The Generalized Bayes Receiver

If we are given a priori probabilities and the cost functions of various types of errors, the receiver that minimizes the average cost of decision is called the *Bayes receiver,* and the decision rule is *Bayes' decision rule*. Note that the receiver that has been discussed so far is the Bayes receiver under the condition that all errors have equal cost (equal weight). We shall now generalize this rule for the case where different errors have different costs. Let

$$C_{kj} = \text{cost of deciding that } \hat{m} = m_k \text{ when } m_j \text{ was transmitted} \tag{7.80}$$

and, as usual,

$$P(m_i|\boldsymbol{q}) = \text{conditional probability that } m_i \text{ was transmitted when } \boldsymbol{q} \text{ is received}$$

If $\boldsymbol{q}$ is received and the receiver decides that $\hat{m} = m_k$, then the probability that m_j was transmitted is $P(m_j|\boldsymbol{q})$ for all $j = 1, 2, \ldots, M$. Hence, the average cost of the decision $\hat{m} = m_k$ is β_k, given by

$$\begin{aligned}\beta_k &= C_{k1}P(m_1|\boldsymbol{q}) + C_{k2}P(m_2|\boldsymbol{q}) + \cdots + C_{kM}(m_M|\boldsymbol{q}) \\ &= \sum_{j=1}^{M} C_{kj}P(m_j|\boldsymbol{q})\end{aligned} \tag{7.81}$$

Obviously, if $\boldsymbol{q}$ is received, the optimum receiver decides that $\hat{m} = m_k$ if

$$\beta_k < \beta_i \qquad \text{for all } i \neq k$$

or

$$\sum_{j=1}^{M} C_{kj}P(m_j|\boldsymbol{q}) < \sum_{j=1}^{M} C_{ij}P(m_j|\boldsymbol{q}) \tag{7.82}$$

for all $i \neq k$. Use of Bayes' mixed rule in Eq. (7.82) yields

$$\sum_{j=1}^{M} C_{kj}P(m_j)p_{\mathbf{q}}(\boldsymbol{q}|m_j) < \sum_{j=1}^{M} C_{ij}P(m_j)p_{\mathbf{q}}(\boldsymbol{q}|m_j) \tag{7.83}$$

for all $i \neq k$. Note the C_{kk} is the cost of setting $\hat{m} = m_k$ when m_k is transmitted. This cost is generally zero. If we assign equal weight to all other errors, then

$$C_{kj} = \begin{cases} 0 & k = j \\ 1 & k \neq j \end{cases}$$

and the decision rule in Eq. (7.83) reduces to the rule in Eq. (7.42), as expected. The generalized Bayes' receiver for $M = 2$, assuming $C_{11} = C_{22} = 0$, sets $\hat{m} = m_1$ if

$$C_{12}P(m_2)p_{\mathbf{q}}(\boldsymbol{q}|m_2) < C_{21}P(m_1)p_{\mathbf{q}}(\boldsymbol{q}|m_1)$$

Otherwise, the receiver decides that $\hat{m} = m_2$.

The Maximum-Likelihood Receiver

The strategy used in the Bayes receiver discussed above is general, except that it can be implemented only when the a priori probabilities $P(m_1), P(m_2), \ldots, P(m_M)$ are known. Frequently this information is not available. Under these conditions various possibilities exist, depending upon the assumptions made. When, for example, there is no reason to expect any one signal to be more likely than any other, we may assign equal probabilities to all the messages:

$$P(m_1) = P(m_2) = \cdots = P(m_M) = \frac{1}{M}$$

Bayes' rule [Eq. (7.42)] in this case becomes: set $\hat{m} = m_k$ if

$$p_{\mathbf{q}}(\boldsymbol{q}|m_k) > p_{\mathbf{q}}(\boldsymbol{q}|m_i) \qquad \text{for all } i \neq k \tag{7.84}$$

Observe that $p_{\mathbf{q}}(\boldsymbol{q}|m_k)$ represents the probability of observing $\boldsymbol{q}$ when m_k is transmitted. Thus, the receiver chooses that signal which, when transmitted, will maximize the likelihood (probability) of observing the received $\boldsymbol{q}$. Hence, this receiver is called the *maximum-likelihood receiver*. Note that the maximum-likelihood receiver is Bayes receiver under the condition that the a priori message probabilities are equal. In terms of geometrical concepts, the maximum-likelihood receiver decides in favor of that signal which is closest to the received data $\boldsymbol{q}$. The practical implementation of the maximum-likelihood receiver is the same as that of the Bayes receiver (Figs. 7.8 and 7.9) under the condition that all a priori probabilities are equal to $1/M$.

If the signal set is geometrically symmetrical, and if all a priori probabilities are equal (maximum-likelihood receiver), then the decision regions for various signals are congruent. In this case, because of symmetry, the conditional probability of a correct decision is the same no matter which signal is transmitted—that is,

$$P(C|m_i) = \text{constant} \qquad \text{for all } i$$

Because

$$P(C) = \sum_{i=1}^{M} P(m_i)P(C|m_i)$$

in this case

$$P(C) = P(C|m_i) \tag{7.85}$$

Thus, the error probability of the maximum-likelihood receiver is independent of the actual source statistics $P(m_i)$ for the case of symmetrical signal sets. It should, however, be realized that if the actual source statistics were known beforehand, one could design a better receiver using Bayes' decision rule.

It is apparent that if the source statistics are not known, the maximum-likelihood receiver proves very attractive for a symmetrical signal set. In such a receiver one can specify the error probability independently of the actual source statistics.

The Minimax Receiver

Designing a receiver with a certain decision rule completely specifies the conditional probabilities $P(C|m_i)$. The probability of error is given by

$$\begin{aligned} P_{eM} &= 1 - P(C) \\ &= 1 - \sum_{i=1}^{M} P(m_i)P(C|m_i) \end{aligned}$$

Thus, in general, for a given receiver (with some specified decision rule) the error probability depends upon the source statistics $P(m_i)$. The error probability is the largest for some source statistics. The error probability in the worst possible case is $[P_{eM}]_{\max}$ and represents the upper bound on the error probability of the given receiver. This upper bound $[P_{eM}]_{\max}$ serves as an indication of the quality of the receiver. Each receiver (with a certain decision rule) will have a certain $[P_{eM}]_{\max}$. The receiver that has the smallest upper bound on the error probability, that is the minimum $[P_{eM}]_{\max}$, is called the *minimax receiver*.

We shall illustrate the minimax concept for a binary receiver with on-off signaling. The conditional PDFs of the receiving-filter output sample r at $t = T_o$ are $p(r|\mathbf{1})$ and $p(r|\mathbf{0})$. These are the PDFs of r for the "on" pulse and the "off" pulse (i.e., no pulse), respectively. Figure 7.21*a* shows these PDFs with a certain threshold a. If we

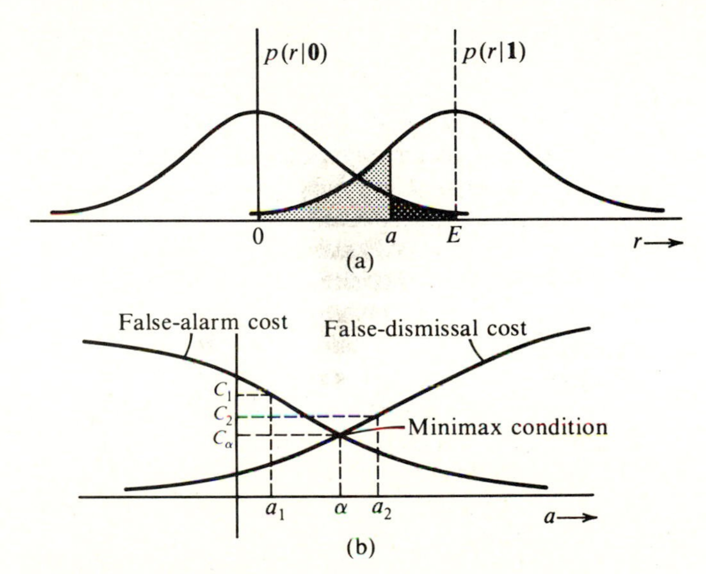

Figure 7.21 Explanation of minimax concept.

receive r $\geq a$, we choose the hypothesis "signal present" (**1**), and the shaded area to the right of a is the probability of *false alarm* (deciding "signal present" when in fact the signal is not present). If r $< a$, we choose the hypothesis "signal absent" (**0**), and the shaded area to the left of a is the probability of *false dismissal* (deciding "signal absent" when in fact the signal is present). It is obvious that the larger is the threshold a, the larger is the false dismissal error and the smaller is the false alarm error (see Fig. 7.21*b*).

We shall now find the minimax condition for this receiver. For the minimax receiver, we consider all possible receivers (all possible values of a in this case) and find the maximum error probability (or cost) that occurs under the worst possible a priori probability distribution. Let us choose $a = a_1$, as shown in Fig. 7.21*b*. In this case the worst possible case occurs when $P(\mathbf{0}) = 0$ and $P(\mathbf{1}) = 1$, that is, when the signal $s(t)$ is always present. The type of error in this case is false alarm. These errors have a cost C_1. On the other hand, if we choose $a = a_2$, the worst possible case occurs when $P(\mathbf{0}) = 1$ and $P(\mathbf{1}) = 0$, that is, when the signal is always absent, causing only the false-dismissal type of errors. These errors have a cost C_2. It is evident that for the setting $a = \alpha$, the cost of false alarm and false dismissal are equal, namely, C_α. Hence, for all possible source statistics the cost is C_α. Because $C_\alpha < C_1$ and C_2, it is obvious that this cost is the *minimum* of the maximum possible cost (because the worst cases are considered) that accrues for all values of a. Hence, $a = \alpha$ represents the minimax setting.

It follows from this discussion that the minimax receiver is rather conservative. It is designed under the pessimistic assumption that the worst possible source statistics exist. The maximum-likelihood receiver, on the other hand, is designed on the assumption that all messages are equally likely. It can, however, be shown that for a symmetrical signal set, the maximum-likelihood receiver is in fact the minimax receiver. This can be proved by observing that for a symmetrical set, the probability of error of a maximum-likelihood receiver (equal a priori probabilities) is independent of the source statistics [Eq. (7.85)]. Hence, for a symmetrical set, the error probability $P_{eM} = \alpha$ of a maximum-likelihood receiver is also equal to its $[P_{eM}]_{\max}$. We will now show that no other receiver exists whose $[P_{eM}]_{\max}$ is less than α of a maximum-likelihood receiver for a symmetrical signal set. This is easily seen from the fact that for equiprobable messages, the maximum-likelihood receiver is optimum by definition. All other receivers must have $P_{eM} > \alpha$ for equiprobable messages. Obviously, $[P_{eM}]_{\max}$ for these receivers can never be less than α. This proves that the maximum-likelihood receiver is indeed the minimax receiver for a symmetrical signal set.

APPENDIX 7.1 Gram-Schmidt Orthogonalization of a Vector Set

We have defined the dimensionality of a vector space as equal to the maximum number of independent vectors in the space. Thus in an N-dimensional space, there can be no more than N vectors that are independent. Alternatively, it is always possible to find a set of N vectors that are independent. Once such a set is chosen, any vector in this space can be expressed in terms of (as a linear combination of) the vectors in this set. This set forms what we commonly refer to as a basis set, which forms the coordinate system. This set of N independent vectors is by no means unique. The reader is familiar with this fact in the physical space of three dimensions, where one can find an infinite number of independent sets of three vectors. This is clear from the fact that we have infinite possible coordinate systems. The orthogonal set, however, is of special interest because it is easier to deal with as compared to nonorthogonal sets. If we are given a set of N independent vectors, it is possible to obtain from this set another set of N independent vectors that is orthogonal. This is done by the *Gram-Schmidt process* of orthogonalization.

In order to get a physical insight into this procedure, we shall consider a simple case of 2-dimensional space. Let $\boldsymbol{x}_1$ and $\boldsymbol{x}_2$ be two independent vectors in a 2-dimensional space (Fig. A7.1). We wish to generate a new set of two orthogonal

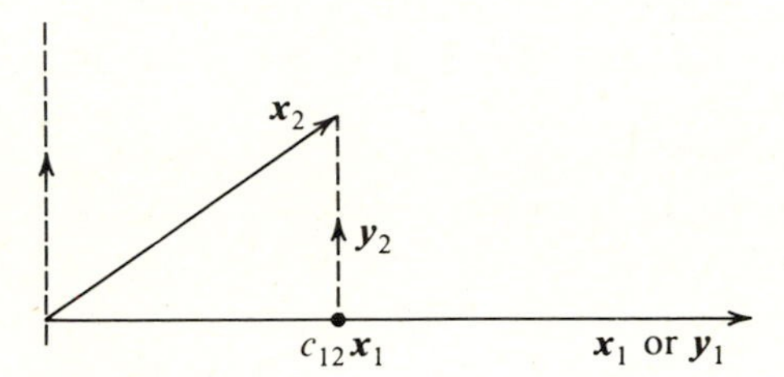

Figure A7.1

vectors $\boldsymbol{y}_1$ and $\boldsymbol{y}_2$ from $\boldsymbol{x}_1$ and $\boldsymbol{x}_2$. For convenience, we shall choose

$$\boldsymbol{y}_1 = \boldsymbol{x}_1 \tag{A7.1}$$

$\boldsymbol{y}_2$ is orthogonal to $\boldsymbol{y}_1$ (and $\boldsymbol{x}_1$), as shown in Fig. A7.1. We now wish to express $\boldsymbol{y}_2$ in terms of $\boldsymbol{x}_1$ and $\boldsymbol{x}_2$. This can be done by observing that the vector $\boldsymbol{x}_2$ is equal to the sum of vector $\boldsymbol{y}_2$ and the projection of $\boldsymbol{x}_2$ upon $\boldsymbol{x}_1$. Thus,

$$\boldsymbol{x}_2 = \boldsymbol{y}_2 + C_{12}\boldsymbol{x}_1 = \boldsymbol{y}_2 + C_{12}\boldsymbol{y}_1 \tag{A7.2}$$

where C_{12} is given by* [see Eq. (7.8)]

$$C_{12} = \frac{\boldsymbol{x}_1 \cdot \boldsymbol{x}_2}{|\boldsymbol{x}_1|^2}$$

But because [by Eq. (A7.1)] $\boldsymbol{y}_1 = \boldsymbol{x}_1$,

$$C_{12} = \frac{\boldsymbol{y}_1 \cdot \boldsymbol{x}_2}{|\boldsymbol{y}_1|^2}$$

Hence,

$$\begin{aligned}\boldsymbol{y}_2 &= \boldsymbol{x}_2 - C_{12}\boldsymbol{y}_1 \\ &= \boldsymbol{x}_2 - \frac{\boldsymbol{y}_1 \cdot \boldsymbol{x}_2}{|\boldsymbol{y}_1|^2}\boldsymbol{y}_1 \end{aligned} \tag{A7.3}$$

Equations (A7.1) and (A7.3) yield the desired orthogonal set. Note that this set is not unique. There are infinite possible orthogonal vector sets ($\boldsymbol{y}_1, \boldsymbol{y}_2$) that can be generated from ($\boldsymbol{x}_1, \boldsymbol{x}_2$). In our derivation, we could as well have started with $\boldsymbol{y}_1 = \boldsymbol{x}_2$ instead of

* This can also be obtained by taking the scalar product of both sides of Eq. (A7.2) with $\boldsymbol{x}_1$.

$\mathbf{y}_1 = \mathbf{x}_1$. This starting point would have yielded an entirely different set. In general,

$$\begin{aligned} \mathbf{y}_1 &= a_{11}\mathbf{x}_1 + a_{12}\mathbf{x}_2 \\ \mathbf{y}_2 &= a_{21}\mathbf{x}_1 + a_{22}\mathbf{x}_2 \end{aligned} \tag{A7.4}$$

and

$$\mathbf{y}_1 \cdot \mathbf{y}_2 = 0 \tag{A7.5}$$

Substituting Eq. (A7.4) in Eq. (A7.5), we get one equation in four unknown coefficients a_{11}, a_{12}, a_{21}, and a_{22}. Hence, infinite possible solutions exist. One can choose any three of the coefficients arbitrarily, and the remaining (fourth) coefficient will be determined from Eqs. (A7.4) and (A7.5).

The reader can extend these results to a 3-dimensional case. If vectors $\mathbf{x}_1$, $\mathbf{x}_2$, $\mathbf{x}_3$ form an independent set in this space, then it can be shown that one possible orthogonal vector set $\mathbf{y}_1$, $\mathbf{y}_2$, $\mathbf{y}_3$ in this space is given by

$$\begin{aligned} \mathbf{y}_1 &= \mathbf{x}_1 \\ \mathbf{y}_2 &= \mathbf{x}_2 - \frac{\mathbf{y}_1 \cdot \mathbf{x}_2}{|\mathbf{y}_1|^2}\mathbf{y}_1 \\ \mathbf{y}_3 &= \mathbf{x}_3 - \frac{\mathbf{y}_1 \cdot \mathbf{x}_3}{|\mathbf{y}_1|^2}\mathbf{y}_1 - \frac{\mathbf{y}_2 \cdot \mathbf{x}_3}{|\mathbf{y}_2|^2}\mathbf{y}_2 \end{aligned} \tag{A7.6}$$

These results can be extended to an N-dimensional space. In general, if we are given N independent vectors $\mathbf{x}_1, \mathbf{x}_2, \ldots, \mathbf{x}_N$, then proceeding along similar lines, one can obtain an orthogonal set $\mathbf{y}_1, \mathbf{y}_2, \ldots, \mathbf{y}_N$ where

$$\mathbf{y}_1 = \mathbf{x}_1$$

and

$$\mathbf{y}_j = \mathbf{x}_j - \sum_{k=1}^{j-1} \frac{\mathbf{y}_k \cdot \mathbf{x}_j}{|\mathbf{y}_k|^2}\mathbf{y}_k \qquad j = 2, 3, \ldots, N \tag{A7.7}$$

Note that this is one of the infinitely many orthogonal sets that can be formed from the set $\mathbf{x}_1, \mathbf{x}_2, \ldots, \mathbf{x}_N$. Moreover, this set is not an orthonormal set. The orthonormal set $\hat{\mathbf{y}}_1, \hat{\mathbf{y}}_2, \ldots, \hat{\mathbf{y}}_N$ can be obtained by normalizing the lengths of the respective vectors.

$$\hat{\mathbf{y}}_k = \frac{\mathbf{y}_k}{|\mathbf{y}_k|}$$

We can apply these concepts to signal space, because one-to-one correspondence exists between the signals and vectors. If we have N independent signals $x_1(t)$, $x_2(t), \ldots, x_N(t)$, we can form a set of N orthogonal signals $y_1(t), y_2(t), \ldots, y_N(t)$ as

$$\begin{aligned} y_1(t) &= x_1(t) \\ y_j(t) &= x_j(t) - \sum_{k=1}^{j-1} C_{kj}y_k(t) \qquad j = 2, 3, \ldots, N \end{aligned} \tag{A7.8}$$

where

$$C_{kj} = \frac{\int y_k(t)x_j(t)\,dt}{\int y_k^2(t)\,dt} \tag{A7.9}$$

Note that this is one of the infinitely many possible orthogonal sets that can be formed from the set $x_1(t), x_2(t), \ldots, x_N(t)$. The set can be normalized by dividing each signal $y_j(t)$ by its energy.

■ EXAMPLE A7.1

The exponential signals

$$g_1(t) = e^{-pt}u(t)$$

$$g_2(t) = e^{-2pt}u(t)$$

$$\vdots$$

$$g_N(t) = e^{-Npt}u(t)$$

form an independent set of signals in N-dimensional space, where N may be any integer. This set, however, is not orthogonal. We can use the Gram-Schmidt process to obtain an orthogonal set for this space. If $y_1(t), y_2(t), \ldots, y_N(t)$ is the desired orthogonal basis set, we choose

$$y_1(t) = g_1(t) = e^{-pt}u(t)$$

From Eqs. (A7.8) and (A7.9) we have

$$y_2(t) = x_2(t) - C_{12}y_1(t)$$

where

$$\begin{aligned} C_{12} &= \frac{\int_{-\infty}^{\infty} y_1(t)x_2(t)\,dt}{\int_{-\infty}^{\infty} y_1^2(t)\,dt} \\ &= \frac{\int_0^{\infty} e^{-pt}e^{-2pt}\,dt}{\int_0^{\infty} e^{-2pt}\,dt} \\ &= \frac{2}{3} \end{aligned}$$

Hence,

$$y_2(t) = (e^{-2pt} - \tfrac{2}{3}e^{-pt})\,u(t)$$

Similarly, we can proceed to find the remaining functions $y_3(t), \ldots, y_N(t)$, and so on. The reader can verify that all this represents a mutually orthogonal set.

REFERENCES

1. H. J. Landau and H. O. Pollak, "Prolate Spheroidal Wave Functions, Fourier Analysis, and Uncertainty, III: The Dimensions of Space of Essentially Time- and Band-Limited Signals," *Bell Syst. Tech. J.*, vol. 41, pp. 1295–1336, July 1962.

2. J. M. Wozencraft and I. M. Jacobs, *Principles of Communication Engineering,* Wiley, New York, 1965.
3. E. Arthurs and H. Dym, "On Optimum Detection of Digital Signals in the Presence of White Gaussian Noise—A Geometric Interpretation and a Study of Three Basic Data Transmission Systems," *IRE Trans. Commun. Syst.,* vol. CS-10, pp. 336–372, Dec. 1962.
4. B. P. Lathi, *An Introduction to Random Signals and Communication Theory,* International Textbook Co., Scranton, Pa., 1968.
5. H. L. Van Trees, *Detection, Estimation, and Modulation Theory,* vols. I, II, and III, Wiley, New York, 1968-71.
6. H. J. Landau and D. Slepian, "On the Optimality of the Regular Simplex Code," *Bell Syst. Tech. J.,* vol. 45, pp. 1247–1272, Oct. 1966.
7. A. V. Balakrishnan, "Contribution to the Sphere-Packing Problem of Communication Theory," *J. Math. Anal. Appl.,* vol. 3, no. 3, pp. 485–506, Dec. 1961.

PROBLEMS

7.1. Find at least two different sets of orthonormal basis signals for a 5-dimensional signal space.

7.2. The basis signals of a 3-dimensional signal space are given by

$$\varphi_1(t) = p(t),\ \varphi_2(t) = p(t - T_o),\ \text{and}\ \varphi_3(t) = p(t - 2T_o)$$

where

$$p(t) = \frac{1}{\sqrt{T_o}}[u(t) - u(t - T_o)]$$

Sketch the waveforms of the signals represented by

$(1, 1, 0)$, $(2, -1, 1)$, $(3, 2, -\frac{1}{2})$, and $(-\frac{1}{2}, -1, 1)$ in this space.

7.3. Repeat Prob. 7.2 if

$$\varphi_1(t) = \frac{1}{\sqrt{T_o}}$$

$$\varphi_2(t) = \sqrt{\frac{2}{T_o}} \sin \frac{2\pi}{T_o} t \qquad 0 \le t \le T_o$$

$$\varphi_3(t) = \sqrt{\frac{2}{T_o}} \cos \frac{2\pi}{T_o} t$$

7.4. If $p(t)$ is as in Prob. 7.2 and

$$\varphi_k(t) = p[t - (k - 1)T_o] \qquad k = 1, 2, 3, 4, 5$$

(a) Sketch the signals represented by $(-1, 2, 3, 1, 4)$, $(2, 1, -4, -4, 2)$, $(3, -2, 3, 4, 1)$, and $(-2, 4, 2, 2, 0)$ in this space.
(b) Find the energy of each signal.
(c) Find the pairs of signals that are orthogonal.

7.5. For a certain stationary gaussian random process x(t), it is given that $R_x(\tau) = e^{-|\tau|}$. Determine the joint PDF of r.v.'s x(t), x$(t + 1)$, and x$(t + 2)$.

7.6. A source emits M equiprobable messages, which are assigned signals $s_1, s_2, \ldots, s_M$, as shown in Fig. P7.6. Determine the optimum receiver and the corresponding error probability for an AWGN channel.

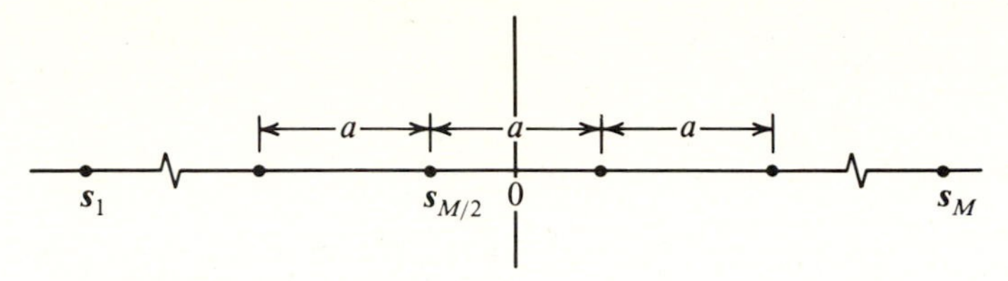

Figure P7.6

7.7. A source emits eight equiprobable messages, which are assigned signals $s_1, s_2, \ldots, s_8$, as shown in Fig. P7.7.

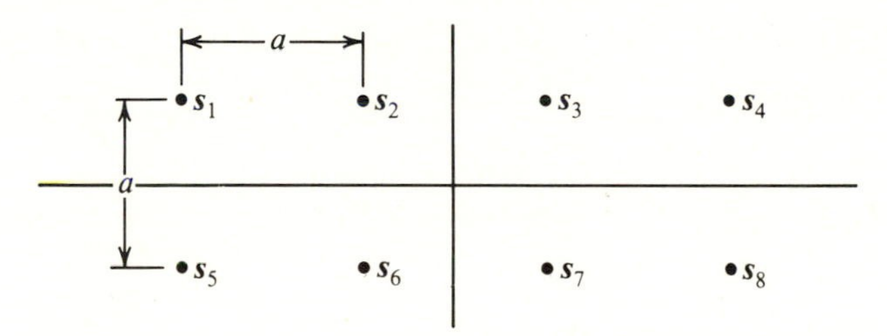

Figure P7.7

(a) Find the optimum receiver for an AWGN channel.
(b) Determine the decision regions and the error probability of the optimum receiver.

7.8. Show that for MPSK, the optimum receiver of the form in Fig. 7.9*a* is equivalent to a phase comparator. Assume all messages equiprobable and an AWGN channel.

7.9. The vertices of a hypercube are a set of 2^N signals

$$s_k(t) = \frac{d}{2} \sum_{j=1}^{N} a_{kj}\varphi_j(t)$$

where $\{\varphi_1(t), \varphi_2(t), \ldots, \varphi_N(t)\}$ is a set of N orthonormal signals, and a_{kj} is either 1 or -1. Note that all the N signals are at a distance of $\sqrt{N}\, d/2$ from the origin and form the vertices of the N-dimensional cube.
(a) Sketch the signal configuration in the signal space for $N = 1$, 2, and 3.
(b) For each configuration in part (a), sketch one possible set of waveforms.
(c) If all the 2^N symbols are equiprobable, find the optimum receiver and determine the error probability P_{eM} of the optimum receiver.

7.10. A ternary signal configuration is shown in Fig. P7.10.
(a) If $P(m_0) = 0.5$ and $P(m_1) = P(m_{-1}) = 0.25$, determine the optimum decision regions and P_{eM} of the optimum receiver. Assume an AWGN channel.
(b) If we define $\lambda = S_i/\mathcal{N}f_o$ where S_i is the received power and f_o is the symbol transmission rate, find P_{eM} as a function of λ.

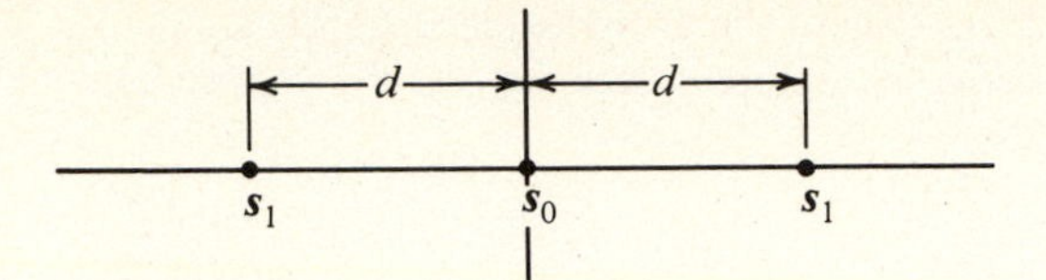

Figure P7.10

7.11. A 16-point QAM signal configuration is shown in Fig. P7.11. Assuming that all symbols are equiprobable, determine the error probability of the optimum receiver for an AWGN channel.

Compare the performance of this scheme with that in Example 7.3.

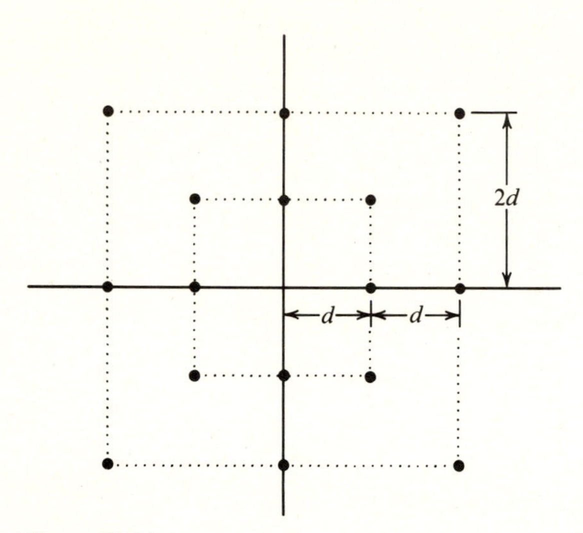

Figure P7.11

7.12. Another 16-ary signal configuration is shown in Fig. P7.12. Write the expression (do not

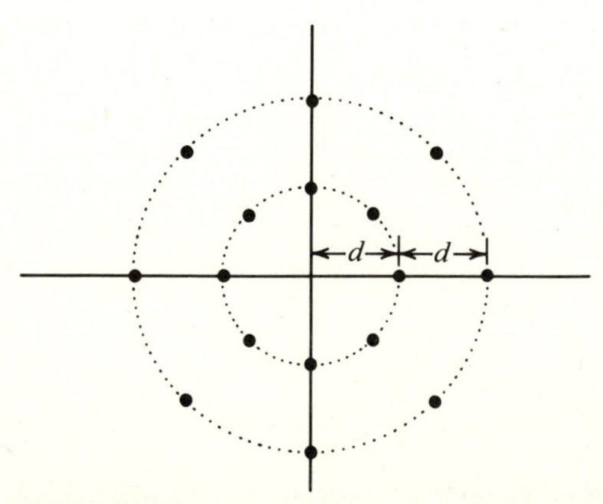

Figure P7.12

evaluate various integrals) for P_{eM} of the optimum receiver, assuming all symbols to be equiprobable. Assume an AWGN channel.

7.13. An orthogonal signal set is given by

$$s_k(t) = \sqrt{E}\ \varphi_k(t) \qquad k = 1, 2, \ldots, N$$

A biorthogonal signal set is formed from the orthogonal set by augmenting it with the negative of each signal. Thus, we add to the orthogonal set another set

$$s_{-k}(t) = -\sqrt{E}\ \varphi_k(t)$$

This gives $2N$ signals in an N-dimensional space. Assuming all signals to be equiprobable and an AWGN channel, obtain the error probability of the optimum receiver. How does the bandwidth of the biorthogonal set compare with that of the orthogonal set?

7.14. A five-signal configuration in a 2-dimensional space is shown in Fig. P7.14.

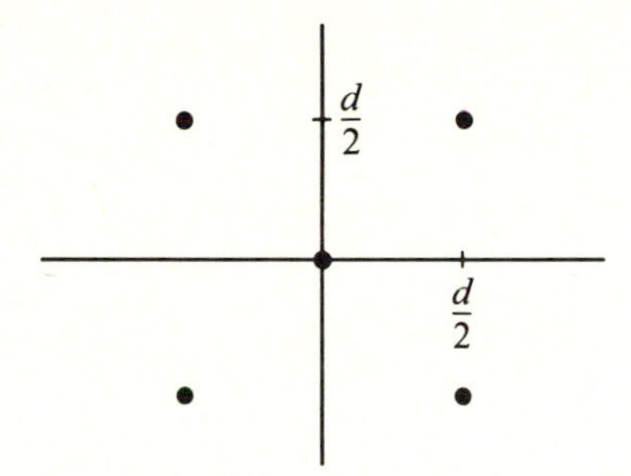

Figure P7.14

(a) Choose the appropriate $\varphi_1(t)$ and $\varphi_2(t)$ and sketch the waveforms of the five signals.
(b) In the signal space, sketch the optimum decision regions.
(c) Determine the error probability of the optimum receiver.
Hint: A rotation of the signal set will be helpful.

7.15. **(a)** What is the minimum-energy equivalent signal set of a binary on-off signal set?
(b) Using geometrical signal space concepts, explain why the binary on-off and the binary orthogonal sets have identical error probabilities and why the binary polar energy requirements are 3 dB lower than those of the on-off or the orthogonal set.

7.16. A quaternary signaling scheme uses for waveforms

$$s_1(t) = (\sqrt{3} - 1)\ \varphi_1(t)$$

$$s_2(t) = -2\varphi_1(t) + (\sqrt{3} - 1)\ \varphi_2(t)$$

$$s_3(t) = -(\sqrt{3} + 1)\ \varphi_1(t) - 2\varphi_2(t)$$

$$s_4(t) = -(\sqrt{3} + 1)\ \varphi_2(t)$$

where $\varphi_1(t)$ and $\varphi_2(t)$ are orthonormal basis signals. All the signals are equiprobable, and the channel noise is white gaussian with PSD $S_n(\omega) = 0.2$.
(a) Represent these signals in the signal space, and determine the optimum decision regions.
(b) Compute the error probability of the optimum receiver.
(c) Find the minimum-energy equivalent signal set.

7.17. A ternary signaling scheme ($M = 3$) uses the three waveforms $s_1(t)$, $s_2(t)$, and $s_3(t)$ shown in Fig. P7.17. The transmission rate is 1000 symbols/second. All three messages are equiprobable, and the channel noise is white gaussian with PSD $S_n(\omega) = 10^{-5}$.
(a) Determine the error probability of the optimum receiver.
(b) Determine the minimum-energy signal set and sketch the waveforms.
(c) Compute the mean energies of the signal set in Fig. P7.17 and its minimum-energy equivalent set, found in part (b).

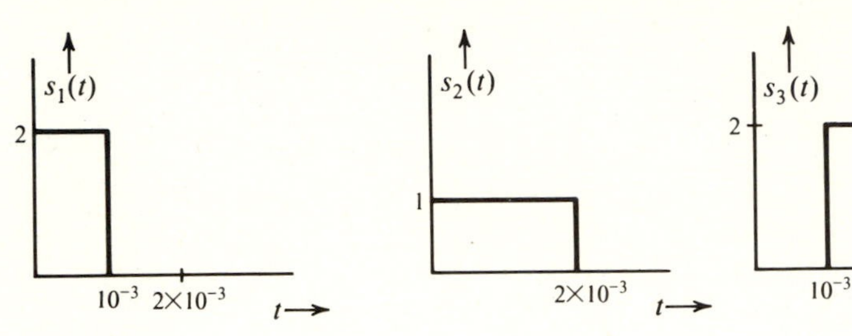

Figure P7.17

7.18. Repeat Prob. 7.17 if $P(m_1) = 0.6$ and $P(m_2) = 0.4$.

7.19. A binary signaling scheme uses the two waveforms $s_1(t)$ and $s_2(t)$ shown in Fig. P7.19. The signaling rate is 1000 pulses/second. Both signals are equally likely, and the channel noise is white gaussian with PSD $S_n(\omega) = 2.5 \times 10^{-6}$.
(a) Determine the error probability of the optimum receiver.
(b) Determine the minimum-energy equivalent signal set.
Hint: Use Gram-Schmidt orthogonalization to determine the appropriate basis signals $\varphi_1(t)$ and $\varphi_2(t)$.

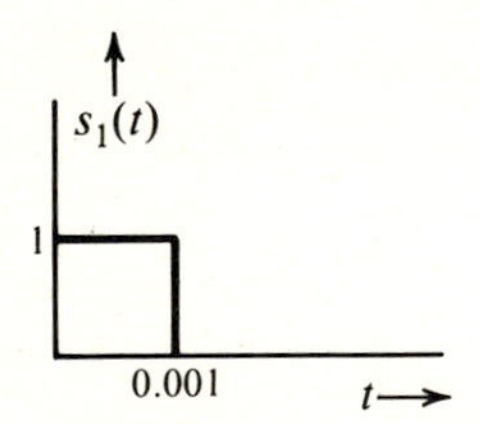

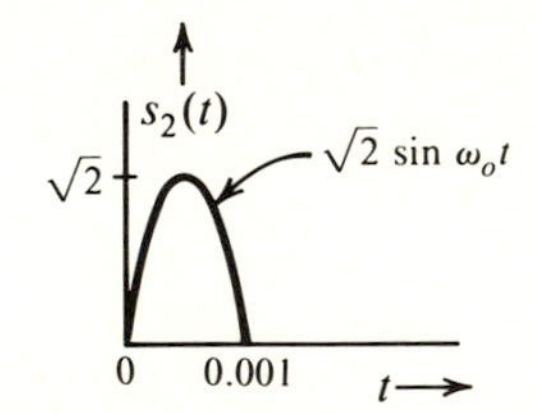

Figure P7.19

8

An Introduction to Information Theory

In all the modes of communication discussed thus far, the communication is not error free. We may be able to improve the accuracy in digital signals by reducing the error probability P_e. But it appears that as long as a channel noise exists, the communication cannot be error free. For example, in all the digital systems discussed thus far, P_e varies as $e^{-k\lambda}$ for $\lambda \gg 1$. By increasing λ, we can reduce P_e to any desired level. But because $\lambda = S_i/\mathcal{N} f_o$, increasing λ means either increasing S_i or decreasing f_o, or both. Because of physical limitations, however, S_i cannot be increased beyond a certain limit. Hence, to reduce P_e further, we must reduce f_o, the rate of transmission of information digits. Thus, the price to be paid for reducing P_e is a reduction in transmission rate. To make $P_e \rightarrow 0$, $f_o \rightarrow 0$. Hence, it appears that in the presence of channel noise it is impossible to achieve error-free communication. Thus thought communication engineers until the publication of Shannon's classical paper[1] in 1948. Shannon showed that for a given channel, as long as the rate of information digits per second to be transmitted is maintained within a certain limit (known as the channel capacity), it is possible to achieve error-free communication. That is, to attain $P_e \rightarrow 0$, it is not necessary to make $f_o \rightarrow 0$. Such a goal ($P_e \rightarrow 0$) can be attained by maintaining $f_o < C$, the channel capacity (per second). The gist of Shannon's paper is that the presence of random disturbance in a channel does not, by itself, set any limit on transmission accuracy. Instead, it sets a limit on the information rate for which arbitrarily small error probability ($P_e \to 0$) can be achieved.

We have been using the phrase "rate of information transmission" as if information can be measured. This is indeed so. We shall now discuss the information content of a message as understood by our "common sense" and also as it is understood in the engineering sense. Surprisingly, both approaches yield the same measure of information in a message.

8.1 MEASURE OF INFORMATION

Common-Sense Measure of Information

Consider the following three hypothetical headlines in a morning paper:

1. Tomorrow the sun will rise in the east
2. United States invades Cuba
3. Cuba invades the United States

The reader will hardly notice the first headline. He will be very, very interested in the second. But what really catches his fancy is the third one. This item will attract much more attention than the previous two headlines. From the point of view of "common sense," the first headline conveys hardly any information, the second conveys a large amount of information, and the third conveys yet a larger amount of information. If we look at the probabilities of occurrence of these three events, we find that the probability of occurrence of the first event is unity (a certain event), that of the second is very low (an event of small but finite probability), and that of the third is practically zero (an almost impossible event). If an event of low probability occurs, it causes greater surprise and, hence, conveys more information than the occurrence of an event of larger probability. Thus, the information is connected with the element of surprise, which is a result of uncertainty, or unexpectedness. The more unexpected the event, the greater the surprise, and hence more information. The probability of occurrence of an event is a measure of its unexpectedness and, hence, is related to the information content. Thus, from the point of view of common sense, the amount of information received from a message is directly related to the uncertainty or inversely related to the probability of its occurrence. If P is the probability of occurrence of a message and I is the information gained from the message, it is evident from the above discussion that when $P \to 1$, $I \to 0$ and when $P \to 0$, $I \to \infty$, and in general a smaller P gives a larger I. This suggests the following model:

$$I \sim \log \frac{1}{P} \tag{8.1}$$

The Engineering Measure of Information

We shall now show that from an engineering point of view, the information content of a message is identical to that obtained on an intuitive basis [Eq. (8.1)]. What do we mean by an engineering point of view? An engineer is responsible for efficient transmission of messages. For this service he will charge a customer an amount proportional to the information to be transmitted. But in reality he will charge the customer in proportion to the time required to transmit his message. In short, from an engineering point of view, the information in a message is proportional to the (minimum) time required to transmit the message. We shall now show that this concept of information also leads to Eq. (8.1). This implies that a message with higher probability can be transmitted in a shorter time than that required for a message with lower probability. This fact may be verified by the example of the transmission of alphabetic symbols in the English language using Morse code. This code is made up of various combinations of two symbols (such as a mark and a space or pulses of height A and $-A$ volts). Each letter is represented by a certain combination of these symbols called the *code word,* which has a certain length. Obviously, for efficient transmission, shorter code words are assigned to the letters e, t, a, and o, which occur more frequently. The longer code words are assigned to letters x, k, q, and z, which occur less frequently. Each letter may be considered as a message. It is obvious that the letters that occur more frequently (with higher probability of occurrence) need a shorter time to transmit (shorter code words) as compared to those with smaller probability of occurrence. We shall now show that on the average, the time required to transmit a symbol (or a message) with probability of occurrence P is indeed proportional to $\log (1/P)$.

For the sake of simplicity, let us begin with the case of binary messages m_1 and m_2, which are equally likely to occur. We may use binary digits to encode these messages. Messages m_1 and m_2 may be represented by digits **0** and **1**, respectively. Clearly, we must have a minimum of one binary digit (which can assume two values) to represent each of the two equally likely messages. Next, consider the case of the four equiprobable messages m_1, m_2, m_3, and m_4. If these messages are encoded in binary form, we need a minimum of two binary digits per message. Each binary digit can assume two values. Hence a combination of two binary digits can form the four code words **00**, **01**, **10**, **11**, which can be assigned to the four equiprobable messages m_1, m_2, m_3, and m_4, respectively. It is clear that each of these four messages takes twice as much transmission time as that required by each of the two equiprobable messages and, hence, contains twice as much information. Similarly, we can encode any one of eight equiprobable messages with a minimum of three binary digits. This is because three binary digits form eight distinct code words, which can be assigned to each of the eight messages. It can be seen that, in general, we need $\log_2 n$ binary digits to encode each of n equiprobable messages.* Because all the messages are equiprobable, P, the probability of any one message occurring, is $1/n$. Hence, each message (with probability P) needs $\log_2 (1/P)$ binary digits for encoding. From the engineering point of view, it is then evident that the information I contained in a message with probability of occurrence P is proportional to $\log_2 (1/P)$.

$$I = k \log_2 \frac{1}{P} \tag{8.2}$$

Once again we come to the conclusion (from the engineering point of view) that the

* Here we are assuming that the number n is such that $\log_2 n$ is an integer. Later on we shall observe that this restriction is not necessary.

information content of a message is proportional to the logarithm of the reciprocal of the probability of the message.

We shall now define the information conveyed by a message according to Eq. (8.2). The constant of proportionality is taken as unity for convenience, and the information is then in terms of binary units, abbreviated *bit* (*bi*nary uni*t*).

$$I = \log_2 \frac{1}{P} \quad \text{bits} \tag{8.3}$$

According to this definition, the information I in a message can be interpreted as the minimum number of binary digits required to encode the message. This is given by $\log_2 (1/P)$, where P is the probability of occurrence of the message. Although here we have shown this result for the special case of equiprobable messages, we shall show in the next section that this is true for nonequiprobable messages, also.

Next, we shall consider the case of r-ary digits instead of binary digits for encoding. Each of the r-ary digits can assume r values $(\mathbf{0}, \mathbf{1}, \mathbf{2}, \ldots, \mathbf{r} - 1)$. Each of n messages (encoded by r-ary digits) can then be transmitted by a particular sequence of r-ary signals.

Because each r-ary digit can assume r values, k r-ary digits can form a maximum of r^k distinct code words. Hence, to encode each of the n equiprobable messages, we need a minimum of $\log_r n$ r-ary digits.* But $n = 1/P$, where P is the probability of occurrence of each message. Obviously, we need a minimum of $\log_r (1/P)$ r-ary digits. The information I per message can be considered as

$$I = \log_r \frac{1}{P} \quad r\text{-ary units} \tag{8.4}$$

Equation (8.4) is a general definition of information, and Eq. (8.3) is the special case of $r = 2$. According to the definition in Eq. (8.4), the information I (in r-ary units) of a message is equal to the minimum number of r-ary digits required to encode the message. This is given by $\log_r (1/P)$, where P is the probability of the message. Although here we have shown this result for the highly special case of equiprobable messages, we shall prove it later for any arbitrary message probabilities.

From Eqs. (8.3) and (8.4) it is evident that

$$I = \log_2 \frac{1}{P} \quad \text{bits} = \log_r \frac{1}{P} \quad r\text{-ary units}$$

Hence,

$$1 \; r\text{-ary unit} = \log_2 r \quad \text{bits} \tag{8.5}$$

In general,

$$1 \; r\text{-ary unit} = \log_s r \quad s\text{-ary units} \tag{8.6}$$

The 10-ary unit of information is called the *hartley* in honor of R. V. L. Hartley,[2] who was one of the pioneers (along with Nyquist[3] and Carson) in the area of information

* Here again we are assuming that n is such that $\log_r n$ is an integer. As we shall see later, this restriction is not necessary.

transmission in the twenties. The rigorous mathematical foundations of information theory, however, were established by C. E. Shannon[1] in 1948:

$$1 \text{ hartley} = \log_2 10 \quad \text{bits}$$
$$= 3.32 \text{ bits}$$

Sometimes the unit *nat* is used:

$$1 \text{ nat} = \log_2 e \quad \text{bits}$$
$$= 1.44 \text{ bits}$$

A Note on the Unit of Information. From the earlier discussion, it follows that a general unit of information is the r-ary unit. Hence, we should use r as the logarithm base everywhere. The binary unit bit ($r = 2$), however, is commonly used in the literature. There is, of course, no loss of generality in using $r = 2$. These units can always be converted into any other units by using Eq. (8.6). Henceforth, we shall always use the binary unit (bit) for information unless otherwise stated. The bases of the logarithm functions will be omitted but will be understood to be 2.

Average Information per Message: Entropy of a Source

Consider a memoriless source m emitting messages $m_1, m_2, \ldots, m_n$ with probabilities $P_1, P_2, \ldots, P_n$, respectively ($P_1 + P_2 + \cdots + P_n = 1$). A *memoriless source* implies that each message emitted is independent of the previous message(s). By the definition in Eq. (8.3) [or Eq. (8.4)], the information content of message m_i is I_i, given by

$$I_i = \log \frac{1}{P_i} \quad \text{bits}$$

The probability of the occurrence of m_i is P_i. Hence, the mean, or average, information per message emitted by the source is given by $\sum_{i=1}^{n} P_i I_i$ bits. The average information per message of a source m is called its *entropy*, denoted by $H(\text{m})$. Hence,

$$H(\text{m}) = \sum_{i=1}^{n} P_i I_i \quad \text{bits}$$

$$= \sum_{i=1}^{n} P_i \log \frac{1}{P_i} \quad \text{bits} \tag{8.7a}$$

$$= -\sum_{i=1}^{n} P_i \log P_i \quad \text{bits} \tag{8.7b}$$

It is evident that the entropy of a source is a function of the message probabilities. It is interesting to find the message probability distribution that yields the maximum entropy. Because the entropy is a measure of uncertainty, the probability distribution that generates the maximum uncertainty will have the maximum entropy. On qualitative grounds, one expects entropy to be maximum when all the messages are equiprobable. We shall now show that this is indeed true.

Because $H(\mathrm{m})$ is a function of $P_1, P_2, \ldots, P_n$, the maximum value of $H(\mathrm{m})$ is found from the equation $dH(\mathrm{m})/dP_i = 0$ for $i = 1, 2, \ldots, n$, with the constraint that

$$P_n = 1 - (P_1 + P_2 + \cdots + P_{n-1}) \tag{8.8}$$

Because

$$H(\mathrm{m}) = -\sum_{i=1}^{n} P_i \log P_i \tag{8.9}$$

It can be seen that in calculating dH/dP_i from Eq. (8.9) we need consider only the terms $-P_i \log P_i$ and $-P_n \log P_n$ [because P_n is a function of P_i, as seen from Eq. (8.8)]. Hence,

$$\begin{aligned}\frac{dH(\mathrm{m})}{dP_i} &= \frac{d}{dP_i}[-P_i \log P_i - P_n \log P_n] \\ &= -P_i\left(\frac{1}{P_i}\right) \log e - \log P_i + P_n\left(\frac{1}{P_n}\right) \log e + \log P_n \\ &= \log \frac{P_n}{P_i}\end{aligned}$$

which is zero if $P_i = P_n$. Because this is true for all i, we have

$$P_1 = P_2 = \cdots = P_n = \frac{1}{n} \tag{8.10}$$

To show that Eq. (8.10) yields $[H(\mathrm{m})]_{\max}$ and not $[H(\mathrm{m})]_{\min}$, we note that when $P_1 = 1$ and $P_2 = P_3 = \cdots = P_n = 0$, $H(\mathrm{m}) = 0$, whereas the probabilities in Eq. (8.10) yield

$$\begin{aligned}H(\mathrm{m}) &= -\sum_{i=1}^{n} \frac{1}{n} \log \frac{1}{n} \\ &= \log n\end{aligned} \tag{8.11}$$

The Intuitive (Common-Sense) and the Engineering Interpretation of Entropy. Earlier, we observed that both the intuitive and the engineering viewpoints lead to the same definition of the information associated with a message. The conceptual bases, however, are entirely different for the two points of view. Consequently, we have two physical interpretations of information. According to the engineering point of view, the information content of any message is equal to the minimum number of digits required to encode the message, and, therefore, the entropy $H(\mathrm{m})$ is equal to the minimum number of digits per message required, on the average, for encoding. From the intuitive standpoint, on the other hand, information is thought of as being synonymous with the amount of surprise, or uncertainty, associated with the event (or message). A smaller probability of occurrence implies more uncertainty about the event. Uncertainty is, of course, associated with surprise. Hence intuitively, the information associated with a message is a measure of the uncertainty (unexpectedness) of the message. Therefore, $\log (1/P_i)$ is a measure of the uncertainty of the message m_i, and $\sum_{i=1}^{n} P_i \log (1/P_i)$ is the average uncertainty (per message) of the source that generates messages $m_1, m_2, \ldots, m_n$ with probabilities $P_1, P_2, \ldots, P_n$. Both these interpretations prove useful in the qualitative understanding of the mathematical definitions and results in information theory. Entropy may also be viewed as a function associated with a random variable m that assumes values $m_1, m_2, \ldots, m_n$ with probabilities $P(m_1), P(m_2), \ldots, P(m_n)$:

$$H(\mathrm{m}) = \sum_{i=1}^{n} P(m_i) \log \frac{1}{P(m_i)}$$

$$= \sum_{i=1}^{n} P_i \log \frac{1}{P_i}$$

Thus, we can associate an entropy with every discrete random variable.

If the source is not memoriless (i.e., a message emitted at any time is not independent of the previous messages emitted), then the source entropy will be less than $H(\mathrm{m})$ in Eq. (8.9). This is because the dependence of a message on previous messages reduces its uncertainty.

8.2 SOURCE ENCODING

The minimum number of binary digits required to encode a message was shown to be equal to the source entropy ($\log 1/P$) if all the messages of the source are equiprobable (each message probability is P). We shall now generalize this result to the case of nonequiprobable messages. We shall show that the average number of binary digits per message required for encoding is given by $H(\mathrm{m})$ [in bits] for an arbitrary probability distribution of messages.

Let a source m emit messages $m_1, m_2, \ldots, m_n$ with probabilities $P_1, P_2, \ldots, P_n$, respectively. Consider a sequence of N messages with $N \to \infty$. Let k_i be the number of times message m_i occurs in this sequence. Then according to the relative frequency interpretation,

$$\lim_{N\to\infty} \frac{k_i}{N} = P_i$$

Thus, the message m_i occurs NP_i times in the whole sequence of N messages (provided $N \to \infty$). Therefore, in a typical sequence of N messages, m_1 will occur NP_1 times, m_2 will occur NP_2 times, . . . , m_n will occur NP_n times. All other compositions are extremely unlikely to occur ($P \to 0$). Thus, any typical sequence (where $N \to \infty$) has the same proportion of the n messages, although in general the order will be different. We shall assume a zero-memory source, that is, the message is emitted from the source independently of the previous messages. Consider now a typical sequence S_N

of N messages from the source. Because the n messages (of probability $P_1, P_2, \ldots, P_n$) occur $NP_1, NP_2, \ldots, NP_n$ times, respectively, and because each message is independent, the probability of occurrence of a typical sequence S_N is given by

$$P(S_N) = (P_1)^{NP_1}(P_2)^{NP_2} \cdots (P_n)^{NP_n} \tag{8.12}$$

Because all possible sequences of N messages from this source have the same composition, all the sequences (of N messages) are equiprobable, with probability $P(S_N)$. We can consider these sequences as new messages (which are now equiprobable). To encode one such sequence we need L_N binary digits, where

$$L_N = \log\left(\frac{1}{P(S_N)}\right) \quad \text{binary digits} \tag{8.13}$$

Substituting Eq. (8.12) in Eq. (8.13), we obtain

$$L_N = N \sum_{i=1}^{n} P_i \log \frac{1}{P_i}$$

$$= NH(\mathrm{m}) \qquad \text{binary digits}$$

Note that L_N is the length (number of binary digits) of the code word required to encode N messages in sequence. Hence, L, the average number of digits required per message, is L_N/N and is given by

$$L = \frac{L_N}{N} = H(\mathrm{m}) \quad \text{binary digits} \tag{8.14}$$

This is the desired result, which states that it is possible to encode the messages emitted by a source using, on the average, $H(\mathrm{m})$ number of binary digits per message, where $H(\mathrm{m})$ is the entropy of the source (in bits). Although it does not prove that, on the average, this is the minimum number of digits required, one can show that $H(\mathrm{m})$ is indeed the minimum. It is not possible to find any uniquely decodable code whose average length is less than $H(\mathrm{m})$.[4,5]

Compact Codes

The source encoding theorem says that to encode a source with entropy $H(\mathrm{m})$, we need, on the average, a minimum of $H(\mathrm{m})$ binary digits per message or $H_r(\mathrm{m})$ r-ary digits per message, where $H_r(\mathrm{m})$ is the entropy in Eq. (8.9) computed with r as the base of the logarithm. The number of digits in the code word is the *length* of the code word. Thus, the average word length of an optimum code is $H(\mathrm{m})$. Unfortunately, to attain this length, in general, we have to encode a sequence of N messages ($N \to \infty$) at a time. If we wish to encode each message directly without using longer sequences, then in general, the average length of the code word per message will be greater than $H(\mathrm{m})$. In practice, it is not desirable to use long sequences, as they cause transmission delay and add to equipment complexity. Hence, it is preferable to encode messages directly, even if the price has to be paid in terms of increased word length. In most cases, the price turns out to be small. The advantages of coding long sequences are only marginal. The following is a procedure, given without proof, for finding the optimum source code (called the Huffman code). The proof that this code is optimum can be found in Refs. 4–6.

We shall illustrate the procedure with an example using the binary code. We first arrange the messages in order of descending probability, as shown in Table 8.1. Here

Table 8.1

Original Source		Reduced Sources			
Messages	Probabilities	S_1	S_2	S_3	S_4
m_1	0.30	0.30	0.30	0.43	0.57
m_2	0.25	0.25	0.27	0.30	0.43
m_3	0.15	0.18	0.25	0.27	
m_4	0.12	0.15	0.18		
m_5	0.10	0.12			
m_6	0.08				

we have six messages with probabilities 0.30, 0.25, 0.15, 0.12, 0.10, and 0.08, respectively. We now combine the last two messages into one message with probability $P_5 + P_6 = 0.18$. This leaves five messages with probabilities, 0.30, 0.25, 0.15, 0.12, and 0.18. These messages are now rearranged in the second column in order of descending probability. We repeat this procedure by combining the last two messages in the second column and rearranging them in order of descending probability. This is done until the number of messages is reduced to 2. These two (reduced) messages are now assigned **0** and **1** as their first digits in the code sequence. We now go back and assign the numbers **0** and **1** to the second digit for the two messages that were combined in the previous step. We keep regressing this way until the first column is reached. The code finally obtained (for the first column) can be shown to be the optimum. The complete procedure is shown in Tables 8.1 and 8.2.

Table 8.2

Original Source			Reduced Sources			
Messages	Probabilities	Code	S_1	S_2	S_3	S_4
m_1	0.30	**00**	0.30 **00**	0.30 **00**	0.43 **1**	0.57 **0**
m_2	0.25	**10**	0.25 **10**	0.27 **01**	0.30 **00**	0.43 **1**
m_3	0.15	**010**	0.18 **11**	0.25 **10**	0.27 **01**	
m_4	0.12	**011**	0.15 **010**	0.18 **11**		
m_5	0.10	**110**	0.12 **011**			
m_6	0.08	**111**				

The optimum (Huffman) code obtained this way is also called a *compact code*. The average length of the compact code in the present case is given by

$$L = \sum_{i=1}^{n} P_i L_i = 0.60 + 0.50 + 0.45 + 0.36 + 0.30 + 0.24$$

$$= 2.45 \text{ binary digits}$$

The entropy $H(\mathrm{m})$ of the source is given by

$$H(\mathrm{m}) = \sum_{i=1}^{n} P_i \log_2 \frac{1}{P_i}$$

$$= 2.418 \text{ bits}$$

Hence, the minimum possible length (attained by an infinitely long sequence of messages) is equal to 2.418 binary digits. Using direct coding (the Huffman code), it is possible to attain an average length of 2.45 bits in the example given. This is a very close approximation of the optimum performance attainable. Thus, little is gained by complex coding using long sequences of messages. It can be shown that the Huffman code is uniquely decodable; that is, a sequence of coded messages can be decoded unambiguously.

The merit of any code is measured by its average length in comparison to $H(\mathrm{m})$ [the average minimum possible length]. We define the *code efficiency* η as

$$\eta = \frac{H(\mathrm{m})}{L}$$

where L is the average length of the code. In our present example,

$$\eta = \frac{2.418}{2.45}$$

$$= 0.976$$

The *redundancy* γ is defined as

$$\gamma = 1 - \eta$$

$$= 0.024$$

A similar procedure is used to find a compact r-ary code. In this case we arrange the messages in descending order of probability, combine the last r messages into one message, and rearrange the new set (reduced set) in order of descending probability. We repeat the procedure until the final set reduces to r messages. Each of these messages is now assigned one of the r numbers $\mathbf{0}, \mathbf{1}, \mathbf{2}, \ldots \boldsymbol{r} - 1$. We now regress in exactly the same way as in the binary case until each of the original messages is assigned a code.

For an r-ary code, we will have exactly r messages left in the last reduced set if, and only if, the total number of original messages is equal to $r + k(r - 1)$, where k is an integer. This is obvious, because each reduction decreases the number of messages by $r - 1$. Hence, if there is a total of k reductions, the total number of original messages must be $r + k(r - 1)$. In case the original messages do not satisfy this condition, we must add some dummy messages with zero probability of occurrence until this condition is fulfilled. As an example, if $r = 4$ and the number of messages n is 6, then we must add one dummy message with zero probability of occurrence to make the total number of messages 7, that is, $[4 + 1(4 - 1)]$, and proceed as usual. The procedure is illustrated in Example 8.1.

■ EXAMPLE 8.1

A zero-memory source emits six messages with probabilities 0.3, 0.25, 0.15, 0.12, 0.1, and 0.08, respectively. Find the 4-ary (quaternary) Huffman code. Determine its average word length, the efficiency, and the redundancy.

The Huffman code is found in Table 8.3. The length L of this code is

$$L = 0.3(1) + 0.25(1) + 0.15(1) + 0.12(2) + 0.1(2) + 0.08(2)$$
$$= 1.3 \quad \text{4-ary units}$$

and

$$H_4(\text{m}) = -\sum_{i=1}^{6} P_i \log_4 P_i$$
$$= 1.209 \quad \text{4-ary units}$$

Table 8.3

Original Source			Reduced Sources	
Messages	Probabilities	Code		
m_1	0.30	**0**	0.30	**0**
m_2	0.25	**2**	→ 0.30	**1**
m_3	0.15	**3**	0.25	**2**
m_4	0.12	**10** ⌉	0.15	**3**
m_5	0.10	**11** ─┘		
m_6	0.08	**12**		
m_7	0.00	**13** ⌋		

The code efficiency η is given by

$$\eta = \frac{1.209}{1.3} = 0.93$$

The redundancy $\gamma = 1 - \eta = 0.07$. ■

To achieve code efficiency $\eta \rightarrow 1$, we need $N \rightarrow \infty$. The Huffman code uses $N = 1$ but its efficiency is, in general, less than 1. A compromise exists between these two extremes of $N = 1$ and $N = \infty$. We can use $N = 2$ or 3. In most cases, the use of $N = 2$ or 3 can yield an efficiency close to 1, as the following example shows.

■ EXAMPLE 8.2

A zero-memory source emits messages m_1 and m_2 with probabilities 0.8 and 0.2, respectively. Find the optimum binary code for this source as well as for its *second- and third-order extensions* (that is, for $N = 2$ and 3). Determine the code efficiencies in each case.

The Huffman code for the source is simply **0** and **1**, giving $L = 1$, and

$$H(\mathrm{m}) = -(0.8 \log 0.8 + 0.2 \log 0.2)$$
$$= 0.72 \text{ bits}$$

Hence,

$$\eta = 0.72$$

For the second-order extension of the source ($N = 2$), there are four possible composite messages, m_1m_1, m_1m_2, m_2m_1, and m_2m_2, with probabilities 0.64, 0.16, 0.16, and 0.04, respectively. The Huffman code is obtained in Table 8.4.

Table 8.4

Original Source			Reduced Sources			
Messages	Probabilities	Code				
m_1m_1	0.64	**0**	0.64	**0**	0.64	**0**
m_1m_2	0.16	**11**	→0.20	**10**	→0.36	**1**
m_2m_1	0.16	**100**	0.16	**11**		
m_2m_2	0.04	**101**				

In this case the average word length L' is

$$L' = 0.64 + 2(0.16) + 3(0.16 + 0.04)$$
$$= 1.56$$

This is the word length for two messages of the original source. Hence L, the word length per message, is

$$L = \frac{L'}{2} = 0.78$$

and

$$\eta = \frac{0.72}{0.78} = 0.923$$

If we proceed with $N = 3$ (the third-order extension of the source), we have in all eight possible messages, and following the Huffman procedure, we find the code given in Table 8.5.

Table 8.5

Messages	Probabilities	Code
$m_1m_1m_1$	0.512	**0**
$m_1m_1m_2$	0.128	**100**
$m_1m_2m_1$	0.128	**101**
$m_2m_1m_1$	0.128	**110**
$m_1m_2m_2$	0.032	**11100**
$m_2m_1m_2$	0.032	**11101**
$m_2m_2m_1$	0.032	**11110**
$m_2m_2m_2$	0.008	**11111**

The word length L'' is

$$L'' = 0.512 + 0.128 \times 9 + 0.032(15) + 0.008 \times 5$$
$$= 2.184$$

and

$$L = \frac{L''}{3} = 0.728$$

and

$$\eta = \frac{0.72}{0.728} = 0.989$$

■

8.3 ERROR-FREE COMMUNICATION OVER A NOISY CHANNEL

As seen in the previous section, messages of a source with entropy $H(\text{m})$ can be encoded by using an average of $H(\text{m})$ digits per message. This encoding has zero redundancy. Hence, if we transmit these coded messages over a noisy channel, some of the information will be received erroneously. There is absolutely no possibility of error-free communication over a noisy channel when messages are encoded with zero redundancy. Use of redundancy, in general, helps combat noise. This can be seen from a simple example of a *single parity-check code,* in which an extra binary digit is added to each code word to ensure that the total number of **1**'s in the resulting code word is always even (or odd). If a single error occurs in the received code word, the parity is violated, and the receiver requests retransmission. This is a rather simple example to demonstrate the utility of redundancy. More complex coding procedures, which can correct up to n digits, will be discussed in Chapter 9.

The addition of an extra digit increases the average word length to $H(\text{m}) + 1$, giving $\eta = H(\text{m})/[H(\text{m}) + 1]$, and the redundancy is $1 - \eta = 1/[H(\text{m}) + 1]$. Thus, the addition of an extra check digit increases redundancy, but it also helps combat noise. Immunity against channel noise can be increased by increasing the redundancy. Shannon has shown that it is possible to achieve error-free communication by adding sufficient redundancy. For example, if we have a binary symmetric channel (BSC) with an error probability P_e, then for error-free communication over this channel, messages from a source with entropy $H(\mathrm{m})$ must be encoded by binary codes with a word length of at least $H(\mathrm{m})/C_s$, where (see Sec. 8.4)

$$C_s = 1 - \left[P_e \log \frac{1}{P_e} + (1 - P_e) \log \frac{1}{1 - P_e}\right] \tag{8.15}$$

The parameter $C_s (C_s < 1)$ is called the *channel capacity*.

The efficiency of these codes is never greater than C_s. If a certain binary channel has $C_s = 0.4$, a code that can achieve error-free communication must have at least 2.5 $H(\mathrm{m})$ binary digits per message, which is two-and-one-half times as many digits as required for coding without redundancy. This means there are 1.5 $H(\mathrm{m})$ redundant digits per message. Thus, on the average, for every 2.5 digits transmitted, one digit is the information digit and 1.5 digits are redundant, or check, digits, giving a redundancy of $1 - C_s = 0.6$.

As seen in Chapters 6 and 7, P_e, the error probability of binary signaling, varies as $e^{-k\lambda}$ where $\lambda = S_i/\mathcal{N} f_o$. Thus, to make $P_e \to 0$, we need $\lambda \to \infty$, that is, either $S_i \to \infty$ or $f_o \to 0$. Because S_i must be finite, $P_e \to 0$ only if $f_o \to 0$. But Shannon's results state that it is really not necessary to let $f_o \to 0$ for error-free communication. All that is required is to hold f_o below C, the channel capacity per second ($C = 2BC_s$). Where is the discrepancy? To answer this question let us investigate carefully the role of redundancy in error-free communication. Although the discussion here is with reference to a binary scheme, it is quite general and can be extended to the M-ary case.

Consider a simple method of reducing P_e by repeating a given digit an odd number of times. For example, we can transmit **0** and **1** as **000** and **111**. The receiver uses the majority rule to make the decision; that is, if at least two out of three digits are **1**, the decision is **1**, and if at least two out of three digits are **0**, the decision is **0**. Thus, even if one out of three digits is in error, the information is received error free. This scheme will fail if two out of three digits are in error. In order to correct two errors, we need five repetitions. In any case, repetitions cause redundancy but improve P_e (see Example 5.8). It will be instructive to understand this situation from a graphic point of view. Consider the case of three repetitions. We can show all eight possible sequences of three binary digits graphically as the vertices of a cube (Fig. 8.1). It is convenient to map binary sequences as shown in Fig. 8.1 and to talk in terms of what is called the *Hamming distance* between binary sequences. If two binary sequences of the same length differ in j places (j digits), then the Hamming distance between the sequences is considered to be j. Thus, the Hamming distance between **000** and **010** (or **001** and **101**) is 1, and that between **000** and **111** is 3. In the case of three repetitions, we transmit binary **1** by **111** and binary **0** by **000**. The Hamming distance between these two sequences is 3. Observe that of the eight possible vertices, we are occupying only two (**000** and **111**) for transmitted messages. At the receiver, however, because of channel noise, we are liable to receive any one of the eight sequences. The majority decision rule can be interpreted as a rule that decides in favor of that message (**000** or **111**) which is at the closest Hamming distance from the received sequence. Se-

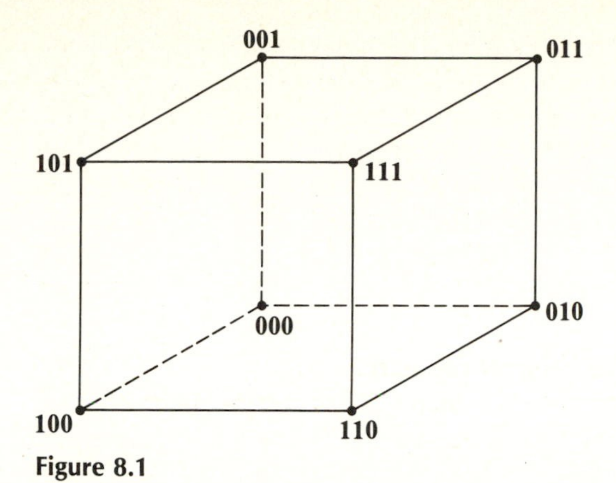

Figure 8.1

quences **000**, **001**, **010**, and **100** are within 1 unit Hamming distance from **000** but are at least 2 units away from **111**. Hence, when we receive any one of the above four sequences, our decision is binary **0**. Similarly, when any one of the sequences **110**, **110**, **011**, or **101** is received, the decision is binary **1**.

We can now see why the error probability is reduced in this scheme. Of the eight possible vertices, we have used only two, which are separated by 3 Hamming units. If we draw a Hamming sphere of unit radius around each of these two vertices (**000** and **111**), the two Hamming spheres* are nonoverlapping. The channel noise can cause a distance between the received sequence and the transmitted sequence, and as long as this distance is equal to or less than 1 unit, we can still detect the message without error. In a similar way, the case of five repetitions can be represented by a hypercube of five dimensions. The transmitted sequences **00000** and **11111** occupy two vertices separated by five units, and the Hamming spheres of 2-unit radius drawn around each of these two vertices would be nonoverlapping. In this case, even if channel noise causes two errors, we can still detect the message correctly. Hence, the reason for reduction in error probability is that we have not used all the available vertices for messages. Had we occupied all the available vertices for messages (as is the case without redundancy, or repetition), then if channel noise causes an error (even one), the received sequence will occupy a vertex assigned to another transmitted sequence, and we are certain to make a wrong decision. Precisely because we have left the neighboring vertices of the transmitted sequence unoccupied are we able to detect the sequence correctly, despite channel errors (within a certain limit). The smaller the fraction of vertices used, the smaller the error probability. It should also be remembered that redundancy (or repetition) is what makes it possible to have unoccupied vertices.

If we continue to increase n, the number of repetitions, we will reduce P_e, but we will also reduce f_o by the factor n. But no matter how large we make n, the error

*Note that the Hamming sphere is not a true geometrical hypersphere because the Hamming distance is not a true geometrical distance (e.g., sequences **001**, **010**, and **100** lie on a Hamming sphere of radius 2 and centered at **111**).

probability never becomes zero. The trouble with this scheme is that we are adding redundant (or check) digits to each information digit. To give an analogy, redundant (or check) digits are like guards protecting the information digit. To hire guards for each information digit is somewhat similar to a case of families living on a certain street hit by several burglaries. Each family panics and hires a guard. This is obviously expensive and inefficient. A better solution would be for all the families on the street to hire one guard and share the expenses. One guard can check on all the houses on the street, assuming a street of reasonable size. If the street is too long, it might be necessary to hire more than one guard. But it is certainly not necessary to hire one guard per house. In using repetitions, we had a similar situation. Redundant (or repeated) digits were used to check on only one transmitted digit. Using the clue from the above analogy, it might be more efficient if we use redundant digits not to check (guard) any one individual transmitted digit but, rather, several transmitted digits. Herein lies the key to our problem. Thus, let us consider a group of information digits over a certain time interval T seconds, and let us add some redundant digits to check on all these digits.

Suppose we need to transmit α binary information digits per second. Then over a period of T seconds, we have a group of αT binary information digits. If to this group of information digits we added $(\beta - \alpha)T$ check digits ($\beta - \alpha$ check digits, or redundant digits, per second), then we need to transmit βT ($\beta > \alpha$) digits for every αT information digits. Therefore over a T second interval, we have

αT information digits

βT ($\beta > \alpha$) total transmitted digits

$(\beta - \alpha)T$ check digits

Thus, instead of transmitting one binary digit every $1/\alpha$ seconds, we let αT digits accumulate over T seconds. Now consider this as a message to be transmitted. There are a total of $2^{\alpha T}$ such supermessages. Thus, every T seconds we need to transmit one of the $2^{\alpha T}$ possible supermessages. These supermessages are transmitted by a sequence of βT binary digits. There are in all $2^{\beta T}$ possible sequences of βT binary digits, and they can be represented as vertices of a βT-dimensional hypercube. Because we have only $2^{\alpha T}$ messages to be transmitted whereas $2^{\beta T}$ vertices are available, we occupy only a $2^{-(\beta-\alpha)T}$ fraction of the vertices of the βT-dimensional hypercube. Observe that we have reduced the transmission rate by a factor of α/β. This rate-reduction factor α/β is independent of T. The fraction of the vertices occupied (occupancy factor) by transmitted messages is $2^{-(\beta-\alpha)T}$ and can be made as small as possible simply by increasing T. In the limit as $T \to \infty$, the occupancy factor $\to 0$. This will make the error probability go to zero, and we have the possibility of error-free communication.

One important question, however, remains to be answered. What must be the rate reduction ratio α/β for this dream to come true? To answer this question, we observe that increasing T increases the length of the transmitted sequence (βT digits). If P_e is the digit error probability, then it can be seen from the relative frequency definition that as $T \to \infty$, the total number of digits in error in a sequence of βT digits ($\beta T \to \infty$) is exactly $\beta T P_e$. Hence, the received sequences will be at a Hamming distance of $\beta T P_e$ from the transmitted sequences. Therefore, for error-free communication, we must leave all the vertices unoccupied within spheres of radius $\beta T P_e$ drawn around each of the $2^{\alpha T}$ occupied vertices. In short, we must be able to pack $2^{\alpha T}$ nonoverlapping spheres, each of radius $\beta T P_e$, into the Hamming space of βT dimensions. This means for a given β, α cannot be increased beyond some limit without causing overlap in the spheres and the consequent failure of the scheme. Shannon's theorem states that for this scheme to work, α/β must be less than the constant C_s (the channel capacity), which is a function of the channel noise and signal power:

$$\frac{\alpha}{\beta} < C_s$$

It must be remembered that such perfect, error-free communication is not practical. In this system we accumulate the information digits for T seconds before encoding them, and because $T \to \infty$, for error-free communication, we must wait until eternity before we start encoding. Hence, there will be an infinite delay at the transmitter and an additional delay of the same amount at the receiver. Secondly, the equipment needed for the storage, encoding, and decoding sequence of infinite digits would be monstrous. Needless to say that in practice the dream of error-free communication cannot be achieved. Then what is the use of Shannon's results? For one thing, they indicate the upper limit on the rate of error-free communication that can be achieved on a channel. This in itself is monumental. Secondly, they indicate the way to reduce error probability with only a small reduction in the rate of transmission of information digits. We can therefore seek a compromise between error-free communication with infinite delay and virtually error-free communication with a finite delay.

8.4 THE CHANNEL CAPACITY OF A DISCRETE MEMORILESS CHANNEL

In this section, discrete memoriless channels will be considered. Let a source emit symbols $x_1, x_2, \ldots, x_r$. The receiver receives symbols $y_1, y_2, \ldots, y_s$. The set of symbols $\{y_k\}$ may or may not be identical to the set $\{x_k\}$, depending upon the nature of the receiver. If we use the types of receivers discussed in Chapter 7, the set of received symbols will be the same as the set transmitted. This is because the optimum receiver, upon receiving a signal, decides which of the r symbols $x_1, x_2, \ldots, x_r$ has been transmitted. Here we shall be more general and shall not constrain the set $\{y_k\}$ to be identical to the set $\{x_k\}$.

If the channel is noiseless, then the reception of some symbol y_j uniquely determines the message transmitted. Because of noise, however, there is a certain amount of uncertainty regarding the transmitted symbol when y_j is received. If $P(x_i|y_j)$ represents the conditional probabilities that x_i was transmitted when y_j is received, then there is an uncertainty of $\log [1/P(x_i|y_j)]$ about x_i when y_j is received. If the channel were noiseless, the uncertainty about x_i when y_j is received would be zero. Obviously, the uncertainty $\log [1/P(x_i|y_j)]$ is caused by channel noise. This, evidently, is the loss of information caused by channel noise. The average loss of information when y_j is received is $H(\mathrm{x}|y_j)$, given by

$$H(\mathrm{x}|y_j) = \sum_i P(x_i|y_j) \log \frac{1}{P(x_i|y_j)} \quad \text{bits/symbol} \tag{8.16}$$

Note that this is the average uncertainty about the transmitted symbol when y_j is received. Next we average the information loss over all the received symbols. This is denoted by $H(\mathrm{x} \mid \mathrm{y})$:

$$H(\mathrm{x} \mid \mathrm{y}) = \sum_j P(y_j) H(\mathrm{x} \mid y_j) \quad \text{bits/symbol} \tag{8.17a}$$

$$= \sum_i \sum_j P(y_j) P(x_i \mid y_j) \log \frac{1}{P(x_i \mid y_j)} \quad \text{bits/symbol} \tag{8.17b}$$

$$= \sum_i \sum_j P(x_i, y_j) \log \frac{1}{P(x_i \mid y_j)} \quad \text{bits/symbol} \tag{8.17c}$$

Thus, $H(\mathrm{x} \mid \mathrm{y})$ represents the average uncertainty about the transmitted symbol averaged over all the received symbols. Evidently, $H(\mathrm{x} \mid \mathrm{y})$ is the average loss of information per received symbol. If the channel is noiseless, then $H(\mathrm{x} \mid \mathrm{y})$ should be zero. This can be easily verified for a binary noiseless channel, where $P(x_i \mid y_j)$ is either **0** or **1**. Therefore, $H(\mathrm{x} \mid y_j)$ as given by Eq. (8.16) must be zero. Hence, $H(\mathrm{x} \mid \mathrm{y})$ in Eq. (8.17a) is also zero. $H(\mathrm{x} \mid \mathrm{y})$ is called the *equivocation* of x with respect to y.

Note that $P(y_j \mid x_i)$ represents the probability that y_j is received when x_i is transmitted. This is a characteristic of the channel and receiver. Thus, a given channel (with its receiver) is specified by the *channel matrix*:

$$\text{Inputs} \quad \begin{array}{c} \\ x_1 \\ x_2 \\ \\ x_r \end{array} \begin{array}{c} \text{Outputs} \\ \begin{array}{cccc} y_1 & y_2 & \cdots & y_s \end{array} \\ \begin{bmatrix} P(y_1 \mid x_1) & P(y_2 \mid x_1) & \cdots & P(y_s \mid x_1) \\ P(y_1 \mid x_2) & P(y_2 \mid x_2) & \cdots & P(y_s \mid x_2) \\ \cdots & \cdots & \cdots & \cdots \\ P(y_1 \mid x_r) & P(y_2 \mid x_r) & \cdots & P(y_s \mid x_r) \end{bmatrix} \end{array}$$

We can obtain the reverse conditional probabilities $P(x_i \mid y_j)$ using Bayes' rule:

$$P(x_i \mid y_j) = \frac{P(y_j \mid x_i) P(x_i)}{P(y_j)} \tag{8.18a}$$

$$= \frac{P(y_j \mid x_i) P(x_i)}{\sum_i P(x_i, y_j)} \tag{8.18b}$$

$$= \frac{P(y_j \mid x_i) P(x_i)}{\sum_i P(x_i) P(y_j \mid x_i)} \tag{8.18c}$$

Thus, if the input symbol probabilities $P(x_i)$ and the channel matrix are known, the reverse conditional probabilities can be computed from Eq. (8.18). The reverse conditional probability $P(x_i \mid y_j)$ is the probability that x_i was transmitted when y_j is received.

If the channel were noise free, the average amount of information received would be $H(\mathrm{x})$ bits (entropy of the source) per received symbol. Note that $H(\mathrm{x})$ is the average information transmitted over the channel per symbol. Because of channel noise, we lose an average of $H(x|y)$ bits of information per symbol. Therefore, in this transaction the amount of information the receiver receives is, on the average, $I(x; y)$ bits per received symbol, where

$$I(x; y) = H(x) - H(x|y) \quad \text{bits/symbol} \tag{8.19}$$

$I(x; y)$ is called the *mutual information* of x and y. We have

$$H(x) = \sum_i P(x_i) \log \frac{1}{P(x_i)} \quad \text{bits}$$

Therefore

$$I(x; y) = \sum_i P(x_i) \log \frac{1}{P(x_i)} - \sum_i \sum_j P(x_i, y_j) \log \frac{1}{P(x_i|y_j)}$$

because

$$\sum_j P(x_i, y_j) = P(x_i)$$

$$I(x; y) = \sum_i \sum_j P(x_i, y_j) \log \frac{1}{P(x_i)} - \sum_i \sum_j P(x_i, y_j) \log \frac{1}{P(x_i|y_j)}$$

$$= \sum_i \sum_j P(x_i, y_j) \log \frac{P(x_i|y_j)}{P(x_i)} \tag{8.20a}$$

$$= \sum_i \sum_j P(x_i, y_j) \log \frac{P(x_i, y_j)}{P(x_i)P(y_j)} \tag{8.20b}$$

Alternatively, using Bayes' rule in Eq. (8.20a), $I(x; y)$ may be expressed as

$$I(x; y) = \sum_i \sum_j P(x_i, y_j) \log \frac{P(y_j|x_i)}{P(y_j)} \tag{8.20c}$$

or we may substitute Eq. (8.18c) in Eq. (8.20a):

$$I(x; y) = \sum_i \sum_j P(x_i)P(y_j|x_i) \log \frac{P(y_j|x_i)}{\sum_i P(x_i)P(y_j|x_i)} \tag{8.20d}$$

Equation (8.20d) expresses $I(x; y)$ in terms of the input symbol probabilities and the channel matrix.

The units of $I(x; y)$ should be carefully noted. $I(x; y)$ is the average amount of information received per symbol transmitted. Hence its units are bits/symbol. If we use binary digits at the input, then the symbol is a binary digit, and the units of $I(x; y)$ are bits per binary digit.

Because $I(x; y)$ in Eq. (8.20b) is symmetrical with respect to x and y, it follows that

$$I(x; y) = I(y; x) \tag{8.21a}$$

$$= H(y) - H(y|x) \tag{8.21b}$$

The quantity $H(\mathrm{y}|\mathrm{x})$ is the equivocation of y with respect to x and is the average uncertainty about the received symbol when the transmitted symbol is known. Equation (8.21a) can be rewritten as

$$H(\mathrm{x}) - H(\mathrm{x}|\mathrm{y}) = H(\mathrm{y}) - H(\mathrm{y}|\mathrm{x}) \tag{8.21c}$$

From Eq. (8.20d) it is clear that $I(\mathrm{x};\mathrm{y})$ is a function of the transmitted symbol probabilities $P(x_i)$ and the channel matrix. For a given channel, $I(\mathrm{x};\mathrm{y})$ will be maximum for some set of probabilities $P(x_i)$. This maximum value is the channel capacity C_s.

$$C_s = \underset{P(x_i)}{\mathrm{Max}}\, I(\mathrm{x};\mathrm{y}) \quad \text{bits/symbol} \tag{8.22}$$

Thus, C_s represents the maximum information that can be transmitted by one symbol over the channel. These ideas will become clear from the following example of a binary symmetric channel (BSC).

■ EXAMPLE 8.3

Find the channel capacity of the BSC shown in Fig. 8.2.

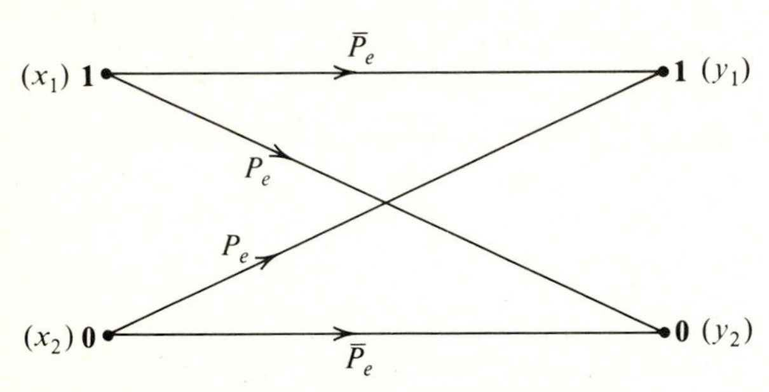

Figure 8.2 A binary symmetric channel (BSC).

☐ **Solution:** Let $P(x_1) = \alpha$ and $P(x_2) = \bar{\alpha} = (1 - \alpha)$. Also,

$$P(y_1|x_2) = P(y_2|x_1) = P_e$$

$$P(y_1|x_1) = P(y_2|x_2) = \overline{P}_e = 1 - P_e$$

Substitution of these probabilities in Eq. (8.20d) gives

$$\begin{aligned} I(\mathrm{x};\mathrm{y}) &= \alpha\overline{P}_e \log\left[\frac{\overline{P}_e}{\alpha\overline{P}_e + \bar{\alpha}P_e}\right] + \alpha P_e \log\left[\frac{P_e}{\alpha P_e + \bar{\alpha}\overline{P}_e}\right] \\ &\qquad + \bar{\alpha}P_e \log\left[\frac{P_e}{\alpha\overline{P}_e + \bar{\alpha}P_e}\right] + \bar{\alpha}\overline{P}_e \log\left[\frac{\overline{P}_e}{\alpha P_e + \bar{\alpha}\overline{P}_e}\right] \\ &= (\alpha P_e + \bar{\alpha}\overline{P}_e) \log\left[\frac{1}{\alpha P_e + \bar{\alpha}\overline{P}_e}\right] + (\alpha\overline{P}_e + \bar{\alpha}P_e) \log\left[\frac{1}{\alpha\overline{P}_e + \bar{\alpha}P_e}\right] \\ &\qquad - \left(P_e \log\frac{1}{P_e} + \overline{P}_e \log\frac{1}{\overline{P}_e}\right) \end{aligned}$$

If we define

$$\Omega(z) = z \log \frac{1}{z} + \bar{z} \log \frac{1}{\bar{z}}$$

with $\bar{z} = 1 - z$, then

$$I(\mathrm{x}; \mathrm{y}) = \Omega(\alpha P_e + \bar{\alpha}\bar{P}_e) - \Omega(P_e) \tag{8.23}$$

The function $\Omega(z)$ vs. z is shown in Fig. 8.3. It can be seen that $\Omega(z)$ is maximum at

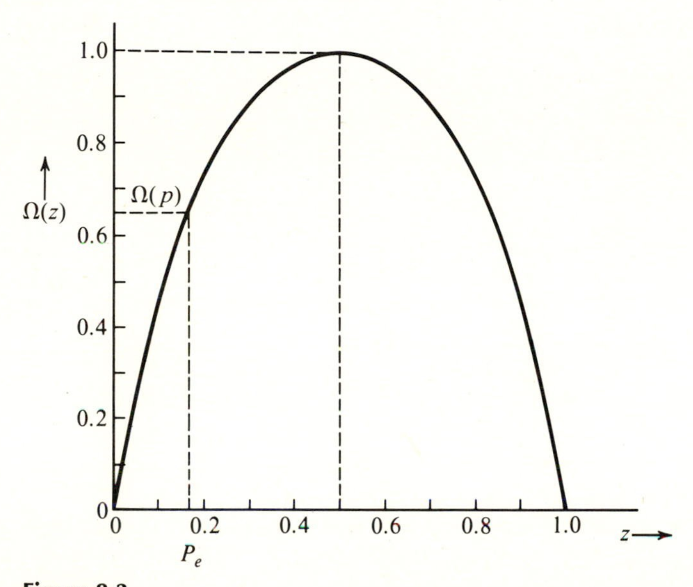

Figure 8.3

$z = \frac{1}{2}$. (Note that we are interested in the region $0 < z < 1$ only.) For a given P_e, $\Omega(P_e)$ is fixed, and $I(\mathrm{x}; \mathrm{y})$ is maximum when $\Omega(\alpha P_e + \bar{\alpha} P_e)$ is maximum. This occurs when

$$\alpha P_e + \bar{\alpha}\bar{P}_e = 0.5$$

or

$$\alpha P_e + (1 - \alpha)(1 - P_e) = 0.5$$

This equation is satisfied when

$$\alpha = 0.5 \tag{8.24}$$

For this value of α, $\Omega(\alpha P_e + \bar{\alpha}\bar{P}_e) = 1$ and

$$C_s = \max_{P(x_i)} I(\mathrm{x}; \mathrm{y}) = 1 - \Omega(P_e)$$

$$= 1 - \left[P_e \log \frac{1}{P_e} + (1 - P_e) \log \left(\frac{1}{1 - P_e}\right)\right] \tag{8.25}$$

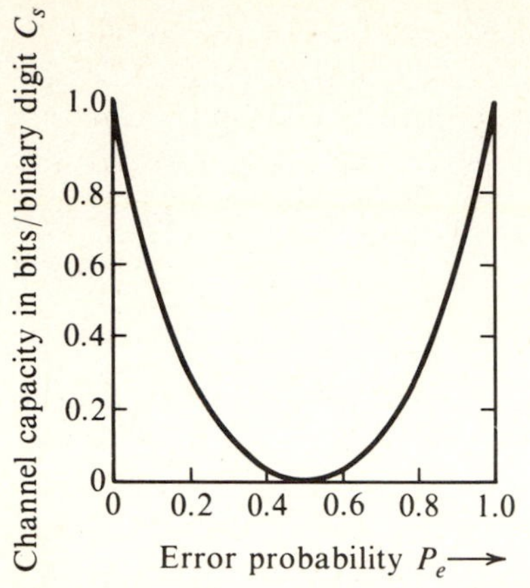

Figure 8.4 Binary symmetric channel capacity as a function of error probability P_e.

C_s vs. P_e is shown in Fig. 8.4. From this figure it follows that the maximum value of C_s is unity. This means we can transmit at most 1 bit of information per binary digit. This is the expected result, because one binary digit can convey one of the two equiprobable messages. The information content of one of the two equiprobable messages is $\log_2 2 = 1$ bit. Secondly, we observe that C_s is maximum when the error probability $P_e = 0$ or $P_e = 1$. When the error probability $P_e = 0$, the channel is noiseless, and we expect C_s to be maximum. But surprisingly, C_s is also maximum when $P_e = 1$. This is easy to explain, because a channel that consistently and with certainty makes errors is as good as a noiseless channel. All we have to do is reverse the decision that is made, and we have error-free reception; that is, if **0** is received, we decide that **1** was actually sent, and vice versa. The channel capacity C_s is zero (minimum) when $P_e = \frac{1}{2}$. If the error probability is $\frac{1}{2}$, then the transmitted symbols and received symbols are statistically independent. If we received **0**, for example, either **1** or **0** is equally likely to have been transmitted, and the information received is zero. ■

Channel Capacity per Second

The channel capacity C_s in Eq. (8.22) gives the maximum possible information transmitted when one symbol (digit) is transmitted. If K symbols are being transmitted per second, then, obviously, the maximum rate of transmission of information per second is KC_s. This is the channel capacity in information units per seconds and will be denoted by C (bits/second):

$$C = KC_s \quad \text{bits/second}$$

A Comment on Channel Capacity. Channel capacity is a property of a particular physical channel over which the information is transmitted. This is true provided the term *channel* is correctly interpreted. A channel means not only the transmission medium but also includes the specifications of the kind of signals (binary, r-ary, etc., or orthogonal, simplex, etc.), and the kind of receiver used (the receiver determines error probability). All these specifications are included in the channel matrix. A channel matrix completely specifies a channel. If we decide to use, for example, 4-ary digits instead of binary digits over the same physical channel, the channel matrix changes (it becomes a 4×4 matrix), as does the channel capacity. Similarly, a change in the receiver or signal power or noise power will change the channel matrix and, hence, the channel capacity.

Magnitude of the Channel Capacity

Because C_s, the channel capacity, is the maximum value of $H(\mathrm{x}) - H(\mathrm{x}|\mathrm{y})$, it is evident that $C_s \leq H(\mathrm{x})$ (because $H(\mathrm{x}|\mathrm{y}) \geq 0$). But $H(\mathrm{x})$ is the average information per input symbol. Hence, C_s is always less than (or equal to) the average information per input symbol. If we use binary symbols at the input, the maximum value of $H(\mathrm{x})$ is one bit, occurring when $P(x_1) = P(x_2) = \frac{1}{2}$. Hence for a binary channel, $C_s \leq 1$ bit per binary digit. If we use r-ary symbols, the maximum value of $H_r(\mathrm{x})$ is 1 r-ary unit. Hence, $C_s \leq 1$ r-ary unit per symbol.

Verification of Error-Free Communication Over a BSC

We have shown that over a noisy channel, C_s bits of information can be transmitted per symbol. If we consider a binary channel, this means that for each binary digit (symbol) transmitted, the received information is C_s bits ($C_s \leq 1$). Thus to transmit one bit of information, we need to transmit at least $1/C_s$ binary digits. This gives a code efficiency C_s and redundancy $1 - C_s$. When the transmission of information is implied, it means error-free transmission, because $I(\mathrm{x};\mathrm{y})$ was defined as the transmitted information minus the loss of information caused by channel noise.

The problem with this derivation is that it is based on a certain speculative definition of information [Eq. (8.1)]. And based on this definition, we defined the information lost during the transmission over the channel. We really have no direct proof that the information lost over the channel will oblige us in this way. Hence, the only way to ensure that this whole speculative structure is sound is to verify it. If we can show that C_s bits of error-free information can be transmitted per symbol over a channel, the verification will be complete. A general case will be discussed later. Here we shall verify the results for a BSC.

Let us consider a binary source emitting messages at a rate of α digits/second. We accumulate these information digits over T seconds to give a total of αT digits. Because αT digits form $2^{\alpha T}$ possible combinations, our problem is now to transmit one of these $2^{\alpha T}$ supermessages every T seconds. These supermessages are transmitted by a code of word length βT digits, with $\beta > \alpha$ to ensure redundancy. Because βT digits can form $2^{\beta T}$ distinct patterns (vertices of a βT dimensional hypercube), and we have only $2^{\alpha T}$ messages, we are utilizing only a $2^{-(\beta-\alpha)T}$ fraction of the vertices. The remaining vertices are deliberately unused in order to combat noise. If we let $T \to \infty$, the fraction of vertices used $\to 0$. Because there are βT digits in each transmitted sequence, the number of digits received in error will be exactly $\beta T P_e$ when $T \to \infty$. We now construct Hamming spheres of radius $\beta T P_e$ each around the $2^{\alpha T}$ vertices used for the messages. When any message is transmitted, the received message will be on the Hamming sphere surrounding the vertex corresponding to that message. We use the following decision rule: If a received sequence falls inside a sphere surrounding message m_i, then the decision is "m_i is transmitted." If $T \to \infty$, the decision will be without error if all the $2^{\alpha T}$ spheres are nonoverlapping.

Of all the possible sequences of βT digits, the number of sequences that differ from a given sequence by exactly j digits is $\binom{\beta T}{j}$ (see Example 5.6). Hence K, the total number of sequences that differ from a given sequence by less than or equal to $\beta T P_e$ digits, is

$$K = \sum_{j=0}^{\beta T P_e} \binom{\beta T}{j} \tag{8.26}$$

Here we use an inequality often used in information theory:[4,7]

$$\sum_{j=0}^{\beta T P_e} \binom{\beta T}{j} \leq 2^{\beta T \Omega(P_e)} \qquad P_e < 0.5$$

Hence,

$$K \leq 2^{\beta T \Omega(P_e)} \tag{8.27}$$

with

$$\Omega(P_e) = P_e \log \frac{1}{P_e} + (1 - P_e) \log \frac{1}{1 - P_e}$$

From the $2^{\beta T}$ possible vertices we choose $2^{\alpha T}$ vertices to be assigned to the supermessages. How shall we select these vertices? From the decision procedure it is clear that if we assign a particular vertex to a supermessage, then none of the other vertices lying within a sphere of radius $\beta T P_e$ can be assigned to another supermessage. Thus, when we choose a vertex for m_1, the corresponding K vertices [Eq. (8.26)] become ineligible for consideration. From the remaining $2^{\beta T} - K$ vertices we choose another vertex for m_2. We proceed this way until all the $2^{\beta T}$ vertices are exhausted. This is a rather tedious procedure. Let us see what happens if we choose the required $2^{\alpha T}$ vertices randomly from the $2^{\beta T}$ vertices. In this procedure there is a danger that we may select more than one vertex lying within a distance $\beta T P_e$. If, however, α/β is sufficiently small, the probability of making such a choice is extremely small as $T \to \infty$. The probability of choosing any particular vertex s_1 as one of the $2^{\alpha T}$ vertices from $2^{\beta T}$ vertices is $2^{\alpha T}/2^{\beta T} = 2^{-(\beta - \alpha)T}$.

Remembering that K vertices lie within a distance of $\beta T P_e$ digits from s_1, the probability that we may also choose another vertex s_2 that is within the distance $\beta T P_e$ from s_1 is

$$P = K 2^{-(\beta - \alpha)T}$$

From Eq. (8.27) it follows that

$$P \leq 2^{-[\beta[1 - \Omega(P_e)] - \alpha]T}$$

Hence, as $T \to \infty$, $P \to 0$ if

$$\beta[1 - \Omega(P_e)] > \alpha$$

that is, if

$$\frac{\alpha}{\beta} < 1 - \Omega(P_e) \tag{8.28a}$$

But $1 - \Omega(P_e)$ is C_s, the channel capacity of a BSC [Eq. (8.25)]. Therefore,

$$\frac{\alpha}{\beta} < C_s \tag{8.28b}$$

Hence, the probability of choosing two sequences randomly within a distance $\beta T P_e$ approaches 0 as $T \to \infty$ provided $\alpha/\beta < C_s$, and we have error-free communication. We can choose $\alpha/\beta = C_s - \epsilon$, where ϵ is arbitrarily small.

8.5 CHANNEL CAPACITY OF A CONTINUOUS CHANNEL*

For a discrete random variable x taking on values $x_1, x_2, \ldots, x_n$ with probabilities $P(x_1), P(x_2), \ldots, P(x_n)$, the entropy $H(\mathrm{x})$ was defined as

$$H(\mathrm{x}) = \sum_{i=1}^{n} P(x_i) \log \frac{1}{P(x_i)} \tag{8.29}$$

For analog data, we have to deal with continuous random variables. It is therefore desirable to extend the definition of entropy to continuous random variables. One is tempted to state that $H(\mathrm{x})$ for continuous random variables is obtained by using the integral instead of discrete summation in Eq. (8.29):†

$$H(\mathrm{x}) = \int_{-\infty}^{\infty} p(x) \log \frac{1}{p(x)}\, dx \tag{8.30}$$

We shall see that Eq. (8.30) is indeed the meaningful definition of entropy for a continuous random variable. We cannot accept this definition, however, unless we show that it has the meaningful interpretation of entropy. To verify this, we shall consider the continuous random variable x as a limiting form of a discrete random variable, which assumes discrete values $0, \pm\Delta x, \pm 2\Delta x, \ldots$, and so on. Let $k\Delta x = x_k$. The continuous random variable x can thus be approximated by a discrete random variable. The random variable x assumes a value in the range $(x_k, x_k + \Delta x)$ with probability $p(x_k)\Delta x$ in the limit as $\Delta x \to 0$. The error in the approximation will vanish in the limit as $\Delta x \to 0$. Hence $H(\mathrm{x})$, the entropy of a continuous random

* The channel is assumed to be memoriless.

† Throughout this discussion, the PDF $p_{\mathrm{x}}(x)$ will be abbreviated as $p(x)$, because it causes no ambiguity and improves the clarity of equations.

variable x, is given by

$$H(\mathrm{x}) = \lim_{\Delta x \to 0} \sum_k p(x_k)\Delta x \log \frac{1}{p(x_k)\Delta x}$$

$$= \lim_{\Delta x \to 0} \left[\sum_k p(x_k)\Delta x \log \frac{1}{p(x_k)} - \sum_k p(x_k)\Delta x \log \Delta x \right]$$

$$= \int_{-\infty}^{\infty} p(x) \log \frac{1}{p(x)}\, dx - \lim_{\Delta x \to 0} \log \Delta x \int_{-\infty}^{\infty} p(x)\, dx$$

$$= \int_{-\infty}^{\infty} p(x) \log \frac{1}{p(x)}\, dx - \lim_{\Delta x \to 0} \log \Delta x \qquad \textbf{(8.31)}$$

In the limit as $\Delta x \to 0$, $\log \Delta x \to -\infty$. It therefore appears that the entropy of a continuous random variable is infinite. This is quite true. The magnitude of uncertainty associated with a continuous random variable is infinite. This fact is also apparent intuitively. A continuous random variable assumes a nonenumerably infinite number of values, and, hence, the uncertainty is of the order of infinity. Does this mean that there is no meaningful definition of entropy for a continuous random variable? On the contrary, we shall see that the first term in Eq. (8.31) serves as a meaningful measure of the entropy (average information) of a continuous random variable x. This may be argued as follows. We can consider $\int p(x) \log [1/p(x)]\, dx$ as a relative entropy with $-\log \Delta x$ serving as a datum, or reference. The information transmitted over a channel is actually the difference between the two terms $H(\mathrm{x})$ and $H(\mathrm{x}|\mathrm{y})$. Obviously, if we have a common datum for both $H(\mathrm{x})$ and $H(\mathrm{x}|\mathrm{y})$, the difference $H(\mathrm{x}) - H(\mathrm{x}|\mathrm{y})$ will be the same as the difference between their relative entropies. We are therefore justified in considering the first term in Eq. (8.31) as the entropy of x. We must, however, always remember that this is a relative entropy and not the absolute entropy. Failure to realize this subtle point generates many apparent fallacies, one of which is given in Example 8.4.

Based on the above argument, we define $H(\mathrm{x})$, the entropy of a continuous random variable x, as

$$H(\mathrm{x}) = \int_{-\infty}^{\infty} p(x) \log \frac{1}{p(x)}\, dx \quad \text{bits} \qquad \textbf{(8.32a)}$$

$$= -\int_{-\infty}^{\infty} p(x) \log p(x)\, dx \quad \text{bits} \qquad \textbf{(8.32b)}$$

EXAMPLE 8.4

A signal amplitude x is a random variable uniformly distributed in the range $(-1, 1)$. This signal is passed through an amplifier of gain 2. The output y is also an r.v., uniformly distributed in the range $(-2, 2)$. Determine entropies $H(\mathrm{x})$ and $H(\mathrm{y})$.

☐ **Solution:**

$$P(\mathrm{x}) = \begin{cases} \frac{1}{2} & |\mathrm{x}| < 1 \\ 0 & \text{otherwise} \end{cases}$$

$$P(\mathrm{y}) = \begin{cases} \frac{1}{4} & |\mathrm{y}| < 2 \\ 0 & \text{otherwise} \end{cases}$$

Hence,

$$H(\mathrm{x}) = \int_{-1}^{1} \tfrac{1}{2} \log 2 \, dx = 1 \text{ bit}$$

$$H(\mathrm{y}) = \int_{-2}^{2} \tfrac{1}{4} \log 4 \, dx = 2 \text{ bits}$$

The entropy of the random variable y is twice that of x. This result may come as a surprise, because a knowledge of x uniquely determines y, and vice versa, because y = 2x. Hence, the average uncertainty of x and y should be identical. Amplification itself can neither add nor subtract information. Why, then, is $H(\mathrm{y})$ twice as large as $H(\mathrm{x})$? This becomes clear when we remember that $H(\mathrm{x})$ and $H(\mathrm{y})$ are relative entropies, and they will be equal if and only if their datum (or reference) entropies are equal. The reference entropy R_1 for x is $-\log \Delta x$, and the reference entropy R_2 for y is $-\log \Delta y$ (in the limit as Δx, $\Delta y \to 0$).

$$R_1 = \lim_{\Delta x \to 0} -\log \Delta x$$

$$R_2 = \lim_{\Delta y \to 0} -\log \Delta y$$

and

$$\begin{aligned} R_1 - R_2 &= \lim_{\Delta x, \Delta y \to 0} \log\left(\frac{\Delta y}{\Delta x}\right) \\ &= \log \frac{dy}{dx} \\ &= \log 2 = 1 \text{ bit} \end{aligned}$$

It is evident that R_1, the reference entropy of x, is higher than the reference entropy R_2 for y. Obviously, if x and y have equal absolute entropies, their relative entropies must differ by 1 bit. ■

Maximum Entropy for a Given Mean Square Value of *x*

For discrete random variables, we observed that the entropy is maximum when all the outcomes (messages) were equally likely (uniform probability distribution). For continuous random variables, there also exists a PDF $p(x)$ that maximizes $H(\mathrm{x})$ in Eq. (8.32). In the case of a continuous distribution, however, we may have additional constraints on x. Either the maximum value of x or the mean square value of x may be given. We shall find here the PDF $p(x)$ that will yield maximum entropy when $\overline{x^2}$ is given to be a constant σ^2. The problem, then, is to maximize $H(x)$:

$$H(x) = \int_{-\infty}^{\infty} p(x) \log \frac{1}{p(x)} \, dx \tag{8.33}$$

with the constraints

$$\int_{-\infty}^{\infty} p(x) \, dx = 1 \tag{8.34a}$$

$$\int_{-\infty}^{\infty} x^2 p(x) \, dx = \sigma^2 \tag{8.34b}$$

To solve this problem, we use a theorem from the calculus of variation. Given the integral I,

$$I = \int_a^b F(x, p) \, dx \tag{8.35}$$

subject to the following constraints:

$$\begin{aligned} \int_a^b \varphi_1(x, p) \, dx &= \lambda_1 \\ \int_a^b \varphi_2(x, p) \, dx &= \lambda_2 \\ &\vdots \\ \int_a^b \varphi_k(x, p) \, dx &= \lambda_k \end{aligned} \tag{8.36}$$

where $\lambda_1, \lambda_2, \ldots, \lambda_k$ are given constants. The result from the calculus of variation states that the form of $p(x)$ that maximizes I in Eq. (8.35) with the constraints in Eq. (8.36) is found from the solution of the equation

$$\frac{\partial F}{\partial p} + \alpha_1 \frac{\partial \varphi_1}{\partial p} + \alpha_2 \frac{\partial \varphi_2}{\partial p} + \cdots + \alpha_k \frac{\partial \varphi_k}{\partial p} = 0 \tag{8.37}$$

The quantities $\alpha_1, \alpha_2, \ldots, \alpha_k$ are adjustable constants (called *undetermined multipliers*), which can be found by substituting the solution of $p(x)$ [obtained from Eq. (8.37)] in Eq. (8.36). In the present case,

$$F(p, x) = p \log \frac{1}{p}$$

$$\varphi_1(x, p) = p$$

$$\varphi_2(x, p) = x^2 p$$

Hence, the solution for p is given by

$$\frac{\partial}{\partial p}\left(p \log \frac{1}{p}\right) + \alpha_1 + \alpha_2 \frac{\partial}{\partial p} x^2 p = 0$$

or

$$-(1 + \log p) + \alpha_1 + \alpha_2 x^2 = 0$$

Solving for p, we have

$$p = e^{(\alpha_1 - 1)} e^{\alpha_2 x^2} \tag{8.38}$$

Substituting Eq. (8.38) in Eq. (8.34a), we have

$$\begin{aligned} 1 &= \int_{-\infty}^{\infty} e^{\alpha_1 - 1} e^{\alpha_2 x^2}\, dx \\ &= 2e^{\alpha_1 - 1} \int_0^{\infty} e^{\alpha_2 x^2}\, dx \\ &= 2e^{\alpha_1 - 1}\left(\frac{1}{2}\sqrt{\frac{\pi}{-\alpha_2}}\right) \end{aligned}$$

provided α_2 is negative, or

$$e^{\alpha_1 - 1} = \sqrt{\frac{-\alpha_2}{\pi}} \tag{8.39}$$

Next we substitute Eqs. (8.38) and (8.39) in Eq. (8.34b):

$$\begin{aligned} \sigma^2 &= \int_{-\infty}^{\infty} x^2 \sqrt{\frac{-\alpha_2}{\pi}} e^{\alpha_2 x^2}\, dx \\ &= 2\sqrt{\frac{-\alpha_2}{\pi}} \int_0^{\infty} x^2 e^{\alpha_2 x^2}\, dx \\ &= -\frac{1}{2\alpha_2} \end{aligned}$$

or

$$\alpha_2 = -\frac{1}{2\sigma^2} \tag{8.40a}$$

and

$$e^{\alpha_1 - 1} = \sqrt{\frac{1}{2\pi\sigma^2}} \tag{8.40b}$$

Substituting Eq. (8.40a and b) in Eq. (8.38), we have

$$p(x) = \frac{1}{\sigma\sqrt{2\pi}} e^{-x^2/2\sigma^2} \tag{8.41}$$

We therefore conclude that for a given mean square value, the maximum entropy (or maximum uncertainty) is obtained when the distribution of x is gaussian. This maximum entropy, or uncertainty, is given by

$$H(\mathrm{x}) = \int_{-\infty}^{\infty} p(x) \log_2 \frac{1}{p(x)}\, dx$$

Note that

$$\log \frac{1}{p(x)} = \log (\sqrt{2\pi\sigma^2}\, e^{x^2/2\sigma^2})$$

$$= \frac{1}{2} \log (2\pi\sigma^2) + \frac{x^2}{2\sigma^2} \log e$$

Hence

$$H(\mathrm{x}) = \int_{-\infty}^{\infty} p(x)\left[\frac{1}{2} \log (2\pi\sigma^2) + \frac{x^2}{2\sigma^2} \log e\right] dx$$

$$= \frac{1}{2} \log (2\pi\sigma^2) \int_{-\infty}^{\infty} p(x)\, dx + \frac{\log e}{2\sigma^2} - \int_{-\infty}^{\infty} x^2 p(x)\, dx$$

$$= \frac{1}{2} \log (2\pi\sigma^2) + \frac{\log e}{2\sigma^2} \sigma^2$$

$$= \frac{1}{2} \log (2\pi e\sigma^2) \tag{8.42a}$$

$$= \frac{1}{2} \log (17.1\sigma^2) \tag{8.42b}$$

To reiterate, for a given mean square value $\overline{\mathrm{x}^2}$, the entropy is maximum for a gaussian distribution, and the corresponding entropy is $\frac{1}{2} \log (2\pi e\sigma^2)$.

The reader can similarly show that if x is constrained to some peak value $M(-M < \mathrm{x} < M)$, then the entropy is maximum when x is uniformly distributed:

$$p(x) = \begin{cases} \frac{1}{2}M & -M < \mathrm{x} < M \\ 0 & \text{otherwise} \end{cases}$$

Entropy of a Bandlimited White Gaussian Noise

Consider a bandlimited white gaussian noise n(t) with PSD $\mathcal{N}/2$. Because

$$R_{\mathrm{n}}(\tau) = \mathcal{N}B \operatorname{sinc} (2B\tau)$$

$$R_{\mathrm{n}}\left(\frac{k}{2B}\right) = 0 \qquad \text{for } k = \pm 1, \pm 2, \pm 3, \ldots$$

Hence,

$$R_{\mathrm{n}}\left(\frac{k}{2B}\right) = \overline{\mathrm{n}(t)\mathrm{n}\left(t + \frac{k}{2B}\right)} = 0 \qquad k = \pm 1, \pm 2, \ldots$$

Because n(t) and n($t + k/2B$) ($k = \pm 1, \pm 2, \ldots$) are Nyquist samples of n(t), it follows that Nyquist samples of n(t) are all uncorrelated. Because n(t) is gaussian, all Nyquist samples are independent. Note that

$$\overline{\mathrm{n}^2} = R_{\mathrm{n}}(0) = \mathcal{N}B$$

Hence, the variance of each Nyquist sample is $\mathcal{N}B$. From Eq. (8.42a) it follows that the entropy $H(\mathrm{n})$ of each Nyquist sample of $\mathrm{n}(t)$ is

$$H(\mathrm{n}) = \tfrac{1}{2}\log(2\pi e\mathcal{N}B) \quad \text{bits/sample} \tag{8.43a}$$

Because $\mathrm{n}(t)$ is completely specified by $2B$ Nyquist samples/second, the entropy/second of $\mathrm{n}(t)$ is the entropy of $2B$ Nyquist samples. Because all the samples are independent, knowledge of one sample gives no information about any other sample. Hence, the entropy of $2B$ Nyquist samples is the sum of the entropies of the $2B$ samples, and

$$H'(\mathrm{n}) = B\log(2\pi e\mathcal{N}B) \quad \text{bits/second} \tag{8.43b}$$

where $H'(\mathrm{n})$ is the entropy/second of $\mathrm{n}(t)$.

From the results derived thus far, we can draw one significant conclusion. Among all signals bandlimited to B Hz and constrained to have a certain mean square value σ^2, the white gaussian bandlimited signal has the largest entropy/second. The reason for this lies in the fact that for a given mean square value, gaussian samples have the largest entropy; moreover, all the $2B$ samples of a gaussian bandlimited process are independent. Hence, the entropy/second is the sum of the entropies of all the $2B$ samples. In processes that are not white, the Nyquist samples are correlated, and, hence, the entropy/second is less than the sum of the entropies of the $2B$ samples. If the signal is not gaussian, then its samples are not gaussian, and, hence, the entropy per sample is also less than the maximum possible entropy for a given mean square value. To reiterate, for a class of bandlimited signals constrained to a certain mean square value, the white gaussian signal has the largest entropy/second, or the largest amount of uncertainty. This is also the reason why white gaussian noise is the worst possible noise in terms of interference with signal transmission.

Mutual Information $I(x; y)$

The ultimate test of any concept is its usefulness. We shall now show that the relative entropy defined in Eq. (8.32) does lead to meaningful results when we consider $I(\mathrm{x}; \mathrm{y})$, the mutual information of continuous random variables x and y. We wish to transmit a random variable x over a channel. Each value of x in a given continuous range is now a message that may be transmitted, for example, as a pulse of height x. The message recovered by the receiver will be a continuous random variable y. If the channel were noise free, the received value y would uniquely determine the transmitted value x. But channel noise introduces a certain uncertainty about the true value of x. Consider the event that at the transmitter, a value of x in the range $(x, x + \Delta x)$ has been transmitted $(\Delta x \to 0)$. The probability of this event is $p(x)\Delta x$ in the limit $\Delta x \to 0$. Hence, the amount of information transmitted is $\log[1/p(x)\Delta x]$. Let the value of y at the receiver be y and $(p(x|y)$ be the conditional probability density of x when $\mathrm{y} = y$. Then $p(x|y)\Delta x$ is the probability that x will lie in the interval $(x, x + \Delta x)$ when $\mathrm{y} = y$ (provided $\Delta x \to 0$). Obviously, there is an uncertainty about the event that x lies in the interval $(x, x + \Delta x)$. This uncertainty, $\log[1/p(x|y)\Delta x]$, arises because of channel noise and therefore represents a loss of information. Because $\log[1/p(x)\Delta x]$ is the information transmitted and $\log[1/p(x|y)\Delta x]$ is the information lost over the channel, the net information received is $I(x; y)$ given by

$$I(x; y) = \log \frac{p(x|y)}{p(x)} \tag{8.44}$$

Note that this relation is true in the limit $\Delta x \to 0$. $I(x; y)$, therefore, represents the information transmitted over a channel when we receive y (y $= y$) when x is transmitted (x $= x$). We are interested in finding the average information transmitted over a channel when some x is transmitted and a certain y is received. We must therefore average $I(x; y)$ over all values of x and y. The average information transmitted will be denoted by $I(\mathrm{x}; \mathrm{y})$, where

$$I(\mathrm{x}; \mathrm{y}) = \int_{-\infty}^{\infty}\int_{-\infty}^{\infty} p(x, y) I(x; y)\, dx\, dy \tag{8.45a}$$

$$= \int_{-\infty}^{\infty}\int_{-\infty}^{\infty} p(x, y) \log \frac{p(x|y)}{p(x)}\, dx\, dy \tag{8.45b}$$

$$= \int_{-\infty}^{\infty}\int_{-\infty}^{\infty} p(x, y) \log \frac{1}{p(x)}\, dx\, dy + \int_{-\infty}^{\infty}\int_{-\infty}^{\infty} p(x,y) \log p(x|y)\, dx\, dy$$

$$= \int_{-\infty}^{\infty}\int_{-\infty}^{\infty} p(x)p(y|x) \log \frac{1}{p(x)}\, dx\, dy + \int_{-\infty}^{\infty}\int_{-\infty}^{\infty} p(x, y) \log p(x|y)\, dx\, dy$$

$$= \int_{-\infty}^{\infty} p(x) \log \frac{1}{p(x)}\, dx \int_{-\infty}^{\infty} p(y|x)\, dy + \int_{-\infty}^{\infty}\int_{-\infty}^{\infty} p(x, y) \log p(x|y)\, dx\, dy$$

Note that

$$\int_{-\infty}^{\infty} p(y|x)\, dy = 1$$

and

$$\int_{-\infty}^{\infty} p(x) \log \frac{1}{p(x)}\, dx = H(\mathrm{x})$$

Hence,

$$I(\mathrm{x}; \mathrm{y}) = H(\mathrm{x}) + \int_{-\infty}^{\infty}\int_{-\infty}^{\infty} p(x, y) \log p(x|y)\, dx\, dy \tag{8.46a}$$

$$= H(\mathrm{x}) - \int_{-\infty}^{\infty}\int_{-\infty}^{\infty} p(x, y) \log \frac{1}{p(x|y)}\, dx\, dy \tag{8.46b}$$

The integral on the right-hand side is the average over x and y of $\log [1/p(x|y)]$. But

$\log [1/p(x|y)]$ represents the uncertainty about x when y is received. This, as we have seen, is the information lost over the channel. The average of $\log [1/p(x|y)]$ is the average loss of information when some x is transmitted and some y is received. This, by definition, is $H(\mathrm{x}|\mathrm{y})$, the equivocation of x with respect to y.

$$H(\mathrm{x}|\mathrm{y}) = \int_{-\infty}^{\infty} \int_{-\infty}^{\infty} p(x, y) \log \frac{1}{p(x|y)} dx\, dy \tag{8.47}$$

Hence

$$I(\mathrm{x}; \mathrm{y}) = H(\mathrm{x}) - H(\mathrm{x}|\mathrm{y}) \tag{8.48}$$

Thus, when some value of x is transmitted and some value of y is received, the average information transmitted over the channel is $I(\mathrm{x}; \mathrm{y})$, given by Eq. (8.47). We can define the channel capacity C_s as the maximum amount of information that can be transmitted, on the average, per sample or per value transmitted:

$$C_s = \operatorname{Max} I(\mathrm{x}; \mathrm{y}) \tag{8.49}$$

For a given channel, $I(\mathrm{x}; \mathrm{y})$ is a function of the input probability density $p(x)$ alone. This can be shown as follows:

$$p(x, y) = p(x)p(y|x) \tag{8.50}$$

$$\begin{aligned} \frac{p(x|y)}{p(x)} &= \frac{p(y|x)}{p(y)} \\ &= \frac{p(y|x)}{\int_{-\infty}^{\infty} p(x, y)\, dx} \\ &= \frac{p(y|x)}{\int_{-\infty}^{\infty} p(x)p(y|x)\, dx} \end{aligned} \tag{8.51}$$

Substituting Eqs. (8.50) and (8.51) in Eq. (8.45b), we obtain

$$I(\mathrm{x}; \mathrm{y}) = \int_{-\infty}^{\infty} \int_{-\infty}^{\infty} p(x)p(y|x) \log \left(\frac{p(y|x)}{\int_{-\infty}^{\infty} p(x)p(y|x)\, dx} \right) dx\, dy \tag{8.52}$$

The conditional probability density $p(y|x)$ is characteristic of a given channel. Hence for a given channel, $I(\mathrm{x}; \mathrm{y})$ is a function of the input probability density $p(x)$ alone. Thus,

$$C_s = \operatorname*{Max}_{p(x)} I(\mathrm{x}; \mathrm{y})$$

If the channel allows transmission of K values per second, then C, the channel capacity per second, is given by

$$C = KC_s \text{ bits/second} \tag{8.53}$$

It should be noted that just as in the case of discrete variables, $I(\mathrm{x}; \mathrm{y})$ is symmetrical with respect to x and y for continuous random variables. This can be seen by rewriting Eq. (8.45b) as

$$I(\mathrm{x}; \mathrm{y}) = \int_{-\infty}^{\infty} \int_{-\infty}^{\infty} p(x, y) \log \frac{p(x, y)}{p(x)p(y)} \, dx \, dy \tag{8.54}$$

It is evident from this equation that $I(\mathrm{x}; \mathrm{y})$ is symmetrical with respect to x and y. Hence,

$$I(\mathrm{x}; \mathrm{y}) = I(\mathrm{y}; \mathrm{x})$$

From Eq. (8.48) it now follows that

$$I(\mathrm{x}; \mathrm{y}) = H(\mathrm{x}) - H(\mathrm{x}|\mathrm{y}) = H(\mathrm{y}) - H(\mathrm{y}|\mathrm{x}) \tag{8.55}$$

The Capacity of a Bandlimited AWGN Channel

The channel capacity C is, by definition, the maximum rate of information transmission over the channel. The mutual information $I(\mathrm{x}; \mathrm{y})$ is given by Eq. (8.55):

$$I(\mathrm{x}; \mathrm{y}) = H(\mathrm{y}) - H(\mathrm{y}|\mathrm{x}) \tag{8.56}$$

The channel capacity C is the maximum value of the mutual information $I(\mathrm{x}; \mathrm{y})$ per second. Let us first find the maximum value of $I(\mathrm{x}; \mathrm{y})$ per sample. We shall find here the capacity of a channel bandlimited to B Hz and disturbed by a white gaussian noise of PSD $\mathcal{N}/2$. In addition, we shall constrain the signal power (or its mean square value) to S. The disturbance is assumed to be additive—that is, the received signal $\mathrm{y}(t)$ is given by

$$\mathrm{y}(t) = \mathrm{x}(t) + \mathrm{n}(t) \tag{8.57}$$

Because the channel is bandlimited, both the signal $\mathrm{x}(t)$ and the noise $\mathrm{n}(t)$ are bandlimited to B Hz. Obviously, $\mathrm{y}(t)$ is also bandlimited to B Hz. All these signals can therefore be completely specified by samples taken at the uniform rate of $2B$ samples per second. Let us find the maximum information that can be transmitted per sample. Let x, n, and y represent samples of $\mathrm{x}(t)$, $\mathrm{n}(t)$, and $\mathrm{y}(t)$, respectively. The information $I(\mathrm{x}; \mathrm{y})$ transmitted per sample is given by Eq. (8.56):

$$I(\mathrm{x}; \mathrm{y}) = H(\mathrm{y}) - H(\mathrm{y}|\mathrm{x})$$

We shall now find $H(\mathrm{y}|\mathrm{x})$. By definition [Eq. (8.47)],

$$\begin{aligned} H(\mathrm{y}|\mathrm{x}) &= \int_{-\infty}^{\infty} \int_{-\infty}^{\infty} p(x, y) \log \frac{1}{p(y|x)} \, dx \, dy \\ &= \int_{-\infty}^{\infty} p(x) \, dx \int_{-\infty}^{\infty} p(y|x) \log \frac{1}{p(y|x)} \, dy \end{aligned}$$

Because

$$y = x + n$$

for a given x, y is equal to n plus a constant (x). Hence, the distribution of y when x has a given value is identical to that of n except for a translation of x. If $p_{\mathrm{n}}(\cdot)$ represents the PDF of noise sample n, then

$$p(y|x) = p_n(y - x) \tag{8.58}$$

$$\int_{-\infty}^{\infty} p(y|x) \log \frac{1}{p(y|x)}\, dy = \int_{-\infty}^{\infty} p_n(y - x) \log \frac{1}{p_n(y - x)}\, dy$$

Letting $y - x = z$, we have

$$\int_{-\infty}^{\infty} p(y|x) \log \frac{1}{p(y|x)}\, dy = \int_{-\infty}^{\infty} p_n(z) \log \frac{1}{p_n(z)}\, dz$$

The right-hand side is the entropy $H(\mathrm{n})$ of the noise sample n. Hence,

$$\begin{aligned} H(\mathrm{y}|\mathrm{x}) &= H(\mathrm{n}) \int_{-\infty}^{\infty} p(x)\, dx \\ &= H(\mathrm{n}) \end{aligned} \tag{8.59}$$

In deriving Eq. (8.59), we made no assumptions about the noise. Hence, Eq. (8.59) is very general and applies to all types of noise. The only condition is that the noise disturbs the channel in an additive fashion. Thus,

$$I(\mathrm{x}; \mathrm{y}) = H(\mathrm{y}) - H(\mathrm{n}) \quad \text{bits/sample} \tag{8.60}$$

We have assumed that the mean square value of the signal $\mathrm{x}(t)$ is constrained to have a value S, and the mean square value of the noise is N. We shall also assume that the signal $\mathrm{x}(t)$ and noise $\mathrm{n}(t)$ are independent. In such a case, the mean square value of y will be the sum of the mean square values of x and n. Hence,

$$\overline{\mathrm{y}^2} = S + N$$

For a given noise [given $H(\mathrm{n})$], $I(\mathrm{x}; \mathrm{y})$ is maximum when $H(\mathrm{y})$ is maximum. We have seen that for a given mean square value of y ($\overline{\mathrm{y}^2} = S + N$), $H(\mathrm{y})$ will be maximum if y is gaussian, and the maximum entropy $H_{\max}(\mathrm{y})$ is then given by

$$H_{\max}(\mathrm{y}) = \tfrac{1}{2} \log [2\pi e(S + N)] \tag{8.61}$$

Because

$$\mathrm{y} = \mathrm{x} + \mathrm{n}$$

and n is gaussian, y will be gaussian only if x is gaussian. As the mean square value of x is S, this implies that

$$p(x) = \frac{1}{\sqrt{2\pi S}} e^{-x^2/2S}$$

and

$$\begin{aligned} I_{\max}(\mathrm{x}; \mathrm{y}) &= H_{\max}(\mathrm{y}) - H(\mathrm{n}) \\ &= \tfrac{1}{2} \log [2\pi e(S + N)] - H(\mathrm{n}) \end{aligned}$$

For a white gaussian noise with mean square value N,

$$H(\text{n}) = \tfrac{1}{2} \log 2\pi e N \qquad (N = \mathcal{N}B)$$

and

$$C_s = I_{\max}(\text{x}; \text{y}) = \frac{1}{2} \log \left(\frac{S + N}{N}\right) \tag{8.62a}$$

$$= \frac{1}{2} \log \left(1 + \frac{S}{N}\right) \tag{8.62b}$$

The channel capacity per second will be the maximum information that can be transmitted per second. Equation (8.62) represents the maximum information transmitted per sample. If all the samples are statistically independent, the total information transmitted per second will be $2B$ times C_s. If the samples are not independent, then the total information will be less than $2BC_s$. Because the channel capacity C represents the maximum possible information transmitted per second,

$$C = 2B\left[\frac{1}{2} \log \left(1 + \frac{S}{N}\right)\right]$$

$$= B \log \left(1 + \frac{S}{N}\right) \quad \text{bits/second} \tag{8.63}$$

The samples of a bandlimited gaussian signal are independent if and only if the signal PSD is uniform over the band. Obviously, to transmit information at the maximum rate [Eq. (8.63)], the PSD of signal $\text{y}(t)$ must be uniform. The PSD of y is given by

$$S_{\text{y}}(\omega) = S_{\text{x}}(\omega) + S_{\text{n}}(\omega)$$

Because $S_{\text{n}}(\omega) = \mathcal{N}/2$, the PSD of $\text{x}(t)$ must also be uniform. Thus, the maximum rate of transmission (C bits/second) is attained when $\text{x}(t)$ is also a white gaussian signal.

To recapitulate, when the channel noise is additive and is white gaussian with mean square value $N (N = \mathcal{N}B)$, the channel capacity C of a bandlimited channel under the constraint of a given signal power S is given by

$$C = B \log \left(1 + \frac{S}{N}\right) \quad \text{bits/second}$$

where B is the channel bandwidth in Hertz. The maximum rate of transmission (C bits/second) can be realized only if the input signal is a white gaussian signal.

The Capacity of a Channel of Infinite Bandwidth. Superficially, Eq. (8.63) seems to indicate that the channel capacity goes to ∞ as the channel's bandwidth B goes to ∞. This, however, is not true. For white noise, the noise power $N = \mathcal{N}B$. Hence as

B increases, N also increases. It can be shown that in the limit as $B \to \infty$, C approaches a limit:

$$C = B \log \left(1 + \frac{S}{N}\right)$$

$$= B \log \left(1 + \frac{S}{\mathcal{N}B}\right)$$

$$\lim_{B\to\infty} C = \lim_{B\to\infty} B \log \left(1 + \frac{S}{\mathcal{N}B}\right)$$

$$= \lim_{B\to\infty} \frac{S}{\mathcal{N}} \left[\frac{\mathcal{N}B}{S} \log \left(1 + \frac{S}{\mathcal{N}B}\right)\right]$$

This limit can be found by noting that

$$\lim_{x\to\infty} x \log_2 \left(1 + \frac{1}{x}\right) = \log_2 e = 1.44$$

Hence,

$$\lim_{B\to\infty} C = 1.44 \frac{S}{\mathcal{N}} \quad \text{bits/second} \tag{8.64}$$

Thus, for a white gaussian channel noise, the channel capacity C approaches a limit of $1.44\ S/\mathcal{N}$ as $B \to \infty$. The variation of C with B is shown in Fig. 8.5. It is evident

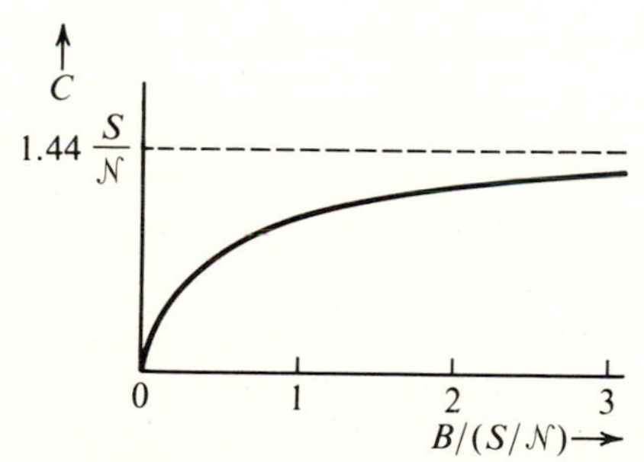

Figure 8.5 Channel capacity vs. bandwidth for a channel with white gaussian noise and fixed power.

that the capacity can be made infinite only by increasing the signal power S to infinity. For finite signal and noise powers, the channel capacity always remains finite.

Verification of Error-Free Communication Over a Continuous Channel

Using the concepts of information theory, we have shown that it is possible to transmit information error free at a rate of $2B \log (1 + S/N)$ bits/second over a channel bandlimited to B Hz. The signal power is S, and the channel noise is white gaussian with power N. This theorem can be verified in a way similar to that used for the verification of the channel capacity of a discrete channel. This verification using signal space is so general that it is in reality an alternate proof of the capacity theorem.

Let us consider M-ary communication with M equiprobable messages m_1, $m_2, \ldots, m_M$ transmitted by signals $s_1(t), s_2(t), \ldots, s_M(t)$. All signals are time limited with duration T and have an essential bandwidth B. Their powers are less than or equal to S. The channel is bandlimited to B, and the channel noise is white gaussian with power N.

All the signals and noise waveforms have $2BT + 1$ dimensions. In the limit we shall let $T \to \infty$. Hence $2BT \gg 1$, and the number of dimensions will be taken as $2BT$ in our future discussion. Because the noise power is N, the energy of the noise waveform of T-seconds duration is NT. The maximum signal energy is ST. Because the signals and noise are independent, the maximum received energy is $(S + N)T$. Hence, all the received signals will lie in a $2BT$-dimensional hypersphere of radius $\sqrt{(S + N)T}$ (Fig. 8.6a). A typical received signal $s_i(t) + n(t)$ has an energy

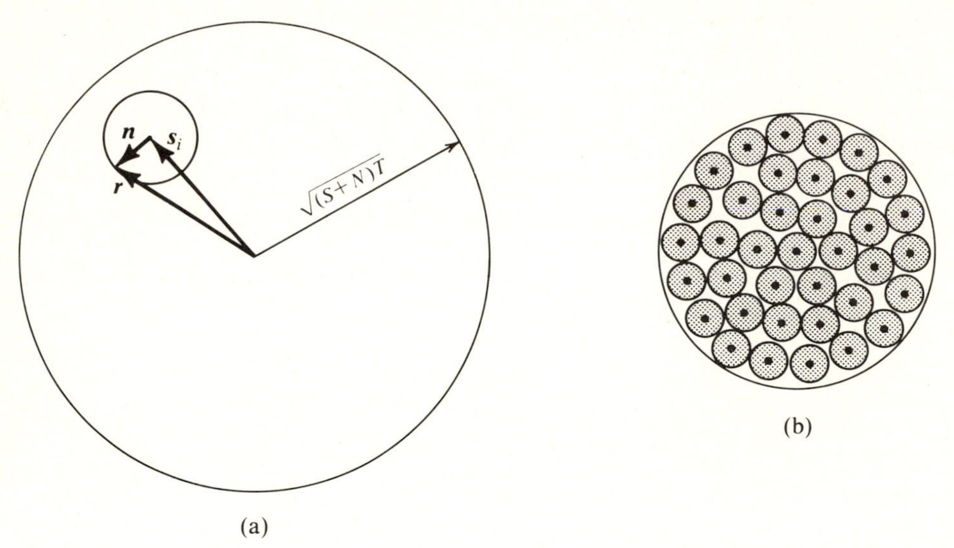

Figure 8.6 (a) Signal space representation of transmitted and received signal and noise signal. (b) Choice of signals for error-free communication.

$(S_i + N)T$, and the point $\boldsymbol{r}$ representing this signal lies at a distance of $\sqrt{(S_i + N)T}$ from the origin (Fig. 6.8a). The signal vector $\boldsymbol{s}_i$, the noise vector $\boldsymbol{n}$, and the received vector $\boldsymbol{r}$ are shown in Fig. 6.8a. Because

$$|\boldsymbol{s}_i| = \sqrt{S_i T}, \; |\boldsymbol{n}| = \sqrt{NT}, \text{ and } |\boldsymbol{r}| = \sqrt{(S_i + N)T},$$

it follows that vectors $\boldsymbol{s}_i$, $\boldsymbol{n}$, and $\boldsymbol{r}$ form a right triangle. Also, $\mathbf{n}$ lies on the sphere of radius $\sqrt{NT}$, centered at $\boldsymbol{s}_i$. Note that because $\mathbf{n}$ is random, it can lie anywhere on the sphere centered at $\boldsymbol{s}_i$.*

*Because N is the average noise power, the energy over an interval T is $NT + \epsilon$, where $\epsilon \to 0$ as $T \to \infty$. Hence, we can assume that $\mathbf{n}$ lies on the sphere.

We have M possible transmitted vectors located inside the big sphere. For each possible $\boldsymbol{s}$, we can draw a sphere of radius $\sqrt{NT}$ around $\boldsymbol{s}$. If a received vector $\boldsymbol{r}$ lies on one of the small spheres, the center of that sphere is the transmitted waveform. If we pack the big sphere with M nonoverlapping and nontouching spheres, each of radius $\sqrt{NT}$ (Fig. 8.6*b*), and use the centers of these M spheres for the transmitted waveforms, we will be able to detect all these M waveforms correctly at the receiver simply by using the maximum-likelihood (ML) receiver. The ML receiver looks at the received signal point $\boldsymbol{r}$ and decides that the transmitted signal is that one of the M possible transmitted points which is closest to $\boldsymbol{r}$ (smallest error vector). Every received point $\boldsymbol{r}$ will lie on the surface of one of the M nonoverlapping spheres, and using the ML criterion, the transmitted signal will be correctly chosen as the point lying at the center of the sphere on which $\boldsymbol{r}$ lies.

To compute the number of small spheres that can be packed into the big sphere, we must determine the volume of a sphere of D dimensions.

The Volume of a D-Dimensional Sphere. A D-dimensional sphere is described by the equation

$$x_1^2 + x_2^2 + \cdots + x_D^2 = R^2$$

where R is the radius of the sphere. We can show that the volume $V(R)$ of a sphere of radius R is given by

$$V(R) = R^D V(1) \tag{8.65}$$

where $V(1)$ is the volume of D-dimensional sphere of unit radius and thus is constant. To prove this, we have by definition

$$V(R) = \underset{x_1^2+x_2^2+\cdots+x_D^2 \le R^2}{\iint \cdots \int} dx_1\, dx_2 \cdots dx_D$$

Letting $y_j = x_j/R$, we have

$$\begin{aligned} V(R) &= R^D \underset{y_1^2+y_2^2+\cdots+y_D^2 \le 1}{\iint \cdots \int} dy_1\, dy_2 \cdots dy_n \\ &= R^D V(1) \end{aligned}$$

A direct consequence of Eq. (8.65) is that when D is large, almost all of the volume of the sphere is concentrated at the surface. To show this, consider the volume of a D-dimensional sphere of radius $R - \Delta$ (Fig. 8.7):

$$\begin{aligned} V(R - \Delta) &= (R - \Delta)^D V(1) \\ &= R^D V(1)\left(1 - \frac{\Delta}{R}\right)^D \end{aligned}$$

and

$$\frac{V(R - \Delta)}{V(R)} = \left(1 - \frac{\Delta}{R}\right)^D$$

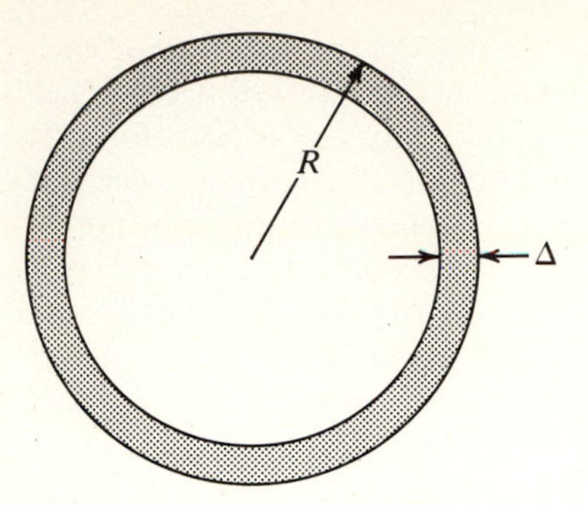

Figure 8.7 Volume of a shell of a D-dimensional hypersphere.

The left-hand side represents the ratio of the volume within radius $R - \Delta$ to the volume within radius R. Because $(1 - \Delta/R)^D \to 0$ as $D \to \infty$, this ratio approaches zero even if Δ/R is very small. This means no matter how small Δ is, the volume within radius $R - \Delta$ is a negligible fraction of the total volume within radius R if D is large enough. Hence, for a large D, almost all of the volume of a D-dimensional sphere is concentrated at the surface. This result sounds strange, but a little reflection will show that it is reasonable. This is because the volume is proportional to Dth power of the radius. Thus for a large D, a small increase in R can increase the volume tremendously, and all the increase comes from a tiny increase in R at the surface of the sphere. This means most of the volume must be concentrated at the surface.

The number of nonoverlapping spheres of radius $\sqrt{NT}$ that can be packed into a sphere of radius $\sqrt{(S + N)T}$ is bounded by the ratio of the volume of the signal sphere to the volume of the noise sphere. Hence,

$$M \leq \frac{(\sqrt{(S + N)T})^{2BT}\,V(1)}{(\sqrt{NT}\,)^{2BT}\,V(1)} = \left(1 + \frac{S}{N}\right)^{BT} \tag{8.66}$$

Each of the M-ary signals carries the information of $\log_2 M$ binary digits. Hence, the transmission of one of the M signals every T seconds is equivalent to the information rate C given by

$$C = \frac{\log M}{T} \leq B \log\left(1 + \frac{S}{N}\right) \quad \text{bits/second} \tag{8.67}$$

This equation gives the upper limit of C. To show that we can actually receive error-free information at a rate of $B \log(1 + S/N)$, we use the argument proposed by Shannon.[8] Instead of choosing the M transmitted messages at the centers of nonoverlapping spheres (Fig. 8.6b), Shannon proposed to select the M points randomly located in the signal sphere I_s of radius $\sqrt{ST}$ (Fig. 8.8). Consider one particular transmitted signal s_k. Because the signal energy is assumed to be $\leq S$, point s_k will lie somewhere inside the signal sphere I_s of radius $\sqrt{ST}$. Because all the M signals are randomly picked from this sphere I_s, however, the probability of finding a signal within a volume ΔV is $M\,\Delta V/V_s$, where V_s is the volume of I_s. But because for large D all of the volume of the sphere is concentrated at the surface, all M signal points

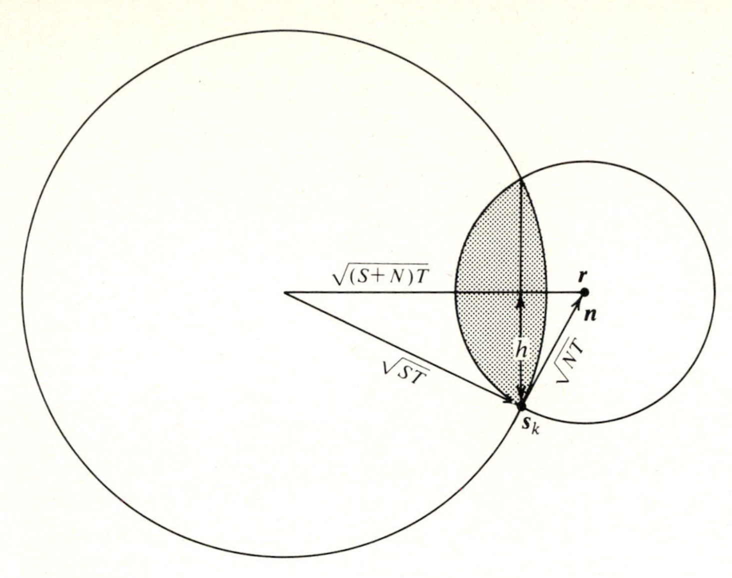

Figure 8.8

selected randomly would lie near the surface of I_s. Figure 8.8 shows the transmitted signal $\boldsymbol{s}_k$, the received signal $\boldsymbol{r}$, and the noise $\boldsymbol{n}$. We draw a sphere of radius $\sqrt{NT}$ with $\boldsymbol{r}$ as the center. This sphere intersects the sphere I_s and forms a common lens-shaped region. The signal $\boldsymbol{s}_k$ lies on the surface of both spheres. We shall use a maximum-likelihood receiver. This means that when $\boldsymbol{r}$ is received, we shall make the decision that "$\boldsymbol{s}_k$ was transmitted" provided none of the remaining $M - 1$ signal points are closer to $\boldsymbol{r}$ than $\boldsymbol{s}_k$. The probability of finding any one signal in the lens is V_{lens}/V_s. Hence P_e, the error probability in the detection of $\boldsymbol{s}_k$ when $\boldsymbol{r}$ is received, is

$$P_e = (M - 1)\frac{V_{\text{lens}}}{V_s}$$

$$< M\frac{V_{\text{lens}}}{V_s}$$

From Fig. 8.8, we observe that $V_{\text{lens}} < V(h)$, where $V(h)$ is the volume of the D-dimensional sphere of radius h. Because $\boldsymbol{r}$, $\boldsymbol{s}_k$, and $\boldsymbol{n}$ form a right triangle,

$$h\sqrt{(S + N)T} = \sqrt{(ST)(NT)} \qquad \text{and} \qquad h = \sqrt{\frac{SNT}{S + N}}$$

Hence,

$$V(h) = \left(\frac{SNT}{S + N}\right)^{BT} V(1)$$

Also,

$$V_s = (ST)^{BT} V(1)$$

and

$$P_e < M\left(\frac{N}{S+N}\right)^{BT}$$

If we choose

$$M = \left[k\left(1 + \frac{S}{N}\right)\right]^{BT}$$

Then

$$P_e < [k]^{BT}$$

If we let $k = 1 - \Delta$ where Δ is a positive number chosen as small as we wish, then

$$P_e \to 0 \quad \text{as} \quad BT \to \infty$$

This means P_e can be made arbitrarily small by increasing T, provided M is chosen arbitrarily close to $(1 + S/N)^{BT}$. This means

$$C = \frac{1}{T} \log_2 M$$

$$= \left[B \log\left(1 + \frac{S}{N}\right) - \epsilon\right] \quad \text{bits/second} \tag{8.68}$$

where ϵ is a positive number chosen as small as we please. This proves the desired result. A more rigorous derivation of this result can be found in Wozencraft and Jacobs.[9]

Because the M signals are selected randomly from the signal space, they tend to acquire the statistics of white noise[9] (i.e., a white gaussian random process).

Comments on Channel Capacity. According to the derived result, theoretically we can communicate error free up to C bits per second. There are, however, practical difficulties in achieving this rate. In proving the capacity formula, we assumed that communication is effected by signals of interval T. This means we must wait T seconds to accumulate the input data and then encode it by one of the waveforms of duration T. Because the capacity rate is achieved only in the limit as $T \to \infty$, we must wait $2T$ seconds (twice eternity) at the receiver to get the information. Secondly, because the number of possible messages that can be transmitted over interval T increases exponentially, the transmitter, as well as the receiver, increases in complexity beyond imagination as $T \to \infty$.

The channel capacity indicated by Shannon's equation [Eq. (8.68)] is the maximum error-free communication rate achievable on an optimum system without any restrictions (except for bandwidth B, signal power S, and gaussian white channel-noise power N). If we have any other restrictions, this maximum rate will not be achieved. For example, if we consider a binary channel (a channel restricted to transmit only binary signals), we will not be able to attain Shannon's rate, even if the channel is optimum. The channel-capacity formula [Eq. (8.68)] indicates that the transmission rate is a monotonically increasing function of the signal power S. If we use a binary channel, however, we know that increasing the transmitted power beyond a certain point buys very little advantage (see Fig. 6.14). Hence on a binary channel, increasing S will not increase the error-free communication rate beyond some value. This does not mean that the channel-capacity equation has failed. It simply means when we have a large amount of power (with a finite bandwidth) available, the binary scheme is not the optimum communication scheme.

We have shown that if we insist on using the binary scheme [Eq. (8.15)],

$$C_s = 1 - \left[P_e \log \frac{1}{P_e} + (1 - P_e) \log \frac{1}{1 - P_e}\right]$$

where P_e is the error probability of detecting a binary digit at the receiver. As the signal power is increased, $P_e \to 0$ and $C_s \to 1$. Thus, over a binary channel, the maximum value of C_s is 1 when $S \to \infty$, and

$$C = 2BC_s = 2B \quad \text{bits/second}$$

One last comment: Shannon's results tell us the upper theoretical limit of error-free communication. But they do not tell us precisely how this can be achieved. To quote the words of Abramson, "[This is one of the problems] which has persisted to mock information theorists since Shannon's original paper in 1948. Despite an enormous amount of effort spent since that time in quest of this Holy Grail of information theory, a *deterministic* method of generating the codes promised by Shannon is still to be found."[4]

8.6 PRACTICAL COMMUNICATION SYSTEMS IN LIGHT OF SHANNON'S EQUATION

It would be instructive to determine the ideal law for the exchange between the SNR and the transmission bandwidth using the channel-capacity equation. Consider a message of bandwidth B that is used for modulation (or coding), with the resulting modulated signal of bandwidth B_T. This signal is received at the input of an ideal demodulator with signal and noise powers of S_i and N_i, respectively* (Fig. 8.9). The demodulator output bandwidth is B, and the SNR is S_o/N_o. Because an SNR

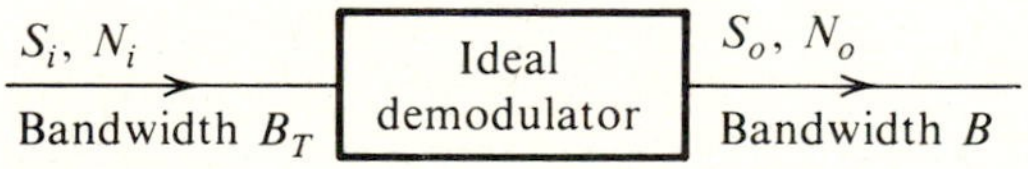

S_i, N_i → Bandwidth B_T → [Ideal demodulator] → S_o, N_o → Bandwidth B

Figure 8.9

of S/N and a bandwidth B can transmit ideally $B \log (1 + S/N)$ bits of information, the ideal information rates of the signals at the input and the output of the demodulator are $B_T \log (1 + S_i/N_i)$ bits and $B \log (1 + S_o/N_o)$ bits, respectively. Because the demodulator neither creates nor destroys information, the two rates should be equal,

* An additive white gaussian channel noise is assumed.

that is,

$$B_T \log\left(1 + \frac{S_i}{N_i}\right) = B \log\left(1 + \frac{S_o}{N_o}\right)$$

and

$$\left(1 + \frac{S_o}{N_o}\right) = \left(1 + \frac{S_i}{N_i}\right)^{B_T/B} \tag{8.69a}$$

In practice, for the majority of systems, S_o/N_o as well as $S_i/N_i \gg 1$, and

$$\frac{S_o}{N_o} \simeq \left(\frac{S_i}{N_i}\right)^{B_T/B} \tag{8.69b}$$

Also,

$$\frac{S_i}{N_i} = \frac{S_i}{\mathcal{N} B_T}$$

$$= \left(\frac{S_i}{\mathcal{N} B}\right)\left(\frac{B}{B_T}\right) = \frac{B}{B_T}\gamma$$

Hence, Eq. (8.69a and b) become

$$\frac{S_o}{N_o} = \left(1 + \frac{\gamma}{B_T/B}\right)^{B_T/B} - 1 \tag{8.70a}$$

$$\simeq \left(\frac{\gamma}{B_T/B}\right)^{B_T/B} \tag{8.70b}$$

Equations (8.69) and (8.70) give the ideal law of exchange between the SNR and the bandwidth. The output SNR S_o/N_o is plotted in Fig. 8.10 as a function of γ for various values of B_T/B.

The output SNR increases exponentially with the bandwidth expansion factor B_T/B. This means to maintain a given output SNR, the transmitted signal power can be reduced exponentially with the bandwidth expansion factor. Thus for a small increase in bandwidth, we can reduce the transmitted power considerably. On the other hand, for a small reduction in bandwidth, we need to increase the transmitted power considerably. Hence in practice, the trade is in the sense of reducing the transmitted power at the cost of increased transmission bandwidth and rarely the other way.

Let us now investigate various systems studies thus far and see how they fare in comparison with the ideal system.

AM

For baseband and SSB-SC systems, $B_T/B = 1$, and Eq. (8.70) yields

$$\frac{S_o}{N_o} = \gamma \tag{8.71}$$

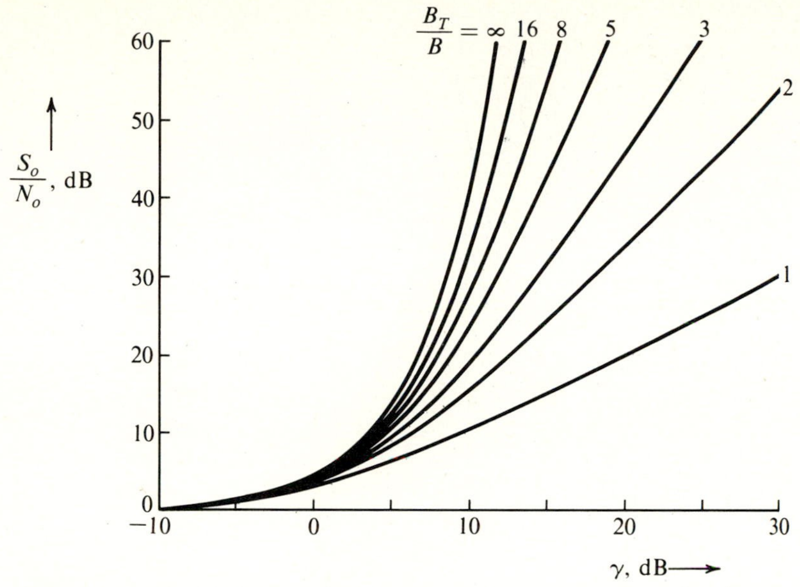

Figure 8.10 Ideal behavior of SNR versus γ for various ratios of B_T/B.

which is exactly the performance of these systems [see Eqs. (6.3) and (6.12)]. For DSB-SC, $B_T/B = 2$, and Eq. (8.70) predicts

$$\frac{S_o}{N_o} \simeq \frac{\gamma^2}{4} \tag{8.72}$$

Thus DSB-SC, for which $S_o/N_o = \gamma$, falls short of ideal performance. If, however, we consider the fact that quadrature multiplexing can be used to simultaneously transmit two DSB signals, with effective bandwidth per signal as B rather than $2B$, DSB-SC has ideal performance. Because for AM, quadrature multiplexing is not used,* AM performance [Eq. (6.14)] falls considerably short of ideal performance [Eq. (8.72)].

The ideal performance of the baseband and SSB-SC or DSB-SC systems is really an empty boast, because these systems really do not exchange SNR for bandwidth.

FM

For FM, we have [Eq. (6.37)]

$$\frac{S_o}{N_o} = 3\beta^2\gamma\left(\frac{\overline{m^2}}{m_p^2}\right)$$

Because $B_T/B = 2(\beta + 2)$,

$$\frac{S_o}{N_o} = 3\left[\frac{1}{2}\left(\frac{B_T}{B}\right) - 2\right]^2\left(\frac{\overline{m^2}}{m_p^2}\right)\gamma \tag{8.73}$$

* Because this requires a phase reference at the receiver, it defeats the purpose of AM.

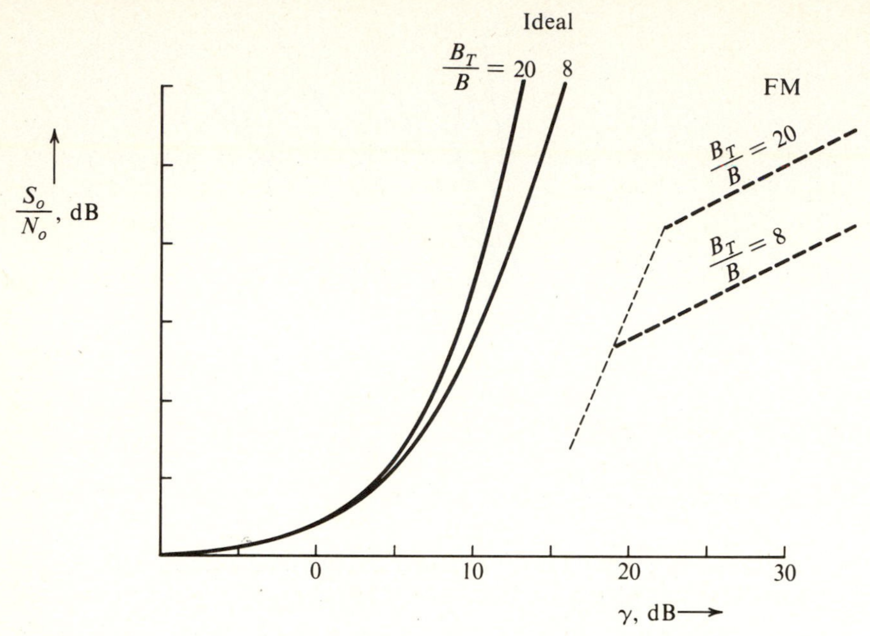

Figure 8.11 Comparison of ideal behavior with FM system behavior.

Figure 8.11 shows the plots of S_o/N_o for $B_T/B = 8$ and 20, assuming $m_p^2/\overline{m^2} = 2$. The ideal S_o/N_o plots for $B_T/B = 8$ and 20 are also shown in the figure for comparison. FM performance falls far below ideal performance. Even if a gain of 13 dB resulting from preemphasis and deemphasis is added to these plots, FM is several dB inferior to ideal curves. The comparison between FM and the ideal system gets progressively worse as γ increases. If we observe the behavior of FM at the threshold, however, it does not fare as badly. It can be shown that for FM, when optimum demodulation (phase-lock loop) is used, the behavior of the SNR at the threshold (the dotted line* in Fig. 8.11) is close to ideal.[10]

PCM

As seen earlier, M-ary PCM shows a saturation effect unless we go to higher values of M as γ increases. If the message signal is quantized in L levels, then each sample can be encoded by $\log_M L$ M-ary pulses. If B is the bandwidth of the message signal, we need to transmit $2B$ samples per second and, hence, $2B \log_M L$ M-ary pulses per second. The transmission bandwidth B_T is

$$B_T = B \log_M L \tag{8.74a}$$

For the M-ary case, the power S_i is given by [Eq. (6.126a)]

$$S_i = \frac{M^2 - 1}{3} E_p f_o \tag{8.74b}$$

* The dotted threshold line shown in Fig. 8.11 is for a frequency discriminator. For optimum demodulation using a PLL, the threshold line is shifted to left by 3 to 5 dB.

Also,

$$N_i = \mathcal{N} B_T = \frac{\mathcal{N} f_o}{2} \tag{8.75}$$

Each of the M-ary pulses carries the information of $\log_2 M$ bits, and we are transmitting $2B \log_M L$ M-ary pulses per second. Hence, we are transmitting information at a rate R:

$$\begin{aligned} R &= (2B \log_M L)(\log_2 M) \\ &= 2B_T \log_2 M \\ &= B_T \log_2 M^2 \quad \text{bits/second} \end{aligned}$$

Substitution of Eqs. (8.74b) and (8.75) in the above equation yields

$$R = B_T \log_2 \left(1 + \frac{3\mathcal{N}}{2E_p} \frac{S_i}{N_i}\right) \quad \text{bits/second} \tag{8.76}$$

We are transmitting the information equivalent of R binary digits per second over the M-ary PCM channel. The reception is not error free, however. The pulses are detected with an error probability P_{eM} given in Eq. (6.127b). If P_{eM} is of the order of 10^{-6}, we could consider the reception to be essentially error free. From Eq. (6.127b),

$$P_{eM} \simeq 2Q\left(\sqrt{\frac{2E_p}{\mathcal{N}}}\right) = 10^{-6} \qquad M \gg 1$$

This gives

$$\frac{2E_p}{\mathcal{N}} = 24$$

Substitution of this value in Eq. (8.76) gives

$$R = B_T \log_2 \left(1 + \frac{1}{8} \frac{S_i}{N_i}\right) \quad \text{bits/second} \tag{8.77}$$

Thus, over a channel of bandwidth B_T with an SNR of S_i/N_i, a PCM system can transmit information at a rate of R in Eq. (8.77). The ideal channel with bandwidth B_T and SNR S_i/N_i transmits information at a rate C:

$$C = B_T \log_2 \left(1 + \frac{S_i}{N_i}\right) \quad \text{bits/second} \tag{8.78}$$

It follows that PCM uses roughly eight times (9 dB) as much power as the ideal system. This performance is still much superior to that of FM. Figure 8.12 shows R/B_T as a function of S_i/N_i. For the ideal system

$$\frac{R}{B_T} = \frac{C}{B_T} = \log_2 \left(1 + \frac{S_i}{N_i}\right)$$

PCM at the threshold is 9 dB inferior to the ideal curve.

When PCM is in saturation, the detection error probability → 0. Each M-ary pulse transmits $\log_2 M$ bits and there are $2B_T$ pulses/second. Hence,

$$R = 2B_T \log_2 M$$

or

$$\frac{R}{B_T} = 2 \log_2 M$$

This is clearly seen in Fig. 8.12.

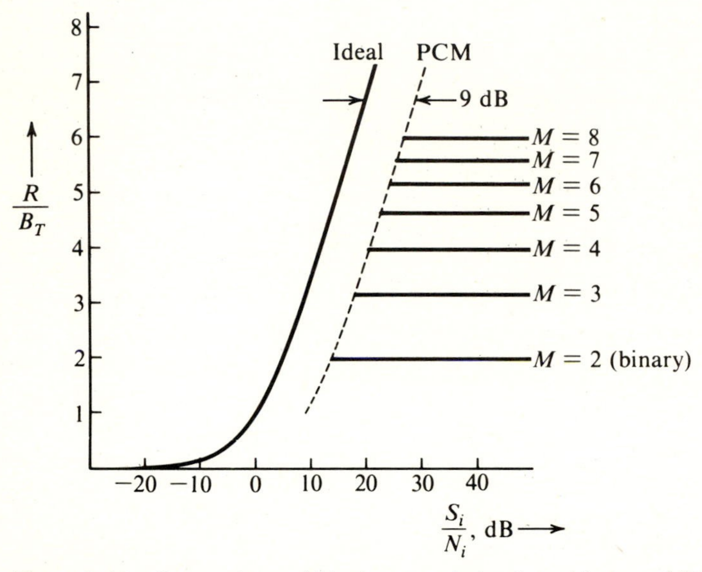

Figure 8.12 Comparison of ideal system behavior with that of PCM.

Orthogonal Signaling

We shall now show that the performance of M-ary orthogonal signaling approaches that of the ideal system in the limit as $M \to \infty$.

The error probability P_{eM} of the optimum receiver was derived in Eq. (7.71b). This expression can be rewritten in terms of λ_b, defined as

$$\lambda_b = \frac{S_i}{\mathcal{N}R} \tag{8.79}$$

where R, the bit rate, is

$$R = f_o \log_2 M$$

with f_o as the M-ary pulse rate. If E is the pulse energy, then

$$\lambda_b = \frac{S_i}{\mathcal{N}f_o \log_2 M} = \frac{E}{\mathcal{N} \log_2 M} \tag{8.80}$$

Equation (7.71b) can now be expressed as

$$P_{eM} = 1 - \frac{1}{\sqrt{2\pi}} \int_{-\infty}^{\infty} [1 - Q(y + \sqrt{2\lambda_b \log_2 M})]^{M-1} e^{-y^2/2}\, dy \tag{8.81}$$

The plot of P_{eM} versus λ_b is shown in Fig. 8.13. The case of $M = \infty$ shows an

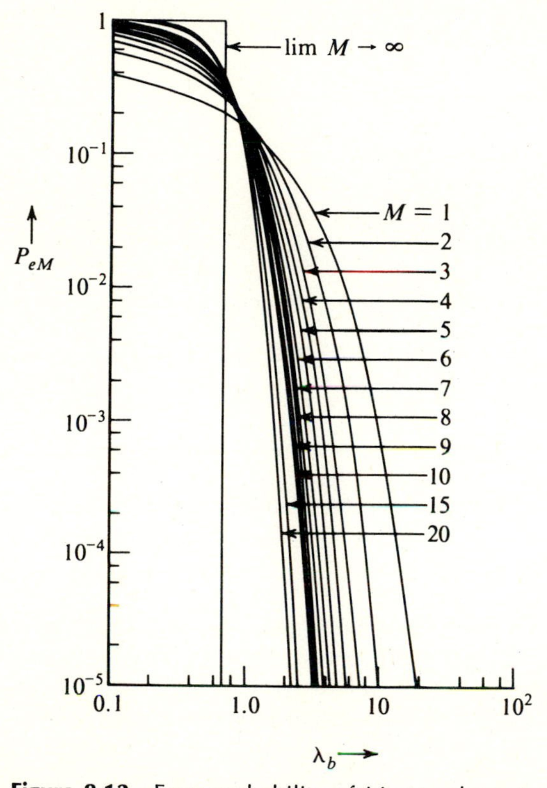

Figure 8.13 Error probability of M-ary coherent orthogonal scheme.

interesting behavior. By properly taking the limit of P_{eM} in Eq. (8.81) as $M \to \infty$, it can be shown that[11]

$$\lim_{M\to\infty} P_{eM} = \begin{cases} 1 & \lambda_b < \log_e 2 \\ 0 & \lambda_b \geq \log_e 2 \end{cases} \tag{8.82}$$

Hence, for error-free communication,

$$\lambda_b \geq \log_e 2 = \frac{1}{1.44} \qquad \text{or} \qquad \frac{S_i}{\mathcal{N}R} \geq \frac{1}{1.44}$$

or

$$R \leq 1.44 \frac{S_i}{\mathcal{N}} \quad \text{bits/second} \tag{8.83}$$

Note that for M-ary orthogonal signals, the dimensionality is M and is infinite for $M = \infty$. The ideal channel with infinite bandwidth can transmit error-free information at a rate of up to $1.44\, S_i/\mathcal{N}$ [Eq. (8.64)]. This is precisely the error-free rate attained by orthogonal signaling.

REFERENCES

1. C. E. Shannon, "Mathematical Theory of Communication," *Bell Syst. Tech. J.*, vol. 27, pp. 379–423, July 1948 and pp. 623–656, Oct. 1948.
2. R. V. L. Hartley, "Transmission of Information," *Bell Syst. Tech. J.*, vol. 7, no. 3, pp. 535–563, July 1928.
3. H. Nyquist, "Certain Factors Affecting Telegraph Speed," *Bell Syst. Tech. J.*, vol. 3, no. 2, pp. 324–346, April 1924.
4. N. Abramson, *Information Theory and Coding*, McGraw-Hill, New York, 1963.
5. R. G. Gallager, *Information Theory and Reliable Communication*, Wiley, New York, 1968.
6. D. A. Huffman, "A Method for Construction of Minimum Redundancy Codes," *Proc. IRE*, vol. 40, no. 10, pp. 1098–1101, Sept. 1952.
7. J. M. Wozencraft and B. Reiffen, *Sequential Decoding*, Wiley, New York, 1961.
8. C. E. Shannon, "Communication in the Presence of Noise," *Proc. IRE*, vol. 37, no.1, pp. 10–21, Jan. 1949.
9. J. M. Wozencraft and I. A. Jacobs, *Principles of Communication Engineering*, Wiley, New York, 1965, Chap. 5.
10. A. J. Viterbi, *Principles of Coherent Communication*, McGraw-Hill, New York, 1966.

PROBLEMS

8.1. A television picture is composed of approximately 300,000 basic picture elements (about 600 picture elements in a horizontal line and 500 horizontal lines per frame). Each of these elements can assume 10 distinguishable brightness levels (such as black and shades of gray) with equal probability. Find the information content of a television picture frame.

8.2. A radio announcer describes a television picture orally in 1,000 words out of his vocabulary of 10,000 words. Assume that each of the 10,000 words in his vocabulary is equally likely to occur in the description of this picture (a crude approximation, but good enough to give the idea). Determine the amount of information broadcast by the announcer in describing the picture. Would you say the announcer can do justice to the picture in 1,000 words? Is the old adage "a picture is worth a thousand words" an exaggeration or an underrating of the reality? Use data in Prob. 8.1 to estimate the information of a picture.

8.3. A source emits six messages with probabilities 1/2, 1/4, 1/8, 1/16, 1/32, and 1/32, respectively. Find the entropy of the source. Obtain the compact binary code and find the average length of the code word. Determine the efficiency and the redundancy of the code.

8.4. A source emits five messages with probabilities 1/3, 1/3, 1/9, 1/9, and 1/9, respectively. Find the entropy of the source. Obtain the compact 3-ary code and find the average length of the code word. Determine the efficiency and the redundancy of the code.

8.5. For the messages in Prob. 8.3, obtain the compact 3-ary code and find the average length of the code word. Determine the efficiency and the redundancy of this code.

8.6. For the messages in Prob. 8.4, obtain the compact binary code and find the average length of the code word. Determine the efficiency and the redundancy of this code. Find the optimum binary code and its efficiency for the second extension of the source.

8.7. For the binary channel in Fig. P8.7, $P_x(1) = 1/3$ and $P_x(0) = 2/3$.
(a) Determine the probabilities $P_y(0)$ and $P_y(1)$.
(b) Determine $H(\mathrm{x})$, $H(\mathrm{x}\,|\,\mathrm{y})$, $H(\mathrm{y})$, $H(\mathrm{y}\,|\,\mathrm{x})$, and $I(\mathrm{x};\,\mathrm{y})$.

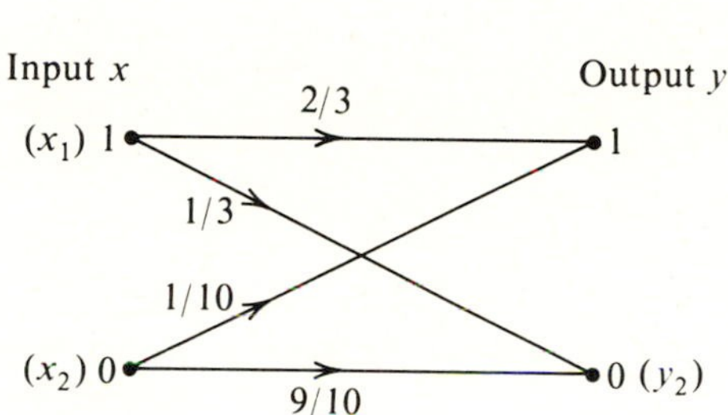

Figure P8.7

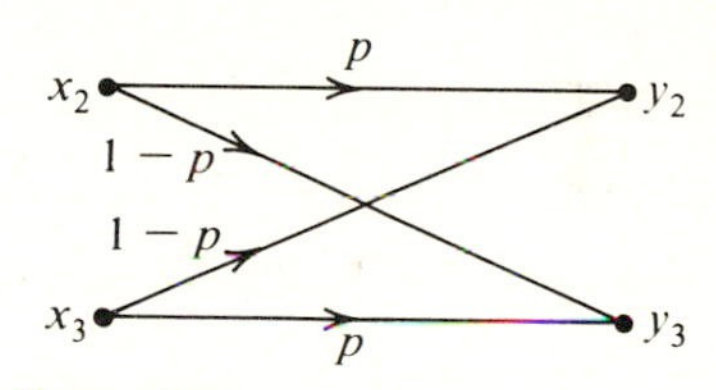

Figure P8.8

8.8. For the ternary channel in Fig. P8.8, $P_x(x_1) = P$, $P_x(x_2) = p_x(x_3) = Q$ (note: $P + 2Q = 1$).
(a) Determine $H(\mathrm{x})$, $H(\mathrm{x}\,|\,\mathrm{y})$, $H(\mathrm{y})$, and $I(\mathrm{x};\,\mathrm{y})$.
(b) Show that the channel capacity C_s is given by:

$$C_s = \log\left(\frac{\beta + 2}{\beta}\right)$$

where $\beta = 2^{-[p\log p + (1-p)\log(1-p)]}$.

8.9. A cascade of two channels is shown in Fig. P8.9. The symbols at the source, at the output of the first channel, and at the output of the second channel are denoted by x, y, and z.

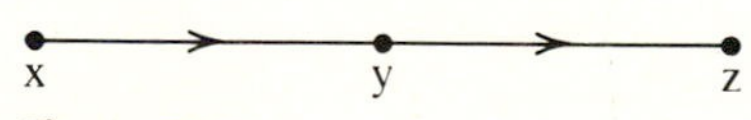

Figure P8.9

Show that

$$H(\mathrm{x}\,|\,\mathrm{z}) \geq H(\mathrm{x}\,|\,\mathrm{y})$$

and

$$I(\mathrm{x};\,\mathrm{y}) \geq I(\mathrm{x};\,\mathrm{z})$$

This shows that the information that can be transmitted over a cascaded channel can be no greater than that transmitted over one link. In effect, information channels tend to leak information. *Hint:* For a cascaded channel, observe that

$$P(\mathrm{z}_k | \mathrm{y}_j, \mathrm{x}_i) = P(\mathrm{z}_k | \mathrm{y}_j)$$

Hence, by Bayes' rule,

$$P(\mathrm{x}_i | \mathrm{y}_j, \mathrm{z}_k) = P(\mathrm{x}_i | \mathrm{y}_j)$$

8.10. For a continuous random variable x constrained to a peak magnitude $M(-M < x < M)$, show that the entropy is maximum when x is uniformly distributed in the range $(-M, M)$ and has zero probability density outside this range. Show that the maximum entropy is given by $\log 2M$.

8.11. For a continuous random variable x constrained to only positive values $0 < x < \infty$ and a mean value A, show that the entropy is maximum when

$$p_{\mathrm{x}}(x) = \frac{1}{A} e^{-x/A}\, u(x)$$

Show that the corresponding entropy is

$$H(\mathrm{x}) = \log eA$$

8.12. A television transmission requires 30 frames of 300,000 picture elements each to be transmitted per second. Using the data in Prob. 8.1, estimate the bandwidth of the AWGN channel if the SNR at the receiver is required to be at least 50 dB.

8.13. Show that the channel capacity of a bandlimited channel disturbed by a colored gaussian noise under the constraint of a given signal power is

$$C = B \log [S_s(\omega) + S_n(\omega)] - \int_{f1}^{f2} \log S_n(\omega)\, df$$

where B is the channel bandwidth (in Hz) over the frequency range (f_1, f_2) $[f_2 - f_1 = B]$. $S_s(\omega)$ and $S_n(\omega)$ are the signal and the noise power densities, respectively.

Show that this maximum rate of information transmission is attained if the desired signal is gaussian and its PSD satisfies the condition.

$$S_s(\omega) + S_n(\omega) = \alpha \text{ (a constant)}$$

Hint: Consider a narrowband Δf in the range (f_1, f_2). The maximum rate of transmission over this band is given by

$$\Delta f \log \left[\frac{S_s(\omega)\Delta f + S_n(\omega)\,\Delta f}{S_n(\omega)\,\Delta f}\right] = \Delta f \log \left[\frac{S_s(\omega) + S_n(\omega)}{S_n(\omega)}\right]$$

provided the signal over this band is gaussian. The rate of transmission over the entire band is given by

$$\int_{f1}^{f2} \log \left[\frac{S_s(\omega) + S_n(\omega)}{S_n(\omega)}\right] df$$

Now maximize this under the constraint

$$2 \int_{f1}^{f2} S_s(\omega)\, df = S$$

8.14. Using the results in Prob. 8.13, show that the worst kind of gaussian noise is a white gaussian noise that is constrained to a given mean square value. *Hint:* Use the expression for channel capacity in Prob. 8.13. The first term in this expression is a constant. Now show that the second term attains a maximum value when $S_n(\omega)$ is a constant under the constraint

$$2\int_{f_1}^{f_2} S_n(\omega)\, df = N$$

9

Error-Correcting Codes

As seen from the discussion in Chapter 8, the key to realizing error-free communication is the use of appropriate redundancy. Addition of a single parity-check digit to detect an odd number of errors is a good example of this. Since Shannon's pioneering paper, a great deal of work has been carried out in the area of error-correcting codes.

9.1 INTRODUCTION

In this chapter we shall discuss two important types of codes: *Block codes* and *convolutional* (or *recurrent*) *codes*. The information coming from the data or message source will be assumed to be in binary form (a sequence of binary digits).

In *block codes,* a block of k data digits is encoded by a code word of n digits ($n > k$). For each sequence of k data digits, there is a distinct code word of n digits.

In *convolutional codes,* the coded sequence of n digits depends not only on the k data digits but also on the previous $N - 1$ data digits ($N > 1$). Hence, the coded sequence for a certain k data digits is not unique but depends on $N - 1$ earlier data digits. In block codes, k data digits are accumulated and then encoded into an n-digit code word. In convolutional codes, the coding is done on a continuous, or running, basis rather than by accumulating k data digits.

If k data digits are transmitted by a code word of n digits, the number of check digits is $m = n - k$. The *code efficiency* (also known as the *code rate*) is k/n. Such a code is known as an (n, k) code. The data digits $(d_1, d_2, \ldots, d_k)$ are a k-tuple and, hence, a k-dimensional vector $\boldsymbol{d}$. Similarly, the code word $(c_1, c_2, \ldots, c_n)$ is an n-dimensional vector $\boldsymbol{c}$. As a preliminary, we shall determine the minimum number of check digits required to detect or correct t number of errors in an (n, k) code.

A total of 2^n code words (or vertices of an n-dimensional hypercube) are available to assign to 2^k data words. Suppose we wish to find a code that will correct up to t number of wrong digits. In this case, if we transmit a data word $\boldsymbol{d}_j$ by one of the code words (or vertices) $\boldsymbol{c}_j$, then because of channel errors the received word will not be $\boldsymbol{c}_j$ but will be $\boldsymbol{c}_j'$. If the channel noise causes errors in t or less digits, then $\boldsymbol{c}_j'$ will lie somewhere in the Hamming sphere of radius t centered at $\boldsymbol{c}_j$. If the code is to correct up to t errors, then the code must have a property that all of the Hamming spheres of radius t centered at the code words are nonoverlapping. This means we may not use vertices (or words) that are within a Hamming distance of t from any code word. If a received word lies within a Hamming sphere of radius t centered at $\boldsymbol{c}_j$, then we decide that the transmitted code word was $\boldsymbol{c}_j$. This scheme is capable of correcting up to t errors, and d_{min}, the minimum distance between t error-correcting code words, is

$$d_{\text{min}} = 2t + 1 \tag{9.1}$$

Next, in order to find relationship between n and k, we observe that 2^n vertices, or words, are available for 2^k data words, and $2^n - 2^k$ are redundant vertices. How many vertices, or words, can lie within a Hamming sphere of radius t? The number of sequences (of n digits) that differ from a given sequence by j digits is the number of the combination of n things taken j at a time and is given by $\binom{n}{j}$ [see Eq. (5.12)]. Hence, the number of ways in which up to t errors can occur is given by $\sum_{j=1}^{t} \binom{n}{j}$. Thus for each code word, we must leave $\sum_{j=1}^{t} \binom{n}{j}$ number of words unused. Because we have 2^k code words, we must leave $2^k \sum_{j=1}^{t} \binom{n}{j}$ words unused. Hence, the total number of words must be at least $2^k + 2^k \sum_{j=1}^{t} \binom{n}{j} = 2^k \sum_{j=0}^{t} \binom{n}{j}$. But the total number of words, or vertices, available is 2^n. Hence,

$$2^n \geq 2^k \sum_{j=0}^{t} \binom{n}{j}$$

or

$$2^{n-k} \geq \sum_{j=0}^{t} \binom{n}{j} \tag{9.2a}$$

Observe that $n - k = m$ is the number of check digits. Hence, Eq. (9.2a) can be expressed as

$$2^m \geq \sum_{j=0}^{t} \binom{n}{j} \tag{9.2b}$$

This is known as the *Hamming bound*. It should also be remembered that the Hamming bound is a necessary but not a sufficient condition. If some m satisfies the Hamming bound, it does not necessarily mean that a t-error-correcting code of n digits can be constructed. Table 9.1 shows some examples of error-correction codes and their efficiencies.

Table 9.1 Some Examples of Error-Correcting Codes

	n	k	Code	Code Efficiency (or Code Rate)
Single-error correcting, $t = 1$, Minimum code separation 3	3	1	(3, 1)	0.33
	4	1	(4, 1)	0.25
	5	2	(5, 2)	0.4
	6	3	(6, 3)	0.5
	7	4	(7, 4)	0.57
	15	11	(15, 11)	0.73
	31	26	(31, 26)	0.838
Triple-error correcting, $t = 3$, Minimum code separation 7	10	4	(10, 4)	0.4
	15	8	(15, 8)	0.533
Double-error correcting, $t = 2$, Minimum code separation 5	10	2	(10, 2)	0.2
	15	5	(15, 5)	0.33
	23	12	(23, 12)	0.52

Another way of correcting errors is to design a code to detect (not to correct) up to t errors. When the receiver detects an error, it requests retransmission. Because error detection requires fewer check digits, these codes operate at a higher efficiency.

To detect t errors, code words must be separated by a Hamming distance of not more than $t + 1$. Suppose a transmitted code word $\mathbf{c}_j$ has α number of errors ($\alpha \leq t$); then the received codeword $\mathbf{c}_j'$ is at a distance of α from $\mathbf{c}_j$. Because $\alpha \leq t$, $\mathbf{c}_j'$ can never be any other valid code word, because all code words are separated by at least $t + 1$. Thus, the reception of $\mathbf{c}_j'$ immediately indicates that an error has been made.

Thus, the minimum distance $d_{\min}$ between t error-correcting code words is

$$d'_{\min} = t + 1 \tag{9.3}$$

In presenting the theory, we shall use modulo-2 addition, defined in Chapter 3:

$$\mathbf{1 \oplus 1 = 0 \oplus 0 = 0}$$

$$\mathbf{0 \oplus 1 = 1 \oplus 0 = 1}$$

Note that the modulo-2 sum of any binary digit with itself is always zero. All the additions in the mathematical development of binary codes presented henceforth are modulo 2.

9.2 LINEAR BLOCK CODES

A code word consists of n digits $c_1, c_2, \ldots, c_n$, and a data word consists of k digits $d_1, d_2, \ldots, d_k$. Because the code word and data word are an n-tuple and k-tuple, respectively, they are n- and k-dimensional vectors. We shall use row matrices to represent these words.

$$\mathbf{c} = (c_1, c_2, \ldots, c_n) \qquad \mathbf{d} = (d_1, d_2, \ldots, d_k)$$

For the general case of linear block codes, all the n digits of $\mathbf{c}$ are formed by linear combinations (modulo-2 additions) of k data digits. A special case where $c_1 = d_1$, $c_2 = d_2, \ldots, c_k = d_k$ and the remaining digits from c_{k+1} to c_n are linear combinations of $d_1, d_2, \ldots, d_k$ is known as a *systematic code.** Thus in a systematic code, the first k digits of a code word are the data digits and the last $m = n - k$ digits are the *parity-check digits,* formed by linear combinations of data digits $d_1, d_2, \ldots, d_k$:

$$\begin{aligned}
c_1 &= d_1 \\
c_2 &= d_2 \\
&\vdots \\
c_k &= d_k \\
c_{k+1} &= h_{11}d_1 \oplus h_{12}d_2 \oplus \cdots \oplus h_{1k}d_k \\
c_{k+2} &= h_{21}d_1 \oplus h_{22}d_2 \oplus \cdots \oplus h_{2k}d_k \\
&\cdots\cdots\cdots\cdots \\
c_n &= h_{m1}d_1 \oplus h_{m2}d_2 \oplus \cdots \oplus h_{mk}d_k
\end{aligned} \tag{9.4a}$$

or

$$\mathbf{c} = \mathbf{d}G \tag{9.4b}$$

* It can be shown that the performance of systematic block codes is identical to that of nonsystematic block codes.

where

$$G = \underbrace{\begin{bmatrix} 1 & 0 & 0 & \cdots & 0 \\ 0 & 1 & 0 & \cdots & 0 \\ \cdot & \cdot & \cdot & \cdots & \cdot \\ 0 & 0 & 0 & \cdots & 1 \end{bmatrix}}_{I_k(k \times k)} \underbrace{\begin{bmatrix} h_{11} & h_{21} & \cdots & h_{m1} \\ h_{12} & h_{22} & \cdots & h_{m2} \\ \cdot & \cdot & \cdots & \cdot \\ h_{1k} & h_{2k} & \cdots & h_{mk} \end{bmatrix}}_{P(k \times m)} \tag{9.5}$$

The k by n matrix G is called the *generator matrix,* which can be partitioned into a k by k identity matrix I_k and a k by m matrix P. All the elements of P are either 0 or 1. The code word can be expressed as

$$\begin{aligned} \mathbf{c} &= \mathbf{d}G \\ &= \mathbf{d}[I_k, P] \\ &= [\mathbf{d}, \mathbf{d}P] \\ &= [\mathbf{d}, \mathbf{c}_P] \end{aligned} \tag{9.6}$$

where $\mathbf{c}_P$ is the row matrix of m parity-check digits:

$$\mathbf{c}_P = \mathbf{d}P \tag{9.7}$$

Thus knowing the data digits, we can calculate the check digits from Eq. (9.7).

■ EXAMPLE 9.1

For a (6, 3) code, the generator matrix G is

$$G = \begin{bmatrix} 1 & 0 & 0 & 1 & 0 & 1 \\ 0 & 1 & 0 & 0 & 1 & 1 \\ 0 & 0 & 1 & 1 & 1 & 0 \end{bmatrix}$$

with the first three columns forming I_k and the last three forming P.

For all eight possible data words, find the corresponding code words, and verify that this code is a single-error-correcting code.

□ **Solution:** Table 9.2 shows that eight data words and the corresponding code words found from $\mathbf{c} = \mathbf{d}G$.

Table 9.2

Data Word	Code Word
111	**111000**
110	**110110**
101	**101011**
100	**100101**
011	**011101**
010	**010011**
001	**001110**
000	**000000**

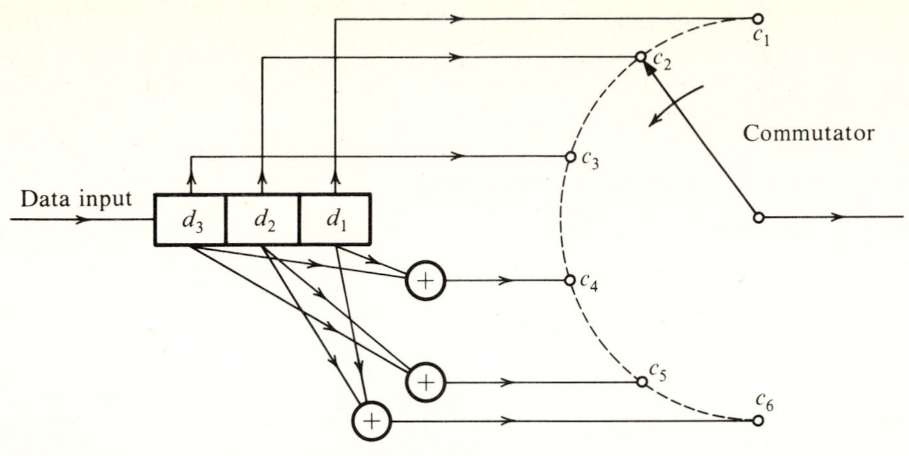

Figure 9.1 An encoder for linear block codes.

Note that the distance between any two code words is at least 3. Hence, the code can correct at least one error. Figure 9.1 shows one possible encoder for this code, using a three-digit shift register and three modulo-2 adders. ■

Decoding

Let us consider some code-word properties that could be utilized for the purpose of decoding. From Eq. (9.7), and the fact that the modulo-2 sum of any sequence with itself is zero, we get

$$\boldsymbol{d}P \oplus \boldsymbol{c}_P = \underbrace{[\boldsymbol{d} \quad \boldsymbol{c}_P]}_{\boldsymbol{c}} \begin{bmatrix} P \\ I_m \end{bmatrix} = 0 \tag{9.8}$$

where I_m is the identity matrix of the order $m \times m$ $(m = n - k)$. Thus,

$$\boldsymbol{c}H^T = 0 \tag{9.9a}$$

where

$$H^T = \begin{bmatrix} P \\ I_m \end{bmatrix} \tag{9.9b}$$

and its transpose

$$H = [P^T \quad I_m] \tag{9.9c}$$

is called the *parity-check matrix*. Every code word must satisfy Eq. (9.9a). This is our clue to decoding. Consider the received word $\boldsymbol{r}$. Because of possible errors caused by channel noise, $\boldsymbol{r}$ in general differs from the transmitted code word $\boldsymbol{c}$.

$$\boldsymbol{r} = \boldsymbol{c} \oplus \boldsymbol{e}$$

where $\boldsymbol{e}$, the error word (or error vector) is also a row vector of n elements. For example, if the data word **100** in Example 9.1 is transmitted as a code word **100101** (see Table 9.2), and the channel noise causes a detection error in the third digit, then

$$\boldsymbol{r} = \mathbf{101101}$$

$$\boldsymbol{c} = \mathbf{100101}$$

and

$$\boldsymbol{e} = \mathbf{001000}$$

Thus, an element **1** in $\boldsymbol{e}$ indicates an error in the corresponding position, and **0** indicates no error. The Hamming distance between $\boldsymbol{r}$ and $\boldsymbol{c}$ is simply the number of **1**'s in $\boldsymbol{e}$.

Suppose the transmitted code word is $\boldsymbol{c}_i$ and the channel noise causes and error $\boldsymbol{e}_i$, making the received word $\boldsymbol{r} = \boldsymbol{c}_i + \boldsymbol{e}_i$. If there were no errors, that is, if $\boldsymbol{e}_i = \mathbf{000000}$, then $\boldsymbol{r}H^T = 0$. But because of possible channel errors, $\boldsymbol{r}H^T$ is in general a nonzero row vector $\boldsymbol{s}$, called the *syndrome*.

$$\boldsymbol{s} = \boldsymbol{r}H^T \tag{9.10a}$$

$$= (\boldsymbol{c}_i \oplus \boldsymbol{e}_i)H^T$$

$$= \boldsymbol{c}_i H^T \oplus \boldsymbol{e}_i H^T$$

$$= \boldsymbol{e}_i H^T \tag{9.10b}$$

Knowing $\boldsymbol{r}$, we can compute $\boldsymbol{s}$ [Eq. (9.10a)] and presumably we can compute $\boldsymbol{e}_i$ from Eq. (9.10b). Unfortunately, knowledge of $\boldsymbol{s}$ does not allow us to solve uniquely for $\boldsymbol{e}_i$. This is because $\boldsymbol{r}$ can also be expressed in terms of code words other than $\boldsymbol{c}_i$. Thus,

$$\boldsymbol{r} = \boldsymbol{c}_j \oplus \boldsymbol{e}_j \qquad j \neq i$$

Hence,

$$\boldsymbol{s} = (\boldsymbol{c}_j \oplus \boldsymbol{e}_j)H^T = \boldsymbol{e}_j H^T$$

Because there are 2^k possible code words,

$$\boldsymbol{s} = \boldsymbol{e}H^T$$

is satisfied by 2^k error vectors. To give an example, if a data word $\boldsymbol{d} = \mathbf{100}$ is transmitted by a code word **100101** in Example 9.1, and if a detection error is caused in the third digit, then the received word is **101101**. In this case we have $\boldsymbol{c} = \mathbf{100101}$ and $\boldsymbol{e} = \mathbf{001000}$. But the same word could have been received if $\boldsymbol{c} = \mathbf{101011}$ and $\boldsymbol{e} = \mathbf{000110}$, or if $\boldsymbol{c} = \mathbf{010011}$ and $\boldsymbol{e} = \mathbf{111110}$, and so on. Thus, there are eight possible error vectors (2^k error vectors) that satisfy Eq. (9.10b). Which vector shall we choose? For this, we must define our decision criterion. One reasonable criterion is the maximum-likelihood rule where, if we receive $\boldsymbol{r}$, then we decide in favor of that $\boldsymbol{c}$ for which $\boldsymbol{r}$ is most likely to be received. In other words, we decide "$\boldsymbol{c}_i$ transmitted" if

$$P(\boldsymbol{r}|\boldsymbol{c}_i) > P(\boldsymbol{r}|\boldsymbol{c}_k) \qquad \text{all } k \neq i$$

For a BSC, this rule gives a very simple answer. Suppose the Hamming distance between $\boldsymbol{r}$ and $\boldsymbol{c}_i$ is d, that is, the channel noise causes errors in d digits. Then if P_e is the digit error probability of a BSC,

$$P(\boldsymbol{r}|\boldsymbol{c}_i) = P_e^d(1 - P_e)^{n-d}$$

If $P_e < 0.5$, then $P(\boldsymbol{r}|\boldsymbol{c}_i)$ is a monotonically decreasing function of d. Hence, to maximize $P(\boldsymbol{r}|\boldsymbol{c}_i)$, we must choose that $\boldsymbol{c}_i$ which is closest to $\boldsymbol{r}$; that is, we must choose the error vector with the smallest number of **1**'s. A vector with the smallest number of **1**'s is called the *minimum-weight vector*.

EXAMPLE 9.2

A (6, 3) code is generated according to the generating matrix in Example 9.1. The receiver receives $\boldsymbol{r} = \mathbf{100011}$. Determine the corresponding data word if the channel is a BSC and the maximum-likelihood decision is used.

Solution:

$$\boldsymbol{s} = \boldsymbol{r}H^T$$

$$= [\mathbf{1\ 0\ 0\ 0\ 1\ 1}]\begin{bmatrix} 1 & 0 & 1 \\ 0 & 1 & 1 \\ 1 & 1 & 0 \\ 1 & 0 & 0 \\ 0 & 1 & 0 \\ 0 & 0 & 1 \end{bmatrix}$$

$$= [\mathbf{1\ 1\ 0}]$$

The correct transmitted code word $\mathbf{c}$ is given by

$$\boldsymbol{c} = \boldsymbol{r} \oplus \boldsymbol{e}$$

where $\mathbf{e}$ satisfies

$$\boldsymbol{s} = [\mathbf{1\ 1\ 0}] = \boldsymbol{e}H^T$$

$$= [e_1\ e_2\ e_3\ e_4\ e_5\ e_6]\begin{bmatrix} 1 & 0 & 1 \\ 0 & 1 & 1 \\ 1 & 1 & 0 \\ 1 & 0 & 0 \\ 0 & 1 & 0 \\ 0 & 0 & 1 \end{bmatrix}$$

We see that $\boldsymbol{e} = \mathbf{001000}$ satisfies this equation. But so does $\boldsymbol{e} = \mathbf{000110}$, or **010101**, or **011011**, or **111110**, or **110000**, or **101101**, or **100011**. The suitable choice, the minimum-weight $\boldsymbol{e}$, is **001000**. Hence,

$$\boldsymbol{c} = \mathbf{100011} \oplus \mathbf{001000} = \mathbf{101011}$$ ■

The decoding procedure just described is quite disorganized. A systematic procedure would be to consider all possible syndromes and for each syndrome associate a minimum-weight error vector. For instance, the single-error-correcting code in Example 9.1 has a syndrome with three digits. Hence, there are eight possible syndromes. We prepare a table of minimum-weight error vectors corresponding to each syndrome [see Table 9.3]. This table can be prepared by considering all possible minimum-weight error vectors and computing s for each of them using Eq. (9.10b). The first minimum-weight error vector **000000** is a trivial case that has the syndrome **000**. Next, we consider all possible unit weight error vectors. There are six such vectors: **100000**, **010000**, **001000**, **000100**, **000010**, **000001**. Syndromes for these can readily be calculated from Eq. (9.10b), and tabulated (Table 9.3). This still leaves one syndrome, **111**, that is not matched with some error vector. Since all unit weight error vectors are exhausted, we must look for error vectors of weight 2.

We find that for the first seven syndromes (Table 9.3), there is a unique minimum-weight vector $\boldsymbol{e}$. But for s = **111**, the error vector $\boldsymbol{e}$ has a minimum weight of 2, and it is not unique. For example, $\boldsymbol{e}$ = **100010** or **010100** or **001001** all have s = **111**, and all the three $\boldsymbol{e}$'s are minimum weight (weight 2). In such a case, we can pick any one of these $\boldsymbol{e}$'s as a *correctable* error pattern. In Table 9.3, we have picked $\boldsymbol{e}$ = **100010** as the double-error correctable pattern. This means the present code can correct all six single-error patterns and one double-error pattern (**100010**). For instance, if $\boldsymbol{c}$ = **101011** is transmitted and the channel noise causes the double error **100010**, the received vector $\boldsymbol{r}$ = **001001**, and

$$s = rH^T = [\mathbf{111}]$$

From Table 9.3 we see that corresponding to s = **111**, $\boldsymbol{e}$ = **100010**, and we immediately decide $\boldsymbol{c} = \boldsymbol{r} \oplus \boldsymbol{e}$ = **101011**. Note, however, that this code will not correct double-error patterns other than **100010**. Thus, this code not only corrects all single errors but one double-error pattern as well. This extra bonus of one double-error correction occurs because n and k oversatisfy the Hamming bound [Eq. (9.2b)]. In case n and k were to satisfy the bound exactly, we would have only single-error-correction ability. This is the case for the (7, 4) code, which can correct all single-error patterns only.

Table 9.3 Decoding Table for Code in Table 9.2

e	s
000000	**000**
100000	**101**
010000	**011**
001000	**110**
000100	**100**
000010	**010**
000001	**001**
100010	**111**

Thus for systematic decoding, we prepare a table of all correctable error patterns and corresponding syndromes. For decoding, we need only calculate $\boldsymbol{s} = \boldsymbol{r}H^T$ and, from the decoding table, find the corresponding $\boldsymbol{e}$. The decision is $\boldsymbol{c} = \boldsymbol{r} \oplus \boldsymbol{e}$.

Because $\boldsymbol{s}$ has $n - k$ digits, there are a total of 2^{n-k} syndromes, each of $n - k$ digits. There are the same number of correctable error vectors, each of n digits. Hence, for the purpose of decoding, we need a storage of $(2n - k)2^{n-k} = (2n - k)2^m$. This storage requirement grows exponentially with m, the number of parity-check digits, and can be enormous, even for moderately complex codes.

Because the maximum-likelihood decision is the same as choosing the code word closest to the received word, we could just as well compare the received word with each of the 2^k possible code words of n digits each. This involves a storage of $n2^k$, which can be much larger than the $(2n - k)2^m$ storage required earlier.

It is still not clear how to choose coefficients of the generator or parity-check matrix. Unfortunately, there is no systematic way to do this, except for the case of single-error-correcting codes (also known as *Hamming codes*). Let us consider a single-error-correcting (7, 4) code. This code satisfies the Hamming bound exactly, and we shall see that a proper code can be constructed. In this case $m = 3$, and there are seven nonzero syndromes, and because $n = 7$, there are exactly seven single-error patterns. Hence, we can correct all single-error patterns and no more. Consider the single-error pattern $\boldsymbol{e} = \mathbf{1000000}$. Because

$$\boldsymbol{s} = \boldsymbol{e}H^T$$

$\boldsymbol{e}H^T$ will be simply the first row of H^T. Similarly, for $\boldsymbol{e} = \mathbf{0100000}$, $\boldsymbol{s} = \boldsymbol{e}H^T$ will be the second row of H^T, and so on. Now for unique decodability, we required that all seven syndromes corresponding to the seven single-error patterns be distinct. Conversely, if all the seven syndromes are distinct, we can decode all the single-error patterns. This means that the only requirement on H^T is that all seven of its rows be distinct and nonzero. Note that H^T is an $(n \times n - k)$ matrix (i.e., 7×3 in this case). Because there exist seven nonzero patterns of three digits, it is possible to find seven nonzero rows of three digits each. There are many ways in which these rows can be ordered. But we must remember that the three bottom rows must form identity matrix I_m [see Eq. (9.9a)].

One possible form of H^T is:

$$H^T = \begin{bmatrix} 1 & 1 & 1 \\ 1 & 1 & 0 \\ 1 & 0 & 1 \\ 0 & 1 & 1 \\ 1 & 0 & 0 \\ 0 & 1 & 0 \\ 0 & 0 & 1 \end{bmatrix} = \begin{bmatrix} P \\ I_m \end{bmatrix}$$

The corresponding generator matrix G is

$$G = [I_k \quad P] = \begin{bmatrix} 1 & 0 & 0 & 0 & 1 & 1 & 1 \\ 0 & 1 & 0 & 0 & 1 & 1 & 0 \\ 0 & 0 & 1 & 0 & 1 & 0 & 1 \\ 0 & 0 & 0 & 1 & 0 & 1 & 1 \end{bmatrix}$$

Thus when $\boldsymbol{d} = \mathbf{1011}$, the corresponding code word $\boldsymbol{c} = \mathbf{1011001}$, and so forth.

A general (n, k) code has m-dimensional syndrome vectors ($m = n - k$). Hence, there are $2^m - 1$ distinct nonzero syndrome vectors that can correct $2^m - 1$ single-error patterns. Because in an (n, k) code there are exactly n single-error patterns, all these patterns can be corrected if

$$2^m - 1 \geqslant n$$

or

$$2^{n-k} \geqslant n + 1$$

This is precisely the condition in Eq. (2.9a) for $t = 1$. Thus, for any (n, k) satisfying this condition, it is possible to construct a single-error-correcting code by the procedure discussed above.

More discussion on block coding can be found in Peterson and Weldon[1] and Lin.[2]

9.3 CYCLIC CODES[1,2]

Cyclic codes are a subclass of linear block codes. As seen before, a procedure for selecting a generator matrix is relatively easy for single-error-correcting codes. This procedure, however, cannot carry us very far in constructing higher-order error-correcting codes. Cyclic codes have a fair amount of mathematical structure that permits the design of higher-order correcting codes. Secondly, for cyclic codes, encoding and syndrome calculations can be easily implemented using simple shift registers.

Cyclic codes are such that code words are simple lateral shifts of one another. For example, if $\boldsymbol{c} = (c_1, c_2, \ldots, c_{n-1}, c_n)$ is a code word, then so is $(c_2, c_3, \ldots, c_n, c_1)$ and $(c_3, c_4, \ldots, c_n, c_1, c_2)$ and so on. We shall use the following notation. If

$$\boldsymbol{c} = (c_1, c_2, \ldots, c_n) \tag{9.11a}$$

is a code vector of a code C, then $\boldsymbol{c}^{(i)}$ denotes $\boldsymbol{c}$ shifted cyclically i places to the left—that is,

$$\boldsymbol{c}^{(i)} = (c_{i+1}, c_{i+2}, \ldots, c_n, c_1, c_2, \ldots, c_i) \tag{9.11b}$$

Cyclic codes can be described in a polynomial form. This property is extremely useful in analysis and implementation of these codes. The code vector $\boldsymbol{c}$ in Eq. (9.11a) can be expressed as the $(n - 1)$-degree polynomial

$$c(x) = c_1 x^{n-1} + c_2 x^{n-2} + \cdots + c_n \tag{9.12a}$$

The coefficients of the polynomial are either **0** or **1**, and they obey the following properties.

$$\begin{array}{ll} \mathbf{0 + 0 = 0} & \mathbf{0 \times 0 = 0} \\ \mathbf{0 + 1 = 1 + 0 = 0} & \mathbf{0 \times 1 = 1 \times 0 = 0} \\ \mathbf{1 + 1 = 0} & \mathbf{1 \times 1 = 1} \end{array}$$

The code polynomial $c^{(i)}(x)$ for the code vector $\mathbf{c}^{(i)}$ in Eq. (9.11b) is

$$c^{(i)}(x) = c_{i+1}x^{n-1} + c_{i+2}x^{n-2} + \cdots + c_n x^i + c_1 x^{i-1} + \cdots + c_i \tag{9.12b}$$

One of the interesting properties of the code polynomials is that when $x^i c(x)$ is divided by $x^n + 1$, the remainder is $c^{(i)}(x)$. We can verify this property as follows:

$$xc(x) = c_1 x^n + c_2 x^{n-1} + \cdots + c_n x$$

$$\begin{array}{r} c_1 \qquad\qquad\qquad\qquad \\ x^n + 1 \overline{)c_1 x^n + c_2 x^{n-1} + \cdots + c_n x} \\ \underline{c_1 x^n + c_1} \qquad\qquad\qquad \\ c_2 x^{n-1} + c_3 x^{n-2} + \cdots + c_n x + c_1 \quad \leftarrow \text{remainder} \end{array}$$

The remainder is clearly $c^{(1)}(x)$. In deriving this result, we have used the fact that subtraction amounts to summation when modulo-2 operations are involved. Continuing in this fashion, we can show that the remainder of $x^i c(x)$ divided by $x^n + 1$ is $c^{(i)}(x)$.

We shall now prove an important theorem in cyclic codes. It says that a cyclic code polynomial $c(x)$ can be generated by the data polynomial $d(x)$ of degree $k - 1$ and a generator polynomial $g(x)$ of degree $(n - k)$ as

$$c(x) = d(x)\, g(x) \tag{9.13}$$

where the generator polynomial $g(x)$ is an $(n - k)$ order factor of $(x^n + 1)$.

For a data vector $(d_1, d_2, \ldots, d_k)$, the data polynomial is

$$d(x) = d_1 x^{k-1} + d_2 x^{k-2} + \cdots + d_k \tag{9.14}$$

☐ **Proof:** Consider a polynomial

$$\begin{aligned} c(x) &= d(x)\, g(x) \\ &= d_1 x^{k-1} g(x) + d_2 x^{k-2} g(x) + \cdots + d_k g(x) \end{aligned} \tag{9.15}$$

This is a polynomial of degree $n - 1$ or less. There are a total of 2^k such polynomials corresponding to 2^k data vectors. Thus, we have a linear (n, k) code generated by Eq. (9.13). To prove that this code is cyclic, let

$$c(x) = c_1 x^{n-1} + c_2 x^{n-2} + \cdots + c_n$$

be a code polynomial in this code [Eq. (9.15)]. Then

$$\begin{aligned} xc(x) &= c_1 x^n + c_2 x^{n-1} + \cdots + c_n x \\ &= c_1(x^n + 1) + (c_2 x^{n-1} + c_3 x^{n-2} + \cdots + c_n x + c_1) \\ &= c_1(x^n + 1) + c^{(1)}(x) \end{aligned}$$

Because $xc(x)$ is $xd(x)\,g(x)$, and $g(x)$ is a factor of $x^n + 1$, $c^{(1)}(x)$ must also be a multiple of $g(x)$ and can also be expressed as $d(x)\,g(x)$ for some data vector $\mathbf{d}$. Therefore, $c^{(1)}(x)$ is also a code polynomial. Continuing this way, we see that $c^{(2)}(x)$, $c^{(3)}(x)$, . . . are all code polynomials generated by Eq. (9.15). Thus, the linear (n, k) code generated by $d(x)\,g(x)$ is indeed cyclic.

■ EXAMPLE 9.3

Find a generator polynomial $g(x)$ for a (7, 4) cyclic code, and find code vectors for the following data vectors: **1010**, **1111**, **0001**, and **1000**.

□ **Solution:** In this case $n = 7$ and $n - k = 3$.

$$x^7 + 1 = (x + 1)(x^3 + x + 1)(x^3 + x^2 + 1)$$

For a (7, 4) code, the generator polynomial must be of the order $n - k = 3$. In this case, there are two possible choices for $g(x)$: $x^3 + x + 1$ or $x^3 + x^2 + 1$. Let us choose the latter, that is,

$$g(x) = x^3 + x^2 + 1$$

as a possible generator polynomial. For

$$d = [\mathbf{1\ 0\ 1\ 0}]$$

$$d(x) = x^3 + x$$

and the code polynomial is

$$\begin{aligned} c(x) &= d(x)\,g(x) \\ &= (x^3 + x)(x^3 + x^2 + 1) \\ &= x^6 + x^5 + x^4 + x \end{aligned}$$

Hence,

$$\mathbf{c} = (\mathbf{1110010})$$

Similarly, code words for other data words can be found (see Table 9.4).

Table 9.4

d	c
1010	**1110010**
1111	**1001011**
0001	**0001101**
1000	**1101000**

Note the structure of the code words. The first k digits are not necessarily the data digits. Hence, this is not a systematic code. In a systematic code, the first k digits are data digits, and the last $m = n - k$ digits are the parity-check digits. Systematic codes are a special case of general codes. Our discussion thus far applies to general cyclic codes, of which systematic cyclic codes are a special case. We shall now develop a method of generating systematic cyclic codes. ■

Systematic Cyclic Codes

We shall show that for a systematic code, the code word polynomial $c(x)$ corresponding to the data polynomial $d(x)$ is given by

$$c(x) = x^{n-k}\, d(x) + \rho(x) \tag{9.16a}$$

where $\rho(x)$ is the remainder from dividing $x^{n-k}\, d(x)$ by $g(x)$.

$$\rho(x) = \text{Rem}\, \frac{x^{n-k}\, d(x)}{g(x)} \tag{9.16b}$$

To prove this we observe that

$$\frac{x^{n-k}\, d(x)}{g(x)} = q(x) + \frac{\rho(x)}{g(x)} \tag{9.17a}$$

where $q(x)$ is of degree $(k - 1)$ or less. We add $\rho(x)/g(x)$ to both sides of Eq. (9.17a), and because $f(x) + f(x) = 0$ under modulo-2 operation, we have

$$\frac{x^{n-k}\, d(x) + \rho(x)}{g(x)} = q(x) \tag{9.17b}$$

or

$$q(x)\, g(x) = x^{n-k}\, d(x) + \rho(x) \tag{9.17c}$$

Because $q(x)$ is of the order of $k - 1$ or less, $q(x)\, g(x)$ is a code word. Because $x^{n-k}\, d(x)$ represents $d(x)$ shifted to the left by $n - k$ digits, the first k digits of this code word are precisely $\boldsymbol{d}$, and the last $n - k$ digits corresponding to $\rho(x)$ must be parity-check digits. This will become clear by considering a specific example.

■ EXAMPLE 9.4

Construct a systematic (7, 4) cyclic code using a generator polynomial (see Example 9.3)

$$g(x) = x^3 + x^2 + 1$$

Consider a data vector $\boldsymbol{d} = \mathbf{(1010)}$

$$d(x) = x^3 + x$$

and

$$x^{n-k}\, d(x) = x^6 + x^4$$

Hence,

$$
\begin{array}{r}
x^3 + x^2 + 1 \qquad q(x) \\
x^3 + x^2 + 1 \overline{)x^6 + x^4} \\
x^6 + x^5 + x^3 \\
\hline
x^5 + x^4 + x^3 \\
x^5 + x^4 + x^2 \\
\hline
x^3 + x^2 \\
x^3 + x^2 + 1 \\
\hline
1 \leftarrow \rho(x)
\end{array}
$$

Hence,

$$
\begin{aligned}
c(x) &= q(x)\, g(x) \\
&= (x^3 + x^2 + 1)(x^3 + x^2 + 1) \\
&= x^6 + x^4 + 1
\end{aligned}
$$

and

$$\mathbf{c} = \mathbf{1010001}$$

Table 9.5 shows the complete code. Note that $d_{\min}$, the minimum distance between two code words, is 3. Hence, this is a single-error-correcting code.

Cyclic codes can also be described by a generator matrix G (see Probs. 9.12 and 9.13). It can be shown that Hamming codes are cyclic codes. ■

Table 9.5

d	**c**
1111	1111111
1110	1110010
1101	1101000
1100	1100101
1011	1011100
1010	1010001
1001	1001011
1000	1000110
0111	0111001
0110	0110100
0101	0101110
0100	0100011
0011	0011010
0010	0010111
0001	0001101
0000	0000000

Cyclic Code Generation

One of the advantages of cyclic codes is that coding and decoding can be implemented using such simple elements as shift registers and modulo-2 adders. A systematic code generation is described in Eq. (9.16a and b). This involves division of $x^{n-k}\,d(x)$ by $g(x)$ and can be implemented by a dividing circuit, which is a shift register with feedback connections according to the generator polynomial* $g(x) = x^{n-k} + g_1x^{n-k-1} + \cdots + g_{n-k-1}x + 1$. The gain g_k's are either 0 or 1. An encoding circuit with $(n - k)$ shift registers is shown in Fig. 9.2. An understanding of this dividing

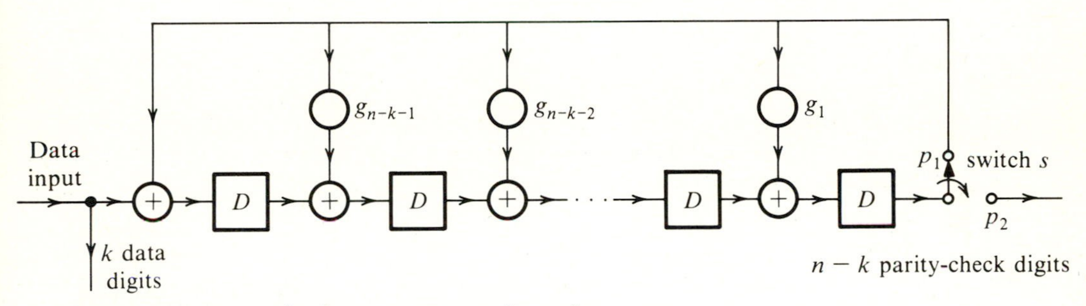

Figure 9.2 An encoder for systematic cyclic code.

circuit requires some background in linear sequential networks. An explanation of its functioning can be found in Peterson and Weldon.[1] The k data digits are shifted in one at a time at the input with the switch s held at position p_1. The symbol D represents a one-digit delay. As the data digits move through the encoder, they are also shifted out onto the output line, because the first k digits of code word are the data digits themselves. As soon as the last (or kth) data digit clears the last $(n - k)$ register, all the registers contain the parity-check digits. The switch s is now thrown to position p_2, and the $n - k$ parity-check digits are shifted out one at a time onto the line.

Decoding

Every valid code polynomial $\mathbf{c}(x)$ is a multiple of $\mathbf{g}(x)$. If an error occurs during the transmission, the received word polynomial $\mathbf{r}(x)$ will not be a multiple† of $\mathbf{g}(x)$. Thus,

$$\frac{r(x)}{g(x)} = m_1(x) + \frac{s(x)}{g(x)} \tag{9.18}$$

and

$$s(x) = \text{Rem}\,\frac{r(x)}{g(x)} \tag{9.19}$$

where the syndrome polynomial $s(x)$ has a degree $n - k - 1$ or less.

---

* It can be shown that for cyclic codes, the generator polynomial must be of this form.

† This assumes that the number of errors in $\mathbf{r}$ is correctable.

If $e(x)$ is the error polynomial, then

$$r(x) = c(x) + e(x)$$

Remembering that $c(x)$ is a multiple of $g(x)$,

$$\begin{aligned} s(x) &= \text{Rem}\,\frac{r(x)}{g(x)} \\ &= \text{Rem}\,\frac{c(x) + e(x)}{g(x)} \\ &= \text{Rem}\,\frac{e(x)}{g(x)} \end{aligned} \tag{9.20}$$

Again, as before, a received word $\boldsymbol{r}$ could result from any one of the 2^k code words and a suitable error. For example, for the code in Table 9.5, if $\boldsymbol{r}$ = **0110010**, this could mean $\boldsymbol{c}$ = **1110010** and $\boldsymbol{e}$ = **1000000**, or $\boldsymbol{c}$ = **1101000** and $\boldsymbol{e}$ = **1011010**, or 14 more such combinations. As seen earlier, the most likely error pattern is the one with the minimum weight (or minimum number of **1**'s). Hence, here $\boldsymbol{c}$ = **1110010** and $\boldsymbol{e}$ = **1000000** is the correct decision.

It is convenient to prepare a decoding table, that is, to list the syndromes for all correctable errors. For any $\boldsymbol{r}$, we compute the syndrome from Eq. (9.19) and from the table determine the corresponding correctable error $\boldsymbol{e}$, and we decide $\boldsymbol{c} = \boldsymbol{r} \oplus \boldsymbol{e}$. Note that computation of $s(x)$ [see Eq. (9.18)] involves exactly the same operation as that required to compute $\rho(x)$ in coding [see Eq. (9.17a)]. Hence, the circuit in Fig. 9.2 can also be used to compute $s(x)$.

■ EXAMPLE 9.5

Construct the decoding table for the single-error-correcting (7, 4) code in Table 9.5. Determine the data vectors transmitted for the following received vectors $\boldsymbol{r}$: (1) **1101101** (2) **0101000**, and (3) **0001100**.

☐ **Solution:** The first step is to construct the decoding table. Because $n - k - 1 = 2$, the syndrome polynomial is of the second order, and there are seven possible nonzero syndromes. There are also seven possible correctable single-error patterns because $n = 7$. Using Eq. (9.20), we compute the syndrome for each of the seven correctable error patterns. For example, for $\boldsymbol{e}$ = **1000000**, $e(x) = x^6$, and for this code $g(x) = x^3 + x^2 + 1$ (see Example 9.4). Hence,

$$\begin{array}{r|l} & x^3 + x^2 + x \\ x^3 + x^2 + 1 & x^6 \\ & \underline{x^6 + x^5 + x^3} \\ & \quad x^5 + x^3 \\ & \quad \underline{x^5 + x^4 + x^2} \\ & \qquad x^4 + x^3 + x^2 \\ & \qquad \underline{x^4 + x^3 + x} \\ & \qquad\qquad x^2 + x \quad \leftarrow s(x) \end{array}$$

Hence,

$$s = (110)$$

In a similar way, we compute the syndromes for the remaining error patterns (see Table 9.6)

Table 9.6

e	s
1000000	110
0100000	011
0010000	111
0001000	101
0000100	100
0000010	010
0000001	001

When the received word $\mathbf{r}$ is **1101101**,

$$r(x) = x^6 + x^5 + x^3 + x^2 + 1$$

We now compute $s(x)$ according to Eq. (9.19):

$$\begin{array}{r} x^3 \\ x^3 + x^2 + 1 \overline{\big)x^6 + x^5 + x^3 + x^2 + 1} \\ \underline{x^6 + x^5 + x^3 } \\ x^2 + 1 \quad \leftarrow s(x) \end{array}$$

Hence s = **101**. From Table 9.6, this gives $\mathbf{e}$ = **0001000**, and

$$\mathbf{c} = \mathbf{r} \oplus \mathbf{e} = \mathbf{1101101} \oplus \mathbf{0001000} = \mathbf{1100101}$$

Hence, from Table 9.5 we have

$$\mathbf{d} = \mathbf{1100}$$

In a similar way, we determine: For $\mathbf{r}$ = **0101000**, s = **110** and $\mathbf{e}$ = **1000000**; hence $\mathbf{c} = \mathbf{r} \oplus \mathbf{e}$ = **1101000**, and $\mathbf{d}$ = **1101**. For $\mathbf{r}$ = **0001100**, s = **001** and $\mathbf{e}$ = **0000001**; hence $\mathbf{c} = \mathbf{r} \oplus \mathbf{e}$ = **0001101**, and $\mathbf{d}$ = **0001**. ■

Bose-Chaudhuri-Hocquenghen (BCH) Codes. The BCH codes are perhaps the most powerful of the random-error-correcting cyclic codes. Moreover, their decoding procedure can be implemented simply. These codes are described as follows:

For any positive integers m' and $t(t < 2^{m'-1})$, there exists a t-error-correcting (n, k) code with $n = 2^{m'} - 1$ and $n - k \leq m't$. The minimum distance $d_{\min}$ between code words is $\geq 2t + 1$.

The detailed treatment of BCH codes requires extensive use of modern algebra and is beyond our scope. For further discussion of BCH codes, the reader is referred to Lin[2] or Peterson and Weldon.[1]

9.4 BURST-ERROR-DETECTING AND -CORRECTING CODES

Thus far we have considered detecting or correcting errors that occur independently, or randomly, in digit positions. On some channels, disturbances can wipe out an entire block of digits. For instance, a stroke of lightning or a man-made electrical disturbance can affect several adjacent transmitted digits. On magnetic storage systems, magnetic tape defects usually affect more than one digit. Burst errors are those errors that wipe out some or all of a sequential set of digits. In general, random-error-correcting codes are not efficient for correcting burst errors. Hence, special *burst-error-correcting codes* are used for this purpose.

A burst of length b is defined as a sequence of digits in which the first and the bth digit are in error, with the $b - 2$ digits in between either in error or received correctly. For example, an error vector $\boldsymbol{e}$ = **0010010100** has a burst length of 6.

It can be shown that for detecting all burst errors of length b or less with a linear block code of length n, b parity-check bits are necessary and sufficient.[1] We shall prove the sufficiency part of this theorem by constructing a code of length n with b parity-check digits that will detect a burst of length b.

To construct such a code, let us group k data digits into segments of b digits in length (Fig. 9.3). To this we add a last segment of b parity-check digits, which are

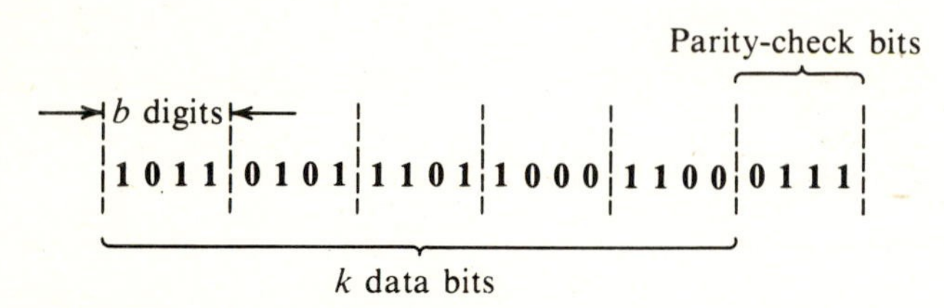

Figure 9.3 Burst-error detection.

determined as follows. The modulo-2 sum of the ith digit in each segment (including the parity-check segment) must be zero. For example, the first digits in the five data segments are **1**, **0**, **1**, **1**, and **1**. Hence, we must have **0** as the first parity-check digit in order to obtain a modulo-2 sum zero. We continue in this way with the second digit, the third digit, and so on, to the bth digit. Because parity-check digits are a linear combination of data digits, this is a linear block code. Moreover, it is a systematic code.

It is easy to see that if a digit sequence of length b or less is in error, parity will be violated and the error will be detected (but not corrected), and the receiver can request retransmission of the digits lost. One of the interesting properties of this code is that b, the number of parity-check digits, is independent of k (or n), which makes it a very useful code for such systems as packet switching, where the data digits may vary from packet to packet. It can be shown that a linear code with b parity bits detects not only all bursts of length b or less, but also a high percentage of longer bursts, as well.[1]

If we are interested in correcting rather than detecting burst errors, we require twice as many parity-check digits. A theorem says: In order to correct all burst errors of length b or less, a linear block code must have at least $2b$ parity-check digits.[1]

In order to correct all burst errors of length b or less, and simultaneously detect all bursts of length $l \geq b$ or less, the code must have at least $b + l$ parity-check digits.[1]

9.5 INTERLACED CODES FOR BURST- AND RANDOM-ERROR CORRECTION

In general, random-error-correcting codes are not efficient for burst-error correcting, and burst-error-correcting codes are not efficient for random-error correcting. Unfortunately, in most practical systems, we have errors of both kinds. Out of the several methods proposed to simultaneously correct random and burst errors, the *interlaced code* is the most effective.

x_1	x_2	x_3	. . .	x_{14}	x_{15}
y_1	y_2	y_3	. . .	y_{14}	y_{15}
z_1	z_2	z_3	. . .	z_{14}	z_{15}

Figure 9.4 Random and burst-error correction.

For an (n, k) code, if we interlace λ code words, we have what is known as a $(\lambda n, \lambda k)$ *interlaced code*. Instead of transmitting code words one by one, we group λ code words and interlace them. Consider, for example, the case of $\lambda = 3$ and a two-error-correcting (15, 8) code. Each code word has 15 digits. We group code words to be transmitted in groups of three. Suppose the first three code words to be transmitted are $\mathbf{x}(x_1, x_2, \ldots, x_{15})$, $\mathbf{y}(y_1, y_2, \ldots, y_{15})$, and $\mathbf{z}(z_1, z_2, \ldots, z_{15})$, respectively. Then instead of transmitting $\mathbf{xyz}$ in sequence as $x_1, x_2, \ldots, x_{15}, y_1, y_2, \ldots, y_{15}, z_1, z_2, \ldots, z_{15}$, we transmit $x_1, y_1, z_1, x_2, y_2, z_2, x_3, y_3, z_3, \ldots, x_{15}, y_{15}, z_{15}$. This can be explained graphically by Fig. 9.4, where λ code words (3 in this case) are arranged in rows. In usual transmission, we transmit one row after another. In the interlaced case, we transmit columns (of λ elements) in sequence. When all the 15 (n) columns are transmitted, we repeat the procedure for next λ code words to be transmitted.

To explain the error-correcting capabilities of this code, we observe that the decoder will first remove the interlacing and regroup the received digits as $x_1, x_2, \ldots, x_{15}, y_1, y_2, \ldots, y_{15}, z_1, z_2, \ldots, z_{15}$. Suppose the shaded digits in Fig. 9.4 were in error. Because the code is a two-error-correcting code, two or less errors in each row will be corrected. Hence, all the errors in Fig. 9.4 are correctable. We see that there are two random, or independent, errors and one burst of length 4 in all the 45 digits transmitted. In general, if the original (n, k) code is t-error correcting, the interlaced code can correct any combination of t bursts of length λ or less.

9.6 CONVOLUTIONAL CODES

Convolutional (or recurrent) codes, first introduced by Elias in 1955,[3] differ from block codes as follows. In a block code, the block of n code digits generated by the encoder in any particular time unit depends only on the block of k input data digits within that time unit. In a convolutional code, on the other hand, the block of n code digits generated by the encoder in a particular time unit depends not only on the block of k message digits within that time unit but also on the block of data digits within a previous span of $N - 1$ time units $(N > 1)$. For convolutional codes, k and n are usually small. Convolutional codes can be devised for correcting random errors, burst errors, or both. Encoding is easily implemented by shift registers. As a class, convolutional codes invariably outperform block codes of the same order of complexity.

A convolutional coder with constraint length N consists of an N-stage shift register and v modulo-2 adders. Figure 9.5 shows such a coder for the case of $N = 3$ and

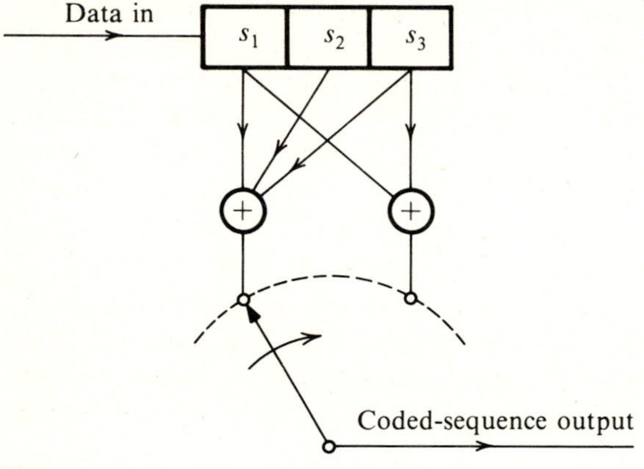

Figure 9.5 A convolutional coder.

$v = 2$. The message digits are applied at the input of the shift register. The coded digit stream is obtained at the commutator output. The commutator samples the v modulo-2 adders in sequence, once during each input-bit interval. We shall explain this operation with reference to the input digits **11010**. Initially, all the stages of the register are clear; that is, they are in a **0** state. When the first data digit **1** enters the register, the stage s_1 shows **1** and all the other stages (s_2 and s_3) are unchanged; that is, they are in a **0** state. The two modulo-2 adders show $v_1 = \mathbf{1}$ and $v_2 = \mathbf{1}$. The commutator samples this output. Hence the coder output is **11**. When the second message bit **1** enters the register, it enters the stage s_1, and the previous **1** in s_1 is shifted to s_2. Hence, s_1 and s_2 both show **1**, and s_3 is still unchanged; that is, it is in a **0** state. The modulo-2 adders now show $v_1 = \mathbf{0}$ and $v_2 = \mathbf{1}$. Hence, the decoder output is **01**. In the same way, when the third message digit **0** enters the register, we have $s_1 = \mathbf{0}$, $s_2 = \mathbf{1}$, and

s_3 = **1**, and the decoder output is **01**. Observe that each data digit influences N groups of v digits in the output (in this case three groups of two digits). The process continues until the last data digit enters the stage s_1.* We cannot stop here, however. We continue adding a sufficient number of **0**'s (N to be exact) to the input stream to make sure that the last data digit (**0** in this case) proceeds all the way through the shift register in order to influence the N groups of v digits. Hence, when the input digits are **11010**, we actually apply **11010000** (the digits augmented by N zeros) to the input of the shift register. It can be seen that when the last digit of the augmented message stream enters s_1, the last digit of the message stream has passed through all the N stages of the register, and the register is in clear state. The reader can verify that the coder output is given by **11 01 01 00 10 11 00 00**. Thus, there are in all $n = (N + k)v$ digits in the coded output for every k data digits. In practice, $k \gg N$, and, hence, there are approximately kv coded output digits for every k data digits, giving an efficiency† $\eta \simeq 1/v$.

It can be seen that unlike the block coder, the convolutional coder operates on a continuous basis, and each data digit influences N groups of v digits in the output.

The Code Tree

Coding and decoding is considerably facilitated by what is known as the *code tree,* which shows the coded output for any possible sequence of data digits. The code tree for the coder in Fig. 9.5 with $k = 5$ is shown in Fig. 9.6. When the first digit is **0**, the coder output is **00**, and when it is **1**, the output is **11**. This is shown by the two tree branches that start at the initial node. The upper branch represents **0**, and the lower branch represents **1**. This convention will be followed throughout. At the terminal node of each of the two branches, we follow a similar procedure, corresponding to the second data digit. Hence, two branches initiate from each node, the upper one for **0** and the lower one for **1**. This continues until the kth data digit. From there on, all the input digits are **0** (augmented digit), and we have only one branch until the end. Hence, in all there are 32 (or 2^k) outputs corresponding to 2^k possible data vectors. The coded output for input **11010** can be easily read from this tree (the path shown dotted in Fig. 9.6).

Figure 9.6 shows that the code tree becomes repetitive after the third branch. This can be seen from the fact that the two blocks enclosed inside the dotted lines are identical. This means that the output from the fourth input digit is the same whether the first digit was **1** or **0**. This is not surprising in view of the fact that when the fourth input digit enters the shift register, the first input digit is shifted out of the register, and it ceases to influence the output digits. In other words, the data vector $\mathbf{1}x_1x_2x_3x_4\ldots$ and the data vector $\mathbf{0}x_1x_2x_3x_4\ldots$ generate the same output after the third group of output digits. It is convenient to label the four third-level nodes (the nodes appearing at the beginning of the third branch) as nodes a, b, c, and d (Fig. 9.6). The repetitive structure begins at the fourth level nodes and continues at the fifth-level

---

* For a systematic code, one of the output digits must be the data digit itself.

† In general, instead of shifting one digit at a time, b digits may be shifted at a time. In this case $\eta \simeq b/v$.

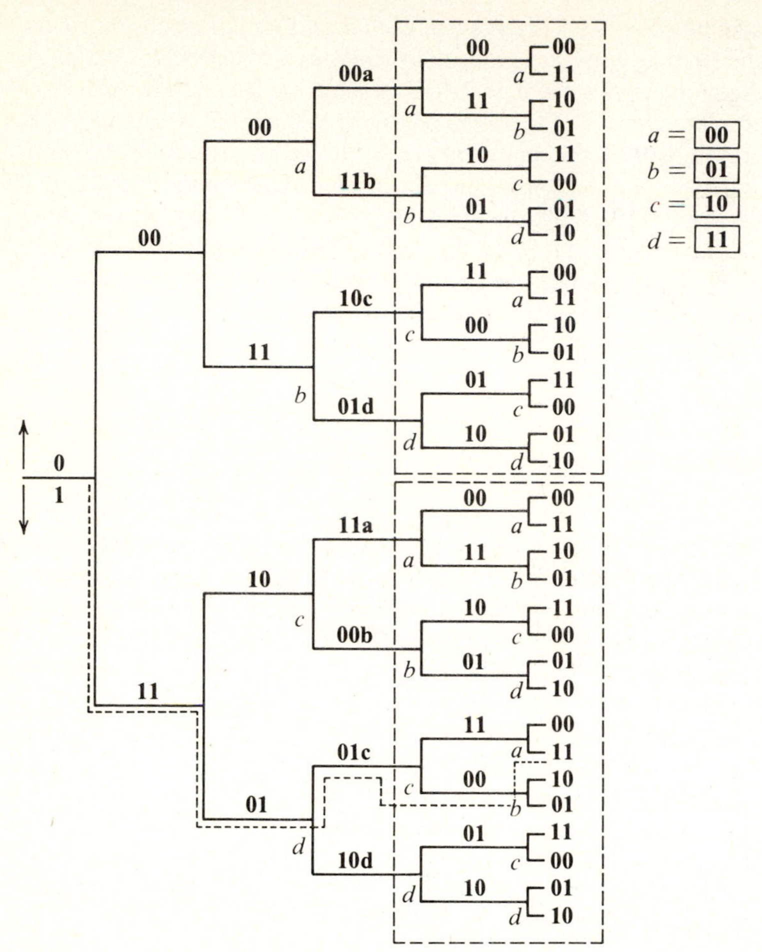

Figure 9.6 Code tree for the coder in Fig. 9.5.

nodes, whose behavior is similar to that of nodes a, b, c, and d at the third level. Hence, we label the fourth- and fifth-level nodes also as either a, b, c, or d. What this means is that at the fifth-level nodes, the first two data digits have become irrelevant; that is, any of the four combinations (**11** **10** **01** or **00**) for the first two data digits will give the same output after the fifth node.

This behavior can be seen from another point of view. The encoder output depends only on the two previous data digits and the one present data digit. It can be seen that all nodes labeled a are the nodes where the two previous data digits are **00**. Similarly, all nodes labeled b represent the previous data digits **01**, and nodes c and d represent the previous data digits **10** and **11**, respectively. Thus, nodes a, b, c, and d may be considered as states of the encoder. If the encoder is in state a (previous two data digits **00**) and the present data digit is **0**, the encoder output is **00**, and if the present data digit is **1**, the encoder output is **11**. This is clearly seen from any node labeled a. This entire description can be concisely expressed by the state diagram shown in Fig. 9.7. This is a four-state directed graph used to uniquely represent the

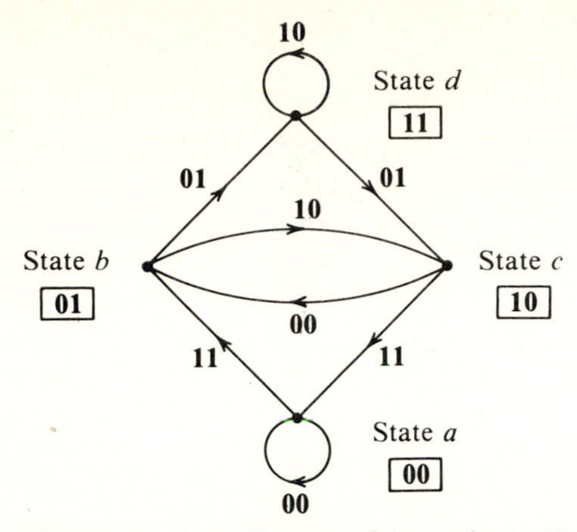

Figure 9.7 State diagram of the coder in Fig. 9.5.

input-output relation of the eight-state machine. Note that the three-digit encoder is an eight-state machine. This graph is interpreted as follows: If the encoder goes from state a to state b, the transition branch gives the encoder output as **11**. Similarly if it goes from a to a, the encoder output is **00**, and so on. The encoder cannot go from a to c in one step. It must go from a to b to c, or from a to b to d to c, and so on. We can verify these facts from the code tree. If the encoder is in state a (i.e., the previous two data digits are **00**), then if the present data digit is **1**, the encoder output will be **11**. But in so doing, the state has changed from a to b, because after the data digit **1** is shifted in, the previous two digits become **01** (state b). Thus, Fig. 9.7 contains the complete information of the code tree.

Another useful way of representing the code tree is the trellis diagram in Fig. 9.8. The diagram starts from scratch (all **0**'s in the shift register) and makes transitions corresponding to each input data digit. These transitions are denoted by a solid line for the next data digit **0** and by a dotted line for the next data digit **1**. Thus when the

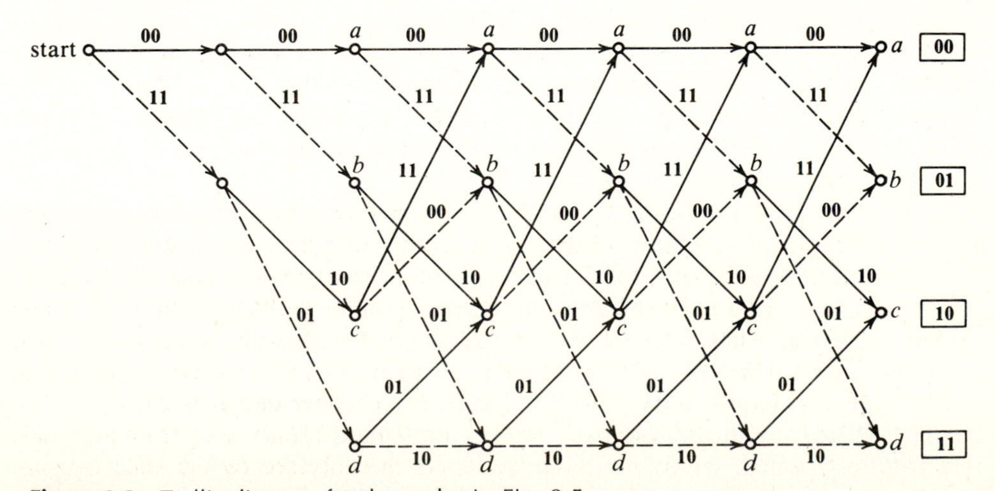

Figure 9.8 Trellis diagram for the coder in Fig. 9.5.

first input data is **0** the encoder output is **00** (solid line), and when the input digit is **1**, the encoder output is **11** (the dotted line). We continue this way with the second input digit. After the first 2 input digits, the encoder is in one of the four states a, b, c, or d as shown in Fig. 9.8. If the encoder is in the state a (previous 2 data digits **00**), it goes to the state b if the next input bit is **1** and remains in the state a if the next input bit is **0**. In so doing, the encoder output is **11** (a to b) or **00** (a to a). Note that the structure of the trellis diagram is completely repetitive, as expected.

Decoding. We shall consider two important techniques: (1) maximum-likelihood decoding (Viterbi's algorithm) and (2) sequential decoding.

Maximum-Likelihood Decoding: Viterbi's Algorithm

Among various decoding methods for convolutional codes, Viterbi's maximum-likelihood algorithm[4] is one of the best techniques yet evolved for digital communications where energy efficiency dominates in importance. It permits major equipment simplification while obtaining the full performance benefits of maximum-likelihood decoding. The decoder structure is relatively simple for short constraint length, making decoding feasible at relatively high rates of up to 100 Mbits/second.

The maximum-likelihood receiver implies selecting a code word closest to the received word. Because there are 2^k code words, the maximum-likelihood decision involves storage of 2^k words and their comparison with the received word. This calculation is extremely difficult for large k and would result in an overly complex decoder.

A major simplification was made by Viterbi in the maximum-likelihood calculation by noting that each of the four nodes (a, b, c, and d) has only two predecessors; that is, each node can be reached through two nodes only (see Fig. 9.7 or 9.8), and only the path that agrees most with the received sequence (the minimum-distance path) need be retained for each node. This can be explained with reference to the trellis diagram in Fig. 9.8. Our problem is as follows: Given a received sequence of bits, we need to find a path in the trellis diagram with output digit sequence that agrees most with the received sequence.

Suppose the first six received digits are **01 00 01**. We shall consider two paths of three branches (for six digits) leading to each of the nodes a, b, c, and d. Out of the two paths reaching into each node, we shall retain only the one that agrees most with the received sequence **01 00 01** (the minimum-distance path). The retained path is called the *survivor* at that node. There are two paths (**00 00 00** and **11 10 11**) that arrive at the third-level node a. These paths are at distances of 2 and 3, respectively, from the received sequence **01 00 01**. Hence, the survivor at the third-level node a is **00 00 00**. We repeat the procedure for nodes b, c, and d. For example, the two paths reaching to the third-level node c (the node after three branches) are **00 11 10** and **11 01 01**, at distances of 5 and 2, respectively, from the received sequence **01 00 01**. Hence, the survivor at the third-level node c is **11 01 01**. Similarly, we find survivors at the third-level nodes b and d. With four paths eliminated, the four survivor paths are the only contenders. The reason behind eliminating the other four paths is as follows. The two paths merging at the third-level node a, for example, imply that the previous two data digits are identical (viz., **00**). Hence regardless of what the future data digits are, both paths must merge at this node a and follow a common path in the future. Clearly, the survivor path is the minimum-distance path between the two, regardless of future data digits. What we need to remember is the four survivor paths and their distances from the received sequence.

Once we have survivors at all the third-level nodes, we look at the next two received digits. Suppose these are **11** (i.e., the received sequence is **01 00 01 11**). We now compare the two survivors that merge into the fourth-level node a. These are the survivors at nodes a and c of the third-level, with paths **00 00 00 00** and **11 01 01 11**, respectively, and distances of 4 and 2, respectively, from the received sequence **01 00 01 11**. Hence, the path **11 01 01 11** is the survivor at the fourth-level node a. We repeat this procedure for nodes b, c, and d and continue in this manner until the end. Note that only two paths merge in a node and there are only four contending paths (the four survivors at nodes a, b, c, and d) until the end. The only remaining problem is how to truncate the algorithm and ultimately decide on one path rather than four. This is done by forcing the last two data digits to be **00**. This forces the final state of the code to be a (remember that "the last two data digits **00**" is state a). Consequently, the ultimate survivor is the survivor at node a after insertion into the coder of two dummy zeros and transmission of the corresponding four code digits. In terms of the trellis diagram, this means that the number of states is reduced from four to two (a and c) by insertion of the first zero and to a single state (a) by insertion of the second zero.

With the Viterbi algorithm, storage and computational complexity are proportional to 2^N and are very attractive for constraint length $N < 10$. To achieve very low error probabilities, longer constraint lengths are required, and sequential decoding (to be discussed next) becomes attractive.

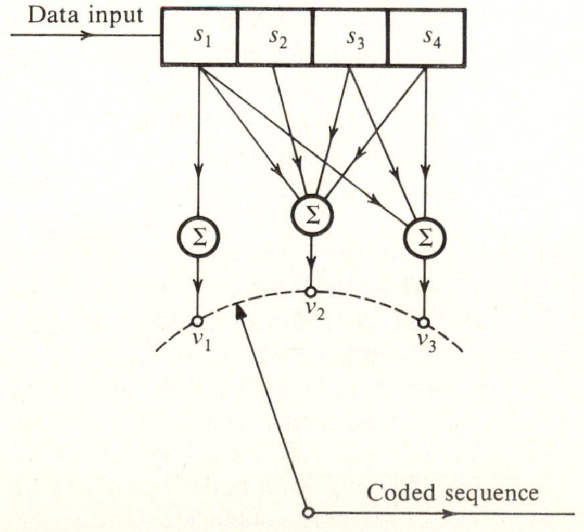

Figure 9.9 A convolution coder.

Sequential Decoding. In this technique, proposed by Wozencraft, decoder complexity increases linearly rather than exponentially. To explain this technique, let us consider a coder with $N = 4$ and $v = 3$ (Fig. 9.9). The code tree for this coder is shown in Fig. 9.10. Each data digit generates three ($v = 3$) output digits but affects four groups of three digits (12 digits) in all.

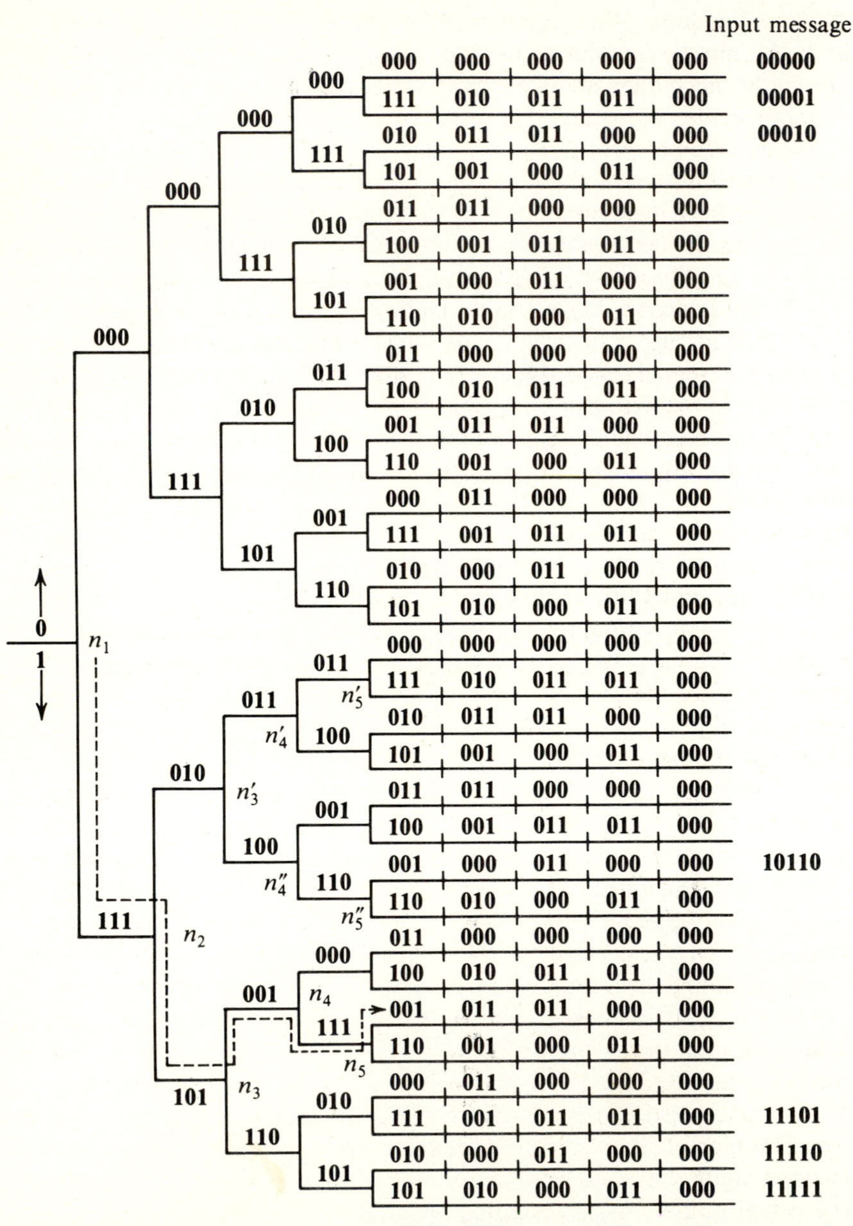

Figure 9.10 Code tree for the coder in Fig. 9.9.

In this decoding scheme, we observe only three (or v) digits at a time to make a tentative decision, with readiness to change our decision if it creates difficulties later. A sequential detector acts much like a driver who occasionally makes a wrong choice at a fork in the road, but quickly discovers his error (because of road signs), goes back and tries the other path.

Applying this insight to our decoding problem, the analogous procedure would be as follows. We look at the first three received digits. There are only two paths of three digits from the initial node n_1. We choose that path whose sequence is at the shortest Hamming distance from the first three received digits. We thus progress to the most likely node. From this node there are two paths of three digits. We look at the second group of the three received digits and choose that path whose sequence is closest to these received digits. We progress this way until the fourth node. If we are unlucky enough to have large number of errors in a certain received group of v digits, we will take a wrong turn, and from there on we will find it more difficult to match the received digits with those along the paths available from the wrong node. This is the clue to the realization that an error has been made. Let us explain this by an example. Suppose a data sequence **11010** is encoded by the coder in Fig. 9.9. The coded sequence will be (see the code tree in Fig. 9.10) **111 101 001 111 001 011 011 000 000**. Let the received sequence be **101 011 001 111 001 011 011 000 000** (three errors: one in the first group and two in the second group). We start at the initial node n_1. The first received group **101** (one error) being closer to **111**, we make a correct decision to go to node n_2. But the second group **011** (two errors) is closer to **010** than to **101** and will lead us to wrong node n_3' rather than to n_3. From here on we are on the wrong track, and, hence, the received digits will not match any path starting from n_3'. The third received group is **001** and does not match any sequence starting at n_3' (viz., **011** and **100**). But it is closer to **011**. Hence, we go to node n_4'. Here again the fourth received group **111** does not match any group starting at n_4' (viz., **011** and **100**). But it is closer to **011**. This takes us to node n_5'. It can be seen that the Hamming distance between the sequence of 12 digits along the path $n_1\ n_2\ n_3'\ n_4'\ n_5'$ and the first 12 received digits is 4, indicating four errors in 12 digits (if our path is correct). Such a high number of errors should immediately make us suspicious. If P_e is the digit error probability, then the expected number of errors n_e in d digits is $P_e d$. Because P_e is of the order of 10^{-4} to 10^{-6}, four errors in 12 digits is unreasonable. Hence, we go back to node n_3' and try the lower branch, leading to n_5''. This path n_1 $n_2\ n_3'\ n_4''\ n_5''$ is even worse than the previous one, because it gives five errors in 12 digits. Hence, we go back even farther to node n_2 and try the path leading to n_3 and farther. We find the path $n_1\ n_2\ n_3\ n_4\ n_5$, giving three errors. If we go back still farther to n_1 and try alternate paths, we find that none yields less than five errors. Thus, the correct path is taken as $n_1\ n_2\ n_3\ n_4\ n_5$. This enables us to decode the first transmitted digit as **1**. Next, we start at node n_2, discard the first three received digits, and repeat the procedure to decode the second transmitted digit. We repeat it until all the digits are decoded.

The next important question concerns the criterion for deciding when the wrong path is chosen. The plot of the expected number of errors n_e as a function of the number of decoded digits d is a straight line ($n_e = P_e d$) with slope P_e, as shown in Fig. 9.11. The actual number of errors along the path is also plotted. If the errors remain within a limit (the discard level), the decoding continues. If at some point it

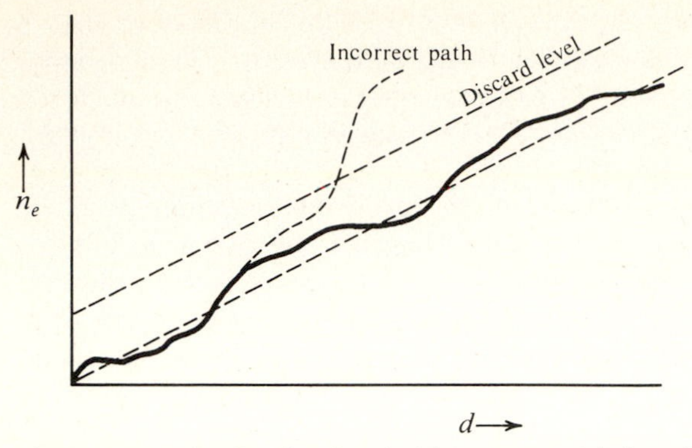

Figure 9.11 Setting the threshold in sequential decoding.

is found that the errors exceed the discard level, we go back to the nearest decision node and try an alternate path. If errors still increase beyond the discard level, we then go back one more node along the path and try an alternate path. The process continues until the errors are within the set limit. By making the discard level very stringent (close to expected error curve), we reduce the average number of computations. On the other hand, if the discard level is made too stringent, the decoder will discard all possible paths in some extremely rare cases where the noise may cause an unusually large number of errors. This difficulty is usually resolved by starting with a stringent discard level. If on rare occasions decoder rejects all paths, the discard level can be relaxed bit by bit until one of the paths is acceptable.

It can be shown that the error probability in this scheme decreases exponentially as N, whereas the system complexity grows only linearly with k.

The disadvantages of sequential decoding are: 1) The number of incorrect path branches, and consequently the computation complexity, is a random variable depending on the channel noise. The efficiency η is $\simeq 1/v$. It can be shown that for $\eta < \eta_o$ (computational cutoff rate), the average number of incorrect branches searched per decoded digit is bounded, whereas for $\eta > \eta_o$ it is not; hence η_o is called the *computational cutoff rate*. 2) To make storage requirements easier, the decoding speed has to be maintained at 10 to 20 times faster than the incoming data rate. This limits the maximum data-rate capability. 3) The average number of branches can occasionally become vary large and may result in a storage overflow, causing relatively long sequences to be erased.

A third technique for decoding convolutional codes is *feedback decoding*, with threshold decoding[5] as a subclass. Threshold decoders are easily implemented. Their performance, however, does not compare favorably with the previous two methods.

9.7 COMPARISON OF CODED AND UNCODED SYSTEMS

It is instructive to compare the error probability of coded and uncoded schemes under similar constraints of power and information rate.

Let us consider a t-error-correcting (n, k) code. In this case, k information digits are coded into n digits. For a proper comparison, we shall assume that k information digits are transmitted in the same time interval over both the systems and that the transmitted power S_i is also maintained the same for both systems. Because only k digits are required to be transmitted over the uncoded system (versus n over the coded one), the digital transmission rate f_o is lower for uncoded system by a factor of k/n as compared to the coded system. Hence, $\lambda = S_i/\mathcal{N}f_o$ is higher for the uncoded case as compared to the coded case by a factor n/k. This tends to reduce the digit error probability for the uncoded case. Let P_{eu} and P_{ec} represent the digit error probabilities in the uncoded and coded cases, respectively.

For the uncoded case, a word of k digits will be received wrong if any one of the k digits is in error. *If* P_{Eu} and P_{Ec} represent the word error probabilities of the uncoded and coded systems, respectively, then

$$P_{Eu} = 1 - P\text{ (all } k \text{ digits received correct)}$$

$$= 1 - (1 - P_{eu})^k \tag{9.21a}$$

$$\simeq kP_{eu} \qquad P_{eu} \ll 1 \tag{9.21b}$$

For a t-error-correcting (n, k) code, the received word will be in error if more than t errors occur in n digits. If $P(j, n)$ is the probability of j errors in n digits, then

$$P_{Ec} = \sum_{j=t+1}^{n} P(j, n)$$

Because there are $\binom{n}{j}$ ways in which j errors can occur in n digits (Example 5.6),

$$P(j, n) = \binom{n}{j} P_{ec}^j (1 - P_{ec})^{n-j}$$

and

$$P_{Ec} = \sum_{j=t+1}^{n} \binom{n}{j} P_{ec}^j (1 - P_{ec})^{n-j} \tag{9.22}$$

For $P_{ec} \ll 1$, the first term in the summation in Eq. (9.22) dominates all the other terms, and we are justified in ignoring all but the first term. Hence,

$$P_{Ec} = \binom{n}{t+1} P_{ec}^{t+1} (1 - P_{ec})^{n-(t+1)} \tag{9.23a}$$

$$\simeq \binom{n}{t+1} P_{ec}^{t+1} \qquad P_{ec} \ll 1 \tag{9.23b}$$

For further comparison, we must assume some specific transmission scheme. Let us consider a coherent PSK scheme. In this case for an AWGN channel,

$$P_{eu} = Q(\sqrt{2\lambda}) \tag{9.24a}$$

and because λ for the coded case is k/n times that for the uncoded case,

$$P_{ec} = Q\left(\sqrt{\frac{2k\lambda}{n}}\right) \tag{9.24b}$$

Hence,

$$P_{Eu} = kQ(\sqrt{2\lambda}) \tag{9.25a}$$

$$P_{Ec} = \binom{n}{t+1}\left[Q\left(\sqrt{\frac{2k\lambda}{n}}\right)\right]^{t+1} \qquad P_{ec} \ll 1 \tag{9.25b}$$

To compare the coded and uncoded systems, we could plot P_{Eu} and P_{Ec} as functions of λ. Because Eq. (9.25a and b) involve parameters t, n, and k, a proper comparison requires families of plots. For the case of a (7, 4) single-error-correcting code ($t = 1$, $n = 7$, and $k = 4$), P_{Ec} and P_{Eu} in Eq. (9.25a and b) are plotted in Fig. 9.12 as a

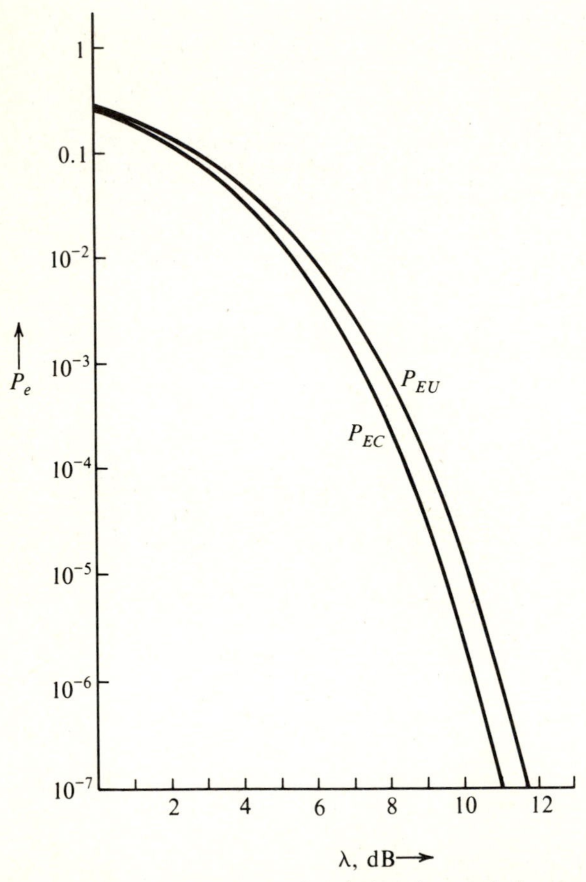

Figure 9.12 Comparison of coded and uncoded systems.

function of λ. Observe that the coded scheme is superior to the uncoded scheme, but the improvement (about 1 dB) is not too significant. For large n and k, however, the coded scheme can become significantly superior to the uncoded one. For practical channels plagued by fading and impulse noise, coding can yield substantial gains.

■ EXAMPLE 9.6

Compare the performance of an AWGN BSC using a single-error-correcting (15, 11) code with that of the same system using uncoded transmission, given that $\lambda = 9.12$ for the uncoded scheme and coherent PSK is used to transmit the data.

□ **Solution:** From Eq. (9.25a),

$$P_{Eu} = 11Q(\sqrt{18.24}) = 1.1 \times 10^{-4}$$

and [Eq. (9.25b)]

$$P_{Ec} = \binom{15}{2}\left[Q\left(\sqrt{\frac{11}{15}(18.24)}\right)\right]^2$$

$$= 105(1.37 \times 10^{-4})^2 = 1.96 \times 10^{-6}$$

Note that the word error probability of the coded system is reduced by a factor of 56. On the other hand, if we wish to achieve the error probability of the coded transmission (1.96×10^{-6}) using the uncoded system, we must increase the transmitted power. If λ' is the new value of λ to achieve $P_{Eu} = 1.96 \times 10^{-6}$,

$$P_{Eu} = 11Q(\sqrt{2\lambda'}) = 1.96 \times 10^{-6}$$

This gives $\lambda' = 10.7$.

This is an increase over the old value of 9.12 by a factor of 1.17, or 0.7 dB. ■

REFERENCES

1. W. W. Peterson and E. J. Weldon, Jr., *Error Correcting Codes,* 2nd ed., Wiley, New York, 1972.
2. S. Lin, *An Introduction to Error Correcting Codes,* Prentice-Hall, Englewood Cliffs, N.J., 1970.
3. P. Elias, "Coding For Noisy Channels," *IRE Nat. Conv. Rec.*, vol. 3, part 4, pp. 37–46, 1955.
4. A. J. Viterbi, "Convolutional Codes and their Performance in Communication Systems," *IEEE Trans. Commun. Technol.*, vol. CT-19, pp. 751–771, Oct. 1971.
5. J. L. Massey, *Threshold Decoding,* M.I.T. Press, Cambridge, Mass., 1963.

PROBLEMS

9.1. Golay's (23, 12) codes are three-error-correcting codes. Verify that $n = 23$ and $k = 12$ satisfies the Hamming bound exactly for $t = 3$.

9.2. **(a)** Determine the Hamming bound for a ternary code.
(b) A ternary (11, 6) code exists that can correct up to two errors. Verify that this code satisfies the Hamming bound exactly.

9.3. Confirm the possibility of a (18, 7) binary code that can correct up to three errors. Can this code correct up to four errors?

9.4. Consider a generator matrix G for a nonsystematic (6, 3) code:

$$G = \begin{bmatrix} 1 & 0 & 1 & 1 & 0 & 0 \\ 0 & 0 & 1 & 0 & 1 & 0 \\ 1 & 1 & 0 & 0 & 0 & 1 \end{bmatrix}$$

Construct the code for this G, and show that $d_{\min}$, the minimum distance between code words, is 3. Consequently, this code can correct at least one error.

9.5. Given a nonsystematic generator matrix

$G = [1\ \ 1\ \ 1]$

construct a (3, 1) code. How many errors can this code correct? Find the code word for data vectors $\mathbf{d} = \mathbf{0}$ and $\mathbf{d} = \mathbf{1}$. Comment.

9.6. Find a generator matrix G for a (15, 11) single-error-correcting linear block code. Find the code word for the data vector **01001010111**.

9.7. For a (6, 3) systematic linear block code, the three parity-check digits c_4, c_5, and c_6 are

$$c_4 = d_1 + d_2 + d_3$$

$$c_5 = d_1 + d_2$$

$$c_6 = d_2 + d_3$$

(a) Construct the appropriate generator matrix for this code.
(b) Construct the code generated by this matrix.
(c) Determine the error-correcting capabilities of this code.
(d) Prepare a suitable decoding table.
(e) Decode the following received words: **101100**, **000110**, **101010**.

9.8. Construct a single-error-correcting (7, 4) linear block code (Hamming code) and the corresponding decoding table.

9.9. For the (6, 3) code in Example 9.1, the decoding table is Table 9.3. Show that if we use this decoding table, and a two-error pattern **010100** or **001001** occurs, it will not be corrected. If it is desired to correct a single two-error pattern **010100** (along with six single-error patterns), construct the appropriate decoding table and verify that it does indeed correct one two-error pattern **010100** and that it cannot correct any other two-error patterns.

9.10. **(a)** Given $k = 8$, find the minimum value of n for a code that can correct at least one error.
(b) Choose a generator matrix G for this code.
(c) How many double errors can this code correct?
(d) Construct a decoding table (syndromes and corresponding correctable error patterns).

9.11. **(a)** Construct a systematic (7, 4) cyclic code using the generator polynomial $g(x) = x^3 + x + 1$.
(b) What are the error-correcting capabilities of this code?
(c) Construct the decoding table.
(d) If the received word is **1101100**, determine the transmitted data word.

9.12. Equation (9.15) suggests a method of constructing a generator matrix G' for a cyclic code:

$$G' = \begin{bmatrix} x^{k-1}g(x) \\ x^{k-2}g(x) \\ \cdots \\ g(x) \end{bmatrix} = \begin{bmatrix} g_1 & g_2 & \cdots & g_{n-k+1} & 0 & 0 & \cdots & 0 \\ 0 & g_1 & g_2 & \cdots & g_{n-k+1} & 0 & \cdots & 0 \\ \cdot & \cdot & \cdot & \cdot & \cdot & \cdot & \cdot & \cdot \\ 0 & 0 & 0 & \cdots & g_1 & g_2 & \cdots & g_{n-k+1} \end{bmatrix}$$

where $g(x) = g_1x^{n-k} + g_2x^{n-k-1} + \cdots + g_{n-k+1}$ is the generator polynomial. This is, in general, a nonsystematic cyclic code.

(a) For a single-error-correcting (7, 4) cyclic code with a generator polynomal $g(x) = x^3 + x^2 + 1$, find G' and construct the code.

(b) Verify that this code is identical to that derived in Example 9.3 (Table 9.4).

9.13. The generator matrix G for a systematic cyclic code (see Prob. 9.12) can be obtained by realizing that adding any row of a generator matrix to any other row yields another valid generator matrix, because the code word is formed by linear combinations of data digits. Also, a generator matrix for a systematic code must have an identity matrix I_k in the first three columns. Such a matrix is formed step by step as follows. Observe that each row in G' in Prob. 9.12 is a left shift of the row below it, with the last row being $g(x)$. Start with the kth (last) row $g(x)$. Because $g(x)$ is of the order $n - k$, this row has the element **1** in the kth column, as required. For the $(k - 1)$th row, use the last row with one left shift. We require a 0 in the kth column of the $(k - 1)$th row to form I_k. If there is a 0 in the kth column of this $(k - 1)$th row, we accept it as a valid $(k - 1)$th row. If not, then we add the kth row to the $(k - 1)$th row to obtain 0 in its kth column. The resulting row is the final $(k - 1)$th row. This row with a single left shift serves as the $(k - 2)$th row. But if this newly formed $(k - 2)$th row does not have a 0 in its kth column, we add the kth (last) row to it to get the desired 0. We continue this way until all k rows are formed. This gives the generator matrix for a systematic (n, k) cyclic code.

(a) For a single-error-correcting (7, 4) systematic cyclic code with a generator polynomial $g(x) = x^3 + x^2 + 1$, find G and construct the code.

(b) Verify that this code is identical to that in Table 9.5 (Example 9.4).

9.14. **(a)** Find the generator matrix G' for a nonsystematic (7, 4) cyclic code using the generator polynomial $g(x) = x^3 + x + 1$.

(b) Find the code generated by this matrix G'.

(c) Determine the error-correcting capabilities of this code.

9.15. Find the generator matrix G for a systematic (7, 4) cyclic code using the generator polynomial $g(x) = x^3 + x + 1$ (see Prob. 9.13).

9.16. The simple burst-error-detecting code in Fig. 9.3 can also be used as a single-error-correcting code with a slight modification. The k data digits are divided into groups of b digits in length, as in Fig. 9.3. To each group we add one parity-check digit, so that each segment now has $b + 1$ digits (b data digits and one parity-check digit). The parity-check digit is chosen to ensure that the total number of **1**'s in each segment of $b + 1$ digits is even. Now we consider these digits as our new data and augment them with the last segment of $b + 1$ parity-check digits, as was done in Fig. 9.3. The data in Fig. 9.3 will be transmitted thus:

10111 01010 11011 10001 11000 01111

Show that this (30, 20) code is capable of single-error correction as well as the detection of a single burst of length 5.

9.17. Discuss the error-correcting capabilities of an interlaced $(\lambda n, \lambda k)$ cyclic code with $\lambda = 10$ and using a three-error correcting (31, 16) BCH code.

9.18. Draw the code tree for the convolutional encoder shown in Fig. P9.18 and determine the output digit sequence for the data digits **1101011000**.

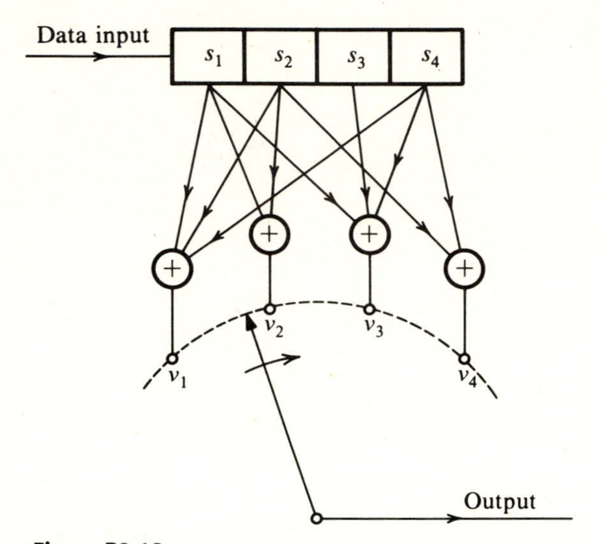

Figure P9.18

9.19. Draw the code tree, the trellis diagram, and the state diagram for the convolutional encoder shown in Fig. P9.19.

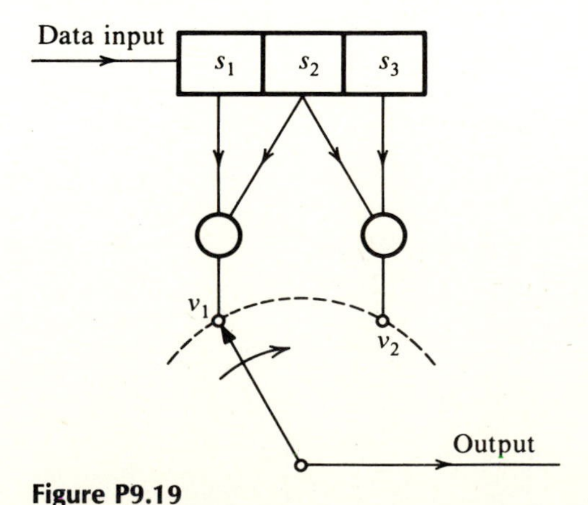

Figure P9.19

9.20. Uncoded data is transmitted using PSK over an AWGN channel with $\lambda = 9.12$. This data is now coded using a three-error-correcting (23, 12) Golay code (see Prob. 9.1) and transmitted over the same channel at the same data rate and with the same transmitted power.

(a) Determine the word error probabilities P_{Eu} and P_{Ec} for the coded and the uncoded systems.

(b) If it is decided to achieve the error probability P_{Ec} computed in part (a) using the uncoded system by increasing the transmitted power, determine the required value of λ.

Appendix A

A.1 TRIGONOMETRIC IDENTITIES

$$e^{\pm jx} = \cos x \pm j \sin x$$

$$\cos x = \tfrac{1}{2}[e^{jx} + e^{-jx}]$$

$$\sin x = \frac{1}{2j}[e^{jx} - e^{-jx}]$$

$$\sin^2 x + \cos^2 x = 1$$

$$\cos^2 x - \sin^2 x = \cos 2x$$

$$\cos^2 x = \tfrac{1}{2}(1 + \cos 2x)$$

$$\sin^2 x = \tfrac{1}{2}(1 - \cos 2x)$$

$$\cos^3 x = \tfrac{1}{4}(3 \cos x + \cos 3x)$$

$$\sin(x \pm y) = \sin x \cos y \pm \cos x \sin y$$

$$\cos(x \pm y) = \cos x \cos y \mp \sin x \sin y$$

$$\tan(x \pm y) = \frac{\tan x \pm \tan y}{1 \mp \tan x \tan y}$$

$$\sin x \sin y = \tfrac{1}{2}[\cos(x - y) - \cos(x + y)]$$

$$\cos x \cos y = \tfrac{1}{2}[\cos(x - y) + \cos(x + y)]$$

$$\sin x \cos y = \tfrac{1}{2}[\sin(x - y) + \sin(x + y)]$$

$$A \cos x + B \sin x = C \cos(x + \theta)$$

where $C = \sqrt{A^2 + B^2}$ and $\theta = -\tan^{-1} \dfrac{B}{A}$

A.2 SERIES EXPANSION

$$(1 + x)^n = 1 + nx + \frac{n(n-1)}{2!}x^2 + \frac{n(n-1)(n-2)}{3!}x^3 + \cdots$$

$$\simeq 1 + nx \qquad |x| \ll 1$$

$$e^x = 1 + x + \frac{x^2}{2!} + \frac{x^3}{3!} + \cdots$$

$$Q(x) = \frac{e^{-x^2/2}}{x\sqrt{2\pi}}\left(1 - \frac{1}{x^2} + \frac{1\cdot 3}{x^4} - \frac{1\cdot 3\cdot 5}{x^6} + \cdots\right)$$

$$\sin x = x - \frac{1}{3!}x^3 + \frac{1}{5!}x^5 - \frac{1}{7!}x^7 + \cdots$$

$$\cos x = 1 - \frac{1}{2!}x^2 + \frac{1}{4!}x^4 - \frac{1}{6!}x^6 + \cdots$$

$$\tan x = x + \frac{1}{3}x^2 + \frac{2}{15}x^3 + \frac{17}{315}x^7 + \cdots$$

$$\sin^{-1} x = x + \frac{1}{2\cdot 3}x^3 + \frac{1\cdot 3}{2\cdot 4\cdot 5}x^5 + \frac{1\cdot 3\cdot 5}{2\cdot 4\cdot 6\cdot 7}x^7 + \cdots$$

$$\cos^{-1} x = \frac{\pi}{2} - \sin^{-1} x$$

$$\tan^{-1} x = \begin{cases} x - \dfrac{x^3}{3} + \dfrac{x^5}{5} - \dfrac{x^7}{7} + \cdots & |x| < 1 \\ \dfrac{\pi}{2} - \dfrac{1}{x} + \dfrac{1}{3x^2} - \dfrac{1}{5x^5} + \dfrac{1}{7x^7} & x > 1 \end{cases}$$

A.3 SUMMATIONS

$$\sum_{k=0}^{N} k = \frac{N(N+1)}{2}$$

$$\sum_{k=0}^{N} k^2 = \frac{N(N+1)(2N+1)}{6}$$

$$\sum_{k=M}^{N} r^k = \frac{r^{N+1} - r^M}{r - 1}$$

A.4 INDEFINITE INTEGRALS

$$\int \sin ax\, dx = -\frac{1}{a}\cos ax \qquad \int \cos ax\, dx = \frac{1}{a}\sin ax$$

$$\int \sin^2 ax\, dx = \frac{x}{2} - \frac{\sin 2ax}{4a}$$

$$\int x \sin ax\, dx = \frac{1}{a^2}(\sin ax - ax\cos ax)$$

$$\int x^2 \sin ax\, dx = \frac{1}{a^3}(2ax \sin ax + 2 \cos ax - a^2x^2 \cos ax)$$

$$\int \cos^2 ax\, dx = \frac{x}{2} + \frac{\sin 2ax}{4a}$$

$$\int x \cos ax\, dx = \frac{1}{a^2}(\cos ax + ax \sin ax)$$

$$\int x^2 \cos ax = \frac{1}{a^3}(2ax \cos ax - 2 \sin ax + a^2x^2 \sin ax)$$

$$\int \sin ax \sin bx\, dx = \frac{\sin(a-b)x}{2(a-b)} - \frac{\sin(a+b)x}{2(a+b)} \qquad a^2 \neq b^2$$

$$\int \sin ax \cos bx\, dx = -\left[\frac{\cos(a-b)x}{2(a-b)} + \frac{\cos(a+b)x}{2(a+b)}\right] \qquad a^2 \neq b^2$$

$$\int \cos ax \cos bx\, dx = \frac{\sin(a-b)x}{2(a-b)} + \frac{\sin(a+b)x}{2(a+b)} \qquad a^2 \neq b^2$$

$$\int \epsilon^{ax}\, dx = \frac{1}{a}\epsilon^{ax}$$

$$\int x\, \epsilon^{ax}\, dx = \frac{\epsilon^{ax}}{a^2}(ax - 1)$$

$$\int x^2\, \epsilon^{ax}\, dx = \frac{\epsilon^{ax}}{a^3}(a^2x^2 - 2ax + 2)$$

$$\int \epsilon^{ax} \sin bx\, dx = \frac{\epsilon^{ax}}{a^2 + b^2}(a \sin bx - b \cos bx)$$

$$\int \epsilon^{ax} \cos bx\, dx = \frac{\epsilon^{ax}}{a^2 + b^2}(a \cos bx + b \sin bx)$$

Supplementary Reading

The following books are listed for readers desiring alternate treatment of the topics covered in this book.

1. Carlson, A. B., *Communication Systems,* 2nd ed., McGraw-Hill, New York, 1975.
2. Gagliardi, R., *Introduction to Communications Engineering,* Wiley, New York, 1978.
3. Gregg, W. D., *Analog and Digital Communication,* Wiley, New York, 1977.
4. Haykin, S., *Communication Systems,* Wiley, New York, 1978.
5. Roden, M. S., *Analog and Digital Comminication Systems,* Prentice-Hall, Englewood Cliffs, N.J., 1979.
6. Sakrison, D. J., *Communication Theory: Transmission of Waveforms and Digital Information,* Wiley, New York, 1968.
7. Schwartz, M., *Information Transmission, Modulation, and Noise,* McGraw-Hill, New York, 1980.
8. Shanmugam, S., *Digital and Analog Communication Systems,* Wiley, New York, 1979.
9. Simpson, R. S., and R. C. Houts, *Fundamentals of Analog and Digital Communication Systems,* Allyn & Bacon, Boston, 1972.
10. Stanley, W. D., *Electronic Communication Systems,* Reston Publishing Co., Reston, Va., 1982.
11. Stark, H., and F. B. Tuteur, *Modern Electrical Communications,* Prentice-Hall, Englewood Cliffs, N.J., 1979.
12. Stremler, F., *Introduction to Communication Systems,* Addison-Wesley, Reading, Mass., 1982.
13. Taub H., and D. L. Schilling, *Principles of Communication Systems,* McGraw-Hill, New York, 1971.

Index